生 物 信 息 学
序列与基因组分析

（第二版）

Bioinformatics
Sequence and Genome Analysis
（Second Edition）

David W. Mount

曹志伟　编译

科 学 出 版 社

北 京

图字:01-2005-3665

内 容 简 介

当前生物信息学研究重点是对基因组序列、蛋白质组学和数组技术所产生的大量数据的计算分析。本书对 DNA、RNA 和蛋白质数据的计算提供了丰富的演算方法,并指出了在解决生物学问题中这些方法的优缺点及应用策略。

本书的第一版是在 Mount 博士讲稿的基础上进行整理出版的,在全球范围内用作教材。第二版对内容进行了全面的修订,由专业教师提供导读,最大程度地适用本科生和研究生教学。

本书为高等院校生物信息学专业本科生和研究生提供理想的学习材料。同时,本书也适宜科研人员、信息专家自学使用。

Bioinformatics:Sequence and Genome Analysis
Second Edition
David W. Mount
ISBN:0-87969-712-1

©2004 by Cold Spring Harbor Laboratory Press, Cold Spring Harbor, New York All rights reserved.

Translation rights arranged with the permission of Cold Spring Harbor Laboratory Press

图书在版编目(CIP)数据

生物信息学/(美)芒特(Mount, D. W.)著;曹志伟编译. —2 版. —北京:科学出版社,2006
　(国外生命科学优秀教材)
　ISBN 978-7-03-017640-0

　Ⅰ. 生… Ⅱ. ①芒…②曹… Ⅲ. 生物信息论-教材 Ⅳ. Q811.4

　中国版本图书馆 CIP 数据核字(2006)第 078363 号

责任编辑:刘　丹　周　辉
责任印制:赵　博/封面设计:卢秋红

科 学 出 版 社 出版
北京东黄城根北街 16 号
邮政编码: 100717
http://www.sciencep.com

北京华宇信诺印刷有限公司印刷
科学出版社发行　各地新华书店经销

*

2006 年 10 月第 二 版　开本:890×1240　1/16
2025 年 1 月第十一次印刷　印张:37
字数:1 102 000

定价:**168.00**元
(如有印装质量问题,我社负责调换)

The book is dedicated to my family:

Jendy,

Lisa,

Jonathan,

and Melissa

第二版前言

第二版的生物信息学（序列与基因组分析）比第一版的读者更广泛，它不仅面向想学计算和统计方法的生物学家，而且也适用于想学生物学的、尤其是遗传学和基因组学的计算生物学家。章节指南介绍了每一章所需的基本计算学（统计学）和生物学背景，接着是本章要学习内容的提纲，后面提供网络资源（表格和正文中仍会显示相关 URL），习题部分用于强化本章概念和相关技术。最后，所有网络资料都均用文字形式表述出来，放在一处。所有原来的章节都经过了相应的更新、修改和重写。

第二版里增加了三章新的内容。原来第三章里的序列比对的概率和统计分析现在单列为第四章，并增添了以序列分析为基础的假设检验和预测准确性检验。第 12 章和第 13 章涵盖了原来没有的 Perl 语言编程和芯片分析。这些比较高级的内容需要读者有一定的专业背景，但可增加各生物信息学最相关内容。增添的内容代表了国际前沿进展，是本书第二版宝贵的补充。在此，我非常感谢 Arizona 大学的同事们贡献了这些章节的内容。一遍遍修改润色非常耗时、艰苦，可他们总是非常合作，不吝时间。

第 12 章由 Nirav Merchant 和 Susan Miller 提供，他们是经验丰富的计算机系统和软件专家。本章具体叙述了使用和编写 Perl 脚本和模块满足不同任务，也包含了数据格式和建立关系数据库等内容。这章里列举的许多 Perl 脚本例子都可以从此书网站上直接下载作为项目模板使用。我们希望第 12 章可以作为很多实用 Perl 程序的起点以支持大规模基因组项目。

第 13 章由统计学家 David Henderson 博士提供。他擅长 QTL 分析、试验设计和统计分析，并在芯片试验方面具有丰富的经验。本章旨在帮助生物学家从芯片的基因表达数据中提取重要的信息。通常生物学家不习惯于设计涉及大量数据的试验以及从这些试验中提取信息。芯片试验麻烦的原因有两种：一是表达数据有各种来源的背景噪音，在复杂噪音中寻找哪个基因表达有差异是很困难的事；二是芯片试验的结果往往是一长串混乱的难以分类的基因。我们为解决这两种问题提供思路：一是为去除噪音达到特定科研目标来设计试验提供指导；二是叙述了寻找显著性表达差异基因的方法；三是介绍了相关分析方法，包括基于不同标准寻找、验证不同分类标志（生物标志）的聚类方法。最后提供了实现这些目标的程序资源。基于这些背景知识，第 13 章旨在指导试验设计使其产出尽可能多的有用信息。

大家对第一版的建设性评论也有益于第二版的改进。第一个是日本的 Yasushi Okazaki 先生，他在将此书译成日文的同时提了很多建议和修改意见。在不同场合提供帮助的还有 John Clark，Gabriel Dorado，Dan Flath，Toni Kusalik 和 Etsuko Moriyama。

此书离不开我生物信息学的同事 Ritu Pandey 和 Rob Klein 的支持以及 Arizona 大学的经济支持，尤其是 Vicki Chandler，Gene Gerner Rich Hoff。谢谢 Walt Klimecki 在图 11.2 提供的帮助和 Roger Miesfeld 在图 9.13A 的协助。我也很感激 Beck Nickerson 给许多章节提出的批评和建议。Pick Weil Lau 帮助我们校对了图片库，Eric Shen 从本书网站上收集整理了意见。

最后，要感谢冷泉港实验室出版社的员工的支持。没有他们的努力此项工作不可能如此成功。Judy Cuddihy 改进了章节的格式，对全文做了极其有益的建议，并给予了大量的鼓励、支持使此书能按时完成。Mary Cozza 指导我整理了参考书目，Patricia Barker，Kathleen Bubbeo，Daniel Debruin 和 Susan Schaefer 在时间紧迫的情况下高效工作，Jan Argentine 和 Denise Weiss 见证了他们为这本书付出的努力。

D. W. M.

2004 年 5 月

图森

第一版前言

此书主要献给那些想理解序列和结构分析方法的科学家。我深信一个使用计算机程序的人应该理解这些程序的原理，所以我的主要目的之一就是让生物学家在使用程序时领会程序背后的算法、假设、局限性以及使用策略。我尽量避免使用复杂的数学公式和注解，而是尽可能地给以简单的数字化例子。希望此书会对那些志在多学一点生物信息学中生物学问题的计算生物学家仍然有用。此书可以用作实验室参考书，或者生物信息学教科书，而不仅仅是一套特定序列分析程序的使用指南。

大部分章节都包含了一个流程图来建议如何有序使用本章中所讨论的方法。这种图表的例子比较稀少，因为需要的假设和过分简化不一定合理。希望这些图表会对那些本领域的新手有用，而对那些较有经验的人，我期望他们能以其他更好的方式达到相同的目的。

有很多参考文献也提供了获取程序或应用方法的网站和 FTP 站点。有时候我对一些常用并重要的 blast 和 clastalW 程序提供大量如何使用的说明以及如何分析结果。然而还有许多其他重要的关于生物序列、基因组分析的工具和方法，虽然时间和篇幅限制，我仍尽量包括进来。我对比较简单的序列分析给予了特别关注，比如寻找限制性酶切位点、翻译序列和组成分析。做这些分析的商业和免费的软件包很多，也有用于基因组分析的商用软件包。

在撰写此书（也是我的第一本书）的过程中，我发现已发表的文献中的信息远比我想像的多。我尽量详尽地把序列和基因组分析中的重要问题包括进来，仍有很多优秀的文章限于时间和篇幅没有引用到，在此我向那些作出重要贡献却未提及到的同事表示歉意。由于篇幅限制及生物信息学本身日新月异，没包括在此文中的资料等和引用的站点链接、例子以及问题都将收录到本书网站（http://www.bioinformaticsonline.org）

此学科令我感受深刻的一个特点就是大多数研究者，尤其是本领域的前沿学者都愿意与同事分享他们的研究成果。我有幸结识了几位这样的先驱，尤其是 David Lipman，Hugo Martinez（我与其共度了学术休假年）和 Temple Smith。这些学者取得的巨大成就因为愿意和同行免费共享他们努力的成果而更加备受称赞。正是因为这样，他们为学术界和商业界在序列分析领域的成功起到了非常重要的作用。

这个大项目需要许多的支持和帮助。本书的一部分来自于 Arizona（亚利桑那）大学 1999 至 2000 学年的"生物信息和基因组分析"课程的课堂笔记。许多学生提出了非常有意义的建议，并且对错误的查找很有帮助；我想要特别感谢 Bryan Zeitler，他做了很多校正工作。书中其他错误将会在本书网站上修正。我要衷心感谢 Bill Pearson 提供了 FASTA 软件包的信息，感谢 Julie Thompson 和 John Kececioglu 对第 4 章提出的意见，感谢 Steve Henikoff 帮助我阅读了第 3 章内容，感谢 Michael Zuker 对第 5 章提出的意见。Bill Montfort 为第 9 章提供了 PDB 文件的信息，另外，在第 8 章中，Foger Miesfeld 提供了复杂基因调控的例子。Jun Zhu 帮助解答了第 3 章中关于贝叶斯块联配校对器的问题。在过去的三年中，为了完成或修改其他的章节，我不得不放弃一些会议和讨论会，我所在的系给了我最大的理解和支持。这段时间里，Rob Han 和 Juwon Kim 每次都用非常短的时间为我定期地提供了大量的文献和相关书本的章节，使我有更多的时间消化这些信息。冷泉港实验室出版社的编辑 Judy Cuddihy 指导了整个写作过程，耐心地督促我遵循合理的写作进度，并给予我鼓励，帮助我完成这个项目。Elisabeth Cuddihy 确认了大部分的网站，仔细检查了书中的方程和数字示例，同时帮助撰写了部分术语表。我还要感谢出版社开发部门的 Joan Ebert 和 Jan Argentine 和生产部门的 Pat Barker 和 Denise Weiss。

最后，我还要感谢我的妻子 Jennifer Hall，感谢她在本书写作过程中所给予我家庭方面的支持、耐心和理解。

David W. Mount

目 录

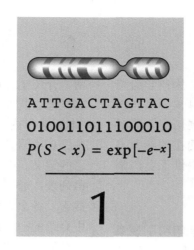

ATTGACTAGTAC
010011011100010
$P(S < x) = \exp[-e-x]$

1

历史简介和概论

目录

简　　介

本章介绍了生物信息学是如何演变为科学研究的一个新领域，描述了生物学和计算科学研究在该领域所扮演的角色，并简要记述了发展历史。同时还提供第二版各章节的概述。早期和现在的参考书籍、文章、综述以及杂志也为该领域提供了更广阔的视角。

本书各章节结构

每一章节以介绍段落开篇，简要地阐明了该章节的目的，接着是分别给生物和计算科学家的指南，介绍了在他们各自的研究领域中可能并不熟悉的概念，同时也提供该章节主题的介绍性指导。"本章学习内容"列出章节中重要的概念和实际的主题。术语表包括各章节中介绍部分以外的主要术语的定义。在各个章节最后的"本章检索词"是术语列表，通过它们能够对本章引用的网址进行引擎搜索。各个章节还总结了习题集，用于探究该章节中介绍的一些概念和步骤。因为是介绍和概述，第一章的内容比其他各章更总括，也没有习题集。

生物学家指南

在这章中，我们描述了实验室中获得的 DNA 序列在计算机文件中通常如何保存，这些文件和普通的文本文件非常相似。序列文件包括序列的信息，如生物体、实验室来源、文献引用、序列名（如果是已知基因），一个或多个唯一数字，这是特定序列的主记录号或其他标识号。由于序列的数量非常大，序列文件就能够组成数据库，如 GenBank 数据库，经过索引，它可以基于文件序列信息非常容易地获取特定的序列；比如获取特定生物体的所有序列。数据库的格式，称为关系数据库，由相互索引的或主键表格组成。我们使用数据管理系统构建数据库，可以将数据存储在里面，也可以从中提取数据。虽然序列是文本格式保存，常用的商业文本编辑器却不能检查或处理这些文档，因为这些编辑器在文本中引入了一些控制符反而破坏了序列文件。在网络和本地计算机系统上有许多计算机程序可用于数据存取的工作。在本书中，这些程序能够从数据库中显示或获取部分或所有的序列，并以适当的格式保存，通常是有一个新的数据的库。

本章的其余部分介绍了比较和分析序列的方法，同时也为本书其他各章的做了简要的概述。主要介绍的概念是序列联配，这一方法是设法比较不同句子之间的每个字母或词，确定它们是否有相同的顺序，以表明它们是否有相关的主题。能够以这一方式联配的 DNA 和蛋白序列有相同生物学功能和共同的进化起源。序列联配是非常困难的计算问题，因为要进行非常多的比对，并且在包括词间空位的情况下有很多可能的不同的联配方式。动态规划是解决这一问题的一种计算方法，它将问题分解成很小的单元，只对序列开始部分做比对分析。一旦在序列开始部分搜索到最好的匹配，比对就逐步延伸到整条序列。相同的方法也被用于搜索 RNA 序列中可形成碱基对的互补区域，或是比较蛋白质的三维结构。

在基因组时代，一个主要的研究目的是为了分析基因的功能，阐述他们在特定生物体中的相互作用。在实验室中，模式生物的遗传学操作，测量基因表达的基因芯片，以及用于蛋白分析的蛋白质组都被用于收集新的生物学数据。来源于这些实验的信息需要组织成合适的数据格式，并且需要开发计算机程序，最好能通过网络来存取相关信息并进行数据分析。生物信息学家已经编写了不少这类程序用于序列的处理和分析。

目前，编写程序已经被简化成编写一些小的，被称为"对象"的模块程序用于处理各类简单的任务。许多对象组成的程序库能够处理几乎所有的任务，这样的库被大程序所使用就能够处理更具体的序列分析任务。比如，一个模块可用于从 GenBank 数据库中提取序列，而另一个模块可以将序列转化成为特定的序列格式。BioPerl 对象库就是这类对象编程的一个例子。BioPerl 的对象可用 Perl 语言编写。使用这些已经开发的模块，只要接受一些训练或是在学生或专业程序员的帮助下就能够很容易开发任何序列处理或分析的应用程序。但是，要真正的精通这些知识，有生物学背景的学生需要通过一些课程加强自己的生物学基础，如分子生物学和生物化学的实验方法课程、种群和进化遗传学的进化分析课程。他们还必须学习数学、概率和统计、数据处理、数据挖掘和建模工具、计算机算法设计和编程，以及在可能的情况下学习基因组分析和生物信息学专题的高等课程。

计算科学家指南

生命的基本单元是细胞，细胞外侧是保护性的细胞膜，它包围着一些细胞器（亚细胞结构）以及能够支持细胞结构、提供细胞能量和进行自我复制的大而复杂的生物分子。在植物和动物中，独立的细胞相互合作形成多细胞组织和器官系统，实现生物体的生物学功能。本书主要涉及在细胞和生物体中调控生物学过程的生物序列的分析，以及指导生物体发育过程中细胞组织的指令信息的分析。

序列存贮在长化学链中，称为 DNA，由 A、G、C、T（分别代表腺嘌呤、鸟嘌呤、胞嘧啶和胸腺嘧啶）四种不同的信息字符或碱基以及糖和磷酸骨架组成。DNA 紧密地缠绕成染色体，在显微镜下就能够看见这样的亚细胞结构。除了如单细胞细菌这样简单的生物体外，染色体位于可见的细胞核中，细胞核周围包裹着细胞质。组织细胞有两套染色体，这是被称为二倍体的一种遗传结构，两条染色体分别来自于母本和父本。在有性发育过程中，形成称为配偶子的性细胞（精子细胞或卵细胞），它拥有一套染色体（单体型）。单体型细胞中的染色体组成了生物体的基因组。在有性复制过程中，来自于两个亲本的染色体重新分类配对。接着，拥有一套新配对染色体的配偶子重新生成，并最终传递到子代中去。

染色体的 DNA 序列编码了制造细胞蛋白质的指令。蛋白质是有化学活性的 20 个氨基酸组成的线性链。每个蛋白质特定的氨基酸序列由染色体上的 DNA 序列确定。蛋白质的形成使生物体能够形成结构以及行使生物学功能。在转录的生物学过程中，细胞通过搜索特定的序列模式（如启动子）"阅读" DNA 序列，这些特定的序列模式标志着遗传信息单位——基因的起始位置。从该点开始，在生物分子上沿着一定的化学方向读序列，直到遇到标志着基因结束的序列模式才停止。转录生成的化学长链称为信使 RNA 或 mRAN，对应于将要生成的蛋白质的氨基酸序列。mRNA 分子在结构和化学上和 DNA 分子非常相似，但是他们通常是单链，并且以新的碱基尿嘧啶（U）代替了胸腺嘧啶（T），在主链上也连接着不同的糖。mRNA 分子也有特定的序列模式，标记了蛋白质编码开始的位置。细胞质中称为核糖体的大细胞器能够结合在蛋白质编码开始的位置，按定义的化学方向移动，每次读三个碱基位置（一个密码子）确定一个氨基酸。接着，该密码子对应的氨基酸就添加到形成蛋白质的氨基酸链上。就这样，氨基酸逐个添加，直至碰到几个终止密码子之一。每个蛋白质的序列就这样基本与染色体上的原始编码序列线性一致。

蛋白质一旦形成，将迅速由线性字串折叠成简单的螺旋和折叠单元（即二级结构），随后这些单元将形成特异的三级结构。产生的蛋白质分子作为组成构成单元形成组织或者参与特定的化学活动。一个生物体的结构及其生物学功能取决于蛋白质组，及一个生物体产成的所有蛋白。并非所有的基因都被翻译成为蛋白质——某些在 RNA 水平上仍保持着 4 个字母序列的基因，调控着许多重要的细胞进程。简单的生物体拥有数以千量的基因，而较为复杂的生物体则拥有 12 000～35 000 左右的基因。在一些生物体中，尤其是植物类生物体，可能会存在巨大数量的重复基因或重复染色体，这些多余的基因逐代复制下去，有时生成包含 100 000 甚至更多基因的基因组。

有个概念对了解生物学家如何看待基因非常重要，就是所有生物体看起来都通过进化过程相互关联。即使是完全不同的生物体，譬如单细胞细菌和多细胞动植物，他们之间也可能有一些共同基因。这些基因通常并不是每次由新的生物起点产生，而是从先前的基因复制得来，在复制中 DNA 序列在有限程度上随机地产生突变。生物体可以通过二元树划分成组，在外面的分支代表联系更加紧密、进化时代较近的生物体，在内部的分支则代表更加原始，通常比较简单的生物体。这些树的构建可以基于生命信息（生物体的结构和行为），但是最近随着模式生物基因组的测序，基因组上基因的互补信息可以用来构建二元树。因此，很多的生物信息学研究关注对比基因和被翻译的蛋白质，将他们视为由 4 个字母（DNA）或 20 个字母（蛋白质）组成的线性字串。当某个生物体编码的一个蛋白质与另一生物体编码的一个蛋白质很容易比对上的话，这就告诉生物学家们这些基因来自共同祖先并且会有相同的功能。

　　学习基因和蛋白质还有一个重要概念。为产生一项新的生物学功能生物体似乎运用三个策略而不是构造一个全新的基因。第一个策略是，复制已有的基因，其中的某个基因通过随机突变逐渐改变而发展出一个新的生物功能。在极少数情况下，这个新功能对生物体本身有利，自然选择就在这个生物中扮演了基因催生的角色。这个过程可以建立起一个相互关联的、功能重要的、可以通过进化传递的基因家族。第二个策略是，生物体将现有的基因部件（功能域）结合在一起从而形成新的基因，这些部件本身代表蛋白质三维结构或者生物功能的基本单位。生物信息通过序列比对，运用统计方法寻找序列的共有模式，做了大量的基因、蛋白质家族和序列功能域发现的工作。生物体内已有基因功能分化的第三个策略是非常少见的从不同、不相关的生物体内转移基因。通常情况下基因都是由父系向子系传递，生物学家把这个过程称作垂直传递，因为遗传沿着树状家谱的分支传递。另一种遗传方式是一段外源 DNA 随机地转入细胞并掺入已有染色体从而增加了基因。因为贡献基因的生物来自通常并不发生基因交换的不同种群，所以这种遗传被称作横向转移。在远古生命中，甚至认为整个细胞都能融合来创造新的物种。横向转移仍在单细胞细菌的进化中起主要的作用，尤其是在发展对特殊抗生素的抵抗力方面。所以，所有生物体都共有许多基因或基因的组成部分，现在这些都能在生物体基因组中被识别。

　　目前生物学家面临的两大挑战是如何发现基因的生物功能以及理解基因相互之间的作用是如何调节生命活动的。如果得不到其他物种的同源基因的信息，可以使用以下实验方法：①破坏基因序列（对序列进行改变，或插入，或删除）。②引入改变形式的基因的 mRNA 拷贝，它能导致细胞 mRNA 拷贝的降解，以一种半持久的、可遗传的方式（外成性改变）改变基因的结构，使得基因不能转录，从而关闭基因的表达。通过一整套生物学实验，可从这些遗传性或表观性改变的影响来推断基因的功能。除了遗传分析以外，生物学家还通过 DNA 微阵列技术和蛋白质分析实验（蛋白质组学）来跟踪许多细胞基因和蛋白质的表达。这些技术帮助发现与生物功能相关的基因表达和蛋白质发生的模式，比如癌细胞中的非正常模式。

生物信息学专业学生的基本要求

- 知道序列、蛋白质和基因组数据的存贮位置以及保存方式。
- 具备编程技术能从网络，数据库中提取数据，将它们整理，重新储存成合适的数据库格式。
- 掌握足够的生命科学指示，熟悉序列和基因组信息与基因功能及蛋白质结构功能之间关系，并知道如何适当保存这些数据。
- 了解在不同物种找到功能相似基因的最新序列基因组分析方法，达到基因调控分析，蛋白质结构功能预测的目的。
- 懂得如何处理海量的数据如基因表达微阵列，蛋白质组数据，种群中的序列变异。
- 能够整合分散的数据，并利用已有的工具来进行数据挖掘，在数据中发现新的关系。
- 与计算机科学家、数学家、统计学家沟通，发展所需的新的分析工具和模型。

术语表

alignment（联配）　通过查找两个甚至更多序列中相同次序的字符串或字符串模式来对序列进行比较。

algorithm（算法）　一种数据分析的方法，强调时间及空间效应，可从数学上证明其正确性。例如用于序列比对的动态规划算法。

annotation（注释）　基因组序列上的基因定位，包括编码蛋白质，RNA 的基因，同时提供了编码蛋白质和 RNA 分子的序列和位置。

codon（密码子）　DNA 中的三核苷酸，被细胞翻译成蛋白质中的氨基酸。在 64 个可能的密码子中，有 61 个通常被翻译成 20 个氨基酸中的一种，其余 3 个是终止子。mRNA 以一系列密码子的形式携带蛋白质序列信息。

comparative genomics（比较基因组学）　比较不同物种的基因组中基因数目、位置、基因的功能，目的在于识别在特定生物中起特殊功能的家族基因。

database（数据库）　存储数据的组织系统，通常是一系列相互关联的表格，被称为关系数据库。

distance score（距离分值）　指相关序列的联配中，比对字母不同或很少找到的那些位点，通常（但并不总是）不考虑缺口位点。

DNA　由 4 个核苷酸 A，T，G，C 组成的双链螺旋分子，A 总是与另一条链的 T 相配对，而 G 总是和 C 配对，核苷酸之间的化学相互作用将两条链连接在一起。两条链分开后，其中每一条链作模板都可以基于 A/T，G/C 配对原理合成新链。这条适用于所有生物繁殖的分子机理是由 James D. Watson 和 Francis Crick 发现的。DNA 分子序列可以用 DNA 测序机以每次 500～800 个的速度阅读出来。

dot matrix analysis（点矩阵分析）　以图形化的方式比较两条序列。一条序列从图的顶部（或底部）开始水平书写，另一条沿左侧从上至下，在两条序列中都出现的核苷酸的交叉点被标记，对角线意味相似性。

dynamic programming（动态规划）　允许配对，错配和空缺的序列联配算法。算法首先从序列开始位置找到最佳匹配，然后在每条序列上逐渐增加最佳匹配直至完成。

extreme value distribution（极值分布）　一些诸如序列比对分值之类的数据服从一种"长尾"分布，在高值端衰减得比正态分布慢得多。其中一种缓慢衰减的分布称为极值分布。不相关序列或随机序列之间的比对分值就是一个例子。这些分值能达到很高的值，尤其在进行数据库相似性搜索进行了大量比对时。特定分值的概率可通过 Gumbel 型的双负指数函数的极值分布准确预测。

functional genomics（功能基因组学）　评估从基因组比较中发现的基因功能。新识别的基因常通过引入突变并观测后代生物性状改变的方法来判定功能。

gene（基因）　与某生物功能相关的一段 DNA，通常是蛋白质的氨基酸序列。与合成 DNA 新链相同，DNA 运用碱基互补原理复制到 mRNA 分子。

gene expression microarrays（基因表达微阵列）代表一个物种中大量基因的显微镜载玻片上的 DNA 微点阵列。微阵列用来比较一个生物样本中整套基因的 mRNA 水平，通过合成化学性质不稳定、带有荧光标记的与 mRNA 互补的 DNA（cDNA）来探测。

genome（基因组）　一个生物全部的遗传信息，包括特殊的蛋白质，RNA 分子以及其他的序列。他包含体细胞中一半的染色体，性染色体的全部。

global sequence alignment（全局序列联配）　一种包含所有序列的比对方法。

homologous（同源性）　基因因相似的序列被证实来自相同的祖先基因。

local sequence alignment（局部同源联配）　通过最高密度匹配方法进行序列区域比对的方法。Maximum parsimony tree－最节省树：多序列比对中已观察到的变化的图形表示，使得在树分支上变化数量的和达到最小。

model organism（模式生物）　可被遗传操纵，从后代生物的变化中找到具有特殊功能的基因的物种。从这样的模式生物研究中得到的结论也可应用于其他生物。模式生物有果蝇，酵母，斑马鱼，老鼠以及植物阿拉伯芥。

motif（protein）　可以代表蛋白质结构中的活性中心，功能区域的一小段氨基酸。

multiple sequence alignment（多序列联配）　排列三个或更多的序列，试图将功能或进化相关的序列位点放在同一纵栏，允许错配，空缺。

orthologous（同源）　形容来自不同生物体，两个甚至更多的基因，因为相似被预测拥有相同的生物功能。

orthologs　同源的基因。

paralogous（旁系同源）　形容一个生物体内一群相似基因，被预测来自于同一祖先的基因复制。

paralogs　旁系同源的基因。

phylogenetic analysis of sequences（序列系统进化分析）　试图应用进化树来找出一系列相似序列的进化关系。

position-specific scoring matrix（PSSM）（位置特异记分矩阵）　表示从一些列关联基因的多序列排列实验中找到的差异的表格。表格纵列对应排列中的纵列，行对应每列序列特征的出现频率。这个频率通常除以序列中特征频率以得到奇数，为了方便，奇数用其对数形式表示。

protein（蛋白质）　由一条长链氨基酸组成的分子，它的序列由基因的密码子序列决定，长链折叠形成三级结构，唯一对应具有某生物功能的蛋白质。

什么是生物信息学？

过去，生物信息学被定义为包括生物学、计算机科学、数学、统计学的交叉学科，它研究生物序列数据、基因组内容及排列、预测生物大分子的结构与功能。随着基因组时代的来临，现在生物信息学在生物学和医学研究中起着更重要的作用，论文发表量逐年上升（Luscome et al. 2001）。

生物信息学与信息学及计算生物学等研究领域有交叉。过去，信息学是这样的学科：数学家、计算机科学家、统计学家、工程师共同发展在诸如健康等领域的信息管理的支持技术。现在，生物信息学也参与进来，它组织与基因组相关的数据，旨在将这些信息应用到农业、制药业和其他商业中。有两种类型的生物信息：序列信息，以及来自基因组和由实验获得的基因产物结构-功能分析等内容。针对人类基因组，生物信息学的作用是收集约35 000个的人类基因的生物信息，以发现那些基因在人类疾病中起着最显著的作用。利用现代技术，例如基因表达芯片，细胞和组织中基因遗传操作，以及蛋白质结构与功能的快速评估，可以收集到许多基因功能方面的新数据。

对于上述问题，密切相关领域的计算生物学也提供了计算方面的支持，但与生物信息学所提供的支持是有所不同的。一方面，计算生物学一般关注发展一些能解决诸如多序列比对、基因组碎片拼接等困难问题的新型高效算法。而生物信息学更多地注重开发数据管理与分析的实用工具，如基因组信息的显示和序列分析，而不注重于对高效性和可证明的正确性的研究。在很多情况诸如进行多序列比对和数据库相似性搜索时，得不到一个适当的序列改变的模型，或者合理定义的序列分析问题太复杂而不能在合理的时间内解决。在这种情况下，生物信息学能提供满足当前需求但没有理论证明基础的计算方法。因此，今天的生物信息学领域支持包括确定数据的生物学意义、提供组织数据的专门技术、发展用于新数据挖掘的实用计算工具等很大范围的研究工作。当信息具有实用的重要性时，生物信息学协助实际应用诸如识别药物疗法的新蛋白药靶等。然而不可否认地，生物信息学、信息学和计算生物学都在快速发展并将在基因组数据的利用方面起着不同的作用。

首先被收集的蛋白质序列

由于蛋白质测序方法的发展（Sanger 和 Tuppy 1951），一些具有代表性的常见蛋白质家族得以测序，如不同物种的细胞色素蛋白。早在 20 世纪 60 年代，Margaret Dayhoff（1972，1978）和她在位于华盛顿的国立生物医学研究基地（National Biomedical Research Foundation，NBRF）的合作者们首先建立了蛋白序列数据库，这一数据中心最终发展成了著名的蛋白信息资源库（Protein Information Resource，PIR；http://watson. gmu. edu:8080/pirwww/index. html）。NBRF 自 1984 年起维护这些数据库。1988 年，作为 NBRF 的合作伙伴，PIR -国际蛋白序列数据库（PIR-International Protein Sequence Database，ht-tp://www-nbrf. georgetown. edu/pir）、慕尼黑蛋白质序列中心（Munich Center for Protein Sequences，MIPS）、日本国际蛋白质信息数据库（Japan International Protein Information Database，JIPID）相继成立。

Dayhoff 和她的合作者们基于序列的相似程度将蛋白质组织成家族和超家族。由此衍生出了反映一群紧密相关的蛋白质序列的氨基酸频度变化的表格。为了保证观察到的氨基酸变化是一次性变化而非两次连续的变化，仅那些序列差异性小于 15% 的蛋白质被选来作研究，这样可以避开多次改变的嫌疑。根据序列比对的结果可生成系统发育树，以图形的方式显示出那些由于密切相关而位于系统发育树同一分支上的序列。这些树可用于对来源于不同物种的蛋白质序列在进化中形成的氨基酸改变打分（图 1.1）。

Margaret Dayhoff

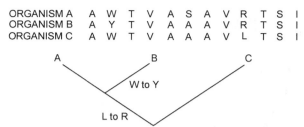

```
ORGANISM A   A W T V A S A V R T S I
ORGANISM B   A Y T V A A A V R T S I
ORGANISM C   A W T V A A A V L T S I
```

图 1.1　对进化上相关的蛋白质序列预测系统发育关系
和可能的氨基酸改变的方法

图中显示了来自 3 个不同物种的相同蛋白质的 3 条高度保守的序列（A，B，C）。这些序列非常相似，在进化中每个位置只有一次改变。蛋白质序列只有一个或两个位置的替换，由此可构建如图所示的树。一旦树形成后对应的氨基酸改变也就确定了。图中显示的特定氨基酸改变的发生概率要高于随机替换过程出现的改变

评估系统发育树分析结果的规则是：两条序列中相同和保守的氨基酸越多，它们进化自共同祖先的可能性越大。如果两条序列非常相似，蛋白质可能有相同的生化功能和三维立体结构。因此，可以构建一组称为 PAM（percent accepted mutation）表的矩阵，来描述进化选择所接受的氨基酸突变百分比。这些表格给出了系统发育树中由一个氨基酸变成另一个氨基酸的概率，因此能显示出两个序列中哪些对应位置的氨基酸是最保守的。PAM 表还可以用来测度蛋白质序列间的相似性以及在数据库中搜索匹配序列。

通过建立第一个蛋白质序列数据库以及建立蛋白质序列比较的 PAM 表，Dayhoff 和她的同事们在许多方面对现代生物序列分析做出贡献。氨基酸置换表成为序列比对和数据库相似性搜索的常规工具，它们在这方面的应用将在第 3 章和第 6 章讨论。

始于 20 世纪 80 年代早期的 DNA 序列数据库

新墨西哥州的 Los Alamos 国立实验室的 George I. Bell 于 1974 年建立的理论生物学与生物物理研究组最先收集了 DNA 序列，并将其存放到 GenBank 数据库中。这些物理学家试图为实验室的工作，主要是免疫学方面的研究提供理论背景。该小组的研究扩展到了计算生物学和生物信息学的领域，并且在 1982～1992 年期间由 Walter Goad 和其同事开发了 GenBank 的第一个版本。Goad 早在 1979 年就已经开始对 GenBank 数据库原型进行构思。由 DNA 翻译的蛋白质序列也被华盛顿的国立生物医学研究基地的蛋白质信息资源数据库（PIR）收录。其他相关的数据库，如欧洲分子生物学

Walter Goad

实验室（EMBL）数据库于 1980 年成立（http://www.ebi.ac.uk），日本 DNA 数据库（DDBJ，http://www.ddbj.nig.ac.jp）于 1984 年成立。GenBank 现在由国家生物工程信息中心（NCBI，http://www.ncbi.nlm.nih.gov）管理。GenBank，EMBL 和 DDBJ 现在已形成国际核酸序列数据库联盟（http://www.ncbi.nlm.nih.gov/collab），旨在进行每日的数据交换。PIR 也进行着同样的工作。

每年第一期的《核酸研究》杂志会刊登许多序列数据库类型。GenBank 数据库中序列数目的增长可在 http://www.ncbi.nlm.nih.gov/GenBank/genebankstats.html 查到。

最初，这些数据库中的一条序列记录包括其计算机文件名以及 DNA 或蛋白质序列文件。最终，这些记录扩展了更多的序列信息，如功能、突变、编码蛋白质、调节位点和参考文献。这些注释信息与序列一起存放成一种易于搜索的数据格式。类似的数据库和格式将在第 2 章讨论。

核酸序列数据库 GenBank 和 EMBL 的记录每天都有大量更新。注释所有这些新序列是耗时、艰苦的过程，有时还易出错。近年来，这个过程变得更自动化，但是又引起了准确性和可靠性的问题。GenBank 在 1997 年 12 月有 1.26×10^9 碱基，到 2004 年 4 月增长到 39×10^9 碱基。尽管储存的序列数成指数增长，有效的搜索方法的应用保证对这些序列快速公开的检索。

为减少数据库搜索所得到的匹配数目，建立了非冗余数据库对于相同序列只列出一条记录。NCBI Refseq 就是这样的数据库。然而，许多序列数据库，如 NCBI 用于 BLAST 搜索的非冗余 NR 数据库，仍然包括了大量相同的基因和蛋白质记录，它们可能来源于不同数据库的序列碎片、专利、重复序列以及其他类似的序列记录。

从公众数据库中方便的获取序列

提供序列数据库访问的重要一步是开发查询页面，这一工作形成了一些重要的序列数据库（GenBank，EMBL 等）。这项技术一个例子是在 NCBI 早期由 D. Benson、D. Lipman 及其同事开发了称为 GENINFO 的菜单驱动程序。这个程序快速搜索已建立索引的序列数据库，为生物学家的查询提供相匹配的记录。NCBI 随后派生了有简单窗口界面的 Entrez 程序，并最终开发成网页界面，形成 NCBI 网站（http://www.ncbi.nlm.nih.gov/Entrez）。

David Lipman

这些程序的想法就是用关键字搜索标准记录字段，以灵活的方式为序列数据库提供易于使用的界面。一些主要数据库中的序列还记录了序列的附加信息，如索引号、序列名称和别名，相关基因名称，调节序列类型，物种来源，参考文献，已知突变。Entrez 程序能访问这些信息，因而能够快速搜索整个序列数据库，查对到一个或几个特定匹配的术语。

这些程序也能根据预先的相似性比较定位相似序列（Entrez 称为"邻居"）。当对一项或多项术语执行查询时，简单的模式搜索程序只能找到精确匹配结果。相反，Entrez 能够容易的进行相似或相关术语的搜索，或是进行多个组合选项的复杂搜索，并将搜索结果按与原查询的相似程度排序。Entrez 最初允许直接访问 DNA 和蛋白质序列数据库及它们的参考文献，甚至允许索引不同或同一个数据库中的相关记录或相似序列。最后，Entrez 还提供了对所有 Medline 数据库的访问，这是一个位于华盛顿区的国家医学图书馆全文文献数据库。另外也提供对许多其他数据库的访问，如物种系统发育数据库、基因组数据库和蛋白质结构数据库。NCBI 任何用户—个人，政府、工业或是研究机构提供免费访问。他们所做出的这一决定对生物学研究的促进作用是不可低估的。NCBI 每天都要处理数百万个访问事件。

注意事项：Entrez 这种数据库查询程序推动了数据库和日益增长的序列数据以及生物化学杂志保持同步。然而，如同任何自动化方法一样，必须注意数据库搜索也许不能够提取出所有的相关材料，重要的记录可能会遗漏。每条数据库记录在某个阶段都需要手工编辑，这会造成少部分不可避免的拼写错误和其他问题。有时在数据库中应有的记录没有被检索到，可能因为搜索术语在相关数据库记录中被误拼，或这条记录未出现在数据库中，或是存在其他更复杂的问题。如果各种搜索尝试都失败了，将此问题报告给程序编写者或系统管理员将有助于解决这一问题。

序列联配程序的发展

DNA 的测序是通过排列一套由测序仪检测的来自测序凝胶的峰值（A，G，C 或 T），这是一个很易出错的过程，主要依赖于数据的质量。为了能够正确地收集数据，华盛顿大学的 Phil Green 和他的同事编写了 phred（Ewing and Green 1998, Ewing et al. 1998）和 phrap 程序，用以帮助读取和处理测序数据。phred 和 phrad 现在由 CardonCode Corporation（http://www.codoncode.com）发布。这些序列读取工具对更准确地收集序列数据有很大帮助。

20 世纪 70 年代末，随着序列数据的增长，人们也增加了使用各种方式开发计算机程序分析序列的兴趣。在 1982 和 1984 年，《核酸研究》出版了两期专辑刊登计算机在序列分析方面的应用，包括大型机和微机程序。不久后，J. Devereux 在威斯康星组建了遗传学计算机研究组（Genetics Computer Group，GCG），提供了一套在 VAX 计算机上运行的分析程序。同一时期还创立了其他的公司，包括 Intelligenetics，DNAstar 等，他们为序列分析提供可在微机上运行的程序。实验室也开发并在免费或低费用的基础上共享计算机程序。

1977 年 Maxam、Gilbert 及 Sanger 等发明了

DNA 序列分析的方法。Sanger 方法在第 2 章开头将详细介绍。

这些商业化及非商业化程序仍被广泛使用。而且，有许多网站为各种序列分析提供平台，它们对学术机构免费，商业用户可以适当的价格购买。以下是对序列分析方法发展的简单综述。

序列比较的点阵法或图示法

1970 年，A. J. Gibbs 和 G. A. McIntyre（1970）提出了氨基酸和核酸序列比较的新方法。可以画图说明这一方法，将其中一条序列横排在页面中，另一条序列竖排在左下方的位置。当相同的字母在两条序列中同时出现时，在序列对应的两个字母的交点位置上放置一个点（图 1.2）。通过扫描图中一系列的点，能够在两条序列中找到揭示相似性或有相同字符的字串的对角线。长序列的比较通过使用较小的点也可以在一页纸内实现。

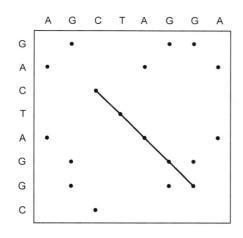

图 1.2　两条 DNA 序列 AGCTAGGA 和
GACTAGGC 的点阵比较图
对角线上的点表明在两条序列中同时出现的一连串序列
CTAGG

由于点阵的方法找到的对角线在水平和垂直方向上的平移是根据序列间差异而得到的，所以很容易反映序列间的缺失和插入。单条序列的自比较能够揭示序列中的相同（正向重复）或相反方向（反向重复或回文）的重复区域。序列自比较的方法能够揭示一些特征，如染色体之间的相似区域、串联基因、蛋白质序列中的重复区域、有重复碱基组成的低复杂度序列，或是 RNA 中能够配对形成双链结构的互补序列。

由于序列间的弱相似性对角线可能无法在图中找到，Gibbs 和 McIntyre 为了找到显著的匹配，计算了图中所有可能出现的对角线，同时与一些随机序列比对时出现的对角线数目相比较。这套随机序列提供了不相关序列之间联配分值的分布范围。如果两个测试序列之间对角线分值有显著性，即对角线分值应该超过随机序列比对的对角线分值，就可以证明序列已成功联配。同时，高于随机序列的对角线分值的程度提供了一个统计结果，该结果能决定测试序列对角线的显著性。随后，Maizel 和 Lenk（1981）开发了各种过滤和彩色显示配置，极大增加了点阵方法的有效性。以点阵展示序列比对在 DNA 和蛋白质序列相似性分析以及基因和染色体上重复序列的分析中一直起着重要的作用，这一内容将在第 3 章介绍（第 79 页）。

动态规划方法在序列联配中的运用

虽然点阵方法可以检测序列中的相似性区域，但是不能很好的解决由于序列间的弱匹配或缺失插入区域的存在而中断序列相似性匹配的问题。因此，人们寻找新的方法希望通过点阵找到一条曲折路径，在两条序列中发现最有可能的联配，或称为最优联配。如图 1.3 所示，将其中一条序列连续的横排在页面中，另一条序列排列在其下方，相同碱基排列在同一栏中，如果同一栏中有不匹配的碱基则被视为错配或加入空位表示插入（在另一条序列中是缺失）。对于一条长度为 300 的蛋白序列，如果要在所有可能的匹配中找到最优的联配，同时考虑插入和缺失的匹配情况，需要进行 10^{88} 比对，这在计算上有很大的困难（Waterman 1989）。

```
SEQUENCEA  A G △ △ C D E V I G
SEQUENCEB  A G E Y C D △ I I G
```

图 1.3　双序列比对显示匹配、不匹配和空缺（▲）
最好的或最合适的比对需要允许所有这三种变化

为了简化上述工作，Needleman 和 Wunsch（1970）将这个问题分解为逐个氨基酸的比较，以渐进的方式建立起序列联配。序列比对从两条序列的结尾开始，而后同时向前移动一个氨基酸，允许匹配、错配或是在一条序列中出现插入或缺失氨基酸的各种组合形式。这一方法在计算机科学中称为动态规划。Needleman 和 Wunsch 的方法产成了以下结果：①可能存在的序列联配，即包括了匹配、错配或是单碱基插入或缺失的各种可能组合的序列联配。②序列联配的打分系统。其目标就是通过判断最高分以确定最好的联配结果。其中每对匹配的碱基记 1 分，错配记 0 分，空位给予一定的罚分。而后沿着序列将分值相加得到联配的总分，拥有最高分值的联配被定义为最优联配（图 1.4）。

```
      G A T C T A
G     1
A       2             1
T         3       1
C           4
A     1               5   (minus gap penalty)
```

Deduced alignment with gap △

```
G A T C T A
G A T C △ A
```

图 1.4　使用 Needleman-Wunsch 的方法联配
GATCTA 和 GATCA 序列

首先，在本例中我们定义所有匹配的碱基记 1 分，错配记 0 分（未显示）。接着将对角线上的分值连续相加，在本例中总分值为 4。这时横行因无法延伸到另一对相匹配的碱基对而不能得到 5 分。但是，如果在 GATCA 中插入空位生成 GATC△A 就可以继续延伸，在这里 △ 表示一个空位。引入空位后，需要从目前行列匹配的总分值 5 分中减去空位罚分值。最优联配的寻找开始于最高分值对应的碱基，并回溯对这一最高分值有贡献的所有位置。

如图 1.4 所示，所有可能的联配产生的过程开始于其中一条序列的末尾位置，并在一个与点阵非常相似的矩阵中有序地移动（图 1.2）。允许从序列任意位置开始，并且包含匹配、错配、插入和缺失的情况，寻找最高分值并放置在矩阵的每个位置中。而后通过寻找图中的高分位置并回溯这些路径就能找到最优序列联配。对应于这些路径的碱基相互匹配最终序列也得以联配。

搜索序列间的局部联配

上面提及的方法是在整条序列中寻找最优联配，被称为全局联配。Smith 和 Waterman（1981a,b）意识到在 DNA 和蛋白质的序列中许多有生物学意义的区域是匹配较好的子区域，而其他弱相关的部分匹配不显著。因此，他们对 Needleman-Wunsch 算法做了重要的修改，开发了称为局部联配或

Smith-Waterman（或 Waterman-Smith）的算法用于类似局部区域的比对。他们还认识到序列之间不同长度的插入和缺失与序列进化上的变化有关，因此他们调整了该方法以适应这样的变化。最后，他们还从数学上证明了动态规划能够保证在序列间找到最优的匹配结果。第 3 章有相关算法的具体介绍（第 83 页）。

Mike Waterman　　　*Temple Smith*

研究者开发了相似性分值和距离分值两种互补的方法用于两条序列联配打分。如图 1.3 所示，在联配序列中有三种情况——一致匹配，错配和空位。为了获得相同的分值，使用简单的打分机制将匹配设定为 1；匹配的分值累加除以匹配以及错配残基的总数（空位通常是忽略的）。这一序列相似性打分机制是生物学家所熟知的，由 Needleman 和 Wunsch 设计，Smith 和 Waterman 也采用这一方法。另一种打分机制，即距离分值，是一条序列转变成另一条序列所需要的替换次数的累加值。这一分值在系统进化分析时对于蛋白质或基因序列进化距离的预测十分有用。该方法是数学家特别是 P. Sellers 的贡献。距离分值计算联配序列中的错配残基总数，并除以匹配和错配残基的总数。距离分值代表了在不考虑空位的情况下，从一条序列转化成为另一条序列所需要的替换次数。在图 1.3 的例子中，有 6 个匹配的残基，1 个错配。对于这一联配的相似性分值是 6 个匹配除以 7＝0.86，距离分值是 1 个错配除以 7＝0.14。一般来说，相似性分值和距离分值相加为 1。如果没有空位出现，联配的长度应该和序列的长度一致。如果出现空位，联配的长度要比序列的长度更长。可以注意到序列的长度等于两倍的匹配和错配的和值，再加上插入或缺失的数目。这样，在我们的例子中，可以这样计算 8＋9＝2×（6＋1）＋3＝17。通常更复杂的打分机制用于生成有意义的联配，联配用似然值和几率分值（第 3 章）评估，但是相似性分值和距离分值之间

的相反关系还是保持着的。

在序列联配中存在一个困难问题，既判断特定的序列联配是否有显著性。我们需要知道序列联配的分值真能够揭示两条序列是相似的吗？或是这样的分值在两条不相关的序列中（或是计算机生成有相同的组分的随机序列）也能容易的找到吗？这一问题由 S. Karlin 和 S. Altschul（1990，1993）提出，在第 4 章中做了具体的描述（第 136 页）。

Karlin-Altschul 分析了不相关或是随机序列的分值，发现这一分值通常比常规分布的期望值要高。分值的分布符合正偏态，也称为极值分布，对随机或不相关序列联配分值的分析提供了一种方法，该方法可以用于评估一段联配的分值在相同长度的不相关或随机序列中出现的概率。正如第 6 章中讨论的，这一发现对于评估查询序列和数据库中序列的比对特别有用。另外，一段特定联配的分值的评估必须考虑数据库搜索时相似序列的数目。假设使用随机序列联配分值的极值分布来评估目标蛋白质序列和数据库中蛋白质序列匹配的分值，有 10^{-7} 的概率和两条不相关序列的联配分值一样高。如果数据库中有 80 000 条蛋白质序列，将进行 80 000 个不同的比较。每次比对都会有不相关序列联配的分值与 10^{-7} 概率分布下的分值一样高。如果做80 000次比对，就增加了 80 000 次这样的机会。这些序列中的任意一条能够得到 10^{-7} 概率分布下的比对分值的概率是 $10^{-7} \times 8 \times 10^{-4} = 8 \times 10^{-3} = 0.008$（称为期望值）。该值介于 0.02～0.05 之间被认为有显著性。但是，在基因组比较时，当数据库中的序列与查询序列进行联配时，低期望值（如 10^{-20}）更有意义。即便在这样的期望值下找到了联配，也需要仔细检查序列联配中的缺陷，如不切实际的氨基酸匹配，连串重复的氨基酸都会降低序列联配的置信度。

多序列联配

多序列联配是用于同时比较三条或多条序列的方法（早期的例子见 Johnson 和 Doolittle 1986）。它采用运算密集的方法，通常基于连续联配最相似的成对序列。常用的工具有 GCG 软件中的 PILE-UP, CLUSTALW（Thompson et al. 1994），以及 T-COFFEE（Notredame et al. 2000）。软件的最新发展在第 5 章中有具体描述。一旦生成一套相关的生物大分子序列（一个家族）的多序列联配，可以确定特定家族中高度保守的区域（Gribskov et al. 1987），也可以用于确定家族新成员。序列谱和位置特异性矩阵（PSSM）是两种替代多序列联配的重要的计算工具。

多序列联配也是系统发育建模的第一步。检查序列联配的每一列，接着构建最可能的系统发育关系或系统发育树，这样的系统发育关系或树基于在序列联配中观察到的改变的残基。

多序列联配的另一种用途是在一套蛋白质或 DNA 序列中的搜索模式（Stormo et al. 1982；Staden 1984，1989；Stormo 和 Hartzell 1989；Lawrence 和 Reilly 1990）。对于蛋白质序列，模式定义为结构或功能的保守组件。对于 DNA 序列，模式可能特指调控蛋白质在启动子区域的结合位点，或是 RNA 分子上的修饰信号。统计或非统计的模式搜索方法在这一领域有广泛的应用。实际上，这些方法都是排列序列，而后当序列联配后，试图在序列上定位一系列能够有最高匹配数目的连续的字符。神经网络、隐马尔科夫模型、期望极大化以及吉布斯采样的方法（Stormo et al. 1982；Lawrence et al. 1993；Krogh et al. 1994；Eddy et al. 1995）都可用于该分析。第 5 章有这些方法相关的解释和示例。

预测 RNA 二级结构的几种方法

使用计算机预测 RNA 二级结构的方法发展得也很早。在 RNA 分子中，如果在 RNA 序列的下游重复并且有化学方向相反的互补序列，这一区域可能组成碱基对，形成发卡结构，如图 1.5 所示。Tinoco（1971）等人在寡核苷酸分子中合成这样的对称结构，并且尝试着用一些方法预测其的稳定性，这些方法用能量值表估算模型中堆积的碱基对、自由能的相关性以及环结构的失稳效应值（Tinoco et al. 1971；Salser 1987）。单链环以及其他不匹配的区域都会降低预测的能量。

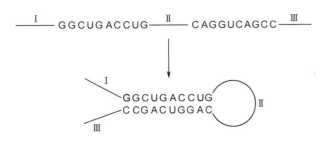

图 1.5　单链 RNA 分子折叠成发夹二级结构

图中显示了序列互补区域：它们可以通过碱基配对形成双链区域。G/C 碱基对之间由于有 3 个氢键，碱基对能量高，A/U 碱基对之间有 2 个氢键而能量次之，G/U 碱基对之间只有 1 个氢键，能量最低

随后，Nussinov 和 Jacobson（1980）设计了快速预测 RNA 分子的计算机方法，该方法与 DNA 和蛋白质序列联配的动态规划算法相同，寻找能够最多配对的碱基对。Zuker 和 Stiegler（1981）又发展这一方法，增加了生物分子限制和热力学信息以预测能量上最稳定的结构。

RNA 结构模型另一个重要的作用在于构建 RNA 分子数据库。其中 C. Woese 实验室的数据库（1987，http://www.cme.msu.edu/RDP.html/index.html）是非常有名的 RNA 二级结构数据库。第 8 章讨论了 RNA 二级结构的预测。

对 RNA 序列的比对、结构模拟和系统发育分析可帮助发现生物之间的进化关系。RNA 研究的一个重要领域是探索 RNA 在细胞中所起的重要的调节作用（Eddy　2001），以及利用特定 RNA 基因的抑制性复制产物处理基因的表达。生物信息学工具能帮助发现这些调节分子，设计抑制性功能的 RNA 分子。如果一组 RNA 分子在序列上存在差异，RNA 序列分析工具也可以根据生物分子采用的最小能量的方法预测其二级结构。生物信息分析特殊价值还在于能够分析共变序列。通过检查序列联配的每一列可以定位可形成二级结构的碱基对，以及以共同变化的方式维持二级结构的碱基对。比如，序列联配后预测得到在 RNA 分子三维结构中能够配对的区域，一条序列在第 10 列是碱基 G，在第 20 列是碱基 C，而另一条序列在这些位置可能有相同的碱基，或是其他能够配对的碱基。这些预测 RNA 结构的方法在第 8 章有非常具体的介绍。

从序列中发现进化关系

一个家族的核苷酸或蛋白质序列的差异为进化生物学提供了非常重要的信息。随着大量序列信息的出现，我们有能通过如核糖体 RNA 和一些蛋白提供的信息追溯祖先基因，重构生命之树以及基于序列发现新的物种（Barns et al. 1996）。分化的基因可能有不同的进化历史，反映了物种之间遗传物质的横向传递。进化的其他分析可以用于确定一个家族中有共同祖先的基因，即直系同源。基因复制事件创造了两个基因，称为并系同源，许多这样的事件可以形成一个基因家族，序列之间有略微不同，或可能有新的功能。一旦生成序列联配，得到联配分值，找到最相似的序列对，他们将被放置在系统发育树的叶分支上，正如图 1.1 中的序列 A 和序列 B。图 1.1 中的次相似序列 C 将被放置在系统发育树的另一分支上。不断重复这一过程可预测特定基因的系统发育模式。找到系统发育树后就可以推测位于各分支的序列之间的差异。

系统发育树构建的第一步是多序列联配。对于每一对序列，相似性分值用于指示哪些序列是最密切相关的。而后生成系统发育树，它以最优的方式解决这些相似序列对之间变化（距离）（Fitch 和 Margoliash　1987）。邻接法是基于这一分析最常用的方法（Saitou 和 Nei　1987），在第 7 章中有相关介绍。从另一方面来看，如果已经构建可信的多序列联配，在某些序列联配列中观察到的残基变化和系统发育树是一致的，所以也可以使用残基的变化构建系统发育树。这样的树选择最少的位点分析其残基改变数（最大简约法树）（Felsensten 1988）。

系统发育分析时，必须考虑不同的树有几乎相同的结果。因此，显著性检验被用于决定序列的变化在多大程度上支持已经存在的特定的树的分支（Felsenstein　1988）。这些方面的发展也在第 7 章中有所讨论。

通过搜索数据库中的相似序列发现基因功能

随着 DNA 测序技术在实验室的普及，许多有重要生物学功能的基因被测序，科学家还期望能够了解这些基因产物的生化特性。反转录病毒编码的 v-sis 和 v-src 致癌基因是其中的一个例子，它们能够在动物体内引起癌症。R. Doolittle 和他的同事（1983）以及 W. Barker 和 M. Dayhoff（1982）共同找到了令人吃惊的发现，通过将病毒产物的预测序列和已知功能的蛋白质序列相比较，这些基因似乎来源于细胞基因。Sis 蛋白质在序列上与哺乳动物的血小板源生长因子（PDGF）非常相似。Src 蛋白质与哺乳动物 cAMP-dependent 激酶的催化链相似。因此，似乎是反转录病毒通过遗传物质交换的事件从宿主细胞中获得了致癌基因，而后生成突变蛋白，当病毒感染其他动物时，这些蛋白能够损害正常蛋白质的功能。随着生物学家分析越来越多的基因序列，他们发现许多生物有相同的基因，可以通过序列的相似性来识别。

来自于模式生物已知的遗传或生化信息极大的方便了搜索工作，如大肠杆菌和芽殖酵母。对这些生物全面的遗传分析揭示了基因的功能，也确定了这些基因的序列。在一个新的生物中（如农作物植物）可以通过和模式生物（如酵母）序列的相似性寻找基因，预测新基因与模式生物的基因是否有相同功能。序列搜索已成为非常平常的方法，FASTA（Pearson 和 Lipman 1988）和 BLAST（Altschul et al. 1990）工具的使用使得这一方法更为方便。

FASTA 和 BLAST 加快了数据库搜索

随着实验室中收集的新序列的数目的增长，对序列比对的计算机程序的需求也不断增长。当时，Needleman 和 Wunsch 的动态规划的方法由于计算速度太慢而无法使用；然而现在高速的计算机已经能够实现这一方法。W. Pearson 和 D. Lipman（1988）开发了名为 FASTA 的程序，它能在非常短的时间内扫描数据库搜索相似序列。

FASTA 提供了快速搜索的方法，它在新序列和数据库的序列中寻找相似的短延伸片段。每条序列被分解成短的"字"，只有几个残基的长度，这些字被放置在表格中，并标明他们在序列中的位置。如果一个或多个字在查询序列和数据库的序列中都出现了，特别是当几个字能够相连时，序列在这一区域必定相似。Pearson（1990，1996）已经进一步发展了 FASTA 的方法以用于序列数据库的相似性搜索。

另一个更快的序列数据库相似性搜索程序是 BLAST，由 S. Altschul 等人（1990）开发。这一方法通过访问设立在美国国立医学图书馆的国家生物技术信息中心的网站（http://www.ncbii.nlm.nih.gov/BLAST）而被广泛使用。BLAST 的服务器可能是世界上使用最多的序列分析工具，它能够对最新的数据库进行相似性搜索。与 FASTA 一样，

BLAST 也为每条序列准备序列字段的列表，同时确定显著出现的字用于建立两条序列相似性的索引。正如图 1.6 中所描述的那样，BLAST 只搜索这些字。BLAST 有不同的版本用于搜索核苷酸和蛋白质序列库，可以在 DNA 序列与蛋白质序列数据库比较前首先翻译 DNA 序列

Bill Pearson

（Altschul et al. 1997）。BLAST 的最近的发展包括 GAPPED-BLAST，在速度上比最初的 BLAST 程序快 3 倍，但是会在数据库中找到更多匹配片断，另一个是 PSI-BLAST（位置特异性迭代 BLAST），它对查询序列和每次匹配的序列不断的重复搜索以增加额外搜索到的序列，该方法能够找到远缘的序列匹配。这些方法将在第 6 章讨论。

```
PORTION OF SEQUENCE A  — — W I V — —
PORTION OF SEQUENCE B  — — W I V — —
```

图 1.6　使用 FASTA 和 BLAST 快速识别序列相似性
FASTA 搜索两条氨基酸序列中的短的匹配片断，然后向两边延伸。在这里，程序通过快速简单的索引方法找到 W，I 和 V 在两条序列中以相同的顺序出现，为联配提供了一个好的开端。BLAST 以相同的原理工作，但是只检索长度为 3，更显著出现的片断，减少了随机匹配的概率

通过翻译的 DNA 序列预测蛋白质序列

蛋白质序列可以通过翻译 DNA 序列来预测，这些 DNA 序列来自于已经预测了起始和结束开放读码框的 mRNA 序列的 cDNA 拷贝。不幸的是，在数据库中 cDNA 序列不如基因组序列普遍。许多研究者在不考虑 mRNA 来源的情况下随机的对一个细胞中所有 mRNA 测序，结果形成了许多生物的序列片段（表达序列标签，ESTs）库，但是这只提供了蛋白质序列的碳末端序列，通常也只有大约 99% 的精确度。对于在基因组 DNA 序列中含有很少或没有内含子的生物（比如细菌基因组），基因组序列可以直接翻译。对于大多数真核生物，它们的基因中有内含子，必须预测蛋白编码的外显子区域后翻译，相关的方法在第 9 章介绍。基于基因组的预测并不总是精确的，因此有蛋白编码区的 cDNA 序列非常重要的。生物信息界使用大量的 EST 数据库开展这方面的工作。一个生物的 EST 序列可以定位到基因组序列上，EST 通过交迭相互联接并聚集成簇。EST 簇对于预测基因和组织中 mRNA 的选择性剪接的分析非常有用。NCBI 为每个基因组构建了一个通用的 EST 簇数据库，称为 UNIGENE 数据库。随着新数据的加入，该数据库定期进行修订，对于研究团体这是一个非常用资源。

基因组中的启动子序列可以通过模式分析反映普遍的调控特征。首先，启动子区域可以用于搜索与已知转录因子结合位点相似的模式。通过多序列联配，已知结合位点的序列变化可以用记分矩阵表示，基于与已知结合位点的相似性原理，该矩阵可为潜在的结合位点打分。TFBind（Tsunoda 和 Takagi　1999）就是这类启动子分析的工具。第二，有相同调控反应的启动子序列能够搜索到相同的模式，代表着保守的转录因子结合位点。这类分析需要复杂的方法，第 9 章中有相关讨论（Hertz et al. 1990）。

预测蛋白结构

大量的蛋白质序列是已知的，但只有很少的结构被解析。分析蛋白质结构传统上包括如 X 射线晶体衍射和核磁共振（NMR）的方法，这些方法耗费时间并且高度专用，一些自动化的方法已经应用于这些过程，并用于大量蛋白的结构分析。但是，人们有更多的兴趣尝试着基于序列预测蛋白的结构。蛋白质有氨基酸链合成；而后沿着链形成二级结构，如通过临近侧链的相互作用形成 α 螺旋。而后，生物分子中有这些二级结构区域自己前后折叠，形成 α 螺旋三级结构，由相互作用的 β 串组成的 β 折叠以及环区结构（图 1.7）。这样的折叠通常将有疏水侧链的氨基酸面向折叠分子的内部，与水以及生物分子相互作用的极性氨基酸面向外侧。氨基酸序列引导了蛋白折叠的过程，折叠有时需要分子伴侣的帮助。折叠后的蛋白有特定的生物学功能，可能在细胞中有结构上的功能或是有特定的生化活性。

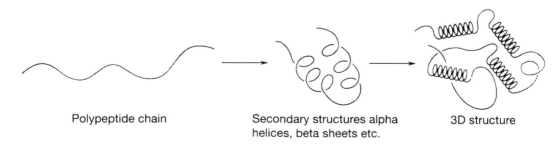

Polypeptide chain　　　Secondary structures alpha helices, beta sheets etc.　　　3D structure

图 1.7　蛋白质从线性的氨基酸链折叠成三级结构

折叠的过程包括氨基酸的相互作用。许多不同的氨基酸模式在相同的折叠结构中出现，这使得从氨基酸预测蛋白结构存在着困难为了能够发现氨基酸和各种二级结构——α 螺旋、转角、β 串之间的相关性，Chou 和 Fasman（1978）以及 Garnier 等人（1978）搜索了蛋白质结构数据库。而后，通过扫描未知结构的蛋白序列，确定每个区域的氨基酸是否与已知某类结构的序列相似。比如，脯氨酸在 α 螺旋中不会出现，因为它的侧链不利于形成螺旋结构。这一方法对于一些蛋白结构预测很好，但是，总体看来，预测正确率在 50% 左右。

随着越来越多的蛋白结构通过实验得以解决，相似的折叠结构可以用计算方法搜索（由相同的环结构连接着的二级结构的排列）。研究发现新解析的蛋白结构中经常有折叠结构是在其他序列中已知的。在蛋白中只发现大约 500 种有限数目的折叠（Chothia 1992），这可能是由于蛋白质折叠中的化学限制或蛋白结构存在单独的进化路径（Gibrat et al. 1996）。但是，没有任何序列相似性的蛋白质序列也可能采用相同的折叠，因此极大的增加了从序列预测结构的复杂性。现在也发展了基于氨基酸的特性方法（Bowie et al. 1991）来判断给定的蛋白序列是否和其他蛋白采用相同的三维构象。结构家族数据库可以通过网络访问，并且在第 10 章有相关介绍。

Amos Bairoch（Bairoch et al. 1997）开发了另一种方法，基于序列预测未知蛋白的生化活性。他收集了有相同生化活性的蛋白序列，比如具有 ATP-结合位点的蛋白，推断负责这一活性的氨基酸模式，并且允许有序列上的变化。这些模式或模体被收录在 Prosite 数据库中（http://www.expasy.ch/prosite）。

第一个全基因组序列——嗜血杆菌（*Hemophilus influenzae*）

虽然很多病毒的测序已经完成，但是首次有计划的对非寄生或共生物种的测序是由 Fred Blattner 和他的同事（Blattner et al. 1997）对大肠杆菌进行的尝试。然而，当时对于有约 4×10^6 碱基对的大基因组能否用当时现有的测序手段实现还有所疑虑。第一个发布的全基因组序列是由 Craig Vemter 研究员建立的美国马里兰州基因组研究院（TIGR，http://www.tigr.org）所测的另外一个具有单一环状染色体的细菌——流行性嗜血杆菌（*Hemophilus influenzae*）（Fleischmann et al. 1995）。这一计划获得了研究此物种多年的微生物学家 Hamilton Smith 的协助。鸟枪法的应用加快了测序的速度，该方法将染色体敲成随机片段，从 DNA 测序胶上自动读取通过染色标记的碱基对这些片段进行快速测序而不用知道这些片段在染色体上的定位。

对鸟枪法的计算机分析和测序技术在之前就已经由剑桥大学的 R. Staden 和其他工作人员提出了，不过相比之下 TIGR 的工作更具有挑战性。在这个基因组计划中，最新获得的序列立即进入计算机数据库中，借助计算机程序，相互之间进行比对，找出重复区域并在两条或多条序列之间生成重叠克隆群。这一过程避免了大量子克隆数据的产生和保存的必要。虽然相同的序列常常会被重复多达十次，但除去一些缺失区域外整个染色体的序列（1.8×10^6 bp）在计算机里迅速拼接完成，用时 9 个多月，耗资大约一百万美元。

这次测序的成功预示着大量其他真核及原核微生物的测序计划将陆续进行（最新的列表，见 genomes，网址 http://www.ncbi.nlm.nih.gov，或者 http://www.ensembl.org），就这些物种有用的基因产物及进化信息来说，测序工作具有巨大的潜在价值。迄今为止，已完成的计划包括大量原核生物（http://www.tigr.org/），如芽殖酵母 *Saccharomyces cerevisiae*（Cherry et al. 1997）、线虫 *Caenorhabditis elegans*（*C. elegans* Sequencing Consortium 1998）、果蝇 *Drosophila*（Adam et al. 2000）、拟南芥（*Arabidopsis* Genome Initiative 2000）和人类（Lander et al. 2001；Venter et al. 2001）及小鼠（http://www.ensembl.org/Mus_musculus/和 Waterston et al. 2002b）基因组。许多与人类疾病有关的物种的基因组都已测序完成，包括肺结核病原体（细菌 *Mycobacteria tuberculosis*）、梅毒（细菌 *Treponema pallidum*）和疟疾（寄生虫 *Plasmodium falciparum*）。作为疟疾的载体的蚊子的基因组测序也已经完成。

人类基因组计划

Craig Venter　　*Francis Collins*

人类基因组计划是一个由联邦基金资助的重大合作项目，于 2001 年完成了人类整个基因组的测序工作。这一计划源于 1984 和 1985 年在科学会议上讨论的一个想法，1986 年美国能源部（DOE）启动了人类基因组提案。美国国家健康研究院（NIH）在基因组研究办公室的领导下于 1987 年开始资助这一计划。当前，这一计划已经组建了国家人类基因组研究协会（NHGRI）。1998 年，Craig Venter 成立商业机构 Celera，并于 2001 年完成了对人类基因组大部分序列的测序工作。Celera 使用了计算机专家 Eugene W. Myers 的技术，使用全基因组的鸟枪克隆以及高密集计算机处理数据方法完成了果蝇和人的基因组测序。在 2002 年 NHGRJ 和 Celera 这两个组织几乎同时宣布人类基因组测序工作的完成。而后，人们一直在争论究竟 Venter 等人对公共数据的使用是否对他们绘制人类基因组草图具有显著的影响（Green 2002；Myers et al. 2002；Waterston et al. 2002a）。

第一个基因组数据库——AceDB

随着有关模式生物遗传学和序列的信息越来越多，人们开始关注建立特定的数据库来查询及获得相关信息。这项工作需要实验室之间数据与资源共享达到一个新的水平。尽管最初关于版权、信誉、准确性、编辑检查等方面还存在担忧，但随着资源在互联网上的逐步建立，这些担忧也随之消失或是解决了。AceDB 是一个基因组数据库管理系统，1989 年由 Jean Thierry-Mieg（CNRS, Montpellier）和 Richard Durbin（Sanger 研究院）开发，另外有人将其进一步开发出了 X-Window（Cherry 和 Cartinhour 1993）和 Web 交互界面（http://www.acedb.org/）。

第一个基因组数据库 AceDB（线虫数据库）、AAtDB（拟南芥数据库，现在被 TAIR［the Arabidopsis Information Resource］取代）和 SaccDB（酵母数据库，即现在的 SGB）都运用了 AceDB 系统。这些数据库可通过网络访问并且允许用户查询序列以及基因、突变、研究者联系方式及参考文献等相关信息。目前，互联网上已有相当大数量的公共数据库，在 11 章将做进一步讨论。

基因组分析方法的开发

当一个生物基因组装配完成后，首先用基因模型定位调控位点，如 RNA 剪切位点，并预测基因数目和类型。基因组的注释可以用计算机图形显示每条染色体的方式实现。如下讨论，计算机处理可以在网页上产生质量非常高的这种图片。预测基因在染色体上定位后被翻译成蛋白序列，从而得到该物种的蛋白质组。每个预测的蛋白再和其他物种的蛋白质组比较就可以地位到对应的蛋白家族（直系同源），这些蛋白家族来源于祖先基因的复制。这些蛋白还可以作为种子在蛋白数据库中检索，从而找出它是否和其他物种的某些基因同源。同源性很强的蛋白往往具有相似的三维结构和生物功能。如果在数据库中通过蛋白搜索找到的基因具有某些已知功能，那么检索蛋白也预测有相似功能。当一个物种中的蛋白往往非常特异的类似与其他物种的某个蛋白时，该蛋白具有类似的功能和结构的置信度非常高。这样的蛋白称为直系同源。

基因和蛋白关系的研究更为详尽。基因组和基因组之间的比较包括一个蛋白质组中的每条蛋白序列和另外一个蛋白质组中每条蛋白序列的比对。这种分析可以预测在物种间的同源蛋白。如果不考虑基因复制和重组的因素，这些蛋白对经过长时间的进化仍然保持很好的序列相似性，而且可以推断具有相似甚至相同的功能。在相近物种中（如哺乳类）进一步对基因功能和染色体位点的分析显示在染色体上功能基因的排列顺序是保守的，从而也提示这些基因组有共同的进化起源。有关基因组分析的详细描述见第 11 章。

通过基因芯片分析基因表达

另外一种基因组分析是在一张特制的显微片子上做一个组织或物种中大量基因的微阵列，监测一次实验某个特定时间点或者样品中所有基因的表达情况。可以将5000～10 000个基因的 mRNA 片段点在玻璃片上（spotted array）或者直接合成在上面（Affymetix 方法）。点样法中，从组织或细胞中提出 mRNA，然后合成与这些 mRNA 相对应的 DNA 并用荧光燃料染色，这样每个基因的染色量就和相应的细胞 mRNA 的量成正比。一般要准备两个不同染色的样品，一个是信号样品，另一个是背景样品。而后将两种样品混合，再杂交到具有探针的玻璃片上，这样两个样品的 RNA 就都杂交到了互补的探针上。接着对这张片子扫描就可以测量每个 DNA 点上杂交样的品量有多少。扫描设备将每个 DNA 点看很多像素点的集合并检验每个像素点上的染料的总量。如果不同点的比率均一并且样点显示了合理的变化，这些数据就认为可以用作后续的分析。与参照样本相比就以分析出那些基因的表达水平升高或者降低了，或是维持在相同表达水平（详细信息可参考 Bowtell 和 Sambrook 2003）。图 1.8 就表示这种分析。

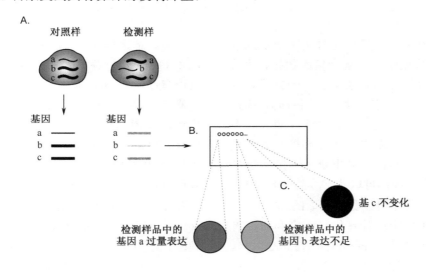

图 1.8 DNA 点样芯片

（A）把参考样本和检测样本里的 RNA 抽提出来，反转录成 cDNA，然后每张芯片都用黑色和红色的荧光染料标记，使得染料标记量差不多跟 RNA 量成正比。（B）这些 DNA 与显微芯片上的基因陈列杂交。（C）芯片扫描后，测量与每个基因杂交的每种标记的量，并随后转化成两种标记的比值。这个比值一般用颜色来表示。在本例中，测试标本中表达上调的 *a* 基因是红色的，而表达下调的基因 *b* 呈现绿色。粗线表示较多的 mRNA

Affymetix 方法与 DNA 点样法略有不同。该方法直接在玻片上合成来自每个基因的一些短片段，同时在相同的对照片段中部加入一个突变以减弱杂交程度。只有一个染料标记的样品能够杂交。因此，要比较样品和对照，必须比较各自的杂交量。

在精心的实验设计下微阵列技术可以产生非常有用的信息。例如，它们曾用于显示芽殖酵母在细胞分裂时期的基因表达模式，或者用于显示不同人类肿瘤发生组织的基因表达模式。但是这些数据的分析非常困难。首先，即使人们在预算内可以重复多次实验，每个基因仍会由于有很多因素可以产生统计上变异从而导致不确定性。其次，因为很多基因的功能及基因之间的互相作用还不清楚，所以很难对得到的表达模式做出合理的解释。再次，分析过程往往要处理大量的数据集。微阵列技术给统计学提出了很多有意思的挑战，这将在第 13 章详细讨论。

对蛋白质的生化分析也可以提供一些关于细胞中基因表达的信息。相比较微阵列技术，蛋白分析可以发现哪些基因翻译成了活性蛋白，这些蛋白分布在哪里，以及他们是如何进行翻译后修饰的。现阶段一般通过电泳或者质谱分析来分离鉴定蛋白。

大量生物数据中的数据存储和挖掘技术

运用基因技术对基因组分析，蛋白质分析，微阵列分析以及遗传学方法检测到的基因功能能够搜集大量的数据。但是系统的保存这些数据对于大部分实验室来说都是非常困难。一般来说，为了方便使用，数据都是保存在电子表格中。但是这些数据应该用一种比较合适的数据库格式来统筹安排。在很多应用领域如药学、农学、药剂学等研究上数据库的存储格式是非常重要的，因为统筹的存储格式非常有利于检验大规模数据的相互关系。统筹的数据也有利于通过数据挖掘发现新的数据间的未知关系，例如分析人群发病群体中变异和多态性的关系。数据可以用数据库技术组织成"多维立方体"的形式，从而可以方便的找出哪些因素的组合能够产生特定的结果，如疾病发生的概率等。这些技术已经开发并且成功的用于商业领域，同样我们也可以把他们用于大规模生物数据的挖掘。

通过 BioPerl 和互联网资源实现自动序列分析

许多实验室面临的一个巨大困难是需要从网络的各个站点及序列数据库中收集大量的数据，然后再重复的分析这些数据。如果采用手工的方式，科研人员必须先访问一个站点，输入序列号码，选择序列格式，再拷贝这些序列到另外一个文件中，去除序列其中的一部分，翻译这些序列等等。大量重复这些过程费时费力，而且容易出错。不过幸运的是，对序列的这些重复操作可以非常容易的自动化，而且有些必要的工具还是可以免费获取的。

这类工具来源于由很多计算机科学家和研究者组成的资源开放团体（open-source community），他们遵循一个共同的协议（the GNU General Public Licese, http://www. gnu. org），研发并分享很多计算机软件。另外是由国际生物信息学科研团体研发的 BioPerl 工程。这个团体开发了一系列小的程序模块，称为 BioPerl 模块，它们可用于各种实验室常见的序列分析（http://www. bioperl. org）。例如，在本地运行 perl 程序中的 BioPerl 模块 Bio∷DB∷GenBank，并输入一系列序列号，这个模块就可以自动从 GenBank 中检索这些序列并把他们按一定格式储存。Perl 程序还可以调用其他的 BioPerl 模块，如 BLAST 或者序列分析等模块。EMBOSS（http://www. hgmp. mrc. ac. uk/Software/EMBOSS）开发出了一系列可以用 Perl 中运行的开放程序（Rice et al. 2000）。通过 CGI. pm（公共网关交互的 perl 模块，http://stein. cshl. org/WWW/software/CGI/）就可以简单的通过网站来使用 perl 程序，只要输入序列信息并提供一些选项它就可以给出规则的输出结果。这些科研机构还用 BioPerl 的模块生成漂亮的互动性的网页。这些特点在 linux、Unix 或者 Mac OS X 系统中得到很好的体现。资源开放团体（open-source community）还提供了完善的技术支持及相关介绍文档。第 12 章对这些方法在实验室的应用有更加详细的描述。

网络上的资源

对于实验室中序列的分析，互联网上大量的优秀站点提供了非常好的资源。除了一些主要的序列和基因组分析站点外，其他很多的站点会随着时间发生改变，这样网上的一些链接就需要经常变化。不过网络搜索引擎可以解决这个问题。在每章的最后都提供了关于网络生物信息资源的搜索引擎的术语以用于搜索引擎。在搜索引擎中输入检索词就可以找到相应可用的连接。

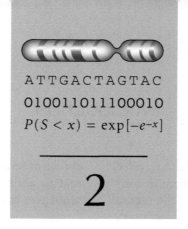

ATTGACTAGTAC
010011011100010
$P(S < x) = \exp[-e^{-x}]$

2

Collecting and Storing Sequences in the Laboratory

C O N T E N T S

INTRODUCTION

This chapter summarizes methods used to collect sequences of DNA molecules and store them in computer files. Procedures ranging from the actual sequencing, through determination of accuracy, choice of sequence format, conversions from one format to another, storage in databases, and accessing sequences in databases are described.

CHAPTER GUIDE FOR BIOLOGISTS

Biologists are familiar with collecting DNA sequences in the laboratory but have less familiarity with the storage and organization of the sequences into databases. A rational storage scheme is particularly important when collecting sequences of large genomes such as plant and animal genomes. Each laboratory sequence has a unique identifier or accession number. Computer programs are then used to join these sequences into longer contigs. When overlapping sequences are identified, the sequences can be combined into longer contigs, e.g., the sequence of a cDNA, a BAC clone or a chromosome, and are given a new identifier. Assembling sequences can involve editing of any variations that may be present in the individual sequences to reach a reliable consensus sequence. The assembled sequences are then annotated to show the positions of promoter sequences, open reading frames, intron and exon start positions, and other important features of the sequences. This information is integrated into each sequence entry in the sequence databases, such as GenBank.

Web sites provide access to mouse-clickable genome maps that can provide links to the genome and cDNA and original clone sequences. An individual sequence entry includes annotation information, literature references, identifiers of related sequences, and mouse-clickable links to this information. This information is organized in a specific format comprising specified fields so that a computer search tool can search one field for a particular word or words. Tools like NCBI's Entrez provide a way to search for combinations of terms in separate specific fields. These tools are important, and familiarity with them is an important step in student training.

Instead of using Web sites involving human access to sequence or genome databases, an automated search method may be used. In this case, a computer program that goes to the sequence database Web site and retrieves one or more sequences is run on a local computer. Programming modules that greatly simplify this process are described in Chapter 12.

CHAPTER GUIDE FOR COMPUTATIONAL SCIENTISTS

This chapter describes the collection of DNA sequences in the laboratory and their subsequent use by biologists. DNA is composed of four chemical bases—A (adenine), G (guanine), C (cytosine), and T (thymine)—joined in long strings up to hundreds of millions of bases long. The DNA molecule itself is made up of two strings wound around each other in a double helix. These strings are complementary, in that A in one strand is always paired with T in the other, and G is always paired with C. As a result, it does not matter which strand is sequenced because the sequence of the other strand can be derived. This complementarity is also the basis for the reproduction, or replication, of the DNA molecule in organisms in which the double-stranded molecule separates and a complementary copy of each strand is then made from the template strand. Sequencing projects involve obtaining multiply redundant, overlapping sequences about 500 bases long from both strands of a given DNA molecule. Computer scientists are primarily concerned with assembling these fragments into longer sequences called contigs.

Most often the DNA sequences to be manipulated are those of the genome or the expressed genes of an organism. The genome is the collection of DNA sequences of an organism that is packaged into microscopically visible objects called chromosomes. Chromosomes are the genetic material of an organism that is passed on from one generation to the next. Certain regions on the chromosome are read by the cell from one DNA strand into messenger RNA (mRNA), a single-stranded molecule also composed of four bases—A, G, C, and U (uracil)—where U pairs as T does in DNA. In many organisms, some sequence regions called introns are spliced out of the mRNA leaving behind the exons, which carry sequence information for encoding a protein. Three consecutive positions in the spliced exons (64 possible groups of three, called codons) specify an amino acid sequence for a colinear protein molecule. Consecutive sets of codons that encode amino acids are called an open reading frame and are read until one of three stop codons indicating the end of the gene sequence is reached. More information on this process is provided in the introduction to Chapter 1.

mRNA molecules are collected from cells by biologists, and their sequences are copied biochemically into more chemically stable DNA molecules (complementary DNA or cDNA) using base-pairing similar to that in the DNA molecule. Often a random, redundant set of these molecules is sequenced to give a set of sequence fragments called expressed sequence tags (ESTs). Full-length cDNA sequences are then obtained by locating the overlapping, similar, or complementary sequences and joining them into contiguous sequences. The protein sequence is then readily obtained by reading the codon sequence from the cDNA sequence using a table of known codon–amino acid matches.

Once the genome and protein sequences have been collected, biologists then attempt to discover the biological functions of the genes. This necessitates organizing the data into annotation databases that can be readily queried on Web sites. It is very useful to provide interactive maps that show relative positions of genes on chromosomes with links to the sequence fragments and supportive biological data.

WHAT SHOULD BE LEARNED IN THIS CHAPTER?

- How DNA sequences of organisms are collected in the laboratory.
- The biological sources and relationships among genome, mRNA, cDNA, EST, and protein sequences.
- The process of assembling gene and chromosome sequences from laboratory data.
- How sequences are organized in databases and annotated with information on the location and function of genes.
- The location and use of the major sequence databases.
- How to use computer search tools for retrieving a particular sequence from the sequence databases.

Glossary Terms

BAC refers to a bacterial artificial chromosome containing one chromosomal fragment about 100 kb in length from the target organism in a cloning project. Enough BACs are made to produce a redundant set of overlapping fragments from which the original chromosomal sequence may be derived.

cDNAs, or complementary DNAs, are DNA copies of mRNA molecules that are made in the laboratory using the mRNA sequences as templates. cDNAs are then sequenced to reveal the mRNA sequence, which in turn is translated to reveal the protein sequence.

Check sum value is a number placed within a sequence file that is uniquely obtained by adding the values of each of the sequences. The values (ASCII) are the binary values of each sequence character. If the sequence has been transmitted and stored correctly to the local computer, and a local program adds up the sequence values, then the check sum value for the sequence should be the same as originally recorded in the file.

Clone, in sequence collecting, refers to inserting a DNA fragment isolated from the chromosome of an organism that is being sequenced into another molecule called a cloning vector, whose sequence behaves as a type of artificial chromosome that is capable of replicating into many copies, usually in a bacterial cell. Many identical copies of the cloned fragment are then produced for sequencing or other laboratory experiments.

Cloning vector is a molecule that is capable of independently multiplying in a cell, in much the same manner as a chromosome. When a fragment of foreign DNA is inserted into the vector, the foreign DNA then is copied in the cell along with the vector at each round of multiplication, thus giving rise to a clone of the foreign DNA.

Contigs are collections of overlapping sequences that are obtained in a sequencing project. They can be found by automatically aligning random sequence fragments to look for overlaps or by using positional information from cloning experiments as a guide to alignment. Contigs may be that of a chromosome or an individual gene, depending on the type of sequence.

Controlled vocabulary is a set of terms used in a database to describe a particular biological object or process. Use of these terms avoids confusion when describing the same type of biological object or process in two different databases.

CORBA, or Common Object Request Broker Architecture, is an industry open standard for working with data and computer programs that are used for analysis of these data (called distributed data objects) on different machines, different computer languages, or in different locations. These distributed data objects are developed by the Object Management Group.

Cosmid is an extra bacterial chromosome into which fragments of foreign DNA of about 40 kb in size from various organisms can be conveniently inserted by recombinant DNA methods. The cosmid also has sequences that allow it to be packaged into the protein coat of a bacterial virus (a phage) which greatly facilitates the introduction of the cosmid into bacteria. Thereafter, the cosmid is replicated each time the bacterium divides, thus creating new copies of the cosmid, including the foreign DNA clone.

DNA is a double-stranded, helical molecule made up of sequences of four nucleotides—A, G, C, and T in each strand—and backbones of sugars and phosphates. A in one strand is always paired with T in the other, and G is always paired with C; the interaction of these nucleotide pairs holds the strands together. The sequence of nucleotides in DNA molecules is read by DNA sequencing machines 500–800 at a time.

Entrez is a highly interactive database search program that is available on the National Center for Biotechnology Information(NCBI) Web site and may be used for retrieving DNA and protein sequences as well as for literature searches on PubMed.

ESTs, or expressed sequence tags, are partial sequences of cDNA copies of mRNA molecules. They are often sequenced randomly from different sequence locations in a large collection of cDNA sequences from one biological source. ESTs are assembled into contiguous sequences that represent individual cDNA sequences and the mRNA sequences from which the cDNAs were derived.

Genome is the entire set of DNA sequence information for an organism that is passed along from one generation to the next.

mRNA is a single-stranded, chemically unstable molecule comprising A, G, C, and U (like T) that is read by the cell from one strand of the genome sequence in a particular chemical (5′ to 3′) direction. The molecule is often spliced to remove introns, leaving the exons that carry the information for the amino acid sequence of a protein in the form of an open reading frame comprising three-base words (codons). mRNAs also include regulatory sequence information at the ends of the molecule.

Object-oriented database is a database in which data are not stored in tables, as in relational databases; rather, data are stored as objects with space reserved for the data but with additional information as to how the data can be analyzed. This type of database facilitates the modeling and sharing of complex data between different locations. It is related to object-oriented programming, which creates computer program modules that perform a specific function in data analysis.

PCR, or polymerase chain reaction, is a method for amplifying a specific region of a DNA fragment using DNA polymerase and short primer sequences to delimit the amplified region. The amplified region can be cloned or the amplification process can produce a sequence of the region, thus circumventing the need for cloning.

Phage, or bacteriophage, is a virus that attacks bacteria by injecting its nucleic acid molecule into the cell. In recombinant DNA technology and sequencing, foreign DNA is inserted into phage λ DNA, which then behaves as a cloning vector. As the phage replicates itself in bacterial cells, the foreign DNA is also replicated.

Plasmid is a small extra chromosome in bacteria that can carry up to 1000 bases of foreign DNA in a convenient size for sequencing.

Relational database is a database that organizes information into tables in which each column represents one type of information that can be stored in a single record. Each row in the table corresponds to a single record. A single database can have many tables, and a query language, such as SQL, is used to access the data.

Shotgun sequencing is the procedure used to obtain the

sequence of long DNA molecules by breaking them into random fragments, sequencing the fragments without regard to their position in the original sequence, and then assembling the original sequence based on overlaps among the fragments.

YAC, or yeast artificial chromosome, is a type of cloning vector into which large fragments of foreign DNA of megabase size can be inserted using recombinant DNA technology. The foreign DNA is then amplified as the yeast cells containing the YAC divide.

DNA SEQUENCING IS AN AUTOMATED PROCEDURE

DNA sequences with an approximate length of 500 bases are produced in the laboratory. Because most DNA molecules in organisms are much longer than 500 bases, a strategy is needed to fragment these molecules into shorter lengths, read these shorter sequences, and then assemble the sequence of the original molecule from these fragments. The large, linear DNA sequences of plant and animal chromosomes are usually first broken down into more manageable lengths of approximately 100,000 bases and stored as inserts in a redundant set of artificial bacterial chromosomes known as BACs. Each BAC sequence is a separate clone of a fragment of the original chromosome. BAC sequences are then further subcloned into even smaller fragments using cosmid, plasmid, and phage cloning vectors prior to sequencing. Hundreds of overlapping sequence reads from each cosmid are then obtained and assembled into contiguous sequence lengths called contigs. Overlapping sequences are then further assembled into full-length chromosome sequences of up to hundreds of millions of bases. The frequently single, circular chromosomes of one-celled microorganisms, such as bacteria, with approximate lengths of millions of bases, are assembled in a similar manner. Once all of the chromosomes of an organism have been sequenced and assembled in this manner, the resulting entire collection of sequences constitutes the genome of that organism.

Because it is difficult to identify accurately the specific chromosomal sequences that code for proteins, and because these sequences may vary in different tissues of an organism, sequences of individual genes that specify the amino acid sequence of individual proteins are also obtained in the laboratory. The method involves making DNA sequence copies called complementary DNA or cDNA copies of the messenger RNA (mRNA) molecules that are present in cells or tissues and then sequencing these cDNA

molecules. In this instance, individual overlapping sequence reads called expressed sequence tags (ESTs) are made from cDNA molecules and then assembled into contigs. Additional sequences may be obtained in the laboratory to fill in gaps. Each resulting cDNA sequence will include a series of codons in the 5′ to 3′ chemical direction that specify the sequence of a protein with extra regulatory sequences on the upstream 5′ and downstream 3′ ends of the sequence. Any tissue-to-tissue variation should also become apparent provided that the individual tissue cDNA sequences are available. The cDNA sequence can then be aligned with the chromosomal sequence to locate the gene on the chromosome.

Assembled sequences are edited to a level of accuracy of 99.99% (1 error or less per 10,000 positions) and stored in a computer file along with identifying features, such as DNA source (organism), gene name, and investigator. Sequences and accessory information are then entered into a database. This procedure organizes the sequences so that specific ones can be retrieved by a database query program for subsequent use. There are a number of different sequence formats in use. Most sequence analysis programs require that the information in a sequence file be stored in a particular format. To use these programs, it is necessary to be aware of these formats and to be able to convert one format to another. Once entered into a computer, DNA sequences can be analyzed for function by a variety of methods that are described in the remainder of the chapters that follow, the most common of these being to identify the function of each encoded protein.

The stored DNA sequences of an organism usually represent a single sequence averaged over several individuals, strains, or races. But as most of us realize, individual-to-individual variation within a population in biological features such as physical appearance, stress resistance, and disease susceptibility is due to sequence variation within the genome. Thus, the final stage of sequence collection is to locate and identify those regions that determine these types of genetic variabilities.

Sequencing DNA has become a routine, automated task in the molecular biology laboratory. Purified fragments of DNA cut from clones of DNA fragments or amplified by polymerase chain reaction (PCR) from clones are denatured to single strands, and one of the strands is hybridized to an oligonucleotide primer. In an automated procedure, new strands of DNA are synthesized from the end of the primer by heat-resistant enzyme *Taq* polymerase from a pool of deoxyribonucleotide triphosphates (dNTPs) that includes a small amount of one of four chain-terminating nucleotides (ddNTPs). For example, if ddATP is used, the resulting synthesis creates a set of nested DNA fragments, each one ending at one of the As in the sequence through the substitution of a fluorescent-labeled ddATP, as shown in Figure 2.1. A similar set of fragments is made for each of the other three bases, but each is labeled with a different fluorescent ddNTP.

The combined mixture of all labeled DNA fragments is electrophoresed to separate the fragments by size, and the ladder of fragments is scanned for the presence of each of the four labels, producing data similar to those shown in Figure 2.2. A computer program is then used to determine the probable order of the bands and to predict the sequence. Depending on the actual procedure being used, one run may generate a reliable sequence of as many as 500 nucleotides. For accurate work, a printout of the scan is usually examined for abnormalities that may decrease the quality of the sequence, and the sequence may then have to be edited manually. The sequence can also be verified by making an oligonucleotide primer complementary to the distal part of the readable sequence and using it to obtain the sequence of the complementary strand on the original DNA template. The first sequence can also be extended by making a second oligonucleotide matching the distal end of the readable sequence and using this primer to read more of the original template. When the process is fully automated, a number of priming sites may be used to obtain sequencing results that give optimal separation of bands in each region of the sequence. By repeating this procedure, both strands of a DNA fragment several kilobases in length can be sequenced (Fig. 2.3).

GENOME SEQUENCING IS DONE USING TWO MAIN METHODS

The first genome sequencing method is an approach based on the known order of DNA fragments on a chromosome. To sequence larger molecules, such as human chromosomes, individual chromosomes are broken into random fragments of approximately 150 kb using an enzyme that cuts DNA. The fragments

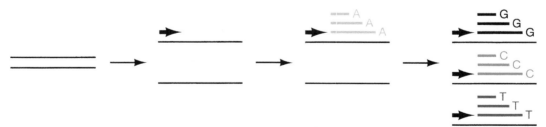

FIGURE 2.1. Method used to synthesize a nested set of DNA fragments, each ending at a base position complementary to one of the bases in the template sequence. To the left is a double-stranded DNA molecule several kilobases in length. After denaturation, the DNA is annealed to a short primer oligonucleotide primer (*black arrow*), which is complementary to an already sequenced region on the molecule. New DNA is then synthesized in the presence of a fluorescently labeled chain-terminating ddNTP for one of the four bases. ddNTP can be added to a growing DNA molecule but the molecule then stops growing. The reactions produce a nested set of labeled molecules. The resulting fragments are separated in order by length to give the sequence display shown in Fig. 2.2.

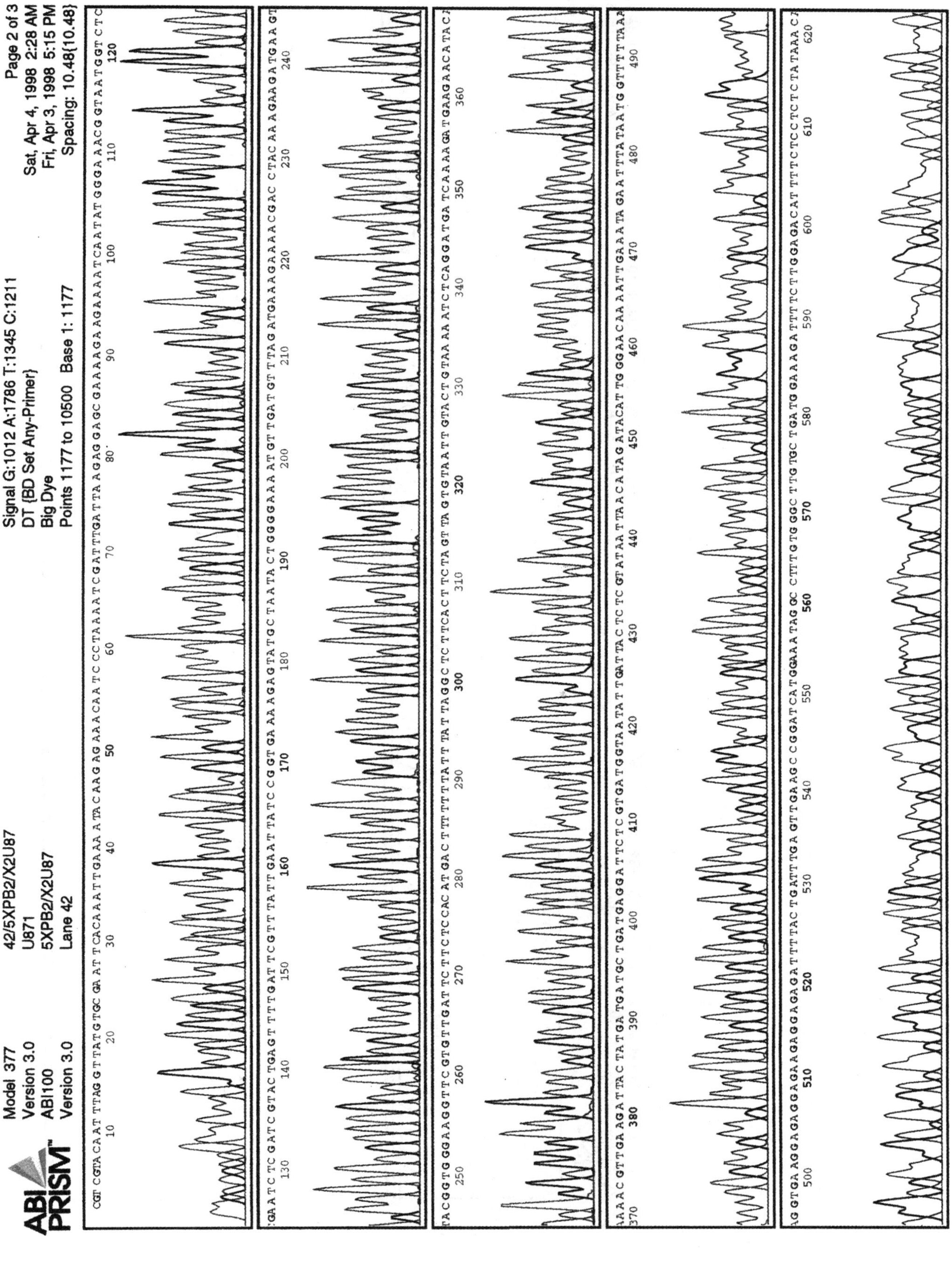

FIGURE 2.2. *See legend on p. 37.*

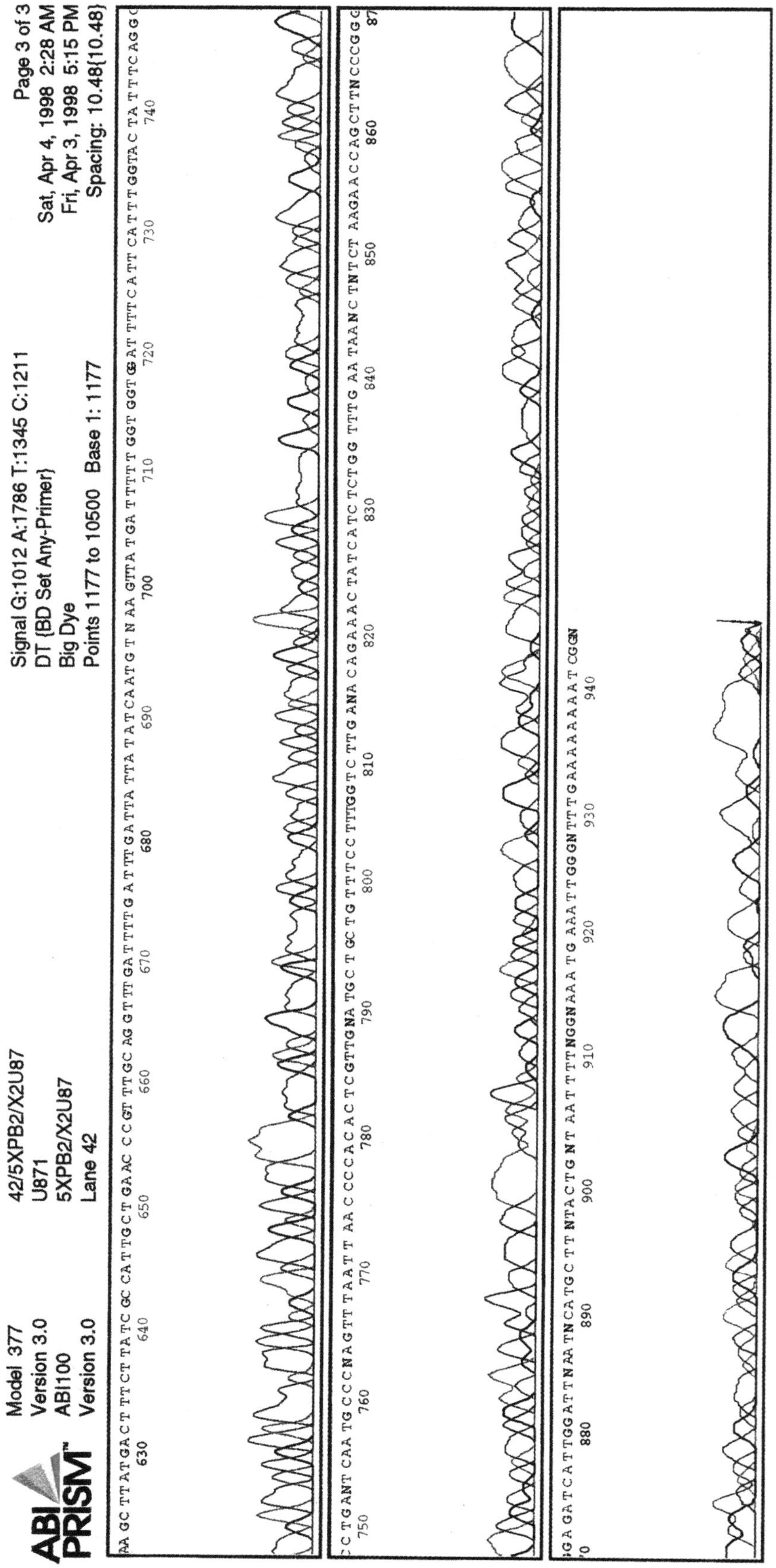

Model 377
Version 3.0
ABI100
Version 3.0

42/5XPB2/X2U87
U871
5XPB2/X2U87
Lane 42

Signal G:1012 A:1786 T:1345 C:1211
DT {BD Set Any-Primer}
Big Dye
Points 1177 to 10500   Base 1: 1177

Page 3 of 3
Sat, Apr 4, 1998 2:28 AM
Fri, Apr 3, 1998 5:15 PM
Spacing: 10.48{10.48}

AA GC TT AT GA CT TT CT T ATC GC CA TT GC T GA AC C C GT TT GC AG GT TT GA T TTT G A T TT GA TTA TTA TA TC AAT G T N AA GTTA T GA T TTT T GGT GGT GG AT TTT C ATT CA TT T GGTA C TA T TT C AGG C
630   640   650   660   670   680   690   700   710   720   730   740

C C TG ANT CAA TG C CC NA GTT TAAT T AA C C CC AC AC T C GTT GNA TGC T GC TG T TT C C TTT GGT C TTG ANA C AG AA AC T ATC ATC TC TGG TTT G AA TAAN C TNT C T AAG AAC C AG CT TN C CC GG G
750   760   770   780   790   800   810   820   830   840   850   860   87

GGA GA TC A TT GG A TT NA A TN CA TG C TT NT ACT GN T AAT TTT NGG NA AA TG AAA TT GGG N TTT GAA AA A AA AT C GG GN
0   880   890   900   910   920   930   940

```
  1  CGTCGTACAA TTTAGGTTAT GTGCGAATTC ACAAATTGAA AATACAAGAG AAACAATCCC TAAAAATCGAT TTGATTAAGA GGAGCGAAAA   90
 91  GAAGAAAATC AATATGGGAA ACGGTAATGG TCTCGAATCT CGATCGTACT GAGTTTTGAT TCGTTTATTG AATTATCCGG TGAAAAGAGT  180
181  ATGCTAATAC TGGGAAAAT  GTTGATGTTT AGATGAAAGA AAACGACCTA CAAAGAAGAT GAAGTACGGT GGGAAGGTTC GTGTTGATTC  270
271  TTCTCCACAT GACTTTTTTT ATTTATTAGG CTCTTCACTT CTAGTTAGTG TAATTGTACT GTAAAAATCT CAGGATGATC AAAAGATGAA  360
361  GAACATACAA AACGTTGAAG ATTACTATGA TGATGCTAAT GTGATGGTAA TATTGATTAC TCTCGTATAA TTAACATAGA  450
451  TACATTGGGA ACAAAATTGA ATAATGGTTT TTAAAGGTGA AGGAGAGAG  ATTTTTACTGA TTTGAGTTGA  540
541  AGCCGGATCA TGGAAATAGG CCTTTGTGGG CTTGTGCTGA TGGAAAGATT TTCTTGGAGA CATTTTCTCC TCTCTATAAA CAAGCTTATG  630
631  ACTTTCTTAT CGCCATTGCT GAACCCGTTT GCAGGTTTGA TTTTGATTTG ATTATTATAT CAATGTNAAG TTATGATTTT TGGTGGTGGA  720
721  TTTTCATTCA TTTGGTACTA TTTCAGGCCT GANTCAATGC CCNAGTTTAA TTAACCCCAC ACTCGTTGNA TGCTGCTGTT TCCTTTGGTC  810
811  TTGANACAGA AACTATCATC TCTGGTTTGA ATAANCTNTC TAAGAACCAG CTTNCCCGGG GAGATCATTG GATTNAATNC ATGCTTNTAC  900
901  TGNTAATTTT NGGNAAATGA AATTGGGNTT TCGGN  990
```

FIGURE 2.2. (*Continued*) Example of a DNA sequence obtained on an ABI-Prism 377 automated sequencer. The target DNA is denatured by heating and then annealed to a specific primer. Sequencing reactions are carried out in a single tube containing Amplitaq (Perkin-Elmer), dNTPs, and four ddNTPs, each base labeled with a different fluorescent dichloro-rhodamine dye. The polymerase extends synthesis from the primer, until a ddNTP is incorporated instead of dNTP, terminating the molecule. The denaturing, reannealing, and synthesis steps are recycled up to 25 times, excess labeled ddNTPs are removed, and the remaining products are electrophoresed on one lane of a polyacrylamide gel. As the bands move down the gel, the rhodamine dyes are excited by a laser within the sequencer. Each of the four ddNTP types emits light at a different wavelength band that is detected by a digital camera. The sequence of changes is plotted as shown in the figure, and the sequence is read by a base-calling algorithm. More recently developed machines allow sequencing of 96 samples at a time by capillary electrophoresis using more automated procedures. The accuracy and reliability of high-throughput sequencing have been much improved by the development of the phred, phrap, and consed systems for base-calling, vector sequence removal, sequence assembly, and assembled sequence editing (Ewing and Green 1998; Gordon et al. 1998).

FIGURE 2.3. Sequential sequencing of a DNA molecule using oligonucleotide primers. One of the denatured template DNA strands is primed for sequencing by an oligonucleotide (*gold*) complementary to a known sequence on the molecule. The resulting sequence may then be used to produce two more oligonucleotide primers downstream in the sequence, one to sequence more of the same strand (*pink*) and a second (*blue*) that hybridizes to the complementary strand and produces a sequence running backward on this strand, thus providing a way to confirm the first sequence obtained.

are then cloned into vectors designed for large molecules, such as BACs. In an intensive, but largely automated, laboratory procedure, the resulting library is screened for clusters of fragments called contigs, which have overlapping or common sequences. These contigs are then joined to produce an integrated physical map of the genome based on the order of the BAC fragments. Many levels of clone redundancy may be required to build a consensus map because individual clones can have rearrangements, deletions, or two separate fragments. These changes are spurious and do not reflect the correct map, and thus have to be eliminated. Once the correct map has been obtained, unique overlapping clones are chosen for sequencing. However, these molecules are too large for direct sequencing. One procedure for sequencing these clones is to subclone them further into smaller fragments that are of sizes suitable for sequencing, make a map of these clones, and then sequence overlapping clones (Fig. 2.4). This method is expensive because it requires a great deal of time to keep track of all the subclones. This method of creating physical maps of genomes and then using this map to guide the sequencing of individual sequence regions was used by the Public Human Genome Consortium to create a draft of the human genome. The method had also been used earlier to sequence the yeast and *Caenorhabditis elegans* (worm) genomes (Olson 2001). This carefully crafted but laborious procedure was designed to produce a sequence of the human genome that is based on a top-down approach, at each stage using a physical map to guide the placement of sequences. The reasoning behind this strategy was avoidance of sequence repeats that might otherwise confound obtaining the correct genome sequence.

A contrasting bottom-up method in which the genome sequence was derived solely from overlaps in large numbers of random sequence without using the physical map as a guide has been devised. This alternative method, called shotgun sequencing, attempts to assemble a linear map from subclone sequences without knowing their order on the chromosome. In the shotgun method, contigs are assembled based on alignments of all possible sequence pairs in the computer. Up to ten levels of redundancy are used to get around the problem of a small fraction of abnormal clones. This procedure was first used to obtain the sequence of the 1.8-megabase-pair chromosome of the bacterium *Haemophilus influenzae* by The Institute of Genetics Research (TIGR) team (Fleischmann et al. 1995). Only a few regions could not be joined because of a problem subcloning those regions into plasmids, requiring manual sequencing of these regions from another library of phage subclones. This method is now routinely used to sequence microbial genomes and the cloned fragments of larger genomes.

Shotgun sequencing has since been used to obtain the genome sequences of many bacterial genomes. However, a controversy arose as to whether or not shotgun sequencing could be applied to genomes with repetitive sequences such as those likely to be encountered in the human genome (Green 1997; Myers 1997). The reason for the controversy is that when DNA fragments derived from different chromosomal regions have repeats of the same sequence, they will appear to overlap, thus giving an incorrect genome sequence. The shotgun method was used by Celera Genomics to assemble the genome of the fruit fly *Drosophila melanogaster* after removal of the most highly repetitive regions (Myers et al. 2000).

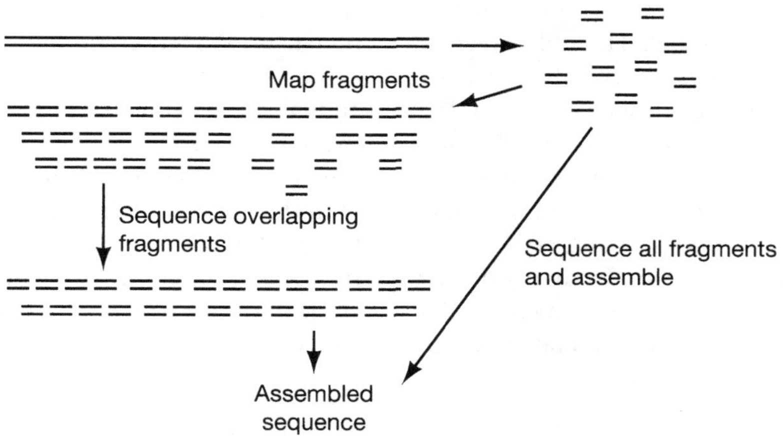

FIGURE 2.4. Methods for large-scale sequencing. A large DNA molecule 150 kb to several megabases in size is randomly sheared by a DNA cutting enzyme and cloned into a cloning vector, usually a bacterial artificial chromosome (BAC). In one method, overlapping fragments are identified. The resulting map of the fragments is called a physical map, which is used to guide the sequencing of the whole genome. This method was used by the public international Human Genome Consortium to produce a draft of the human genome (Lander et al. 2001; Olson 2001). In the whole-genome shotgun method, DNA fragments of different size ranges are sequenced without knowledge of their chromosomal location, and the sequence of the parent molecule is assembled in several stages from any overlaps found. This method was used by Celera Genomics to produce a draft of the human genome (Venter et al. 2001).

In another approach for avoiding problems with repeats, Celera Genomics sequenced both ends of DNA fragments of short (2 kb), medium (10 kb), and long (~100 kb) BAC lengths and then used overlaps between longer fragments to assemble contigs that circumvent repeats.

In 2001, both the Public Human Genome Consortium, which used the physical map as an assembly guide, and Celera Genomics, which used whole-genome shotgun sequencing, published drafts of the human genome (Lander et al. 2001; Venter et al. 2001). There has since been controversy as to whether or not use of the public data by the Venter group contributed significantly to their draft of the human gene sequence (Green 2002; Myers et al. 2002; Waterston et al. 2002b).

SEQUENCING cDNA LIBRARIES OF EXPRESSED GENES IDENTIFIES PROTEIN-ENCODING REGIONS

Two common goals in sequence analysis are to identify sequences that encode proteins, which determine all cellular metabolism, and to discover sequences that regulate the expression of genes or other cellular processes. Genomic sequencing as described above meets both goals. However, only a small percentage of the genomic sequence of many organisms actually encodes proteins because of the presence of introns within coding regions and other noncoding sequences in the genome. Although there has been a great deal of progress in developing computational methods for analyzing genome sequences to find these protein-encoding regions (see Chapter 9), these methods are not completely reliable and, furthermore, such genomic sequences are often not available. Therefore, cDNA libraries have been prepared that have the same sequences as the mRNA molecules produced by organisms, or else cDNA copies are sequenced directly by reverse transcriptase (RT)-PCR (copying of mRNA by reverse transcriptase followed by sequencing of the cDNA copy by PCR).

By using cDNA sequences with the introns

removed, it is much simpler to locate protein-encoding sequences in these molecules. The only possible difficulty is that a gene of interest may be developmentally expressed or regulated in such a way that the mRNA is not present except under unusual circumstances. This problem has been circumvented by pooling mRNA preparations from tissues that express a large proportion of the genome, from a variety of tissues and developing organs, or from organisms subjected to several environmental influences. An important development for computational purposes

was the decision by Craig Venter to prepare databases of partial sequences of the expressed genes, called expressed sequence tags or ESTs, which have just enough DNA sequence to give a pretty good idea of the protein sequence. The translated sequence can then be compared to a database of protein sequences with the hope of finding a strong similarity to a protein of known function, and hence to identify the function of the cloned EST. The corresponding cDNA clone of the gene of interest can then be obtained and the gene completely sequenced.

SUBMITTING SEQUENCES TO THE DATABASES IS EASY

Investigators are encouraged to submit their newly obtained sequences directly to a member of the International Nucleotide Sequence Database Collaboration, such as the National Center for Biotechnology Information (NCBI), which manages GenBank (http://www.ncbi.nlm.nih.gov); the DNA Databank of Japan (DDBJ; http://www.ddbj.nig.ac.jp); or the European Molecular Biology Laboratory (EMBL)/EBI Nucleotide Sequence Database (http://www.embl-heidelberg.de). NCBI reviews new entries and updates existing ones, as requested. A database accession number, which is required to publish the sequence, is provided. New sequences are exchanged daily by the GenBank, EMBL, and DDBJ databases.

The simplest way of submitting sequences is through the Web site http://www.ncbi.nlm.nih.gov/ on a Web form page called BankIt. The sequence can also be annotated with information about the sequence, such as mRNA start and coding regions. The submitted form is transformed into GenBank format and returned to the submitter for review before it is added to GenBank. The other method of submission is to use Sequin (formerly called Authorin), which runs on personal computers and UNIX machines. The program provides an easy-to-use graphic interface and can manage large submissions such as genomic sequence

information. It is described and demonstrated on http://www.ncbi.nlm.nih.gov/Sequin/index.html and may be obtained by anonymous FTP (file transfer program) from ncbi.nlm.nih.gov/sequin/. Completed files can also be E-mailed to gb-sub@ncbi.nlm.nih.gov or can be mailed on diskette to GenBank Submissions, National Center for Biotechnology Information, National Library of Medicine, Bldg. 38A, Room 8N-803, Bethesda, Maryland 20894.

Submitted sequences can include well-characterized sequences, but they also can be batches of first-pass sequences intended for assembly and subsequent editing into larger molecules. For example, random GSS (genome survey sequences), HTG (high-throughput genome), or EST (expressed sequence tag) sequences may be submitted. These sequences are stored in specific divisions of GenBank. STS (sequence tagged site) sequences are another class of unique genome sequences that serve as specific genome markers. All of these high-throughput sequences become publicly available. Once sufficiently characterized, sequences are placed in organism-based GenBank divisions—BCT (bacteria), PRI (primate), ROD (rodent), MAM (other mammals), VRT (vertebrate), INV (invertebrate), PLN (plants and fungi), PHG (phage), VRL (virus), RNA, SYN (synthetic), and UNA (unannotated sequence).

SEQUENCE ACCURACY VARIES

It should be apparent from the above description of sequencing projects that the higher the level of accu-

racy required in DNA sequences, the more time-consuming and expensive the procedure. There is no detailed check of sequence accuracy prior to submission to GenBank and other databases. Often, a sequence is submitted at the time of publication of the sequence in a journal article, providing a certain level of checking by the editorial peer-review process. However, many sequences are submitted without being published or prior to publication. In laboratories performing large sequencing projects, such as those engaged in genome projects of organisms, the granting agency requires a final level of accuracy of the order of 1 possible error per 10 kb. This level of accuracy should be sufficient for most sequence analysis applications such as sequence comparisons, pattern searching, and translation.

GenBank does permit submission of raw sequence data that have a high error rate. GSS or EST sequences, referred to in the above section, represent first-pass sequences with an error rate as high as 1 in 100, including incorrectly identified bases and inserted or deleted bases. Thus, in translating EST sequences in GenBank and other databases, incorrect bases may translate to the wrong amino acid. The worst problem, however, is that base insertions/deletions will cause frameshifts in the sequence, thus making alignment with a protein sequence very difficult. Another type of database sequence that is error-prone is a fragment of sequence from the immunological variant of a pathogenic organism, such as the regions in the protein coat of the human immunodeficiency virus (HIV). Although this low level of accuracy may be suitable for some purposes, such as classification of the organism, the accuracy of such sequence fragments should be verified for more detailed analyses, e.g., evolutionary analyses.

The accuracy of the first-pass sequences is greatly increased as they are curated through increasing phases of editing involving assembly into contigs, removal of gaps, and further editing into derivative sequences that have 99.99% accuracy. These derived, curated sequences are useful resources for the scientific community. There are two GenBank resources that provide these more refined sequence collections (see http://www.ncbi.nlm.nih.gov/) for many research organisms. First, the RefSeq resource provides an integrated, nonredundant database of sequences including genomic DNA, cDNA, and protein sequences for the major research organisms. RefSeq sequences include sequences derived from high-throughput sequencing projects and sequences that are representative of multiple entries of the same sequence in GenBank. Second, the UniGene resource is composed of thousands of clusters of GenBank sequences for an organism, each of which appears to represent the same gene based on alignments between EST sequences with each other and with genome sequences of the organism. This set of apparent gene sequences changes as more sequences are added to the database and hence cannot be used for publication. For the human genome, there are over 100,000 UniGene clusters compared to a predicted 35,000 genes based on gene prediction methods. As better overlapping sequences are added, the human UniGene number will presumably decrease.

SEQUENCES ARE STORED IN COMPUTERS USING SPECIFIED FORMATS

Before using a sequence file in a sequence analysis program, it is important to ensure that computer sequence files contain only sequence characters and not special characters used by text editors. Editing a sequence file with a word processor can introduce such changes if one is not careful to work only with text or so-called ASCII files (those on the typewriter keyboard), and, even then, additional characters may be introduced. Most text editors normally create text files that include control characters in addition to standard ASCII characters. These control characters will only be recognized correctly by the text editor program. Sequence files that contain such control characters may not be analyzed correctly, depending on whether or not the sequence analysis program filters them out. Editors usually provide a way to save files with only standard ASCII characters, and these are the files that will be suitable for most sequence analysis programs. In general, it is best to avoid commercial text editors normally used for writing reports, papers, and grants. Instead, use simple text editors that are designed for writing computer programs.

Sequence and other data files that contain non-ASCII characters also may not be transferred correctly from one machine to another and may cause unpredictable behavior of communications software. Some communications software can be set to ignore such control characters. For example, the file transfer program (FTP) has ASCII and binary modes, which may be set by the user. The ASCII mode is useful for

ASCII AND HEXADECIMAL

Computers store sequence information as simple rows of sequence characters called strings, which are similar to the sequences shown on the computer terminal. Each character is stored in binary code in the smallest unit of memory, called a byte. Each byte comprises 8 bits, with each bit having a possible value of 0 or 1, producing 255 possible combinations. By convention, many of these combinations have a specific definition, called their ASCII equivalent. Some ASCII values are defined as keyboard characters and others as special control characters, such as signaling the end of a line (a line feed and a carriage return) or the end of a file full of text (end-of-file character). A file with only ASCII characters is called an ASCII file. For convenience, all binary values may be written in a hexadecimal format, which corresponds to our decimal format 0, 1, ... 9 plus the letters A, B, ... F. Thus, hexadecimal 0F corresponds to binary 0000 1111 and decimal 15, and FF corresponds to binary 1111 1111 and decimal 255. A DNA sequence is usually stored and read in the computer as a series of 8-bit words in this binary format. A protein sequence appears as a series of 8-bit words comprising the corresponding binary form of the amino acid letters.

transferring text files, and the binary mode is useful for transferring compressed data files, which also contain non-ASCII characters.

Most sequence analysis programs require not only that a DNA or protein sequence file be a standard ASCII file, but also that the file be in a particular format such as the FASTA format (see below). The use of windows on a computer has simplified such problems, since one merely has to copy a sequence from one window, for example, a window that is running a Web browser on the Entrez Web site, and paste it into another, for example, that of a translation program.

In addition to the standard four base symbols, A, T, G, and C, the Nomenclature Committee of the International Union of Biochemistry has established a standard code to represent bases in a nucleic acid sequence that are uncertain or ambiguous. The codes are listed in Table 2.1.

For computer analysis of proteins, it is more convenient to use single-letter than three-letter amino acid codes. For example, GenBank DNA sequence entries contain a translated sequence in single-letter code. The standard, single-letter amino acid code was established by a joint international committee, and is shown in Table 2.2. When the name of only one amino acid starts with a particular letter, then that letter is used; e.g., C, cysteine. In other cases, the letter chosen is phonetically similar (R, arginine) or close in the alphabet (K, lysine).

One major difficulty encountered in running sequence analysis software is the use of differing sequence formats by different programs. These formats all are standard ASCII files, but they may differ in the presence of certain characters and words that

indicate where different types of information and the sequence itself are to be found. The more commonly used sequence formats are discussed below. A more detailed list of sequence formats may be found at http://www.hgmp.mrc.ac.uk/Software/EMBOSS/.

GenBank DNA Sequence Entry

The format of a database entry in GenBank, the NCBI nucleic acid and protein sequence database, is as fol-

TABLE 2.1. *Base–nucleic acid codes*

Symbol	Meaning	Explanation
G	G	Guanine
A	A	Adenine
T	T	Thymine
C	C	Cytosine
R	A or G	puRine
Y	C or T	pYrimidine
M	A or C	aMino
K	G or T	Keto
S	C or G	Strong interactions 3 h bonds
W	A or T	Weak interactions 2 h bonds
H	A, C or T not G	H follows G in alphabet
B	C, G or T not A	B follows A in alphabet
V	A, C or G not T (not U)	V follows U in alphabet
D	A, G or T not C	D follows C in alphabet
N	A,C,G or T	Any base

Adapted from NC-IUB (1984).

TABLE 2.2. *Table of standard amino acid code letters*

1-letter code	3-letter code	Amino acid
A[a]	Ala	alanine
C	Cys	cysteine
D	Asp	aspartic acid
E	Glu	glutamic acid
F	Phe	phenylalanine
G	Gly	glycine
H	His	histidine
I	Ile	isoleucine
K	Lys	lysine
L	Leu	leucine
M	Met	methionine
N	Asn	asparagine
P	Pro	proline
Q	Gln	glutamine
R	Arg	arginine
S	Ser	serine
T	Thr	threonine
V	Val	valine
W	Trp	tryptophan
X	Xxx	undetermined amino acid
Y	Tyr	tyrosine
Z[b]	Glx	either glutamic acid or glutamine

Adapted from IUPAC-IUB (1969, 1972, 1983).

[a] Letters not shown are not commonly used.

[b] Note that sometimes when computer programs translate DNA sequences, they will put a "Z" at the end to indicate the termination codon. This character should be deleted from the sequence.

lows: Information describing each sequence entry is given, including literature references, information about the function of the sequence, locations of mRNAs and coding regions, and positions of important mutations. This information is organized into fields, each with an identifier, shown as the first text on each line. In some entries, these identifiers may be abbreviated to two letters, e.g., RF for reference, and some identifiers may have additional subfields. The information provided in these fields is described in Figure 2.5 and the database organization is described in Figure 2.6. The CDS subfield in the field FEATURES gives the amino acid sequence, obtained by translation of known and potential open reading frames, i.e., a consecutive set of three-letter words that could be codons specifying the amino acid sequence of a protein. The sequence entry is assumed by computer programs to lie between the identifiers "ORIGIN" and "//."

The sequence includes numbers on each line so that sequence positions can be located by eye. Because the sequence count or a sequence checksum value may be used by the computer program to verify the sequence composition, the sequence count should not be modified except by programs that also modify the count. The GenBank sequence format often has to be changed for use with sequence analysis software.

European Molecular Biology Laboratory Data Library Format

The European Molecular Biology Laboratory (EMBL) maintains DNA and protein sequence data-

```
LOCUS        name of locus, length and type of sequence,
             classification of organism, data of entry
DEFINITION   description of entry
ACCESSION    accession numbers of original source
KEYWORDS     key words for cross referencing this entry
SOURCE       source organism of DNA
ORGANISM     description of organism
REFERENCE
COMMENT      biological function or database information
FEATURES     information about sequence by base position or range of positions
     source          range of sequence, source organism
     misc_signal     range of sequence, type of function or signal
     mRNA            range of sequence, mRNA
     CDS             range of sequence, protein coding region
     intron          range of sequence, position of intron
     mutation        sequence position, change in sequence for mutation
BASE COUNT   count of A, C, G, T and other symbols
ORIGIN       text indicating start of sequence
        1 gaattcgata aatctctggt ttattgtgca gtttatggtt ccaaaatcgc
       51 atatactcac agcataactg tatatacacc caggggggcgg aatgaaagcg
//               database symbol for end of sequence
```

FIGURE 2.5. GenBank DNA sequence entry.

Accession no	Organism	Reference	Name	Keywords	Sequence
..123	Escherichia. coli	Medline1,.	LexA protein	SOS regulon, repressor, transcriptional regulator,..	ATG..
..124	Escherichia coli	Medline2,.	UmuD protein	SOS regulon,..	GTA..
..125	Saccharomyces. cerevisiae	Medline3,.	GAL4 protein	transcriptional regulator,..	CAT..
..125	Homo. sapiens	Medline4,.	gluco-corticoid receptor	transcriptional regulator,..	TGT..

FIGURE 2.6. Organization of the GenBank database and the search procedure used by Entrez. In this database format, each row is another sequence entry and each column another GenBank field. When one sequence entry is retrieved, all of these fields will be displayed, as in Fig. 2.5. Only a few fields and simple examples are shown for illustration. A search for the term "SOS regulon and coli" in all fields will find two matching sequences. Finding these sequences is simple because indexes have been made listing all of the sequences that have any given term, one index for each field. Similarly, a search for transcriptional regulator will find three sequences.

bases. The format for each entry in these databases is shown in Figure 2.7. As with GenBank entries, a large amount of information describing each sequence entry is given, including literature references, information about the function of the sequence, locations of mRNAs and coding regions, and positions of important mutations. This information is organized into fields, each with an identifier, shown as the first text on each line. The meaning of each of these fields is explained in Figure 2.7. These identifiers are abbreviated to two letters, e.g., RF for reference, and some identifiers may have additional subfields. The sequence entry is assumed by computer programs to lie between the identifiers "SEQUENCE" and "//" and includes numbers on each line to locate parts of the sequence visually. The sequence count or a checksum value for the sequence may be used by computer programs to make sure that the sequence is complete and accurate. For this reason, the sequence part of the entry usually should not be modified except with programs that also modify this count. This EMBL sequence format is very similar to the GenBank format. The main differences are in the use of the term ORIGIN in the GenBank format to indicate the start

```
ID              identification code for sequence in the database
AC              accession number giving origin of sequence
DT              dates of entry and modification
KW              key cross-reference words for lookup up this entry
OS, OC          source organism
RN, RP, RX, RA, RT, RL  literature reference or source
DR              i.d. in other databases
CC              description of biological function
FH, FT          information about sequence by base position or range of positions
                source range of sequence, source organism
                misc_signal range of sequence, type of function or signal
                mRNA range of sequence, mRNA
                CDS range of sequence, protein coding region
                intron range of sequence, position of intron
                mutation sequence position, change in sequence for mutation
SQ              count of A, C, G, T and other symbols
gaattcgata aatctctggt ttattgtgca gtttatggtt ccaaaatcgc cttttgctgt 60
atatactcac agcataactg tatatacacc caggggcgg aatgaaagcg ttaacggcca 120
.
.
// symbol to indicate end of sequence
```

FIGURE 2.7. EMBL sequence entry format.

of sequence; also, the EMBL entry does not include the sequences of any translation products, which are shown instead as a different entry in the database. This EMBL sequence format often has to be changed for use with sequence analysis software. The output of a DDBJ DNA sequence entry is almost identical to that of GenBank.

SwissProt Sequence Format

The format of an entry in the SwissProt protein sequence database is very similar to the EMBL format, except that considerably more information is provided about the physical and biochemical properties of the protein.

FASTA Sequence Format

The FASTA sequence format includes three parts, as shown in Figure 2.8: (1) a comment line identified by a ">" character in the first column followed by the name and origin of the sequence; (2) the sequence in standard one-letter symbols; and (3) an optional "*" indicating the end of the sequence that may or may not be present. The presence of "*" may be essential for reading the sequence correctly by some sequence analysis programs. The FASTA format is the one most often used by sequence analysis software. This format provides a very convenient way to copy just the sequence part from one window to another because there are no numbers or other nonsequence characters within the sequence. The FASTA sequence format is similar to the protein information resource National Biomedical Research Foundation (NBRF) format except that the NBRF format includes a first line with a ">" character in the first column followed by information about the sequence, a second line containing an identification name for the sequence, and the third to last lines containing the sequence, as described below.

National Biomedical Research Foundation/Protein Information Resource Sequence or CODATA Format

This sequence format, which is sometimes also called the PIR format, has been used by the National Biomedical Research Foundation/Protein Information Resource (NBRF/PIR) and also by other sequence analysis programs. Note that sequences retrieved from the PIR database on their Web site (http://www-nbrf.georgetown.edu) are not in this compact format, but in an expanded format with much more information about the sequence, as shown below. The NBRF format is similar to the FASTA sequence format but with significant differences. An example of a PIR sequence format is given in Figure 2.9. The first line includes an initial ">" character followed by a two-letter code such as P for complete sequence or F for fragment, followed by a 1 or 2 to indicate type of sequence (linear or circular sequence, respectively), then a semicolon, then a four- to six-character unique name for the entry. There is also an essential second line with the full name of the sequence, a hyphen, then the species of origin. In FASTA format, the second line is the start of the sequence and the first line gives the sequence identifier after a ">" sign. The sequence terminates with an asterisk.

Genetics Computer Group Sequence Format

Earlier versions of the Genetics Computer Group (GCG) programs require a unique sequence format and include programs that convert other sequence formats into the GCG format. Later versions of GCG accept several sequence formats. A converted GenBank file is illustrated in Figure 2.10. Information about the sequence in the GenBank entry is first included, followed by a line of information about the sequence and

```
>YCZ2_YEAST protein in HMR 3' region
MKAVVIEDGKAVVKEGVPIPELEEGFV
GNPTDWAHIDYKVGPQGSILGCDAAGQ
IVKLGPAVDPKDFSIGDYIYGFIHGSS
VRFPSNGAFAEYSAISTVVAYKSPNEL
KFLGEDVLPAGPVRSLEGAATIPVSLT*
```

FIGURE 2.8. FASTA sequence entry format.

```
>P1;ILEC
lexA repressor - Escherichia coli
MKALTARQQEVFDLIRDHISQTGMPPTRAE
IAQRLGFRSPNAAEEHLKALARKGVIEIVS
GASRGIRLLQEEEEGLPLVGRVAAGEPLLA
QQHIEGHYQVDPSLFKPNADFLLRVSGMSM
KDIGIMDGDLLAVHKTQDVRNGQVVVARID
DEVTVKRLKKQGNKVELLPENSEFKPIVVD
LRQQSFTIEGLAVGVIRNGDWL
```

FIGURE 2.9. NBRF sequence entry format.

```
BASE COUNT 215 A 224 C 263 G 250 T
ORIGIN
Filename, Length of sequence, Date, Checksum value, ..
   1  GAATTCGATA AATCTCTGGT TTATTGTGCA GTTTATGGTT CCAAAATCGC
  51  CTTTTGCTGT ATATACTCAC AGCATAACTG TATATACACC CAGGGGGCGG
```

FIGURE 2.10. GCG sequence entry format.

a checksum value. This value (not shown) is provided as a check on the accuracy of the sequence by the addition of the ASCII values of the sequence. If the sequence has not been changed, this value should stay the same. If one or more sequence characters become changed through error, a program reading the sequence will be able to determine that the change has occurred because the checksum value in the sequence entry will no longer be correct. Lines of information are terminated by two periods, which mark the end of information and the start of the sequence on the next line. The rest of the text in the entry is treated as sequence. Note the presence of line numbers. Because there is no symbol to indicate the end of sequence, no text other than sequence should be added beyond this point. The sequence should not be altered except by programs that will also adjust the checksum score for the sequence. The GCG sequence format may have to be changed for use with other sequence analysis software. GCG also includes programs for reformatting sequence files.

Format of Sequence File Retrieved from the National Biomedical Research Foundation/Protein Information Resource

This file format has approximately the same information as a GenBank or EMBL sequence file but is formatted slightly differently, as in Figure 2.11. This format is presently called the PIR/CODATA format.

Plain/ASCII Staden Sequence Format

This sequence format is a computer file that includes only the sequence with no other accessory information. This particular format is used by the Staden Sequence Analysis programs (http://www/.mrc lmb.com.ac.uk/pubseq) produced by Roger Staden at

Cambridge University (Staden et al. 2000). The sequence must be further formatted to be used for most other sequence analysis programs.

Abstract Syntax Notation Sequence Format

Abstract Syntax Notation (ASN.1) is a formal data description language that has been developed by the computer industry. ASN.1 (http://www-sop.inria.fr/ rodeo/personnel/hoschka/asn1.html; NCBI 1993) has been adopted by the National Center for Biotechnology Information (NCBI) to encode data such as sequences, maps, taxonomic information, molecular structures, and bibliographic information. These data sets may then be easily connected and accessed by computers. The ASN.1 sequence format is a highly structured and detailed format especially designed for computer access to the data. All the information found in other forms of sequence storage, e.g., the GenBank format, is present. For example, sequences can be retrieved in this format by Entrez (see below). However, the information is much more difficult to read by eye than a GenBank formatted sequence. One would normally not need to use the ASN.1 format except when running a computer program that uses this format as input.

XML Sequence Format

XML is a standard data format that is becoming widely used for sharing genome data between computer locations. The format consists of series of successively generated fields in a tree-like structural arrangement. The tags used to define these fields are very similar to those used in HTML formats that are used for displaying Web pages on a browser. For example, the sequence in a GenBank file in XML format is between the XML tags <Seq-data> and <\Seq-data>. As with the ASN.1 format, all the information in the GenBank entry is present between defined sets

```
ENTRY ILEC
      #type complete
TITLE lexA repressor - Escherichia coli
ORGANISM
      #formal_name Escherichia coli
DATE 29-Jul-1981
      #sequence_revision 01-Sep-1981
      #text_change 14-Nov-1997
ACCESSIONS A90808; A93734; S11945; B65212; A03569
REFERENCE A90808
      #authors Horii, T.; Ogawa, T.; Ogawa, H.
      #journal Cell (1981) 23:689-697
      #title Nucleotide sequence of the lexA gene of Escherichia coli.
      #cross-references MUID:81186269
      #contents lexA
      #accession A90808
          ##molecule_type DNA
          ##residues 1-202
          ##label HOR
REFERENCE
.

.
COMMENTS
GENETICS
      #gene lexA
      #map_position 92 min
CLASSIFICATION
      #superfamily lexa repressor
KEYWORDS DNA binding, repressor, transcription regulator
SUMMARY
      #length 202
      #molecular_weight 22358
SEQUENCE
                5       10      15      20      25      30
        1 M K A L T A R Q Q E V F D L I R D H I S Q T G M P P T R A E
```

FIGURE 2.11. Protein Information Resource sequence format.

of tags so that the information may be readily retrieved by a computer program. Sequences can be retrieved from GenBank in XML format.

Genetic Data Environment Sequence Format

Genetic Data Environment (GDE) format is used by a sequence analysis system called the Genetic Data Environment, which was designed by Steven Smith and collaborators (Smith et al. 1994) around a multiple sequence alignment editor that runs on UNIX machines. The GDE features are incorporated into the SEQLAB interface of the GCG software, version 9. GDE format is a tagged-field format similar to ASN.1 that is used for storing all available information about a sequence, including residue color. The file consists of various fields (Fig. 2.12), each enclosed by brackets, and each field has specific lines, each with a given name tag. The information following each tag is placed in double quotes or follows the tag name by one or more spaces.

AceDB Sequence Format

This format indicates the type of sequence followed by a colon and then by an accession number. The sequence then begins on the next line as shown in Figure 2.13.

General Feature Format

The General Feature format (GFF) is used by the EMBOSS suite of sequence analysis programs and is

```
{
name              "Short name for sequence"
longname          "Long (more descriptive) name for sequence"
sequence-ID       "Unique ID number"
creation-date     "mm/dd/yy hh:mm:ss"
direction         [-1|1]
strandedness      [1|2]
type              [DNA|RNA|PROTEIN|TEXT|MASK]
offset            (-999999,999999)
group-ID          (0,999)
creator           "Author's name"
descrip           "Verbose description"
comments          "Lines of comments about a sequence"
sequence          "gctagctagctagctagctcttagctgtagtcgtagctgatgctag
                   ctgatgctagctagctagctagctgatcgatgctagctgatcgtag
                   ctgacggactgatgctagctagctagctagctgtctagtgtcgtag
                   tgcttattgc"
}
```

FIGURE 2.12. The Genetic Data Environment format.

illustrated in Figure 2.14. This format also has additional features for storing information about the sequence, e.g., gene start and stop positions that, in combination with the programming modules described in Chapter 12, may be used for graphical display of these features. The home of these GFF formats may be found at http://www.hgmp.mrc.ac.uk/Software/EMBOSS/.

ONE SEQUENCE FORMAT CAN BE CONVERTED TO ANOTHER

READSEQ is an extremely useful sequence formatting program developed by D.G. Gilbert at Indiana University, Bloomington (gilbertd@bio.indiana.edu). READSEQ can recognize a DNA or protein sequence file in any of the formats shown in Table 2.3, identify the format, and write a new file with an alternative format. Some of these formats are used for special types of analyses such as multiple sequence alignment and phylogenetic analysis. The appearance of these formats for two sample DNA sequences, seq1 and seq2, is shown in Table 2.4. READSEQ may be accessed at the Baylor College of Medicine site at http://dot.imgen.bcm.tmc.edu:9331/seq-util/readseq.html and also by anonymous FTP from ftp.bio.indiana.edu/molbio/readseq or ftp.bioindi ana.edu/molbio/mac to obtain the appropriate files.

Data files that have multiple sequences, such as those required for multiple sequence alignment and phylogenetic analysis using parsimony (PAUP), are also converted. Examples of the types of files produced are shown in Table 2.4. Options to produce a reverse-complement sequence and to remove gaps from sequences are included.

SEQIO, another sequence conversion program for a UNIX machine, is described at http://bioweb.pasteur.fr/docs/seqio/seqio.html and is available for download at http://www.cs.ucdavis.edu/gusfield/seqio.html. There are also BioPerl programming modules that may be used for changing between sequence formats that are described in Chapter 12.

```
DNA : "UVH1"

gccaatatgcagctctttgtccgcgcccaggagctacacaccttcgaggt

gatcaaggctcatgtagcctcactgggcggtagctgggaccgccgttcagtc

gaccggccaggaaacggtcgcccattcctctttctcgactccatcttc
```

FIGURE 2.13. AceDB format.

```
##gff-version 2
##source-version EMBOSS 2.2.1
##date 2002-01-22
##DNA HSFAU
##ttcctctttctcgactccatcttcgcggtagctgggaccgccgttcagtcgccaatatgc
##agctctttgtccgcgcccaggagctacacaccttcgaggtgaccggccaggaaacggtcg
##end-DNA
```

FIGURE 2.14. The GFF sequence format.

The "from" programs in the GCG suite of programs convert sequence files from GCG format into the named format, and the "to" programs convert the alternative format into GCG format. The following are the actual program names, with no spaces included: FROMEMBL, FROMFASTA, FROMGENBANK, FROMIG, FROMPIR, FROMSTADEN, TOFASTA, TOIG, TOPIR, TOSTADEN. There are no programs to convert to GenBank and EMBL formats.

In addition, the GCG programs include the following sequence formatting programs: (1) GETSEQ, which converts a simple ASCII file being received from a remote PC to GCG format; (2) REFORMAT, which will format a GCG file that has been edited, and will also perform other functions; and (3) SPEW, which sends a GCG sequence file as an ASCII file to a remote PC.

TABLE 2.3. *Major sequence formats recognized by format conversion program READSEQ*

1. Fasta/Pearson
2. Intelligenetics/Stanford
3. GenBank
4. National Biomedical Research Foundation (NBRF)
5. European Molecular Biology Laboratory (EMBL)
6. Genetics Computer Group (GCG)[a]
7. DNA Strider
8. Fitch (for phylogenetic analysis)
9. Phylogenetic Inference package (PHYLIP v3.3, v3.4)
10. Protein Information Resource (PIR or CODATA)
11. Multiple sequence format (MSF)
12. Abstract Syntax Notation (ASN.1)
13. Phylogenetic Analysis Using Parsimony (PAUP) NEXUS format

[a] For conversion of single sequence files only. The other conversions can be performed on files with single or multiple sequences.

TABLE 2.4. *Multiple sequence format conversions by READSEQ*

1. Fasta/Pearson format

```
>seq1
agctagct agct agct
>seq2
aactaact aact aact
```

(*Continued on following pages.*)

TABLE 2.4. (*Continued*)

2. Intelligenetics format	

```
;seq1, 16 bases, 2688 checksum.
seq1
agctagctagctagct1
;seq2, 16 bases, 25C8 checksum.
seq2
aactaactaactaact1
```

3. GenBank format	

```
LOCUS       seq1         16 bp
DEFINITION seq1, 16 bases, 2688 checksum.
ORIGIN
        1 agctagctag ctagct
//
LOCUS       seq2         16 bp
DEFINITION seq2, 16 bases, 25C8 checksum.
ORIGIN
  1  aactaactaa ctaact
//
```

4. NBRF format	

```
>DL;seq1
seq1, 16 bases, 2688 checksum.
 agctagctag ctagct*

>DL;seq2
seq2, 16 bases, 25C8 checksum.
 aactaactaa ctaact*
```

5. EMBL format	

```
ID seq1
DE seq1, 16 bases, 2688 checksum.
SQ          16 BP
   agctagctag ctagct
//
ID seq2
DE seq2, 16 bases, 25C8 checksum.
SQ          16 BP
   aactaactaa ctaact
//
```

6. GCG format	

```
seq1
    seq1  Length: 16  Check: 9864 ..
  1  agctagctag ctagct

seq2
    seq2  Length: 16  Check: 9672 ..
  1  aactaactaa ctaact
```

TABLE 2.4. (*Continued*)

7.	Format for the Macintosh sequence analysis program DNA Strider

```
; ### from DNA Strider ;-)
; DNA sequence  seq1, 16 bases, 2688 checksum.
;
agctagctagctagct
//
; ### from DNA Strider ;-)
; DNA sequence  seq2, 16 bases, 25C8 checksum.
;
aactaactaactaact
//
```

8.	Format for phylogenetic analysis programs of Walter Fitch

```
seq1,  16 bases, 2688 checksum.
 agc tag cta gct agc t
seq2,  16 bases, 25C8 checksum.
 aac taa cta act aac t
```

9.	Format for phylogenetic analysis programs PHYLIP of J. Felsenstein v 3.3 and 3.4.

```
2 16
seq1        agctagctag ctagct
seq2        aactaactaa ctaact
```

10.	Protein International Resource PIR/CODATA format

```
\\\
ENTRY           seq1
TITLE           seq1, 16 bases, 2688 checksum.
SEQUENCE
                    5        10       15       20
25        30
      1 a g c t a g c t a g c t a g c t
///
ENTRY           seq2
TITLE           seq2, 16 bases, 25C8 checksum.
SEQUENCE
                    5        10       15       20
25        30
      1 a a c t a a c t a a c t a a c t
///
```

11.	GCG multiple sequence format (MSF)

```
/tmp/readseq.in.2449  MSF: 16 Type: N January 01,
1776  12:00  Check: 9536 ..

 Name: seq1           Len:     16 Check:   9864
Weight:  1.00
 Name: seq2           Len:     16 Check:   9672
Weight:  1.00

 //

            seq1  agctagctag ctagct
            seq2  aactaactaa ctaact
```

TABLE 2.4. (*Continued*)

12.	Abstract Syntax Notation (ASN.1) format	

```
Bioseq-set ::= {
seq-set {
  seq {
    id { local id 1 },
    descr { title "seq1" },
    inst {
      repr raw, mol dna, length 16, topology linear,
      seq-data
        iupacna "agctagctagctagct"
    } } ,
  seq {
    id { local id 2 },
    descr { title "seq2" },
    inst {
      repr raw, mol dna, length 16, topology linear,
      seq-data
        iupacna "aactaactaactaact"
    } } ,
} }
```

13.	NEXUS format used by the phylogenetic analysis program PAUP by David Swofford	

```
#NEXUS
[/tmp/readseq.in.2506 — data title]

[Name: seq1                  Len: 16 Check: 2688]
[Name: seq2                  Len: 16 Check: 25C8]


begin data;
 dimensions ntax=2 nchar=16;
 format datatype=dna interleave missing=-;
  matrix
        seq1   agctagctagctagct
        seq2   aactaactaactaact
```

Two sequences in FASTA multiple sequence format (1) were used as input for the remainder of the format options (2–14).

SEVERAL FORMATS ARE AVAILABLE FOR MULTIPLE SEQUENCES

Most of the sequence formats listed above can be used to store multiple sequences in tandem in the same computer file. Exceptions are the GCG and raw sequence formats, which are designed only for single sequences. GCG has an alternative multiple sequence format, which is described below. In addition, there are formats especially designed for multiple se-quences that can also be used to show their align-ments or to perform types of multiple sequence analyses such as phylogenetic analysis. In the case of PAUP, the program will accept the MSA format and convert it to the NEXUS format. These formats are illustrated below using several short sequence fragments.

1. Aligned sequences in FASTA format. The aligned sequence characters occupy the same line and column and gaps are indicated by a dash.

```
>gi|730305|
MATHHTLWMGLALLGVLGDLQAAPEAQVSVQPNFQQDKFL
RTQTPRAELKEKFTAFCKAQGFTEDTIVFLPQTDKCMTEQ
>gi|404390|
--------------------APEAQVSVQPNFQPDKFL
RTQTPRAELKEKFTAFCKAQGFTEDSIVFLPQTDKCMTEQ
>gi|895868
MAALRMLWMGLVLLGLLGFPQTPAQGHDTVQPNFQQDKFL
RTQTLKDELKEKFTTFSKAQGLTEEDIVFLPQPDKCIQE-
```

represents the same alignment as:

```
MATHHTLWMGLALLGVLGDLQAAPEAQVSVQPNFQQDKFL
--------------------APEAQVSVQPNFQPDKFL
MAALRMLWMGLVLLGLLGFPQTPAQGHDTVQPNFQQDKFL
RTQTPRAELKEKFTAFCKAQGFTEDTIVFLPQTDKCMTEQ
RTQTPRAELKEKFTAFCKAQGFTEDSIVFLPQTDKCMTEQ
RTQTLKDELKEKFTTFSKAQGLTEEDIVFLPQPDKCIQE-
```

2. GCG multiple sequence format (MSF) produced by the GCG multiple sequence alignment program PILEUP. The gap symbol is "~". The length indicated is the length of the alignment, which is the length of the longest sequence including gaps.

```
Symbol comparison table: GenRunData:blosum62.cmp CompCheck: 6430

   GapWeight:                12
   GapLengthWeight:           4

list4.msf  MSF: 883  Type: P  February 28, 1997 16:42  Check: 482

Name:   haywire      Len:   883   Check:   3979   Weight:  1.00
Name:   xpb-human    Len:    83   Check:   9129   Weight:  1.00
Name:   rad25        Len:   883   Check:   5359   Weight:  1.00
Name:   xpb-ara      Len:   883   Check:   2015   Weight:  1.00

//

   1                                                               50
   haywire     ~~~~~~~~~~  ~~~~~~~~~~  ~~~~~~~~~~  ~~~~~~~~~~  ~~~~~MGPPK
   xpb-human   ~~~~~~~~~~  ~~~~~~~~~~  ~~~~~~~~~~  ~~~~~~~~~~  ~~~~~~~~~~
   rad25       MTDVEGYQPK  SKGKIFPDMG  ESFFSSDEDS  PATDAEIDEN  YDDNRETSEG
   xpb-ara     ~~~~~~~~~~  ~~~~~~~~~~  ~~~~~~~~~~  ~~~~~~~~~~  ~~~~~~~~~~

   51                                                              100
   haywire     KSRKDRSG..  GDKFGKKRRA  EDEAFTQLVD  DNDSLDATES  EGIPGAASKN
   xpb-human   MGKRDRAD..  RDKKKSRKRH  YED...EEDD  EEDAPGNDPQ  EAVPSAAGKQ
   rad25       RGERDTGAMV  TGLKKPRKKT  KSSRHTAADS  SMNQMDAKDK  ALLQDTNSDI
   xpb-ara     ~~~~~~~~~~  ~~~~~~~~~~  ~~~~~~~~~~M  KYGGKDDQKM  KNIQNAEDYY
```

3. ALN form produced by multiple sequence alignment program CLUSTALW (Thompson et al. 1994). In addition to the alignment position, the program also shows the current sequence position at the end of each row.

4. Blocked alignment used by GDE and GCG SEQLAB (Fig. 2.15). Unlike the other examples shown, which are all simple text files of an alignment, the following figure is a screen display of an alignment, using GDE and SEQLAB display programs. The underlying alignment in text format would be similar to the GCG multiple sequence alignment file shown above.

5. Format used by Fitch phylogenetic analysis programs.

```
seq1, 16 bases, 2688 checksum.
 agc tag cta gct agc t
seq2, 16 bases, 25C8 checksum.
 aac taa cta act aac t
```

6. Formats used by Felsenstein phylogenetic analysis programs PHYLIP (phylogenetic inference package): 2 for two sequences, 16 for length of alignment.

```
a. version 3.2

2 16 YF
seq1          agctagctag ctagct
  seq2          aactaactaa ctaact


b. versions 3.3 and 3.4

2 16
  seq1          agctagctag ctagct
  seq2          aactaactaa ctaact
```

7. Format used by phylogenetic analysis program PAUP. ntax is number of taxa, nchar is the length of the alignment, and interleave allows the alignment to be shown in readable blocks. The other terms describe the type of sequence and the character used to indicate gaps.

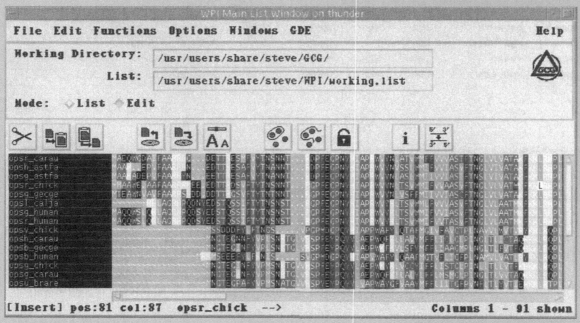

FIGURE 2.15. A multiple sequence alignment editor for GCG MSF files. For information on using multiple sequence alignment editors and for examples of other editors, see Chapter 4.

```
#NEXUS

[ comments ]

begin data;
      dimensions ntax=4 nchar=100;
      format datatype=protein interleave gap=-;
      matrix
[             1                                                                        50]
      haywire    ----------    ----------    ----------    ----------    -----
MGPPK

   xpb-human    ----------    ----------    ----------    ----------    ----------
-

       rad25    MTDVEGYQPK    SKGKIFPDMG    ESFFSSDEDS    PATDAEIDEN    YDDNRETSEG
     xpb-ara    ----------    ----------    ----------    ----------    ----------
-


[            51
                                                                              100]
     haywire    KSRKDRSG--    GDKFGKKRRA    EDEAFTQLVD    DNDSLDATES    EGIPGAASKN
   xpb-human    MGKRDRAD--    RDKKKSRKRH    YED---EEDD    EEDAPGNDPQ    EAVPSAAGKQ
       rad25    RGERDTGAMV    TGLKKPRKKT    KSSRHTAADS    SMNQMDAKDK    ALLQDTNSDI
     xpb-ara    ----------    ----------    ---------M    KYGGKDDQKM    KNIQNAEDYY

  ;
endblock;
```

8. The Selex format used by hidden Markov program HMMER by Sean Eddy has been used to keep track of the alignment of small RNA molecules.

```
# Example selex file

seq1    ACGACGACGACG.
seq2    ..GGGAAAGG.GA
seq3    UUU..AAAUUU.A

seq1    ...ACG
seq2    AAGGG
seq3    AA...UUU
```

Each line contains a name, followed by the aligned sequence. A gap is indicated by a period in this example but a gap can also be denoted by a dash or underscore. Long alignments are split into multiple blocks and interleaved or separated by blank lines. The number of sequences, their order, and their names must be the same in every block, and every sequence must be represented even though there are no residues present.

9. The block multiple sequence alignment format (see http://www.blocks.fhcrc.org/).

Identification (ID) starts contain a short identifier for the group of sequences from which the block was made and often is the original Prosite group ID. The identifier is terminated by a semicolon, and "BLOCK" indicates the entry type. AC contains the block number, a seven-character group number for sequences from which the block was made, followed by a letter (A–Z) indicating the order of the block in the sequences. The block number is a 5-digit number preceded by BL (BLOCKS database) or PR (PRINTS database). min,max is the minimum,maximum number of amino acids from the previous block or from the sequence start. DE describes sequences from which the block was made. BL contains information about the block: xxx is the amino acids in the spaced triplet found by MOTIF upon which the block is based. w is the width of the sequence segments (columns) in the block. s is the number of sequence segments (rows) in the block. Other values (n1, n2) describe statistical features of the block. Sequence_id is a list of sequences. Each sequence line contains a sequence identifier, the offset from the beginning of the sequence to the block in parentheses, the sequence segment, and a weight for the segment.

```
ID    short_identifier; BLOCK
AC    block_number; distance from previous block = (min,max)
DE    description
BL    xxx motif; width=w; seqs=s; 99.5%=n1; strength=n2
sequence_id   (offset) sequence_segment   sequence_weight.

//

ID    GLU_CARBOXYLATION; BLOCK
    AC    BL00011; distance from previous block=(1,64)
    DE    Vitamin K-dependent carboxylation domain proteins.
    BL    ECA motif; width=40; seqs=34; 99.5%=1833; strength=1412
    FA10_BOVIN (   45) LEEVKQGNLERECLEEACSLEEAREVFEDAEQTDEFWSKY 31
    FA10_CHICK (   45) LEEMKQGNIERECNEERCSKEEAREAFEDNEKTEEFWNIY 46
    FA10_HUMAN (   45) LEEMKKGHLERECMEETCSYEEAREVFEDSDKTNEFWNKY 33
     FA7_BOVIN (    5) LEELLPGSLERECREELCSFEEAHEIFRNEERTRQFWVSY 57
     FA7_HUMAN (   65) LEELRPGSLERECKEEQCSFEEAREIFKDAERTKLFWISY 42
    OSTC_CHICK (    6) SGVAGAPPNPIEAQREVCELSPDCNELADELGFQEAYQRR 94
```

SEQUENCE DATABASES HAVE SPECIFIC STORAGE REQUIREMENTS

As shown by the above examples, each DNA or protein sequence database entry has much information, including an assigned accession number(s); source organism; name of locus; reference(s); keywords that apply to sequence; features in the sequence such as coding regions, intron splice sites, and mutations; and finally the sequence itself. The above information is organized into a tabular form very much like that found in a relational database. (Additional information about databases is given in the box "Database Types," below.) If one imagines a large table with each sequence entry occupying one row, then each column will include one of the above types of information for each sequence, and each column is called a FIELD (see Fig. 2.6). The last column contains the sequences themselves. It is very easy to make an index of the information in each of these fields so that a search query can locate all the occurrences through the index. Even related sequences can be cross-referenced. In addition, the information in one database can be cross-referenced to that in another database. The DNA, protein, and reference databases have all been cross-referenced so that moving between them is readily accomplished (see Entrez section below, p. 57).

DATABASE TYPES

There are several types of databases; the two principal types are the relational and object-oriented databases. The relational database orders data in tables made up of rows giving specific items in the database, and columns giving the features as attributes of those items. These tables are carefully indexed and cross-referenced with each other, sometimes using additional tables for cross-referencing, so that each item in the database has a unique set of identifying features. A relational model for the GenBank sequence database has been devised at the National Center for Genome Resources.

The object-oriented database structure has been useful in the development of biological databases. The objects, such as genetic maps, genes, or proteins, each have an associated set of utilities for analysis and display of the object and a set of attributes such as identifying name or references. In developing the database, relationships among these objects are identified. To standardize some commonly arising objects in biological databases, e.g., maps, the Object Management Group (http://www.omg.org) has formed a Life Science Research Group. The Life Science Research Group is a consortium of commercial companies, academic institutions, and software vendors that is trying to establish standards for displaying biological information from bioinformatics and genomics analyses (http://www.omg.org/homepages/lsr). The Common Object Request Broker Architecture (CORBA) is the Object Management Group's interface for objects that allows different computer applications to communicate with each other through a common language, Interface Definition Language (IDL). To plan an object-oriented database by defining the classes of objects and the relationships among these objects, a specific set of procedures called the Unified Modeling Language (UML) has been devised by the OMG group.

DNA sequence analysis software packages often include sequence databases that are updated regularly. The organizations that manage sequence databases also provide public access through the Internet. Using a browser such as Netscape or Internet Explorer on a local personal computer, these sites may be visited through the Internet and a form can be filled out with the sequence name. Once the correct sequence has been identified, the sequence is delivered to the browser and may be saved as a local computer file, cut-and-pasted from the browser window into another window of an analysis program or editor, or even pasted into another browser page for analysis at a second Web site. A useful feature of browser programs for sequence analysis is the capability of having more than one browser window running at a time. Hence, one browser window may retrieve sequences from a database, and a second may analyze these sequences. At the time

of retrieving the sequence, several sequence formats may be available. The FASTA format, which is readily converted into other formats and also is smaller and simpler, containing just a line of sequence identifiers followed by the sequence without numbers, is very useful for this purpose. A list of sequence databases accessible through the Internet is provided in Table 2.5.

The growth of the sequence databases since their inception has been dramatic. In April, 2004, there were approximately 34 million sequence records including 39 billion nucleotides in 114 Gigabytes of storage space in GenBank. Tens of thousands of new nucleotides are added each day. New releases of GenBank are produced every 2 months. Moreover, there are approximately 200,000 users making 4 million database queries per day, making this site second only to the U.S. Internal Revenue Service in amount of use.

DATABASES CAN BE ACCESSED EASILY USING ENTREZ

One straightforward way to access the sequence databases is through Entrez, a resource prepared by the staff of the National Center for Biotechnology Information, National Library of Medicine, Bethesda, Maryland, and available through their Web site at http://ncbi.nlm.nih.gov/Entrez. Entrez provides a series of forms that can be filled out to retrieve a DNA or protein sequence, or a Medline reference related to

the molecular biology sequence databases. After a choice of database is made from the drop-down "search" list on the upper right (e.g., PubMed for references, Protein, and Nucleotide), then search terms are added to the "for" window (e.g., the gene name ERCC1), and "go" clicked with the mouse. The resulting search looks for all nucleotide entries in GenBank that have this word in any GenBank field (Fig. 2.5).

TABLE 2.5. *Major sequence databases accessible through the Internet*

1. GenBank at the National Center for Biotechnology Information, National Library of Medicine, Washington, D.C. accessible from:

   http://www.ncbi.nlm.nih.gov/Entrez

2. European Molecular Biology Laboratory (EMBL) Outstation at Hixton, England

   http://www.ebi.ac.uk/embl/index.html

3. DNA DataBank of Japan (DDBJ) at Mishima, Japan

   http://www.ddbj.nig.ac.jp/

4. Protein International Resource (PIR) database at the National Biomedical Research Foundation in Washington, D.C. (see Barker et al. 1998), an annotated protein database

   http://www-nbrf.georgetown.edu/pirwww/

5. The SwissProt protein sequence database at ISREC, Swiss Institute for Experimental Cancer Research in Epalinges/ Lausanne, an annotated protein database

   http://www.expasy.ch/cgi-bin/sprot-search-de

6. The Sequence Retrieval System (SRS) at the European Bioinformatics Institute allows both simple and complex concurrent searches of one or more sequence databases. The SRS system may also be used on a local machine to assist in the preparation of local sequence databases.

   http://srs6.ebi.ac.uk

The databases are available at the indicated addresses and return sequence files through an Internet browser. Many of the sites shown provide access to multiple databases. The first three database centers are updated daily and exchange new sequences daily, so that it is only necessary to access one of them. Additional Web addresses of databases of protein families and structure, and genomic databases, are given in Chapters 10 and 11. These databases can also provide access to sequences of a protein family or organism.

The annotated protein data banks traditionally examine the scientific literature for physical evidence that the protein is actually produced in cells. The presence of mRNA sequences reveals that the gene is expressed but do not reveal whether or not the mRNA is translated into a protein. However, some proteins may be difficult to detect because they are made in small quantities, in specific cells or tissues, or at a particular time in development. Codon use by the mRNA of suspect genes can be examined for consistency with codon use by other genes that are known to be translated, as discussed in Chapter 9.

Usually many sequences will be shown (in this example, 78) if the query is not restricted to a particular GenBank field (e.g., gene name) and database (e.g., RefSeq). These search restrictions can be made by clicking the mouse on the "Limits" link and by then by choosing the field to search and database, and by eliminating all draft sequences, for example. When these choices are made, the number of matches is reduced to 4. The 'history" link will show the number of matches as the search criteria are changed and "preview/index" link allows extra terms to be added and for combinations of search terms to be used. Biological databases are beginning to use "controlled vocabularies" for entering data so that these defined terms can confidently be used for database subsequent searches.

The found sequences can be readily scanned to find one of interest and search for additional information such as structural information, sequence domains, and related sequences. Boolean logic can also be used to search for database entries that have combinations of terms as commonly provided by most public search engines. Thus, using the search phrase <mouse AND ERCC1> and searching all GenBank fields, Entrez will look for sequence entries with both of these terms and using the phrase <"mouse ERCC1"> where the quotation marks are included, Entrez will search for an entry with the phrase "mouse ERCC1" in one of the fields. While visiting the site, note that Entrez has been adapted to search through a number of other biological databases, and also through PubMed, and these searches are available from the Entrez Web page.

Entrez has a number of additional features, including a batch feature that supports retrieval of a set of sequences given a list of their accession numbers, the ability to save a search term for later searches, and the ability to create data links to external Web sites. The problems at the end of the chapter provide some examples using the sophisticated search routines of Entrez to retrieve sequences. In Chapter 12, methods for accessing GenBank to retrieve a sequence when the accession number is known without using a Web page are described.

RETRIEVING A SPECIFIC SEQUENCE

Even following the above instructions, it can be difficult to retrieve the sequence of a specific gene or protein simply because of the sheer number of sequences in the GenBank database and the complex problem of indexing them. For projects that require the most currently available sequences, the nonredundant (nr) databases should be searched. Other projects may benefit from the better-curated and annotated protein sequence databases, including RefSeq, PIR, and SwissProt. The genome databases described in Chapter 11 can also provide the sequence of a particular gene or protein. Another valuable resource is genome display Web sites such as NCBI's LocusLink, which provides a query Web page for searching a genome for a particular feature such as the *ERCC1* gene used in the above example. LocusLink provides links to all sequence, reference, molecular, and other information, in addition to providing a display of the chromosomal location in which the gene is located in the human genome. There are many additional genome Web sites that offer similar facilities; these sites are discussed further in Chapter 11.

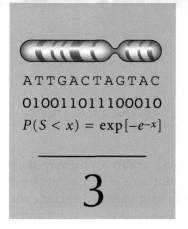

ATTGACTAGTAC

010011011100010

$P(S < x) = \exp[-e^{-x}]$

3

Alignment of Pairs of Sequences

CONTENTS

INTRODUCTION

Knowing how to align a pair of nucleic acid or protein sequences is a fundamentally important area of bioinformatics. For very similar sequences with runs of identical or commonly found substitutions, this is quite readily done; but as the sequences become more divergent, they also become more difficult to align. Thus, a method is required to find the very best possible or optimal alignment given an expected pattern of variation in sequences that are related but have also diverged over evolutionary time. Even if such requirements are met, human judgment may have to be used because there may be more than one possible alignment and some regions may align much better than others, leaving the poorly aligning regions in doubt. This chapter discusses how to perform pair-wise sequence alignments and how to score these alignments. The chapter following this one describes how to evaluate the significance of alignment scores.

CHAPTER GUIDE FOR BIOLOGISTS

This chapter is a very important one for biologists because sequence alignment is one type of sequence analysis they need to perform well. The chapter explores methods for aligning a significant fraction or all of two protein or DNA sequences, usually protein sequences. If two sequences can be aligned, they share an evolutionary relationship. Proteins may also share short sequence patterns that represent functional domains. Similarly, nucleic acid sequences, e.g., promoter sequences, may have short patterns that represent binding sites for transcription factors. Finding these patterns in protein or DNA sequences is another form of sequence alignment that is discussed in Chapters 5 and 10.

For aligning two protein sequences, the beginner often chooses a program without much of an idea as to how it works and then takes for granted that the alignment produced is correct. If sequences are very much alike, the approach probably does not matter. But biologists usually want to push sequence alignment methods to find more distant sequence relationships as evidence of functional or evolutionary diversity. For genome analysis, very long sequences need to be compared to identify similar regions. It is important to learn which approaches are best in each situation. However, it is also important to recognize that some problems, e.g., the alignment of long sequences that have variable degrees of sequence similarity, are difficult computationally, with no ready solutions.

There are many ways to align two sequences, especially when there are insertions/deletions in the alignment. To find the best or optimal alignment of the sequences, one must first devise a system for scoring alignments and then use a computer program that finds the highest-scoring alignment by utilizing this scoring system. For proteins, the scoring system in common use consists of a table of positive and negative scores (scoring matrix) for all possible amino acid substitutions based on those found in many related protein sequences and a gap penalty score for introducing gaps into the alignment. For DNA sequences, a simple positive match score and mismatch/gap penalty scores are used. Varying the scoring system can sometimes produce a better alignment with the commonly used sequence alignment programs.

Fortunately, we have reliable information as to which programming tools are most useful for aligning protein sequences. Experience has proven that what is usually needed is a local alignment of sequences, an alignment that emphasizes finding the most-alike regions in the sequences without necessarily including all of the sequences in the alignment. A local alignment of a query sequence with each matching database sequence is produced by BLASTP. A program such as LALIGN uses a proven algorithm, the Smith–Waterman algorithm, for finding the highest-scoring or optimal localized region of similarity. Use of such programs along with standard amino acid substitution scores and gap penalties will provide an optimal alignment for most pairs of sequences. Global alignments between sequences that include all of the sequences are sometimes needed. Programs that employ the Needleman–Wunsch algorithm are used in this case.

As more sequences have been obtained and analyzed, they have been grouped into classes and families based on patterns of conserved sequence similarity, e.g., a functional domain in proteins or a transcription-factor-binding site in DNA. Rather than using the standard scoring matrices described above for all proteins, more specific scoring systems based on substitutions in these groups can be used to analyze and align these groups. These methods depend on producing a multiple sequence alignment of the related sequences, as described in Chapter 5.

Suitable programs for pair-wise sequence alignment may be found on Web sites with windows into which the sequences can be copied and pasted, variables set, and an output alignment produced. Alternatively, a local copy of alignment programs may be used. The EMBOSS project includes alignment programs and is ideally suited for this purpose. Local copies can be easier to use if a Web interface is made, a relatively straightforward task using the programming tools described in Chapter 12.

CHAPTER GUIDE FOR COMPUTATIONAL SCIENTISTS

As already discussed in other chapters, one of the most important recent discoveries in biology is that organisms use similar genes to provide their biological functions. To discover the similarities, gene sequences comprising the four DNA bases (A, G, C, and T) are usually translated into protein sequences made up of 20 amino acids using the genetic code. Recall that when reading DNA sequences which code for a protein sequence, 3 contiguous bases (a codon) specify 1 amino acid and that up to 6 codons may specify the same amino acid. The use of codons varies among organisms, making the alignment of the DNA sequences that encode the same protein in different organisms a difficult task. Also, the larger alphabet size for proteins facilitates their alignment because few random alignments in sequences that are not related will occur.

A goal of biologists is to perform comparisons between the genes found in one organism and those of another to predict those that have the same function based on sequence similarity. In many cases, the biologist knows what the gene does in one of the organisms, and if a protein from the other can be aligned with the known protein, a functional prediction can be made. Proteins that align very well will also have the same three-dimensional structure. Hence, the biologist can use the alignment to formulate new experiments, design drugs, etc., based on a structure prediction.

Why does sequence variation occur? During evolution, gene sequences undergo spontaneous change or mutation. The following generations will then have a different sequence. Most of the changes are deleterious to the organism and are never observed, and, of the ones that are not deleterious, most have no effect at all. Hence, when we compare proteins, we see substitutions that are compatible with keeping the same structure and function. Rarely, a change occurs that is of benefit to the organism by providing some kind of a biological advantage. These rare changes, which will also be observed as sequence variation, are the basis of evolutionary change.

Another type of sequence variation during evolutionary change is the movement of blocks of sequence to create new genes and proteins. Sequences that are the result of such complex rearrangements are more difficult to align except in the conserved domains. Analysis of these types of rearrangements is discussed in Chapters 5 and 10.

WHAT SHOULD BE LEARNED IN THIS CHAPTER?

- What is a sequence alignment?
- The difference between a global and a local alignment and what the uses of each are.
- How to use the dot matrix methods to analyze genes and chromosomes.
- The steps performed by the Needleman–Wunsch and Smith–Waterman algorithms to produce a sequence alignment.
- How to use scoring matrix values and gap penalties to produce a sequence alignment.

Glossary Terms

Affine gap penalty is a gap penalty score that is a linear function of gap length, consisting of a gap opening penalty and a gap extension penalty multiplied by the length of the gap. Using this penalty scheme greatly enhances the performance of dynamic programming methods for sequence alignment and makes sense for scoring deletions of diverse length as single evolutionary steps with a slight extra weight for being longer.

Alignment score is a computed score based on the number of matches, substitutions, and insertions/deletions (gaps) within an alignment. For proteins, scores for matches and substitutions are often derived from a scoring matrix such as the BLOSUM and PAM matrices, and gap penalties suitable for the matrix are chosen. Alignment scores are in log odds units, often bit units (log to the base 2). Higher scores denote better alignments. For DNA sequences, usually a match and mismatch score is chosen along with a gap penalty that will produce the most reasonable alignment.

BLOSUM scoring matrices are commonly used to align protein sequences. The BLOSUM scoring matrices are based on the substitutions found in alignments of a large number of protein families of variable sequence similarity. The alignments are found in the BLOCKS database. Overrepresented sequences in these alignments are grouped together to different extents to reduce their contribution. The BLOSUM62 matrix, which has 62% of the alike sequences grouped, is the most commonly used matrix for scoring protein sequence alignments.

Convergent evolution refers to the evolution of two genes to the same biological function. However, because they have different genetic starting points, the resulting sequences are not similar.

Distance score between aligned sequences is a measure of the evolutionary distance between the sequences. A scoring system opposite to the commonly used similarity scores is used. For example, a mismatch may be given a positive score as it represents a greater amount of change between sequences, and a match may be given no score or a negative penalty score.

Dot matrix diagrams provide a graphical method for comparing two sequences. One sequence is written horizontally across the top of the graph and the other along the left-hand side. Dots are placed within the graph if the same letter appears at the corresponding positions in the sequences. A diagonal series of dots appearing as lines on the graph indicates an alignment of a series of positions in the sequences. The matrix may be filtered to reveal the most-alike regions by scoring a minimal threshold number of matches (stringency) within a sliding sequence window.

Dynamic programming algorithm solves the problem of finding the optimal alignment between two sequences by breaking the alignment down into a series of sequential subalignments that can be readily computed. A table or matrix is built with one sequence running across the top and the other down the side. The object is to calculate the best possible score for each position in the matrix, by considering every possible combination of matches, mismatches, and gaps up to that position. Starting at the upper-left matrix positions corresponding to the sequence ends, the algorithm scores the initial positions, including any possible gap combination. Longer alignments are then built starting with these initial ones by filling in the remaining positions in the matrix. Eventually, all possible combinations of sequence matches and gaps are taken into account. The highest-scoring matrix position is the best possible score between the sequences. The route followed through the matrix to achieve this score gives the corresponding sequence alignment.

Filtered dot matrix During pair-wise sequence alignment using the dot matrix method, random matches can be filtered out by using a sliding window to compare the two sequences. Rather than comparing a single pair of sequence positions at a time, a window of adjacent positions in the two sequences is compared and a dot, indicating a match, is generated only if a certain threshold number of matches (stringency) is found.

Gap is a blank position in the alignment of two sequences caused by an insertion or deletion in one of the sequences.

Genome is the entire complement of genetic material of an organism, including all the genes that specify proteins and RNA molecules, and any other sequences that are present. This is composed of half the paired chromosomes in a somatic (body) cell and all the chromosomes in a germ (sex) cell.

Global alignment is an alignment of two or more sequences that matches as many characters as possible in all of the sequences.

Heuristic method for sequence alignment is a tried-and-proven-by-experience method that tries to find a reasonable alignment. In contrast, an algorithm for sequence alignment is generally proven to produce a given result, e.g., an optimal alignment between two sequences. An example of a heuristic alignment method is starting with a search for similar sequence patterns and then building a longer alignment based on these patterns.

Homologous sequences refers to two or more sequences that can be quite readily aligned such that they must have originated from a common ancestor sequence in earlier evolutionary time.

Horizontal transfer is the transfer of genetic material between two distinct species that do not ordinarily exchange genetic material. The transferred DNA becomes established in the recipient genome and can be detected by a novel phylogenetic history or codon content compared to the rest of the genome.

Indel is a term used to represent either insertions or deletions in sequence alignments.

Local alignment is an alignment that includes only the best-matching, highest-scoring regions in two or more sequences. For two sequences, a local alignment is obtained by using the Smith–Waterman algorithm and suitable choices for match, mismatch, and gap penalty scores.

Log odds score is the logarithm of an odds score. In sequence alignment, the log odds score for aligning each pair of sequence characters is added to the score for all other aligned pairs and gap penalties are subtracted. The result is a log odds score for the alignment.

Markov chain, as used in sequence analysis, describes an idealized model of DNA or protein sequence change over evolutionary time in which each sequence position changes independently of the previous history of the position and of the other positions in the sequence. These models are called percent accepted mutation (PAM) models of sequence change.

Needleman–Wunsch algorithm is a dynamic programming algorithm for producing a global alignment of sequences.

Natural selection is a process originally conceived by the father of evolutionary theory, Charles Darwin. This process refers to the influence of the environment on an organism. Species constantly undergo random mutations that change their DNA sequences, often with deleterious or no effect and rarely with a beneficial effect for survival of the organism. As the environment changes, a selection force develops so that only the most genetically fit organisms survive and reproduce.

Odds score is the ratio of the likelihoods of two events or outcomes, e.g., the chance of winning to the chance of losing. In sequence alignments, the odds score for matching two sequence characters is the ratio of the frequency with which the characters are found to be aligned in related sequences divided by the frequency with which those same two characters are expected to align by chance alone, calculated from the frequency of occurrence of each character in sequences. Odds scores for aligning a set of individually aligned positions are obtained by multiplying the odds scores of the individual positions. Odds scores are often converted to logarithms to create log odds scores that can be added to obtain the log odds score of a sequence alignment.

Optimal alignment is the highest-scoring of all possible alignments between two sequences usually found by the dynamic programming algorithm for a given scoring system, usually a proven combination of match/mismatch scores and gap penalties.

Orthologous sequences are usually protein sequences found in two or more organisms that are so much alike that they almost certainly have a similar three-dimensional structure, domain structure, and biological function. Orthology is found by searching for all members of protein families in two organisms and identifying the most alike.

Pair-wise sequence alignment is an alignment of two sequences.

PAM scoring matrix, or percent accepted mutation scoring matrix, is a table or matrix that describes the odds that a sequence position, e.g., an amino acid, has changed into a second one during a period of evolutionary time. Amino acid PAM matrices are derived from an analysis of observed (accepted by the evolutionary process) amino acid substitutions in families of closely related sequences. A PAM1 matrix describes the pattern of changes expected over a period of time when an average of 1% of the positions change. Higher-value PAM matrices to predict changes over longer periods of time are derived by multiplying the PAM1 matrix by itself. These matrices are commonly used for proteins, but may also be used for DNA sequence alignments.

Paralogous sequences are usually protein sequences that are related through gene duplication events. These events may lead to the production of a family of related proteins with similar sequence but also variable biological function within a species.

Parametric sequence alignment is an algorithm that finds a range of possible alignments based on varying the parameters of the scoring system for matches, mismatches, and gap penalties.

Percent identity in sequence alignment describes the percentage of aligned positions in an alignment in which the sequence characters are identical. Gapped positions are usually not counted.

Percent similarity in sequence alignment is commonly used to describe the percentage of aligned positions in a protein sequence alignment in which the amino acids are identical or commonly occurring substitutions. Positions aligned with gaps are usually not counted.

Scoring matrix, as used in this chapter, refers to a scoring matrix that lists all 20 amino acids across the top and down the side of the matrix, and then gives log odds scores for amino acid substitutions at the intersection of the rows and columns for each amino acid pair. These scores may be based on observed evolutionary changes or alignments of many sequence families. The matrix may be used to score an alignment of any pair of protein sequences. For comparison, there is another type of scoring matrix, the position-specific scoring matrix (see Chapter 5), that represents the substitutions found in each column of an alignment of a specific family of proteins.

Sequence alignment is the comparison of two or more sequences by searching for a series of individual characters or character patterns that are in the same order in the sequences.

Sequence alignment score is the sum of the individual log odds scores for each pair of aligned sequence characters in an alignment less a penalty for each gap of one or more positions in the alignment.

Sequence blocks, or BLOCKS, are conserved ungapped patterns approximately 3–60 amino acids in length in a set of related proteins that were used to produce the BLOSUM amino acid scoring matrices.

Sequence similarity. Two sequences are similar if the order of sequence characters is recognizably the same in the sequences, and is usually found by showing that they can be aligned.

Sequence similarity score (sequence alignment) is the sum of the number of identical matches and conservative (high scoring) substitutions in a sequence alignment divided by the total number of aligned sequence characters. Gaps are usually, but not always, ignored.

Smith–Waterman algorithm is a dynamic programming algorithm for locating the highest-scoring local alignments of sequences. The key feature is that all negative scores calculated in the dynamic programming matrix are changed to zero to avoid extending poorly scoring alignments and to assist in identifying local alignments starting and stopping anywhere in the matrix.

Space or time complexity of an algorithm is the amount of computer memory or time (number of steps) required by the algorithm to solve a problem. For sequence alignment algorithms, complexity is a function of the sequence lengths.

Stringency is used in describing the dot matrix method of sequence alignment in which sliding windows of sequence in the two sequences are compared and a specified number of matches are required for producing a dot in the matrix.

WHAT IS A SEQUENCE ALIGNMENT?

Sequence alignment is the procedure of comparing two (pair-wise alignment) or more (multiple sequence alignment) DNA or protein sequences by searching for a series of individual characters or character patterns that are in the same order in the sequences. Two sequences are aligned by writing them across a page in two rows. Identical or similar characters are placed in the same column, and nonidentical characters can be placed either in the same column as a mismatch or opposite a gap in the other sequence. In an optimal alignment, nonidentical characters and gaps are placed to bring as many identical or similar characters as possible into vertical register. Sequences that can be readily aligned in this manner are said to be similar.

There are two types of pair-wise sequence alignments, global and local, and they are illustrated below in Figure 3.1. In global alignment, an attempt is made to align the entire sequence, using all sequence characters, up to both ends of each sequence. Sequences that are quite similar and approximately the same length are suitable candidates for global alignment. In local alignment, stretches of sequence with the highest density of matches are aligned, thus generating one or more islands of matches or subalignments in the aligned sequences. Local alignments are more suitable for aligning sequences that are similar along some of their lengths but dissimilar in others, sequences that differ in length, or sequences that share a conserved region or domain.

Global Alignment

The Needleman–Wunsch algorithm is used to produce global alignments between pairs of DNA or protein sequences. For the two hypothetical protein sequence fragments in Figure 3.1, the global alignment is stretched over the entire sequence length to include as many matching amino acids as possible up to and including the sequence ends. A global alignment is made possible by including gaps either within the middle of the alignment or at either end of one or both sequences. Vertical bars between the sequences indicate the presence of identical amino

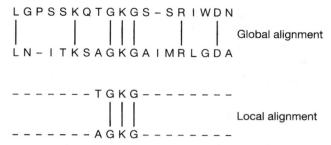

FIGURE 3.1. Distinction between global and local alignments of two protein sequences.

acids. Although there is an obvious region of identity in this example (the sequence GKG preceded by a commonly observed substitution of T for A), a global alignment will misalign this area so that more

amino acids along the entire sequence lengths can be matched.

Local Alignment

The Smith–Waterman algorithm is used to produce local alignments between pairs of DNA or protein sequences. In a local alignment, the alignment stops at the ends of regions of strong similarity, and a much higher priority is given to finding these local regions (Fig. 3.1) than to extending the alignment to include more neighboring amino acid pairs. Dashes in the figure indicate sequence not included in the alignment. This type of alignment favors finding conserved nucleotide patterns in DNA sequences or amino acid domains in protein sequences.

SEQUENCE ALIGNMENT REVEALS FUNCTION, STRUCTURE, AND EVOLUTIONARY INFORMATION

Sequence alignment is useful for discovering functional, structural, and evolutionary information in DNA or protein sequences, particularly for protein sequences. It is important to obtain the best possible, so-called "optimal," alignment to discover this information. Sequences that are very much alike, or "similar" in the parlance of sequence analysis, probably have the same function, be it a regulatory role in the case of similar DNA molecules or a similar biochemical function and three-dimensional structure in the case of proteins. Additionally, if two sequences from different organisms are similar, there may have been a common ancestor sequence, and the sequences are then defined as being homologous. An alignment indicates the changes that could have occurred between the two homologous sequences and a common ancestor sequence during evolution, as shown in Figure 3.2.

With the advent of genome analysis and large-scale sequence comparisons, it becomes important to recognize that sequence similarity may be an indicator of several possible types of ancestor relationships, or there may be no ancestor relationship at all, as illustrated in Figure 3.3. For example, new gene evolution is often thought to occur by gene duplication, creating two tandem copies of the gene at the same chromosomal location, followed by independent mutational events in these two copies. In rare cases, new mutations in one of the copies provide an advantageous change in function. The two copies may then

evolve along separate pathways. Although the resulting separation of function will generate two related sequence families, sequences among both families will still be similar due to the single gene ancestor. In addition to gene duplication events, genetic rearrangements can reassort domains in proteins, leading to more complex proteins with an evolutionary history that is difficult to reconstruct (Henikoff et al. 1997). Sequence alignment methods can be used to sort out such rearrangements.

Evolutionary theory provides terms that may be used to describe sequence relationships. Homologous genes that share a common ancestry and function in the absence of any evidence of gene duplication are called orthologs. When there is evidence for gene duplication, the genes in an evolutionary lineage derived from one of the copies and with the same function are also referred to as orthologs. The two copies of the duplicated gene and their progeny in the evolutionary lineage are referred to as paralogs. In other cases, similar regions in sequences may not have a common ancestor but may have arisen independently by two evolutionary pathways converging on the same function, called convergent evolution. There are some remarkable examples of convergent evolution in protein structures. For instance, although the enzymes chymotrypsin and subtilisin have totally different three-dimensional structures and folds, the active sites show similar structural features, including histidine (H), serine (S), and aspartic acid (D) in the

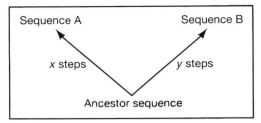

FIGURE 3.2. The evolutionary relationship between two similar sequences and a possible common ancestor sequence that would make the sequences homologous. The number of steps required to change one sequence to the other is the evolutionary distance between the sequences, and is also the sum of the number of steps to change the common ancestor sequence into one of the sequences (x) plus the number of steps required to change the common ancestor into the other (y). The common ancestor sequence is not available, such that x and y cannot be calculated; only $x + y$ is known. By the simplest definition, the distance $x + y$ is the number of mismatches in the alignment (gaps are not usually counted), as illustrated in Fig. 1.3. In a phylogenetic analysis of three or more similar sequences, the separate distances from the ancestor can be estimated, as discussed in Chapter 7.

catalytic sites of the enzymes (for discussion, see Branden and Tooze 1991). Additional examples are given in Chapter 11 (p. 529). Such sequences are referred to as analogous (Fitch 1970). A closer examination of alignments can help to sort out possible evolutionary origins among similar sequences (Tatusov et al. 1997). The sequence analysis methods that are used are discussed further in Chapter 11.

As pointed out by Fitch and Smith (1983), sequences can be either homologous or nonhomologous, but not in between. Sequences that are similar to each other, i.e., can be readily aligned, are homologous and derived from a common ancestor. In contrast, sequences that do not align well are nonhomologous, since there is no evidence of common ancestry. The genetic rearrangements referred to above can give rise to chimeric genes, in which only certain regions of the sequences can be aligned and others cannot. In such cases, the regions of similarity do appear to have a common ancestry. Referring to sequences as 30% homologous in such situations leads to an inaccurate and incomplete description of the sequence lineage.

It is important to describe these relationships accurately in publications. A common error in the molecular biology literature is to refer to sequence "homology" when one means sequence "similarity." Sequence similarity is a measurement of the matching characters in an alignment, whereas homology is a statement of common evolutionary origin.

Another complication in tracing the origins of similar sequences is that individual genes may not share the same evolutionary origin as the rest of the genome in which they presently reside. Genetic events such as symbioses and virus-induced transduction can cause horizontal transfer of genetic material between unrelated organisms. Genetic variation in bacteria is thought to be largely due to the transfer of DNA between unrelated species. In such cases, the evolutionary history of the transferred sequences and that of the recipient organisms will be different. Again, with the capability of detecting such events in the genomes of organisms comes the responsibility to describe these changes with the correct evolutionary terminology. In this case, the resulting sequences are xenologous (Gray and Fitch 1983). Lawrence and Ochman (1997) have shown that horizontal transfer of genes between species is as common in enteric bacteria, if not more common, than mutation. Describing such changes requires a careful description of the sequence origins. As discussed in Chapters 7 and 11, phylogenetic and other types of sequence analyses help to uncover such events. (See Bushman [2002] for a fuller discussion of lateral DNA transfer.)

THERE ARE THREE PRINCIPAL METHODS OF PAIR-WISE SEQUENCE ALIGNMENT

Alignment of two sequences is performed using the following methods:

1. Dot matrix analysis.

2. The dynamic programming (or DP) algorithm.

3. Word or *k*-tuple methods, such as used by the programs FASTA and BLAST, which are described in detail in Chapter 6.

The following overview discussion briefly describes the pros and cons of these methods and their use. Each method is described in detail in the sections below.

Unless the two sequences are known to be very much alike, the dot matrix method should be considered as a first choice for pair-wise sequence alignment, because this method displays graphically any possible

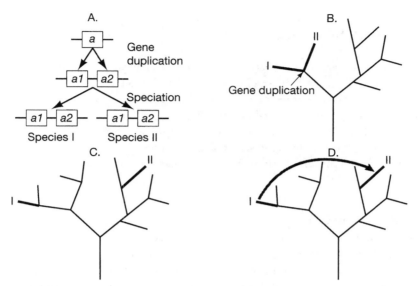

FIGURE 3.3. Origins of genes having a similar sequence. Shown are illustrative examples of gene evolution. In *A*, a duplication of gene *a* to produce tandem genes *a1* and *a2* in an ancestor of species I and II has occurred. Species I and II are then compared to a number of other species, and a binary tree is made whose outer branches (or leaves) represent species and whose inner branches represent evolutionary relationships among the species (for more details, see Chapter 7). Separation of the duplicated region by speciation gives rise to two separate branches in the evolutionary tree, shown in *B* as blue and red. *a1* in species I and *a1* in species II are orthologous because they have not diverged from each other very much (they are side by side in the tree) and are therefore likely to have maintained the same biological function. Similarly, *a2* in species I and *a2* in species II are orthologous. However, the *a1* genes are paralogous to the *a2* genes because they arose from a gene duplication event, indicated in *A*. If two or more copies of a gene family have been separated by speciation in this fashion, they tend to all undergo change as a group, due to gene conversion-type mechanisms (Li and Graur 1991). In *C*, a gene in species I and a different gene in species II have converged on the same function by separate evolutionary paths. Such analogous genes, or genes that result from convergent evolution, include proteins that have a similar active site but within a different backbone sequence. In *D*, genes in species I and II are related through the transfer of genetic material between species, even though the two species are separated by a long evolutionary distance. Although the transfer is shown between outer branches of the evolutionary tree, it could also have occurred in lower-down branches, thus giving rise to a group of organisms with the transferred gene. Such genes are known as xenologous or horizontally transferred genes. Transfer of the P transposable elements between *Drosophila* species is a prime example of such horizontal transfer (Kidwell 1983). Horizontal transfer also is found in bacterial genomes and can be traced as a regional variation in base composition within chromosomes. A similar type of transfer is that of the small ribosomal RNA subunits of mitochondria and chloroplasts, which originated from early prokaryotic organisms. Symbiotic relationships between organisms may be a precursor event leading to such exchanges. Other rearrangements within the genome (not shown) may produce chimeric genes comprising domains of genes that were evolving separately.

sequence alignments as diagonals on the matrix. Dot matrix analysis can readily reveal the presence of insertions/deletions and direct and inverted repeats that are more difficult to find by the other, more automated methods. The major limitation of the method is that most dot matrix computer programs do not show an actual alignment of the sequences.

The dynamic programming algorithm, first used for global alignment of protein sequences by Needleman and Wunsch (1970) and for local alignment by Smith and Waterman (1981a), provides one

or more alignments of the sequences. The algorithm generates an alignment by starting at the ends of two sequences and attempting to match all possible pairs of characters between the sequences and by following a scoring scheme for matches, mismatches, and gaps. This procedure generates a matrix of numbers that represents all possible alignments between the sequences. The highest set of sequential scores in the matrix defines an optimal alignment. For proteins, an amino acid substitution matrix, such as the Dayhoff percent accepted mutation matrix 250 (PAM250) or

A NOTE ABOUT MULTIPLE SEQUENCE ALIGNMENT

Specific computational methods are also available for producing alignments of three or more protein or DNA sequences. The highly conserved residues that define structural and functional domains in protein families can be identified from a multiple alignment of three or more protein sequences. New members of such families can then sometimes be found by searching sequence databases for other sequences with these same domains. Alignment of three or more DNA sequences can assist in finding conserved regulatory patterns in DNA sequences. Despite the great value of multiple sequence alignments, obtaining such alignments presents a very difficult algorithmic problem. The methods that have been devised to optimize multiple sequence alignment are discussed in Chapter 5.

blosum substitution matrix 62 (BLOSUM62), is used to score matches and mismatches. Similar matrices are available for aligning DNA sequences (see Tables 3.4–3.6) although a simple combination of a positive match score, a negative mismatch score, and a negative gap penalty score is often used.

The dynamic programming algorithm is guaranteed in a mathematical sense to provide the optimal (very best, or highest-scoring) alignment for a given set of user-defined variables, including choice of scoring matrix and gap penalties. Fortunately, experience in the computational community with the dynamic

programming algorithm has provided much help for the user to make the best choices of parameters, and dynamic programming has become widely used. However, the dynamic programming method can be slow due to the very large number of computational steps required, which increase approximately as the square or cube of the sequence lengths. The computer memory requirement also increases as the square of the sequence lengths. Thus, it has been difficult to use the method for very long sequences. Fortunately, computer scientists have greatly reduced these time and space requirements to near-linear relationships

Notes for Flowchart (p. 75)

1. This chart assumes that both sequences are protein sequences or that both are DNA sequences. If only one is a DNA sequence, that sequence should be translated and then aligned with the second protein sequence.

2. The local alignment program, e.g., LALIGN in the FASTA suite or WATER in the EMBOSS suite, usually has a recommended scoring matrix and gap penalty combination. It is important to make sure that the combination is one that is known to produce a confined, local alignment with random (or scrambled) sequences. A global alignment program such as NEEDLE in the EMBOSS suite may also be used with sequences that are similar enough to be readily aligned and that are approximately the same length.

3. For protein sequences, a high-quality alignment, e.g., 50%, is one that includes most of each sequence, a significant proportion of identities throughout the alignment, multiple examples of conservative substitutions (chemically and structurally similar amino acids), and relatively few gaps that are confined to specific regions of the alignment. A poor-quality alignment of protein sequences includes only a portion, e.g., less than 20%, of the sequences, has few and widely dispersed identities and conservative substitutions, tends to include regions of low complexity (repeats of the same amino acid), and includes gaps that are obviously necessary to obtain the alignment. For DNA sequences, a significant alignment

must include long runs of identity, few gaps in the aligned regions, and an overall high degree of identity (greater than 80%). For two random or unrelated DNA sequences of length 100 and normal composition (0.25 of each base), the longest run of identical matches that can be expected is 6 or 7 (see text). A clue as to the significance of an alignment may also be obtained by using an alignment program that gives multiple alternative alignments, e.g., LALIGN. The first alignment found should have a highly significant score, which should be much greater than the following ones, which are designed so that the same sequence positions will not be aligned a second time. Usually, these subsequent alignments should be random and hence low scoring. If the subsequent alignments give a high-scoring alignment, then there are alternative ways of aligning the sequences, possibly due to low-complexity regions, repeated domains, or some other feature. Protein sequences are more readily aligned because it is relatively difficult to align more than 1 or 2 amino acids in unrelated sequences without introducing a gap.

4. The result of this analysis can be a guide for the test of significance that follows. In the test described in this chapter, the second sequence is scrambled and realigned with the first sequence. Scrambling can be done at the level of the individual nucleotide or amino acid, or at the level of words by keeping the composition of short stretches of sequence intact.

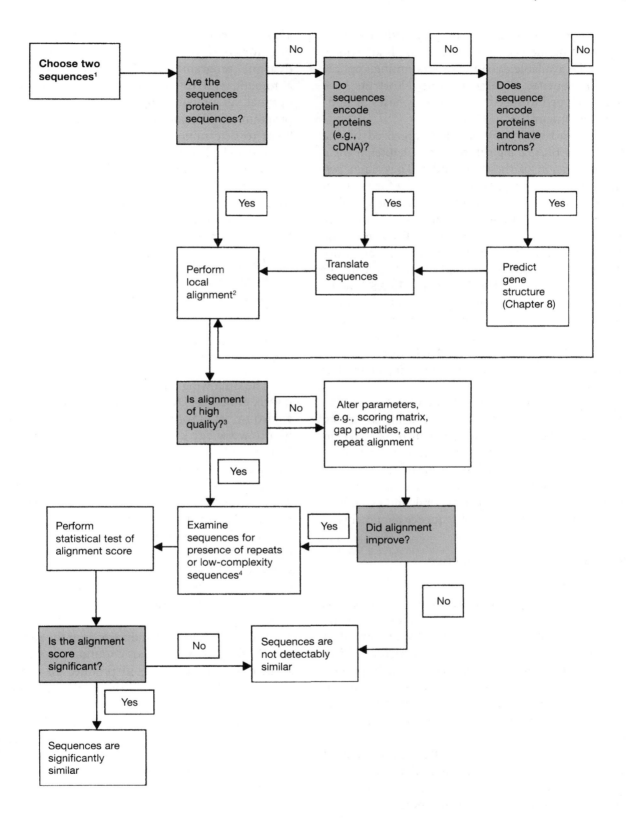

without compromising the reliability of the dynamic programming method, and these methods are widely used in the available dynamic programming applications to sequence alignment. Other shortcuts have been developed to speed up the early phases of finding an alignment (see p. 234).

The word or *k*-tuple methods are used by the FASTA and BLAST algorithms (see Chapter 6 for a fuller discussion of these methods). They align two sequences very quickly, by first searching for identical short stretches of sequences (called words or *k*-tuples) and by then joining these words into an alignment by the dynamic programming method. These methods are fast enough to be suitable for searching an entire database for the sequences that align best with an input query sequence. The FASTA and BLAST methods are heuristic; i.e., an empirical method of computer programming in which rules of thumb are used to find solutions and feedback is used to improve performance. These methods are reliable in a statistical sense, and usually provide the best scoring alignment that is possible.

Optimal alignments provide useful information to biologists concerning sequence relationships by giving the best possible information as to which characters in a sequence should be in the same column in an alignment, and which are insertions in one of the sequences (or deletions on the other). This information is important for making functional, structural, and evolutionary predictions on the basis of sequence alignments.

Dot Matrix Pair-wise Sequence Comparison

A dot matrix analysis, first described by Gibbs and McIntyre (1970), is primarily a method for comparing two sequences to look for possible alignment of characters between the sequences. The method is also used for finding direct or inverted repeats in protein and DNA sequences, and for predicting regions in RNA that are self-complementary and that, therefore, have the potential of forming secondary structure through base-pairing.

The major advantage of the dot matrix method for finding sequence alignments is that all possible matches of residues between two sequences are found, leaving the investigator the choice of identifying the most significant ones by examination of the dot matrix for long runs of matches, which appear as diagonals. Then, sequences of the actual regions that

align can be detected by using one of two other methods for performing sequence alignments, e.g., dynamic programming. These latter methods are automatic and usually show a highest-scoring, or optimal, alignment. Alignments generated by these programs can be compared to the dot matrix alignment to determine whether the longest regions are being matched and whether insertions and deletions are located in the most reasonable places.

Every laboratory that does sequence analysis should have at least one dot matrix program available. In choosing a program, look for as many of the features described below as possible.

1. The dot matrix should be visible on the computer terminal, thus providing an interactive environment so that different types of analyses may be tried.

2. Use of colored dots can enhance the detection of regions of similarity (Maizel and Lenk 1981).

The following are some commonly used dot matrix programs. DOTTER, for the UNIX X-Window environment (Sonnhammer and Durbin 1995; http://www.cgr.ki.se/cgr/groups/sonnhammer/ Dotter.html), has interactive features. The Genetics Computer Group programs COMPARE and DOT-PLOT also perform a dot matrix analysis. Although not a dot matrix method, the program PLALIGN in the FASTA suite may be used to display the alignments found by the dynamic programming method between two sequences on a graph (http://fasta.bioch.vir ginia.edu/; Pearson 1990). The EMBOSS suite (http://www.hgmp.mrc.ac.uk/Software/EMBOSS/) includes three dot matrix programs: (1) dotmatcher, which aligns sequences using a scoring matrix; (2) dottup, which finds common words in the sequences; and (3) dotplot, which finds common patterns in the sequences and uses these to produce an alignment (Rice et al. 2000). A dot matrix program that may be used with a Web browser is described in Junier and Pagni (2000) (http://www.isrec.isb-sib.ch/java/dotlet/ Dotlet.html). The examples given in the next section use the dot matrix feature of DNA Strider (version 1.3) on a Macintosh computer and the EMBOSS dotmatcher program on a Mac OS X or Linux machine. To the best of our knowledge, there are no MS Windows-based programs. (Additional descriptions of the dot matrix method have appeared in Doolittle [1986] and States and Boguski [1991].)

The Dot Matrix Method

In the dot matrix method of sequence comparison, one sequence (A) is listed across the top of a page and the other sequence (B) is listed down the left side, as illustrated in Figure 3.4. Starting with the first character in B, the comparison then moves across the page in the first row and places a dot in any column where the character in A is the same. The second character in B is then compared to the entire A sequence, and a dot is placed in row 2 wherever a match occurs. This process is continued until the page is filled with dots representing all the possible matches of A characters with B characters. Any region of similar sequence is revealed by a diagonal row of dots. Isolated dots not on the diagonal represent random matches that are probably not related to any significant alignment.

Detection of matching regions may be improved by filtering out random matches in a dot matrix. Filtering is achieved by using a sliding window to compare the two sequences. Instead of comparing single-sequence positions, windows of adjacent posi-

tions in the two sequences are compared at the same time, and a dot is printed on the page only if a certain minimal number of matches occur when comparing these windows. The window starts at the positions in A and B to be compared and includes characters in a diagonal line going down and to the right, comparing each pair in turn, as in making an alignment. A larger window size is generally used for DNA sequences than for protein sequences because the number of random matches expected between unrelated sequences is much greater due to the use of only four DNA symbols as compared to 20 amino acid symbols. A typical window size for DNA sequences is 15 bases and a suitable stringency or match requirement in this window is 10, meaning that if there are 10–15 matches within the window, a dot is printed at the first base in the window. For protein sequences, the matrix is often not filtered, but a window size of 2 or 3 amino acids and a stringency of 2 identities will highlight matching regions. If two proteins are expected to be related but to have long regions of dissimilar sequence with only a small proportion of identities, such as similar active sites, a large window, e.g., 20, and small stringency, e.g., 5, should be useful for seeing any similarity. Identification of sequence alignments by the dot matrix method can be aided by performing a count of dots in all possible diagonal lines through the matrix to determine statistically which diagonals have the most matches, and by comparing these match scores with the results of random sequence comparisons (Gibbs and McIntyre 1970; Argos 1987).

An example of a dot matrix analysis between the DNA sequences that encode the *Escherichia coli* phage λ *c*I and phage P22 *c*2 repressor proteins is shown in Figure 3.4. With a window of 1 and stringency of 1, there is so much noise that no diagonals can be seen, but, as shown in the figure, with a window of 11 and a stringency of 7, diagonals appear in the lower right. The analysis reveals that there are regions of similarity in the 3′ ends of the coding regions, which, in turn, suggests similarity in the carboxy-terminal domains of the encoded proteins. Note that sequential diagonals in the matrix do not line up exactly, indicating the presence of extra nucleotides in one sequence (the λ *c*I gene on the vertical scale). The diagonals shown in the lower part of the matrix (D) reveal a region of sequence similarity in the carboxy-terminal domains of the proteins. A small insertion in the *c*I gene that is approximately in the middle of this region (C) shifts

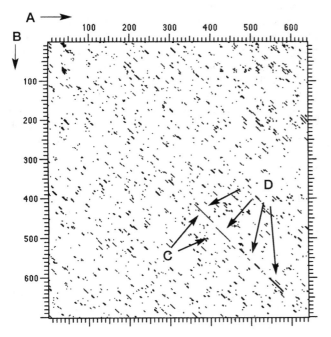

FIGURE 3.4. Dot matrix analysis of DNA sequences encoding phage λ *c*I (*vertical sequence, B*) and phage P22 *c*2 (*horizontal sequence, A*) repressors. This analysis was performed using the dot matrix display of the Macintosh DNA sequence analysis program DNA Strider, vers. 1.3. The window size was 11 and the stringency 7, meaning that a dot is printed at a matrix position only if 7 out of the next 11 positions in the sequences are identical.

the diagonal slightly downward, accounting for this pattern.

An example of a dot matrix analysis between the amino acid sequences of the same two *E. coli* phage λ *cI* and phage P22 *c2* repressor proteins is shown in Figure 3.5. This matrix was filtered differently: A window of 10 was chosen and the BLOSUM62 scoring matrix values were summed within the window. Also, the sequence runs up instead of down the page. A dot was printed only if a threshold score of 23 BLOSUM62 units (half-bits) was found in the window. As found with the DNA sequence alignment of the corresponding genes, diagonals going to the upper right reveal a region of sequence similarity in the carboxy-terminal domains of the proteins. The small insertion in the *cI* protein approximately in the middle of this region, which shifts the diagonal slightly and which is also observed in the DNA alignment of these corresponding genes, is also visible. Note that when comparing amino acids with a simple scoring system of 1 for identities and 0 for nonidentities (not shown), the window size is much smaller, e.g., 1, than required for DNA sequence comparisons due to the greater number of possible symbols (20 amino acids) and therefore fewer expected random matches.

In summary, for DNA sequence dot matrix comparisons, use long windows and high stringencies, e.g., 7 and 11, 11 and 15. For protein sequences, use short windows, e.g., 1 and 1, for window size and stringency, respectively, except when looking for a short domain of partial similarity in otherwise not-similar sequences. In this case, use a longer window size and a small stringency, e.g., 15 and 5, for window and stringency, respectively.

Variations in protein analysis by the dot matrix method. There are three types of variations in the analysis of two protein sequences by the dot matrix method:

1. Chemical similarity of the amino acid R group or some other feature for distinguishing amino acids may be used to score similarity in protein sequences.

2. A symbol comparison table such as the PAM250 or BLOSUM62 table may be used (States and Boguski 1991) for protein sequences. An example is given in Figure 3.5. These tables provide scores for matches based on their occurrence in aligned protein families, and they are discussed later in this

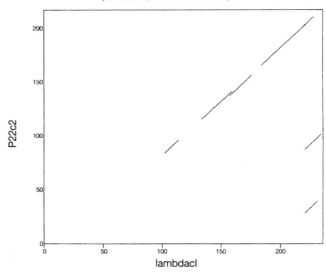

FIGURE 3.5. Dot matrix analysis of the amino acid sequences of the phage λ *cI* (*horizontal sequence*) and phage P22 *c2* (*vertical sequence*) repressors performed as described in Fig. 3.4. Note that the sequence on the left side runs up the *y* axis instead of down as in Fig. 3.4. Hence, similarity diagonals will go from lower left to upper right. The program used in this case was the DotMatcher program in the EMBOSS suite at http://www.hgmp.mrc.ac.uk/Software/EMBOSS/. Instead of using a value of 1 for a match and 0 for no match in the dot matrix, scores were based on the BLOSUM62 values shown in Fig. 3.16. The window size was 10 and a dot was printed at the first window position only if a score of at least 23 was obtained in the window. It is possible to produce and use a simple scoring matrix with match score = 1 and mismatch score = 0 and to score the number of matches with windows as short as 3 with this software. DotMatcher uses an X-Window graphical display that can be implemented on a computer running UNIX, Linux, or a Macintosh OS X operating system or can be used on any PC running an X terminal interface (e.g., visit http://www.uk.research.att.com/vnc/) to a remote UNIX or Linux server that is running the program.

chapter (pp. 95 and 102). When these tables are used, a dot is placed in the matrix only if a minimum similarity score is found. These table values may also be used in a sliding window option, which averages the score within the window and prints a dot only above a certain average score.

3. Several different matrices can be made, each with a different scoring system, and the scores can be averaged. The scores of each possible diagonal through

the matrix are then calculated, and the most significant ones are identified and shown on a computer screen (Argos 1987). This method should be useful for aligning more distantly related proteins.

Finding Sequence Repeats

Dot matrix analysis can also be used to find direct and inverted repeats within DNA or protein sequences. Repeated regions in whole chromosomes may be detected by a dot matrix analysis, and an interactive Web-based program has been designed for showing these regions at increasing levels of detail (http://genome-www.stanford.edu/Saccharomyces/SSV/viewer_start.html). Direct repeats may also be found by performing sequence alignments with dynamic programming methods (see next section). When used to align a sequence with itself, the program LALIGN will show alternative possible alignments between the repeated regions; PLALIGN will plot these alignments on a graph similar in appearance to a dot matrix (see http://fasta.bioch.virginia.edu/; Pearson 1990). Using these programs, the sequence is analyzed against itself and the presence of repeats is revealed by diagonal rows of dots that are offset from the main diagonal of sequence identity. A Bayesian method for finding direct repeats is described on page 151. Inverted repeats require special handling and are discussed in Chapters 8 and 9.

An example of a dot matrix analysis for locating direct repeats in the amino acid sequence of the human low-density lipoprotein (LDL) receptor is shown in Figure 3.6. A list of additional proteins with direct repeats is given in Doolittle (1986, p. 50), and repeats are also discussed in States and Boguski (1991, p. 109). As discussed in Chapters 10 and 11, there are many examples of proteins composed of multiple copies of a single domain.

Finding Repeats of a Single Sequence Symbol

A dot matrix analysis can also reveal the presence of repeats of the same sequence character many times. These repeats become apparent on the dot matrix of a protein sequence against itself as horizontal or vertical rows of dots that sometimes merge into rectangular or square patterns. Such patterns are particularly apparent in the right and lower regions of the dot matrix of the human LDL receptor shown in Figure 3.6A, but they are also seen throughout the rest of the

matrix. The occurrence of such repeats of the same sequence character increases the difficulty of aligning sequences because they can create alignments with artificially high scores. A similar problem occurs with regions in which only a few sequence characters are found, called low-complexity regions. Programs that automatically detect and remove such regions from the analysis so that they do not interfere with database similarity searches are discussed in Chapter 6.

Dynamic Programming Method for Sequence Alignment

Dynamic programming is a computational method that is used to align two protein or nucleic acid sequences. The method is very important for sequence analysis because it provides the highest-scoring, or optimal, alignment between two sequences. Programs that perform this type of analysis on sequences are readily available, and there are Web sites that will perform the analysis. However, the method requires the intelligent use of several variables in the programs. Thus, it is important to understand how the programs work so that informed choices of these variables can be made.

The dynamic programming method encompasses the following features:

1. *Optimal alignment.* The dynamic programming algorithm provides a reliable computational method for aligning DNA and protein sequences. This method compares every pair of characters in the two sequences and generates an alignment. This alignment will include matched and mismatched characters and gaps in the two sequences that are positioned so that the number of matches between identical or related characters gives the maximum possible score. The method has been proven mathematically to produce the best-scoring, or optimal, alignment between two sequences for a given scoring system.

2. *Global and local alignments.* Both global and local types of alignments may be made by simple changes in the basic dynamic programming algorithm. A global alignment program is based on the Needleman–Wunsch algorithm, and a local alignment program on the Smith–Waterman algorithm, described below (p. 83). The predicted alignment will be given an odds score that gives

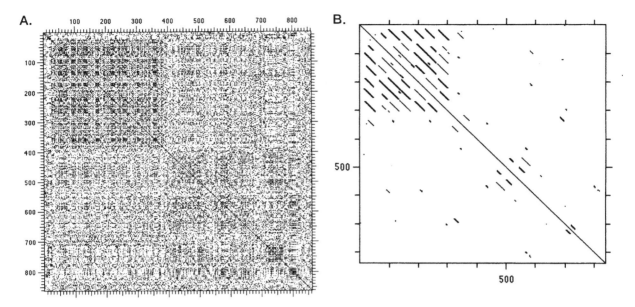

FIGURE 3.6. Dot matrix analysis of the human LDL receptor against itself using DNA Strider, vers. 1.3, on a Macintosh computer. (*A*) Window 1, Stringency 1. There is a diagonal line from upper left to lower right due to the fact that the same sequence is being compared to itself. The rest of the graph is symmetrical about this line. Other (quite hard to see) lines on either side of this diagonal are also present. These lines indicate repeated sequences perhaps 50 amino acids or so long. Patches of high-density dots, e.g., at the position corresponding to position 800 in both sequences representing short repeats of the same amino acid, are also seen. (*B*) Window 23, Stringency 7. The occurrence of longer repeats may be found by using this sliding window. In this example, a dot is placed on the graph at a given position only if 7/23 of the residues are the same. These choices are arbitrary, and several combinations may need to be tried. Many repeats are seen in the first 300 positions. A pattern of approximate length 20 and at position 30 is repeated at least six times at positions 70, 100, 140, 180, 230, and 270. Two longer, overlapping repeats of length 70 are also found in this same region starting at positions 70 and 100, and repeated at position 200. Since few of these diagonals remain in a new analysis at 11/23 (stringency/window) and all disappear at 15/23, they are not repeats of exactly the same sequence but they do represent an average of about 7/23 matches with no deletions or insertions. The information from the above dot matrix may be used as a basis for listing the actual amino acid repeats themselves by one of the other methods for sequence alignment described below.

the ratio of the chance of obtaining the score between related sequences to the chance of obtaining the same score between unrelated sequences. There is a method to calculate whether or not an alignment obtained this way is statistically significant. One of the sequences may be scrambled many times, and each randomly generated sequence may be realigned with the second sequence to demonstrate that the original alignment is unique. The statistical significance of alignment scores is discussed in Chapter 4.

3. *Choices in dynamic programming.* Another feature of the dynamic programming algorithm is that

the alignments obtained depend on the choice of a scoring system for comparing character pairs and on penalty scores for gaps. For protein sequences, the simplest system of comparison is one based on identity. A match in an alignment is only scored if the two aligned amino acids are identical. However, one can also examine related protein sequences that can be aligned easily and find which amino acids are commonly substituted for each other. The chance of a substitution between any pair of the 20 amino acids may then be used to produce alignments. Recent improvements and experience with the dynamic programming programs and the scoring systems have greatly simplified their use. For example, computational biologists have found that optimal alignments are most readily obtained by using log odds scores for matches, mismatches, and substitutions and by including gaps in the alignment that have an associated, size-dependent gap penalty score.

4. *Alignments, alignment scores, and alternative alignments.* The dynamic programming algorithm is guaranteed to produce an optimal alignment that has the highest possible alignment score between two sequences for a given scoring system. However, it is important to recognize that several different alignments of two protein or DNA sequences may have the highest possible alignment score, and an even greater number of alignments may have alignment scores that are almost as high as the optimal one. These suboptimal alignments may sometimes be more biologically meaningful than optimal ones. When two sequences are aligned using dynamic programming, the best score is calculated at each scoring matrix position based on scores that were previously calculated at certain other matrix positions, as prescribed by the dynamic programming algorithm. A trace back to the previously used score is kept to be used later for constructing a path of best scores through the matrix that represents the best scoring alignment. Sometimes, there are two previous positions that give the same score in the current location. This result means that two different combinations of matches, mismatches, and gaps can provide the best alignment score up to that position. This effect can eventually lead to a family of possible alignments, each giving the highest possible alignment score. The number of alignments can be increased even further by keeping track of alignments that have alignment scores that are almost as high as that of an optimal alignment.

Most sequence alignment programs report only one optimal alignment. In some cases, the alternative alignments are not very different from the optimal one and are found in regions that do not align well, so they are not very significant. In other cases, repeated patterns or domains or other sequence features may be present in one or both sequences. Some alignment programs, e.g., LALIGN, can help to sort out these possibilities.

LALIGN first computes an optimal alignment of two sequences. The program then finds a second alignment of the sequences in which the sequence positions that matched in the first alignment are not repeated. Similarly, a third sequence alignment is found, and so on. If there is a single unique way of aligning the sequences, then the first alignment will give a score that reflects this similarity. The subsequent alignments will then give a much lower score because they are matching unrelated sequence positions. Conversely, if there is more than one way of aligning the sequences, e.g., if there are repeated sequence domains or low-complexity regions comprising just a few amino acids or bases, the subsequent alignments may have a high score. Hence, there will be less confidence that the first alignment is unique and optimal. In addition to such features, alignment programs have been greatly improved in algorithmic design and performance. With the advent of faster machines, it is possible to do a dynamic programming alignment between a query sequence and an entire sequence database and to find the similar sequences in several minutes. Dynamic programming has also been used to perform multiple sequence alignments, but only for a small number of sequences, because the complexity of the calculations increases substantially for more than two sequences (see Chapter 5).

Sequence alignment programs that use dynamic programming are available as a component of most commercial sequence analysis packages. Sequences can also be pasted into a text area on a guest Web page on a remote host machine that will perform a dynamic programming alignment, and there are also versions of alignment programs that will run on a microcomputer (Table 3.1).

5. *Analysis goals.* In deciding to perform a sequence alignment, it is important to keep the goal of the analysis in mind. Is the investigator interested in trying to find whether two proteins have similar domains or structural features, whether they are in the same family with a related biological function, or whether they share a common ancestor relationship? The desired objective will influence the way the analysis is done. In addition, there are several decisions to be made along the way, including the type of program, whether to pro-

TABLE 3.1. *Web sites for alignment of sequence pairs*

Name of site	Web address	References
Bayes block aligner[a]	http://www.wadsworth.org/resnres/bioinfo	Zhu et al. (1998)
Likelihood-weighted sequence alignment[b]	http://stateslab.bioinformatics.med.umich.edu/service	see Web site
PipMaker (percent identity plot), a graphical tool for assessing long alignments	http://bio.cse.psu.edu	Schwartz et al. (2000)
BCM Search Launcher[c]	http://searchlauncher.bcm.tmc.edu/	see Web site
SIM—Local similarity program for finding alternative alignments	http://us.expasy.org/	Huang et al. (1990); Huang and Miller (1991); Pearson and Miller (1992)
Global alignment programs (GAP, NAP)	http://genome.cs.mtu.edu/align/align.html	Huang (1994)
FASTA program suite[d]	http://fasta.bioch.virginia.edu/	Pearson and Miller (1992); Pearson (1996)
Pairwise BLAST[e]	http://www.ncbi.nlm.nih.gov/gorf/bl2.html	Altschul et al. (1990)
AceView[f] shows alignment of mRNAs and ESTs to the genome sequence	http://www.ncbi.nlm.nih.gov/IEB/Research/Acembly	see Web site
BLAT[f] Fast alignment for finding genes in genome	http://genome.ucsc.edu	Kent (2002)
GeneSeqer[f] predicts genes and aligns mRNA and genome sequences	http://www.bioinformatics.iastate.edu/bioinformatics2go/	Usuka et al. (2000)
SIM4[f]	http://globin.cse.psu.edu	Floria et al. (1998)

[a]See Chapter 4 for description and examples.

[b]A description of the probabilistic method of aligning two sequences is described in Durbin et al. (1998) and Chapter 4. A related topic, hidden Markov models for multiple sequence alignments, is discussed in Chapter 5.

[c]This server provides access to a number of Web sites offering pair-wise alignments between nucleic acid sequences, between protein sequences, or between a nucleic acid and a protein sequence.

[d]The FASTA algorithm normally used for sequence database searches (see Chapter 6) provides an alternative method to dynamic programming for producing an alignment between sequences. Briefly, all short patterns of a certain length are located in both sequences. If multiple patterns are found in the same order in both sequences, these provide the starting point for an alignment by the dynamic programming algorithm. Older versions of FASTA performed a global alignment, but more recent versions perform a local alignment with statistical evaluations of the scores. The program PLFASTA in the FASTA program suite provides a plot of the best-matching regions, much like a dot matrix analysis, and thus gives an indication of alternative alignments. The FASTA suite is also available from Genestream at http://vega.igh.cnrs.fr/. Programs include ALIGN (global, Needleman–Wunsch alignment), LALIGN (local, Smith–Waterman alignment), LALIGNO (Smith–Waterman alignment, no end gap penalty), FASTA (local alignment, FASTA method), and PRSS (local alignment with scrambled copies of second sequence to do statistical analysis). Versions of these programs that run with a command-line interface on MS-DOS and Macintosh microcomputers are available by anonymous FTP from ftp.virginia.edu/pub/fasta.

[e]The BLAST algorithm normally used for database similarity searches (Chapter 6) can also be used to align two sequences.

[f]Program useful for aligning expressed gene sequences (ESTs or mRNA) to genomic DNA.

duce a global or local alignment, the type of scoring matrix, and the value of gap penalties to be used. There are a very large number of amino acid scoring matrices to choose from, some much more popular than others, and these scoring matrices are designed for different purposes. Some, such as the Dayhoff PAM matrices, are based on an evolutionary model of protein change, whereas others, such as the BLOSUM matrices, are designed to identify members of the same family.

Alignments between DNA sequences using dynamic programming require similar kinds of con-

siderations. It is often worth the effort to try several approaches to find out which choice of scoring system and gap penalty gives the most reasonable result. Fortunately, most alignment programs come with a recommended scoring matrix and gap penalties that are useful for most situations. A more recent development (see Bayesian methods discussed in Chapter 4) is the simultaneous use of a set of scoring matrices by a method that generates the most probable alignments (see Table 3.1). The final choice as to the most believable alignment is up to the investigator, subject to the condition that reasonable decisions have been made regarding the methods used.

Sequence alignment by dynamic programming

can give an alignment of strongly or weakly similar sequences. For sequences that are very similar, e.g., 95%, the optimal sequence alignment is usually quite obvious, and a computer program may not even be needed to produce the alignment. As the sequences become less and less similar, the alignment becomes more difficult to produce, and one is less confident of the result. For protein sequences, similarity can still be detected down to a level of approximately 20% amino acid identity. At this level of identity, the relative numbers of mismatched amino acids and gaps in the alignment have to be decided empirically and a decision made as to which gap penalties work the best for a given scoring matrix. Alignment of sequences at this level of identity is called the "twilight zone" of sequence alignment by Doolittle (1981). The alignment program may provide a quite convincing alignment at 20% identity, which suggests that the two sequences are homologous. However, in such a case, the statistical significance of the alignment score should be evaluated, as described in Chapter 4.

Description of the Dynamic Programming Algorithm

Alignment of two sequences without allowing gaps requires an algorithm that performs a number of comparisons roughly proportional to the square of the average sequence length, as in a dot matrix comparison. If the alignment is to include gaps of any length at any position in either sequence, the number of comparisons that must be made becomes astronomical and is not achievable by direct comparison methods. Dynamic programming is a method of sequence alignment that can take gaps into account but requires a manageable number of comparisons.

The method of sequence alignment by dynamic programming and the proof that the method provides an optimal (highest-scoring) alignment are illustrated in Figures 3.7 and 3.8. To understand how the method works, we must first recall what is meant by an alignment, using the two protein sequences shown in Figure 3.7 as an example. The two sequences are written across the page, one under the

other, with the object being to bring as many amino acids as possible into register. In some regions, amino acids in one sequence are placed directly below identical amino acids in the second. In other regions, this process may not be possible, and nonidentical amino acids may have to be placed next to each other, or else gaps must be introduced into one of the sequences. Gaps are added to the alignment in a manner that increases the matching of identical or similar amino acids at subsequent portions in the alignment. Ideally, when two similar protein sequences are aligned, the alignment should have long regions of identical or related amino acid pairs and very few gaps. As the sequences become more distant, more mismatched amino acid pairs and gaps should appear.

The quality of the alignment between two protein sequences is computed using a scoring system that favors the matching of related or identical amino acids and penalizes for poorly matched amino acids and gaps. To decide how to score these regions, information on the types of changes found in related protein sequences is needed. These changes may be expressed by the following probabilities:

1. that a particular amino acid pair is found in alignments of related proteins;

2. that the same amino acid pair is aligned by chance in the sequences, given that some amino acids are abundant in proteins and others rare; and

3. that the insertion of a gap of one or more residues in one of the sequences (the same as an insertion of the same length in the other sequence), thus forcing the alignment of each partner of the amino acid pair with another amino acid, would be a better choice.

The ratio of the first two probabilities is usually provided in an amino acid substitution matrix. Each table entry gives the ratio of the observed frequency of substitution between each possible amino acid pair in related proteins to that expected by chance, given the frequencies of the amino acids in proteins. These ratios are called odds scores. The ratios are transformed to logarithms of odds scores, called log odds

```
sequence 1   V   D   S   -   C   Y
sequence 2   V   E   S   L   C   Y
SCORE        4   2   4  -11  9   7    SCORE = SUM OF AMINO ACID PAIR SCORES
(26)                                  MINUS SINGLE GAP PENALTY (11) = 15
```

FIGURE 3.7. Example of scoring a sequence alignment with a gap penalty. The individual alignment scores are taken from an amino acid substitution matrix.

```
I.  SCORE OF NEW      =    SCORE OF PREVIOUS +   SCORE OF NEW
    ALIGNMENT              ALIGNMENT (A)         ALIGNED PAIR

    V  D  S  -  C  Y       V  D  S  -  C             Y
    V  E  S  L  C  Y       V  E  S  L  C             Y

          15          =          8          +       7


II. SCORE OF          =    SCORE OF PREVIOUS +   SCORE OF NEW
    ALIGNMENT (A)          ALIGNMENT (B)         ALIGNED PAIR

    V  D  S  -  C          V  D  S  -                C
    V  E  S  L  C          V  E  S  L                C

          8           =          -1         +       9
```

III. REPEAT REMOVING ALIGNED PAIRS UNTIL END OF ALIGNMENT IS REACHED.

FIGURE 3.8. Derivation of the dynamic programming algorithm.

scores. These odds and log odds scores for amino acid pairs may be used to score a sequence alignment in the following ways. The odds score of a sequence alignment is the product of the individual odds scores for aligning each of the amino acid pairs. If the score of each pair is a log odds score, then the log odds score of the sequence alignment can be obtained by adding the log odds scores of the individual amino acid pairs, which is a much faster calculation (see box below). Examples of log odds scores for matching amino acid pairs may be found in the Dayhoff PAM250 and BLOSUM62 substitution matrices described below (pp. 95 and 102). These matrices contain positive and negative log odds values, reflecting the occurrence of amino acid pairs that are found more often than expected by chance and those expected less often by chance, respectively, in related proteins. A more detailed explanation of the use of odds and log odds scores may be found in the discussion of position-specific scoring matrices (PSSMs) in Chapter 5.

Using these scoring matrices, an alignment of sequential amino acid pairs with no gaps receives an overall score that is the sum of the positive and negative log odds scores for each individual amino acid pair in the alignment. The higher this score, the more significant is the sequence alignment and the more it resembles sequence alignments in related proteins. The score given for gaps in aligned sequences is negative, because insertions and deletions are expected to be rare events in the evolution of related proteins. The penalty for a gap will reduce the score obtained from an adjacent, matching region in the sequences, thus lowering the overall score for the alignment. The score of the alignment in Figure 3.7, using values from the BLOSUM62 amino acid substitution matrix and a gap penalty score of –11 for a gap of length 1, is 26 (the sum of amino acid pair scores) –11 = 15. The value of –11 as a penalty for a gap of length 1 is used because this value is already known from experience to favor the alignment of similar regions when the BLOSUM62 matrix is used. Choice of the gap penalty is discussed further below, where a table giving suitable choices is presented (see Table 4.3 on p. 142). As shown in Figure 3.7, the presence of the gap decreases significantly the overall score of the alignment.

CALCULATING THE ODDS SCORE OF AN ALIGNMENT FROM THE ODDS SCORES OF INDIVIDUAL AMINO ACID PAIRS

Protein sequence alignment scores are based on the individual scores of all amino acid pairs found in the alignment. The odds score for an amino acid pair is the ratio of the observed frequency of occurrence of that pair in alignments of related proteins over the expected frequency based on the proportion of amino acids in proteins. Alignments are built by making possible lists of amino acid pairs and by finding the most likely list using odds scores.

To calculate the odds score for an alignment, the odds scores for the individual pairs are multiplied. This calculation is similar to finding the probability of one event AND also of a second independent event by multiplying the probabilities (in contrast, if one event OR another is the choice, the probabilities are added). Thus, if the odds score of C/C is 7/1 and that of W/W is 50/1, then the probability of C/C and W/W being in the alignment is 7/1 x 50/1 = 350/1 (note that the order or position of the pairs in the alignment does not matter). If a less likely match is present in this same alignment, e.g., P/R with an odds score of 1/25, then the probability of C/C, W/W, and P/R being in the alignment is much lower, i.e., 7/1 x 50/1 x 1/20 = 17.5/1. Commonly, log odds scores are used in these calculations, and these scores are added to produce an overall log odds score for the alignment.

To perform this optimal alignment using odds scores, the method assumes that the odds score for matching a given pair of sequence positions is not influenced by the odds score of any other matching pair; i.e., there are no correlations expected among the amino acids found at various sequence positions. Another way of describing this assumption is that the sequences are each being modeled as a Markov chain, with the amino acid found at each position not being influenced by other amino acids in the sequence. Although correlations among positions in some sequence are expected, since they give rise to structure and function in large molecules like nucleic acids and proteins, this simplifying assumption allows the determination of a reasonable alignment between the sequences.

Even though one may be able to align the two short sequences in Figure 3.7 and to place the gap by eye, the dynamic programming algorithm will automatically place gaps in much longer sequence alignments to achieve the optimal, highest-scoring alignment. The derivation of the dynamic programming algorithm is illustrated in Figure 3.8, using the above alignment as an example. Consider building this alignment in steps, starting with an initial matching aligned pair of characters from the sequences (V/V) and then sequentially adding a new pair until the alignment is complete, at each stage choosing a pair from all the possible matches that provides the highest score for the alignment up to that point. If the full alignment finally reached on the left side of Figure 3.8 (part I) has the highest possible or optimal score, then the old alignment from which it was derived (A) by addition of the aligned Y/Y pair must also have been optimal up to that point in the alignment. If this were incorrect, and a different preceding alignment other than A was the highest-scoring one, then the alignment on the left would also not be the highest-scoring alignment,

although we started with that as a known condition. Similarly, in Figure 3.8 (part II), alignment A must also have been derived from an optimal alignment (B) by addition of a C/C pair. In this manner, the alignment can be traced back sequentially to the first aligned pair that was also an optimal alignment. One concludes that the building of an optimal alignment in this stepwise fashion can provide an optimal alignment of the entire sequences.

The example in Figure 3.8 also illustrates two of the three choices that can be made in adding to an alignment between two sequences: Match the next two characters in the next positions in each sequence, or match the next character to a gap in the upper sequence. The last possibility, not illustrated, is to add a gap to the lower sequence. This situation is analogous to performing a dot matrix analysis of the sequences, and of either continuing a diagonal or shifting the diagonal sideways or downward to produce a gap in one of the sequences. An example of using the dynamic programming algorithm to align two short protein sequences is illustrated in Figure 3.9.

MATHEMATICAL FORMULATION OF THE DYNAMIC PROGRAMMING ALGORITHM

The dynamic programming algorithm shown in Figure 3.9 may be written in mathematical form, using the diagram shown in Figure 3.10. The diagram indicates the moves that are possible to reach a certain matrix position (i, j) starting from the previous row and column at position $(i - 1, j - 1)$ or from any position in the same row and column.

The following equation describes the algorithm that is illustrated in Figures 3.9 and 3.10. There are three paths in the scoring matrix for reaching a particular position i, j: (1) a diagonal move from position $i - 1, j - 1$ to position i, j with no gap penalties, (2) a move from any position in column j to i, j, with a gap penalty, or (3) a move from any position in row i to i,j with a gap penalty. For two sequences $\mathbf{a} = a_1a_2 \ldots a_n$ and $\mathbf{b} = b_1 b_2 \ldots b_n$, where $S_{ij} - S(a_1a_2 \ldots a_i, b_1b_2 \ldots b_j)$ then (Smith and Waterman 1981a,b)

$$S_{ij} = \max \left\{ \begin{array}{l} S_{i-1,j-1} + s(a_ib_j), \\[2mm] \displaystyle\max_{x \geq 1}(S_{i-x,j} - w_x), \\[2mm] \displaystyle\max_{y \geq 1}(S_{i,j-y} - w_y) \end{array} \right.$$

(1)

where S_{ij} is the score at position i in sequence $\mathbf{a}$ and position j in sequence $\mathbf{b}$, $s(a_ib_j)$ is the score for aligning the characters at positions i and j, w_x is the penalty for a gap of length x in sequence $\mathbf{a}$, and w_y is the penalty for a gap of length y in sequence $\mathbf{b}$. Note that S_{ij} is a type of running best score as the algorithm moves through every position in the matrix. Eventually, all of the matrix positions (all S_{ij} values) are filled. If a global alignment that involves all of the sequences is required, the matrix score in the last row and column is used as the alignment score. The global score takes into account any end gap penalties that are required to include all of the sequences in the global alignment. Sometimes, end gap penalties are not included, in which case the highest score in the last row and column will represent the global alignment score, with the rest of one of the sequences not being included or scored in the alignment. If a local alignment that requires only the best-scoring local matches is required, then the highest-scoring matrix position will represent the score of an optimal alignment. To determine an optimal alignment of the sequences from the scoring matrix, a second matrix called the trace-back matrix is used (Fig. 3.9). The trace-back matrix keeps track of the positions in the scoring matrix that contributed to the highest overall score found. The sequence characters corresponding to these high-scoring positions may align or may be next to a gap, depending on the information in the trace-back matrix.

Use of the dynamic programming method requires a scoring system for the comparison of symbol pairs (bases for DNA sequences and amino acids for protein sequences), and a scheme for insertion/deletion (GAP) penalties. Once those parameters have been set, the resulting alignment for two sequences should always be the same. Scoring matrices are described below (see p. 94). The most commonly used matrices for protein sequence alignments are the log odds form of the PAM250 matrix and the BLOSUM62 log odds matrix. However, a number of other choices are available.

Dynamic Programming Can Provide Global or Local Sequence Alignments

The dynamic programming algorithm as described above is guaranteed to provide an optimal or highest-scoring alignment between two sequences. However, the alignment found will still depend on the scores chosen for matches, mismatches, substitutions, and gaps. Furthermore, there are two types of alignments that can be computed: a global one that includes most or all of the sequences, and a local one that includes only those parts of the sequences that produce a high-scoring alignment. The dynamic programming algorithm can be varied slightly to provide either a global or a local sequence alignment. The Needleman–Wunsch algorithm provides a global alignment of sequences, whereas the Smith–Waterman algorithm provides a local alignment. The difference between the global and local algorithms lies in what kinds of scores are kept in the dynamic programming scoring matrix. In the Needleman–Wunsch algorithm, all scores are kept no matter whether positive or negative. In the Smith–Waterman algorithm, only positive values are kept. Any negative values are converted to zero.

Examples of Global and Local Alignments

An example of global and local alignments using dynamic programming between two phage repressor proteins using the EMBOSS needle (Needleman–Wunsch) and water (Smith–Waterman) programs is shown in Figure 3.11. Note that the local alignment program revealed that the proteins have 65.6% similarity in the carboxy-terminal domain (see Fig. 3.11B), which is the region required for protein–protein interactions and a self-cleavage function that leads to phage induction. In these implementations of the Needleman–Wunsch and Smith–Waterman algorithms, the alignments found in the carboxy-terminal domain are identical. However, the Smith–Waterman method (Fig. 3.11B) only reports the most-alike regions, as expected by the focus on a local alignment strategy. In contrast, the Needleman–Wunsch method (Fig. 3.11A) shows the entire alignment of the sequences but reports a lower score of similarity due to the longer global alignment.

LALIGN (Fig. 3.12) is an implementation of the SIM algorithm for finding multiple unique (nonintersecting) alignments in DNA and protein sequences (Huang and Miller 1991) distributed in the FASTA package from W. Pearson. The program is also available on Web sites (see Table 3.1). Two features of these alignments are noteworthy: (1) The highest-scoring alignment is similar to that found by the needle program using a different amino acid substitution matrix and different gap penalties, with some minor variations in the more dissimilar regions and extension of the alignment farther into the amino-terminal domains. (2) By design, the alternative alignments never align the same amino acids and, in this example, the second and third alignments score much lower than the first one. These observations that strongly aligning regions are not significantly influenced by the scoring system, and that alternative high-scoring alignments are not possible, add convincing support that the initial alignment represents true similarity between these sequences. Another example of an alignment of these same sequences with a different scoring system is given on page 157.

Global alignment: Needleman–Wunsch algorithm. The dynamic programming method, as described above and in Figure 3.9, can be used to give a global alignment of sequences, as described by Needleman and Wunsch (1970). The global alignment algorithm has been proven mathematically and extended to include an improved scoring system by Smith and Waterman (1981a,b). The optimal score at each matrix position is calculated by adding the current match score to previously scored positions and subtracting gap penalties, if applicable. Each matrix position may have a positive or negative score, or 0. The Needleman–Wunsch algorithm will maximize the number of matches between the sequences along the entire length of the sequences. Gaps may also be present at the ends of sequences, in case there is extra sequence left over after the alignment. These end gaps

are often, but not always, given a gap penalty. The effect of penalties is illustrated later in this chapter.

To produce a global sequence alignment from the scoring matrix, a second matrix, called a trace-back matrix, is produced. This matrix keeps track of how the highest score was produced in each matrix position. For example, if the highest score was made by moving diagonally into the current position, an arrow will be placed between those two positions in the trace-back matrix. As the scoring matrix is filled, the trace-back matrix is filled with arrows to show how each scoring matrix position was calculated. The alignment is then produced from the trace-back matrix. An example of a trace-back matrix is given in Figure 3.9, part 4.

Local alignment: Smith–Waterman algorithm. A modification of the dynamic programming algorithm for sequence alignment provides the ability to create a local sequence alignment giving the highest-scoring local match between two sequences (Smith and Waterman 1981a,b). Local alignments are usually more meaningful than global matches because they identify conserved local sequence domains that are present in both sequences. They can also be used instead of the Needleman–Wunsch algorithm to match two sequences that may have a matched region that is only a fraction of their lengths, that have different lengths, that overlap, or where one sequence is a fragment or subsequence of the other. As discussed below, the Smith–Waterman algorithm is not particularly suitable for finding the highest-scoring alignment when the sequences include several regions that align locally but are separated by other regions that align poorly. In such cases, one would like to know the alignment that has the highest density of matches per unit length (e.g., number of identities in the alignment per hundred residues). An example would be an alignment of two genomic sequences that included matching and nonmatching regions. Finding this type of local alignment is a difficult computational problem and an area of current research. The Smith–Waterman algorithm is guaranteed to find the best-aligning regions between any pair of DNA or protein sequences.

When using the Smith–Waterman algorithm, the rules for calculating scoring matrix values are slightly different from the Needleman–Wunsch algorithm. The most important differences are (1) the scoring system must include negative scores for mismatches,

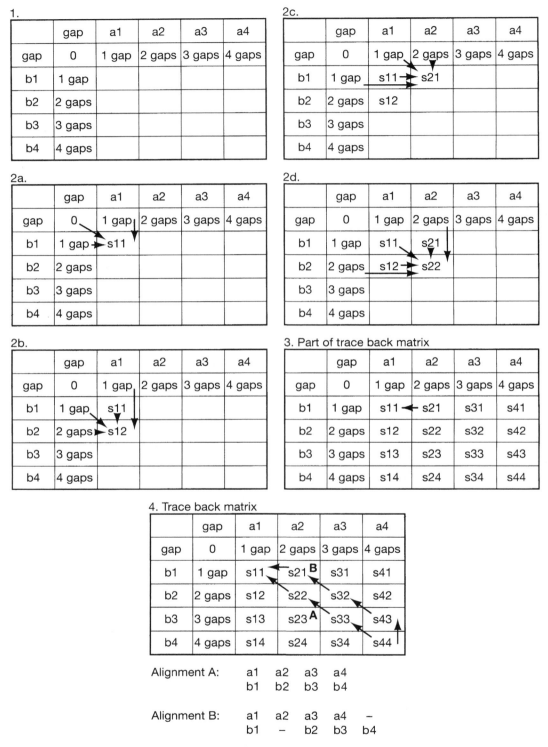

FIGURE 3.9. (*See facing page for legend.*)

and (2) when a dynamic programming scoring matrix value becomes negative, that value is set to zero, which has the effect of terminating any align-ment up to that point. As with a global alignment, in the local alignment a trace-back matrix is used to keep track of the moves in the scoring matrix.

Alignments are produced by starting at the highest-scoring positions in the scoring matrix and following a trace path from those positions up to a box that scores zero. The mathematical formulation of the dynamic programming algorithm is revised to include a choice of zero as the minimum value at any

FIGURE 3.9. Example of using the dynamic programming algorithm to align sequences a1 a2 a3 a4 and b1 b2 b3 b4 in a global alignment.

1. The sequences are written across the top and down the left side of a matrix, respectively, similar to that done in the dot matrix analysis, except that an extra row and column labeled "gap" are added to allow the alignment to begin with a gap of any length in either sequence. The gap rows are filled with penalty scores for gaps of increasing lengths, as indicated. A zero is placed in the upper left box corresponding to no gaps in either sequence.

2. Maximum possible values are calculated for all other boxes below and to the right of the top row and left column, taking into account any sized gap or no gap, using the steps listed in *a* through *d* below. The scores for individual matches a1-b1, a1-b2, etc., are obtained from a scoring matrix (symbol comparison table). To calculate the value for a particular matrix position, trial values are calculated from all moves into that position allowed by the algorithm. The allowed moves are from any position above or to the left of the current position, in the same column or row, or from the upper left diagonal position. The diagonal move attempts to align the sequence characters without introducing a gap. Thus, there is no gap penalty in this case. However, moves from above and to the left will introduce gaps, and thus will require one or more gap penalties to be used. (*a*) s11 is the score for an a1-b1 match added to 0 in the upper left position. According to the algorithm, there are two other possible paths to this position shown by the vertical and horizontal arrows, but they would probably have to give a lower score because they start at a gap penalty and must include an additional gap penalty. (*b*) Trial values for s12 are calculated and the maximum score is chosen. Trial 1 is to add the score for the a1-b2 match to the one gap penalty from the upper left diagonal position. Trial 2 is to subtract a single gap penalty from s11 above. The other two trials include only gap penalties and so likely cannot yield a higher score than trials 1 or 2. (*c*) All possible scores for s21 are calculated by the trial moves indicated. The best score should be obtained by either adding the score of an a2-b1 match to a single gap penalty from an upper left diagonal move or by subtracting a single gap penalty from s11 to the left. All other moves include only gap penalties. (*d*) Trial values of s22 are calculated by considering moves from s11, s21, and s12, and from the top row and left end column. s22 will be the best score of several possible choices, including adding the score for an a2-b2 match to s11, or subtracting a single gap penalty from s12 or s21. Other trials will normally be attempted from other positions above and to the left of this position, but in this case, they will probably not provide a higher score for s22 because they include multiple gap penalties. This procedure is followed to fill in the rest of the matrix positions until the lowest, rightmost position is reached. That position is calculated the same way as the others, except that any introduced gap will now be at the end of one of the sequences.

3. As the maximum scores for each matrix position are calculated, a record of the paths that produced the highest scores to reach each matrix position is kept. These short paths, which represent extending the alignment to another matching pair, with or without gaps, are recorded in another matrix called the trace-back matrix, illustrated below. For example, if moving from s11 to s21 gave the highest score of all moves to s21, then the corresponding region of the matrix will appear as shown.

4. The paths in the trace-back matrix are joined to produce an alignment. In the example shown, the highest-scoring matrix position in the sequence comparison matrix is located, in this case s44, and the arrows are then traced back as far as possible, generating the path shown. The corresponding alignment A is shown below the matrix. More than one alignment may be possible if there is more than one path from the highest-scoring matrix position. As an example, s43 could also be a high-scoring position, generating trace-back alignment B, an alignment that includes a gap opposite a2. Another gap may also be placed opposite b4, which has no matching symbol. Scoring end gaps is optional in the alignment programs. If included in this case, alignment B would be disfavored by an additional gap penalty. In addition to this series of alignments, or so-called clump of alignments starting from the highest-scoring position, there will be other possible alignments starting from other high-scoring matrix positions, and these may also have multiple pathways through the scoring matrix, each representing a different alignment. Note that these alignments are global alignments because they include the entire sequences.

matrix position. For two sequences $\mathbf{a} = a_1 a_2 \ldots a_n$ and $\mathbf{b} = b_1 b_2 \ldots b_n$, where $H_{ij} = H(a_1 a_2 \ldots a_i, b_1 b_2 \ldots b_j)$, then (Smith and Waterman 1981a):

$$
H_{ij} = \max \left\{
\begin{array}{l}
H_{i-1,j-1} + s(a_i b_j), \\
\max_{x \geq 1} (H_{i-x,j} - w_x), \\
\max_{y \geq 1} (H_{i,j-y} - w_y), \\
0
\end{array}
\right\} \tag{2}
$$

where H_{ij} is the score at position i in sequence $\mathbf{a}$ and position j in sequence $\mathbf{b}$, $s(a_i b_j)$ is the score for aligning the characters at positions i and j, w_x is the penalty for a gap of length x in sequence $\mathbf{a}$, and w_y is the penalty for a gap of length y in sequence $\mathbf{b}$.

The procedure is similar to that used for a global alignment in Figure 3.9. The major differences are that (1) there will not be an initial row and column for gaps because all negatively scoring positions in the matrix are replaced by zeros, (2) negative scores at all other matrix positions are also changed to zero, and (3) the trace-back matrix will start the alignment from the highest-scoring position in the matrix and thus may include only a portion of each sequence. Terminal gaps play no role in a local alignment, since they would only decrease the best score.

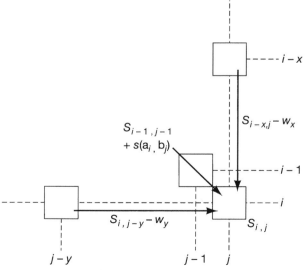

FIGURE 3.10. Formal description of the dynamic programming algorithm.

A. needle (Needleman Wunsch) global alignment.

Percent Identity: 75/238 (31.5%)
Percent Similarity: 118/238 (49.6%)

```
  1 MSTKKKPLTQEQLEDARRLKAIYEKKKNELGLSQESVADKMGMGQSGVGA    50
    ...||..  ..|::|     ::.:|.:.|.:::..:|:....:.
  1       MNTQLM-GERIRA----RRKKLKIRQAALGKMVGVSNVAISQ    37

 51 LFNGINALNAYNAALLAKILKVSVEEFSPSIAREIYEMYEAVSMQPSLRS   100
    ........|..|....|:|.|:.|.:..  :..::.:...|....:..|.
 38 WERSETEPNGENLLALSKALQCSPDYL---LKGDLSQTNVAYHSRHEPRG    84

101 EYEYPVFSHVQAGMFSPELRTFTKGDAERWVSTTKKASDSAFWLEVEGNS   150
    .||:.|.|.||.:..:...:..:.|...|.|..||...|::|||:|:|:|
 85 --SYPLISWVSAGQWMEAVEPYHKRAIENWHDTTVDCSEDSFWLDVQGDS   132

151 MTAPTGSKPSFPDGMLILVDPEQAVEPGDFCIARLGGD-EFTFKKLIRDS   199
    ||||.|.  |.|:||:||||||.....:|:|.|: |.||||:.|:
133 MTAPAGL--SIPEGMIILVDPEVEPRNGKLVVAKLEGENEATFKKLVMDA   180

200 GQVFLQPLNPQYPMIPCNESCSVVGKVIASQWPEETFG             237
    |:.||:|||||||||||..|.:|.::|.|.:::..:.
181 GRKFLKPLNPQYPMIEINGNCKIIGVVVDAKLANLP              216
```

B. water (Waterman Smith) local alignment.

Percent Identity: 60/125 (48.0%)
Percent similarity: 82/125 (65.6%)

```
104 YPVFSHVQAGMFSPELRTFTKGDAERWVSTTKKASDSAFWLEVEGNSMTA   153
    ||:.|.|.||.:..:...:..:.|...|.|..||...|::|||:|:|:|||
 86 YPLISWVSAGQWMEAVEPYHKRAIENWHDTTVDCSEDSFWLDVQGDSMTA   135

154 PTGSKPSFPDGMLILVDPEQAVEPGDFCIARLGGD-EFTFKKLIRDSGQV   202
    |.|.  |.|:||:||||||.....:|:|.|:  |.||||:.|:||:.
136 PAGL--SIPEGMIILVDPEVEPRNGKLVVAKLEGENEATFKKLVMDAGRK   183

203 FLQPLNPQYPMIPCNESCSVVGKVI             227
    ||:||||||||||..|.:|.::|.|:
184 FLKPLNPQYPMIEINGNCKIIGVVV             208
```

FIGURE 3.11. Example of alignment of phage λ *cI* and phage P22 *c2* repressors by dynamic programming using the EMBOSS(http://www.hgmp.mrc.ac.uk/Software/EMBOSS/) needle (*A*, Needleman–Wunsch algorithm) and water (*B*, Smith–Waterman algorithm) programs. These programs were run by typing the name of the program on the command lines of a UNIX or Linux host machine or a Mac OS X (UNIX) terminal window. A Web interface may be built as described in Chapter 12. The BLOSUM62 amino acid substitution matrix (Fig. 3.16) was used with a gap opening penalty of –11 and a gap extension penalty of –1, which are the values commonly used for alignments produced by BLAST searches (Chapter 6). In the program output, percent identity is based on the number of identical amino acid pairs '|' in the alignment, and percent similarity, on the number of identical (|) or similar ':' amino acid pairs divided by the length of the alignment including gaps. Note that gaps are not always counted in the denominator in this manner by other alignment programs. Similar amino acids are defined by high-scoring matches between the amino acid pairs in the substitution matrix. The most similar pairs were indicated by a ':', less similar pairs by a '.', and unrelated pairs by a space ' ' between the aligned amino acids. Although these dynamic programming programs provide a single optimal alignment, it is important to realize that a series of alignments with the same or a slightly lower score is usually possible. Note that the local alignment program reports a shorter alignment with stronger similarity than the global one and that the same amino acids are aligned by both methods in the aligned region they have in common.

```
LALIGN finds the best local alignments between two sequences
version 2.0u64 March 1998
Please cite:
X. Huang and W. Miller (1991) Adv. Appl. Math. 12:373-381

Comparison of:
(A) lamc1.pro LAMC1  REFORMAT of: cipro.pro  from: 1 - 237
(B) p22c2.pro P22C2  REFORMAT of: p22c2.pro  from: 1 - 216
using matrix file: pam250.mat, gap penalties: -12/-2

  34.0% identity in 206 aa overlap; score:  338

          30        40        50        60        70        80
LAMC1 KKNELGLSQESVADKMGMGQSGVGALFNGINALNAYNAALLAKILKVSVEEFSPSIAREI
      ....: .::.....:.... .... .. ..: .: :.: :. : ...  ...
P22C2 RRKKLKIRQAALGKMVGVSNVAISQWERSETEPNGENLLALSKALQCSPDYLLKGDLSQT
         20        30        40        50        60        70

          90       100       110       120       130       140
LAMC1 YEMYEAVSMQPSLRSEYEYPVFSHVQAGMFSPELRTFTKGDAERWVSTTKKASDSAFWLE
      : ..   ...   .::..: : :: .... :.: .:: . :....:::
P22C2 NVAYHSRHEPRG-----SYPLISWVSAGQWMEAVEPYHKRAIENWHDTTVDCSEDSFWLD
          80           90       100       110       120

         150       160       170       180       190       200
LAMC1 VEGNSMTAPTGSKPSFPDGMLILVDPEQAVEPGDFCIARLGGD-EFTFKKLIRDSGQVFL
      :.:.:::::.:  :.:.::.::::::  . :..  .:.:.  : :::::::...: ::
P22C2 VQGDSMTAPAG--LSIPEGMIILVDPEVEPRNGKLVVAKLEGENEATFKKLVMDAGRKFL
         130       140         150       160       170       180

         210       220       230
LAMC1 QPLNPQYPMIPCNESCSVVGKVIASQ
      .:::::::::: :..:...: :....
P22C2 KPLNPQYPMIEINGNCKIIGVVVDAK
         190       200       210

----------

  17.8% identity in 90 aa overlap; score:   37

          20        30        40        50        60        70
LAMC1 RRLKAIYEKKKNELGLSQESVAD-KMGMGQSGVGALFNGINALNAYNAALLAKILKVSVE
      ..::   .  . .:.:. .... ... ... ..:. ..::. . :: :. .
P22C2 KKLKIRQAALGKMVGVSNVAISQWERSETEPNGENLLALSKALQCSPDYLLKGDLSQTNV
         20        30        40        50        60        70

          80        90       100
LAMC1 EF-SPSIAREIYEMYEAVSMQPSLRSEYEY
      .. :. .:..:  :.: ::...: :.:
P22C2 AYHSRHEPRGSYPLISWVSAGQWMEAVEPY
          80        90       100

----------

  40.0% identity in 15 aa overlap; score:   36

         220       230
LAMC1 SCSVVGKVIASQWPE
      :..... : :.:: :
P22C2 SYPLISWVSAGQWME
          90
```

FIGURE 3.12. Example of LALIGN program (Table 3.1) for finding multiple, independent local alignments of two protein sequences in which the same amino acids are not aligned. Three independent alignments of the phage λ and P22 repressors are shown. The amino acid substitution matrix used was the log odds form of the Dayhoff PAM250 matrix provided with the program, with a gap opening penalty of −12 and a gap extension penalty of −2.

Does a Local Alignment Program Always Produce a Local Alignment and a Global Alignment Program Always Produce a Global Alignment?

Although a computer program that is based on the above Smith–Waterman local alignment algorithm is used for producing an optimal local alignment, this feature alone does not assure that a local alignment

will be produced. The scoring matrix or match and mismatch scores and the gap penalties chosen also influence whether or not a local alignment is obtained. Similarly, a program based on the Needleman–Wunsch algorithm, typically used to produce a global alignment, can also return a local alignment depending on the weighting of end gaps and on other scoring parameters. Often, one can simply inspect the alignment obtained by either method to see how many gaps are present. If the matched regions are long and cover most of the two aligning sequences and obviously depend on the presence of many gaps, the alignment is global. A local alignment, on the other hand, will tend to be shorter and not include many gaps. However, these tests are quite subjective, and a more precise method of knowing whether a given program and set of scoring parameters will provide a local or global alignment is required. Looking ahead in the chapter for a moment, the best way of knowing whether you have a global or local alignment is by looking at what happens when many random or completely unrelated sequences are aligned under the chosen conditions. As the length of the random sequences being aligned increases, the score of a global alignment will just increase proportionally. This is easy to see. Because a global alignment matches most of the sequences, and the negative mismatch score and gap penalties are deliberately chosen to be small in comparison to match scores in order to provide a long alignment, only matches count, and the score has to be proportional to the length.

If using a scoring matrix, a matrix that gives on the average a positive score to each aligned position, combined with a small enough gap penalty to allow extension of the alignment through poorly matched regions, will give a global alignment. Conversely, for the local alignment, a negative mismatch score and gap penalties are chosen to balance the positive score of a match and to prevent the alignment from growing into regions that do not match very well. The scoring matrix in this case will on average give a negative value to the mismatched positions, and the gap penalty will be large enough to prevent gaps from extending the alignment. The local alignment score of random sequences does not increase proportionally to sequence length, because the positive score of matches is offset by the mismatch and penalty scores. In this case, it may be shown by theory and experiment that the score of local random alignments increases much more slowly and proportionally to the logarithm of the product of the sequence lengths. It is this different behavior of the alignment score of random sequences with length that distinguishes global and local alignments.

One may well ask, Does it really matter whether I use a sequence alignment program based on the global alignment algorithm or one based on the local alignment algorithm? The answer is that sometimes both methods will provide the same alignment with the same scoring system and sometimes they will not. The most reasonable approach is to use a program based on the appropriate algorithm for the analysis at hand, and then to choose the scoring system carefully. Small changes in the scoring system can abruptly change an alignment from a local to a global one. There are examples in the bioinformatics literature where this feature of alignment scoring systems has been overlooked. The rest of this chapter and the following chapter are designed to provide a suitable guide for making the right choices.

Additional Use and Enhancements of the Dynamic Programming Algorithm for Sequence Alignments

The dynamic programming algorithm has been used for sequence alignment by a totally different scoring scheme based on distances instead of similarity scores between sequences. The intense computational requirements of sequence alignment by dynamic programming have also been reduced by enhancements to the way scores are stored in the scoring matrix. These improvements greatly speed up the computation time.

Use of distance scores for sequence alignment. As originally designed by Needleman and Wunsch and Smith and Waterman, the dynamic programming algorithm was used for sequence alignments scored on the basis of the similarity or identity of sequence characters with negative scores for unlikely mismatches and gaps. An alternative method is to score alignments based on differences between sequences and sequence characters; i.e., how many changes are required to change one sequence into another. Using this measure, the greater the distance between sequences, the greater the evolutionary time that has elapsed since the sequences diverged from a common ancestor. Hence, distance scores provide a more bio-

logically natural way to compare sequences than do similarity scores. Using a distance scoring scheme, Sellers (1974, 1980) showed that the dynamic programming method could be used to provide an alignment that highlighted the evolutionary changes. Smith et al. (1981) and Smith and Waterman (1981b) showed that alignments based on a similarity scoring scheme could give a similar alignment. Conversion between distance and similarity scores for sequence alignments is discussed in Chapter 7.

Regarding the equivalency of similarity and distance algorithms, Smith and Waterman (1981a,b) showed that the Needleman–Wunsch algorithm based on maximizing similarity scores and the Sellers algorithm based on minimizing distance scores were equivalent, if the following is true:

$$\text{sim}(a_i, b_j) + \text{dist}(a_i, b_j) = s_{\text{MAX}} \tag{3}$$

$$d_k = s_{\text{MAX}} \times k/2 + w_k \tag{4}$$

where $\text{sim}(a_i, b_j)$ and $\text{dist}(a_i, b_j)$ are the respective distance and similarity scores between characters a_i in one sequence and b_j in the second; s_{MAX} is the maximum possible similarity score for any pair of different characters; d_k and w_k are the respective gap scores used in the distance and similarity methods for a gap of length k.

In written form, Equation 3 states that the sum of the distance and similarity scores for two different characters should be equal to the maximum similarity score used for any pair of characters. In Equation 4, the weight assigned to gaps in the distance score should be larger than the Needleman–Wunsch alignment gap score by one-half the length of the gap times the largest similarity score used. For two sequences $\mathbf{a} = a_1 a_2 ... a_n$ and $\mathbf{b} = b_1 b_2 ... b_n$ of lengths m and n, respectively, the resulting relationship is also true,

$$S(\mathbf{a}, \mathbf{b}) + D(\mathbf{a}, \mathbf{b}) = s_{\text{MAX}}(n + m)/2 \tag{5}$$

where $S(\mathbf{a}, \mathbf{b})$ and $D(\mathbf{a}, \mathbf{b})$ are the maximum similarity and minimum similarity scores obtained by the Needleman–Wunsch and Smith–Waterman dynamic programming algorithms, respectively.

When simple scoring systems are used for matching individual sequence characters, the above equations provide a way to convert similarity to distance scores, and vice versa, for individual sequence characters and alignment scores. The conversions are more complicated for more sophisticated scoring schemes, such as a PAM scoring matrix for scoring amino acid similarities. For now, it is important to realize that similarity and distance methods produce approximately the same alignments, provided that equivalent similarity and distance scoring schemes are used. If one is to draw the correct evolutionary inferences, however, similarity scores should be converted to the equivalent distance scores.

Improvement in the dynamic programming algorithm for sequence alignments. The dynamic programming methods for sequence alignments originally required between $n \times m$ and $n \times m^2$ steps and storage in several matrices of size $n \times m$, where n is the length of the shorter sequence (Needleman and Wunsch 1970; Waterman et al. 1976; Smith and Waterman 1981a). A series of improvements in this algorithm that reduced the number of steps and amount of memory required have been described. These steps include: (1) a decreased number of steps in the alignment algorithm by Gotoh (1982); (2) a reduction in the amount of memory required to a linear function of sequence length (Myers and Miller 1988); (3) ability to find near-optimal alignments (Chao et al. 1994) and to align long sequences (Schwartz et al. 1991); and (4) ability to find the best-scoring alternative alignments that do not include alignments of the same sequence positions (Waterman and Eggert 1987; Huang et al. 1990; Huang and Miller 1991).

In producing either global or local alignments, a concern is that there may be more than one way to align the sequences. For example, if the sequences have repeated domains or low-complexity regions, it may be possible to produce alignments in which different sequence positions are matched. To address this problem, both global and local alignment methods have been improved to allow finding of alternative alignments. A different global alignment that does not match the same sequence characters is found by giving the matrix position that begins an alignment with a score of zero, and then all matrix positions that are affected by this change are recalculated. The next highest matrix score and the path leading to it provide an alternative alignment of the sequences that does not include the same sequence matches as were present in the original alignment (Waterman and Eggert 1987). Alternative local alignments are found by a more complex algorithm (the SIM algorithm) that includes improvements similar to those listed above (Huang et al. 1990; Huang and Miller 1991).

A LIMITATION OF THE SMITH–WATERMAN LOCAL ALIGNMENT ALGORITHM FOR FINDING DISPERSED LOCAL ALIGNMENTS

The Smith–Waterman algorithm was designed to reveal highly conserved regions in sequences that align well and to disregard regions that do not. A problem arises when the sequences being aligned include two or more matching regions that have intervening regions which do not match very well. After traversing the first matching region, the running alignment score generated by the Smith–Waterman algorithm will drop in the next poorly matching region. If a second matching region is to be included in the overall alignment, the running alignment score (1) must stay positive in the poorly matching region and (2) must reach a new maximum value in the second matching region. Otherwise, the alignment will only include the first matching region as required by the Smith–Waterman algorithm. However, if Smith–Waterman requirements are met and several dispersed highly scoring alignments are included in the local alignment, a further problem arises. The alignment will have a maximal score based on the high-scoring local alignments, but the score will have been subjected to the negative influence of the intervening poorly aligning regions. The resulting alignment will not represent what one would like to have, i.e., an alignment over the entire region that has the maximum possible percentage of matches.

This type of problem becomes particulary significant when comparing long sequences, as in the case of genome sequences with cDNA sequences. To get around this problem, Schwartz et al. (2000) have a developed a dot matrix-like tool called Pip-Maker (percent identity plot maker; see Table 3.1) for showing all of the high-scoring regions in alignments. Arslan et al. (2001) have introduced a new sequence comparison algorithm that finds regions with a maximum degree of similarity (normalized local alignment). This new algorithm uses a new scoring system in which the similarity score of an alignment between two sequences is divided by the sum of the lengths of the sequences in the alignment. To the sum of the sequence lengths is added a parameter L to facilitate computation and find longer alignments. The algorithm then optimizes this new length-normalized score. The method involves repeatedly running the Smith–Waterman algorithm and including updated information on the current length-normalized score until the best normalized alignment is found.

Word and k-tuple Methods

The word or k-tuple methods align two sequences very quickly by first searching for identical short stretches of sequences (called words or k-tuples) and by then joining these words into an alignment using the dynamic programming method. See Chapter 6, pp. 234–258, for a complete discussion of these methods.

HOW TO USE SCORING MATRICES AND GAP PENALTIES IN SEQUENCE ALIGNMENTS

The choice of a scoring system including scores for matches, mismatches, substitutions, insertions, and deletions influences the alignment of both DNA and protein sequences. Both global alignments by the Needleman–Wunsch algorithm and local alignments by the Smith–Waterman algorithm are influenced by the scoring system used. For aligning DNA sequences, a simple positive score for matches and a negative score for mismatches and gaps are most often used. To score matches and mismatches in alignments of proteins, it is necessary to know how often one amino acid is substituted for another in related proteins. The substitutions found are counted, divided by the frequency expected for each type of substitution to give an odds score, and the logarithm of the odds score is placed in a scoring matrix of substitution values, which can then be used to score those pairs in sequence alignments. In addition, a method is needed to account for insertions and deletions that sometimes appear in related DNA or protein sequences. To accommodate such sequence variations, gaps that appear in sequence alignments are given a negative penalty score reflecting the fact that they are not expected to occur very often. Mathematically speaking, it is very difficult to produce the best-possible alignment, either global or local, unless gaps are included in the alignment.

Amino Acid Substitution Matrices

Protein chemists discovered early on that certain amino acid substitutions commonly occur in related proteins from different species. Because a protein still functions with these substitutions, the substituted amino acids are compatible with protein structure and function. Often, these substitutions are for a chemically similar amino acid, but are relatively rare. Knowing the types of changes that are most and least common in a large number of proteins can assist with predicting alignments for any set of protein sequences, as illustrated in Figure 3.13. If related protein sequences are quite similar, they are easy to align, and one can readily determine the single-step amino acid changes. If ancestor relationships among a group of proteins are assessed, the most likely amino acid changes that occurred during evolution can be predicted. This type of analysis was pioneered by Margaret Dayhoff (1978) and used by her to produce a type of scoring matrix called a percent accepted mutation (PAM) matrix (see below).

Amino acid substitution matrices (or symbol comparison tables, as they are sometimes called) are used for such analyses. Although the most common use of such tables is for comparison of protein sequences, other tables of nucleic acid symbols are

	Alignment			
sequence A	Tyr	Cys	Asp	Ala
sequence B	Phe	Met	Glu	Gly
BLOSUM62 matrix value	3	−1	2	0

Total score for alignment of sequence A with sequence B
= 3 − 1 + 2 + 0 = 4

FIGURE 3.13. Use of amino acid substitution matrix to evaluate an alignment of two protein sequences. The score for each amino acid pair (Tyr/Phe, etc.) is looked up in the BLOSUM62 matrix. Each value represents an odds score, the likelihood that the two amino acids will be aligned in alignments of similar proteins divided by the likelihood that they will be aligned by chance in an alignment of unrelated proteins. In a series of individual matches in an alignment, these odds scores are multiplied to give an overall odds score for the alignment itself. For convenience, odds scores are converted to log odds scores so that the values for amino acid pairs in an alignment may be summed to obtain the log odds score of the alignment. In this case, the logarithms are calculated to the base 2 and multiplied by 2 to give values designated as half-bits (a bit is the unit of an odds score that has been converted to a logarithm to the base 2). The value of 4 indicates that the 4 amino acid alignment is $2^{(4/2)} = 4$-fold more likely than expected by chance.

also used for comparison of nucleic acid sequences to accommodate ambiguous nucleotide characters or models of expected sequence changes during different periods of evolutionary time that vary scoring of transitions and transversions (see p. 105).

In amino acid substitution matrices, amino acids are listed both across the top of a matrix and down the side, and each matrix position is filled with a score that reflects how often one amino acid would have been paired with the other in an alignment of related protein sequences. The probability of changing amino acid A into B is always assumed to be identical to the reverse probability of changing B into A. This assumption is made because, for any two sequences, the ancestor amino acid in the phylogenetic tree is usually not known. Additionally, the likelihood of replacement should depend on the product of the frequency of occurrence of the two amino acids, which can be used to predict how often they would be matched by random chance in an alignment, and on their chemical and physical similarities. A prediction of this model is that amino acid frequencies in related proteins will not change over evolutionary time (Dayhoff 1978).

Dayhoff Amino Acid Substitution Matrices (Percent Accepted Mutation or PAM Matrices)

This family of matrices lists the likelihood of change from one amino acid to another in homologous protein sequences during evolution and thus is focused on tracking the evolutionary origins of proteins. Presently, there is no other type of scoring matrix that is based on such sound evolutionary principles as the PAM matrices. Even though they were originally based on a relatively small data set, the PAM matrices remain a useful tool for sequence alignment. Each matrix gives the changes expected for a given period of evolutionary time, evidenced by decreased sequence similarity as genes that encoded the same protein diverge with increased evolutionary time. Thus, one matrix gives the changes expected in homologous proteins that have diverged only a small amount from each other in a relatively short period of time, so that they are still 50% or more similar. Another gives the changes expected of proteins that have diverged over a much longer period, leaving only 20% similarity. These predicted changes are used to produce optimal alignments between two protein sequences and to score the alignment. The assumption in this evolutionary model is that the amino acid substitutions observed over short periods of evolutionary history can be extrapolated to longer distances. In contrast, another type of matrix, the BLOSUM matrix (see below), is based on scoring substitutions found over a range of evolutionary periods. The BLOSUM matrices reveal that substitutions are not always as predicted by the PAM model.

In deriving the PAM matrices, each change in the current amino acid at a particular site is assumed to be independent of previous mutational events at that site (Dayhoff 1978). Thus, the probability of change of any amino acid **a** to amino acid **b** is the same, regardless of the previous changes at that site and also regardless of the position of amino acid **a** in a protein sequence. Amino acid substitutions in a protein sequence are thus viewed as a Markov model (see also hidden Markov models in Chapter 5), which is characterized by a series of changes of state in a system such that a change from one state to another does not depend on the previous history of the state. Use of this model makes it possible to extrapolate amino acid substitutions observed over a relatively short period of evolutionary time to longer periods of evolutionary time.

To prepare the Dayhoff PAM matrices, amino acid substitutions that occurred in a group of evolving proteins were estimated using 1572 changes in 71 groups of protein sequences that were at least 85% similar. Because these changes were observed in closely related proteins, they represented amino acid substitutions that do not significantly change the function of the protein. Hence they are called "accepted mutations," defined as amino acid changes "accepted" by natural selection. Similar sequences were first organized into a phylogenetic tree, as illustrated in Figure 1.1 in Chapter 1. The number of changes of each amino acid into every other amino acid was then counted. To make these numbers useful for sequence analysis, information on the relative amount of change (relative mutabilities) for each amino acid was needed.

Relative mutabilities were evaluated by counting, in each group of related sequences, the number of changes of each amino acid and by dividing this number by a factor, called the exposure to mutation of the amino acid. This factor is the product of the frequency of occurrence of the amino acid in that group of sequences being analyzed and the total number of all amino acid changes that occurred in that group per 100 sites. This factor normalizes the data for variations in amino acid composition, mutation rate, and sequence length. The normalized frequencies were then summed for all sequence groups. By these scores, Asn, Ser, Asp, and Glu were the most mutable amino acids, and Cys and Trp were the least mutable. An example for changing Phe to any other amino acid is shown in Table 3.2 and boxed text.

The above amino acid exchange counts and mutability values were then used to generate a 20 x 20 mutation probability matrix representing all possible amino acid changes. Because amino acid change was modeled by a Markov model, in which the probability of mutation at each site is independent of the previous history of mutations, the changes predicted for more distantly related proteins that have undergone N percent mutations could be calculated. A PAM1 matrix showing the relative frequencies of change for each amino acid into any other adding up to a total frequency of 1% change was first calculated. According to the Markov model, the PAM1 matrix could be multiplied by itself N times (see Table 3.2B), to give transition matrices for comparing sequences with lower and lower levels of similarity due to separation over longer periods of evolutionary history. Thus, the commonly used PAM250 matrix represents

a level of 250% of change expected in 2500 my (million years). Although this amount of change seems very large, sequences at this level of divergence still have about 20% similarity. For example, alanine will be matched with alanine 13% of the time and with another amino acid 87% of the time.

The percentage of remaining similarity for any PAM matrix can be calculated by summing the percentages for amino acids not changing (Ala versus

TABLE 3.2. (A) *Normalized probability scores for changing Phe to any other amino acid (or of not changing) at PAM1 and PAM250 evolutionary distances*

Amino acid change	PAM1	PAM250
Phe to Ala	0.0002	0.04
Phe to Arg	0.0001	0.01
Phe to Asn	0.0001	0.02
Phe to Asp	0.0000	0.01
Phe to Cys	0.0000	0.01
Phe to Gln	0.0000	0.01
Phe to Glu	0.0000	0.01
Phe to Gly	0.0001	0.03
Phe to His	0.0002	0.02
Phe to Ile	0.0007	0.05
Phe to Leu	0.0013	0.13
Phe to Lys	0.0000	0.02
Phe to Met	0.0001	0.02
Phe to Phe	0.9946	0.32
Phe to Pro	0.0001	0.02
Phe to Ser	0.0003	0.03
Phe to Thr	0.0001	0.03
Phe to Trp	0.0001	0.01
Phe to Tyr	0.0021	0.15
Phe to Val	0.0001	0.05
SUM[a]	1.0000	1.00

[a]Approximate since scores are rounded off.

(B) *The multiplication of two PAM1 matrices to give a PAM2 matrix*

$A = a^2 + bd + cg + \ldots$
$B = ab + be + ch + \ldots$
$C = ac + bf + ci + \ldots$
$D = da + ed + fg + \ldots$, etc.

Only three rows and columns are shown for illustrative purposes.

Ala, etc.) after multiplying each by the frequency of that amino acid pair in the database (e.g., 0.089 for Ala) (Dayhoff 1978). The PAM120, PAM80, and PAM60 matrices should be used for aligning sequences that are 40%, 50%, and 60% similar, respectively (also see p. 107). Simulations by George et al. (1990) have shown that, as predicted, the PAM250 matrix provides a better-scoring alignment than lower-numbered PAM matrices for distantly related proteins of 14–27% similarity. *Note:* Do not confuse this mutation probability form of the PAM250 matrix with the log odds form of the matrix described below.

PAM matrices are usually converted into another form, called log odds matrices. Since the direction of mutation is not known, the log odds score in the scoring matrix for changing amino acid **a** into amino acid **b** is the same as that changing **b** into **a**. These scores are the average of the observed changes of **a** to **b**, and of **b** to **a**. The odds score represents the ratio of the chance of amino acid substitution by two different hypotheses—one that the change actually represents an authentic evolutionary variation at that site (the numerator), and the other that the change occurred because of random sequence variation of no biological significance (the denominator). Odds ratios are converted to logarithms to give log odds scores for convenience in multiplying odds scores of amino acid pairs in an alignment by adding the logarithms (see Fig. 3.13).

EXAMPLE: CALCULATIONS FOR OBTAINING THE LOG ODDS SCORE FOR CHANGES BETWEEN PHE AND TYR AT AN EVOLUTIONARY DISTANCE OF 250 PAMs

1. Of 1572 observed amino acid changes, there were 260 changes between Phe and Tyr. These numbers were multiplied by (1) the relative mutability of Phe (see text) and (2) the fraction of Phe to Tyr changes over all changes of Phe to any other amino acid (since Phe to Tyr and Tyr to Phe changes are not distinguished in the original mutation counts, sums of changes are used to calculate the fraction) to obtain a mutation probability score of Phe to Tyr. A similar score was obtained for changes of Phe to each of the other 18 amino acids, and also for the calculated probability of not changing at all. The resulting 20 scores were summed and divided by a normalizing factor such that their sum represented a probability of change of 1%, as illustrated in Table 3.2.

   In this matrix, the score for changing Phe to Tyr was 0.0021, as opposed to a score of Phe not changing at all of 0.9946, as shown in Table 3.2A. These calculations were repeated for Tyr changing to any other amino acid. The score for changing Tyr to Phe was 0.0028, and that of not changing Tyr was 0.9946 (not shown). These scores were placed in the PAM1 matrix, in which the overall probability of each amino acid changing to another is ~1%, and that of each not changing is ~99%.

2. The above PAM1 matrix was multiplied by itself 250 times to obtain the distribution of changes expected for 250 PAMs of evolutionary change, as illustrated in Table 3.2B for the first of these multiplications. These changes can include both forward changes to another amino acid and reverse changes to a former one. At this distance, the probability of change of Phe to Tyr was 0.15 as opposed to a probability of 0.32 of no change in Phe. The corresponding probabilities for Tyr to Phe at 250 PAMs, obtained through the matrix multiplication described above, were 0.20 and 0.31 for no change.

3. The log odds values for changes between Phe and Tyr were then calculated. The Phe to Tyr score in the 250 PAM matrix, 0.15, was divided by the frequency of Phe in the sequence data, 0.040, to give the relative frequency of change. This ratio, 0.15/0.04 = 3.75, was converted to a logarithm to the base 10 ($\log_{10}3.75 = 0.57$) and multiplied by 10 to remove fractional values ($0.57 \times 10 = 5.7$). Similarly, the Tyr-to-Phe score is 0.20/0.03 = 6.7, and the logarithm of this number is $\log_{10}6.7 = 0.83$, and multiplied by 10 is ($0.83 \times 10 = 8.3$). The average of 5.7 and 8.3 is 7, the number entered in the log odds table for changes between Phe and Tyr at 250 PAMs of evolutionary distance.

   The log odds form of the PAM250 matrix, which is sometimes referred to as the mutation data matrix (MDM) at 250 PAMs and also as MDM78, is shown in Figure 3.14. The log odds scores in this matrix lie within the range of –8 to +17. A value of 0 indicates that the frequency of the substitution between a matched pair of amino acids in related proteins is as expected by chance; a value less than 0 or greater than 0 indicates that the frequency is less than or greater than that expected by chance, respectively. Using such a matrix, a high positive score between two amino acids means that the pair is more likely to be found aligned in sequences that are related than in unrelated sequences. The highest-scoring replacements are for amino acids whose side chains are chemically similar, as might be expected if the amino acid substitution is not to impede function. In the original data, the largest number of observed changes (83) was between the acidic amino acids Asp (D) and Glu (E), which have similar chemical properties. This number is reflected as a log odds score of +3 in the MDM. Many changes were not observed. For example, there were no changes between Gly (G) and Trp (W), resulting in a score of –7 in the table.

	C	S	T	P	A	G	N	D	E	Q	H	R	K	M	I	L	V	F	Y	W	
C	12																				C
S	0	2																			S
T	-2	1	3																		T
P	-3	1	0	6																	P
A	-2	1	1	1	2																A
G	-3	1	0	-1	1	5															G
N	-4	1	0	-1	0	0	2														N
D	-5	0	0	-1	0	1	2	4													D
E	-5	0	0	-1	0	0	1	3	4												E
Q	-5	-1	-1	0	0	-1	1	2	2	4											Q
H	-3	-1	-1	0	-1	-2	2	1	1	3	6										H
R	-4	0	-1	0	-2	-3	0	-1	-1	1	2	6									R
K	-5	0	0	-1	-1	-2	1	0	0	1	0	3	5								K
M	-5	-2	-1	-2	-1	-3	-2	-3	-2	-1	-2	0	0	6							M
I	-2	-1	0	-2	-1	-3	-2	-2	-2	-2	-2	-2	-2	2	5						I
L	-6	-3	-2	-3	-2	-4	-3	-4	-3	-2	-2	-3	-3	4	2	6					L
V	-2	-1	0	-1	0	-1	-2	-2	-2	-2	-2	-2	-2	2	4	2	4				V
F	-4	-3	-3	-5	-4	-5	-4	-6	-5	-5	-2	-4	-5	0	1	2	-1	9			F
Y	0	-3	-3	-5	-3	-5	-2	-4	-4	-4	0	-4	-4	-2	-1	-1	-2	7	10		Y
W	-8	-2	-5	-6	-6	-7	-4	-7	-7	-5	-3	2	-3	-4	-5	-2	-6	0	0	17	W
	C	S	T	P	A	G	N	D	E	Q	H	R	K	M	I	L	V	F	Y	W	

FIGURE 3.14. The log odds form (the mutation data matrix or MDM) of the PAM250 scoring matrix. Amino acids are grouped according to the chemistry of the side group: (C) sulfhydryl; (STPAG) small hydrophilic; (NDEQ) acid, acid amide, and hydrophilic; (HRK) basic; (MILV) small hydrophobic; and (FYW) aromatic. Each matrix value is calculated from an odds score, the probability that the amino acid pair will be found in alignments of homologous proteins divided by the probability that the pair will be found in alignments of unrelated proteins by random chance. The logarithm of these odds scores to the base 10 is multiplied by 10 and then used as the table value (see text for details). Thus, +10 means the ancestor probability is greater, 0 that the probabilities are equal, and –4 that the alignment is more often a chance one than due to an ancestor relationship. Because these numbers are logarithms, they may be added to give a combined probability of two or more amino acid pairs in an alignment. Thus, the probability of aligning two Ys in an alignment YY/YY is 10 + 10 = 20, a very significant score, whereas that of YY/TP is –3 –5 = –8, a rare and unexpected alignment between homologous sequences.

At one time, the PAM250 scoring matrix was modified in an attempt to improve the alignments obtained when PAM250 was used as the scoring matrix with dynamic programming. All scores for matching a particular amino acid were normalized to the same mean and standard deviation, for example, by summing the scores and choosing the mean, and all amino acid identities were given the same score to provide an equal contribution for each amino acid in a sequence alignment (Gribskov and Burgess 1986). These modifications were included as the default matrices for the GCG sequence alignment programs in versions 8 and earlier; they are optional in later versions. Their use is not recommended because these modifications will not give an optimal alignment that is in accord with the evolutionary model.

Choosing the Best PAM Scoring Matrices for Detecting Sequence Similarity

The ability of PAM scoring matrices to distinguish statistically between chance and biologically meaningful alignments has been analyzed using a statistical theory for sequences (Altschul 1991) that is discussed in Chapter 4. As discussed above, each PAM matrix is designed to score alignments between sequences that have diverged by a particular degree of evolutionary distance. Altschul (1991) has examined how well the PAM matrices actually can distinguish proteins that have diverged to a greater or lesser extent, when these proteins are subjected to a local alignment.

Initially, when using a scoring matrix to produce an alignment, the amount of similarity between sequences may not be known. However, the ungapped alignment scores obtained are maximal when the correct PAM matrix, i.e., the one corresponding to the degree of similarity in the target sequences, is used (Altschul 1991). One approach is to use a series of PAM matrices and then to choose the best alignment score (for a Bayesian statistical method, see p. 152). Altschul (1991) has also examined the ability of PAM matrices to provide a reliable enough indication of an ungapped local alignment score between sequences on an initial attempt of alignment. For sequence alignments, the PAM200 matrix is able to detect a significant ungapped alignment of 16–62 amino acids whose score is within 87% of the optimal one. Alternatively, several combinations, such as PAM80 and PAM250 or PAM120 and PAM350, can also be used. Altschul (1993) has also proposed using a single matrix and adjusting a statistical parameter in the scoring system to reach more distantly related sequences, but this change would primarily be for database searches.

In addition to the aforementioned differences among PAM scoring matrices for scoring alignments of more- or less-related proteins, the ability of each PAM matrix to discriminate real local alignments from chance alignments also varies. To calculate the ability of the entire matrix to discriminate related from unrelated sequences (*H*, the relative entropy), the score for each amino acid pair s_{ij} (in units of $\log_2$, called bits) is multiplied by the probability of occurrence of that pair in the original data set, q_{ij} (Altschul 1991). This weighted score is then summed over all the amino acid pairs to produce a score that represents the ability of the average amino acid pair in the matrix to discriminate actual from chance alignments. When the correct PAM matrix is chosen for a pair of sequences, as discussed above, this discriminatory ability is best utilized.

$$H = \sum_{i=1}^{20} \sum_{j=1}^{i} q_{ij} \times s_{ij} \qquad (6)$$

In information theory, this score is called the average mutual information content per pair, and the sum over all pairs is the relative entropy of the matrix (termed *H*). The relative entropy will be a small positive number. For the PAM250 matrix the number is +0.36; for PAM120, +0.98; and for PAM160, +0.70. In general, all other factors being equal, the higher the value of *H* for a scoring matrix, the more likely it is to be able to distinguish real from chance alignments. The practical application for using *H* is to choose a scoring matrix that has the highest value of *H*. The lower value of *H* in the higher PAM matrix is a reflec-

USING SCORING MATRICES FOR DATABASE SEARCHES

Scoring matrices are also used in database searches for similar sequences. The optimal matrices for these searches have also been determined. (See Chapter 6 for a more detailed discussion.) It is important to remember that these predictions assume that the amino acid distributions in the set of protein families used to make the scoring matrix are representative of all families that are likely to be encountered. The original PAM matrices represent only a small number of families, although larger data sets have been obtained. Scoring matrices obtained more recently, such as the BLOSUM matrices, are based on a much larger number of protein families. BLOSUM matrices are not based on a PAM evolutionary model in which changes at large evolutionary distance are predicted by extrapolation of changes found at small distances. Matrix values in BLOSUM matrices are based on the observed frequency of change in a large set of diverse proteins. The BLOSUM scoring matrices (especially BLOSUM62) appear to capture more of the distant types of variations found in protein families. Hence, extrapolating the substitutions observed over short evolutionary periods does not appear to provide a good prediction of changes over longer distances. Several more recent scoring matrices are described in Chapter 6.

tion of the increased uncertainty that arises as to the number of changes that have occurred in sequences that are more divergent. Note also that the value of H goes down as the PAM number increases. Thus, using the PAM250 for sequences that are very similar is not a useful way to distinguish real from chance alignments, because this matrix will not take advantage of the higher expected level of identity.

Analysis of the Dayhoff Model of Protein Evolution as Used in PAM Matrices

As outlined above, the Dayhoff model of protein evolution is a Markov process. In this model, each amino acid site in a protein can change at any time to any of the other 20 amino acids with probabilities given by the PAM table, and the changes that occur at each site are independent of the amino acids found at other sites in the protein and depend only on the current amino acid at the site. The assumptions that underlie the method of constructing the Dayhoff scoring matrix have been challenged (for discussion, see George et al. 1990; States and Boguski 1991). First, it is assumed that each amino acid position is equally mutable, whereas, in fact, sites vary considerably in their degree of mutability. Mutagenesis hot spots are well known in molecular genetics, and variations in mutability of different amino acid sites in proteins are also well known.

The more conserved amino acids in similar proteins from different species are ones that play an essential role in structure and function, and the less conserved are in sites that can vary without having a significant effect on function. Thus, there are many factors that influence both the location and types of amino acid changes that occur in proteins. Wilbur (1985) has tested the Markov model of evolution (see box, below) and has shown that it can be valid if certain changes are made in the way that the PAM matrices are calculated.

TEST OF MARKOV MODEL OF EVOLUTION IN PROTEINS

To test the model, Wilbur (1985) addressed a major criticism of the PAM scoring matrix, namely, that the frequency of amino acid changes that require two nucleotide changes is higher than would be expected by chance. About 20% of the observed amino acid changes require more than a single mutation for the necessary codon changes. This fraction is far greater than would be expected by chance.

To correct for changes that require at least two mutations, Wilbur recalculated the PAM1 matrix using only amino acid substitution data from 150 amino acid pairs that are accountable by single mutations. To accomplish this calculation, he used a refined mathematical model that provided a more precise measure of the rate of substitution. He then estimated frequencies of the other 230 amino acid substitutions reachable only by at least two mutations, and compared these frequencies to the values calculated by Dayhoff, who had assumed these were single-step changes. If these numbers agreed, argued Wilbur, then the PAM model used to produce the Dayhoff matrix is a reliable one. In fact, the Dayhoff values exceeded the two-step model values by a factor of about 117. One source of discrepancy was the assumption that the two-step changes were a linear function of evolutionary time over short evolutionary periods of 1 PAM (average time of 1 PAM = 10 my), whereas, because two mutations are required to make the change, a quadratic function is expected. With this correction made to the Dayhoff calculations for amino acid substitutions requiring two mutations, agreement with the two-step model improved about 10-fold, leaving another 11.7-fold unaccounted for.

Wilbur analyzed the remainder by the covarion hypothesis (Fitch and Markowitz 1970; Miyamoto and Fitch 1995), in which it is assumed that only a certain fraction of amino acid sites in a protein are variable and that one site influences another. Thus, a change in one site may influence the variability of others. This model seems to be reasonable from many biological perspectives. For example, amino acids at different regions of a protein sequence will interact to give a three-dimensional structure; thus, a change in one of these without some kind of compensating change in the other region could be detrimental to protein structure and function. The prediction of this hypothesis is that the frequency of two-step changes would be overestimated because we did not take into account the failure of many sites to be mutable. Using a reasonable estimate of 0.3 for the fraction of the sites that could change, the effect on the Dayhoff calculations for frequencies of two-step changes would be 3.3-fold. The remaining discrepancy in the 11.7-fold ratio between Dayhoff values and two-step values may be attributable to variations in mutation rates from site to site, or to the exclusion of certain amino acids at a particular site.

In conclusion, Wilbur (1985) has shown that the Dayhoff model for protein evolution appears to give predictable and consistent results, but that frequencies of change between amino acids that require two mutational steps must be calculated as a two-step process. Failure to do so generates errors due to variations in site-to-site mutability. George et al. (1990) have counterargued that it has never been demonstrated that two independent mutations must occur, each becoming established in a population before the next appears. In fact, double mutations can occur following DNA damage by ultraviolet light. However, the issue has never been resolved.

In addition to the questions raised by Wilbur (see box above), a further criticism of the PAM scoring matrices is that they are not more useful for sequence alignment than simpler matrices, such as one based on a chemical grouping of amino acid side chains. Although alignment of related proteins is straightforward and quite independent of the symbol comparison scoring scheme, alignments of less-related proteins are much more speculative (Feng et al. 1985). PAM and BLOSUM matrices have both been very useful for finding more distantly related sequences (George et al. 1990).

There have been recent changes in the way that members of protein families are identified (see Chapters 5 and 11). Once a family has been identified, family-specific scoring matrices can be produced, and there is no point in using these general matrices. As described in Chapter 5, a scoring matrix representing a section of aligned sequences with no gaps, or a matrix representing a section of aligned sequences with matches, mismatches, and gaps (a profile), is the best tool to search for more family members.

Another criticism of the PAM matrix is that constructing phylogenetic relationships prior to scoring mutations has limitations, due to the difficulty of determining ancestral relationships among sequences, a topic discussed in Chapter 7. Early on in the Dayhoff analysis, the evolutionary trees were estimated by a voting scheme for the branches in the tree, each node being estimated by the most abundant amino acid in distal parts of the tree. Once available, the PAM matrices were used to estimate the evolutionary distance between proteins, given the amount of sequence similarity. Such data can be used to produce a tree based on evolutionary distances (Chapter 7). This circular analysis of using alignments to score amino acid changes and then to use the matrices to

produce new alignments has also been criticized. However, no method has yet been devised in any type of sequence analysis for completely circumventing this problem. Evidence that the values in the scoring matrix are insensitive to changes in the phylogenetic relationships has been provided (George et al. 1990).

Finally, the Dayhoff PAM matrices have been criticized because they are based on a small set of closely related proteins. The Dayhoff data set has been augmented to include the 1991 protein database (Gonnet et al. 1992; Jones et al. 1992) as shown below. The Dayhoff matrices have also been extensively compared to other scoring matrices, as discussed below (p. 111). More recent scoring matrices are described in Chapter 6 (p. 238).

Updates to PAM Amino Acid Scoring Matrices

New PAM amino acid scoring matrices that include the 1991 protein database (Gonnet et al. 1992; Jones et al. 1992) have been developed. The ability of the original Dayhoff PAM250 scoring matrix to identify distantly related sequences has also been extensively compared to that of more recently developed scoring matrices. Gonnet et al. (1992) produced a new scoring matrix known as the Gonnet92 matrix that was intended to be an updated version of the Dayhoff PAM250 scoring matrix (see also Benner et al. 1993). However, since widely divergent sequences were used to produce this matrix, it is more comparable to the BLOSUM62 scoring matrix described below than to the Dayhoff PAM250 matrix.

Another substitute for the Dayhoff PAM250 matrix is an amino acid substitution matrix named PET91 (pair exchange table for year 1991) and also called JTT250 (Jones et al. 1992). The new data set for the initial JTT matrix corresponding to the original

Dayhoff PAM1 matrix included 59,190 accepted point mutations in 16,130 protein sequences, a much smaller number than the more divergent set used by Gonnet et al. (1992). The JTT PAM1 matrix was also based on sequences that were >85% similar, and in this respect was identical to the original Dayhoff analysis. Hence, the JTT PAM1 matrix could be used to produce PAM matrices corresponding to any evolutionary distance. An example of using a set of JTT PAM matrices to align two protein sequences by a Bayesian statistical method and to estimate the evolutionary distance between the sequences is given in Chapter 4, page 155.

There are significant differences both between the new Gonnet92 and JTT250 matrices and the old Dayhoff matrix at 250 PAMs and also between the Gonnet92 and JTT250 matrices. The most striking differences with the Dayhoff matrix are in the substitutions between C (Cys) and other amino acids and W (Trp) and other amino acids. C changes to other amino acids and W changes to other amino acids quite rarely, but the later data sets used to make the Gonnet92 and JTT250 matrices have examples. In the new protein comparisons, C (Cys) and W (Trp) were exchanged for each other, whereas they were not exchanged in the original Dayhoff analysis. There are also some significant differences between the Gonnet92 and JTT250 matrices, attributable to the overall number of sequences compared and their relatedness.

On the basis of these and other differences in these scoring matrices, the best recommendation that can be made is to use the JTT PAM-N matrices as a more modern substitute for the Dayhoff PAM-N matrices (where N is the PAM distance) and the Gonnet92 matrix as a substitute for the Dayhoff PAM250 matrix for scoring amino acid changes between proteins that are more distantly related. Gonnet92 behaves similarly to the BLOSUM62 scoring matrix described below (see also p. 111).

Blocks Amino Acid Substitution Matrices (BLOSUM)

The original Dayhoff PAM matrices were developed based on a small number of protein sequences and an evolutionary model of protein change. By extrapolating from the observed changes at small evolutionary distances to large ones, they were able to establish a PAM250 scoring matrix for sequences that were high-ly divergent. Another approach to finding a scoring matrix for divergent sequences is to start with a more divergent set of sequences and produce a scoring matrix from the substitutions found in those less-related sequences. The blocks amino acid substitution matrices (BLOSUM) scoring matrices were prepared this way.

The BLOSUM62 substitution matrix (Henikoff and Henikoff 1992) is widely used for scoring protein sequence alignments. The matrix values are based on the observed amino acid substitutions in a large set of approximately 2000 conserved amino acid patterns, called blocks. These blocks have been found in a database of protein sequences representing more than 500 families of related proteins (Henikoff and Henikoff 1992) and act as signatures of these protein families. The BLOSUM matrices, which are designed to find the conserved domains of proteins, are thus based on an entirely different type of sequence analysis and a much larger data set than the Dayhoff PAM matrices.

The protein families used for making the BLO-SUM matrices were originally identified by Amos Bairoch in the Prosite catalog (Bairoch 1991). This catalog provides lists of proteins that are in the same family because they have a similar biochemical function. For each family, a pattern of amino acids that are characteristic of that function is provided. Henikoff and Henikoff (1991) examined each Prosite family for the presence of ungapped amino acid patterns (blocks) that were present in each family and that could be used to identify members of that family. To locate these patterns, the sequences of each protein family were searched for similar amino acid patterns by the MOTIF program of H. Smith (Smith et al. 1990), which can find patterns of the type aa1 d1 aa2 d2 aa3, where aa1 and aa2 are conserved amino acids and d1 and d2 are stretches of intervening sequence up to 24 amino acids long located in all sequences. These initial patterns were organized into larger ungapped patterns (blocks) between 3 and 60 amino acids long by the Henikoffs' PROTOMAT program (http://www.blocks.fhcrc.org). Because these blocks were present in all of the sequences in each family, they could be used to identify other members of the same family. Thus, the family collections were enlarged by searching the available sequence databases for more proteins with these same conserved blocks.

The columns of the aligned blocks that character-ized each family indicated the types of amino acid substitutions that occurred. The amino acid changes that were observed in each column of the alignment could then be counted. The types of substitutions were then scored for all aligned patterns in the database and used to prepare a scoring matrix, the BLOSUM matrix, indicating the frequency of each type of substitution. As previously described for the PAM matrices, BLOSUM matrix values were given as logarithms of odds scores of the ratio of the observed frequency of amino acid substitutions divided by the frequency expected by chance. An example of the calculations is shown in Figure 3.15.

This procedure of counting all of the amino acid changes in the blocks, however, can lead to an over-representation of amino acid substitutions that occur in the most closely related members of each family. To reduce this dominant contribution from the most-alike block sequences, these block sequences were grouped together into one sequence before scoring the amino acid substitutions in the aligned blocks. The amino acid changes within these grouped sequences were then averaged. Patterns that were 60% identical were grouped together to make one substitution matrix called BLOSUM60, and those 80% alike to make another matrix called BLOSUM80, and so on. As with the PAM matrices, these matrices differ in the degree to which the more common amino acid pairs are scored relative to the less common pairs. Thus, when used for aligning protein sequences, they provide a different level of distinction between the more common and less common amino acid pairs. Experience has shown that BLOSUM62 generally provides the best alignment of proteins over a range of sequence similarity and that other BLOSUM matrices may provide a better alignment when dealing with more closely related proteins.

The ability of these different BLOSUM matrices to distinguish real from chance alignments and to identify as many members as possible of a protein family has been determined (Henikoff and Henikoff 1992). Two types of analyses were performed: (1) an information content analysis of each matrix, as was described above for the PAM matrices, and (2) an actual comparison of the ability of each matrix to find members of the same families in a database search, discussed below. As the grouping percentage was increased, the ability of the resulting matrix to distinguish actual from chance alignments, defined as the relative entropy of the matrix or the average information content per residue pair (see above), also increased. As grouping increased

from 45% to 62%, the information content per residue increased from approximately 0.4 to 0.7 bits per residue, and was approximately 1.0 bits at 80% grouping. However, at the same time, the number of blocks that contributed information decreased by 25% between no grouping and 62% grouping. BLOSUM62

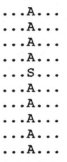

FIGURE 3.15. Derivation of the matrix values in the BLOSUM62 scoring matrix. As an example of the calculations, if a column in one of the blocks consisted of 9 A and 1 S amino acids, the following is true for this data set (see Henikoff and Henikoff 1992).

1. Since the original sequence from which the others were derived is not known, each column position has to be considered a possible ancestor of the other nine positions. Hence, there are $8+7+6+\ldots+1 = 36$ possible AA pairs (f_{AA}) and 9 possible AS pairs (f_{AS}) to be compared.

2. There are $20+19+18+\ldots+1 = 210$ possible amino acid pairs.

3. The frequency of occurrence of an AA pair, $q_{AA} = f_{AA}/(f_{AA} + f_{AS}) = 36/(36+9) = 0.8$, and that of an AS pair, $q_{AS} = f_{AS}/(f_{AA} + f_{AS}) = 9 / (36+9) = 0.2$.

4. The expected frequency of A being in a pair, $p_A = (q_{AA} + q_{AS}/2) = 0.8 + 0.2/2 = 0.9$, and that of $p_S = q_{AS}/2 = 0.1$.

5. The expected frequency of occurrence of AA pairs, $e_{AA} = pA \times pA = 0.9 \times 0.9 = 0.81$, and that of AS, $e_{AS} = 2 \times p_S \times p_A = 2 \times 0.9 \times 0.1 = 0.18$.

6. The matrix entry for AA will be calculated from the ratio of the occurrence frequency to the expected frequency. For AA, the ratio = $q_{AA}/e_{AA} = 0.8/0.81 = 0.99$, and for AS, the ratio = $q_{AS}/ e_{AS} = 0.2/0.18 = 1.11$.

7. Both ratios are converted to logarithms to the base 2 called bits (logs to the base 2 may be calculated from log to the base 10 by dividing by 0.693) and then multiplied by 2 to give units of half-bits. Matrix entry for AA, $s_{AA} = 2 \times \log_2 (q_{AA}/e_{AA}) = -0.04$, and for AS, $s_{AS} = 2 \times \log_2(q_{AS}/ e_{AS}) = 0.30$. These logarithms are both rounded to the nearest half-bit unit, which in this case would be 0 and 0, respectively. The entire BLOCKS multiple sequence alignment database was then used in this same manner to obtain the values shown in Fig. 3.16.

represents a balance between information content and data size. The BLOSUM62 matrix is shown in Figure 3.16.

Henikoff and Henikoff (1993) have prepared a set of interval BLOSUM matrices that represent the changes observed between more closely related or more distantly related representatives of each block. Rather than representing the changes observed in very-alike sequences up to sequences that were n% alike to give a BLOSUM-n matrix, the new BLOSUM-nm matrix represented the changes observed in sequences that were between n% alike and m% alike. The idea behind these matrices was to have a set of matrices corresponding to amino acid changes in sequence blocks that are separated by different evolutionary distances.

Comparison of the PAM and BLOSUM Amino Acid Substitution Matrices

There are important differences in the ways that the PAM and BLOSUM scoring matrices were derived, and these differences should be appreciated when interpreting the results of protein sequence alignments obtained with these matrices.

1. The PAM matrices are based on a mutational model of evolution which assumes that amino acid changes occur as a Markov process, with each amino acid change at a site being independent of previous changes at that site. Changes are scored in sequences that are 85% similar after predicting a phylogenetic history of the changes in each family. Thus, the PAM matrices are based on prediction of the first changes that occur as proteins diverge from a common ancestor during evolution of a protein family. Matrices that may be used to compare more distantly related proteins are then derived by extrapolation from these short-term changes, assuming that these more distant changes are a reflection of the short-term changes occurring over and over again. For each longer evolutionary interval, each amino acid can change to any other with the same frequency as observed in the short term.

2. In contrast, the BLOSUM matrices are not based on an explicit evolutionary model. They are derived from considering all amino acid changes observed in an aligned region from a related family of proteins, regardless of the overall degree of

similarity between the protein sequences. However, these proteins are known to be related biochemically and, hence, should share common ancestry. The evolutionary model implied in such a scheme is that the proteins in each family share a common origin, but closer versus distal relationships are ignored, as if they all were derived equally from the same ancestor, called a starburst model of protein evolution (see Chapters 5 and 7).

3. The PAM matrices are based on scoring all amino acid positions in related sequences, whereas the BLOSUM matrices are based on substitutions and conserved positions in blocks, which represent the most-alike common regions in related sequences.

The PAM model is thus designed to track the evolutionary origins of proteins, whereas the BLOSUM model is designed to find their conserved domains. The choice of which matrix to use depends on the goals of the investigator.

Other Amino Acid Scoring Matrices

In addition to the Dayhoff PAM, and the related Gonnet et al. (1992), Benner et al. (1994), and Jones et al. (1992) matrices, and the BLOSUM matrices, a number of other amino acid substitution matrices have been used for producing protein sequence alignments, and several representative ones are listed in Table 3.3. For a more complete list and comparison, see Vogt et al. (1995). These tables vary from a comparison of simple chemical properties of amino acids to a complex analysis of the substitutions found in secondary structural domains of proteins. Because most of these tables are designed to align proteins on the basis of some such feature of the amino acids, and not on an evolutionary model, they are not particularly suitable for evolutionary analysis. They should be useful, however, for discovering structural and functional relationships, or family relationships among proteins.

A sequence alignment program that uses a combination of these tables has been found to be particularly useful for detecting distant protein relationships (Argos 1987; Rechid et al. 1989). There have been extensive comparisons (Henikoff and Henikoff 1993; Pearson 1995, 1996, 1998) of the usefulness of various amino acid substitution matrices for aligning sequences, for finding similar sequences in a protein

	C	S	T	P	A	G	N	D	E	Q	H	R	K	M	I	L	V	F	Y	W	
C	9																				C
S	-1	4																			S
T	-1	1	5																		T
P	-3	-1	-1	7																	P
A	0	1	0	-1	4																A
G	-3	0	-2	-2	0	6															G
N	-3	1	0	-2	-2	0	6														N
D	-3	0	-1	-1	-2	-1	1	6													D
E	-4	0	-1	-1	-1	-2	0	2	5												E
Q	-3	0	-1	-1	-1	-2	0	0	2	5											Q
H	-3	-1	-2	-2	-2	-2	1	-1	0	0	8										H
R	-3	-1	-1	-2	-1	-2	0	-2	0	1	0	5									R
K	-3	0	-1	-1	-1	-2	0	-1	1	1	-1	2	5								K
M	-1	-1	-1	-2	-1	-3	-2	-3	-2	0	-2	-1	-1	5							M
I	-1	-2	-1	-3	-1	-4	-3	-3	-3	-3	-3	-3	-3	1	4						I
L	-1	-2	-1	-3	-1	-4	-3	-4	-3	-2	-3	-2	-2	2	2	4					L
V	-1	-2	0	-2	0	-3	-3	-3	-2	-2	-3	-3	-2	1	3	1	4				V
F	-2	-2	-2	-4	-2	-3	-3	-3	-3	-3	-1	-3	-3	0	0	0	-1	6			F
Y	-2	-2	-2	-3	-2	-3	-2	-3	-2	-1	2	-2	-2	-1	-1	-1	-1	3	7		Y
W	-2	-3	-2	-4	-3	-2	-4	-4	-3	-2	-2	-3	-3	-1	-3	-2	-3	1	2	11	W
	C	S	T	P	A	G	N	D	E	Q	H	R	K	M	I	L	V	F	Y	W	

FIGURE 3.16. The BLOSUM62 amino acid substitution matrix. The amino acids in the table are grouped according to the chemistry of the side group: (C) sulfhydryl, (STPAG) small hydrophilic; (NDEQ) acid, acid amide, and hydrophilic; (HRK) basic; (MILV) small hydrophobic; and (FYW) aromatic. Each entry is the logarithm of the odds score, found by dividing the frequency of occurrence of the amino acid pair in the BLOCKS database (after sequences 62% or more in similarity have been clustered) by the likelihood of an alignment of the amino acids by random chance. The denominator in this ratio is calculated from the frequency of occurrence of each of the two individual amino acids in the BLOCKS database and provides a measure of a chance alignment of the two amino acids. The actual/expected ratio is expressed as a log odds score in so-called half-bit units, obtained by converting the odds ratio to a logarithm to the base 2, and then multiplying by 2. A zero score means that the frequency of the amino acid pair in the database is as expected by chance, a positive score that the pair is found more often than by chance, and a negative score that the pair is found less often than by chance. The accumulated score of an alignment of several amino acids in two sequences may be obtained by adding up the respective scores of each individual pair of amino acids. As with the PAM250-derived matrix, the highest-scoring matches are between amino acids that are in the same chemical group, and the very highest-scoring matches are for cysteine–cysteine matches and for matches among the aromatic amino acids. Compared to the PAM160 matrix, however, the BLOSUM62 matrix gives a more positive score to mismatches with the rare amino acids, e.g., cysteine, a more positive score to mismatches with hydrophobic amino acids, but a more negative score to mismatches with hydrophilic amino acids (Henikoff and Henikoff 1992).

sequence database, or for aligning similar sequences based on structure. The use of these scoring matrices depends also on the appropriate choice for gap penalties, and on a proper statistical evaluation of local alignment scores, discussed later in this chapter and in Chapter 4.

Nucleic Acid PAM Scoring Matrices

Just as amino acid scoring matrices have been used to score protein sequence alignments, nucleotide scoring matrices for scoring DNA sequence alignments have also been developed. The DNA matrix can incorporate ambiguous DNA symbols (see Table 2.1) and information from mutational analysis, which reveals that transitions (substitutions between the purines A and G or between the pyrimidines C and T) are more probable than transversions (substitutions from purine to pyrimidine or pyrimidine to purine) (Li and Graur 1991). These substitution matrices may be used to produce global or local alignments of DNA sequences.

TABLE 3.3. *Criteria used in amino acid scoring matrices for sequence alignments*

1. Simple identity, which scores only identical amino acids as a match and all others as a mismatch.
2. Genetic code changes, which score the minimum number of nucleotide changes to change a codon for one amino acid into a codon for another, due to Fitch (1966), and also with added information based on structural similarity of amino acid side chains (Feng et al. 1985). A similar matrix based on the assumption that genetic code is the only factor influencing amino acid substitutions has been produced (Benner et al. 1994).
3. Matrices based on chemical similarity of amino acid side chains, molecular volume, and polarity and hydrophobicity of amino acid side chains (see Vogt et al. 1995).
4. Amino acid substitutions in structurally aligned three-dimensional structures (Risler et al. 1988; matrix JO93, Johnson and Overington 1993). A similar matrix was described by Henikoff and Henikoff (1993). Sander and Schneider (1991) prepared a similar matrix based on these same substitutions but augmented by substitutions found in proteins which are so similar to the structure-solved group that they undoubtedly have the same three-dimensional structure.
5. Gonnet et al. (1994) have prepared a 400 × 400 dipeptide substitution matrix for aligning proteins based on the possibility that amino acid substitutions at a particular site are influenced by neighboring amino acids, and thus that the environment of an amino acid plays a role in protein evolution.
6. Jones et al. (1994) have prepared a scoring matrix specifically for transmembrane proteins. This matrix was prepared using an analysis similar to that used for preparing the original Dayhoff PAM matrices, and therefore provides an estimate of evolutionary distances among members of this class of proteins.

States et al. (1991) have developed a series of nucleic acid PAM matrices based on a Markov transition model similar to that used to generate the Dayhoff PAM scoring matrices. Although designed to improve the sensitivity of similarity searches of sequence databases, these matrices also may be used to score nucleic acid alignments. The advantage of using these matrices is that they are based on a defined evolutionary model, and the statistical significance of alignment scores obtained by local alignment programs may be evaluated, as described in Chapter 4.

To prepare these DNA PAM matrices, a PAM1 mutation matrix representing 99% sequence conservation and one PAM of evolutionary distance (1% mutations) were first calculated. For a model in which all mutations from any nucleotide to any other are equally likely, and in which the four nucleotides are present at equal frequencies, the four diagonal elements of the PAM1 matrix representing no change are 0.99, whereas the six other elements representing change are 0.00333 (Table 3.4). The values are chosen so that the sum of all possible changes for a given nucleotide in the PAM1 matrix is 1% (3 × 0.00333 = 0.00999). For a biased mutation model in which a given transition is threefold more likely than a transversion (Table 3.4), the off-diagonal matrix elements corresponding to the one possible transition for each nucleotide are 0.006 and those for the two possible

transversions are 0.002, and the sum for each nucleotide is again 1% (0.006 + 0.002 + 0.002 = 0.01).

As with the amino acid matrices, the above matrix values are then used to produce log odds scoring matrices that represent the frequency of substitutions expected at increasing evolutionary distances. In terms of an alignment, the probability (s_{ij}) of obtaining a match between nucleotides i and j, divided by the random probability of aligning i and j, is given by

$$s_{ij} = \log (p_i M_{ij} / p_i p_j) \qquad (7)$$

where M_{ij} is the value in the mutation matrix given in Table 3.4, and p_i and p_j are the fractional composition of each nucleotide, assumed to be 0.25. The base of the logarithm can be any value, corresponding to multiplying every value in the matrix by the same constant. With such scaling variations, the ability of the matrix to distinguish among significant and chance alignments will not be altered. The resulting tables with s_{ij} expressed in units of bits (logarithm to the base 2) and rounded off to the nearest whole integer are shown in Table 3.5.

From these PAM1 matrices, additional log odds matrices at an evolutionary distance of n PAMs may be derived by multiplying the mutation matrix in Table 3.4 by itself n times, as illustrated in Table 3.1B. The ability of each matrix to distinguish real from

TABLE 3.4. *Nucleotide mutation matrix for an evolutionary distance of 1 PAM, which corresponds to a probability of a change at each nucleotide position of 1%*

A. Model of uniform mutation rates among nucleotides

	A	G	T	C
A	0.99			
G	0.00333	0.99		
T	0.00333	0.00333	0.99	
C	0.00333	0.00333	0.00333	0.99

B. Model of threefold higher transitions than transversions

	A	G	T	C
A	0.99			
G	0.006	0.99		
T	0.002	0.002	0.99	
C	0.002	0.002	0.006	0.99

Values are frequency of change at each site, or of no change for all base combinations.

random nucleotide matches in an alignment, designated H, measured in bit units ($\log_2$) can be calculated using the equation

$$H = \sum_{i,j} p_i p_j s_{ij} 2^{s_{ij}}$$

(8)

where the s_{ij} scores are also expressed in bit units. Table 3.6 shows the log odds values of the match and mismatch scores for PAM matrices at increasing evolutionary distances, assuming a uniform rate of mutation among all nucleotides. Also shown is the percentage of nucleotides that will be changed at that distance. The identity score will be 100 minus this value. This percentage is not as great as the PAM score due to expected back-mutation over longer time periods. Also shown are the H scores of the matrices at each PAM value.

The following points may be made regarding matrix choice:

1. If comparing sequences that are quite similar, it is better to use a lower-scoring matrix because the information content of the small PAM matrices is relatively higher. As discussed earlier for lower-numbered Dayhoff PAM matrices for more-alike protein sequences, a higher scoring alignment will be obtained.

2. As the PAM distance increases, the mismatch scores in the biased mutational model in Table 3.7 become positive and appear as conservative substitutions. Thus, the bias model can provide considerably more information than the uniform mutation model when aligning sequences that are distantly related (>30% different) and may be used for this purpose (States et al. 1991).

3. The scoring matrices at large evolutionary distances provide very little information per aligned nucleotide pair. In sequences aligned with such scoring matrices, the alignment score increases much more slowly as the sequences become longer so that longer sequences are necessary to achieve the same odds score for the alignment.

TABLE 3.5. *Nucleotide substitution matrix at 1 PAM of evolutionary distance*

A. Model of uniform mutation rates among nucleotides

	A	G	T	C
A	2			
G	−6	2		
T	−6	−6	2	
C	−6	−6	−6	2

B. Model of threefold higher transitions than transversions

	A	G	T	C
A	2			
G	−5	2		
T	−7	−7	2	
C	−7	−7	−5	2

Units are log odds scores obtained as described in the text.

TABLE 3.6. *Properties of nucleic acid substitution matrices assuming a uniform rate of mutation among nucleotides*

PAM distance	Percentage difference	Match score (bits)	Mismatch score (bits)	Average information per position (bits)
10	9.4	1.86	−3.00	1.40
25	21.3	1.66	−1.82	0.92
50	36.5	1.34	−1.04	0.47
100	55.2	0.84	−0.44	0.13
125	60.8	0.65	−0.30	0.07

As with amino acid scoring matrices, the average information content shown is only achieved by using the scoring matrix that matches the percentage difference between the sequences. For example, for sequences that are 21% different (79% identical), the matrix at 25 PAM distance should be used. One cannot know ahead of time what the percentage similarity or difference between two sequences actually is until an alignment is done, thus a trial alignment must first be done. States et al. (1991) have calculated how efficient a given scoring matrix is at achieving the highest possible score in aligning two sequences that vary in their levels of similarity. Once the initial similarity score has been obtained with these matrices, a more representative score can be obtained by using another PAM matrix designed specifically for sequences at that level of similarity. In Chapter 4, page 155, a Bayesian statistical method for finding the most probable PAM distance between two DNA or protein sequences is described.

Gap Penalties

Both including gaps in sequence alingments and using gap penalties are necessary to obtain the best possible alignment between two sequences. A gap opening penalty for any gap (g) and a gap extension penalty for each element in the gap (r) is most often used, to give a total gap score w_x, according to the equation

$$w_x = g + rx \qquad (9)$$

where x is the length of the gap. Note that in some formulations of the gap penalty, the equation $w_x = g + r(x-1)$ is used. Thus, the gap extension penalty is not added to the gap opening penalty until the gap size is 2. Although this difference does not affect the alignment obtained, one needs to distinguish which method is being used by a particular computer program if the correct results are to be obtained. In the former case, the penalty for a gap of size 1 is $g + x$, whereas in the latter case this value is g. The values for these penalties have to be chosen to balance the scores in the scoring matrix that is used. Thus, the Dayhoff log odds matrix at PAM250 is expressed in units of $\log_{10}$, which is approximately 1/3 bits, but if this matrix were converted to 1/2 bits, the same gap penalties would no longer be appropriate.

If too high a gap penalty is used relative to the range of scores in the substitution matrix, gaps will never appear in the alignment. Conversely, if the gap penalty is too low compared to the matrix scores, gaps will appear everywhere in the alignment to align as many of the same characters as possible. Fortunately, most alignment programs will suggest

TABLE 3.7. *Properties of nucleic acid substitution matrices assuming transitions are threefold more frequent than transversions*

PAM distance	Percentage difference	Match score (bits)	Transition score (bits)	Transversion score (bits)	Average information per position (bits)
10	9.3	1.86	−2.19	−3.70	1.42
25	21.0	1.66	−1.06	−2.46	0.96
50	35.8	1.36	−0.37	−1.60	0.54
100	53.7	0.89	0.06	−0.86	0.19
150	62.9	0.57	0.16	−0.52	0.08

gap penalties that are appropriate for a given scoring matrix in most situations. In the GCG and FASTA program suites, the scoring matrix itself is formatted in a way that includes default gap penalties. When deciding gap penalties for local alignment programs, another consideration is that the penalties should be large enough to provide a local alignment of the sequences. Examples of suitable values are given in Table 4.3 on page 142. Altschul and Gish (1996) and Pearson (1996, 1998) have found that use of appropriate gap penalties will provide an improved local alignment based on statistical analysis. These studies are described in detail in Chapter 4.

Mathematician Peter Sellers (1974) showed that if sequence alignment is formulated in terms of distances instead of similarity between sequences, a biologically more appealing interpretation of gaps is possible. The distance is the number of changes that must be made to convert one sequence into the other and represents the number of mutations that will have occurred following separation of the genes during evolution; the greater the distance, the more distantly related are the sequences in evolution. In this case, substitution produces a positive score of 1 and conservation of the position a score of 0. Notice that the distance score plus the similarity score for an alignment is equal to 1. Sellers proved that this distance formulation of sequence alignment has a desirable mathematical property that also makes evolutionary sense. If three sequences, **a**, **b**, and **c**, are compared using the above scoring scheme, the distance score as defined above is described as a metric that satisfies the triangle inequality relationship

$$d(\mathbf{a}, \mathbf{b}) + d(\mathbf{b}, \mathbf{c}) \geq d(\mathbf{a}, \mathbf{c}) \qquad (10)$$

where $d(\mathbf{a}, \mathbf{b})$ is the distance between sequences **a** and **b**, and likewise for the other two d values. Expressed another way, if the three possible distances between three sequences are obtained, then the distance between any first pair plus that for any second pair cannot underscore the third pair. Violating this rule would not be consistent with the expected evolutionary origin of the sequences. To satisfy the metric requirement, the scoring of individual matches, mismatches, and gaps must be such that in an alignment of two identical sequences **a** and **a**′, $d(\mathbf{a}, \mathbf{a}')$ must equal 0, and for two totally different sequences **b** and **b**′, $d(\mathbf{b}, \mathbf{b}')$ must equal 1. For any other two sequences **a** and **b**, $d(\mathbf{a}, \mathbf{b}) = d(\mathbf{b}, \mathbf{a})$. Hence, it is important that the distance score for changing one sequence charac-

ter into a second is the same as the converse score for changing the second into the first, if the distance score of the alignment is to remain a metric and to make evolutionary sense. The above relationships were shown by Sellers to be true for gaps of length 1 in a sequence alignment. He also showed that the smallest number of steps required to change one sequence into the other could be calculated by the dynamic programming algorithm. The method was similar to that discussed above for the Needleman–Wunsch global and Smith–Waterman local alignments, except that these former methods found the maximum similarity between two sequences, as opposed to the minimum distance found by the Sellers analysis.

Subsequently, Smith et al. (1981) and Smith and Waterman (1981a,b) showed that gaps of any length could also be included in an alignment and still provide a distance metric for the alignment score. In this formulation, the gap penalty was required to increase as a function of the gap length. The argument was made that a single mutational event involving a single gap of n residues should be more likely to have occurred than n single gaps. Thus, to increase the likelihood of such gaps of length >1 being found, the penalty for a gap of length n was made smaller than the score for n individual gaps. The simplest way of implementing this feature of the gap penalty was to have the gap score w_x be a linear function of gap length by consisting of two parts, a larger gap opening penalty (g) and a smaller gap extension penalty (r) for each extra position in the gap, or $w_x = g + rx$, where x is the length of the gap, as described above. This type of gap penalty is referred to as an affine gap penalty in the literature. Any other formula for scoring gap penalties should also work, provided that the score increases with length of the gap but that the score is less than x individual gaps. Scoring of gaps by the above linear function of gap length has now become widely used in sequence alignment. However, more complex gap penalty functions have also been used (Miller and Myers 1988).

Penalties for Gaps at the Ends of Alignments

Some programs produce global sequence alignments that include penalties for gaps opposite nonmatching characters at the ends of an alignment, whereas others do not. In practice, local sequence alignments should not have such gaps because they would decrease the score of the alignment. End gaps may be given the same penalty score as gaps inside of the alignment or, alternatively, they may not be given any penalty score. End gaps were an important component in the mathematical formulation of both the similarity and distance methods of sequence alignment for producing both global and local alignments. Failure to include them in distance calculations can result in a failure to obtain distance scores that make evolutionary sense (Smith et al. 1981). If gaps at the ends of alignments are not given a penalty score, gaps may then be liberally placed at the ends of alignments by the dynamic programming algorithm to increase the matching of internal characters. In general, it is usually best to include end gap penalties when using Needleman–Wunsch global alignments, but they will not affect Smith–Waterman local alignments.

If comparing sequences that are homologous and of about the same length, it makes a great deal of sense to include end gap penalties to achieve the best overall alignment. For sequences that are of unknown homology or of different lengths, it may be better to use an alignment that does not include end gap penalties (States and Boguski 1991). If one sequence is expected to be contained within the other, it is reasonable to include end gap penalties only for the shorter sequence. However, for any test alignment, these end penalties should be included in at least one alignment to assure that they do not have an effect. It is also important to use alignment programs that include them as an option.

Parametric Sequence Alignments

Computer methods that find a range of possible alignments in response to varying the scoring system used for matches, mismatches, and gaps are called parametric sequence comparisons (Waterman et al. 1992; Waterman 1994 and references therein). There is also an effort to use scores such that the results of global and local types of sequence alignments provide consistent results. For example, if two sequences are similar along their entire lengths, both global and local methods should provide the same alignment. The program Xparal (Gusfield and Stelling 1996) performs this type of analysis and is available from http://www.cs.ucdavis.edu/~gusfield. The program runs on a UNIX environment under X-Window. When provided with two sequences and some of the alignment parameters, such as gap score, the program graphically displays the types of possible alignments

when the remaining parameters are varied. Another sequence alignment program that performs parametric sequence alignment is the Bayes block aligner, which is discussed in detail in Chapter 4 (p. 152).

Effects of Varying Mismatched Gap Penalties on Local Alignment Scores

Vingron and Waterman (1994) have reviewed the effect of varying the parameters of the scoring system on the alignment of random DNA and protein sequences. To simplify the number of parameters, a constant penalty for any size gap was used. If a very high mismatch penalty is used relative to a positive score for a match, with zero gap penalty, the local alignment of these sequences will not include any gaps and is defined as the longest common subsequence. The global alignment with the same scoring parameters will have no mismatches but will have many gaps placed to maximize the matches, and the score will be positive. In this case, the score of the local alignment of the sequences is predicted to increase linearly with the length of the sequences being compared.

Another case of varying alignment is penalizing gaps heavily. Then the best-scoring local alignment between the sequences will be one that optimizes the score between matches and mismatches, without any gaps. If both mismatches and gaps are heavily penalized, the resulting alignment will also be a local alignment that contains the longest region of exact matches. In the above two cases, the alignment score of the highest-scoring local alignment will increase as the logarithm of the length of the sequences. Under these same conditions, the score of the corresponding global alignment between the sequences will be negative. The transition between a linear and logarithmic dependence of the local similarity score on sequence length occurs when the score of the corresponding global alignment is zero. When both the mismatch and gap penalties are varied between zero and a high negative score, the number of possible alignments of random DNA sequences becomes very large.

Three general conclusions can be drawn from this theoretical study of random sequence alignments:

1. Use of high mismatch and gap penalties that are greater than a match score will find local alignments, of which there are relatively few.

2. When the penalty for a mismatch is greater than twice the score for a match, the gap penalty becomes the decisive parameter in the alignment.

3. For a mismatch penalty less than twice the score of a gap and a wide range of gap penalties, there are a large number of possible alignments that depend on both the mismatch and gap penalty scores.

Distinguishing local from global alignments has an important practical application. A local alignment is rarely produced between random sequences. Accordingly, the significance of a local alignment between real sequences may be readily calculated, as described in Chapter 4. In contrast, the significance of a global alignment is difficult to determine because a global alignment is readily produced between random sequences.

Optimal Combinations of Scoring Matrices and Gap Penalties for Finding Related Proteins

The usefulness of scoring matrix–gap penalty combinations for identifying related proteins, including distantly related ones, has been compared (Feng et al.

1985; Doolittle 1986; Henikoff and Henikoff 1993; Pearson 1995, 1996, 1998; Agarwal and States 1998; Brenner et al. 1998). The method generally used is to start with a database of protein sequences organized into families, based on either sequence similarity or structural similarity (described in Chapters 5 and 10, respectively). A member of a family is then selected and used as a query sequence in a search of the entire database from which the sequence came, using a database similarity search method (FASTA, BLAST, SSEARCH), as described in Chapter 6. The measure of success is the number of known family members found (true positives) while keeping false-negative predictions of non-family members to a minimum.

Another earlier method for testing amino acid matrix–gap penalty combinations was to determine how well the combination reproduced a known alignment based either on analysis of families of similar proteins or on structural alignments. The results are summarized in the following box.

STUDIES OF VARYING ALIGNMENT ALGORITHM, AMINO ACID SCORING MATRIX, AND GAP PENALTIES

Comparison of amino acid scoring matrix–gap penalty combinations poses several problems for the reader. One major problem is that the analysis often overlooks the purposes of different matrices; e.g., protein family or domain searching, evolutionary analysis, or structural alignment. In the past, gap penalties were usually not published or well known, thus throwing a level of uncertainty into the results. More recently, when investigators publish a new scoring matrix, they usually provide suitable choices for gap penalties that may be used for comparisons with other matrices.

Many extremely useful scoring matrices, e.g., the BLOSUM62, Gonnet92, and JO93 matrices, are not based on an explicit model of protein evolution. Instead, they are designed to locate sequence similarity as a basis for some other common feature between proteins, such as structure, biochemical function, or a family relationship, because they were generated by examining similarity in sequences that share such features in common. Recent research (Henikoff and Henikoff 1993; Pearson 1995; 1996; 1998) has shown that, for the majority of sequence analysis projects, the above three matrices are the most likely to identify a relationship or to find related sequences in a database search, provided that appropriate gap penalties are chosen. At the same time, these matrices are not as useful for discovering evolutionary relationships, because an explicit model of protein evolution was not used in their generation. If evolutionary relationships are to be emphasized, the Dayhoff PAM and more recent JTT PAM scoring matrices are the best analysis tools, provided that the matrix which corresponds to the evolutionary distance between the sequences and appropriate gap penalties for that matrix are selected (see Altschul et al. 1994).

Pearson (1995) has performed the most detailed analysis to date of the ability of combinations of alignment algorithm, scoring matrix, and gap penalties to assist in the recognition of proteins that are known to be in the same superfamily. He used the computer programs SSEARCH, FASTA, and BLAST (see Chapter 6 for detailed description), which are designed to be used in a database search and to identify proteins that are related to an input query sequence. Several different scoring matrices were tested, each with a range of gap penalties. The best combination of program and scoring system was the SSEARCH program (using the Smith–Waterman algorithm) in combination with the BLOSUM55 scoring matrix and gap penalties of −12, −2. The PAM250 matrix performed significantly worse at every gap penalty tested. The JO93 and JTT160/JTT200 (–14, –2) combinations performed as well as BLOSUM55 (–12, –2). The higher PAM JTT matrices (JTT250 and 320) did not perform as well as BLOSUM55 (–12, –2). BLOSUM62 (–6, –4 and –8, –2), which is scaled in 1/2-bit units and thus requires smaller gap penalties, was almost, but not quite, as good as BLO-SUM55 (–12, –2). Pearson later performed a detailed statistical analysis of alignment scores found in a database search by SSEARCH, using methods described in Chapters 4 and 6. This more refined analysis produced slightly different results; BLOSUM50 (–12, –2) was now the best combination (Pearson 1998).

Most investigators use the BLAST programs for searching sequence databases with a query sequence. Following extensive types of analyses similar to those described above, BLOSUM62 (–11, –1) is the standard (default) scoring matrix–gap penalty combination. This combination optimizes the identification of related sequences (Altschul and Gish 1996).

There have been several analyses of the effectiveness of scoring matrices to produce alignments that match alignments based on three-dimensional structures. Johnson and Overington (1993) prepared a scoring matrix (JO93) based on structurally aligned proteins. They then compared the ability of this matrix with 12 other published matrices with a series of simple (nonaffine) gap penalties to predict the previously observed structural alignments. Of the matrices examined, JO93 performed the best, but Gonnet92 and BLOSUM62 performed almost as well.

Vogt et al. (1995) performed a similar study of scoring matrix accuracy also using known alignments based on structural information as a measure of success. The authors first performed an extensive analysis of the optimum values of the gap opening and extension penalties for a large number of amino acid substitution matrices with only small positive scores that should produce a global alignment. They found that these positive scoring matrices could predict the known alignments after optimizing the gap penalties.

Abagyan and Batalov (1997) also performed a study of the value of scoring matrices in predicting global alignments between structurally related protein sequences. Different gap opening and extension penalties were tried with a series of matrices with a similar range of scores. The Gonnet92 and BLOSUM50 scoring matrices performed the best global alignments. This method is worth considering as a way to perform a global alignment of proteins when the objective is to predict the a structural fold (see Chapter 10).

In summary, the following general observations can be made regarding scoring matrices and gap penalties:

1. Some scoring matrices are superior to others at finding related proteins based on either sequence or structure. For example, matrices prepared by examining the full range of amino acid substitutions in families of related proteins, such as the BLOSUM62 matrix, perform better than matrices based on variations in closely related proteins that are extrapolated to produce matrices for more distantly related sequences, such as the Dayhoff PAM250 matrix.

2. Gap penalties that for a given scoring matrix are adjusted to produce a local alignment are generally the most suitable.

3. To identify related sequences, the significance of the alignment scores should be estimated, as described in Chapter 4.

The main conclusion to be drawn from this chapter is that the dynamic programming algorithm can be used to provide alignment of DNA or protein sequences that includes either all of the sequences (Needleman–Wunsch) or just localized regions that represent conserved domains by a local alignment (Smith–Waterman). Finding the best alignment depends on appropriate choices of a scoring matrix and gap penalties that provide an optimal alignment with the highest score. The methods discussed in this chapter provide the means to demonstrate sequence similarity in even the most distantly related proteins. It is very important to validate the sequence alignment by showing that there are not alternative alignments that are almost as good in terms of alignment score and probability. This chapter has described methods for obtaining alternative alignments of sequences as a method for testing validity. Chapter 4 describes the statistical analysis of the local alignment scores in which the probability that two random or unrelated sequences can achieve as good a score as one found between two sequences can be determined.

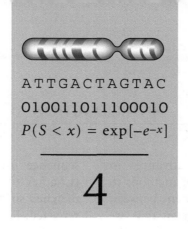

ATTGACTAGTAC
010011011100010
$P(S < x) = \exp[-e^{-x}]$

4

Introduction to Probability and Statistical Analysis of Sequence Alignments

C O N T E N T S

INTRODUCTION

One of the most important recent advances in sequence analysis is the development of methods to assess the significance of a local alignment between DNA or protein sequences. For sequences that are obviously related—two proteins that are clearly in the same family, or two matching or overlapping DNA fragments—such an analysis is hardly necessary. The question of significance arises when comparing two sequences that are not so clearly similar but are shown to align in a promising way. In such a case, a significance test can help the biologist to decide whether an alignment found by the computer program is one that would be expected between related sequences or would just as likely be found if the sequences were not related. A significance test is also critical for evaluating the results of a database search for sequences that are found to be similar to a query

sequence using the BLAST and FASTA programs (Chapter 6). The test is applied to every sequence matched so that the most significant matches can be reported. Finally, a significance test can also help to identify regions in a single sequence that have an unusual composition suggestive of an interesting function.

Our goal here is to examine the significance of sequence alignment scores obtained by the dynamic programming method. Adequate theory has been developed and supportive experimental data have been obtained that together provide a reliable evaluation of local sequence alignments. This chapter outlines some of the major features of statistical testing and probability calculations and shows how to use these features to evaluate the significance of a sequence alignment.

CHAPTER GUIDE FOR BIOLOGISTS

Many biologists have at some time been exposed to the bell-shaped curve of a normal probability distribution and statistical analyses such as t and χ^2 tests that are needed to determine whether experimental data support a certain hypothesis. Sequence analysis introduces several new levels of statistical analysis and probability. First, as discussed in Chapter 3, sequence alignment scores are calculated using odds and log odds scores that provide an indication as to how good an alignment actually is on the basis of alignments of individual sequence positions. We learned that an odds scoring system with scores for matches and gaps in the correct proportions is essential for obtaining a meaningful sequence alignment. When mathematicians began to think about alignments and computational biologists examined alignment scores between computer-generated random sequences, however, they discovered that the scores often far exceeded those expected based on a normal curve. The results gave rise to some guiding principles for evaluating alignment scores that are summarized in the flowchart on page 75 of Chapter 3. The distribution of scores in local alignments between unrelated sequences followed a curve with a long tail at high alignment score values. This result meant that, to be believed, an alignment score that was to be tested for significance had to be much higher than would be the case for a normal distribution. A similar distribution had been observed previously with other types of data and was known as the extreme value distribution after Gumbel.

The formula needed to evaluate the statistical significance of an alignment score is a double exponential function with two constants and two sequence lengths. The constants depend on the scoring system used and can be calculated quite readily. The result of applying the equation is a value giving the probability that two random or unrelated sequences of the same length can give a score as high as the one found between two test sequences. The probability is usually expressed as an E value, which is defined as follows: For a large number of random alignments, e.g., 10,000, E is the number of alignments that will exceed the test score. For a BLAST search of a database of 10,000 sequences, E has the same meaning—it is the number of unrelated database sequences of the same length that are expected to produce as good a score with the query sequence as the related sequences that are found. E values frequently appear in sequence analysis data, and they all have the same meaning as just described.

Another application of probability to sequence analysis is Bayesian statistics, which introduces a new kind of probability called conditional probability. For example, in sequence analysis, we can think of an alignment score between two sequences as depending on the scoring system used. This dependence is a type of conditional probability. If the scoring system is changed, the alignment score will also change. Bayesian statistical analysis uses the alignment scores found by using a range of scoring systems to determine the most probable scoring system and the most probable alignments. A Bayesian analysis also has an updating feature in which new data are used in a special way. Rather than averaging new data with the old to produce a new model of the data, the old data are used to produce a model against which the new data are tested. This analysis results in an updated model with a new probability of being correct. Some people have an intuitive block against using this kind of approach to data analysis. However, simple examples, such as approaches used in the "Let's Make a Deal" game show (outlined on p. 149), can be proven to give a more reliable prediction. This updating feature approach is a recurring theme in bioinformatics.

CHAPTER GUIDE FOR COMPUTATIONAL SCIENTISTS

Biologists are interested in comparing DNA and protein sequences from different organisms to discover similar sequences that can be aligned. Underlying this analysis is the well-supported theory that similar sets of genes carry out the fundamental processes of living organisms such as metabolism, reproduction, and development. These genes have been inherited from ancestor organisms that lived in prehistoric times. By studying variations between similar sequences, biologists can acquire an appreciation as to which sequence positions are important for biological function. The variations are also used to produce binary trees of historical information on the ancestral changes in the sequences.

These biological goals underscore the need to assess the significance of sequence alignments. Sequences are not random permutations of building blocks but instead follow patterns that are compatible with biological function and evolutionary change. Insertions, deletions, repeats, and uneven distribution of sequence characters in different genes and different organisms can make aligning sequences a difficult undertaking. Sequences can also diverge sufficiently over time to make finding any remaining similarity a difficult task. Fortunately, mathematicians and computer scientists have developed methods for evaluating sequence alignment scores when the alignment is a local alignment. These methods are now routinely used by the biological community to screen through large sets of genes to discover similar ones. (However, care must still be taken, because some kinds of sequence variation can give false information.) These methods have introduced both probability and statistics into biological sequence analysis. Bayesian methods that use a range of possible models of sequence change to find sequence alignments have also been developed. These methods provide posterior predictions of the most probable alignments and supporting models.

As an example of the application of these methods, in the introduction to Bayesian statistics and conditional probability on page 148, a genetic example is used that stems from using inbred strains of an organism that are almost identical genetically to the peas used by the famous geneticist Gregor Mendel. The geneticist looks for changes in the strain that are inherited and locates the gene on a chromosome with respect to other genes by genetic experiments. These experiments provide two copies of each gene called alleles—the original one and a new one that gives the new genetic trait—and locates genes with respect to each other. The point of the example is that alleles of genes that are close together on the chromosome stay together in the offspring, whereas genes that are not together on the chromosome or are located on different chromosomes are separated. These genetic linkages are reflected in the conditional probabilities that are reported in the example.

WHAT SHOULD BE LEARNED IN THIS CHAPTER?

- Basic concepts of probability and statistical testing.
- Appreciation for the determination of statistical reliability.
- How to evaluate the statistical significance of a local alignment score.
- An understanding of the meaning of an E value of a sequence alignment score.
- Familiarity with applications of Bayesian statistics to sequence analysis.

Glossary Terms

Bayes' rule Forms the basis of conditional probability by calculating the likelihood of an event occurring based on the history of the event as updated with new information. In terms of two parameters A and B, the theorem is stated in an equation: The conditional probability of A, given B, $P(A|B)$, is equal to the probability of A, $P(A)$, times the conditional probability of B, given A, $P(B|A)$, divided by the probability of B, $P(B)$. $P(A)$ is the historical or prior distribution value of A, $P(B|A)$ is a new prediction for B for a particular value of A, and $P(B)$ is the sum of the newly predicted values for B. $P(A|B)$ is a posterior probability, representing a new prediction for A given the prior knowledge of A and the newly discovered relationships between A and B.

Bayesian analysis A statistical procedure used to estimate the conditional probability of an event based on better and better estimates of the probability using prior conditions from the previous estimate and updating these with new data using Bayes' rule.

Bernoulli trial A repeated number of tests in which the outcome is one of two choices (heads or tails of a coin), the probability in each test does not change, and each measurement is independent of the others.

Binomial distribution A probability distribution of a measurement with two possible outcomes such as tossing a coin to obtain heads or tails. For a large number of trials, approximately 1/2 will be heads and 1/2 tails. For a smaller number, say 10, the number of heads may vary from 1 to 10 and the distribution of repeated trials is given by the binomial distribution. The distribution is useful for determining the reliability of measurements and the power of statistical tests.

Conditional probability The probability of a particular result (or of a particular value of a variable) given one or more events or conditions (or values of other variables). $P(A|B)$ is read as the probability of A, given condition B, which simply may be a value of a variable B.

Erdös and Rényi law In a toss of a "fair" coin, the number of heads in a row that can be expected is the logarithm of the number of tosses to the base 2. The law may be generalized for more than two possible outcomes by changing the base of the logarithm to the number of outcomes. This law was used to analyze the number of matches and mismatches that can be expected between random sequences as a basis for scoring the statistical significance of a sequence alignment.

Expect value (E) In a database similarity search with a query sequence, the number of local alignment scores that are expected between the query sequence and random or unrelated sequences that are as good as the score found between the query sequence and a matched database sequence. E depends on the sequence lengths, the number of sequences in the database, and the type of scoring system used. E is calculated using the extreme value probability distribution corrected for the number of sequences in the database. In an evaluation of a local alignment between two sequences, one sequence may be scrambled many times, and each scrambled sequence realigned with the other sequence to determine the range of scores expected between unrelated sequences. In other types of sequence analysis, E has a similar meaning.

Extreme value distribution Some measurements are found to follow a distribution that has a long tail which decays at high values much more slowly than that found in a normal distribution. One slow-falling type of distribution due to Gumbel is called the extreme value distribution. Local sequence alignment scores between unrelated or random sequences are an example. These scores can reach very high values, particularly when a large number of comparisons are made, as in a database similarity search. The probability of a particular alignment score may be accurately predicted by the extreme value distribution, which follows a double negative exponential function.

False negative In the test of a hypothesis, the calculated probability that the answer obtained will not support the hypothesis, when in fact it should have supported the hypothesis. It is also called a type II or β error in a statistical test.

False positive In the test of a hypothesis, the calculated probability that the answer obtained will support the hypothesis, when in fact it should not have supported the hypothesis. Also called a type I or α error. The power of a statistical test of the hypothesis can be expressed as a value of α.

K In the evaluation of sequence alignment scores, a statistical parameter of the extreme value distribution that depends on the scoring matrix used and the gap penalty scores. For most scoring matrices, $K \approx 0.1$.

Log odds score The logarithm of an odds score. See also Odds score.

Moments The mean or expected value of a variable is the first moment of the values of the variable around the mean, defined as that number from which the sum of deviations to all values is zero. The standard deviation is the second moment of the values about the mean, and so on.

Normal distribution The distribution found for many types of data such as body weight, size, and exam scores.

The distribution is a bell-shaped curve that is described by a mean and standard deviation of the mean. Normal distributions are expected when the data points are affected by many unrelated factors. Local sequence alignment scores between unrelated or random sequences do not follow the normal distribution but instead they follow extreme value distribution, which has a much extended tail for higher scores. See also Extreme value distribution.

Null hypothesis In a statistical test, a specific hypothesis about the value of a variable or outcome of the test is chosen. The objective of the test is to obtain sufficient data that will lead to acceptance or rejection of this hypothesis. The null hypothesis is that there will be no change as a result of the test or that a test condition will be met or not met.

Odds score The ratio of the likelihoods of two events or outcomes. In sequence alignments and scoring matrices, the odds score for matching two sequence characters is the ratio of the frequency with which the characters are aligned in related sequences divided by the frequency with which those same two characters align by chance alone, given the frequency of occurrence of each in the sequences. Odds scores for a set of individually aligned positions are obtained by multiplying the odds scores for each position. Odds scores are often converted to logarithms to create log odds scores that can be added to obtain the log odds score of a sequence alignment.

Poisson distribution Used to predict the occurrence of infrequent events over a long period of time or of a large number of trials, each with a small chance of success. In sequence analysis, it is used to calculate the chance that one pair of a large number of pairs of unrelated sequences may give a high local alignment score.

Posterior probability (Bayesian analysis) A conditional probability based on prior knowledge and newly evaluated relationships among variables based on new information using Bayes' rule. See also Bayes' rule.

Prior probability (Bayesian analysis) The expected distribution of a variable based on previous data.

Probability A score between 0 and 1 that describes the likelihood of an event.

Probability distribution A plot of theoretical or actual values of some variable in which a range of values on the x axis is plotted against the frequency of scores in each range on the y axis. The curve may be normalized by scaling to make the area under the curve (AUC) equal to 1. The probability of measurements in a certain range of values may then be calculated from the AUC between the values. In sequence analysis, a local alignment score of randomly aligned sequences follows the extreme value probability distribution.

Receiver operator characteristic The receiver operator characteristic (ROC) curve describes the probability that a test will correctly declare the condition present against the probability that the test will declare the condition present when actually absent. This is shown through a graph of the test's sensitivity against one minus the test's specificity for different possible threshold values.

Selectivity (in database similarity searches) The ability of a search method to locate members of a protein family without making a false-positive classification of members of other families.

Sensitivity (in database similarity searches) The ability of a search method to locate as many members of a protein family as possible, including distant members of limited sequence similarity.

Significance A significance measurement is the probability that the result of a statistical test or measurement is due to random chance as opposed to being a positive result in support of a model or hypothesis. In sequence analysis, the significance of a local alignment score may be calculated as the chance that such a score would be found between random or unrelated sequences. See Expect value.

Statistical test A test to determine whether or not a measurement or result supports or rejects a particular hypothesis. For sequence alignment scores, the object is to determine the probability that a local alignment score between two sequences would be found between unrelated sequences expressed as an E value.

True negative In the test of a hypothesis, this is the calculated probability that the answer obtained will not support the hypothesis, which was the expected result.

True positive In the test of a hypothesis, this is the calculated probability that the answer obtained will support the hypothesis, which was the expected result.

λ In the evaluation of sequence alignment scores, a statistical parameter of the extreme value distribution that is primarily a scaling factor for the scoring matrix values. For the BLOSUM62 matrix and various gap penalties, $\lambda \approx 0.25$.

WHAT ROLE DOES PROBABILITY PLAY IN SEQUENCE ALIGNMENT?

The goals of this chapter are to outline some of the major features of statistical testing and probability calculations and show how to use these features to evaluate the significance of a sequence alignment. In the previous chapter, the need for odds scores in sequence alignment was discussed. Recall that an odds score is the ratio of previous wins to losses, as in horse racing. In sequence alignments of proteins, for example, an odds score is the chance of an aligned amino acid pair being found in alignments of related sequences compared to the chance of that pair being found in random alignments of unrelated sequences. However, there are questions of probability and uncertainty as to whether or not the amino acids should actually align that an odds score does not address. As an alignment between two protein sequences grows to include alignments of more related pairs, the score for the alignment increases, making the alignment more probable. The statistical significance of the resulting sequence local alignment score between these sequences can be expressed by the probability that random or unrelated sequences could be aligned to produce the same score. This probability should be a very small value, e.g., <0.05, but the smaller, the better. In Chapter 3 (p. 75), a flowchart for performing and evaluating sequence alignments is shown. The later steps in this chart are for determining the statistical significance of the local sequence alignment score.

A task related to evaluating the significance of an alignment score is evaluating data reliability and how well those data support a particular hypothesis or prediction. For example, genes under study may be divided into two classes based on some property that they have, such as expression or function. A test that can discriminate between these two classes based on a statistical or probabilistic model is then devised.

Finally, the test is used to classify an unknown. Examples in which this approach is used in bioinformatics are gene predictions in genomic sequences, protein structure or domain classifications, and sample classification based on gene expression data on microarrays.

The next issue that needs to be addressed is how to evaluate the reliability of the model, because a positive prediction can be either correct or incorrect. The correct prediction is referred to as a true positive and the incorrect one as a false positive. Clearly, we want the evaluation method to give as many true positives as possible. But the method may also fail to predict some cases where a positive prediction should have been made. Consequently, there are true negative cases where the method correctly does not make a prediction and false negatives where a prediction should have been made, but was not. The sensitivity and selectivity of the method may be evaluated using these four measurements of accuracy as described on page 384. The receiver operator characteristic (pp. 192 and 194) is a test that has been used to evaluate the predictive value of scoring matrices for identifying sequence domains in proteins. An overriding difficulty with these methods is that it is very difficult to devise an unbiased evaluation that does not favor the test because the same study objects that are used to devise the test (or related ones) are often used to evaluate the test.

A new application of statistical analysis in bioinformatics is the use of data mining to discover relationships between sequence and genetic variation in genomes and biological effects. All of the examples introduced above stem from basic applications of probability theory that are discussed throughout this chapter. The following box gives a review of some of the basic concepts in probability.

WORKING WITH PROBABILITY AND ODDS SCORES

Odds scores and probabilities may be either multiplied or added, depending on the type of analysis. If the purpose of the analysis is to calculate the probability of one event AND a second event, the odds scores for the events are multiplied. An example is the calculation of the odds of an alignment of two sequences from the odds scores for each of the matched pairs of bases or amino acids in the alignment. The odds scores for the pairs are multiplied, the result being the odds score of the alignment. Usually, the odds score for each pair is transformed into a log odds score, in which case the log odds score for the alignment will be found by adding the logs odds score of the individual pairs.

A second type of analysis is to calculate the odds score for one event OR a second event, or of a series of events (event 1 OR event 2 OR event 3). In this case, the odds scores are added. An example is scoring an alignment of two protein sequences using a set of amino acid substitution matrices (PAM matrices). Each scoring matrix will give a different odds score for the alignment. The most suitable scoring matrix is the one that gives the highest odds score. The probability that this scoring matrix is the best one may be found by adding all the odds scores found using each matrix. The proportion of the sum that is due to the highest scoring matrix is then calculated. In this kind of odds summation, each proportion gives a measure of the probability that the scoring matrix provides the best score. There are also a number of other uses of odds score addition for locating common patterns in a set of sequences by statistical methods that are discussed in Chapter 5.

PROBABILITY IS THE BASIC ELEMENT OF TESTS FOR STATISTICAL SIGNIFICANCE

One of the basic elements of probability theory is the probability of a particular event—for example, What is the probability that a coin toss will yield a head? A series of independent trials with the same outcome probabilities and number of choices (e.g., a head or a tail) are called Bernoulli trials. For n trials, the proportion of heads h will not be exactly 0.5 but will vary about 0.5. The number of heads H in n trials is given by the binomial distribution in which the probability of H heads, designated $P(H)$, is given by $P(H) = n! \, p^H (1 - p)^{n-H} / H! \, (n - H)!$. The mean or expected value of H, $E(H)$, for n trials is given by $E(H) = np$ and the variance σ^2 is given by $\sigma^2 = np(1 - p)$. For larger numbers of trials, plotting different values of $P(H)$ versus H gives an approximately bell-shaped curve of area n that is similar to the normal distribution shown in Figure 4.1. The normal distribution is described by Equation 14 (p. 137).

The relationship between the binomial and normal distributions has led to the theory that many types of data can be due to a series of unrelated effects making the data simulate accumulative Bernoulli trials. If each individual effect could be isolated from the rest, then Bernoulli trials based on this effect would follow the binomial distribution. However, if a number of unrelated effects are present, Bernoulli trials will produce the sum of a series of uncorrelated binomial distributions that appears as a normal distribution. A more detailed and entertaining explanation of the historical origin of these relationships is available

in Gonick and Smith (1993). When the probability of success is small but a large number of trials have been conducted, then another distribution, the Poisson distribution, comes into play. See page 135 for application of the Poisson distribution to sequence alignments.

When data from a series of experimental trials follow a particular probability distribution such as the binomial distribution, it is possible to define the statistical power of the experiment. For a given number of trials, statistical power refers to the ability of the experiment to provide a correct answer subject to

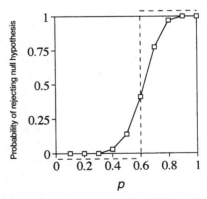

FIGURE 4.1. Use of the binomial distribution for determining the probability that, for a sample size and condition for rejection of the null hypothesis, the null hypothesis will be accepted or rejected. The example used and calculation of the curve are described in the text. The dotted line represents an ideal test result.

a hypothesis about the outcome of the trials; e.g., that a certain proportion of the trials will produce some hypothetical result. An example is guessing the proportion of black marbles p in a bag containing a mixture of a large number of black and white marbles. Drawing n marbles will produce a variable number B of black marbles that is binomially distributed with the mean pn. The larger the value of n, the closer the ratio B/n will be to p. In this example, the predictive power of the test depends on p. For example, if the bag contains mostly black marbles or mostly white marbles, a relatively small sample will give a reliable estimate of the proportion p. Conversely, if the numbers of white and black marbles are more nearly equal, a larger sample will be needed to obtain an estimate of p. To determine the statistical power of a test, we start with a hypothesis that the value of p is less than or equal to some value, e.g., $p \leq 0.6$, and we call this test the null hypothesis. The objective is to perform a test that will lead to acceptance or rejection of this hypothesis. For any sample size n, a number of black marbles r sufficient to lead to rejection of the null hypothesis is chosen. If n marbles are sampled, B or more black marbles are found, and $B \geq r$, then the null hypothesis is rejected; otherwise, the null hypothesis is accepted.

Once these choices of p, n, and r have been made, the binomial distribution will reveal how well the sample of n marbles will support acceptance or rejection of the null hypothesis. Figure 4.1 illustrates the use of the binomial distribution for this purpose. For different values of p, the proportion of black marbles actually in the bag, the probability that the null hypothesis will be rejected if at least r black marbles are found in the sample of n marbles is calculated. This probability of rejecting the null hypothesis is calculated by summing all of the binomial terms, $P(\text{NullHypothesis}) = n!\, p^k (1-p)^{n-k} / k!\, (n-k)!$ summed from $k = r$ to $k = n$. For an ideal test, the probability would follow the dotted line remaining zero until $p = 0.6$ and then rising to 1 for $p > 0.6$. Instead, because of the uncertainty of the test, the curve starts to rise at $p = 0.3$ and does not reach 1 until beyond $p = 0.9$. For $p < 0.6$, the difference between the ideal (dotted) result and the calculated result provides a measure of what is called a type I error, α error, or false-positive result (that is, the hypothesis has been rejected when it should not have been). When $p = 0.6$, the chance of a false-positive result is seen to be $\alpha = 0.42$. For $p > 0.6$, the difference between the calculated proba-

bility and the higher ideal one is a measure of a type II or β error, or a false-negative result (that is, the hypothesis is not rejected when it should have been rejected). When $p = 0.6$, the probability of a false-negative result is given by $\beta = 0.58$.

For this example, similar to one from Goldberg (1960), Figure 4.2 has been drawn for $n = 20$ and $B = 13$. There are two ways of improving the test. First, it may be shown that if a choice of a larger value of r is used to reject the null hypothesis of $p \leq 0.6$, the curve in Figure 4.2 will move to the left, leading to a decrease in the probability of a false-positive result, but with an accompanying increase in the probability of a false-negative result. Second, as the sample size n is increased, the probability curve in Figure 4.2 becomes steeper and more closely resembles the ideal curve, thus decreasing the probability of both types of errors for most values of p. By varying these parameters in the test, it is possible to devise a statistical test that will provide a specified level of reliability or statistical power, e.g., $\alpha = 0.05$. There are related examples in bioinformatics in which the choice of a suitable scoring system or model can lead to a more powerful test for identifying particular classes of genes or proteins.

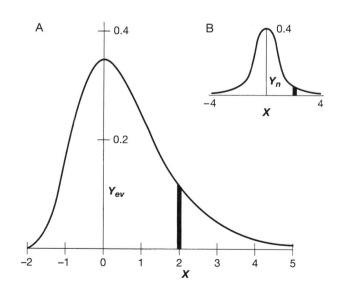

FIGURE 4.2. Probability values for the extreme value distribution (*A*) and the normal distribution (*B*). The area under each curve is 1.

ASSESSING THE SIGNIFICANCE OF SEQUENCE ALIGNMENT SCORES

As discussed in Chapter 3, sequence alignment has two requirements: (1) choice of a suitable scoring system for matches, mismatches, and gaps in the alignment and (2) an algorithm that will find the highest-scoring alignment between the sequences for the chosen scoring system. The first requirement is usually met by using log odds scores based on observed substitutions in known alignments compared to frequencies in random alignments and a suitable gap penalty score comprising a gap opening penalty and a smaller gap extension penalty for gaps longer than 1. The balance between the substitution and gap penalty scores is usually chosen to favor a local alignment of the best-matching regions in the sequences. Increasing the gap penalty favors local alignments with few gaps. Reducing the gap penalty favors global alignments with many gaps that can include all of the sequences. The dynamic programming algorithm described in Chapter 3 is used to find the highest-scoring alignment between the sequences, either local (Smith–Waterman algorithm) or global (Needleman–Wunsch algorithm) alignments.

This chapter discusses how to determine whether or not the alignment found between two sequences is reasonable. This task boils down to asking how often two unrelated sequences or two different random sequences of about the same length as the test sequences can be aligned and achieve as high an alignment score as the two test sequences using the same scoring system. A computer experiment can generate a distribution of thousands of random alignment scores. If the random alignment scores hardly ever are as high as the score between the test sequences, the test score is significant.

The E value is a score that is often used as a measure of significance of an alignment score between two sequences. Consider aligning a test sequence with a library of 10,000 other sequences in which only one of the library sequences matches the test sequence. A list of 10,000 alignment scores is obtained. One of these will be that score between the test sequence and the related sequence, and the other 9999 will be between the test sequences and all of the unrelated sequences. The higher the score between the two related sequences, the fewer the unrelated sequence scores that will reach that high a score. The E value gives an estimate of the number of unrelated sequence alignment scores that will be as high as the related sequence score. The lower this number, the better the alignment between the related sequences. $E = 0.05$ would correspond to the 95% confidence level normally associated with statistical analysis, but the corresponding alignment may be marginally believable. Truly related sequences produce alignment scores where E will be $<10^{-10}$ for partial alignments or as low as 10^{-100} (essentially 0) for sequences matching along their entire lengths. In other words, at $E = 0.05$, the probability that one unrelated sequence will reach the alignment score found between the related sequences is 0.05, and for $E = 10^{-10}$ this probability is 10^{-10}.

Development of Statistical Methods for Sequence Alignments

Originally, the significance of sequence alignment scores was evaluated on the assumption that alignment scores follow a normal statistical distribution. If the sequences to be aligned are randomly generated in a computer by a Monte Carlo or sequence-shuffling method, as in generating a sequence by picking marbles representing four bases or 20 amino acids out of a bag (the number of each type is proportional to the frequency found in sequences), the distribution of alignment scores may look normal at first glance. However, further analysis of the local alignment scores of random sequences will reveal that the scores follow a different distribution than the normal distribution. The actual distribution is called the Gumbel extreme value distribution (see p. 136). In this section, we review some of the earlier methods used for assessing the significance of alignments, then describe the extreme value distribution, and finally, discuss some useful programs for this type of analysis with some illustrative examples.

The statistical analysis of alignment scores is much better understood for local alignments than for global alignments. Recall that the Smith–Waterman alignment algorithm and the scoring system used to produce a local alignment are designed to reveal regions of closely matching sequence with a positive alignment score. In random or unrelated sequence alignments, these regions are rarely found. Hence, their presence in real sequence alignments is significant, and the probability of their occurring by chance alignment of unrelated sequences can be readily calculated. The significance of the scores of global alignments, on the other hand, is more difficult to determine. Using the Needleman–Wunsch algorithm and a suitable scoring system, there are many ways to produce a global alignment between any pair of sequences, and the scores of many different alignments may be quite similar. When random or unrelated sequences are compared using a global alignment method, they can have very high scores, reflecting the tendency of the global algorithm to match as many characters as possible. Thus, assessment of the statistical significance of a global alignment is a much more difficult task. Rather than being used as a strict test for sequence homology, a global alignment is more appropriately used to align sequences that are of approximately the same length and already known to be related. The method will conveniently show which sequence characters align. One can then use this information to perform other types of analyses, such as structural modeling or an evolutionary analysis.

Significance of Global Alignments

In general, global alignment programs use the Needleman–Wunsch alignment algorithm and a scoring system that scores the average match of an aligned nucleotide or amino acid pair as a positive number. Hence, global alignment scores of unrelated or random sequences grow proportionally to the length of the sequences. These scores can quite readily reach values as high as those found between related sequences, thus creating a high background against which the related sequence score must be compared. In addition, there are many possible different global alignments for any two sequences depending on the scoring system chosen, and small changes in the scoring system can produce a different alignment. Thus, finding the best global alignment of two sequences and knowing how to assess its significance are not simple tasks, as reflected by the absence of studies in the literature.

Waterman (1989) provided a set of means and standard deviations of global alignment scores between random DNA sequences, using mismatch and gap penalties that produce a linear increase in score with sequence length, a distinguishing feature of global alignments. However, these values are of limited use because they are based on a simple gap scoring system. Abagyan and Batalov (1997) suggested that global alignment scores between unrelated protein sequences followed the extreme value distribution, similar to local alignment scores. However, because the scoring system they used favored local alignments, the alignments they produced may not be global but local (see below). Unfortunately, there is no equivalent theory on which to base an analysis of global alignment scores as there is for local alignment scores. For zero mismatch and gap penalties, which is the most extreme condition for a global alignment giving the longest subsequence common to two sequences, the score between two random or unrelated sequences P is proportional to sequence length n, such that $P \simeq cn$ (Chvátal and Sankoff 1975), but it has not proven possible to calculate the proportionality constant c (Waterman and Vingron 1994a) because of the computational complexity of the problem.

To evaluate the significance of a Needleman–Wunsch global alignment score, Dayhoff (1978) and Dayhoff et al. (1983) evaluated Needleman–Wunsch alignment scores for a large number of randomized and unrelated but real protein sequences, using their log odds scoring matrix at 250 PAMs and a constant gap penalty. The distribution of the resulting random scores matched a normal distribution. On the basis of this analysis, the significance of an alignment score between two apparently related sequences A and B was determined by obtaining a mean and standard deviation of the alignment scores of 100 random permutations or shufflings of A with 100 of B, conserving the length and amino acid composition of each. If the score between A and B is significant, the authors specify that the real score should be at least 3–5 standard deviations (s.d.'s) greater than the mean of the random scores. This level of significance means that the probability that two unrelated sequences would give such a high score is 1.35×10^{-3} (3 s.d.'s) and 2.87×10^{-6} (5 s.d.'s). In evaluating an alignment, two parameters were varied to maximize the alignment score: First, a constant called the matrix bias was added to each value in the scoring matrix and, second, the gap penalty was varied. The statistical analysis was then performed after the score between A and B had been maximized. It was these optimized alignments on which the significance test was performed. Feng et al. (1985) used the same method to compare the significance of alignment scores obtained by using different scoring matrices. They used 25–100 pairs of randomized sequences for each test of an alignment.

There are several potential problems with this approach for determining statistical significance, some of which apply to other methods as well.

1. The method is expensive in terms of the number of computational steps, which increase at least as much as the square of sequence length because many Needleman–Wunsch alignments must be done. However, this problem is much reduced with the faster computers and more efficient algorithms of today.

2. If the amino acid composition is unusual, and if there is a region of low complexity (for example, many occurrences of one or two amino acids), the analysis will be oversimplified.

3. When natural sequences were compared more closely, the patterns found did not conform to a random set of the basic building blocks of sequences but rather to a random set of sequence segments that were varying.

Consider use of the 26-letter alphabet in English sentences. Alphabet letters do not appear in any random order in these sentences but rather in a vocabulary of meaningful words. What happens if sentences, which are made up of words, are compared? On the one hand, if just the alphabet composition of many sentences is compared, not much variation is seen. On the other hand, if words are compared, much greater variation is found because there are many more words than alphabet characters. If random sequences are produced from segments of sequences, rather than from individual residues, more variation is observed, more like that observed when unrelated natural sequences are compared. The increased variation found among natural sequences is not surprising when one thinks of DNA and proteins as sources of information. For example, protein-encoding regions of DNA sequences are constrained by the genetic code and by amino acid patterns that produce functional domains in proteins.

Lipman et al. (1984) analyzed the distribution of local alignment scores among 100 vertebrate nucleic acid sequences and compared these scores with randomized sequences prepared in different ways. When the randomized sequences were prepared by shuffling the sequence to conserve base composition, as was done by Dayhoff and others, the standard deviation was approximately one-third less than the distribution of scores of the natural sequences. Thus, natural sequences are more variable than randomized ones because of the use of sequence words and greater variation in these words than found in randomized sequences at the single residue level. Using such randomized sequences for a significance test may lead to an overestimation of the significance. If, instead, the random sequences were prepared in a way that maintained the local base composition by producing them from overlapping fragments of sequence, the distribution of scores has a higher standard deviation that is closer to the distribution of the natural sequences. The conclusion is that the presence of conserved local patterns can influence the score in statistical tests such that a test alignment score between two sequences can appear to be more significant than it actually is. Although this study was done using the Smith–Waterman algorithm with nucleic acids, the same cautionary note applies for other types of alignments. The final problem with the above methods is that the correct statistical model for alignment scores was not used. However, these earlier types of statistical analysis methods set the stage for later ones.

The GCG alignment programs have a RANDOMIZATION option, which shuffles the second sequence and calculates similarity scores between the unshuffled sequence and each of the shuffled copies. If the new similarity scores generated with the shuffled sequences are significantly smaller than the real alignment score, the real alignment is considered significant. This analysis is only useful for providing a rough approximation of the significance of an alignment score and can easily be misleading because it does not take into account the greater local variation in overlapping sequence fragments than in randomized sequences found above by Lipman et al. (1984), as well as the statistical variation observed in local alignment scores discussed below.

Dayhoff (1978) and Dayhoff et al. (1983) devised a further method for evaluating global alignment scores between sequences based on scores between many random sequence fragments derived from the sequences. This method is also useful for finding repeated regions within a sequence, similar regions that are in a different order in two sequences, or a small conserved region such as an active site. As used in a computer program called RELATE (Dayhoff 1978), all possible segments of a given length of one sequence are compared with all segments of the same length from another. An alignment score using a scoring matrix is obtained for each comparison to give a score distribution among all the segments. A segment comparison score in standard deviation units is calculated as the difference between the value for real sequences minus the average value for random sequences divided by the standard deviation of the scores from the random sequences. A version of the program RELATE that runs on many computer platforms is included with the FASTA distribution package by W. Pearson. An example of the output of the RELATE program for the phage λ and P22 repressor sequences is shown in Table 4.1. This program also calculates a distribution based on the normal distribution, thus it provides only an approximate indication of the significance of an alignment.

Modeling the Best Possible Alignment between Two Random DNA Sequences

The above types of analyses assume that alignment scores between random sequences follow a normal distribution that can be used to test the significance of a score between two test sequences. For a number of reasons, mathematicians were concerned that this statistical model might not be appropriate. Let us start by creating two aligned random DNA sequences by drawing pairs of marbles, one from each of two large bags filled with four kinds of labeled marbles in equal proportions and labeled A, T, G, and C to represent an assumed equal representation of the four nucleotides in DNA. Now consider the probability of removing six identical pairs representing six columns in an alignment between two random sequences. The probability of removing an identical pair (an A and another A) is 1/4 × 1/4, but there are four possible identical pairs (A/A, C/C, G/G, and T/T), so the probability of removing any identical pair is 4 × 1/4 × 1/4 = 1/4 and that for removing six identical pairs is $(1/4)^6 = 2.4 \times 10^{-4}$. The probability of drawing a mismatched pair is 1 − 1/4 = 3/4, and that of drawing 6/6 mismatched pairs is $(3/4)^6 = 0.178$. Most random alignments produced in this manner will have a mixture of a few matches and many mismatches.

The calculations are a little more complex if the four nucleotides are not equally represented, but the results will be approximately the same. The probability of drawing the same pair is p, where $p = p_A^2 + p_C^2 + p_G^2 + p_T^2$, where p_X is the proportion of nucleotide X.

p is an important parameter to remember for the discussion below. An even more complicated situation is when the two random sequences to be aligned

have different nucleotide distributions. One way would be to use an average p for the two sequences. This example illustrates the difficulty of modeling sequence alignments between two different organisms that have a different base composition.

The above marble model is not suitable for predicting the number of sequentially matched positions to expect between random sequences of a given length. To estimate this number, a DNA sequence alignment may be modeled by coin-tossing experiments (Arratia and Waterman 1989; Arratia et al. 1986, 1990). Random alignments will normally comprise mixtures of matches and mismatches, just as a series of coin tosses will produce a mixture of heads and tails. The chance of producing a series of matches in a sequence alignment with no mismatches is similar to the chance of tossing a coin and coming up with a series of only heads. The numbers of interest are the highest possible score that can be obtained and the probability of obtaining such a score in a certain number of trials. In such models, coins are usually considered to be "fair" in that the probability of a head is equal to that of a tail. The coin in this example has a certain probability p of scoring a head (H) and $q = 1 − p$ of scoring a tail (T). The longest run of heads R that can be expected on the average in n trials has been shown by Erdös and Rényi to be given by $\log_{1/p}(n)$. If $p = 0.5$ as for a normal coin, the base of the logarithm is $1/p = 2$. For the example of $n = 100$ tosses, then R = $\log_2 100 = \log_e 100/\log_e 2 = 4.605/0.693 = 6.65$.

To use the coin model, an alignment of two random sequences $\mathbf{a} = a_1, a_2, a_3, \ldots, a_n$ and $\mathbf{b} = b_1, b_2, b_3, \ldots, b_n$, each of the same length n, is converted to a series of heads and tails. If $a_i = b_i$, the equivalent toss result is an H, otherwise the result is a T. The following example illustrates the conversion of an alignment to a series of H and T tosses:

$$a_1 a_2 a_3 \text{ --- } a_n \text{ ----> H T H ---}$$
$$b_1 b_2 b_3 \text{--- } b_n$$

where $a_1 = b_1$ and $a_3 = b_3$ only (1)

The average longest run of matches in the alignment is now equivalent to the longest run of heads in the coin-tossing sequence, and it should be possible to use the Erdös and Rényi law to predict the longest run of matches. This score, however, only applies to one particular alignment of random sequences, such as generated above by the marble draw. In performing a

TABLE 4.1. *Distribution of alignment scores produced by program RELATE*

```
<  -120      0 :
   -115      0 :
   -110      0 :
   -105      0 :
   -100      0 :
    -95      0 :
    -90      0 :
    -85      0 :
    -80      0 :
    -75      0 :
    -70      0 :
    -65      9 :.
    -60     69 :===
    -55    293 :==============
    -50    932 :=================================================
    -45   1868 :=============================================================
    -40   3214 :=============================================================
    -35   4784 :=============================================================
    -30   5858 :=============================================================
0   -25   6091 :=============================================================
    -20   5384 :=============================================================
    -15   4470 :=============================================================
1   -10   2960 :=============================================================
     -5   2076 :=============================================================
      0   1131 :=============================================================
2     5    590 :==============================
     10    288 :===============
3    15    154 :========
     20     67 :===
     25     34 :=
4    30     18 :.
     35     10 :.
5    40      1 :.
     45      0 :
     50      0 :
6    55      0 :
     60      0 :
     65      0 :
7    70      0 :
     75      0 :
8    80      0 :
     85      0 :
     90      0 :
     95      0 :
    100      0 :
    105      0 :
    110      0 :
    115      0 :
    120      0 :
    125      0 :
>   125      0 :
40301 comparisons of window: 25, mean score: -27.3 (13.34)
matrix file: PAM250

29 segments >= 4 sd above mean
```

The sequences of two phage repressors were broken down into overlapping 25-amino-acid segments, and all 40,301 combinations of these segments were compared. The first column gives the approximate location of the number of standard deviations (13.34) from the mean score of −27.3. The second column is increasing ranges of the alignment score, and the third is the number of segment alignment scores that fall within the range. Twenty-nine scores were greater than 3 s.d.s from the mean. Thus, these two sequences share segments that are significantly more related than the average segment, and the proteins share strong regions of local similarity. In such cases of strong local similarity, a local alignment program such as LFASTA, PLFASTA, or LALIGN can provide the alignments and a more detailed statistical analysis, as described below. The graph is truncated on the right side.

sequence alignment, two sequences are in effect shifted back and forth with respect to each other to find regions that can be aligned. In addition, the sequences may be of different lengths. If two random sequences of length m and n are aligned in this same manner, the same law still applies, but the length of the predicted match is $\log_{1/p}(mn)$ (Arratia et al. 1986). If $m = n$, the longest run of matches is doubled. Thus, for a pair of random DNA sequences of length 100 and $p = 0.25$ (equal representation of each nucleotide), the longest expected run of matches is $2 \times \log_{1/p}(n) = 2 \times \log_4 100 = 2 \times \log_e 100 / \log_e 4 = 2 \times 4.605 / 1.386 = 6.65$, the same number as in the coin-tossing experiment. This number corresponds to the average longest sub-alignment that can be expected between two random sequences of this length and composition.

The application of this result for evaluating an alignment of DNA sequences of length 100 is that, for the alignment to indicate sequence relatedness, a series of sequential aligned positions significantly greater than 7 should be present. However, as shown below, random sequences can achieve a greater number of matches than the average predicted above, and multiple series of matches may be present. Thus, a more detailed analysis is required.

A more precise formula for the expectation value or mean of the longest match M, $E(M)$, and its variance $\mathrm{Var}[M(n,m)]$ has been derived (Arratia et al. 1986; Waterman et al. 1987; Waterman 1989):

$$E(M) \cong \log_{1/p}(mn) + \log_{1/p}(q) \\ + \gamma \log(e) - 1/2 \quad (2)$$

$$\mathrm{Var}[M(n,m)] \simeq [\pi \log_{1/p}(e)]^2/6 + 1/12 \quad (3)$$

where $\gamma = 0.577$ is Euler's number and $q = 1 - p$. Note that Equation 2 can be simplified to

$$E(M) \simeq \log_{1/p}(Kmn) \quad (4)$$

where K is a constant that depends on the base composition.

Equation 4 also applies when there are k mismatches in the alignment, except that another term $= k \log_{1/p}\log_{1/p}(qmn)$ appears in the equation (Arratia et al. 1986). K, the constant in Equation 4, depends on k. The log log term is small and can be replaced by a constant (Mott 1992), and simulations also suggest

that it is not important (Altschul and Gish 1996). Altschul and Gish (1996) have found a better match to Equation 4 when the length of each sequence is reduced by the expected length of a match. In the example given above with two random sequences of length 100, the expected length of a match was 6.65. As the sequences slide-align each other, it is not possible to have overlaps on the ends that are shorter than 7 aligned base pairs because there is not enough sequence remaining. Hence, the effective length of the sequences is $100 - 7 = 93$ (Altschul and Gish 1996). This correction is also used for the calculation of statistical significance by the BLAST database search tool discussed in Chapter 5.

Equation 4 is fundamentally important for calculating the statistical significance of local alignment scores. Basically, it states that as the lengths of random or unrelated sequences increase, the mean of the highest possible local alignment scores will be proportional to the logarithm of the product of the sequence lengths, or twice the logarithm of the sequence length if the lengths are equal (since $\log(nn) = 2 \log n$). Equation 3 also predicts a constant variance among scores of random or unrelated sequences, and this prediction is also borne out by experiment. It is important to emphasize once again that this relationship depends on the use of scoring parameters appropriate for a local alignment algorithm, such as 1 for a match and −0.9 for a mismatch, or a scoring matrix that scores the average aligned position as negative, and also on the use of sufficiently large gap penalties. This type of scoring system gives rise to positive scoring regions in random sequences only rarely. The significance of these scores can then be estimated as described herein.

Another way of describing the result in Equation 4 uses a different parameter, λ, where $\lambda = \log_e(1/p)$ (Karlin and Altschul 1990):

$$E(M) = [\log_e(Kmn)] / \lambda \quad (5)$$

Recall that p is the probability of a match between the same two characters, given above as 1/4 for matching a random pair of DNA bases, assuming equal representation of each base in the sequences. p may also be calculated as the probability of a match averaged over scoring matrix and sequence composition values. Instead, it is λ that is more commonly used with scoring matrix values. The calculation of λ and also of K is described later in this chapter.

Using Alignment Scores

It is more useful in sequence analysis to use alignment scores instead of lengths for comparing alignments. The mean alignment length between two random sequences given by Equations 4 and 5 can be easily converted to a mean alignment score just by using match and mismatch or scoring matrix values along with some simple normalization procedures. Thus, in addition to predicting length, Equations 4 and 5 can also be used to predict the mean alignment score $E(S)$ between random sequences of lengths m and n. Assessing statistical significance then boils down to calculating the probability that $E(S)$ between two random or unrelated sequences will actually be greater than an observed alignment score between two test sequences. Using these changes, the expected or mean extreme score between random or unrelated sequences is

$$E(S) = [\log_e(Kmn)] / \lambda \qquad (6)$$

The next step is to determine the probability that an alignment score S between two unrelated (or random) sequences can exceed the mean score $E(S)$ and reach a much greater score x found between two related sequences, i.e., $x \gg E(S)$. This probability should be very small. For analysis of this problem, it is useful to think of x as a score that can seldom be reached or exceeded by S scores between pairs of unrelated sequences. The probability of S reaching or exceeding x can be predicted by the Poisson distribution where the mean of the Poisson distribution is x (Waterman and Vingron 1994b). The Poisson distribution applies when the probability of success in a single trial is small but the number of trials is large, as in comparing many unrelated sequences and finding a rare positive alignment that has a high score. The Poisson distribution gives the probability P_n of the number of successes, i.e., 0, 1, 2, 3, . . . , n when the average number is x and is given by the formula $P_n = e^{-x} x^n / n!$. Using this approximation, the following probabilities have been derived (Arratia et al. 1986; Karlin and Altschul 1990):

$$P(S < x) \simeq \exp(-Kmne^{-\lambda x}) \qquad (7)$$

$$P(S > x) \simeq 1 - \exp(-Kmne^{-\lambda x}) \qquad (8)$$

where x is the score found between two test sequences, and K and λ are parameters derived from the scoring system used. Equation 8 estimates the probability of a score (S) greater than x between two random sequences, and the resulting probability distribution is identical to the extreme value distribution described below. The Poisson approximation provides a very convenient way to estimate K and λ from alignment scores between many random or unrelated sequences by using the fraction of alignments that have a score less than the value of x.

Local Alignments with Gaps

Gaps in local sequence alignments were not included in the above analysis. It was subsequently predicted on mathematical grounds and shown experimentally that a similar type of analysis holds for local sequence alignments that include gaps (Smith et al. 1985). Thus, when Smith et al. (1985) optimally aligned a large number of unrelated vertebrate and viral DNA sequences of different lengths .(n and m) and their complements to each other, using a dynamic programming local alignment method that allowed for a score of +1 for matches, –0.9 for mismatches, and –2 for a single gap penalty (longer gaps were not considered to simplify the analysis), a plot of the similarity score (S) versus the $\log_{1/p}(nm)$ produced a straight line with approximately constant standard deviation. This result is as expected in the above model except that with the inclusion of gaps, the slope was increased and was of the form

$$S_{\text{mean}} = 2.55 \ (\log_{1/p}(mn)) - 8.99 \qquad (9)$$

with constant standard deviation $\sigma = 1.78$. This result was then used to calculate how many standard deviations were between the predicted mean and sequences. If the actual alignment score exceeded the predicted S_{mean} by several standard deviations, then the alignment score should be significant. For example, the expected score between two unrelated sequences of lengths 2948 and 431, average $p = 0.279$, was $S_{\text{mean}} = 2.55 \times \log_{1/0.279}(2948 \times 431) - 8.99 = 2.55 \times (\log_e(2948 \times 431)/\log_e(1/0.279)) - 8.99 = 2.55 \times 14.1 / 1.28 - 8.99 = 28.1 - 8.99 = 19.1$, based on Equation 9. The actual optimal alignment score between the two real sequences of these lengths was 37.20, which exceeds the alignment score expected for random sequences by $(37.20 - 19.1) / 1.78 = 10.2\sigma$. Is this number of standard deviations significant? Smith et al. (1985) and Waterman (1989) suggested the use of a conservative statistic known as Chebyshev's inequality, which is valid for many probability distributions. This statistic states that the probability that a random variable exceeds its mean is less than or equal to the square of 1 over the number of standard deviations from the mean. In this example, where the actual score is 10 standard deviations above the mean, the probability is $(1/10)^2 = 0.01$.

Waterman (1989) has noted that for low mismatch and gap penalties, e.g., 11 for matches, –0.5 for mismatches, and –0.5 for a single gap penalty, the predicted alignment scores between random sequences as estimated above are not accurate because the score will increase linearly with sequence length instead of with the logarithm of the length. The linear relationship arises when the alignment is more global in nature, and the logarithmic relationship when it is local. Waterman (1989) has fitted alignment scores from a large number of randomly generated DNA sequences of varying lengths to either the predicted $\log(n)$ or n linear relationships expected for low- and high-valued mismatch and gap penalties. The results provide the mean and standard deviation of an alignment score for several scoring schemes, assuming a constant gap penalty.

With further mathematical analysis, it became apparent that the expected scores between alignment of random and unrelated sequences follow a distribution called the Gumbel extreme value distribution (see following section) (Arratia et al. 1986; Karlin and Altschul 1990). This type of distribution is typical of values that are the highest or best score of a variable, such as the number of heads only expected in a coin toss discussed previously. Subsequently, S. Karlin and S. Altschul (1990, 1993) further developed the use of this distribution for evaluating the significance of ungapped segments in comparisons between a test sequence and a sequence database using the BLAST program (for review, see Altschul et al. 1994). The method is also used for evaluating the statistical features of repeats and amino acid patterns and clusters in the same sequence (Karlin and Altschul 1990; Karlin et al. 1991). The program SAPS developed by S. Karlin and colleagues at Stanford University and available at http://gnomic.stanford.edu/pub/SAPS. SSPA/ provides this type of analysis. The extreme value distribution is now widely used for evaluating the significance of the score of local alignments of DNA and protein sequence alignments, especially in the context of database similarity searches.

The Gumbel Extreme Value Distribution

When two sequences have been aligned optimally, the significance of a local alignment score can be tested on the basis of the distribution of scores expected by aligning two random sequences of the same length and composition as the two test sequences (Karlin and Altschul 1990; Altschul et al. 1994; Altschul and Gish 1996). These random sequence alignment scores follow a distribution called the extreme value distribution, which is somewhat like a normal distribution with a positively skewed tail in the higher score range. When a set of values of a variable are obtained in an experiment, biologists are used to calculating the mean and standard deviation of the entire set, assuming that the distribution of values will follow the normal distribution. For sequence alignments, this procedure would be like obtaining many different alignments, both good and bad, and averaging all of the scores. However, biologically interesting alignments are those that give the highest possible scores, and lower scores are not of interest. The experiment is one of obtaining a set of values, and then of using only the highest value and discarding the rest. The focus changes from the statistical approach of wanting to know the average of scores of random sequences to one of knowing how high a value will be obtained the next time another set of alignment scores of random sequences is obtained.

Accordingly, the distribution of alignment scores between random sequences follows the extreme value distribution, not the normal distribution. After many alignments, a probability distribution of highest values will be obtained. The goal is to evaluate the probability that a score between random or unrelated sequences will reach the score found between two real sequences of interest. If that probability is very low, the alignment score between the real sequences is significant and the sequence similarity score is significant.

The probability distribution of highest values in an experiment, the extreme value distribution, is compared to the normal probability distribution in Figure 4.2. The equations giving the respective y coordinate values in these distributions, Y_{ev} and Y_n, are

$$Y_{ev} = \exp[-x - e^{-x}]$$

for the extreme value distribution (10)

$$Y_n = 1/\sqrt{(2\pi)} \exp[(-x^2)/2]$$

for the normal distribution (11)

The area under both curves is 1. The normal curve is symmetrical about the expectation value or mean at $x = 0$, such that the area under the curve below the mean (0.5) is the same as that above the mean (0.5) and the variance σ^2 is 1. The probability of a particular value of x for the normal distribution is obtained by calculating the area under curve B, usually between $-x$ and $+x$. For $x = 2$, often used as an indication of a significant deviation from the mean, the area between -2 and $+2$ is 0.9544. For the extreme value distribution, the expectation value or mean of x is the value of the Euler–Mascheroni constant, 0.57722 . . . and the variance of x, σ^2, is the value of $\pi^2 / 6 = 1.6449$. The probability that score S will be less than value x, $P(S < x)$, is obtained by calculating the area under curve A from $-\infty$ to x, by integration of Equation 10 giving

$$P(S < x) = \exp[-e^{-x}]$$ (12)

and the probability of $S \geq x$ is 1 minus this probability,

$$P(S \geq x) = 1 - \exp[-e^{-x}]$$ (13)

For the extreme value distribution, the area below $x = 0$, which represents the peak or mode of the distribution, is $1/e$ or 0.368 of the total area of 1, and the area above the mean is $1 - 0.368 = 0.632$. At a value of $x = 2$, $Y_{ev} = 0.118$ and $P(S < 2) = \exp[-e^{-2}] = 0.873$. Thus, just over 0.87 of the area under the curve is found below $x = 2$. An area of 0.95 is not reached until $x = 3$. The difference between the two distributions becomes even greater for larger values of x. As a result, for a variable whose distribution comes from extreme values, such as random sequence alignment scores, the score must be greater than expected from a normal distribution in order to achieve the same level of significance.

The above equations can be modified for use with scores obtained in an actual analysis. For a variable x that follows the normal distribution, values of x are used to estimate the mean m and standard deviation σ of the distribution, and the probability curve given by Equation 11 then becomes

$$Y_n = 1/(\sigma\sqrt{(2\pi)}) \exp[-(x-m)^2/2\sigma^2]$$ (14)

The probability of a particular value of x can be estimated by using m and s to estimate the number of standard deviations from the mean, Z, where $Z = (x - m)/\sigma$. Similarly, Equations 10 and 13 can be modified to accommodate the extreme values such as sequence alignment scores

$$P(S \geq x) = 1 - \exp[-e^{-\lambda(x-u)}]$$ (15)

where u is the mode (highest point or characteristic value of the distribution), and λ is the decay or scale parameter. As is apparent in Equation 15, λ converts the experimentally measured values into standard values of x after subtraction of the mode from each score.

It is quite straightforward to calculate u and λ. There is an important relationship between u and λ and the mean and standard deviation ($\sqrt{\sigma^2}$) of a set of extreme values. The mean and variance not only apply to the normal distribution, but in fact are mathematically defined for any probability distribution. The mean of any set of values of a variable may always be calculated as the sum of the values divided by their number. The mean m or expected value of a variable x, $E(x)$, is defined as the first moment of the values of the variable around the mean. From this definition, the mean is that number from which the sum of deviations to all values is zero. The variance σ^2 is the second moment of the values about the mean and is the sum of the squares of the deviations from the mean divided by the number of observations less one ($n - 1$). The mean $\underline{x}$ and standard deviation of a set of extreme values can be calculated in the same way, and then u and λ can be calculated using the following equations derived by mathematical evaluation of the first and second moments of the extreme value distribution (Gumbel 1962; Altschul and Erickson 1986):

$$\lambda = \pi / (\sigma\sqrt{6}) = 1.2825 / \sigma$$ (16)

$$u = \underline{x} - \gamma / \lambda = \underline{x} - 0.4500\,\sigma$$ (17)

where γ was already introduced. Equation 16 is derived from the ratio of the variance σ^2 of the two distributions in Figure 4.2, or 1 to $\pi^2/6$. Equation 17 is derived from the observation that the mode or the extreme value distribution (zero in Figure 4.2) has the value of γ less than the mean. However, the value of γ must be scaled by the ratio of the variances. Hence γ/λ is subtracted from the mean. This method of calculating u and λ from means and variances is called the method of moments.

As with the normal distribution, z scores (number of standard deviations above or below the mean) may be calculated for each extreme value x, where $z = (x - m) / \sigma$ is the number of standard deviations from the mean m to each score x. z scores are used by the FASTA, version 3, programs distributed by W. Pearson (1998). Equation 15 may be written in a form that directly uses z scores to evaluate the probability that a particular score Z exceeds a value z,

$$P(Z > z) = 1 - \exp(-e^{-1.2825\ z-\ 0.5772}) \tag{18}$$

For sequence analysis, u and λ depend on the length and composition of the sequences being compared, and also on the particular scoring system being used. They can be calculated directly or estimated by making many alignments of random sequences or shuffled natural sequences, using a scoring system that gives local alignments. The parameters will change when a different scoring system is used. Examples of programs that calculate these values are given below. For alignments that do not include any gaps, u and λ may be calculated from the scoring matrix. The scaling factor λ is calculated as the value of x, which satisfies the condition

$$\Sigma\ p_i p_j e^{s_{ij}x} = 1 \tag{19}$$

where p_i and p_j are the respective fractional representations of residues i and j in the sequences, and s_{ij} is the score for a match between i and j, taken from a log odds scoring matrix. u, the characteristic value of the distribution (as in Equation 17), is given by (Altschul and Gish 1996):

$$u = (\ln Kmn) / \lambda \tag{20}$$

where m and n are the sequence lengths and K is a constant that can also be calculated from the values of

p_i and s_{ij}. Note that this value originates from the coin-toss analysis that gave rise to Equation 7. Combining Equations 15 and 20 eliminates u and gives the following relationship:

$$\begin{aligned} P(S \geq x) &= 1 - \exp[-e^{-\lambda(x-u)}] \\ &= 1 - \exp[-e^{-\lambda\left(x - (\ln Kmn)/\lambda\right)}] \\ &= 1 - \exp[-e^{-\lambda x + \ln Kmn}] \end{aligned} \tag{21}$$

$$= 1 - \exp[-Kmn\,e^{-\lambda x}] \tag{22}$$

To facilitate calculations, a sequence alignment score S may also be normalized to produce a score S'. The effect of normalization is to change the score distribution into the form shown above in Figure 4.2 with $u = 0$ and $\lambda = 1$. From Equation 21, S' is calculated by

$$S' = \lambda S - \ln Kmn \tag{23}$$

The probability of $P(S' \geq x)$ is then given by Equation 20 with $S = S'$,

$$P(S' \geq x) = 1 - \exp[-e^{-x}] \tag{24}$$

The probability of a particular normalized score may then be readily calculated. This capability depends on a determination of the λ and K to calculate the normalized scores S' by Equation 23.

The probability function $P(S' \geq x)$ decays exponentially in x as x increases and $P(S' \geq x) = 1 - \exp[-e^{-x}] \rightarrow e^{-x}$. Consequently, an important approximation for Equations 22 and 24 for the significant part of the extreme value distribution where $x > 2$ is

$$P(S \geq x) \simeq Kmn\,e^{-\lambda x} \tag{25}$$

$$P(S' \geq x) \simeq e^{-x} \tag{26}$$

Note that the replacement equations are single and not double exponentials.

A comparison of probability calculations using this approximation instead of that given in Equation 24 is shown in Table 4.2. For $x > 2$, the estimates differ by less than 2%. The estimate given in Equation 25

TABLE 4.2. *Approximation of* $P(S' \geq x)$ *by* e^{-x}

x	$1-\exp[-e^{-x}]$	e^{-x}
0	0.63	1
1	0.308	0.368
2	0.127	0.125
3	0.0486	0.0498
4	0.0181	0.0183

also provides a quicker method for estimating the significance of an alignment score.

Quick Determination of the Significance of an Alignment Score

Scoring matrices are most useful for statistical work if they are scaled in logarithms to the base 2, called bits. Scaling the matrices in this fashion does not alter their ability to score sequence similarities, and thereby to distinguish good matches from poor ones, but it does allow a simple estimation of the significance of an alignment. The actual alignment may then be calculated by summing the matrix values for each of the aligned pairs, using matrix values in bit units. If the actual alignment score in bits is greater than expected for alignment of random sequences, the alignment is significant.

For a typical amino acid scoring matrix and protein sequence, $K = 0.1$ and λ depends on the values of the scoring matrix. If the log odds matrix is in units of bits as described above, then $\lambda = \log_e 2 = 0.693$, and the following simplified form of Equation 25 may be derived (Altschul 1991) by taking logarithms to the base 2 and setting p as the probability of the scores of random or unrelated alignments reaching a score of S or greater,

$$
\begin{aligned}
\log_2 p &= \log_2(Kmn\, e^{-\lambda S}) \\
&= \log_2(Kmn) + \log_2(e^{-\lambda S}) \\
&= \log_2(Kmn) + (\log_e(e^{-\lambda S}))/\log_e 2 \\
&= \log_2(Kmn) - \lambda S/\log_e 2 \\
&= \log_2(Kmn) - S \qquad (27)
\end{aligned}
$$

then S, the score corresponding to probability P, may be obtained by rearranging terms of Equation 27 as follows:

$$
\begin{aligned}
S &= \log_2(Kmn) - \log_2 P \\
&= \log_2(K/P) + \log_2(nm) \qquad (28)
\end{aligned}
$$

Since for most scoring matrices $K \approx 0.1$ and, when choosing $P = 0.05$, the first term is 1, the second term in Equation 28 becomes the most important one for calculating the score (Altschul 1991), thus giving

$$
S \simeq \log_2(nm) \qquad (29)
$$

EXAMPLE: USING THE EXTREME VALUE DISTRIBUTION TO CALCULATE THE SIGNIFICANCE OF A LOCAL ALIGNMENT

Suppose that two sequences approximately 250 amino acids long are aligned by the Smith–Waterman local alignment algorithm using the PAM250 matrix and a high gap score to omit gaps from the alignment, and that the following alignment within the sequences is found:

```
FWLEVEGNSMTAPTG
FWLDVQGDSMTAPAG
```

1. Using Equation 29, a significant alignment between unrelated or random sequences will have a score of $S \approx \log_2(nm) = \log_2(250 \times 250) = 16$ bits.

2. The score of the above actual alignment is 73 using the scores in the Dayhoff mutation data matrix (MDM) (see Fig. 3.14) that provides log odds scores at 250 PAMs evolutionary distance.

3. A correction to the alignment score must be made because the MDM table at 250 PAMs is not in bit units but in units of logarithm to the base 10, multiplied by 10. These MDM scores actually correspond to units of 1/3 bits ([MDM score in units of $\log_{10}$] $\times$ 10 = [MDM score in bits of $\log_2 \times \log_2 10$] / 10 = [MDM score in units of $\log_{10} \times 10$] $\times$ 0.333). Thus, the score of the alignment in bits is 73/3 = 24.3 or 8.3 bits greater than the 16 expected by chance. Therefore, this alignment score is highly significant.

4. Altschul and Gish (1996) have provided estimates of $K = 0.09$ and $\lambda = 0.229$ for the PAM250 scoring matrix, for a typical amino acid distribution and for an alignment score based on using a very high gap penalty. By Equations 23 and 24, $S' = 0.229 \times 73 - \ln(0.09 \times 250 \times 250) = 16.72 - 8.63 = 8.09$ bits, and $P(S' \geq 8.09) = 1 - \exp[-e^{-8.09}] = 3.1 \times 10^{-4}$. Thus, the chance that an alignment between two random sequences will achieve a score greater than or equal to 73 using the MDM matrix is 3.1×10^{-4}. Note that the calculated S' of 8.09 bits in step 4 is approximately the same as the 8.3 bits calculated by the simpler method in step 3.

5. The probability may also be calculated by the approximation given in Equation 26 where $P(S' > 8.09) \approx e^{-x} = e^{-8.09} = 3.1 \times 10^{-4}$.

Importance of the Type of Scoring Matrix for Statistical Analyses

Using a log odds matrix in bit units simplifies estimation of the significance of an alignment. The Dayhoff PAM matrices, the BLOSUM matrices, and the nucleic acid PAM scoring matrices are examples of this type. Such matrices are also useful for finding local alignments because the matrix includes both positive and negative values. Another important feature of the log odds form of the scoring matrix is that this design is optimal for assessing statistical significance of alignment scores. A set of matrices, each designed to detect similarity between sequences at a particular level, is best for this purpose. Use of a matrix that is designed for aligning sequences that have a particular level of similarity (or evolutionary distance) assures the highest-scoring alignment and therefore the very best estimate of significance. Thus, lower-numbered PAM matrices are most suitable for aligning sequences that are known in advance to be more similar. In the above example, the Dayhoff PAM250 matrix designed for sequences that are 20% similar was used to align sequences that are approximately 20% identical and 50% similar (identities plus common replacements in the alignment). Using a lower PAM120 matrix produces a slightly higher score for this alignment, and thus increases the significance of the alignment score.

Another important parameter of the scoring matrix to consider for statistical purposes is the expected value of the average amino acid pair, calculated as shown in Equation 30. This value should be negative if alignment scores for the matrix are to be used for statistical tests, as performed in the above example. Otherwise, in any aligned pair of sequences the scores will increase with length faster than the logarithm of the length. Not all scoring matrices will meet this requirement that the score not increase with the length faster than the log of the length. To calculate the expected score (E), the score for each amino acid pair (s_{ij}) is multiplied by the fractional occurrences of each amino acid (p_i and p_j). This weighted score is then summed over all of the amino acid pairs. The expected values of the log odds matrices $E(M)$ such as the Dayhoff PAM, BLOSUM, JTT, JO93, PET91, and Gonnet92 matrices all meet this statistical requirement:

$$E(M) = \sum_{i=1}^{20} \sum_{j=1}^{i} p_i p_j s_{ij} \qquad (30)$$

For example, for the PAM120 matrix in one-half bits, $E(M) = -1.64$ and for PAM160 in one-half bits, $E(M) = -1.14$. Thus, scores obtained with these matrices may be used in the above statistical analysis because $E(M) < 0$. Ungapped alignment scores obtained using the BLOSUM62 matrix may also be subject to a significance test, as described above for the PAM matrices. The test is valid because the expect score for a random pair of amino acids is negative ($E = -0.52$). Because the matrix is in half-bit units, the alignment is significant when a score exceeds $16/0.52 \approx 32$ half-bits.

To assist in keeping track of information about the matrix, scoring matrices have appeared in a new format suitable for use by many types of programs. An example is given in Figure 4.3. The matrix includes (1) the scale of the matrix and the value of the statistical parameter λ; (2) $E(M)$, identified as E, the expect score of the average amino acid pair in the matrix, which if negative assures that local alignments will be emphasized (see Equation 30); (3) H, the information content or entropy of the matrix (see Chapter 3) giving the ability of the matrix to discriminate related from unrelated sequence align-

```
#
# This matrix was produced by "pam" Version 1.0.6 [28-Jul-93]
#
# PAM 120 substitution matrix, scale = ln(2)/2 = 0.346574   [1/2 bits]
#
# Expected score = -1.64, Entropy = 0.979 bits
#
# Lowest score = -8, Highest score = 12
#
    A   R   N   D   C   Q   E   G   H   I   L   K   M   F   P   S   T   W   Y   V   B   Z   X   *
A   3  -3  -1   0  -3  -1   0   1  -3  -1  -3  -2  -2  -4   1   1   1  -7  -4   0   0  -1  -1  -8
R  -3   6  -1  -3  -4   1  -3  -4   1  -2  -4   2  -1  -5  -1  -1  -2   1  -5  -3  -2  -1  -2  -8
N  -1  -1   4   2  -5   0   1   0   2  -2  -4   1  -3  -4  -2   1   0  -4  -2  -3   3   0  -1  -8
D   0  -3   2   5  -7   1   3   0   0  -3  -5  -1  -4  -7  -3   0  -1  -8  -5  -3   4   3  -2  -8
C  -3  -4  -5  -7   9  -7  -7  -4  -4  -3  -7  -7  -6  -6  -4   0  -3  -8  -1  -3  -6  -7  -4  -8
Q  -1   1   0   1  -7   6   2  -3   3  -3  -2   0  -1  -6   0  -2  -2  -6  -5  -3   0   4  -1  -8
E   0  -3   1   3  -7   2   5  -1  -1  -3  -4  -1  -3  -7  -2  -1  -2  -8  -5  -3   3   4  -1  -8
G   1  -4   0   0  -4  -3  -1   5  -4  -4  -5  -3  -4  -5  -2   1  -1  -8  -6  -2   0  -2  -2  -8
H  -3   1   2   0  -4   3  -1  -4   7  -4  -3  -2  -4  -3  -1  -2  -3  -3  -1  -3   1   1  -2  -8
I  -1  -2  -2  -3  -3  -3  -3  -4  -4   6   1  -3   1   0  -3  -2   0  -6  -2   3  -3  -3  -1  -8
L  -3  -4  -4  -5  -7  -2  -4  -5  -3   1   5  -4   3   0  -3  -4  -3  -3  -2   1  -4  -3  -2  -8
K  -2   2   1  -1  -7   0  -1  -3  -2  -3  -4   5   0  -7  -2  -1  -1  -5  -5  -4   0  -1  -2  -8
M  -2  -1  -3  -4  -6  -1  -3  -4  -4   1   3   0   8  -1  -3  -2  -1  -6  -4   1  -4  -2  -2  -8
F  -4  -5  -4  -7  -6  -6  -7  -5  -3   0   0  -7  -1   8  -5  -3  -4  -1   4  -3  -5  -6  -3  -8
P   1  -1  -2  -3  -4   0  -2  -2  -1  -3  -3  -2  -3  -5   6   1  -1  -7  -6  -2  -2  -1  -2  -8
S   1  -1   1   0   0  -2  -1   1  -2  -2  -4  -1  -2  -3   1   3   2  -2  -3  -2   0  -1  -1  -8
T   1  -2   0  -1  -3  -2  -2  -1  -3   0  -3  -1  -1  -4  -1   2   4  -6  -3   0   0  -2  -1  -8
W  -7   1  -4  -8  -8  -6  -8  -8  -3  -6  -3  -5  -6  -1  -7  -2  -6  12  -2  -8  -6  -7  -5  -8
Y  -4  -5  -2  -5  -1  -5  -5  -6  -1  -2  -2  -5  -4   4  -6  -3  -3  -2   8  -3  -3  -5  -3  -8
V   0  -3  -3  -3  -3  -3  -3  -2  -3   3   1  -4   1  -3  -2  -2   0  -8  -3   5  -3  -3  -1  -8
B   0  -2   3  -4  -6   0   3   0   1  -3  -4   0  -4  -5  -2   0   0  -6  -3  -3   4   2  -1  -8
Z  -1  -1   0   3  -7   4   4  -2   1  -3  -3  -1  -2  -6  -1  -1  -2  -7  -5  -3   2   4  -1  -8
X  -1  -2  -1  -2  -4  -1  -1  -2  -2  -1  -2  -2  -2  -3  -2  -1  -1  -5  -3  -1  -1  -1  -2  -8
*  -8  -8  -8  -8  -8  -8  -8  -8  -8  -8  -8  -8  -8  -8  -8  -8  -8  -8  -8  -8  -8  -8  -8   1
```

FIGURE 4.3. Example of BLASTP format of the Dayhoff MDM matrix giving log odds scores at 120 PAMs. Note that the matrix has mirror-image copies of the same score on each side of the main diagonal. Besides the standard single-letter amino acid symbols, there are four new symbols: B, Z, X, and *. B is the frequency-weighted average of entries for D and N pairs, Z similarly for Q and E entries, X similarly for all pairs in each row, and * is the lowest score in the matrix for matches with any other sequence character that may be present. The expected score is given by Equation 30, and the entropy is described in Chapter 3.

ments, not shown here; and (4) suitable gap penalties. The BLOSUM matrices are also available in this same format.

Significance of Gapped, Local Alignments

When random sequences of varying lengths are optimally aligned with the Smith–Waterman dynamic programming algorithm using an appropriate scoring matrix and gap penalties, the distribution of scores also matches the extreme value distribution (Altschul and Gish 1996). Similarly, in optimally aligning a given sequence to a database of sequences, the scores of the unrelated sequences also follow this distribution (Altschul et al. 1994; Pearson 1996, 1998). In these and other cases, optimal scores increase linearly with $\log(n)$, where n is the sequence length. Equation 29 (p. 139) predicts that the optimal alignment score (x) expected between two random or unrelated sequences should be proportional to the logarithm of the product of the sequence lengths, $x \approx \log_2(nm)$. If the sequence lengths are approximately equal, $n \approx m$, then x should be proportional to $\log_2(n^2) = 2\log_2(n)$, and the predicted score should also increase linearly with $\log(n)$. $\log_2(n)$ is equivalent to $\log(n)$ because, to change the base of a logarithm, one merely multiplies by a constant. In comparing one sequence of length m to a sequence database of length n, m is a constant and the predicted score should increase linearly as $\log(n)$. This $\log(n)$ relationship has been found in several studies of the distribution of optimal local alignment scores that have included gap penalties (Smith et al. 1985; Arratia et al. 1986; Collins et al. 1988; Pearson 1996, 1998; for additional references, see Altschul et al. 1994). Thus, the same statistical methods described above for assessing the significance of ungapped alignment scores may also be used for gapped alignment scores.

TABLE 4.3 *Statistical parameters for combination of scoring matrices and affine gap penalties*

Scoring matrix	Gap opening penalty[b]	Gap extension penalty[b]	K	λ	H[c]
BLOSUM50	∞[a]	0–∞	0.110	0.232	0.34
BLOSUM50	15	8–15	0.090	0.222	0.31
BLOSUM50	11	8–11	0.050	0.197	0.21
BLOSUM50	11	1	—	—	—
BLOSUM62	∞[a]	0–∞	0.13	0.318	0.40
BLOSUM62	12	3–12	0.100	0.305	0.38
BLOSUM62	8	7–88	0.060	0.270	0.25
BLOSUM62	7	1	—	—	—
PAM250	∞[a]	0–∞	0.09	0.229	0.23
PAM250	15	5–15	0.060	0.215	0.20
PAM250	10	8-10	0.031	0.175	0.11
PAM250	11	1	—	—	—

Dashes indicate that no value can be calculated because the relationship between alignment score and sequence length is linear and not logarithmic, indicating that the alignment is global, not local, in character. Statistical significance may not be calculated for these gap penalty-scoring matrix combinations. The corresponding values for gap penalties define approximate lower limits that should be used.

[a] A value of ∞ for gap penalty will produce alignments with no gaps.

[b] The penalty for a gap opening of length *l* is the value of the gap opening penalty shown. The gap extension penalty is not added until the gap length is 2. Make sure that the alignment program uses this same scheme for scoring gaps. The extension penalty is shown over a range of values; values within this range did not change K and λ.

[c] The entropy in units of the natural logarithm.

METHODS FOR CALCULATING THE PARAMETERS OF THE EXTREME VALUE DISTRIBUTION

In the analysis by Altschul and Gish (1996), 10,000 random amino acid sequences of a given length were aligned using the Smith–Waterman method and a combination of the scoring matrix and a set of gap penalties for local alignments. The scores found by this method followed the same extreme value distribution predicted by the underlying statistical theory described above. Values of K and λ were then estimated for each scoring system by fitting the data to the predicted extreme value distribution. Some representative results are shown in Table 4.3. Readers should consult Tables V–VII in Altschul and Gish (1996) or the current BLAST Web site (http://www.ncbi.nlm.nih.gov/BLAST) for a more detailed list of the gap penalties used.

Method of Moments

In the method of moments used by Altschul and Gish (1996), the mean and standard deviation of random alignment scores of a given length range are calculated. These values are then used to calculate *u* and λ by Equations 16 and 17 (Altschul and Erickson 1986). A similar method used by Pearson (1996, 1998) in the FASTA suite of programs is to align the first or second sequence of interest with a sequence library, noting the length of each library sequence. The lengths of both the test sequence and of the library sequence will influence the range of alignment scores. However, because the length of the test sequence is the same in each comparison, only the length of the library sequence will influence the range of scores found. As with random sequences, scores from alignments with library sequences of approximately the same length provide the range of scores for that length. The only additional requirement is that high scores from alignments of sequences related to the test sequence and of low-complexity sequences must be pruned from the distribution of scores. The mean score and standard deviation for each length range provide an estimate of the *u* and λ.

A better estimate of λ and K is obtained by plotting the calculated values of u against the logarithm of the sequence length $(\log n)$ over the range of lengths found. This procedure should give a straight line $u = (\ln K)/\lambda + \ln(n)/\lambda$. Both λ and K may be calculated by linear regression because the slope of the straight line is $1/\lambda$ and the y intercept is $(\ln K)/\lambda$. A variation of this method is used in the SSEARCH and FASTA vers. 3 programs. Instead of calculating u and $(\ln K)/\lambda$ for each length, the average score over a short range of sequence lengths is plotted against the logarithm of the average length in each range, $\log n$. The resulting points are then fitted to a straight line by linear regression. Each score is then normalized by subtracting the predicted values along the fitted line from the alignment scores, and the variance of these scores from the fitted line is determined. Finally, each score is converted to z scores, where the z score is the number of standard deviations of an alignment score from the fitted line. At this point (Pearson 1998), there are repeated prunings of high scoring, presumably related or less complex sequences, and also of very low scoring alignments that do not fit the straight line, and a recalculation of the z scores with these outliers excluded.

Maximum Likelihood Estimation

A second general method for estimating K and λ is to use maximum likelihood estimation (Mott 1992). In this case, trial but reasonable values for the parameters are used in the equation describing the extreme value equation to predict the distribution found in the data. Based on the goodness of fit, new trial values are adjusted appropriately to provide a better fit to the data and are then used. This procedure is repeated until parameters providing a good fit are found and the fit of the extreme value distribution to the data cannot be further improved. Pruning of unusual low and high scores is also performed in this analysis. The programs are available from Mott (1992).

Poisson Approximation Method

A novel, rapid method for estimating K and λ is the Poisson approximation method. Scores from a series of best-scoring alignments between the same two random sequences are used (Waterman and Vingron 1994a,b). The idea is to align a few pairs of random sequences (say 10), but to find a large number of possible alignments from each pair (say 300). This procedure is much faster than aligning 3,000 different pairs of random sequences, because once the dynamic programming scoring matrix has been established, all of the possible alignments are there. If the positions of the dynamic programming scoring matrix that give the highest scoring alignment are neutralized as described in Chapter 3, then the second best alignment may be easily found, and this procedure may be repeated many times. If the sequences are long enough, i.e., 900-nucleotide sequences, this so-called declumping procedure can be repeated 300 times in a relatively short computer run. These subsequent alignments are also independent because they do not intersect (i.e., do not both align the same residues).

As discussed previously (see p. 90), each locally aligned region between two sequences is a cluster or "clump" of different possible alignments having about the same score and starting at the same scoring matrix position. The highest-scoring alignment is removed, and the next highest scoring "clump" is found. Some of the same sequence included in the first alignment may again be used, but in a different alignment. In the Poisson approximation method, alignment **a** is removed by rescoring the region of the matrix containing the alignment, and other overlapping clumps such as **b** and **c** with **a** are removed from consideration by using a Poisson approximation for the number of scores that is greater than the expected, average score (Waterman and Vingron 1994a,b). Using such methods, independent alignments with different sequence from the same scoring matrix can be used to calculate statistical parameters.

After this procedure is repeated with other sets of random sequences (say ten times), the average number of sequence alignments N_{ave} that exceeds a threshold score t is determined. This number represents the mean value of a Poisson distribution (Waterman and Vingron 1994a,b). Thus, the average number of scores above a threshold t will yield a value for s, and in turn values for K and λ, according to the following derivations: $N_{ave}(S > t) = Kmn\, e^{-\lambda t}$; $\log(N_{ave}) = \log(Kmn\, e^{-\lambda t}) = \log(Kmn) + \log(e^{-\lambda t}) = \log(Kmn) - \lambda t$.

Waterman and Vingron performed many alignments of random sequences of the same lengths, and kept track of the fraction of the aligned sequences that have a score $P(S < t)$. By the Poisson approximation, $P(S < t) = \exp(-Kmne^{-\lambda t})$. A log-log transformation provides a linear equation from which λ and K may be calculated by linear regression. The y intercept is $\log(Kmn)$ and the slope is λ. K may be calculated from the sequence lengths. These methods provide reliable estimates of the statistical parameters (Waterman and Vingron 1994a,b).

Choices of Statistical Parameters

Altschul and Gish (1996) have cautioned users of the statistical parameters λ and K. First, the parameters were generated by alignment of random sequences that were produced assuming a particular amino acid distribution, which may be a poor model for some proteins. Second, the accuracy of λ and K cannot be estimated easily. Finally, for gap costs that give values of $H < 0.15$, the optimal alignment length is a significant fraction of the sequence lengths and produces a source of error called the edge effect. The edge effect is an inability to align the sequence ends because there is not enough sequence to be included in the optimal alignment length. The effect occurs when the expected length of an alignment is a significant fraction of the sequence length. When an edge effect occurs, the expected length is subtracted from the sequence length before λ is estimated. If no such correction is done, λ may be overestimated.

These values for gap penalties should also not be construed to represent the best choice for a given pair of sequences or the only choices, simply because the statistical parameters are available. The process of choosing a gap penalty remains a matter of reasoned choice. In trying the effects of varying the gap penalty, it is important to recognize that as the gap penalty is lowered, the alignments produced will have more gaps and will eventually change from a local to a global type of alignment, even though a local alignment program is being used. In contrast, higher H values are generated by a very large gap penalty and produce alignments with no gaps (Table 4.3), thus suggesting an increased ability to discriminate between related and unrelated sequences. In this respect, Altschul and Gish (1996) note that, beyond a certain point, increasing the gap extension penalty does not change the parameters, indicating that most gaps in their simulations are probably of length 1. However, reducing the gap penalty can also allow an alignment to be extended and create a higher-scoring alignment. Eventually, however, the optimal local alignment score between unrelated sequences will lose the log length relationship with sequence length and become a linear function. At this point, gap penalties are no longer useful for obtaining local alignments and the above statistical relationships are no longer valid.

The higher the H value, the better the matrix can distinguish related from unrelated sequences. The lower the value of H, the longer the expected alignment. These conditions may be better if a longer alignment region is required, such as testing a structural or functional model of a sequence by producing an alignment. Conversely, scoring parameters giving higher values of H should produce shorter, more compact alignments. If $H < 0.15$, the alignments may be very long. In this case, the sequences have a shorter effective length because alignments starting near the ends of the sequences may not be completed. This edge effect can lead to an overestimation of λ, but was corrected for in Table 4.3 (Altschul and Gish 1996).

The above determination of the parameters of the extreme value distribution and the choice of a scoring matrix and gap penalty combination provide a reliable method for determining the significance of a local alignment score between two sequences. Unfortunately, this method for calculating the significance of a local alignment score may not be used for a global alignment score. The theory does not apply when the same substitution matrices are used for global alignments. Transformation of these matrices by adding a fixed constant value to each entry or by multiplying each value by a constant has no effect on the relative scores of a series of global alignments. Hence, there is no theoretical basis for a statistical analysis of such scores as there is for local alignments (Altschul 1991).

As discussed in Chapter 6, two programs are commonly used for database similarity searches: FASTA and BLAST. These programs both calculate the statistical significance of the local alignment scores found upon alignment with similar database sequences, but the types of analyses used to determine the statistical significance of these scores are somewhat different. BLAST uses the value of K and λ found by aligning random sequences and Equation 29, where n and m are shortened to compensate for inability of ends to align. FASTA calculates the statistical significance using the distribution of scores with unrelated sequences found during the database search. In effect, FASTA calculates the mean and standard deviation of the low scores in a given length range. These scores represent the expected range of scores of unrelated sequences for that sequence length (recall that the local alignment scores increase as the logarithm of the sequence length). The number of standard deviations to the high scores of related sequences in the same length range (z score) is then determined. The significance of this z score is then calculated according to the expected extreme value distribution of the z scores, given in Equation 18. Pearson (1996) showed that this method is useful in database similarity searches for detecting sequences more distantly related to the input query sequence.

Pearson (1996) has also determined the influence of scoring matrices and gap penalties on alignment scores of moderately related and distantly related protein sequences in the same family. For two examples of moderately related sequences, the choice of scoring matrix and gap penalties did not matter, i.e., BLOSUM50 −12/−2, BLOSUM62 −8/−2, Gonnet93 −10/−2, and PAM250 −12, −2 all produced statistically significant scores. Distantly related proteins in the same family were more difficult to detect. Trying a combination of different scoring matrix and gap penalties is a reasonable approach. Pearson recommends using caution in evaluating alignment scores using only one particular combination of scoring matrix and gap penalties. He also suggests that using a larger gap penalty, e.g., −14, −2 with BLOSUM50, can increase the selectivity of a database search for similarity (fewer sequences known to be unrelated will receive a significant alignment score).

A difficulty encountered by FASTA in calculating statistical parameters during a database search is that of distinguishing unrelated from related sequences, because only scores of unrelated sequences must be used. As score and sequence length information is accumulated during the search, the scores will include high, intermediate, and sometimes low scores of sequences that are related to the query sequence, as well as low scores and sometimes intermediate and even high scores of unrelated sequences. As an example, a high score with an unrelated database sequence can occur because the database sequence has a region of low complexity, such as a high proportion of one amino acid. Regardless of the reason, these high scores must be pruned from the search if accurate statistical estimates are to be made. Pearson (1998) has devised several such pruning schemes, and then determined the influence of the scheme on the success of a database search at demonstrating statistically significant alignment scores among members of the same protein family or superfamily. However, no particular scheme proved to be better than another.

EXAMPLE: USE OF THE ABOVE PRINCIPLES TO ESTIMATE THE SIGNIFICANCE OF A SMITH–WATERMAN LOCAL ALIGNMENT SCORE

The alignment shown in step 1 in the next example box is a local alignment between the phage λ and P22 repressor protein sequence examples used previously (see p. 90). The alignment is followed by a statistical analysis of the score in steps 2 and 3. To perform this analysis, the second sequence (the P22 repressor sequence) was shuffled 1000 times and realigned with the first sequence to create a set of random alignments. Two types of shuffling are available: (1) a global type of shuffling in which random sequences are assembled based on amino acid composition, and (2) a local one in which the random sequences are assembled by random selection of an amino acid from a sliding window of length n in the original sequence to preserve local amino acid composition. An example of a global analysis is shown in step 2.

The distribution of scores in each case was fitted to the extreme value distribution (Altschul and Gish 1996) to obtain estimates of λ and K to be used in the estimation of significance. The program and parameters used were LALIGN (see Table 3.1, p. 82), which produces the highest-scoring n independent alignments and which was described previously (p. 91), and the scoring matrix BLOSUM50 with a gap opening penalty of −12 and −2 for extra positions in the gap, with end gaps weighted. These programs do not presently have windows or Web page interfaces, and must be run using command line options.

The program PRSS, a sequence scrambling tool that is a part of the FASTA suite of programs, performs a statistical analysis based on the correct statistical distribution of alignment scores, as shown below. PRSS version 3 (PRSS3) gives the results as z scores.

EXAMPLE: ESTIMATION OF STATISTICAL SIGNIFICANCE OF A LOCAL ALIGNMENT SCORE

1. Optimal alignment of phage λ and P22 repressor sequences using the program LALIGN.

   a. The command line used was lalign -f -12 -g -2 lamc1.pro p22c2.pro >results.doc. The -f and -g flags indicate the gap opening and extension parameters to be used, and are followed by the sequence files in FASTA format.

   b. No scoring matrix was specified so that the default BLOSUM50 matrix was used.

   c. Program output is directed to the file results.doc, as indicated by the symbol >.

   The alignment shown is the highest-scoring or optimal one using this scoring matrix and these gap penalties. The next two alignments reported were only 9 and 15 amino acids long and each one had a score of 35 (not shown). As discussed in the text, these alignments are produced by repeatedly erasing the previous alignment from the dynamic programming matrix and then rescoring the matrix to find the next best alignment that does not align the same positions in the two sequences.

   The fact that the first alignment has a much higher score than the next two is an indication that (1) there are no other reasonable alignments of these sequences and (2) the first alignment score is highly significant.

```
LALIGN finds the best local alignments between two sequences
version 2.0u64 March 1998
Please cite:
X. Huang and W. Miller (1991) Adv. Appl. Math. 12:373-381

Comparison of:
(A) lamc1.pro  LAMC1 REFORMAT of: cipro.pro  check: -1 from: 1  - 237 aa
(B) p22c2.pro  P22C2 REFORMAT of: p22 check: 4729 from: 1 to  - 216 aa
using matrix file: blosum50.mat, gap penalties: -12/-2

36.1% identity in 208 aa overlap; score:  401 [1/2 bits]

           30        40        50        60        70        80
LAMC1  KKNELGLSQESVADKMGMGQSGVGALFNGINALNAYNAALLAKILKVSVEEFSPSIAREI
       ....: : ...  .:....  ...    .  . :. :  :.:.:...:... ...
P22C2  RRKKLKIRQAALGKMVGVSNVAISQWERSETEPNGENLLALSKALQCSPDYLLKGDLSQT
           20        30        40        50        60        70

           90       100       110       120       130       140
LAMC1  YEMYEAVSMQPSLRSEYEYPVFSHVQAGMFSPELRTFTKGDAERWVSTTKKASDSAFWLE
       :.. .: : :.::  :.:.... ..... .: . : . :...:.
P22C2  NVAYHS-RHEP--RGSY--PLISWVSAGQWMEAVEPYHKRAIENWHDTTVDCSEDSFWLD
          80         90       100       110       120

          150       160       170       180       190       200
LAMC1  VEGNSMTAPTGSKPSFPDGMLILVDPEQAVEP--GDFCIARLGGD-EFTFKKLIRDSGQV
       :.:.:::::::  :.:.::.::::: :::  :: . :.:: :. : :::::. :.:.
P22C2  VQGDSMTAPAGL--SIPEGMIILVDPE--VEPRNGKLVVAKLEGENEATFKKLVMDAGRK
          130       140       150       160       170       180

          210       220       230
LAMC1  FLQPLNPQYPMIPCNESCSVVGKVIASQ
       ::.::::::::: : .:...: :. ..
P22C2  FLKPLNPQYPMIEINGNCKIIGVVVDAK
          190       200       210
```

2. Statistical analysis with program PRSS using a global shuffling strategy.

   a. The program prompts for input information and requests the name of a file for saving output.

   b. The second sequence has been shuffled 1000 times, conserving amino acid composition, and realigned to the first sequence.

   c. The distribution of scores is shown.

d. Fitting the extreme value distribution to these scores provides an estimate of λ and *K* needed for performing the statistical estimate by Equation 24.

Recent versions of PRSS estimate these parameters by the method of maximum likelihood estimation (Mott
1992; W. Pearson, pers. comm.) as described above.

```
lamc1.pro, 237 aa vs p22c2.pro

            s-w   est
   < 24      0     0:
     26      0     0:
     28      3     1:*==
     30     13     6:=====*=======
     32     27    21:==================*======
     34     68    50:==================================================*
     36     98    84:================================================*
     38    128   111:==============================================*
     40    129   123:=============================================*
     42    105   121:============================================*
     44    110   108:==========================================*
     46     63    91:===================================*
     48     75    72:==================================*
     50     35    56:=============================          *
     52     48    42:==============================*======
     54     30    32:=======================  *
     56     19    23:==================   *
     58     17    16:=============*=
     60      6    13:======        *
     62      7     9:======= *
     64      7     6:=====*=
     66      2     5:==  *
     68      4     3:==*=
     70      0     2: *
     72      1     2:=*
     74      0     1:*
     76      1     1:*
     78      2     1:*=
     80      0     0:
     82      0     0:
     84      0     0:
     86      1     0:=
     88      1     0:=
     90      0     0:
     92      0     0:
     94      0     0:
   > 96      0     0: O
    216000 residues in  1000 sequences,
    BLOSUM50 matrix, gap penalties: -12,-2
    unshuffled s-w score: 401; shuffled score range: 30 - 89
   Lambda: 0.16931 K: 0.020441; P(401)= 3.7198e-27
   For 1000 sequences, a score >=401 is expected 3.72e-24 times
```

The method given in the boxed example does not necessarily ensure that the choice of scoring matrix and gap penalties provides a realistic set of local alignment scores. In the comparable situation of matching a test sequence to a database of sequences, the scores also follow the extreme value distribution. For this situation, Mott (1992) has explained that for local alignments the end point of the alignment should on the average be halfway along the query sequence. Pearson (1996) has pointed out that the presence of known, unrelated sequences in the upper part of the curve where $E > 1$ (see Chapter 6) can be an indication of an inappropriate scoring system.

Statistical Significance of Individual Alignment Scores between Sequences and Significance of Scores Found in a Database Search Are Calculated Differently

When a database search between a query sequence and a sequence database is performed, a new comparison is made for each sequence in the database. FASTA employs alignment scores between unrelated sequences to calculate the parameters of the extreme value distribution. Then, the probability that scores between unrelated sequences could reach as high as those found for matched sequences can be calculated (Pearson 1998). Similarly, the database similarity search program BLAST calculates estimates of the statistical parameters based on the scoring matrix and sequence composition. These parameters are then used to calculate the probability of finding conserved patterns by chance alignment of unrelated sequences (Altschul et al. 1994). When performing such database searches, many trials are made to find the most strongly matching sequences.

As more and more comparisons between unrelated sequences are made, the chance that one of the alignment scores will be the highest one yet found increases. Therefore, the probability of finding a high score in one of a series of unrelated sequence alignments has to be higher than for one sequence pair. The length of the query sequence is about the same as it would be in a normal sequence alignment, but the effective database sequence is very large and represents many different sequences, each one a different test alignment. Theory shows that the Poisson distribution should apply (Karlin and Altschul 1990, 1993;

Altschul et al. 1994), as it did above for estimating the parameters of the extreme value distribution from many alignments between random sequences.

The probability of observing, in a database of D sequences, no alignments with scores higher than the mean of the highest possible local alignment scores s is given by e^{-Ds}, and that of observing at least one score s is $P \approx 1 - e^{-Ds}$. For the range of values of P that are of interest, i.e., $P < 0.1$, $P = Ds$. If two sequences are aligned by PRSS as given in the above example (p. 147), and the significance of the alignment is calculated, two scores must be considered. The probability of the score may first be calculated using the estimates of λ and K. Thus, in the phage repressor alignment, $P(s > 401) = 3.7 \times 10^{-27}$. However, to estimate the extreme value parameters, 1000 shuffled sequences were compared, and the probability that one of those sequences would score as high as 401 is given by Ds, or $1000 \times 3.7 \times 10^{-27} = 3.7 \times 10^{-24}$. These numbers are also shown in the statistical estimates computed by PRSS. Finally, if the score had arisen from a database search of 50,000 sequences, the probability of a score of 401 among this many sequence alignments is 5×10^{-19}, still a small number, but 50,000 larger than that for a single comparison. These probability calculations are used for reporting the significance of scores with database sequences by FASTA and BLAST, as described in Chapter 6.

SEQUENCE ALIGNMENT AND EVOLUTIONARY DISTANCE CAN BE ESTIMATED BY BAYESIAN STATISTICAL METHODS

A recent development in sequence alignment methods is the use of Bayesian statistical methods to produce alignments between pairs of sequences (Zhu et al. 1998) and to calculate distances between sequences (Agarwal and States 1996). This section begins with some introductory comments about Bayesian probability before discussing these methods further.

Introduction to Bayesian Statistics

Bayesian statistical methods differ from other types of statistics by the use of conditional probabilities. These probabilities are used to derive the joint probability of two events or conditions. An example of a conditional probability is $P(B \mid A)$, meaning the probability of

B, given A, whereas $P(B)$ is the probability of B, regardless of the value of A. Suppose that A can have two states, A1 and A2, and that B can also have two states, B1 and B2, as shown in Table 4.4. These states might, for instance, correspond to two allelic states of two genes. Then, $P(B) = P(B1) + P(B2) = 1$ and $P(A) = P(A1) + P(A2) = 1$. Suppose, further, that the probability $P(B1) = 0.3$ is known. Hence $P(B2) = 1 - 0.3 = 0.7$. In our genetic example, each probability might correspond to the frequency of an allele, for which p and q are often used. These probabilities $P(B1)$, etc., can be placed along the right margins of the table as the respective sum of each row or column and are referred to as the marginal probabilities.

Next, the missing data in the middle two columns of the table must be filled in. The probability of A1

TABLE 4.4. *Prior information for a Bayes analysis*

	A1	A2	
B1			0.3
B2			0.7
			1.0

TABLE 4.5. *Completed table of joint and marginal probabilities*

	A1	A2	
B1	0.24	0.06	0.3
B2	0.21	0.49	0.7
	0.45	0.55	1.0

and B1 occurring together (the value to be entered in row B1 and column A1) is called the joint probability, $P(B1 \text{ and } A1)$ (also denoted $P[B1, A1]$). The marginal probability $P(A1)$ is also missing. The available information up to this point, called the prior information, is not sufficient to calculate the joint probabilities. With additional data on the co-occurrence of A1 with B1, etc., these joint probabilities may be derived by Bayes' rule. Suppose that the conditional probabilities $P(A1 \mid B1) = 0.8$ and $P(A2 \mid B2) = 0.70$ are known, the first representing, for example, the proportion of a population with allele B1 that also has allele A1. First, note that $P(A1 \mid B1) + P(A2 \mid B1) = 1$, and hence that $P(A2 \mid B1) = 1.0 - 0.8 = 0.2$. Similarly, $P(A1 \mid B2) = 1.0 - 0.70 = 0.3$. Then the joint probabilities and other conditional probabilities may be calculated by Bayes' rule, illustrated using the joint probability for A1 and B1 as an example:

$$P(A1 \text{ and } B1) = P(B1)\,P(A1 \mid B1) \qquad (31)$$

$$P(A1 \text{ and } B1) = P(A1)\,P(B1 \mid A1) \qquad (32)$$

Thus, $P(A1 \text{ and } B1) = P(B1) \times P(A1 \mid B1) = 0.3 \times 0.8 = 0.24$, and $P(A2 \text{ and } B2) = P(B2) \times P(A2 \mid B2) = 0.7 \times 0.7 = 0.49$. The other joint probabilities may be calculated by subtraction; e.g., $P(A2 \text{ and } B1) = P(B1) - P(A1 \text{ and } B1) = 0.30 - 0.24 = 0.06$. To calculate $P(A1)$ and $P(A2)$, the joint probabilities in each column may be added, thereby completing the additions to the table, and shown in Table 4.5.

However, note that $P(A1)$ may also be calculated in the following manner:

$$
\begin{aligned}
P(A1) &= P(A1 \text{ and } B1) + P(A1 \text{ and } B2) \\
&= P(B1)\,P(A1 \mid B1) + P(B2)\,P(A1 \mid B2) \quad (33)
\end{aligned}
$$

Other conditional probabilities may be calculated from Equations 31 and 32 by rearranging terms and by substituting Equation 33, deriving the following form of Bayes' rule:

$$
\begin{aligned}
P(B2 \mid A1) &= P(A1 \text{ and } B2) / P(A1) \\
&= P(B2)\,P(A1 \mid B2) / P(A1) \\
&= P(B2)\,P(A1 \mid B2) / [P(B1)\,P(A1 \mid B1) \\
&\quad + P(B2)\,P(A1 \mid B2)] \qquad (34)
\end{aligned}
$$

Using Equation 34, $P(B2 \mid A1) = 0.7 \times 0.30 / [0.3 \times 0.80 + 0.7 \times 0.3] = 0.467$, and also $P(B1 \mid A1) = 1.0 - 0.467 = 0.533$. Such calculated probabilities are called posterior probabilities or posteriors, as opposed to the prior probabilities or priors initially available. Thus, based on the priors and additional information, application of Bayes' rule allows the calculation of posterior estimates of probabilities not initially available. This procedure of predicting probability relationships among variables may be repeated as more data are collected, with the existing model providing the prior information and the new data providing the information to derive a new model. The initial beliefs concerning a parameter of interest are expressed as a prior distribution of the parameter, the new data provide a likelihood estimate for the parameter, and the normalized product of the prior and likelihood (Equation 41) forms the posterior distribution.

EXAMPLE: BAYESIAN ANALYSIS

One illustrative example of a Bayesian analysis is the television game show "Let's Make a Deal" hosted by Monty Hall. A prize is placed behind one of three doors by the host. A contestant is then asked to choose a door. The host opens one door (one that he knows the prize is not behind) and reveals that the prize is not behind that door. The contestant is then given the choice of changing to one of the other doors of the three to win. The initial or prior probability for each door is 1/3, but after the new information is provided, these probabilities must be revised. The original door chosen still has a probability of 1/3, but the second door that the prize could be behind now has a probability of 2/3. These new estimates are posterior probabilities based on the new information provided when the host revealed a door that the prize was not behind.

In the above example (Equation 34), note that the joint probability of A1 and B1 [P(A1 and B1)] is not equal to the product of P(A1) and P(B1); i.e., 0.24 is not equal to 0.3 × 0.45 = 0.135. Such would be the case if the states of A and B were completely independent; i.e., if A and B were statistically independent variables as, for example, in a genetic case of two unlinked genes A and B. In the above example, the state of one variable is influencing the state of the other such that they are not independent of each other, as might be expected for linked genes in the genetic example.

A more general application of Bayes' rule is to consider the influence of several variables on the probability of an outcome. The analysis is essentially the same as that outlined above. To see how the method works with three instead of two values of a variable, think first of an example of three genes, each having three alleles, and of deriving the corresponding conditional probabilities. The resulting joint probabilities will depend on the choice made of the three possible values for each variable. To go even further, instead of a small number of discrete sets of alternative values of a variable, Bayesian statistical methods may also be used with a large number of values of variables or even with continuous variables.

For sequence analysis by Bayesian methods, a slightly different approach is taken. The variables may include combinations of possible alignments, gap scoring systems, and log odds substitution matrices. The most probable alignments may then be identified. The scoring system used for sequence alignments is quite readily adapted to such an analysis. In an earlier discussion, it was pointed out that a sequence alignment score in bits is the logarithm to the base 2 of the likelihood of obtaining the score in alignments of related sequences divided by the likelihood of obtaining the score in alignments of unrelated sequences. It was also indicated that the highest alignment score should be obtained if the scoring matrix is used that best represents the nucleotide or amino acid substitutions expected between sequences at the same level of evolutionary distance. Bayesian

methodology carries this analysis one step farther by examining the probabilities of all possible alignments of the sequences using all possible variations of the input parameters and matrices. These selections are the prior information for the Bayesian statistical analysis and provide various estimates of the alignment that allow us to decide on the most probable alignments. The alignment score for each combination of these variables provides an estimate of the probability of the alignment. By using equations of conditional probability such as Equation 34, posterior information on the probability of alignments, gap scoring system, and substitution matrix can be obtained. For further reading, a Bayesian bioinformatics tutorial by C. Lawrence is available at http://www.wadsworth.org/resnres/bioinfo/.

Application of Bayesian Statistics to Sequence Analysis

To use an example from sequence analysis, a local alignment score (s) between two sequences varies with the choice of scoring matrix and a gap scoring system. In the previous sections, an amino acid scoring matrix was chosen on the basis of its performance in identifying related sequences. Gap penalties were then chosen for a particular scoring matrix on the basis of their performance in identifying known sequence relationships and of their keeping a local alignment behavior by the increase in score between unrelated sequences remaining a logarithmic function of sequence length. The alignment score expressed in bit units was the ratio of the alignment score expected between related sequences to that expected between unrelated sequences, expressed as a logarithm to the base 2. The scores may be converted to an odds ratio (r) using the formula $r = 2^s$. The probability of such a score between unrelated or random sequences can then be calculated using the parameters for the extreme value distribution for that combination of scoring matrix and gap penalty. Finally, the above analysis may provide several different alignments, without providing any information as to which is the most likely. With the application of Bayesian statistics, the approach is different.

The application of Bayesian statistics to this problem allows one to examine the effect of prior information, such as the chosen amino acid substitution matrix, on the probability that two sequences are homologous. The method provides a posterior probability distribution of all alignments taking into account all possible scoring systems. Thus, the most likely alignments and their probabilities may be determined. This method circumvents the need to choose a particular scoring matrix and gap scoring system because a range of available choices can be tested. The approach also provides conditional posterior distributions on the gap number and substitution matrix. Another application of Bayes statistics for sequence analysis is to find the PAM DNA substitution matrix that provides the maximum probability of a given level of mismatches in a sequence alignment, and thus to predict the evolutionary distance between the sequences.

Bayesian Evolutionary Distance

Agarwal and States (1996) have applied Bayesian methods to provide the best estimate of the evolutionary distance between two DNA sequences. The examples used here are sequences of the same length that have a certain level of mismatches. Consequently, there are no gaps in the alignment between the sequences. Sequences of this type originated from gene duplication events in the yeast and *Caenorhabditis elegans* genomes. When there are multiple mismatches between such repeated sequences, it is difficult to determine the most likely length of the repeats. With the application of Bayesian methods, the most probable repeat length and evolutionary time since the repeat was formed may be derived.

The alignment score in bits between sequences of this type may be calculated from the values for matches and mismatches in the DNA PAM scoring matrices described earlier (see Table 3.6, p. 105). Recall that a PAM1 evolutionary distance represents a change of one sequence position in 100 and is thought to correspond roughly to an evolutionary distance of 10^7 years. Higher PAM*n* tables are calculated by multiplying the PAM1 scoring matrix by itself *n* times. This Markovian model of evolution assumes that any DNA sequence position can change with equal probability, and subsequent changes at a

site are not influenced by preceding changes at that site. In addition, a changed position can revert to the original nucleotide at that position. The problem is to discover which scoring matrix (PAM50, PAM100, etc.) gives the most likely alignment score between the sequences. This corresponding evolutionary distance will then represent the time at which the sequence duplication event could have occurred.

An approach described earlier was to evaluate the alignment scores using a series of matrices and then to identify the matrix giving the highest similarity score. For example, if there are 60 mismatches between sequences that are 100 nucleotides long, the PAM50 matrix score of the alignment in bits ($\log_2$) is $40 \times 1.34 - 60 \times 1.04 = -8.8$, but the PAM125 matrix score is much higher, $40 \times 0.65 - 60 \times 0.30 = 8$. When these log odds scores in bits are converted to odds scores, the difference is 0.002 versus 256. Thus, the PAM125 matrix provides a much better estimate of the evolutionary distance between sequences that have diverged to this degree. The Bayesian approach continues this type of analysis to discover the probability of the alignment as a function of each evolutionary distance represented by a different PAM matrix. If x is the evolutionary distance represented by the PAM*n* matrix divided by 100, and k is the number of mismatches in a sequence of length n, then by Bayes' rule and related formulas discussed above,

$$
\begin{aligned}
P(x \mid k) &= P(k \mid x)\,P(x)\,/\,P(k) \\
&= P(k \mid x)\,P(x)\,/\,\textstyle\sum_x P(k \mid x)\,P(x) \quad (35)
\end{aligned}
$$

$P(x \mid k)$ is the probability of distance x given the sequence with k mismatches (and $n - k$ matches), $P(k \mid x)$ is the odds score for the sequence with k mismatches using the log odds scores in the DNA PAM100x matrix, and $P(x)$ is the prior probability of distance x (usually estimated by 1 over the number of matrices, thus making each equally possible). The denominator is the sum of the odds scores over the range of x, which is 0.01–4, representing PAM1 to PAM400 (~10 million to 4 billion years) times the prior probability of each value of x. Like the conditional probabilities calculated by Equation 35, this sum represents the area under the probability curve and has the effect of normalizing the probability for each individual scoring matrix used. The shape of the probability curve reveals how $P(x \mid k)$ varies with x. An example is shown in Figure 4.4.

The probability curves have a single mode (highest score) for $k < 3n/4$. Because the curves are not symmetrical about this mode but are skewed toward higher distances, the expected value or mean of the distribution and its standard deviation are the best indication of evolutionary distance. For a sequence 100 nucleotides long with 40 mismatches, the expected value of x is 0.60 with $s = 0.11$ representing a distance of ~600 million years. These estimates are different from the earlier method that was described of finding the matrix that gives the highest alignment score, which would correspond to the mode or highest scoring distance. Other methods of calculating evolutionary distances are described in Chapter 7.

One difficulty with making such calculations is that the estimate depends on the assumption that the mutation rate in sequences has been constant with time (the molecular clock hypothesis) and that the rate of mutation of all nucleotides is the same. Such problems may be solved by scoring different portions of a sequence with a different scoring matrix, and then using the above Bayesian methods to calculate the best evolutionary distance. These partitions may be located where variations occur in the sequence alignment score. Another difficulty is deciding on the length of sequence that was duplicated. In genomes, the presence of repeats may be revealed by long regions of matched sequence positions dispersed among regions of sequence positions that do not match. However, as the frequency of mismatches is increased, it becomes difficult to determine the extent of the repeated region. The application of the above Bayesian analysis allows a determination of the probability distributions as a function of both length of the repeated region and

evolutionary distance. A length and distance that gives the highest overall probability may then be determined. Such alignments are initially found using an alignment algorithm and a particular scoring matrix. Analysis of the yeast and worm genomes for such repeats has underscored the importance of using a range of DNA scoring matrices such as PAM1 to PAM120 if most repeats are to be found (Agarwal and States 1996). One disadvantage of the Bayesian approach is that a specific mutational model is required, whereas other methods, such as the maximum likelihood approach described in Chapter 7, can be used to estimate the best mutational model as well as the distance. Computationally, however, the Bayesian method is much more practical.

Bayesian Sequence Alignment Algorithms

Zhu et al. (1998) have devised a computer program called the Bayes block aligner that, in effect, slides two sequences along each other to find the highest-scoring ungapped regions or blocks. These blocks are then joined in various combinations to produce alignments. There is no need for gap penalties because only the aligned sequence positions in blocks are scored. Instead of using a given substitution matrix and gap scoring system to find the highest-scoring alignment, a Bayesian statistical approach is used. Given a range of substitution matrices and number of blocks expected in an alignment as the prior information, the method provides posterior probability distributions of alignments. The Bayes aligner is available through a licensing agreement from http://www.wadsworth.org/resnres/bioinfo. A graphical interface for X-Window in a UNIX environment and a nongraphical interface for PCs running MS-Windows are available.

The method may be used for both protein and DNA sequences. An alignment block between two sequences is defined as a run of one or more identical characters in the sequence alignment that can include intervening mismatches but no gaps, as shown in the following example. Only the aligned blocks are identified and scored; regions of unaligned sequence and gaps between these blocks are not scored. The probability of a given alignment is given by the product of the probabilities of the individual alignment scores in the blocks, as indicated in the following example. The Bayes block aligner scores every possible combination of blocks to find the best scoring alignment.

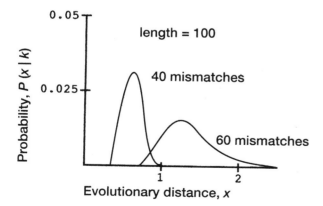

FIGURE 4.4. $P(x \mid k)$ for sequence length $n = 100$ and number of mismatches $k = 40$ or 60. (Redrawn, with permission, from Agarwal and States 1996.)

EXAMPLE: BLOCK ALIGNMENT OF TWO SEQUENCES AND OF THE SCORING OF THE ALIGNMENT AS USED IN THE BAYES BLOCK ALIGNER (ZHU ET AL. 1998)

The score of the sequence alignment is obtained by adding the log odds scores of each block in the alignment, and the alignment score of each block is found by summing the log odds scores of each sequence pair within a block. Any sequence that is not within these blocks and between blocks is not scored and there is no penalty for gaps. Regions of both sequences that are not aligned can be present within the gap. The sequence alignment score is therefore determined entirely by the placement of block boundaries.

```
                       Block 1              Block 2

   Sequence 1     S  G  T  G  K (gap) K  K  R  L  E
   Sequence 2     P  G  S  G  K (gap) K  Q  R  L  T


   BLOSUM62
   score         -1  6  1  6  5        5  1  5  4  -1

   Sum of scores              =     31 half bits
                              =     15.5 bits

   Odds of alignment score

                              =     2^{15.5} to 1
                              =     4.6 x 10^4 to 1
```

Unlike the commonly used methods for aligning a pair of sequences, the Bayesian method does not depend on using a particular scoring matrix or designated gap penalties. Hence, there is no need to choose a particular scoring system or gap penalty. Instead, a number of different scoring matrices and range of block numbers up to some reasonable maximum are examined, and the most probable alignments are determined. The Bayesian method provides a distribution of alignments weighted according to probability and can also provide an estimate of the evolutionary distance between the sequences that is independent of scoring matrix and gaps.

Like dynamic programming methods and the BLAST and FASTA programs, the Bayes block aligner has been used to find similar sequences in a database search. The most extensive comparisons of database searches have shown that the program SSEARCH based on the Smith–Waterman algorithm, with the BLOSUM50 –12,–2 matrix and gap penalty scoring system, can find the most members of protein families previously identified on the basis of sequence similarity (Pearson 1995, 1996, 1998) or structural homology (Brenner et al. 1998). In a similar comprehensive analysis, Zhu et al. have shown that the Bayes block aligner has a slightly better rate than even SSEARCH of finding structurally related sequences at a 1% false-positive level. Hence, this method may be superior over other methods for database similarity searching.

The Bayes block aligner defines blocks by an algorithm due to Sankoff (1972). This algorithm is designed to locate blocks by finding the best alignment between two sequences for any reasonable number of blocks. The example shown in Figure 4.5 illustrates the basic block-finding algorithm.

Following the initial finding of block alignments in protein sequences by the Sankoff method, the Bayes block aligner calculates likelihood scores for these alignments for various block numbers and amino acid or DNA substitution matrices. To be biologically more meaningful by avoiding too many blocks, the number of protein sequence blocks k is limited from 0 to 20 or the length of the shorter sequence divided by 10, whichever is smaller. For a set of amino acid substitution matrices such as the Dayhoff PAM or BLOSUM matrices, the only requirement is that they be in the log odds format to provide the appropriate likelihood scores by additions of rows and columns in the V and W matrices (Fig. 4.5).

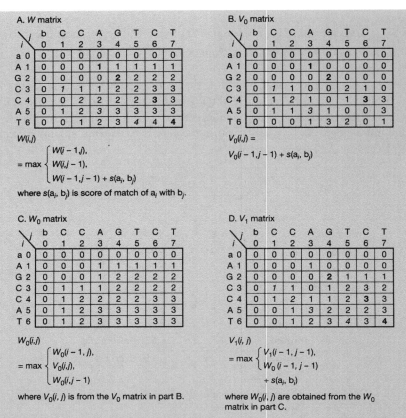

A. W matrix

$i \backslash j$	b 0	C 1	C 2	A 3	G 4	T 5	C 6	T 7
a 0	0	0	0	0	0	0	0	0
A 1	0	0	0	1	1	1	1	1
G 2	0	0	0	1	2	2	2	2
C 3	0	1	1	1	2	2	3	3
C 4	0	0	2	2	2	2	3	3
A 5	0	1	2	3	3	3	3	3
T 6	0	0	1	2	3	4	4	4

$$W(i,j) = \max \begin{cases} W(i-1,j), \\ W(i,j-1), \\ W(i-1,j-1) + s(a_i, b_j) \end{cases}$$

where $s(a_i, b_j)$ is score of match of a_i with b_j.

B. V_0 matrix

$i \backslash j$	b 0	C 1	C 2	A 3	G 4	T 5	C 6	T 7
a 0	0	0	0	0	0	0	0	0
A 1	0	0	0	1	0	0	0	0
G 2	0	0	0	0	2	0	0	0
C 3	0	1	1	0	0	2	1	0
C 4	0	1	2	1	0	1	3	3
A 5	0	1	1	3	1	0	0	3
T 6	0	0	0	1	3	2	0	1

$$V_0(i,j) = V_0(i-1,j-1) + s(a_i, b_j)$$

C. W_0 matrix

$i \backslash j$	b 0	C 1	C 2	A 3	G 4	T 5	C 6	T 7
a 0	0	0	0	0	0	0	0	0
A 1	0	0	0	1	1	1	1	1
G 2	0	0	0	1	2	2	2	2
C 3	0	1	1	1	2	2	2	2
C 4	0	1	2	2	2	2	3	3
A 5	0	1	2	3	3	3	3	3
T 6	0	1	2	3	3	3	3	3

$$W_0(i,j) = \max \begin{cases} W_0(i-1,j), \\ V_0(i,j), \\ W_0(i,j-1) \end{cases}$$

where $V_0(i,j)$ is from the V_0 matrix in part B.

D. V_1 matrix

$i \backslash j$	b 0	C 1	C 2	A 3	G 4	T 5	C 6	T 7
a 0	0	0	0	0	0	0	0	0
A 1	0	0	0	1	0	0	0	0
G 2	0	0	0	0	2	1	1	1
C 3	0	1	1	0	1	2	3	2
C 4	0	1	2	1	1	2	3	3
A 5	0	0	1	3	2	2	2	3
T 6	0	0	1	2	3	4	3	4

$$V_1(i,j) = \max \begin{cases} V_1(i-1,j-1), \\ W_0(i-1,j-1) \end{cases} + s(a_i, b_j)$$

where $W_0(i,j)$ are obtained from the W_0 matrix in part C.

FIGURE 4.5. The Sankoff algorithm for finding the maximum number of identical residues in two sequences without scoring gaps. The example of two DNA sequences shown is taken from Sankoff (1972). A series of scoring matrices called V and W are made according to the matrix scoring scheme shown in parts A–D. (A) The algorithm first examines the maximum number of bases that can match. The scoring scheme used in this example is that a match between two bases is scored as 1 and a mismatch as 0. This number, 4, is shown in the lower right-hand corner of the matrix. To obtain this number, the method does not consider the number of gapped regions between each group of matched pairs, defined as an unconstrained set of matches by Sankoff. For example, a_1 can pair with b_3, and a_2 with b_4, to comprise a group of two sequential pairs, shown in bold. Then there is an unmatched region followed by a match of a_4 with b_6, unmatched base a_5, and finally a match between a_6 and b_7. Thus, two unmatched (gapped) regions will be included in this alignment. A second such set of matches that gives a maximum number of matches is shown as italicized positions. In this case, there is one unmatched region between the groups of matches. (B–D) A slightly different computational method is used to find the maximum possible number of matches given that there are zero gapped regions, one gapped region, two gapped regions, etc. (B) A matrix V_0, where subscript 0 indicates the number of gapped regions permitted, is first calculated. The bold and italicized positions indicate the scores found for the two groups of matches. To simplify the calculation of higher-level V matrices (V_1, V_2, etc.), another set of matrices (W_1, W_2, etc.) is also calculated. (C) The calculation of W_0 is shown. Using the scores calculated in W_0 matrix position and the algorithm shown in D, V_1 is then produced. V_1 shows the same combinations of matches found in the unconstrained case in A, and, therefore, no further calculation of matrices is necessary. In other cases, q, V, and W matrices will be calculated so that alignments with an increased number of unmatched or gapped regions may be found according to the formulas:

$$W_q(i,j) = \max \begin{cases} W_q(i-1,j), \\ V_q(i,j), \\ W_q(i,j-1) \end{cases}$$

$$V_q(i,j) = \max \begin{cases} V_q(i-1,j-1), \\ W_{q-1}(i-1,j-1) \\ + s(a_i, b_j) \end{cases}$$

The number of computational steps required is equal to the product of the sequence lengths times the number of cycles needed to reach the unconstrained alignment, as shown in the lower right-hand corner of the matrix (A). The method may also be used for aligning protein sequences (Zhu et al. 1998) that are distantly related, as described below.

A large number of matrices like those in Figure 4.5 are used, each for a different amino acid substitution matrix and block number. In each of these matrices, a number of alignments of the found block regions are possible. The score in the lower right-hand corner of each matrix is the sum of the odds scores of all possible alignments in that particular matrix. The odds scores calculated in each matrix are then summed to produce a grand total of odds scores. The fraction of this total that is shared by a set of alignments under given conditions (e.g., a given number of blocks or an amino acid substitution matrix) provides the information needed to calculate the most probable scoring matrix, block number, etc., by Bayesian formulas. The joint probabilities equivalent to the interior row and column entries in Tables 4.4 and 4.5 are then calculated. In this case, each joint probability is the likelihood of the alignment given a particular block alignment, number of blocks, and substitution matrix, multiplied by the prior probabilities.

These prior probabilities of particular alignment, block number, and scoring matrix are treated as having an equally likely prior probability. Once all joint probabilities have been computed for every combination of the alignment variables, the conditional posterior information can be obtained by Bayes' rule, using equations similar to Equation 34. As in Equation 34, the procedure involves dividing the sum of all alignment likelihoods that apply to a particular value of a particular variable by the sum of all alignment likelihoods found for all variables.

USE OF THE BAYES BLOCK ALIGNER FOR PAIR-WISE SEQUENCE ALIGNMENT

There are several possible uses of the Bayes block aligner for sequence alignment. The overall probability that a given pair of residues should be aligned may be found by several methods.

1. Alignments may be sampled in proportion to their joint posterior probability, as for example, alignments produced by a particular combination of substitution matrix and gap number. A particular substitution matrix and gap number may be chosen based on their posterior probabilities. An alignment may then be obtained from the alignment matrix in much the same manner as the trace-back procedure used to find an alignment by dynamic programming. Once a number of sample alignments has been obtained, these samples may be used to estimate the marginal distribution of all alignments. This distribution then gives the probability that each pair of residues will align.
2. An alternative method of sampling the joint posterior probability distribution is to identify an average alignment for k blocks by sampling the highest peaks in the marginal posterior alignment distribution and by using each successively lower peak as the basis for another alignment block down to a total of k blocks, concatenating any overlaps. These alignments may then be used to obtain the probability of each aligned residue.
3. The exact marginal posterior alignment distribution of a specific pair of residues may be obtained by summing over all substitution matrices and possible blocks.
4. Optimal alignment and near-optimal alignments for a given number of blocks can also be obtained.
5. The Bayes block aligner provides an indication as to whether or not the sequence similarity found is significant. Bayesian statistics examines the posterior probabilities of all alternative models over all possible priors. The Bayesian evidence that two sequences are related is given by the probability that K, the maximum allowed number of blocks, is greater than 0, as calculated in the following example taken from Zhu et al. (1998). The posterior probability of the number of blocks, the substitution matrices, and the aligned residues can all be calculated as described above.

EXAMPLE: BAYES BLOCK ALIGNER

The same two phage λ and P22 repressor proteins that were aligned by the Smith–Waterman dynamic programming algorithm in Chapter 3 (p. 90) were also aligned by the Bayes block aligner so that the methods could be compared. The prior information provided was an equal chance that the block number k can be any number between 0 and 20, and that a series of new amino acid substitution PAM matrices based on the data of Jones et al. (1992) could be equal-

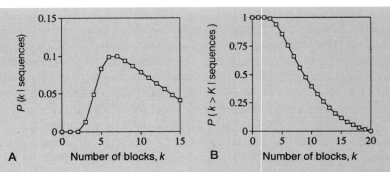

A Number of blocks, *k* **B** Number of blocks, *k*

FIGURE 4.6. Posterior probability distribution of number of blocks from alignment of phage λ and P22 repressor proteins. (*A*) Posterior probability distribution of the block number *k*. (*B*) Cumulative posterior probability distribution. This distribution shows the probability of a block number *K* greater than or equal to the value *k*. Values are derived from the probability distribution of *k* given in A.

ly correct. These matrices were prepared by the author (D. Mount), incorporated into the Bayes block aligner, and are available on request. The posterior probability distribution of the number of blocks, *k*, is shown in Figure 4.6A. The probability rises rapidly from 0 to a maximum of $k = 7$ and falls off slowly at higher values of *k*. This analysis reveals that there are at least six blocks in common in the two proteins, but that there could be more.

The cumulative posterior probability that the block number *K* is greater than a given value *k* is shown in Figure 4.6B. The area under the curve for $k \geq 1$ has the value 1.000000. Although at first glance this number appears to represent the probability that the sequences are related, i.e., that $K > 0$, the probability is actually higher by Bayesian standards. Instead, the maximum value for $P(k \mid \text{sequences})$ in Figure 4.6A, i.e., 0.10000 at $k = 7$, is used. This number times the maximum number of blocks $0.10000 \times 20 = 2.0000$ represents the accumulated best evidence that the blocks are related or that $K > 0$. This calculation assumes that all block numbers are equally likely or that $p(k \mid k>0) = 1/K = 1/20$. The value $P(k = 0 \mid \text{sequences}) = 1.17804 \times 10^{-32}$ (program output values not shown), is the corresponding best evidence that the sequences are not related or that $K = 0$. The probability that the sequences are related is then calculated as $2.0000 / (2.0000 + 1.17804 \times 10^{-32}) = 1.000$. This value is the supremum of $P(k > 0)$ taken over all prior distributions on *k*, where the supremum is a mathematical term that refers to the least upper bound of a set of numbers. This high a Bayesian probability is strong evidence for the hypothesis that the sequences are homologous. Normally, a Bayesian probability of $p > 0.5$ will suffice (Zhu et al. 1998).

The posterior probability distribution of the Jones et al. PAM matrices shown in Figure 4.7 was found. Note that a very clear peak is evident at a PAM value of 70, suggesting a time at which these proteins diverged from a common ancestor. This method provides an excellent phylogenetic comparison of related protein sequences, as described further in Chapter 7. This analysis should provide information as to the evolutionary history of genes, including the possible involvement of duplications, rearrangements, and genetic events producing chimeric sequences.

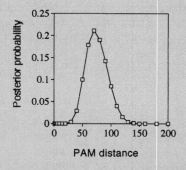

PAM distance

FIGURE 4.7. Posterior probability distribution of the Jones et al. (1992) PAM scoring matrices for alignment of the phage λ and P22 repressor proteins.

A. Bayes block aligner

```
 85 SYPLISWVSAGQWMEAVEPYHKRAIENWHDTTVDCSEDSFWLDVQGDSMTAPAGL  139
                            ...::::::::::::::::.
          .                ..:..::...::::::::::::::::::::::
        ::::::::::          ...::::::::::::::::::::::::::::::
        ::::::::::.        ..:::::::::::::::::::::::::::::::::
        :::::::::::::::::::::::::::::::::::::::::::::::::::::::
103 EYPVFSHVQAGMFSPELRTFTKGDAERWVSTTKKASDSAFWLEVEGNSMTAPTGS  157
    **   *  *  **       *   *  *  **    *    *** * *  ***** *
```

```
140 SIPEGMIILVDPEVEPRNGKLVVAKLEGE  168
       ::::::::::.
     .:::::::::::::         .::::
     ::::::::::::::        ..:::::::
     ::::::::::::::::....:::::::::::
     :::::::::::::::::::::::::::::::
160 SFPDGMLILVDPEQAVEPGDFCIARLGGD  188
    *  *  **  *****       *    *  * *
```

```
170 EATFKKLVMDAGRKFLKPLNPQYPMIEINGNCKIIGVVVDAKLANLP  216
       ::::::::::::::::::::::::::..:::::::::
      ..:::::::::::::::::::::::::::::::::::::::::
      :::::::::::::::::::::::::::::::::::::::::::..
      :::::::::::::::::::::::::::::::::::::::::::::.
      :::::::::::::::::::::::::::::::::::::::::::::::::..
189 EFTFKKLIRDSGQVFLQPLNPQYPMIPCNESCSVVGKVIASQWPEET  235
    *  *****   *  *   **  *********   *   *    * *
```

B. Emboss WATER local alignment.

```
 86 YPLISWVSAGQWMEAVEPYHKRAIENWHDTTVDCSEDSFWLDVQGDSMTA  135
    ||:.|.|.||.:....:..:.|...|.|..||...|:.:|||:|:|||||
104 YPVFSHVQAGMFSPELRTFTKGDAERWVSTTKKASDSAFWLEVEGNSMTA  153
```

```
136 PAGL--SIPEGMIILVDPEVEPRNGKLVVAKLEGENEATFKKLVMDAGRK  183
    |.|.   |.|:||:|||||||....|...:|:|.|: |.|||||:.|:|:.
154 PTGSKPSFPDGMLILVDPEQAVEPGDFCIARLGGD-EFTFKKLIRDSGQV  202
```

```
184 FLKPLNPQYPMIEINGNCKIIGVVV  208
    ||:|||||||||..|.:|.::|.|:
203 FLQPLNPQYPMIPCNESCSVVGKVI  227
```

FIGURE 4.8. The alignment of the phage λ and P22 repressor proteins obtained with the Bayes aligner (*A*), using a new set of PAM matrices, and by the Smith–Waterman local alignment algorithm (EMBOSS WATER program) (B). The highest-scoring sequence positions in the marginal posterior alignment distribution for the sequences for the block number of probability greater than 0.9 were successively sampled with the Bayes block aligner, and are shown in *A*. The accumulated posterior probabilities are shown as vertical columns of dots between the aligned sequence positions. Neighboring aligned positions with scores greater than 0.25 of the peak value were included. Asterisks indicate sequence identities. (*B*) A local alignment of the sequences with WATER is shown. This method is described in Chapter 3 and an example is given on page 89.

Another type of analysis that can be performed with the Bayes block aligner is to examine the probability of the alignments. The procedure is entirely different from other methods of sequence alignment such as dynamic programming. On the one hand, the dynamic programming method finds a single best alignment for a given scoring matrix and gap penalty, and the odds for finding as good a score between random sequences of the same length and complexity is determined. On the other hand, in Bayesian alignment methods, all possible alignments are considered for a reasonable number of blocks and a set of substitution matrices. Rather than a probability of a single alignment, the probabilities of many alignments are provided. Many possible alignments may be examined and compared, and the frequency of certain residues in the sequences in these alignments may be determined.

For the phage repressors, the three most probable block alignments found by the Bayes block aligner are shown in Figure 4.8A. These alignments are compared to a sequence alignment of the same two proteins in Figure 4.8B found by the Smith–Waterman dynamic programming algorithm using the BLOSUM62 substitution matrix and gap penalties of −11/−1. Considering the many differences between these methods, the alignments are remarkably similar with the same identities (*) and gapped regions. Gaps are revealed by the ends of the blocks in A, and by a "-" in B.

Experience with the Bayes block aligner has suggested that a minimum number of blocks that best represents the expected domain structure is the best approach. An average alignment for a number of blocks of probability greater than 0.9 has been found to give good agreement with predicted structural alignments. For other proteins that share little sequence identity than the example shown here, the Bayes aligner correctly predicts many, but not all, features of structural alignments, and does so better than a dynamic programming method that provides local alignments. In other cases, the Bayes aligner may not perform as well as dynamic programming. The prudent choice is to use the Bayes block aligner as one of several computer tools for aligning sequences (Zhu et al. 1998).

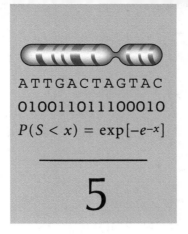

ATTGACTAGTAC
010011011100010

$$P(S < x) = \exp[-e^{-x}]$$

5

Multiple Sequence Alignment

C O N T E N T S

INTRODUCTION

One of the most important contributions of biological sequences to evolutionary analysis is the discovery that sequences of different organisms are often related. Similar genes are conserved across widely divergent species, often performing a similar or even identical function, and at other times, mutating or rearranging to perform an altered function through the forces of natural selection. Thus, many genes are represented in highly conserved forms in a wide range of organisms. Through simultaneous alignment of the sequences of these genes, the patterns of change in the sequences may be analyzed. Because the potential for learning about the structure and function of molecules by multiple sequence alignment (msa) is so great, the necessary computational methods have received a great deal of attention. In msa, sequences are aligned optimally by bringing the greatest number of similar characters into register in the same column of the alignment, just as described in Chapter 3 for the alignment of two sequences.

As with aligning a pair of sequences, the difficulty in aligning a group of sequences varies considerably, being much greater as the degree of sequence similarity decreases. If the amount of sequence variation is minimal, it is quite straightforward to align the sequences, even without the assistance of a computer program. However, if the amount of sequence variation is great, it may be very difficult to find an optimal alignment of the sequences because so many combinations of substitutions, insertions, and deletions, each predicting a different alignment, are possible.

CHAPTER GUIDE FOR BIOLOGISTS

Biologists often find a protein with approximately the same sequence in different species, suggesting that the proteins have a closely related biological function and that the genes encoding these proteins have come from a common genetic source, an evolutionary ancestor gene. If these proteins are aligned with each other in pairs as described in Chapter 3, we usually find that some pairs are more alike than others and some (for example, from two primates) may be almost identical. These pair-wise alignments reveal which amino acids will have to be changed, inserted, or deleted to change one sequence into the other. The changes also represent the sum of all the substitutions going from one sequence to the common ancestor sequence and back to the other sequence.

For three related sequences, many amino acid positions should readily align in a three-sequence alignment, or multiple sequence alignment (msa). For other positions that vary more, the task is to predict what substitutions and insertions/deletions have occurred during the evolution of these three sequences. As with pairs of sequences, we need to predict the changes using a scoring matrix of expected changes (like BLOSUM62) with penalties for insertions/deletions. Using these scores, the msa should thus provide a reasonable prediction as to which positions are changing over time and where insertions/deletions have occurred. The predicted alignment should also make evolutionary sense in that the number of changes from ancestor sequences to each sequence should be a minimum number, referred to in evolutionary biology as the most parsimonious solution.

Making an msa is quite easily done for alike sequences, but the msa then does not have much information about which sequence positions are conserved and which are varying—information that is often of biological interest. Conversely, the more diverse the sequences, the more difficult it is to align them into an msa, and many alignments might be equally probable. The best situation is somewhere in between in which the sequences are similar enough that a believable msa can be produced but variable enough that there is information on amino acid changes. Regardless of the likeness of the sequences, msa editors can be used to adjust the computer-aligned positions manually based on experience with sequence variation.

A situation that often arises is that two or more sequences of the set are alike and the rest of the sequences are much more diverse. Usually, the msa program will align the alike sequences into a consensus; the other less-alike sequences will then be aligned against this consensus. The resulting msa will not provide a good indication of which positions are changing in the evolution of these sequences but will instead show what changes are needed to align the diverse sequences with the consensus of the conserved sequences. One way around this problem is to down-weight the contribution of the alike sequences to the msa. Although weighting the sequences in an alignment makes the most evolutionary sense and is usually implemented in modern msa programs, this is not always done. Computer scientists have also developed programs that base alignments on the most-alike sequences in the set based on the expectation that the sequences are a representative, unbiased sample.

In the genome era, the evolutionary relationships among the sequences can be estimated, allowing a test of this assumption. If the alike genes are from diverse sources, the assumption is a good one (i.e., the conserved sequence has been around a long time), but if sequences are from closely related species, the assumption is not a good one. Thus, careful attention needs to be given to the results of using msa programs that do not weight sequences according to their similarity. An alternative method for obtaining more representative msa's is to avoid focusing on global msa's that include all of the sequences, but instead to concentrate on a local alignment of the sequences in which only conserved domains are aligned, avoiding the more variable regions.

The dynamic programming algorithm described for aligning pairs of sequences in Chapter 3 is also used in most msa programs. However, aligning more than three sequences at the same time, even short sequences, is computationally extremely difficult. Most methods simplify the problem by using dynamic programming to align two sequences or two groups of previously aligned sequences, which is a simple extension of aligning two sequences. The most-alike sequences are aligned first, but then sequence weighting is used to make the msa represent the likely evolutionary changes in all of the sequences. Another computational method used to produce a local msa is to find common patterns in a set of sequences by statistical methods or pattern-searching algorithms. These patterns may then be used as a starting point for alignment using dynamic programming. Such combinations of local and global alignment approaches provide rapid and accurate methods of producing msa's.

A more recently developed method uses a graphical representation of the msa called a partial-order graph. A graph is a mathematical representation of relationships among a set of objects. Nodes in the graph, representing the objects, are joined by edges, which indicate relationships among the objects. A sequence may be represented by a joined set of nodes going from the beginning to the end of the sequence. The msa is represented by collapsing columns of identical sequence positions into a line of nodes and columns with substitutions and gaps indicated by loops along this line. The edges joining these nodes represent the variation found in the sequences going from the beginning to the end of the msa (a partial-order alignment, or POA). The graph, as a mathematical representation of relationships among a set of objects, keeps track of previous alignments as new sequences are added to an existing msa and allows a very rapid alignment of a large number of sequences, e.g., an EST collection using dynamic programming.

Often, conserved patterns or domains in the sequences are located, and the sequences are aligned by these patterns without being concerned about the more variable parts of the sequences. These regions may be found in global msa's that include all of the sequences described above. Another method is to search for conserved patterns in the sequences to produce local msa's. There are well-established computational methods for finding such patterns. However, as with whole sequences, one would like to find patterns that are similar enough to be aligned but also different enough to reveal variable positions in the pattern. One solution has been to apply complex statistical methods to locate such conserved but variable patterns in the sequences. The patterns can then be aligned into a local msa.

A local msa can be conveniently represented by a table (a scoring matrix or position-specific scoring matrix) whose columns represent the columns in the msa and whose rows represent the proportion of each amino acid (or base for nucleic acid sequences) converted into a probability value. Gaps in the alignment can also be represented in the matrix, in which case the matrix is called a profile. The local msa can also be represented by a hidden Markov model (HMM), which has "match states" corresponding to the aligned column and other states that allow for insertions and deletions in the alignment. These matrix and HMM representations of local msa's are very useful in searches for the pattern in other sequences. Understanding how these methods work requires a review of how probabilities are calculated, as discussed in Chapter 4.

There are important applications for msa programs in assembling whole sequences from short fragments (chromosomes from sequence reads or mRNAs from ESTs). In this case, the goal is to find overlaps between sequences that are convincing enough to combine them into a contig. This topic is discussed in greater detail in Chapters 2 and 11.

CHAPTER GUIDE FOR COMPUTATIONAL SCIENTISTS

An important biological concept introduced previously is that different organisms frequently use the same set of proteins to perform the major biological functions of living systems, including synthesis and breakdown of molecules for energy, multicellular development, and reproduction. These proteins are made using sequence information encoded in a four-letter base code in the DNA of the cell's chromosomes that is translated into the 20-letter code for the amino acid sequence of proteins. Using the universal genetic code as a guide, the amino acid sequence of the proteins can be reliably predicted from the DNA sequence. With this information, it is relatively easy for biologists to discover these conserved proteins in diverse organisms.

Starting with the known sequence of a protein in one organism and a prediction as to which codons in the universal code are most commonly used for each amino acid in a second organism, small DNA probes (primers) can be made to sequence genes that encode a similar protein in the second organism using polymerase chain reaction (PCR) technology. Similar proteins in sequenced genomes can readily be found just by translating the predicted genes and searching for protein sequence similarity. Either method reveals the presence of genes in diverse organisms that encode a similar protein.

Finding the same gene in different organisms provides the biologist with several types of useful information. First, the mere presence of a recognizable protein reveals that the organism is using that same protein probably for the same biological function. Second, if the protein sequences can be successfully aligned into an msa, the msa reveals which sequence positions remain the same and which vary. The conserved positions are important for function and the variable ones are probably not; but they also do not interfere with the function of the protein, since changes that did interfere with function were lost when the organism carrying them did not survive. Third, the number of changes between the sequences in the msa can be used to compute which ones are most closely related in an evolutionary sense. The numbers can be used to produce a binary tree that represents the evolutionary relationships among the proteins and often (but not always) reflects the evolutionary relationships among the organisms that are producing them. The use of msa's for producing evolutionary trees is discussed in Chapter 7. msa programs that align sets of related sequences are usually designed to optimize recovery of these types of information.

In addition to aligning entire protein sequences, biologists are also frequently interested in finding proteins that share localized regions of similarity that may include a few to several hundred amino acids. Proteins can be aligned based on the presence of these conserved sequence domains without concern for the rest of the sequence. Frequently, domains correspond to a unit of biological function, biochemical activity, and three-dimensional structure. Proteins may have several copies of the same domain presumably to amplify its function, and new domain combinations often appear that represent a new type of function. Copies of genes are also produced, providing a duplicate that can evolve into a new function. This process can lead to a family of related proteins that have related but somewhat different functions. For example, they may provide the organism with the ability to regulate different cell processes by a similar molecular mechanism.

A final use of a special class of msa programs is to assemble 500-nucleotide-long sequence reads into compatible overlapping sets called contigs. One use arises in genome sequencing projects. Sometimes, after collecting DNA sequences, random fragments are assembled without a guide as to where they are located in the chromosome using an msa program that keeps track of all of the overlaps and assembles a contiguous sequence representing the original genome sequence. Most smaller genomes are now sequenced this way. Another use of msa programs is in the assembly of mRNA sequences from sequence fragments. Recall that mRNAs are copies of the protein-encoding genes that are processed and then translated into proteins. Because of the difficulty of making DNA copies of mRNAs, biologists frequently sequence random DNA fragments of all cellular mRNAs. A special msa program for finding all of the overlaps then is used to assemble a sequence representing part or all of the original mRNA sequence.

WHAT SHOULD BE LEARNED IN THIS CHAPTER?

- Reasons for performing msa.

- The complexity of performing msa's and the computational approaches that are used.

- The difference between msa's that include the whole of each sequence (global msa's) and msa's that include only localized regions of similarity (local msa's).

- Methods used to perform a global msa and their limitations.

- Locating and representing localized regions of similarity in three or more sequences using pattern-searching and statistical methods and knowing these methods.

- Description of the position-specific scoring matrix (PSSM), the profile, and the hidden Markov model (HMM) for representing local msa's and how they are used in sequence similarity searches.

- How to evaluate the ability of PSSMs, profiles, and profile HMMs to discriminate sequences that have a matching pattern from those that do not.

- Basics of uncertainty and information analysis.

Glossary Terms

Algorithm is a computational procedure for solving a problem that has been analyzed thoroughly to ensure that it provides a reasonably correct solution to the problem.

Bacterial artificial chromosomes (BACs) are artificially constructed chromosomes that multiply in bacterial cells as they divide and contain one chromosomal fragment about 100 kb in length from a target organism in a cloning project. Enough BACs are made to produce a redundant set of overlapping fragments from which the original chromosomal sequence may be derived.

Baum–Welch algorithm is a special case of the expectation maximization method that is designed to find the best estimates of parameters during the training of a hidden Markov model.

Blocks are ungapped patterns of amino acids that are present in related proteins.

Contig is an assembled set of overlapping DNA sequence fragments.

Directed acyclic graph is a class of data structures that has an ordered set of vertices (nodes) joined by edges such that no path through the graph starts and ends at the same vertex. A partial-order graph used for producing an msa is an example.

Domain (sequence analysis) generally refers to an extended sequence pattern found by sequence alignment methods. The domain may include all of a given protein sequence or only a portion of the sequence. Some domains are complex and made up of several smaller domains that became joined to form a larger one during evolution.

e value is a parameter used in the multiple sequence alignment program MSA. For a given sequence pair, *e* is the difference between the score of the alignment of that pair in the msa and the score of the optimal pair-wise alignment.

ESTs (expressed sequence tags) are partial sequences of cDNA copies of mRNA molecules.

Expect score (E) is the number of local alignment scores that are expected between a large number of unrelated or random sequences of a given length and are as high as or exceed the score found between two sequences of that same length that are being tested for similarity. E has a similar meaning in other sequence analysis contexts; e.g., when the score for matching a sequence to a scoring matrix is being evaluated.

Expectation maximization (sequence analysis) An algorithm for locating similar sequence patterns in a set of sequences. A guessed alignment of the sequences is first used to generate an expected scoring matrix representing the distribution of sequence characters in each column of the alignment. This pattern is matched to each sequence, and the scoring matrix values are then updated to maximize the alignment of the matrix to the sequences. The procedure is repeated until there is no further improvement.

Forward–backward algorithm The forward algorithm is a method for summing the probabilities of all possible alignments of a sequence with a hidden Markov model in the forward direction. The backward algorithm is used for a similar purpose but starts at the end of the sequence. Together, these algorithms provide a probability that a given part of the model, e.g., a particular match state, is used in alignments with a sequence.

Genetic distance (between sequences) is the number of changes that must be made in one sequence to turn it into another in a sequence alignment. Gaps and commonly observed substitutions are usually not counted.

Genome is the entire set of genetic material encoded in the DNA sequence of an organism including coding sequences for macromolecules and noncoding sequences.

Genome mapping is a determination of the physical gene order in genomes through the identification of overlapping DNA sequence fragments, usually on BACs.

Gibbs sampling method (sequence analysis) is an algorithm for finding conserved patterns within a set of related sequences. A guessed alignment of all but one sequence is made and used to generate a scoring matrix that represents the alignment. The matrix is then matched to the left-out sequence, and a probable location of the corresponding pattern is found. This prediction is then input into a new alignment and another scoring matrix is produced and tested on a new left-out sequence. The process is repeated until there is no further improvement in the matrix.

Global multiple sequence alignment is a multiple sequence alignment that includes most or all of each sequence in the alignment.

Hidden Markov models (HMM for sequence analysis) is a probabilistic model of an msa of proteins, but can also be a model of periodic patterns in DNA sequences, representing, for example, patterns found at the boundaries of introns and exons in genes. In a model of msa, each column of symbols in the alignment is represented by a frequency distribution of the symbols called a state, and insertions and deletions by other states. One then moves through the model along a particular path from state to state in a Markov chain (random choice of next move) trying to match a given sequence. The next matching symbol is chosen from each state, recording its probability (frequency) and also the probability of going to that particular state from a previous one (the transition probability). State and transition probabilities are then multiplied to obtain a probability of the given sequence. The hidden nature of the HMM is due to lack of information about the value of a specific state that is instead represented by a probability distribution over all possible values.

Hill climbing is a nonoptimal search algorithm that works by progressively trying to reach a computationally difficult

goal, e.g., producing an optimal msa by identifying intermediate steps toward that goal (e.g., a progressive or iterative alignment), and then moving to those intermediate steps (or states). The solution may result in a locally best solution that is not a globally best solution. The method works best when there is a good way to score progress toward an optimal solution.

Information theory (in sequence analysis) is an analysis of variation in the columns of a position-specific scoring matrix representing the variation found in columns of an msa. The information content is a measure of how useful that matrix will be for finding another sequence that correctly aligns with the msa as opposed to making a false positive prediction. If one matrix column is strongly biased toward one sequence character, then the information content of that column is high since a matching sequence character may be readily identified, and if a variety of different characters are present, then it is low because there is uncertainty. Information content is determined from the number of questions that must be asked to acquire the correct information. For example, to identify an unknown base in DNA sequences, two questions must be asked; i.e., is it AG or CT, and then is it A or G, or is it C or T, depending on the first answer. If only AG is possible, then only one question need be asked, and if only G, then no question need be asked. A DNA scoring matrix plays the role of reducing the number of questions that must be asked. The amount of information varies from 2 (e.g., just G present) to 0 (e.g., all bases present equally), depending on any residual uncertainty.

Local multiple sequence alignment is a multiple sequence alignment that includes localized regions of sequence similarity in the alignment.

Markov chain is a series of states with a given probability of making a transition from one state to the next that does not depend on previous states. An example is a Markov model of sequence change during evolution in which each sequence position changes independently of previous ones. Another is found in hidden Markov models of an msa in which the next matching state to a sequence is not influenced by what happened in a previous state.

Motif (sequence analysis) refers to a conserved pattern of amino acids that is found in two or more proteins. A motif may be an amino acid pattern that is found in a group of proteins that have a similar biochemical activity, and that is often near the active site of the protein. Sometimes, motif is used in certain computer programs to mean an extended sequence pattern, and in this context is a synonym for domain.

Multiple sequence alignment (msa) is an alignment of three or more sequences such that each column of the alignment is an attempt to represent the evolutionary changes in one sequence position, including substitutions, insertions, and deletions.

Partial-order graph is a graphical representation of sequence and sequence alignments in which sequence positions or aligned columns are presented by a row of nodes joined by edges that extend from the start of the sequence or msa to the end. Insertions and deletions are represented by branches of nodes extending from this line. These branches make the graph deviate from being a one-dimensional, linear-order graph.

Phylogenetic tree is a binary tree representing the ancestor relationships among a group of related nucleic acid or protein sequences. Strongly similar sequences will be represented by adjacent outer branches on the tree joined to a common node, whereas more distantly related sequences will be represented by deeper branches joining additional nodes. Branch lengths are the number of changes between adjacent nodes in the tree.

Position-specific scoring matrix (PSSM) represents the variation found in the columns of an alignment of a set of related sequences. Each subsequent matrix column corresponds to the next column in the alignment and each row corresponds to a particular sequence character (one of four bases in DNA sequences or 20 amino acids in protein sequences). Matrix values are log odds scores obtained by dividing the counts of the residue in the alignment, dividing by the expected number of counts based on sequence composition, and converting the resulting ratio (odds score) to a log (log odds score). The matrix is moved along sequences to find similar regions by adding the matching log odds scores and looking for high values. There is no allowance for gaps. Also called a weight matrix or scoring matrix.

Prior model is used in a Bayesian statistical analysis to represent prior knowledge of some variable. For example, in sequence analysis, a range of possible evolutionary distances between sequences may be possible based on some biological evidence. Another example is an estimate of the amino acid distribution of a match state in a hidden Markov model of an msa. The prior probabilities are then used in Bayes' theorem along with newly acquired information; e.g., new sequence alignments to produce a new estimate of the state probabilities, called posterior probabilities. In this manner, the HMM is trained using a group of related sequences.

Profile is a scoring matrix representation of a conserved region in the msa that allows for gaps in the alignment. The rows include scores for matching sequential columns of the alignment to a test sequence. The columns include substitution scores for amino acids and gap penalties. This use of rows and columns can be reversed. The profile is moved along sequences using dynamic programming to find the optimal alignment at each sequence position, and high-scoring alignments indicate a match of the sequence to the profile.

Program is a series of statements that instruct the computer to perform some computational method.

Progressive multiple sequence alignment is a procedure for generating an msa that reduces the construction of the msa to a series of pair-wise alignments. Initially, a dynamic programming alignment is made between the two most-alike sequences, and the resulting alignment is then extended to include other, less-alike sequences.

Regularization refers to a set of techniques for reducing data overfitting when training a model such as a hidden Markov model with a related family of sequences. A small amount of model variation will often improve the matching of the model to more variable family members.

Sequence weighting (in multiple sequence alignment) refers to the application of an alignment score reducing factor to lower the dominant effect of high scores between related sequences and thus help to make the msa more representative of the evolutionary changes in all of the sequences.

Shotgun cloning refers to obtaining the sequence of long DNA molecules by breaking them into random fragments, joining them to other DNA molecules that act as artificial chromosomes in bacteria by recombinant DNA technology (cloning), sequencing the fragments without regard to their position in the original sequence, and then assembling the original sequence based on overlaps among the fragments.

Shotgun sequencing refers to the procedure used to obtain the sequence of long DNA molecules by breaking them into random fragments, sequencing the fragments without regard to their position in the original sequence, and then assembling the original sequence based on overlaps among the fragments.

Sum of pairs in the msa of a set of sequences is the sum of the alignment scores for each pair of sequences in the msa.

Viterbi algorithm is a computational method similar to the dynamic programming algorithm for sequence alignment. In this case, the most probable path of a test sequence through a hidden Markov model is found. This information shows the best alignment of the sequence with the msa that is represented by the HMM.

z score is a statistical score for a variable, calculated as the difference between the raw score of the variable and the mean of a set of values for the variable, divided by the standard deviation of that set.

HOW MULTIPLE SEQUENCE ALIGNMENT IS USED

Just as the alignment of a pair of nucleic acid or protein sequences can reveal whether there is a functional and evolutionary relationship between the sequences, so can the alignment of three or more sequences reveal similar relationships among a family of sequences. If the structure of one or more members of a protein msa is known, for example, it may be possible to predict which amino acids occupy the same spatial relationship in other proteins through their positions in the alignment. Other uses of an msa include searches for additional sequences and genome sequencing.

Searches for Additional Sequences

Consensus information retrieved from an msa is also used for the design of specific DNA probes to identify other members of the same group or family of similar sequences in different organisms. In the laboratory, the consensus sequence may be used to design PCR primers for amplification of related sequences. The consensus can be used as a query sequence in database-searching programs to find other sequences with a similar pattern.

Genome Sequencing

In genome sequencing projects, large chromosomal fragments are first cloned into bacterial artificial chromosome (BAC) cloning vectors, and then overlapping BAC sequences are identified. The BAC sequences are then subcloned into smaller fragments for sequencing. msa programs can be useful for assembling a set of smaller sequence DNA fragments derived from a BAC into a longer linear sequence (a contig). Alternatively, in the shotgun cloning method, instead of cloning and arranging a very large number of BAC fragments derived from a large DNA molecule, and then moving along the molecule and sequencing the BAC fragments in order, random fragments of the large molecule are sequenced, and those that overlap are found by msa programs designed for this purpose. This approach enables automated assembly of large sequences. Bacterial

genomes have been quite readily sequenced by this method, and it has also been used to assemble a draft of the much larger *Drosophila* and human genomes at Celera Genomics (Weber and Myers 1997, and see Chapter 2).

A related use of msa in genome studies is in predicting the mRNA sequences of the expressed genes in a cell. It is difficult to make a complete DNA copy of an mRNA sequence, but it is quite possible to make a set of partial sequences called expressed sequence tags (ESTs) that represent different parts of the mRNA sequence. ESTs are made from an mRNA preparation without purifying the individual mRNA molecules. msa is then used to assemble partial DNA copies of cell mRNAs into contigs that represent the mRNA sequences of the cell.

The requirements for the msa program for genome projects differ in several respects from those for general sequence analysis. First, the sequences are fragments of the same large sequence molecule, and the sequences of overlapping fragments should be the same except for sequence copying and reading errors, which may introduce the equivalent of substitutions and insertions/deletions between the compared fragments. Thus, there should be one correct alignment that corresponds to that of the genome sequence instead of a range of possibilities. Second, the sequences may be from one DNA strand or the other, and hence the complements of each sequence must also be compared. Third, sequence fragments will usually overlap, but by an unknown amount, and, in some cases, one sequence may be included within another. Finally, all of the overlapping pairs of sequence fragments must be assembled into a large, composite genome sequence, taking into account any redundant or inconsistent information.

Interested readers may wish to consult a description of the type of methodology for genome assembly analysis (Myers 1995, and see Chapter 11) and a comparison of the methods, including several commercial packages that are useful for managing the sequence data from laboratory sequencing projects (Miller and Powell 1994). The Institutue of Genome Research (http://www.tigr.org/) has also developed and made available software and methods for genome assembly and analysis.

MULTIPLE SEQUENCE ALIGNMENTS ARE STARTING POINTS FOR PHYLOGENETIC ANALYSES

Once the msa has been found, the number or types of changes in the aligned sequence residues may be used for a phylogenetic analysis. The alignment provides a prediction as to which sequence characters correspond. Each column in the alignment predicts the mutations that occurred at one site during the evolution of the sequence family, as illustrated in Figure 5.1. Within the column are original characters that were present early, as well as other derived characters that appeared later in evolutionary time. In some cases, the position is so important for function that mutational changes are not observed. It is these conserved positions that are useful for producing an alignment. In other cases, the position is less important, and substitutions are observed. Deletions and insertions may also be present in some regions of the alignment. Thus, starting with the alignment, one can hope to dissect the order of appearance of the sequences during evolution.

MULTIPLE SEQUENCE ALIGNMENT CAN BE GLOBAL OR LOCAL

Just as alignment of two sequences can be a global alignment of the entire sequences or a local alignment of just locally conserved regions in these sequences, these two types of alignment also are used in msa. In a global msa, pair-wise alignment is extended to include three or more related sequences. For example, protein sequences may be conserved in their entirety through evolutionary change, performing some important biological function in closely or distantly related organisms. These conserved sequences can be recognized using a global msa method that attempts to align the sequences. Alternatively, functional domains in protein sequences may be conserved, while the remaining sequence diverges to play new functional roles. Similarly, local conservation of patterns in DNA sequences represents conserved regulatory information in genomes. These patterns may be found by local msa methods designed to search for these conserved domains. For example, aligned promoters of a set of similarly regulated genes may reveal consensus binding sites for regulatory proteins.

Representative examples of programs for producing global or local msa's are listed in Table 5.1. A flowchart illustrating the considerations to be made in choosing a multiple sequence alignment method is also shown on page 173.

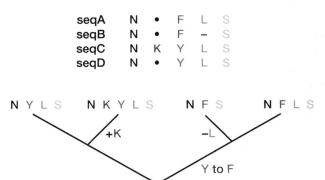

FIGURE 5.1. The close relationship between msa and phylogenetic analysis or evolutionary tree construction. Shown is a short section of one msa of four protein sequences including conserved and substituted positions, an insertion (of K), and a deletion (of L). Below is a hypothetical evolutionary tree that could have generated these sequence changes. Each outer "branch" in the tree represents one of the sequences. The outer branches are also referred to as "leaves." The deepest, oldest branch is that of sequence D, followed by A, then by B and C. The optimal alignment of several sequences can thereby be thought of as minimizing the number of mutational steps in an evolutionary tree for which the sequences are the outer branches or leaves. The mathematical solution to this problem was first outlined by Sankoff (1975). Fast msa programs that are tree-based have since been developed (Ravi and Kececioglu 1998). However, such an approach depends on knowing the evolutionary tree to perform an alignment, and often this is not the case. Usually, pair-wise alignments are generated first and then used to predict the tree. In this example, the alignment could be explained by several different trees, including the one shown, following one of several types of analyses described in Chapter 7. The sequences then become the outer leaves of the tree, and the inner branches are constructed by this analysis.

TABLE 5.1. *Examples of programs for multiple sequence alignment*

Name	Uses	Reference
Global alignments including progressive		
CLUSTALW	standard progressive alignment	Thompson et al. (1994a,1997)
CLUSTALX (graphical interface)	most useful for similar sequences	Higgins et al. (1996)
MAFFT: rapid multiple sequence alignment based on Fourier transform (progressive and iterative programs)	fast, accurate msa alignments with novel scoring systems	Katoh et al. (2002)
MAVID for progressive msa of genome sequences	progressive alignment method for large numbers of DNA sequences with viewer	Bray and Pachter (2003)
MSA	optimal alignment using dynamic programming—limited to few/short sequences	Lipman et al. (1989); Gupta et al. (1995)
MULTIPIPMAKER	produces percent identity plot of multiple DNA sequences	Schwartz et al. (2003)
POA	fast, accurate alignment of large numbers of sequences (ESTs) by partial-order graphs	Lee et al. (2002)
PRALINE	versatile tool kit for producing msa's by different strategies	Heringa (1999); Simossis and Heringa (2003)
T-COFFEE	uses CLUSTALW method but with pair-wise alignments to increase accuracy; flexible	Poirot et al. (2003)
Iterative and other methods		
DIALIGN	segment alignment; very accurate msa method for DNA and protein sequences; aligns based on matching segments without gap penalties	Morgenstern et al. (1998)
PRRP	progressive global alignment method repeatedly improves msa; produced by progressive alignment using command line options	Gotoh (1996)
SAGA	genetic algorithm; user intense method based on biologically relevant method	Notredame and Higgins (1996)
Local alignments of proteins		
Aligned Segment Statistical Evaluation Tool (ASSET)	sophisticated pattern-finding and statistical analysis method—command line	Neuwald and Green (1994)
BLOCKS Web site	finds blocks (ungapped domains) by pattern search or Gibbs sampling	Henikoff and Henikoff (1991, 1992)
eMOTIF Web server	useful analysis of protein families to find most significant patterns in families	Nevill-Manning et al. (1998)
GIBBS, the Gibbs sampler statistical method	finds patterns in unaligned sequences by statistical method—command line	Lawrence et al. (1993); Liu et al. (1995); Neuwald et al. (1995)
HMMER	hidden Markov model software tools for producing a profile hidden Markov model to represent an msa	Eddy (1998)
MACAW, a workbench for multiple alignment construction and analysis	aligner/editor for locating and adjusting local alignment blocks on PC	Schuler et al. (1991)
MEME Web site, expectation maximization method	locates localized sequence blocks "motifs" by statistical method	Bailey and Elkan (1995); Grundy et al. (1996, 1997); Bailey and Gribskov (1998)
Profile analysis at UCSD	produces a sequence profile from an msa	Gribskov and Veretnik (1996)
SAM hidden Markov model Web site	produces an HMM for an msa	Krogh et al. (1994); Hughey and Krogh (1996)

GLOBAL MULTIPLE SEQUENCE ALIGNMENT

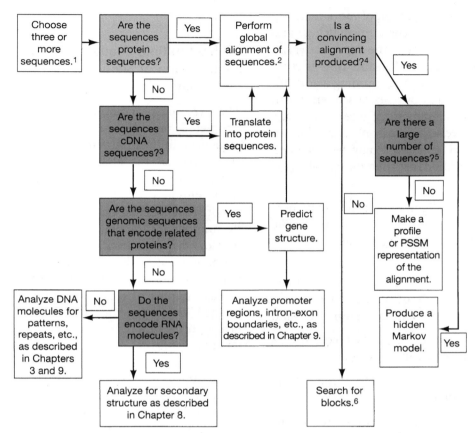

1. Sequences chosen for analysis may have been found by sequence similarity searches (Chapter 6) or from laboratory experiments. Complex features of the sequences, including repeated or low-complexity regions that interfere with alignments, can be analyzed as described in Chapters 3 and 6. The flowchart describes the production of three classes of multiple sequence alignments:
   a. global alignments that include most or all of each sequence in the alignment,
   b. local alignments of common patterns in protein sequences that include matches and mismatches in each column but no gaps are present (a position-specific scoring matrix [PSSM] is produced),
   c. local alignments of common patterns in protein sequences that include matches, mismatches, insertions, and deletions to make a type of scoring matrix called a profile or a probabilistic model called a hidden Markov model (HMM).
   Methods for finding common patterns in DNA sequences are discussed in Chapter 9.

2. Examples of global alignment programs are given in Table 5.1.

3. cDNA sequences of the same gene from a closely related group of organisms may be multiply aligned by a global method so that synonymous (i.e., change the amino acid) and nonsynonymous (i.e., do not change the amino acid) sequences may be analyzed, as described in Chapter 7.

4. A convincing alignment should include a series of columns in which a majority of the sequences have the same amino acid or an amino acid that is a conservative substitution for that amino acid, with relatively few examples of other substitutions or gaps in these columns. These columns of alike amino acids should be found throughout the alignment, often clustered into domains. There may also be variable regions in the alignment that represent sequences that diverged more during evolution of the protein family.

5. This decision rests on whether or not there are enough sequences on which to build an HMM of the entire alignment or of a well-defined region in the alignment (a profile HMM). For sequences that are related but show considerable variations in many columns, as many as 100 sequences may be needed to produce an HMM of the alignment. This number is reduced to ~25–50 if there is less variation among the sequences. A scoring matrix representing the sequence variation found in each column of the alignment may also be made. These matrices may accommodate gaps in the alignment (a profile or HMM profile) or may not include gaps (PSSM).

6. For finding patterns common to a set of sequences, pattern-searching algorithms and statistical methods (expectation maximization and Gibbs sampling) are used.

Global Multiple Sequence Alignment Is Computationally Challenging

Finding a global optimal alignment of more than two sequences that includes matches, mismatches, and gaps and that takes into account the degree of variation in all of the sequences at the same time is especially difficult. The dynamic programming algorithm used for optimal alignment of pairs of sequences can be extended to global alignment of three sequences, but for more than three sequences, only a small number of relatively short sequences may be analyzed. Thus, approximate methods are used for global alignments, including:

1. Progressive global alignment of the sequences starting with an alignment of the most-alike sequences and then building an alignment by adding more sequences.

2. Iterative methods that make an initial global alignment of groups of sequences and then revise the alignment to achieve a more reasonable result.

3. Alignments based on locally conserved patterns found in the same order in the sequences.

4. Statistical methods that generate probabilistic models of the sequences.

5. Producing the msa by using graph-based methods.

Global msa's are also difficult to score. Only when an appropriate scoring system has been identified can an optimal alignment be produced. In aligning pairs of sequences, an alignment score can be computed based on log odds scores for identical characters in the alignment, substitution scores for nonidentical positions, and gap penalties. In an msa, multiple comparisons must be made in each column of the msa including a cumulative score for the substitutions in columns of an msa and the placement and scoring of gaps. The placement and scoring of substitutions and gaps in global msa's are also more difficult for more divergent sequences.

Globally aligning a set of closely related sequences derived from a recent common ancestor sequence is relatively easy. However, if the sequences share a more complex and distant evolutionary relationship, then obtaining an msa that reflects the correct ancestry relationships is challenging. The situation is even more difficult if some of the sequences are alike and others are more divergent. The correct msa should reflect the correct ancestry of the divergent sequences and not be unduly influenced by the closely related ones. Solutions to these problems have been identified and are addressed in the following sections.

Global Multiple Sequence Alignment Is an Extension of Sequence Pair Alignment by Dynamic Programming

The dynamic programming algorithm described in Chapter 2 provides an optimal alignment of two sequences. In the program MSA (Lipman et al. 1989), application of the global alignment algorithm has been extended to provide an optimal alignment of a small number of sequences greater than 2. The number of sequences that can be aligned is limited because the number of computational steps and the amount of memory required grow exponentially with the number of sequences to be analyzed. This limitation means that the program has somewhat limited application to a small number of sequences. For example, MSA 2.1 on the University of Washington server will produce a near-optimal msa of up to eight nucleic acids or protein sequences each 500 amino acids long. The supercomputer center at the University of Pittsburgh can handle up to ten sequences, each 1,000 amino acids long. Caution is needed in using these implementations of the MSA program, since Gupta et al. (1995) have shown that the MSA program rarely produces a provable optimal alignment.

When the dynamic programming method of sequence alignment is used for two sequences, it builds a dynamic programming scoring matrix where each position provides the best alignment up to that point in the sequence comparison. The number of comparisons that must be made to fill this matrix without using any short cuts and excluding gaps is the product of the length of the two sequences. Now imagine extending this analysis to three or more sequences. For three sequences, instead of the two-dimensional matrix for two sequences, we now have to fill the inside lattice of a cube with optimal scores. Scoring positions on three surfaces of the cube represent the alignment values between a pair of the sequences, ignoring the third sequence, as illustrated in Figure 5.2. In the MSA program, positions inside the lattice of the cube are given values based on the sum of the initial scores of the three pairs of sequences.

For three protein sequences each 300 amino acids in length and excluding gaps, the number of compar-

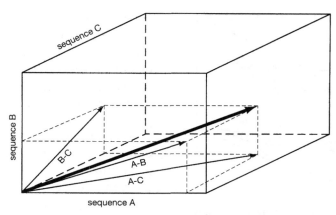

FIGURE 5.2. Alignment of three sequences by dynamic programming. Arrows on the surfaces of the cube indicate the direction for filling in the scoring matrix for pairs of sequences, A with B, etc., performed as previously described. The alignment of all three sequences requires filling in the lattice of the cube space with optimal alignment scores following the same algorithm. The best score at each interior position requires a consideration of all possible moves within the cube up to that point in the alignment. The trace-back matrix will align positions in all three sequences including gaps.

isons to be made by dynamic programming is equal to $300^3 = 2.7 \times 10^7$, whereas only $300^2 = 9 \times 10^4$ comparisons are required for two sequences of this length. This number is sufficiently small that alignment of three sequences by this method is practical. For alignment of more than three sequences, one has to imagine filling an N-dimensional space or hypercube. The number of steps and memory required for a 300-amino-acid sequence (300^N, where N is the number of sequences) then becomes too large for most practical purposes, and it is necessary to find a way to reduce the number of comparisons that must be made without compromising the attempt to find an optimal alignment. Fortunately, Carrillo and Lipman (1988) found such a method, called the sum of pairs, (SP) method. Since the publication of the MSA program, Gupta et al. (1995) have substantially reduced the memory requirements and number of steps required using the SP method. The enhanced version of MSA is available by anonymous FTP from fastlink.nih.gov/pub/msa.

The basic idea of the SP method is that an msa imposes an alignment on each of the pairs of sequences. The heavy arrow in Figure 5.2 represents the path followed inside the cube to find an msa for three sequences, but the msa can be projected onto the sides of the cube, thus defining an alignment for each pair of sequences. The alignments found for each pair of sequences likewise impose bounds on the location of the msa within the cube, and thus define the number of positions within the cube that have to be evaluated. Pair-wise alignments are first computed between each pair of sequences. Next, a trial msa is produced by first predicting a phylogenetic tree for the sequences (Saitou and Nei 1987; see Chapter 7 for the neighbor-joining method of tree construction), and the sequences are then multiply aligned in the order of their relationship on the tree. This method is used by other programs described below (e.g., PILEUP, CLUSTALW, T-COFFEE) and provides a heuristic alignment that is not guaranteed to be optimal. However, the alignment serves to provide a limit to the space within the cube within which optimal alignments are likely to be found. In Figure 5.3, the green area on the left surface of the cube is bounded by the optimal alignment of sequences B and C and a projection of the heuristic alignment for all three sequences. The orange and blue areas are similarly defined for other sequence pairs. The dark gray volume within the cube is bounded by projections from each of the three surface areas. For more sequences, a similar type of analysis of bounds may be performed in the corresponding higher-order space. Thus, the SP method assists in producing a global msa by limiting the number of dynamic programming scores that must be computed.

In practice, MSA calculates the multiple alignment score within the cube lattice by adding the scores of the corresponding pair-wise alignments in the msa. This measure is known as the SP measure (for sum of pairs), and the optimal alignment is based on obtaining the best SP score. These scores may or may not be

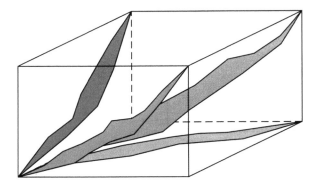

FIGURE 5.3. Bounds within which an optimal alignment will be found by MSA for three sequences. For MSA to find an optimal alignment among three sequences by the DP algorithm, it is only neccessary to calculate optimal alignment scores within the gray volume. This volume is bounded on one side by the optimal alignments found for each pair of sequences, and on the other by a heuristic multiple alignment of the sequences. The colored areas on each cube surface are two-dimensional projections of the gray volume.

weighted so as to reduce the influence of more closely related sequences in the msa. The Dayhoff PAM250 matrix and an associated gap penalty are used by MSA for aligning protein sequences. MSA uses a constant penalty for any size of gap and scores gaps according to the scheme illustrated in Figure 5.4 (Altschul 1989; Lipman et al. 1989). MSA calculates a value ε for each pair of sequences that provides an idea of how much of a role the alignment of those two sequences plays in the msa. ε for a given sequence pair is the difference between the score of the alignment of that pair in the msa and the score of the optimal pair-wise alignment. The bigger the value of ε, the more divergent the msa from the pair-wise alignment and the smaller the contribution of that alignment to the msa. For example, if an extra copy of one of the sequences is added to the alignment project, then ε for sequence pairs that do not include that sequence will increase, indicating a lesser role because the contributions of that pair have been outvoted by the alike sequences (Altschul et al. 1989). Weighting the sequence pairs is designed to get around the common difficulty that some pairs in most sets of sequences are similar. Another score δ is the sum of the εs and gives an indication of the degree of divergence among the sequences—closely related sequences will have low εs and δs and distantly related sequences will have high εs and δs.

The MSA program avoids the bias in an align-

ment due to alike sequences by weighting the pair-wise scores before they are added to give the SP score. These weights are determined by using the predicted tree of the sequences discussed above. The pair-wise scores between all sequence pairs are adjusted to reduce the influence of the more unlike sequence pairs that occupy more distant "leaves" on the evolutionary tree (i.e., by sequences that are joined by more branches) based on the argument that these sequence pairs provide less useful information for computing the msa. This scheme is different from that used by other msa programs (see below), which generally increase the weight of scores from more distant sequences because these sequences represent greater divergence in the evolutionary tree (see Vingron and Sibbald 1993). This increased weighting by the MSA program to the most-alike sequences limits its usefulness to less divergent sequences. However, an optimal global alignment is assured.

In using MSA, several additional practical considerations should be considered (described on MSA Web sites):

1. MSA is a heavy user of machine resources and is limited to a small number of sequences of relatively short lengths.

2. In the UNIX command line mode of the program, there are options that allow users to speci-

		Natural gap cost	Quasi-natural gap cost
sequence 1	x – – – x		
sequence 2	x x – x x	3	4
sequence 3	x x x x x		

FIGURE 5.4. Method of scoring gap penalties by the msa program MSA. x indicates aligned residues, which may be a match or a mismatch, and – indicates a gap. In this example, each gap cost is 1, regardless of length. The "natural" gap cost is the sum of the number of gaps in all pair-wise combinations (sequences 1 and 2, 1 and 3, and 2 and 3). Note that the alignment of a gap of three in sequence 1 with a gap of length one in sequence 2 scores as gap of 1 because the gap in sequence 1 is longer. The quasi-natural gap cost is the natural cost for the gap plus an additional value for any gap that begins and ends within another. In this example, there is an additional penalty score for the presence of a single gap in sequence 2 that falls within a larger gap in sequence 1. The inclusion of this extra cost for a gap has little effect on the alignments produced but provides an enormous reduction in the amount of information that must be maintained in the DP scoring matrix (Altschul 1989), thus making possible the simultaneous alignment of more sequences by MSA.

fy gap costs, force the alignment of certain residues, specify maximum values for *e*, and tune the program in other ways.

3. When the output shows that some ε values are greater than the respective maximum ε, a better alignment usually can be found by increasing the maximum ε in question. However, increasing ε also increases the computational time.

4. If the program bogs down, try dividing the problem into several smaller ones by using fewer sequences or partial sequences.

Below is an example from http://www.psc.edu of using MSA to align a group of phospholipase a2 pro-

teins. Note that the program uses the FASTA sequence format. The following steps are used:

1. Calculate all pair-wise alignment scores (alignment costs).
2. Use the scores (costs) to predict a tree.
3. Calculate pair weights based on the tree.
4. Produce a heuristic msa based on the tree.
5. Calculate the maximum ε for each sequence pair.
6. Determine the spatial positions that must be calculated to obtain the optimal alignment.
7. Perform the optimal alignment.
8. Report the ε found compared to the maximum ε.

EXAMPLE OF MSA

```
MSA release 2.1  (PSC revision b) started on Thu Jun 19 14:55:31 1997
Sequence file format is Fasta.
Calculating pairwise alignments.
..........
**********
Calculating weights.

---------------- Tree given from ancestor ----------------
On the left:    Internal Node   Distance to parent =  278.83
On the left:    Internal Node   Distance to parent =   23.63
On the left:    Internal Node   Distance to parent =  118.62
On the left:    SEQ#01          Distance to parent =  230.50
On the right:   SEQ#04          Distance to parent =  205.50
On the right:   SEQ#05          Distance to parent =  238.37
On the right:   SEQ#02          Distance to parent =  256.17
On the right:   SEQ#03          Distance to parent =    0.00


Calculating epsilons.


Sequence    ID        Description
    1       SEQ#01    P1;1POA Phospholipase a2 (EC 3.1.1.4) - Chinese cobra
    2       SEQ#02    P1;1POD Phospholipase a2 (EC 3.1.1.4) - human
    3       SEQ#03    P1;1PPA Phospholipase a2 (EC 3.1.1.4) lys 49 variant
    4       SEQ#04    P1;1BPQ phospholipase A2 (EC 3.1.1.4) mutant (K56M) -
    5       SEQ#05    P1;1PP2R phospholipase A2 (EC 3.1.1.4) (calcium-free)


            ***  Heuristic Multiple Alignment  ***

**********23541    *****************************35214 **35214       ***
NLYQFKNMIQCTVPSR-SWWDFADYGCYCGRGGSGTPVDDLDRCCQVHDNCYNEAEKISGC-----WPYFKTYSY
NLVNFHRMIK-LTTGKEAALSYGFYGCHCGVGGRGSPKDATDRCCVTHDCCYKRLEK-RGC-----GTKFLSYKF
SVLELGKMIL-QETGKNAITSYGSYGCNCGWGHRGQPKDATDRCCFVHKCCYKKLT---DC-----NHKTDRYSY
ALWQFNGMIKCKIPSSEPLLDFNNYGCYCGLGGSGTPVDDLDRCCQTHDNCYKQAMKLDSCKVLVDNPYTNNYSY
SLVQFETLIM-KIAGRSGLLWYSAYGCYCGWGGHGLPQDATDRCCFVHDCCYGKAT---DC-----NPKTVSYTY
```

```
********35214 ****************14325
ECSQGTLTCKGGNNACAAAVCDCDRLAAICFAG--APYNDNDYNINLKARC-------
SNSGSRITC-AKQDSCRSQLCECDKAAATCFARNKTTYNKKYQYYS-NKHCRGSTPRC
SWKNKAIIC-EEKNPCLKEMCECDKAVAICLRENLDTYNKKYKAYF-KLKCKKPDT-C
SCSNNEITCSSENNACEAFICNCDRNAAICFSK--VPYNKEHKNLD-KKNC-------
SEENGEIIC-GGDDPCGTQICECDKAAAICFRDNIPSYDNKYWLFP-PKDCREEPEPC


Calculating pairwise projection costs.
..........
*********


Calculating multiple alignment.
....1....2....3....4....5....6....7....8....9....0
**************************************************

          ***  Optimal Multiple Alignment  ***


NLYQFKNMIQCTVPSR-SWWDFADYGCYCGRGGSGTPVDDLDRCCQVHDNCYNEAEKISGC-----WPYFKTYSY
NLVNFHRMIK-LTTGKEAALSYGFYGCHCGVGGRGSPKDATDRCCVTHDCCYKRLEK-RGC-----GTKFLSYKF
SVLELGKMIL-QETGKNAITSYGSYGCNCGWGHRGQPKDATDRCCFVHKCCYKKL---TDC-----NHKTDRYSY
ALWQFNGMIKCKIPSSEPLLDFNNYGCYCGLGGSGTPVDDLDRCCQTHDNCYKQAMKLDSCKVLVDNPYTNNYSY
SLVQFETLIM-KIAGRSGLLWYSAYGCYCGWGGHGLPQDATDRCCFVHDCCYGKA---TDC-----NPKTVSYTY


ECSQGTLTCKGGNNACAAAVCDCDRLAAICFAG--APYNDNDYNINLKARC-------
SNSGSRITC-AKQDSCRSQLCECDKAAATCFARNKTTYNKKYQYYS-NKHCRGSTPRC
SWKNKAIIC-EEKNPCLKEMCECDKAVAICLRENLDTYNKKYKAYF-KLKCK-KPDTC
SCSNNEITCSSENNACEAFICNCDRNAAICFSK--VPYNKEHKNLD-KKNC-------
SEENGEIIC-GGDDPCGTQICECDKAAAICFRDNIPSYDNKYWLFP-PKDCREEPEPC


End gaps not penalized.
Costfile:              pam250
Alignment cost:   35132    Lower bound:    34945
Delta:              187    Max. Delta:       285
```

Sequences		Proj. Cost	Pair. Cost	Epsilon	Max. Epsi.	Weight	Weight*Cost
1	2	1864	1825	39	39	1	1864
1	3	1891	1843	48	57	1	1891
1	4	1654	1653	1	5	4	6616
1	5	1814	1787	27	28	2	3628
2	3	1735	1733	2	8	4	6940
2	4	1876	1866	10	10	1	1876
2	5	1713	1712	1	8	2	3426
3	4	1901	1889	12	21	1	1901
3	5	1648	1648	0	11	2	3296
4	5	1847	1842	5	6	2	3694

```
Elapsed time =   0.895


Tree
    A tree is given for the heuristic alignment (not shown).
```

Scoring Global Multiple Sequence Alignments

The scoring method used for producing a multiple sequence alignment can influence which sequence positions are aligned in the msa. The SP method used by the MSA program is one such method. Several additional methods for scoring msa's have been proposed: a tree-based method, star phylogeny, information content, and the trace method. However, many of these methods have not been implemented by msa programs. As discussed above, the SP method provides a way to score the msa by summing the scores of all possible combinations of amino acid pairs in a column of an msa. The method assumes a model for evolutionary change in which any of the sequences could

be the ancestor of the others, as illustrated in Figure 5.5. This figure also illustrates a difficulty with the SP method when a substitution table of log odds scores such as BLOSUM62 is used for protein sequences (see Durbin et al. 1998, pp. 139–140). Shown is the effect of adding a small number of amino acid substitutions to a column that initially has all matching amino acids. Scores in the msa column decrease rapidly as the number of mismatched residue pairs increases. For a larger number of sequences than 5 with all N, or with one or two C substitutions, these decreases should be greater because there will be more N-N matched pairs relative to mismatched N-C pairs.

However, the reverse is true with the SP method of scoring. For n sequences, the number of combinations of pairs in a column is $n(n-1)/2$. If all are amino acids

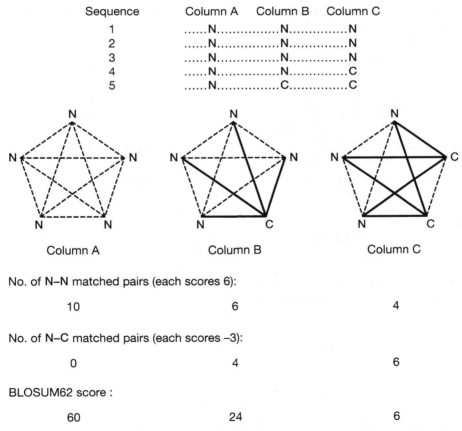

FIGURE 5.5. The SP model for scoring an msa. This model represents one method for optimizing the msa by maximizing the number of matched pairs (or minimizing the cost or number of mismatched pairs) summed over all columns in the msa. Shown first are three columns of a five-sequence msa with all matched (*A*), four matched and one mismatched (*B*), or three matched and two mismatched (*C*) sequence characters. The SP method of calculating the cumulative scores for columns of an msa is then illustrated by a graph with the five sequences as vertices and representing the ten possible sequence pair-wise sequence comparisons. Solid lines represent a mismatched pair and dotted lines a matched pair. Shown are the BLOSUM62 scores for each column calculated by the SP method. (Adapted, with permission, from Altschul 1989 [©1989 Elsevier].)

N, as in column A, then the BLOSUM62 score for the column is $6 \times n(n-1)/2$. If there is one C in the column, as in column B, then $n-1$ matched N-N pairs will be replaced by $n-1$ mismatched N-C pairs, giving a score of $[6-(-3)][n-1] = 9 \times (n-1)$ less. The score for one C in the column divided by that for zero Cs is $9(n-1)/[6n(n-1)/2] = 3/n$. For three sequences, the relative difference is 1, whereas for six sequences, the relative difference is 2. As more sequences are present in the column, the relative difference increases, not in agreement with expectation. Hence, the SP method does not provide a reasonable result when this type of scoring matrix is used.

Two other methods for scoring an msa (Altschul 1989) have been described and are illustrated in Figure 5.6. The first is a tree-based method. Because a phylogenetic tree describing the relationships among the sequences is found by the program MSA, the sum of the lengths of the tree branches can be calculated using the substitutions in the column of the msa. Alternatively, a simplified tree with one of the sequences as the ancestor of all of the others (a star phylogeny) can also be used.

Other scoring methods include information content (see p. 213) and a graph-based method called the trace method (Kececioglu 1993). A novel branch-and-cut algorithm for msa has been developed based on the trace method (Kececioglu et al. 2000). Other methods of scoring and producing an alignment guided by a tree are described below.

Using Progressive Methods for Global Multiple Sequence Alignment

Progressive alignment methods use the dynamic programming method to build an msa starting with the most related sequences and then progressively adding less-related sequences or groups of sequences to the initial alignment (Waterman and Perlwitz 1984; Feng and Doolittle 1987, 1996; Thompson et al. 1994a; Higgins et al. 1996). Relationships among the sequences are modeled by an evolutionary tree in which the outer branches or leaves are the sequences (Fig. 5.7). The tree is based on pair-wise comparisons of the sequences using a phylogenetic method as described in Chapter 7. Progenitor sequences represented by the inner branches of the tree are derived by alignment of the outermost sequences. These inner branches will have uncertainties where positions in the outermost sequences are dissimilar, as illustrated in Figure 5.7.

Three examples of programs that use progressive methods are CLUSTALW, the Genetics Computer Group program PILEUP, and T-COFFEE. These programs are discussed in the following sections.

CLUSTALW

CLUSTAL has been in use since 1988, and the authors have done much to support and improve the program (Higgins and Sharp 1988; Thompson et al. 1994a; Higgins et al. 1996). CLUSTALW is a more recent version of CLUSTAL, with the W standing for "weighting" to represent the ability of the program to provide weights to the sequence and program parameters; CLUSTALX provides a graphic interface (see Table 5.1). These program changes provide more realistic alignments that should reflect the evolutionary changes in the aligned sequences and the more appropriate distribution of gaps between conserved domains. CLUSTALW is designed to provide an adequate alignment of a large number of more closely

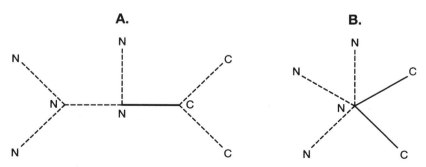

FIGURE 5.6. Alternative methods for scoring a column in the msa (Altschul 1989). The variations in column C of Fig. 5.5 are shown modeled by a phylogenetic tree (*A*) and a simplified phylogenetic tree called a star phylogeny (*B*) where one of the sequences is treated as the ancestor of all the others (instead of treating them as all equally possible ancestors as in the original sum of pairs scoring method).

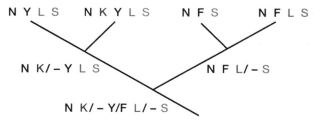

FIGURE 5.7. Progressive sequence alignment. Sequences are represented as the outermost branches (leaves) on an evolutionary tree. The most closely related sequences are first aligned by dynamic programming, providing a representation of ancestor sequences in deeper branches with uncertainties where amino acids have been substituted or positioned opposite a gap. These sequences are the same as those shown in Fig. 5.1. The challenge to the msa method is to utilize an appropriate combination of sequence weighting, scoring matrix, and gap penalties so that the correct series of evolutionary changes may be found.

related sequences and a reliable indication of the domain structure of those sequences. CLUSTALW also has options for adding one or more additional sequences with weights or an alignment to an existing alignment (Higgins et al. 1996). Once an alignment has been made, a phylogenetic tree may be made by the neighbor-joining method, with corrections for possible multiple changes at each counted position in the alignment (see Chapter 6). The predicted trees may also be displayed by various programs described in Chapter 6.

The steps used by CLUSTALW include (1) perform pair-wise alignments of all of the sequences; (2) use the alignment scores to produce a phylogenetic tree (for an explanation of the neighbor-joining method used, see Chapter 7); and (3) align the sequences sequentially using the dynamic programming algorithm, guided by the phylogenetic relationships indicated by the tree. Thus, the most closely related sequences are aligned first, and then additional sequences and groups of sequences are added, guided by the initial alignments to produce an msa that shows in each column the sequence variations among the sequences. The initial alignments used to produce the guide tree may be obtained by a fast *k*-tuple or pattern-finding approach similar to FASTA that is useful for many sequences, or a slower, full dynamic programming method may be used. An enhanced dynamic programming alignment algorithm (Myers and Miller 1988; see Chapter 3) is used to obtain optimal alignment scores. For producing a phylogenetic tree, genetic distances between the

sequences are computed. The genetic distance is the number of mismatched positions in an alignment divided by the total number of matched positions (positions opposite a gap are not scored).

As with the MSA program, contributions of sequences to the msa are weighted according to their relationships on the predicted evolutionary tree. Sequences that are similar to others in the group will be represented by adjacent branches on the tree. The scores for aligning these sequences will be given a small weight so that they have a less significant effect on the alignment. Conversely, sequences that are less similar to others will be represented by more distant branches on the tree. Their scores will be given a larger weight so that they have a greater influence on the alignment. The result of this weighting scheme is that the alignment will more accurately reflect sequence changes in the evolutionary tree. To compute the weights, a rooted tree with known branch lengths of which the sequences are outer branches (leaves) is examined (see Chapter 7). Weights are based on the distance of each sequence from the root, as illustrated in Figure 5.8. The alignment scores between two positions in the msa are then calculated using the resulting weights as multiplication factors.

The scoring of gaps in the msa has to be performed in a different manner from scoring gaps in a pair-wise alignment because as more sequences are added to a profile of an existing msa, gaps accumulate and influence the alignment of further sequences in an unsuitable way (Thompson et al. 1994b; Taylor 1996). CLUSTALW calculates gaps in a novel way designed to place them between conserved domains. When Pascarella and Argos (1992) aligned sequences of structurally related proteins, the gaps were preferentially found between secondary structural elements. These authors also prepared a table of the observed frequency of gaps next to each amino acid in these regions. CLUSTALW uses the information in this table and also attempts to locate what may be the corresponding domains by appropriate gap placement in the msa.

Like other alignment programs, CLUSTAL uses a penalty for opening a gap in a sequence alignment and an additional penalty for extending the gap by one residue. Default values recommended by the progam are provided, but gap penalties may also be used. Gaps found in the initial alignments remain fixed. New gaps introduced as more sequences are added also receive this same gap penalty, even when they occur within an existing gap, but the gap penal-

A. Calculation of sequence weights

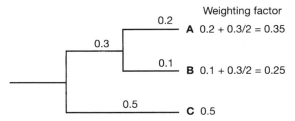

Weighting factor

A 0.2 + 0.3/2 = 0.35

B 0.1 + 0.3/2 = 0.25

C 0.5

B. Use of sequence weights

Column in alignment 1

Sequence A (weight a)   ....K.........

Sequence B (weight b)   ....I.........

Column in alignment 2

Sequence C (weight c)   ....L.........

Sequence D (weight d)   ....V.........

Score for matching these two columns in an msa =

[a x c x score (K,L) +
a x d x score (K,V) +
b x c x score (I,L) +
b x d x score (I,V)] / 4

FIGURE 5.8. Weighting scheme used by CLUSTALW (Higgins et al. 1996). (*A*) Sequences that arise from a unique branch deep in the tree receive a weighting factor equal to the distance from the root. Other sequences that arise from branches shared with other sequences receive a weighting factor that is less than the sum of the branch lengths from the root. For example, the length of a branch common to two sequences will only contribute one-half of that length to each sequence. Once the specific weighting factors for each sequence have been calculated, they are normalized so that the largest weight is 1. As CLUSTALW aligns sequences or groups of sequences, these fractional weights are used as multiplication factors in the calculation of alignment scores. (*B*) Illustration of using sequence weights for aligning two columns in two separate alignments. Note that this sequence weighting scheme is the opposite to that used by MSA, because the more distant a sequence from the others, the higher the weight given. For a comparison of additional weighting schemes, see Vingron and Sibbald (1993).

ties for an alignment are then modified according to the average match value in the substitution matrix, the percent identity between the sequences, and the sequence lengths (Higgins et al. 1996). These changes are attempts to compensate for the scoring matrix, expected number of gaps (alignment with more identities should have fewer gaps), and differences in sequence length (should limit placement of gaps if

one sequence shorter).

Tables of gaps are then calculated by CLUSTALW for each group of sequences to be aligned to confine them to less conserved regions in the alignment. Gap penalties are decreased where gaps already occur (increased in regions adjacent to already gapped regions), decreased within stretches of hydrophilic regions (amino acids DEGKNQPRS), and increased or decreased according to the table in Pascarella and Argos (1992). These rules are most useful when a correct alignment of some of the sequences is already known. Another method for achieving this same result is to enhance the scores of more closely matching regions on the alignment, as described in Taylor (1996). The CLUSTALW algorithm and the results of using the above sequence weighting gap adjustment method are illustrated in Figure 5.9.

PILEUP

PILEUP is the msa program that is a part of the Genetics Computer Group (GCG) package of sequence analysis programs, owned since 1997 by Oxford Communications, and is widely used as a result of its popularity and availability. PILEUP uses a method for msa that is very similar to CLUSTALW. The sequences are aligned pair-wise using the Needleman–Wunsch dynamic programming algorithm, and the scores are used to produce a tree by the unweighted pair-group method using arithmetic averages (UPGMA; Sneath and Sokal 1973, and see Chapter 7). The resulting tree is then used to guide the alignment of the most closely related sequences and groups of sequences. The resulting alignment is a global alignment produced by the Needleman–Wunsch algorithm. Standard scoring matrices and gap opening/extension penalties are used. Unfortunately, there have not been any recent enhancements of this program such as sequence weighting or gap modifications comparable to those introduced for CLUSTALW. Hence, PILEUP does not have the capability of CLUSTALW to reduce the dominating effect of related sequences on the msa and to give emphasis to gaps in more poorly aligning regions, and is of more limited value.

T-COFFEE

T-COFFEE is an advanced progressive alignment program that uses a system of sequence position weights to generate an msa that is most consistent

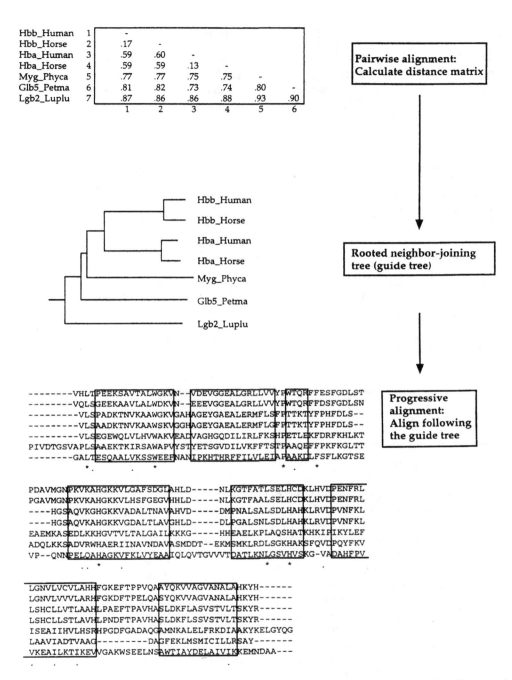

FIGURE 5.9. A multiple sequence alignment of seven globins by CLUSTALW. The protein identifiers are from the SwissProt database. The amino acid substitution matrix was the Dayhoff PAM250 matrix, and gap penalties were varied to emphasize conserved ungapped regions. The approximate and known locations of seven α-helices in the structure of this group are shown in boxes. (Reprinted, with permission, from Higgins et al. 1996 [©1996 Elsevier].)

with pair-wise alignments of all of the component sequences (T-COFFEE stands for Tree-based Consistency based Objective Function For alignmEnt Evaluation). The method, illustrated in Figure 5.10, begins by making an optimal global alignment of all possible sequence pairs using the CLUSTAL program and a set of possible local alignments for each sequence pair using the LALIGN program (see Chapter 3, p. 91). The best alignments found are then used by the program to devise a weighting system that

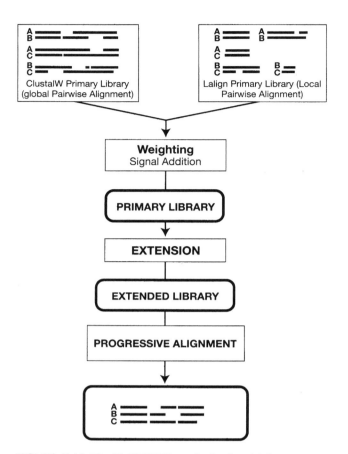

FIGURE 5.10 The T-COFFEE method of multiple sequence alignment. The program starts with both global and local alignments of all sequence pairs and then uses these to produce libraries of alignments that are weighted based on percent identity. These libraries are then extended to weight each position pair in each aligned pair of sequences (A and B). These extended weights are found by examining the pair-wise alignments of A and B with each of the other sequences (ACB, ADB, etc.) to find the most conserved positions. The sequences are then progessively aligned starting with the most-alike sequences by dynamic programming using as a scoring system the weights found above. The alignment of this pair of sequences including gaps is then fixed as other sequences are added to the progressive msa. When including a group of sequences in the alignment, the average scores in each column of the alignment are used. There is no penalty for gaps. (Reprinted, with permission, from Notredame et al. 2000 [©2000 Elsevier].)

can be used as a guide for producing the msa by the progressive method. T-COFFEE is better than CLUSTALW at reproducing known alignments of related proteins (approximately 50% identity) but is much slower. Other programs described below produce a better alignment than T-COFFEE of proteins that are less similar (approximately 25% identity).

Problems with Progressive Global Alignment

The major problem with progressive alignment programs is that the ultimate msa depends on the initial pair-wise sequence alignments. The very first sequences to be aligned using this method are the most closely related ones on the sequence tree. If these sequences align very well, there will be few errors in the initial alignments. However, the more distantly related these sequences are, the more errors will be made, and these errors will be propagated to the msa. There is no simple way to circumvent this problem. T-COFFEE attempts to circumvent this problem by a more judicious weighting of sequence pairs using information from pair-wise alignments. A second problem with the progressive alignment method is the choice of suitable scoring matrices and gap penalties that apply to the set of sequences (Higgins et al. 1996; Lee et al. 2002).

Lee et al. (2002) have pointed out an additional problem with the progressive method. As sequences are added to the msa, the current msa is represented by an alignment profile that shows a combination of matches, substitutions, and gaps in each column. As a new sequence is aligned to a profile column by dynamic programming, the weighted scores for each substitution in the column are used to find the optimal score for matching the sequence to the profile at that position in the profile and sequence. This choice is made without regard to the scores used at neighboring positions in the profile. The result is that the alignments previously found and used to produce the scoring profile are not being used to produce the new alignment. Rather, previous alignment information has been lost in the profile, which cannot be used to regenerate the earlier alignments in the progressive msa. A new graphical method that tracks previous alignments and avoids this problem (Lee et al. 2002) is described below.

Using Iterative Methods for Global Multiple Sequence Alignment

The major problem with the progressive alignment method described above is that errors in the initial alignments of the most closely related sequences are propagated to the msa. This problem is more acute when the starting alignments are between more distantly related sequences. Iterative methods attempt to correct for this problem by repeatedly realigning subgroups of the sequences and then by aligning these subgroups into a global alignment of all of the sequences. The objective is to improve the overall alignment score, such as an SP score. Selection of these groups may be based on the ordering of the sequences on a phylogenetic tree predicted in a manner similar to that of progressive alignment, separation of one or two of the sequences from the rest, or a random selection of the groups. These methods are compared in Hirosawa et al. (1995).

Three programs that use iterative methods are MultiAlin, PRRP, and DIALIGN. MultiAlin (Corpet 1988) recalculates pair-wise scores during the production of a progressive alignment and uses these scores to recalculate the tree, which is then used to refine the alignment in an effort to improve the score. The program PRRP (Table 5.1) uses iterative methods to produce an alignment. An initial pair-wise alignment is made to predict a tree. The tree is then used to produce weights for making alignments in the same manner as the MSA program except that the sequences are analyzed for the presence of aligned regions that include gaps rather than being globally aligned. These regions are iteratively recalculated to improve the alignment score. The best-scoring alignment is then used in a new cycle of calculations to predict a new tree, new weights, and new alignments, as illustrated in Figure 5.11. The program repeats this process until there is no further increase in the alignment score (Gotoh 1994, 1995, 1996).

The program DIALIGN (see Table 5.1) finds an alignment by a different iterative method. Pairs of sequences are aligned to locate aligned regions that do not include gaps, much like continuous diagonals in a dot matrix plot. Diagonals of various lengths are identified. A consistent collection of weighted diagonals that provides an alignment, which is a maximum sum of weights, is then found. The result is an alignment of the sequences based on alignment of these weighted diagonals. Additional methods that use iterative methods—genetic algorithms, partial-order graphs, and hidden Markov models—are described in detail in the following sections.

Genetic Algorithm

The genetic algorithm is a general type of machine-learning algorithm that has no direct relationship to biology and that was invented by computer scientists. The method has been recently adapted for msa by Notredame and Higgins (1996) in a computer program package called SAGA (Sequence Alignment by

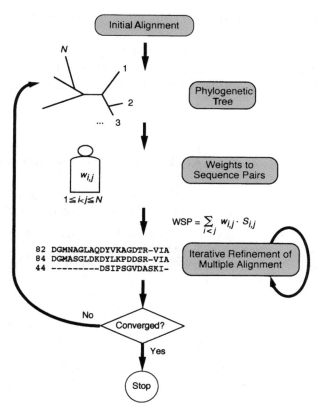

FIGURE 5.11 The iterative procedures used by PRRP to compute a multiple sequence alignment. (Reprinted, with permission, from Gotoh 1996 [© 1996 Elsevier].)

Genetic Algorithm; see Table 5.1). Zhang and Wong (1997) have developed a similar program. The method is of considerable interest because the algorithm can find high-scoring alignments that are as good as those found by other methods such as CLUSTALW. Similar genetic algorithms have been used for RNA sequence alignment (Notredame et al. 1997) and for prediction of RNA secondary structure (Shapiro and Navetta 1994). Although the method is relatively new and not used extensively, it likely represents the first of a series of sequence analysis programs that produce alignments by attempted simulation of the evolutionary changes in sequences. Genetic algorithms are quite complex, and user experience with them is necessary.

The basic idea behind SAGA is to try to generate many different msa's by rearrangements that simulate gap insertion and recombination events during replication to generate a higher and higher score for the msa. The alignments are not guaranteed to be optimal (highest scoring that is achievable). Although SAGA can generate alignments for many sequences, the program is slow for more than about 20 sequences.

A similar approach for obtaining a higher-scoring msa by rearranging an existing alignment uses a probability approach called simulated annealing (Kim et al. 1994). The program MSASA (Multiple Sequence Alignment by Simulated Annealing) starts with a heuristic msa and then changes the alignment by following an algorithm designed to identify changes that increase the alignment score.

STEPS IN THE SAGA GENETIC ALGORITHM FOR GLOBAL SEQUENCE ALIGNMENT

The success of the genetic algorithm may be attributed to the steps used to rearrange sequences, many of which might be expected to have occurred during the evolution of the protein family. The steps in the algorithm are as follows:

1. The sequences to be aligned (up to approximately 20 in number) are written in rows, as on a page, except that they are made to overlap by a random amount of sequence, up to 50 residues long for sequences about 200 in length. The ends are then padded with gaps. A typical population of 100 of these msa's is made, although other numbers may be set. Shown below is an initial msa for the genetic algorithm (1 of approximately 100 in number):

    ```
    XXXXXXXXXX----
    ---XXXXXXXXXXXX
    -XXXXXXXXXX---
    ```

2. The 100 initial msa's are scored by the SP method, using both natural and quasi-natural gap-scoring schemes (see Fig. 5.4). Standard amino acid scoring matrices and gap opening and extension penalties are used by SAGA.

3. These initial msa's are now replicated to give another generation of msa's. The half of the replicates with the lowest SP scores are sent to the next generation unchanged. The remaining half for the next generation are selectively chosen by lot, like picking marbles from a bag, except that the chance for a particular choice is inversely proportional to the msa score (the lower the score, the better the msa, therefore giving that one a greater chance of replicating).

    This latter one-half of the choices for the next generation is now subject to mutation, as described in step 4 below, to produce the children of the next generation. All members of the next-generation msa's undergo recombination to make new child msa's derived from the two parents, as described in step 5 below. The relative probabilities of these separate events are governed by program parameters. These parameters are also adjusted dynamically as the program is running to favor those processes that have been most useful for improving msa scores.

4. In the mutation process, the sequence is not changed (else it would no longer be an alignment), but gaps are inserted and rearranged in an attempt to create a better-scoring msa. In the gap insertion process, the sequences in a given msa are divided into two groups based on an estimated phylogenetic tree, and gaps of random length are inserted into random positions in the alignment. Alternatively, in a "hill-climbing" version of the procedure, the position is so chosen as to provide the best possible score following the change.

    ```
    XXXXXXXXXX              XXX--XXXXXXX
    XXXXXXXXXX              XXX--XXXXXXX
    XXXXXXXXXX    ---->      XXXXXXXXXX--X
    XXXXXXXXXX              XXXXXXXXXX--X
    XXXXXXXXXX              XXXXXXXXXX--X
    ```

    Shown above are random gap insertions into phylogenetically related sequences. The first two and last three sequences comprise the two related groups in this example. X indicates any sequence character.

    Another mutational process is to move common blocks of sequence (overlapping ungapped regions) delineated by a gap, or blocks of gaps (overlapping gaps). Some of the possible moves are illustrated below. These moves may also be tailored to improve the alignment score.

```
XXX--XXXXX        XX--XXXXXX        XXX--XXXXX        XXXXX--XXX
XXXXXXXXXX        XXXXXXXXXX        XXXXXXXXXX        XXXXXXXXXX
XX--XXXXXX        X--XXXXXXX        XXX--XXXXX        XX-XX-XXXX
XXXXXXXXXX        XXXXXXXXXX        XXXXXXXXXX        XXXXXXXXXX
```

Starting block   Whole block   Split block   Split block
                   move   horizontal   vertically
                                       (guided by
                                       phylogenetic grouping)

5. Recombination among next-generation parent msa's is accomplished by one of two mechanisms. The first is not homology-driven. One msa is cut vertically through, and the other msa is cut in a staggered manner that does not lose any sequence after the fragments are spliced. The higher scoring of the two reciprocal recombinants is kept. The second mechanism, illustrated below, is recombination between msa's driven by conserved sequence positions. It is driven by homology expressed as a vertical column of the same residue and is very much like standard homologous recombination.

```
xxGxxxxDxx        xxGxx-xDxx        xxGxx-xDxx
xxGx-xxDxx        xxGxxxxDxx        xxGxxxxDxx
xxGxx-xDxx        xxGxxxxDxx        xxGxxxxDxx
xxGxxxxDxx        xxGx-xxDxx        xxGx-xxDxx
```

Parent A   **Parent B**   **Child**
alignment   alignment   alignment

6. The next generation, an overlapping one of the previous one-half of the best-scoring parental msa's and the mutated children, is now evaluated as in step 2, and the cycle of steps 2–5 is typically repeated as much as 100 times, although as many as 1,000 generations can be run. The best-scoring msa is then kept.

7. The entire process of producing a set of msa's for replication and mutation is repeated several times to obtain several possible msa's, and the best-scoring one is chosen.

Partial-order Graphs

A dramatic improvement in the speed of producing an msa has been achieved by representing sequences and msa's as partial-order graphs, a class of directed acyclic graphs (Lee et al. 2002). An example of a partial-order graph representation of an msa is illustrated in Figure 5.12. The partial-order graph representing an msa can be rapidly aligned with a sequence or with another msa-representing graph by dynamic programming in an amount of time proportional to the average number of branches per node. This method is particularly efficient for sequences that share many identities, as in overlapping sets of EST sequences. Another advantage to this method is that each stage of the alignment can store and use information from previous alignments, in contrast to progressive alignment methods, which use profiles that do not have this information. The method of alignment is illustrated in Figure 5.12C. The program POA (Progressive Order Alignment) for a UNIX or Linux

environment is available from http://www. bioinformatics.ucla.edu/poa/ and a Web page for data input is also available.

Directed acyclic graphs are also used to model gene ontologies, which classify gene functions.

Hidden Markov Models of a Global Multiple Sequence Alignment

The hidden Markov model (HMM) is a probabilistic, statistical model that considers all possible combinations of matches, mismatches, and gaps to generate an alignment of a set of sequences. Both global and local msa's (PROFILE HMM's) may be modeled, and the methods are quite similar. Examples are given below.

Other Programs and Methods for Global Multiple Sequence Alignment

The msa method often used, especially for ten or more sequences, is first to determine sequence simi-

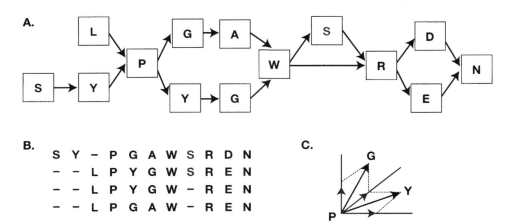

FIGURE 5.12. A partial-order graph representation of an msa (Lee et al. 2002). (*A*) A partial-order graph. The graph has a line of nodes representing columns of conserved sequence positions (*black*) in the msa joined by directed edges representing consecutive sequence letters that run from the start to the end of the graph. Aligned substitutions (*purple*) and an internal insertion (*blue*) in the msa are depicted by loops with edges representing consecutive positions in the divergent sequences. Initial unaligned termini of the sequences are also depicted (*green*). Each pair of nodes is joined by only one edge, but one node can have more than one edge entering or leaving it, thereby serving as a junction. There are no connections between branches. (*B*) The msa represented by the graph. The graph depicts the msa in a compacted form from which the original msa can be derived, since the nodes representing conserved msa positions store information about the location of the character in each sequence. (*C*) Possible moves between cells in the dynamic programming matrix by aligning the graph in *A* at position P with a new sequence, also represented by a graph, by the Smith–Waterman dynamic programming algorithm. Branch joining at this position gives two possible diagonal moves (*purple arrows*, one for each branch), two horizontal moves, and one vertical move (*blue arrows* to each branch). In addition, there is a start move at each cell of score zero that is the graphical equivalent of the scoring system used in the Smith–Waterman local alignment of sequences. In simpler unbranched regions, only the three standard sequence alignment moves are used. Optimal scores are calculated in each cell and the best overall score is found. The optimal alignment between sequence and nodes is then produced by a trace-back procedure similar to that used for sequence alignment until a start move is reached. The existing graph is then updated with the new alignment information according to a set of rules as described in Lee et al. (2002). As more sequences are added to a node, the lists of sequence and aligned sequence positions are also updated. This method does not follow a progressive alignment strategy as do CLUSTALW and T-COFFEE. The order of addition of sequences to the msa can influence the results in regions of low sequence similarity. Although not implemented in the version of POA described here, this information can be used to analyze rearrangements of conserved domains in the sequences.

larity between all pairs of sequences in the set. On the basis of these similarities, various methods are used to cluster the sequences into the most related groups or into a phylogenetic tree. There are several approaches to using these methods.

1. In the group approach, a consensus is produced for each group of sequences and then used to make further alignments between groups. Two examples of programs using the group approach are the program PIMA (Smith and Smith 1992), which uses several novel alignment techniques, and the program MULTAL described by Taylor (1990, 1996; see Table 5.1).

2. The tree method uses the distance method of phylogenetic analysis to arrange the sequences. The two

closest sequences are first aligned, and the resulting consensus alignment is aligned with the next best sequence or cluster of sequences, and so on, until an alignment is obtained that includes all of the sequences. The program CLUSTALW, discussed above, is an example of this approach. The ALIGN set of programs (Feng and Doolittle 1996) and the MS-DOS program by Corpet (1988) use this method. Additional programs for msa are also described in Barton (1994), Kim et al. (1994), and Morgenstern et al. (1996).

3. Another approach (Vingron and Argos 1991) aligns all possible pairs of sequences to create a set of dot matrices, and the matrices are then filtered sequentially to find motifs that provide a

starting point for sequence alignment. A set of programs for interactive msa by dot matrix analysis and other alignment techniques has also been developed (Boguski et al. 1992).

4. The program TREEALIGN takes the approach that msa's should be done in a fashion that simultaneously minimizes the number of changes needed during evolution to generate the observed sequence variation (Hein 1990). TREEALIGN (also named ALIGN in the program versions) performs the alignment and the most parsimonious tree construction at the same time. The initial steps are similar to other msa methods, except that TREEALIGN uses a distance scale: i.e., the sequences are aligned pair-wise and the resulting distance scores are used sequentially to produce a tree, which is rearranged as more sequences are added. The sequences are then realigned so that the same tree can be produced by maximum parsimony. Finally, the tree is rearranged to maximize parsimony. The advantage to this method is the increased use of phylogenetic analysis to improve the msa.

Performance of Global Multiple Sequence Alignment Programs

The performance of global msa programs is commonly assessed by comparing the computed msa with a structural alignment of the proteins and by other objective methods (Notredame et al. 1998). The programs are compared for their ability to reproduce structurally derived alignments from BALiBASE (Thompson et al. 1999b), a database of protein families, each with a known three-dimensional structure and a documented msa alignment based on a scheme of expected sequence changes using the program Rose (Stoye et al. 1998). Reviews on the performance of msa software are given in McClure et al. (1994; progressive alignment methods), Gotoh (1996), and Thompson et al. (1999a). A review of Web sites is given in Briffeuil et al. (1998) and a review on iterative algorithms is given in Hirosawa et al. (1995) and Gotoh (1999).

In a recent set of program comparisons, T-COFFEE slightly outperformed the iterative program PRRP (11–17%) in matching BALiBASE alignments (Notredame et al. 2000). A later review of POA, DIALIGN, T-COFFEE, and CLUSTALW for speed and quality of their alignments was performed (Lassmann and Sonnhammer 2002). Of these, DIALIGN was most accurate for msa's of sequences of low sequence identity, T-COFFEE was best for sequences of high sequence identity, POA was fastest and almost as accurate as DIALIGN and T-COFFEE, and CLUSTALW was only as good as the others for global msa's of sequences with high sequence similarity.

LOCAL MULTIPLE SEQUENCE ALIGNMENTS

The methods of msa described above produce global alignment of the sequences, including most or all parts of all sequences. Local msa methods align the most similar regions in sequences, ignoring dissimilar regions, to produce a local msa using the following approaches.

1. Profile analysis identifies a portion of the alignment produced by a global msa that is highly conserved and produces a type of scoring matrix called a profile. A profile includes scores for amino acid substitutions and gaps in each column of the conserved region. An alignment of the region to a new sequence can be determined using dynamic programming.

2. Block analysis scans a global msa for regions that include only substituted regions without gaps, called blocks, and these blocks are then used in sequence alignments.

3. Pattern-searching or statistical methods find a localized region of sequence similarity in a set of sequences. These methods do not depend on first producing a global msa.

All of these methods for finding localized sequence similarity are discussed below.

Profile Analysis

Profiles are found by performing the global msa of a group of sequences and then removing the more highly conserved regions in the alignment into a smaller msa. A scoring matrix for the msa, called a profile, is then made. The profile is composed of columns much like a mini-msa and includes scores for matches, mismatches, and gap penalties. Once produced, the profile is used to search a target sequence for possible matches to the profile using the scores in the table to evaluate the likelihood at each position. For example, the table value for a profile that is 25 amino acids long will have 25 rows of 20 scores, each score in a row for matching one of the amino acids at the corresponding position in the profile. If a sequence 100 amino acids in length is to be searched, each 25-amino-acid-long stretch of sequence will be examined, 1–25, 2–26, . . . , 76–100. The first 25-amino-acid-long stretch will be evaluated using the profile scores for the amino acids in that sequence, then the next 25-long stretch, and so on. The highest-scoring sections will be the most similar to the profile.

The disadvantage of this method of profile extraction from an msa is that the profile produced is only as representative of the variation in the family of sequences as the msa itself. If several sequences in the msa are similar, the msa and the derived profile will be biased in favor of those sequences. Methods have been devised for partially circumventing this problem with the profile (Gribskov and Veretnik 1996), but the difficulty with the msa itself is not easily reconciled, as discussed at the beginning of this section. Sequence weighting is based on the production of a simple phylogenetic tree by distance methods; more closely related sequences then receive a reduced weight in the profile. Another problem with profiles is that some amino acids may not be represented in a particular column because not enough sequences have been included. Although absence of an amino acid may mean that the amino acid may not occur at that position in the protein family, adding counts to such positions generally increases the usefulness of the profile. This feature is built into the profile method discussed below (Gribskov and Veretnik 1996).

The profile is similar to the log odds form of the amino acid substitution table, such as the PAM250 and BLOSUM62 matrices used for sequence alignments. The profile matrix is 23 columns wide, one column for each of the 20 amino acids, plus one column for an unknown amino acid z and two columns for a gap opening and extension penalty. There is one row for each column in the msa. The consensus sequence, derived from the most common amino acid in each column of the msa, is listed down the left-hand column. The scores on each row reflect the number of occurrences of each amino acid in the aligned sequences. The highest positive score on each row is in the column corresponding to the consensus amino acid, the most negative score for an amino acid not expected at that position. These values are derived from the log odds amino acid substitution matrix that was used to produce the alignment, such as the log odds form of the Dayhoff PAM250 matrix.

Two methods are used to produce profile tables— the average method and the evolutionary method. The evolutionary method seems somewhat better for finding family members. In the average method, the profile matrix values are weighted by the proportion of each amino acid in each column of the msa. For example, if column 1 in the msa has 5 Ile (I), 3 Thr (T), and 2 Val (V), then the frequency of each amino acid in this column is 0.5 I, 0.3 T, and 0.2 V. These amino acids are considered to have arisen with equal probability from any of the 20 amino acids as ancestors. Suppose that they arose from an Ile (I). The profile values in the Ile (I) column of the corresponding row in the profile matrix would then use the amino acid scoring matrix values for I-I, I-T, and I-V, which

SOURCES OF PROFILE ANALYSIS PROGRAMS

A tutorial on preparing profiles, prepared by M. Gribskov, is at Web address http://www.sdsc.edu/projects/profile/profile_tutorial.html, and the Web site at http://www.sdsc.edu/projects/profile/ will perform an analysis on the University of California at San Diego Supercomputer Center. The program Profilemake can be used to produce a profile from an msa (Gribskov et al. 1987, 1990; Gribskov and Veretnik 1996). A version of the Profilesearch program, which performs a database search for matches to a profile, is available at the University of Pittsburgh Supercomputer Center (http://www.psc.edu/general/software/packages/profiless/profiless.html). A special grant application may be needed to use this facility. Profile-generating programs are available by FTP from ftp.sdsc.edu/pub/sdsc/biology and are included in the Genetics Computer Group suite of programs (http://www. gcg.com/), although the more recent features (Gribskov and Veretnik 1996) were not included.

are log odds scores of 5, 0, and 4 in the Dayhoff PAM250 matrix. Then the profile value for column 1 in the profile is the frequency-weighted value, or 0.5 × 5 + 0.3 × 0 + 0.2 × 4 = 3.3. The value 3.3 is converted to an integer (3) to facilitate calculations.

The profile table also includes penalties for matching a gap in the target sequence, shown in the two right columns. All of these table values are multiplied by a constant for convenience so that only the value of a score with one sequence relative to the score with another sequence matters. Once a profile table has been obtained, the table may be used in database searches for additional sequences with the same pattern (program Profilesearch) or as a scoring matrix for aligning sequences (program Profilegap). If several profiles characteristic of a particular protein family can be identified within a global msa of the family, then the chance of a positive identification of additional family members is greatly increased (Bailey and Gribskov 1998; also see http://www.sdsc. edu/MEME).

The evolutionary method for producing a profile table is based on the Dayhoff model of protein evolution (Chapter 2) (Gribskov and Veretnik 1996). The amino acids in each column of the msa are assumed to be evolving at different rates, as reflected in the amount of amino acid variation observed. As with the average method, the object is to consider each of the 20 amino acids as a possible ancestor of the pattern of each column. In the evolutionary model, the evolutionary distance in PAM units required to give the observed amino acid distribution in each column is determined. Each PAM unit represents an overall probability of 1% change in a sequence position. For example, in the original Dayhoff PAM1 matrix for an evolutionary distance of 1 PAM unit (very roughly 10 my), the probability of an I not changing is 0.9872, and the probabilities for I changing to a T or a V are 0.0011 and 0.0057, respectively. All of the probabilities of changing I to any other amino acid add up to 1.0000, for a combined probability of change of 1% for I. For an evolutionary distance of n PAMs, the PAM1 matrix is multiplied by itself n times to give the expected changes at that distance. At a distance of 250 PAMs, the above three probabilities of an I not changing or of changing to a T or V are 0.10, 0.06, and 0.15, respectively, representing a much greater degree of change than for a shorter time (Dayhoff 1978).

Do not confuse these probabilities of one amino acid changing to another in the original Dayhoff PAM250 matrix with scores from the log odds form of the PAM250 matrix, which have been used up to now. The log odds scores are derived from the original Dayhoff matrix by dividing each probability of change with the probability of a chance matching of the amino acids in a sequence alignment. These ratios are then converted to logarithms.

Thus, for the example of the msa column 1 with 5 Ile (I), 3 Thr (T), and 2 Val (V), the object is to find the amount of PAM distance from each of the 20 amino acids as possible ancestors will generate this much diversity. This amount can be found by a formula giving the amount of information (entropy) of the observed column variation given the expected variation in the evolutionary model,

$$H = - \sum_{\text{all } a\text{'s}} f_a \log(p_a) \qquad (1)$$

where f_a is the observed proportion of each amino acid a in the msa column and p_a is the expected frequency of the amino acid when derived from a given ancestor amino acid. For a given column in the msa, H is calculated for each 20 ancestor amino acids and for a large number of evolutionary distances (PAM1, PAM2, PAM4, ...). The distance that gives the minimum value for H for each column–possible ancestor combination is the best estimate of the distance that generates the column diversity from that ancestor. This analysis provides 20 possible models (M_a for a = 1,2,3, . . . , 20) for how the amino acid frequencies in a column (F) may have originated. The next step in the evolutionary profile construction determines the extent to which each M_a predicts F by the Bayes conditional probability analysis:

$$P(M_a|F) = P(M_a) \times P(F|M_a) / \sum_{\text{all } a\text{'s}} [P(M_a)$$
$$\times P(F|M_a)] \qquad (2)$$

where the prior distribution $P(M_a)$ is the given by the background amino acid frequencies and

$$P(F|M_a) = p_{aa1}{}^{faa1} \times p_{aa2}{}^{faa2}$$
$$\times p_{aa3}{}^{faa3} \times \cdots \times p_{aa20}{}^{faa20} \qquad (3)$$

i.e., the product of the expected amino acid frequencies in M_a raised to the power of the fraction observed for each amino acid in the msa column, as defined above. The weights for each of the 20 possible distri-

A

```
rhle_ecoli   GVDVLVATPG  RLLDLEHQNA  ....VKLDQV  EILVLDEADR  MLDMGFIHDI
dbp2_schpo   GVEICIATPG  RLLDMLDSNK  ....TNLRRV  TYLVLDEADR  MLDMGFEPQI
dbp2_yeast   GSEIVIATPG  RLIDMLEIGK  ....TNLKRV  TYLVLDEADR  MLDMGFEPQI
dbpa_ecoli   APHIIVATPG  RLLDHLQKGT  ....VSLDAL  NTLVMDEADR  MLDMGFSDAI
```

B

Cons	A	B	C	D	E	F	G	H	I	K	L	M	N	P	Q	R	S	T	V	W	Y	Z	Gap	Len
G	17	18	0	19	14	-22	31	0	-9	12	-15	-5	15	10	9	6	18	14	1	-15	-22	11	100	100
P	18	0	13	0	0	-12	13	0	8	-3	-3	-1	-2	23	2	-2	12	11	17	-31	-8	1	100	100
H	5	24	-12	29	25	-20	8	32	-9	9	-10	-9	22	7	30	10	0	4	-8	-20	-7	27	100	100
I	-1	-12	6	-13	-11	33	-12	-13	63	-11	40	29	-15	-9	-14	-15	-6	7	50	-17	8	-11	100	100

FIGURE 5.13. An example of an msa (*A*) and the derived evolutionary profile (*B*).

butions that give rise to the msa column diversity are calculated from $P(M_a \mid F)$, as follows:

$$W_a = P(M_a \mid F) - P(M_{random} \mid F) \qquad (4)$$

where W_a is the weight given to M_a and $P(M_{random} \mid F)$ is calculated as above using the background amino acid distribution.

The log odds scores for the profile ($Profile_{ij}$) are given by

$$Profile_{ij} = \log \left[\sum_{all\ a's} (W_{ai} \times p_{aij})/p_{random\ j} \right] \qquad (5)$$

where W_{ai} is the weight of an ancestral amino acid a at row i in the profile, p_{aij} is the frequency of amino acid j in the PAM amino acid distribution that best matches at row i, and $p_{random\ j}$ is the background fre-

quency of amino acid j. An example of an evolutionary profile matrix for the ATP-dependent RNA helicase ("DEAD" box family) from the M. Gribskov laboratory is given in Figure 5.13.

The usefulness of the evolutionary profile is demonstrated by the following: A profile for the 4Fe-4S ferredoxin family was prepared from six sequences. This profile was then used to search the SwissProt database for family members. Success was measured by the so-called receiver operating characteristic test (ROC) plot described in the following box.

The ROC plot is used to measure the accuracy of a profile or scoring matrix for discriminating between members and nonmembers of protein families in a database similarity search. A range of scores is produced by the database search starting with high scores representing strong predictions of family member-

THE ROC PLOT

The ROC plot originated as a method for determining the accuracy of radar detection of enemy aircraft in World War II. The method is now often used to evaluate the accuracy of clinical tests, e.g., deciding whether a person has an illness based on a blood measurement. Suppose that measurements are made of a number of blood samples from 100 normal and 100 diseased individuals and a particular assay is found that discriminates quite well between presence or absence of disease. Suppose further that a low assay value indicates absence of disease and a high value indicates presence of disease. The separation of samples is not complete, however, and some normal samples have a high score for the assay and some disease samples have a low score.

The object of the ROC plot is to calculate how well the assay discriminates diseased from normal individuals. Using the above example, the ability of the test to distinguish these two classes is determined when different cutoff values are used as an indicator of disease. If a very high score is used as an initial cutoff value, then only a few diseased individuals will be scored. As the cutoff value is gradually decreased, then more diseased individuals are added to the initial number found and a few high-scoring normal individuals will start to appear. Eventually, normal individuals will appear in higher numbers and the increase in diseased will slow down. Since the assay value is being used to predict diseased individuals, then all numbers counted are assumed to be from diseased individuals. Numbers of diseased individuals found at each cutoff value are labeled true positive predictions (TP) and the number of normals incorrectly predicted to be diseased are false positive (FP) predictions. The fraction of all known cases that are found at each cutoff value is then calculated by dividing TP and FP by the numbers used in the test; in this example, 100 diseased and 100 normal individuals. A plot is then made of the TP fraction found on the *y* axis against the FP fraction found on the *x* axis. The resulting graph will appear as follows:

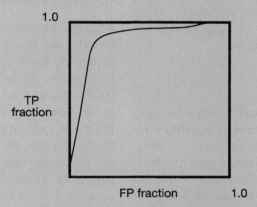

When the chosen scoring system is a good discriminator, the curve defined by the (TP, FP) points will stay close to the *y* axis, turn quite suddenly, and then stay close to the upper horizontal line as the TP fraction increases only slightly more while the FP fraction increases significantly. If the test is not as good, the curve will more closely follow a 45° diagonal across the plot. The overall accuracy of the test is measured by the area between the curve and the *x* axis and gives the probability of a correct result.

In the above example, an intermediate cutoff value at which many TPs but relatively few FPs are included under the curve may be used. Some FPs will be missed, but most diseased individuals should be found if the test is a good one.

The TP fraction is also known as the sensitivity of the test and is also calculated as sensitivity = TP/(TP + FN), where FN is the number of false negatives and TP+FN is the number of diseased individuals (100) in the example used. The FP fraction is related to the specificity of the test, defined as specificity = TN/(TN+FP), where TN is the count of true negative predictions and TN+FP is the number of normal individuals (100) in the example. It is easy to see that FP fraction is 1 – specificity. Sensitivity and specificity of a test and the effect of using different cutoff scores to test a hypothesis are also discussed in Chapter 4 (p. 126).

ship followed by lower scores representing less certain predictions. The first step toward making an ROC plot for these data is to determine how well the profile identifies known protein family members divided by the known number of family members in the searched database (true positive fraction) as opposed to identifying known non-family members incorrectly as family members (false positive fraction) as the scores decrease. A series of cutoff scores are chosen and the number of true and false positives is determined in the range of scores from the highest score to each cutoff score. The scoring of true and false positive fractions for different cutoff scores to produce an ROC plot is illustrated by an example in the neighboring box. The area under the curve and the *x* axis give the probability of correct identification. In a database similarity using a profile to identify known family members, there will be a large number of non-family sequences. Hence, a truncated ROC plot (ROC$_{50}$) is used in which the cutoff score is not reduced any further once 50 known nonfamily members have been found. ROC$_{50}$ is the area under the ROC curve when this point has been reached (Gribskov and Veretnik 1996). For a ferredoxin family search, the ROC$_{50}$, 95.6 ± 0.6% of the known family members, was identified in a search of SwissProt by an evolutionary profile, whereas 93.0 ± 2.0% was identified by the average profile method (Gribskov and Veretnik 1996). The success rate was increased 0.4–0.6% by using 12 training sequences and 2–3% by using 134 training sequences.

The ROC plot is also used to evaluate the accuracy of database similarity searches using scoring matrices produced by the PSI-BLAST program, and the method was further enhanced for statistical comparisons between different search methods (Schäffer et al. 2001). Thus, the ROC plot can be used to evaluate the accuracy of a profile or scoring matrix in finding known protein family members. Profiles take account of matches, mismatches, and gap penalties found in multiple sequence alignments.

Block Analysis

Like profiles, blocks represent a conserved region in an msa. Blocks are different from profiles in that they do not accommodate insertions and deletions in the sequences. Instead, each column includes only matches and mismatches. Blocks, like profiles, may be made by searching for a section of a global msa that is highly conserved. However, in a block analysis, alignable regions may also be found by searching each sequence in turn for similar patterns of the same length. These patterns may include a region with one or a few matching characters followed by a short spacer region of unmatched characters and then by another set of a few matching characters, and so on, until the sequences start to be significantly different. The pattern found in each sequence is the same length, and when the sequences are aligned, the matching sequence characters will appear in columns. The first alignments of this type were performed by computer programs that searched for conserved patterns of the above type in sequences (Henikoff and Henikoff 1991; Neuwald and Green 1994).

Several blocks located in different regions in a set of sequences may be used to produce an msa (Zhang et al. 1994), and blocks may be constructed from a set of aligned sequence pairs (Miller et al. 1994). Statistical and Bayesian statistical methods are also used to locate the most-alike regions of sequences (Lawrence and Reilly 1990; Lawrence et al. 1993). Web sites that perform some of these types of analyses are discussed below and also given in Table 5.1.

Few of these types of analyses presently provide a method for phylogenetic estimates of the sequence relationships so that sequence weighting can be used to make the changes more reflective of the phylogenetic histories among the sequences. Additionally, except where noted, these methods do not use substitution matrices such as the PAM and BLOSUM matrices to score matches. Rather, they are based on

finding exact matches that have the same spacing in at least some of the input sequences, and that may be repeated in a given sequence.

Extraction of Blocks from a Global Multiple Sequence Alignment

A global msa of related protein sequences usually includes no insertions or deletions in any of the sequences. These ungapped patterns may be extracted from these aligned regions and used to produce blocks. Blocks found in this manner are only as good as the msa from which they are derived. Using the BLOCKS program (http://www. blocks.fhcrc.org/ blocks/ process_blocks.html), blocks of width 10–55 are extracted from a protein msa of up to 400 sequences (Henikoff and Henikoff 1991, 1992). The program accepts FASTA, CLUSTAL, or MSF formats, or manually reformatted msa's. Several types of analyses may be performed with such extracted blocks. The BLOCKS server primarily generates blocks from unaligned sequences. The eMOTIFs server at http:// dna.stanford.edu/emotif/ (Nevill-Manning et al. 1998) similarly extracts similar conserved patterns from global msa's, but these tend to be shorter than BLOCKS and are called MOTIFS. These types of analyses are discussed below in greater detail.

Pattern Searching

This type of analysis was performed on groups of related proteins, and the common amino acid patterns were placed in a database known as the Prosite catalog (Bairoch 1991). This catalog groups proteins that have similar biochemical functions on the basis of amino acid patterns such as those in the active site. Subsequently, these families were searched for amino acid patterns by the MOTIF program (Smith et al. 1990), which finds patterns of the type aa1 d1 aa2 d2 aa3, where aa1 and aa2 are conserved amino acids and d1 and d2 are stretches of intervening sequence up to 24 amino acids long. These initial patterns are then organized into blocks between 3 and 60 amino acids long by the Henikoff PROTOMAT program (Henikoff and Henikoff 1991, 1992). The BLOCKS database can be accessed at http://www.blocks. fhcrc.org/, and the server may also be used to produce new blocks by the original pattern-finding method or other methods described below.

Although used successfully for making the

BLOCKS database, the MOTIF program is limited in the pattern sizes that can be found. The MOTIF program distinguishes true motifs from random background patterns by requiring that motifs occur in a number of the input sequences and tend not to be internally repeated in any one sequence. As the length of the motif increases, there are many possible combinations of patterns of a given length where only a few characters match, e.g., >10^9 possible patterns for a 15-amino-acid-long pattern with only five matches. The MOTIF program always provides a motif, even for random sequences, thus making it difficult to decide how significant the found motif really is.

This problem has been circumvented by combining the analysis performed by MOTIF with that of the Gibbs sampler (discussed on p. 201), which is based on sound statistical principles. Another pattern search tool, the TEIRESIAS algorithm, which searches for all short patterns and then progressively builds these into longer and longer sets of unique patterns, has been described (Rigoutsos and Floratos 1998). Another rigorous searching algorithm called Aligned Segment Statistical Evaluation Tool (ASSET) has been devised (Neuwald and Green 1994) that can find patterns in sequence up to 50 amino acids long, group them, and provide a measure of the statistical significance of the patterns. These patterns may also include certain pairs, the 26 positive scoring pairs in the BLOSUM62 scoring matrix. Consideration of all BLOSUM pairs by ASSET or any other pattern-finding method is not possible because this would greatly increase the complexity of the analysis.

The efficiency of ASSET is achieved by a combination of an efficient pattern search strategy called the depth-first method, which assures searching for the same patterns only once, and the use of methods for efficiently organizing the patterns. Low-complexity regions with high proportions of the same residue and use of a group of sequences in which some are more similar than others can interfere with the method. ASSET removes low-complexity regions and redundant sequences from consideration. The program was easily able to find subtle motifs in the DNA methylase, reverse transcriptase, and tRNA ligase families, as previously identified by the MOTIF program. In addition, however, ASSET gave these motifs an expect score (the probability that these are random matches of unrelated sequences) of <0.001. The program also found motifs in families with only a fraction of the sequences sharing a motif (the acyltrans-

ferase family) and in a set of distantly related sequences sharing the helix-turn-helix motif. Finally, the program found several repeat sequences in a prenyltransferase and ankyrin-like repeats in an *E. coli* protein. This source code of the program is available by anonymous FTP from ftp://ftp.ncbi.nih.gov/pub/neuwald/. The European Bioinformatics Institute has a Web page for another complex pattern-finding program (PRATT) at http://www2.ebi.ac.uk/pratt/ (Jonassen et al. 1995).

Another method indentifies common patterns in related sequences and then uses these to produce a global msa (Stoye et al. 1997; Sammeth et al. 2003; see http://bibiserv.TechFak.Uni-Bielefeld.DE/dca/).

The advantage of all pattern-finding methods is that they focus attention on finding short conserved regions in sequences that may represent conserved biological functions. The patterns may then be used as the starting point for a global msa. In programs that produce global msa's without initial pattern searches, the focus is on longer sequence alignments, and thus shorter regions may be overlooked. Two disadvantages of pattern-finding methods are that allowances for gaps and substitutions in the conserved regions are not included and the ability to test the significance of the pattern is often not included in the analysis. Some of the local alignment methods discussed later in the chapter address these issues.

Blocks Produced by the BLOCKS Server from Unaligned Sequences

As described above, the BLOCKS server can extract a conserved, ungapped region from an msa to produce a sequence block. This same server can also find blocks in a set of unaligned, input sequences. The advantage of the latter is that finding the block does not depend on first having a global msa, which is often difficult to produce, especially from more divergent sequences, and may not produce correct local alignments that reveal the blocks. The BLOCKS server maintains a large database of blocks based on an analysis of proteins in the Prosite catalog. Blocks are found by the PROTOMAT program (Henikoff and Henikoff 1991) in a two-step process: First, the program MOTIF (Smith et al. 1990) described on the previous page is used to locate spaced patterns. The second step takes the best and most consistent patterns found in step 1 and uses the program MOTOMAT to merge overlapping triplets and extend them, order the resulting blocks, and choose those that are in the largest subset of sequences. Since 1993, the Gibbs sampler (see below) has been used as an additional tool for finding the initial set of short patterns also by specifying that the sampler search for short motifs. The Gibbs sampler is based on a statistical analysis of the sequences and can identify the most significant common patterns in a set of sequences.

An example of BlockMaker output from Lawrence et al. (1993) is shown below. The program first searches for blocks using either the MOTIF or Gibbs sampler program to identify patterns, then the PROTOMAT program consolidates the patterns into meaningful blocks. The MOTIF program is based on a heuristic method that will always find motifs, even in random sequences, whereas the Gibbs sampler discriminates found motifs based on sound statistical methods. The results of both types of analyses are reported.

In the example below, two blocks identified as LipocalA and LipocalB are reported using both the MOTIF and Gibbs sampler programs for step 1, the initial pattern-finding step. The locations of these blocks in the sequences correspond to the known locations of conserved three-dimensional structures in these same sequences. However, incorrect predictions may be made. In the example below, MOTIF aligned MUP2_MOUSE incorrectly in the B block. The Gibbs sampler predictions may also differ when the same sequences are submitted repeatedly with a different initial alignment (see below). Hence, block analysis can be used for three-dimensional structure prediction based on the presence of common blocks in sequences of known and unknown structure, but false predictions arise.

The eMOTIF Method of Motif Analysis

Another somewhat different but extremely useful method of identifying motifs in protein sequences has been described by Nevill-Manning et al. (1998). Using the BLOCKS database (derived from msa of proteins in the Prosite catalog) and the HSSP database (derived from msa of proteins based on predicted structural similarities), this method found a set of amino acid substitution groups characteristic of each column in all of the alignments. These patterns reflect the higher log odds scores in the amino acid substitution matrices. A blocks search for conserved patterns

---

EXAMPLE: BLOCKMAKER OUTPUT

```
    A.  Motif analysis

                LipocalA, width = 15          LipocalB, width = 11
     BBP_PIEBR   16 NFDWSNYHGKWWEVA (   70)    101 VLSTDNKNYII
    ICYA_MANSE   17 DFDLSAFAGAWHEIA (   73)    105 VLATDYKNYAI
    LACB_BOVIN   25 GLDIQKVAGTWYSLA (   70)    110 VLDTDYKKYLL
    MUP2_MOUSE   27 NFNVEKINGEWHTII (  101)    143 DLSSDIKERFA
    RETB_BOVIN   14 NFDKARFAGTWYAMA (   77)    106 IIDTDYETFAV


    B.  Gibbs sampler analysis

                LipocalA, width = 15          LipocalB, width = 11
     BBP_PIEBR   16 NFDWSNYHGKWWEVA (   70)    101 VLSTDNKNYII
    ICYA_MANSE   17 DFDLSAFAGAWHEIA (   73)    105 VLATDYKNYAI
    LACB_BOVIN   25 GLDIQKVAGTWYSLA (   70)    110 VLDTDYKKYLL
    MUP2_MOUSE   27 NFNVEKINGEWHTII (   68)    110 IPKTDYDNFLM
    RETB_BOVIN   14 NFDKARFAGTWYAMA (   77)    106 IIDTDYETFAV
```

using the BLOCKS server or other pattern-finding methods described above does not take substitutions into account. If substitutions are included, the pattern-finding analysis can be extended to include finding both similar and exact patterns in related sequences. A statistical analysis of eMOTIF results was performed to identify amino acids that are found together in the same msa column at the 0.01 level of significance. Thirty and 51 substitution groups that met this criterion were found in the BLOCKS and HSSP msa's, respectively. For example, the chemically aromatic group of amino acids F, W, and Y were found to define a group often located in the same column of the local msa.

From the msa for a particular group of proteins, each column is examined to see whether these groups are represented in the column, as illustrated in Figure 5.14. In column 1, M is always present, and because M is one group, M is used in column 1 of the motif, as shown in Figure 5.14b. Similarly for column 2, Y and F, which are members of the group FYW, are found, and hence this group is used as column 2 in the motif. The final motif shown in Figure 5.14b describes the variation in all the sequences. A motif may be made for only the first group of 19 sequences, and this situation is shown in Figure 5.14c. This second motif (c) has less variability and greater specificity for the first 19 sequences, and thus motif c would be more likely to find those sequences in a database search (i.e., it is a more sensitive motif for those sequences) than motif b.

The probability of each motif's existence in a protein sequence is estimated from the frequencies of the individual amino acids in the SwissProt database. The probability of the motif b above is given by the product of the probability sums in each column, or $p(\text{Motif}) = p(M) \times 1 \times [p(F)+p(W)+p(Y)] \times [p(Y)+p(R)] \times \cdots$. This value has been found to provide a good estimate of false positives, or of the selectivity of the motif, in a database search. As described above for profiles, both sensitivity and selectivity of a given motif must be taken into account in using the motif for a database similarity search. Ideally, validation of a motif should find only sequences used to generate the motif but none other. In practice, eMOTIF produces a large set of motifs, some more and some less sensitive at finding the original true-positive set of aligned sequences. The more sensitive ones, which are also the most selective based on the value of $p(\text{Motif})$, are then chosen. Some are even

```
a MFRRKAFLHWYTGEGMDEMEFTEAESNMNDPVAEYQQY
  MFKRKAFLHWYTGEGMDEMEFTEAESNMNDLVSEYQQY
  MFKRKAFLHWYTGEGMDEMEFTEVRANMNDLVAEYQQY
  MFKRKAFLHWYTSEGMDELEFSEAESNMNDLVSEYQQY
  MFKRKGFLHWYTGEGMEPVEFSEAQSDLEDLILEYQQY
  MFRRKAFLHWFTGEGMDEMEFSEAEGNTNDLVSEYQQY
  MFRRKAFLHWYTGEGMDEMEFSEAEGNTNDLVSEYQQY
  MFRRKAFLHWYTGEGMDEMEFTEAESNMNDLMSEYQQY
  MFRRKAFLHWYTGEGMDEMEFTEAESNMNDLVAEYQQY
  MFRRKAFLHWYTGEGMDEMEFTEAESNMNDLVHEYQQY
  MFRRKAFLHWYTGEGMDEMEFTEAESNMNDLVSEYQQY
  MFRRKAFLHWYTGEGMDEMEFTEAESNMNELVSEYQQY
  MFRRKAFLHWYTLEGMEELEFTEAESNMNDLVYEYQQY
  MFRRKAFLHWYTNEGMDITEFAEAESNMNDLVSEYQQY
  MFRRKAFLHWYTSEGMDEMEFTEAESNMNDLVSEYQQY
  MFRRKRFLHWYTGEGMDEMEFTEAESNMNDLVSEYQQY
  MFRRNAFLHWYTGEGMDEMEFTEAESNMNDLVSEYQQY
  MFRRQAFLHWYTSEGMDEMEFTEAESNMNDLVSEYQQY
  MFSRKAFLHWYTGEGMEEGDFAEADNNVSDLLSEYQQY

  MFGKRAFVHHYVGEGMEENEFTDARQDLYELEVDYANL
  MFKKRAFVHWYVGEGMEEGEFTEARENIAVLERDFEEV
  MFVKRAFVHWYVGEGMEEGEFAEARDDLLALEKDYESV
  MYAKRAFVHWYVGEGMEEGEFAEAREDLAALEKDYEEV
  MYAKRAFVHWYVGEGMEEGEFSEAREDIAALEKDYEEV
  MYAKRAFVHWYVGEGMEEGEFSEAREDLAALEKDFEEV
  MYAKRAFVHWYVGEGMEEGEFSEAREDLAALEKDYEEV
  MYAKRAFVHWYVGEGMEEGEFSEAREDMAALEKDYEEV
  MYAKRAFVHWYVGEGMEEGEFSEVREDLAALEKDYEEV
  MYAKRAFVHWYVGEGMEEGEFTEAREDLAALEKDYEEV
  MYAKRAFVHWYVGEGMEEGEFTEAREDLAALERDYIEV
  MYAKRAFVHWYVGEGMEEVEFSEAREDLAALEKDYEEV
  MYAKRAFVHWYVSEGMEEGEFAEAREDLAALEKDYDEV
  MYSKRAFVHWYVGEGMEEGEFSEAREDLAALEKDYEEV
  MYSKRAFVHWYVGEGMEEGEFSEAREDLAALERDYEEV

b MF.K..FVH.F..EGMQ..QFPQ...Q......QF...
    YR   L Y      N   N AN  N       NY
    W    I W      E   E GE  E       EW
                  D   D SD  D       D
                          T

c MF.KR.FLHWFT.EGMQ..QFPE...Q..DLI.DYQQY
       R      Y      N   N A    N   L
       W             E   E G    E   M
                     D   D S    D   V
                             T
```

FIGURE 5.14. Aligned block of 34 tubulin proteins used by eMOTIF. (*a*) The sequences are divided into two groups based on the occurrence of R or K in the fourth position and Y or NOT Y in the last position. (*b*) Specific substitution groups found in the columns of the block. If a group cannot be found, then the position is ambiguous and a dot is printed at the position. (*c*) If only the first group of sequences is used, a more specific motif may be found because sequences in this group are more closely related to each other. (Reprinted, with permission, from Nevill-Manning et al. 1998 [©1998 National Academy of Sciences, U.S.A.])

useful for distinguishing subfamilies within a protein superfamily. A database of such motifs called Identify is a useful resource for discovering protein function based on presence of motifs (Nevill-Manning et al. 1998; http://dna.stanford.edu/emotif/).

Using Statistical Methods for Aiding Local Alignments

The above methods utilize pattern identification methods and alignments to locate a conserved sequence pattern in a group of sequences. An alternative method of searching for patterns is to produce trial alignments and then improve these alignments using statistical methods. These methods allow for variation in each column of the alignment and produce a scoring matrix that may be used to search other sequences for the same pattern. These methods as a group are especially useful when the sequence patterns are variable and therefore not as easy to find, thus requiring an exhaustive search of the sequences.

Expectation Maximization Algorithm

Expectation maximization (EM) in sequence analysis is a two-stage process in which an initial guess (expectation step) is made as to the location of a variable sequence pattern in a set of sequences. The sequences are then scanned for this pattern to locate probable matches, and the pattern is adjusted (maximization step) to provide a better match to the sequences. This cycle of sequence scanning and pattern updating is repeated until there is no further change on the pattern. The EM algorithm has been used to identify both conserved domains in unaligned protein sequences and protein-binding sites in unaligned DNA sequences (Lawrence and Reilly 1990), including sites that may include gaps (Cardon and Stormo 1992), and is still used.

The EM approach is as follows. A set of sequences expected to have a common but variable sequence pattern that may not be easily recognizable is initially identified. An initial guess is made by the method as to the location and size of the site of interest in each of the sequences, and these parts of the sequence are aligned. The alignment provides an estimate of the base or amino acid composition of each column in the site. The EM algorithm then consists of two steps, which are repeated consecutively. In step 1, the expectation step, the column-by-column composition of the initial guess of the site in a trial alignment of the sequences is used to estimate the probability of finding that site at any position in each of the sequences. These probabilities are used in turn to provide new information as to the expected base or amino acid distribution for each column in the site. In step 2, the maximization step, the new counts of bases or amino acids for each position in the site found in step 1 are substituted for the previous set. Step 1 is then repeated using these new counts. The cycle is repeated until the algorithm converges on a solution and does not change with further cycles. At that point, the best location of the site in each sequence and the best estimate of the residue composition of each column in the site are given.

As an example, suppose that there are ten DNA sequences having very little similarity with each other, each about 100 nucleotides long and thought to contain a binding site near the middle 20 residues, based on biochemical and genetic evidence. As we will later see when examining the EM program MEME (see p. 201), the size and number of binding sites, the location in each sequence, and whether the site is present in each sequence do not necessarily have to be known. For the present example, the following steps would be used by the EM algorithm to find the most probable location of the binding sites in each of the ten sequences.

INITIAL SETUP FOR THE EM ANALYSIS

A starting guess of the location of the 20-residue-long binding motif patterns in each sequence is made by the program. The base composition of each column in the aligned patterns is also determined. The composition of the flanking sequence on each side of the site provides the surrounding base or amino acid composition for comparison, as illustrated below. For illustration purposes, each sequence is assumed to be the same length and to be aligned by the ends, and each character in the alignment represents five sequence positions (o, not in motif; x, in motif).

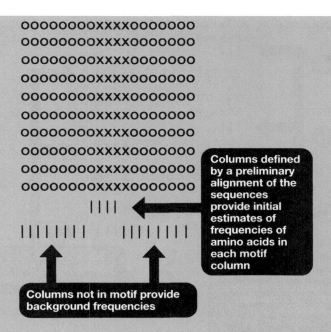

OOOOOOOOOXXXXOOOOOOO
OOOOOOOOOXXXXOOOOOOO
OOOOOOOOOXXXXOOOOOOO
OOOOOOOOOXXXXOOOOOOO
OOOOOOOOOXXXXOOOOOOO
OOOOOOOOOXXXXOOOOOOO
OOOOOOOOOXXXXOOOOOOO
OOOOOOOOOXXXXOOOOOOO
OOOOOOOOOXXXXOOOOOOO
OOOOOOOOOXXXXOOOOOOO

| | | |

| | | | | | | | | | | | | | | |

Columns defined by a preliminary alignment of the sequences provide initial estimates of frequencies of amino acids in each motif column

Columns not in motif provide background frequencies

The number of each base in each column is determined and then converted to fractions. Suppose, for example, that there are four Gs in the first column of the ten sequences, then the frequency of G in the first column of the site, $fs_G = 4/10 = 0.4$. This procedure is repeated for each base and each column. For the rest of the sequences not included in the sites, the background frequency of each base is calculated. For example, let one of these four values for the background frequency, the frequency of G, be $fb_G = 224/800 = 0.28$. These values are now placed in a 4 × 21 matrix, the first column for the background frequencies, and the next 20 columns for the base frequencies in each successive column in the sites. Thus, the counts in the first three columns of the matrix may appear as shown in Table 5.2.

TABLE 5.2. *Column frequencies of each base in the example given*

	Background	Site column 1	Site column 2	...
G	0.27	0.4	0.1	...
C	0.25	0.4	0.1	...
A	0.25	0.2	0.1	...
T	0.23	0.2	0.7	...
	1.00	1.0	1.0	

The first column gives the background frequencies in the flanking sequence. Subsequent columns give base frequencies within the site given in the above example.

The following calculations are performed in the expectation step of the EM algorithm:

1. The estimates described above provide an initial estimate of the composition of the site and the location in each sequence. The object of this step is to improve this estimate by discriminating to the greatest possible extent between sequence within and sequence not within the site. Using the above estimates of base frequencies for (1) background sequences that are not within the site and (2) each column within the site, each sequence is scanned for all possible locations for the site to find the most probable location of the site. For the ten-residue DNA sequence example, there are 100 − 20 + 1 = 81 possible starting sites for a 20-residue-long site, the first one being at position 1 in the sequence ending at 20 and the last beginning at position 81 and ending at 100 (there is not enough sequence for a 20-residue-long site beyond position 81).

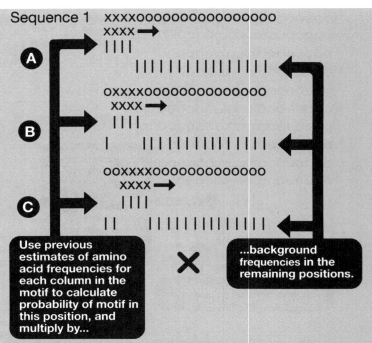

Sequence 1 xxxxooooooooooooooooooo
xxxx →

Use previous estimates of amino acid frequencies for each column in the motif to calculate probability of motif in this position, and multiply by...

×

...background frequencies in the remaining positions.

The resulting score gives the likelihood that the motif matches positions (a) 1–20, (b) 6–25, or (c) 11–30 in sequence Repeat for all other positions and find most likely locator. Then repeat for the remaining sequences.

For each possible site location, the probability that the site starts there is just the product of the probabilities given by Table 5.1. For example, suppose that the site starts in column 1 and that the first two positions in sequence 1 are A and T, respectively. The site will then end at position 20, and the first two flanking background sequence positions are 21 and 22. Suppose that these positions have an A and a T, respectively. Then the probability of this location of the site in sequence 1 is given by $P_{\text{site1,sequence1}} = 0.2$ (for A in position 1) × 0.7 (for T in position 2) × Ps for next 18 positions in site × 0.25 (for A in first flanking position) × 0.23 (for T in second flanking position) × Ps for next 78 flanking positions. Similar probabilities for $P_{\text{site2, sequence1}}$ to $P_{\text{site78, sequence1}}$ are then calculated, thus providing a comparative set of probabilities for the site location. The probability of this best location in sequence 1, say at site k, is the ratio of the site probability at k divided by the sum of all the other site probabilities $P(\text{site } k \text{ in sequence 1}) = P_{\text{site } k, \text{ sequence 1}} / (P_{\text{site 1, sequence 1}} + P_{\text{site 2, sequence 1}} + \ldots + P_{\text{site 78, sequence 1}})$. The probability of the site location in each sequence is then calculated in this manner.

2. The above site probabilities for each sequence are then used to provide a new table of expected values for base counts for each of the site positions using the site probabilities as weights. For example, suppose that P (site 1 in sequence 1) = 0.01 and that P (site 2 in sequence 1) = 0.02. In the above example, the first base in site 1 is an A and the first base for site 2 is a T. Then 0.01 A's and 0.02 T's are added to the accumulated list of bases at site column 1. This procedure is repeated for every other 76 possible first columns in sequence 1. Similarly, site column 2 in the new table of expected values is augmented by counts from the 78 possible column 2 positions in sequence 1, the first, for example, being 0.01 T's. The weighted sequence data from the remaining sequences are also added to the new table, resulting finally in a new estimate of the expected number of each base at each site position and providing a new version of Table 5.2.

In this maximization step, the base frequencies found in the expectation step are used as an updated estimate of the site residue composition. In this case, the data are more complete than the initial estimate because all possible sites in each of the sequences have been evaluated. The expectation and maximization steps are repeated until the estimates of the base frequencies do not change.

AN ALTERNATIVE METHOD OF CALCULATING SITE PROBABILITIES BY THE EM ALGORITHM

The example shown above uses the frequencies of each base in the trial alignment and background base frequencies to calculate the probabilities of each possible location in each sequence. An alternative method is to produce an odds scoring matrix, which is calculated by dividing each base frequency by the background frequency of that base. The probability of each location is then found by multiplying the odds scores from each column. An even simpler method is to use log odds scores in the matrix. The column scores are then simply added. In this case, the log odds scores must be converted to odds scores before position probabilities are calculated.

Multiple EM for Motif Elicitation (MEME)

A Web resource for performing local msa's by the above expectation maximization method is the program Multiple EM for Motif Elicitation (MEME) developed at the University of California's San Diego Supercomputing Center. The Web page for two versions of MEME—ParaMEME, a Web program that searches for blocks by an EM algorithm (described below), and a similar program MetaMEME, which searches for profiles using HMMs (described below)—is found at http://www.sdsc.edu/MEME/meme/website/meme.html. The Motif Alignment and Search Tool (MAST) for searching through databases for matches to motifs is found at http://www.sdsc.edu/MEME/meme/website/mast.html.

MEME will locate one or more ungapped patterns in a single DNA or protein sequence or in a series of DNA or protein sequences. A search is conducted for a range of possible motif widths, and the most likely width for each profile is chosen on the basis of the log-likelihood score after one iteration of the EM algorithm. The EM algorithm then iterates to find the best EM estimate for that width.

Three types of possible motif models may be chosen. The OOPS model is for one expected occurrence of a motif per sequence, the ZOOPS model is for zero or one occurrence per sequence, and the TCM model is for a motif to appear any number of times in a sequence. These models are reflected in the choices on the Web site.

MEME can use prior knowledge about a motif being present in all or only some of the sequences, the length of the motif and whether it is a palindrome (DNA sequences), and the expected patterns in individual motif positions (Dirichlet mixtures, see section on HMMs, p. 207) that provide information as to which amino acids are likely to be interchangeable in a motif (Bailey and Elkan 1995). Once a motif has been found, the motif and its position are effectively erased

to prevent finding the same one twice. An example of the resulting position-specific scoring matrix from a ParaMEME analysis is given in Figure 5.15. Not shown are additional information on the positions of the correeponding motifs in the sequences and the information content given by the scoring matrix.

The Gibbs Sampler

The Gibbs sampler in sequence analysis is a method for finding one or more variable patterns in a set of DNA or protein sequences. A trial pattern is identified in all but one of the sequences and the left-out sequence is evaluated for probable locations of the pattern. The pattern is then updated on the basis of this information. The process is repeated until there is no further change in the pattern. The method is similar in principle to the EM method described above, but the algorithm is different (see below). Like the EM method, given a set of sequences, the Gibbs sampler searches for the statistically most probable motifs and can find the optimal width and number of these motifs in each sequence (Lawrence et al. 1993; Liu et al. 1995; Neuwald et al. 1995). The source code for one implementation of the Gibbs sampler code is available by anonymous FTP from ftp://ftp.ncbi.nih.gov/ pub/neuwald/. A combinatorial approach of the Gibbs sampler and MOTIF may be used to make blocks at the BLOCKS Web site (http://www. blocks.fhcrc.org/). The expected number of blocks in the search is one block for approximately each 40 residues of sequence. The Gibbs sampler is also an option of the msa block-alignment and editing program MACAW (Schuler et al. 1991), which runs on MS-DOS, Macintosh, and other computer platforms and is available by anonymous FTP from ftp://ftp. ncbi.nih.gov/pub/schuler.

To understand the Gibbs sampler algorithm, consider a simple example using this algorithm to locate a single 20-residue-long motif in ten sequences, each

Position-specific scoring matrix

Log odds matrix: alength = 20 w = 9 n = 9732 bayes = 8.36118

```
-2.725  0.818 -5.204 -4.539 -0.082 -4.432 -3.515  1.560 -4.218  1.814  0.701 -4.126 -3.146 -3.848 .
-3.441 -3.841 -4.023 -1.204 -4.313 -2.395 -0.889 -4.226 -4.009 -4.571 -3.882 -0.220 -4.682 -3.547 ·
-0.768 -2.342 -4.756 -4.189 -2.319  0.376 -3.154  1.757 -3.870  0.288  0.918 -3.149 -4.229 -3.492 .
-3.379 -2.600 -5.066 -4.331 -0.586 -5.089 -3.668 -0.081 -4.098  3.045  1.107 -4.393 -4.287 -3.383 ·
-1.373 -1.895 -3.823 -3.574 -1.086 -1.952 -0.466  1.480 -3.565 -2.234 -1.834 -3.701 -3.612 -3.536 ·
-1.879 -0.980 -2.231 -4.187 -3.807 -3.562 -0.892 -3.306 -3.238 -2.753 -3.337  4.193 -2.276 -2.750 ·
-2.460 -0.912 -2.252 -4.176 -3.833 -2.391 -0.968 -3.339 -3.262 -4.256 -3.364  4.217 -4.026 -2.768 ·
-3.475 -1.137 -3.874 -3.535 -3.304 -2.080 -2.080 -2.826 -3.544 -3.127 -2.263 -3.592 -4.599 -3.533 ·
-0.693 -3.833 -3.137 -3.879 -4.963  3.663 -3.647 -3.364 -3.716 -5.287 -4.212 -2.849 -4.518 -4.155 ·
```

FIGURE 5.15. Position-specific scoring matrix. This matrix is a log odds matrix calculated by taking the log (base 2) of the ratio of the observed to expected counts for each amino acid in each column of the profile generated by MEME. Columns and rows in the matrix correspond to the amino acids in each column and positions of the motif, respectively. The counts for each column may have additional pseudocounts added to compensate for zero occurrences of an amino acid in a column or for a small number of sequences, as discussed below for this type of matrix.

200 residues long, as was done above to illustrate the EM algorithm. The method iterates through two steps. In the first step, the predictive update step, one sequence of the set is removed. A trial guess of the pattern location in the rest of the sequences is then made. For example, suppose that sequence 1 is chosen as the outlier and the other nine to find an initial guess of the motif. These other nine sequences are randomly aligned. The following box illustrates how this initial motif is located (an x equals 20 sequence positions and M indicates the random location of the motif chosen for each sequence.

EXAMPLE: LOCATION OF INITIAL MOTIF

A. Estimate the amino acid frequencies in the motif columns of all but one sequence. Also obtain background.

Motif

```
xxxMxxxxx              xxxMxxxxx
xxxxxxMxx              xxxxxxMxx
xxxxxMxxx              xxxxxMxxx
xMxxxxxxx                xMxxxxxxx
xxxxxxxxx          xxxxxxxxx
Mxxxxxxxx                  Mxxxxxxxx
xxxxMxxxx              xxxxMxxxx
xMxxxxxxx                xMxxxxxxx
xxxxxxxxM          xxxxxxxxM
```

Random start   Location of motif in each sequence
positions chosen   provides first estimate of motif composition

B. Use the estimates from A to calculate the ratio of probability of motif to background score at each position in the left-out sequence. This ratio for each possible location in the sequence is the weight of the position.

```
xxxxxxxxx      xxxxxxxxx      xxxxxxxxx      xxxxxxxxx      xxxxxxxxx
M→             M→             M→             M→             M→
```

C. Choose a new location for the motif in the left-out sequence by a random selection using the weights to bias the choice.

```
xxxxxxxMxx        Estimated location of the motif in left-out sequence.
```

D. Repeat steps A to C >100 times.

The objective is to find the most probable pattern common to all of the sequences by sliding them back and forth until the ratio of the motif probability to the background probability is a maximum. This is accomplished by first using the initial alignment shown above to estimate the residue frequencies in each column of the motif, and the sequence residues that are not included in the motif to estimate the background residue frequencies. For example, if these sequences are DNA sequences and the first column of the estimated motif in the nine sequences includes three G's, then the value for $f_{g, \text{column1}}$ = 3/9 = 0.33. Similarly, let $f_{t, \text{column2}}$ = 1/9 = 0.11 for illustration purposes. These frequencies are determined for each of the 20 columns in our example. Similarly, if there are 486 G's among the 9 × 180 = 1620 sequence positions not within the estimated motif, then $f_{g, \text{background}}$ = 486/1620 = 0.30. Also, let $f_{t, \text{background}}$ = 365/1620 = 0.225. If the first two positions in sequence 1 are G and T in that order, then the probability of the motif starting at position 1, Q_1, is calculated as 0.33 × 0.11 × $\cdots$ × $f_{\text{last base, column20}}$. The background probability of this first possible motif, P_1, is also calculated as 0.30 × 0.225 × $\cdots$ × $f_{\text{last base, background}}$. These values are calculated for each possible location of the trial pattern in the left-out sequence. The ratio of these values Q_1/P_1 is a measure of how well the trial pattern matches each sequence position: The higher the ratio, the better the match. The Gibbs sampler method then chooses one probable location for the pattern by giving greater emphasis to the higher ratios. This step amounts to making an informed guess of a better pattern in the sequences. The goal of the algorithm is to converge on a best pattern in the sequences by continuing to make guesses of the pattern until one is found.

Note the difference between the Gibbs sample method and the EM method. EM calculates the probability of the entire sequence using the motif column frequencies within the motif and the background frequencies outside the motif.

To choose a probable location for the pattern in the left-out sequence, the ratio Q_1/P_1 is designated as weight A_1 for motif position 1 in sequence 1. A_1's are then calculated for all other 100 – 20 + 1 = 81 possible locations of the 20-residue-long motif in sequence 1. These weights are then normalized by dividing each weight by its sum to give a probability for each motif position. From this probability distribution, a random start position is chosen for position 1. In so doing, the chance of choosing a particular position is proportional to the weight of that position so that a higher-

scoring position is more likely to be chosen. (You can think of a bag with 81 kinds of balls, with the number of each ball proportional to the weight or probability of that kind. Drawing a random ball will favor the more prevalent ones.) This position in the left-out sequence is then used as an estimate of the location for the motif in sequence 1. The procedure is then repeated by selecting the next sequence to be scanned, aligning the motifs in the other nine sequences with sequence 1 and using the new estimate of a pattern location found above. This process is repeated until the residue frequencies in each column of the motif do not change. For different starting alignments, the number of iterations needed may range from several hundred to several thousand.

As the above cycles are repeated, the more accurate the initial estimate of the motif in the aligned sequences, the more accurate the pattern location in the outlier sequence. The second step in the algorithm tends to move the sequence alignments in a direction that favors a better score, but it also has a random element to search for other possible better locations. When optimal start positions have been selected in several sequences by chance, the compositions of the motif columns begin to reflect a pattern that the algorithm can search for in the other sequences, and the method converges on the optimal motif and the probability distribution of the motif location in each sequence.

Several additional procedures are used to improve the performance of the algorithm.

1. For a correct Bayesian statistical analysis, the amino acid counts in the motif and the background in the outlier sequence are estimated and added to the counts in the remaining aligned sequences. This step is the equivalent of combining prior and updated information to improve the estimation of the motif. These counts may be estimated by Dirichlet mixtures (see discussion of HMMs, p. 207), which give frequencies expected on the basis of prior information from amino acid distributions (Liu et al. 1995). The missing background counts for each residue b_i are estimated by the formula $b_i = f_i \times B$, where B is chosen based on experience with the method as $\sqrt{N}$, the number of sequences in the motif, and f_i is the frequency of residue i in the sequences (Lawrence et al. 1993).

2. Another feature is a procedure to prevent the algorithm from getting locked in a suboptimal

solution. In the HMM method (see below), noise is introduced for this purpose. In the Gibbs sampler, after a certain number of iterations, the current alignments are shifted a certain number of positions to the right and left, and the scores from these shifted positions are found. A probability distribution of these scores is then used as a basis for choosing a new random alignment.

3. The results of a range of motif widths can be investigated. The major difficulty in exploring motif width is to arrive at a criterion for comparing the resulting scores. One suitable measure is to optimize the average information (see below) per free parameter in the motif, a value that can be calculated (Lawrence et al. 1993; Liu et al. 1995). The number of free parameters for proteins is $20 - 1 = 19$ (and for DNA, $4 - 1 = 3$) times the model width.

4. The method can be readily extended to search for multiple motifs in the same set of sequences.

5. The method has been extended to seek a pattern in only a fraction of the input sequences.

The Gibbs sampler was used to align 30 helix-turn-helix DNA-binding domains showing very little recognizable sequence similarity. The information per parameter criterion was used to find the best motif width. Multiple motifs were found in lipocalins, a family with quite dissimilar motif sequences separated by variable spacer regions, and also in protein isoprenyltransferase subunits, which have very large numbers of repeats of several kinds (Lawrence et al. 1993). Thus, the method is widely applicable for discovering complex and variable motifs in proteins.

Hidden Markov Models

The HMM is a statistical model that considers all possible combinations of matches, mismatches, and gaps to generate an alignment of a set of sequences (Fig. 5.16). These models are primarily used for protein sequences to represent protein families or sequence domains, but they are also used to represent patterns in DNA sequences such as RNA splice junctions. Using a computer program designed for producing HMMs, a model of a sequence family that takes into account the lengths of the sequences and accommodating insertions and deletions is first produced and initialized with prior information; i.e., a guess of the expected variation in each position of the multiple

sequence alignment. The program then uses the sequences of a set of 20–100 sequences or more as data to train the model. The trained model may then be used to produce the most probable alignment of the sequences as posterior information. Alternatively, the model may be used to search sequence databases to identify additional members of a sequence family. A different HMM is produced for each set of sequences.

HMMs have been previously used very successfully for speech recognition, and an excellent review of the methodology is available (Rabiner 1989). In addition to their use in producing multiple sequence alignments (Baldi et al. 1994; Krogh et al. 1994; Eddy 1995, 1996), HMMs are used extensively in sequence analysis to produce an HMM that represents a sequence profile (a profile HMM) to analyze sequence composition and patterns (Churchill 1989), to locate genes by predicting open reading frames (Chapter 9), and to produce protein structure predictions (Chapter 10). Pfam, a database of profiles that represent protein families, is based on profile HMMs (Sonhammer et al. 1997).

Advantages and disadvantages of HMM. HMMs often provide an msa as good as, if not better than, other methods such as global alignment and local alignment methods, including profiles and scoring matrices. The approach also has a number of other strong features: It is well grounded in probability theory, no sequence ordering is required, guesses of insertion/deletion penalties are not needed, and experimentally derived information can be used. The disadvantage to using HMMs is that at least 20 sequences and sometimes many more are required to accommodate the evolutionary history of the sequences (see Mitchison and Durbin 1995). The HMM can be used to improve an existing heuristic alignment. The two HMM programs in common use are Sequence Alignment and Modeling Software System (SAM) (Krogh et al. 1994; Hughey and Krogh 1996) and HMMER (see Eddy 1998). The software is available at http://www.cse.ucsc.edu/research/comp bio/sam.html and http://hmmer.wustl.edu/. The algorithms used for producing HMMs are discussed extensively in Durbin et al. (1998). A comparison of HMMs with other methods is given at the end of this section (see p. 210).

The HMM representation of a section of multiple sequence alignment that includes deletions and insertions was devised by Krogh et al. (1994) and is shown in Figure 5.16. This HMM generates sequences with various combinations of matches, mismatches, inser-

A. Sequence alignment

```
N  •  F  L  S
N  •  F  L  S
N  Y  L  T
Q  •  W  –  T
```

RED POSITION REPRESENTS ALIGNMENT IN COLUMN
GREEN POSITION REPRESENTS INSERT IN COLUMN
PURPLE POSITION REPRESENTS DELETE IN COLUMN

B. Hidden Markov model for sequence alignment

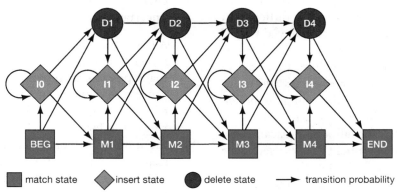

■ match state ◆ insert state ● delete state → transition probability

FIGURE 5.16. Relationship between the sequence alignment and the hidden Markov model of the alignment (Krogh et al. 1994). This particular form for the HMM was chosen to represent the sequence, structural, and functional variation expected in proteins. The model accommodates the identities, mismatches, insertions, and deletions expected in a group of related proteins. (A) A section of an msa. The illustration shows the columns generated in an msa. Each column may include matches and mismatches (*red* positions), insertions (*green* positions), and deletions (*purple* positions). (B) The HMM. Each column in the model represents the possibility of a match, insert, or delete in each column of the alignment in A. The HMM is a probabilistic representation of a section of the msa. Sequences can be generated from the HMM by starting at the beginning state labeled BEG and then by following any one of many pathways from one type of sequence variation to another (states) along the state transition arrows and terminating in the ending state labeled END. Any sequence can be generated by the model and each pathway has a probability associated with it. Each square match state stores an amino acid distribution such that the probability of finding an amino acid depends on the frequency of that amino acid within that match state. Each diamond-shaped insert state produces random amino acid letters for insertions between aligned columns and each circular delete state produces a deletion in the alignment with probability 1. For example, one of many ways of generating the sequence **N K Y L T** in the above profile is by the sequence BEG →M1 →I1 →M2 →M3 →M4 →END. Each transition has an associated probability, and the sum of the probabilities of transitions leaving each state is 1. The average value of a transition would thus be 0.33, since there are three transitions from most states (there are only two from M4 and D4, hence the average from them is 0.5). For example, if a match state contains a uniform distribution across the 20 amino acids, the probability of any amino acid is 0.05. Using these average values of 0.33 or 0.5 for the transition values and 0.05 for the probability of each amino acid in each state, the probability of the above sequence **N K Y L T** is the product of all of the transition probabilities in the path BEG →M →I1 →M2 →M3 →M4 →END, and the probability that each state will produce the corresponding amino acid in the sequences, or $0.33 \times 0.05 \times 0.33 \times 0.05 \times 0.33 \times 0.05 \times 0.33 \times 0.05 \times 0.33 \times 0.05 \times 0.5 = 6.1 \times 10^{-10}$. Since these probabilities are very small numbers, amino acid distributions and transition probabilities are converted to log odds scores, as done in other statistical methods (see pp. 199–204), and the logarithms are added to give the overall probability score. The secret of the HMM is to adjust the transition values and the distributions in each state by training the model with the sequences. The training involves finding every possible pathway through the model that can produce the sequences, counting the number of times each transition is used and which amino acids were required by each match and insert state to produce the sequences. This training procedure leaves a memory of the sequences in the model. As a consequence, the model will be able to give a better prediction of the sequences. Once the model has been adequately trained, of all the possible paths through the model that can generate the sequence **N K Y L T**, the most probable should be the match-insert-3 match combination (as opposed to any other combination of matches, inserts, and deletions). Likewise, the other sequences in the alignment would also be predicted with highest probability as they appear in the alignment; i.e., the last sequence would be predicted with highest probability by the path match-match-delete-match. In this fashion, the trained HMM provides a multiple sequence alignment, such as shown in A. For each sequence, the objective is to infer the sequence of states in the model that generate the sequences. The generated sequence is a Markov chain because the next state is dependent on the current one. Because the actual sequence information is hidden within the model, the model is described as a hidden Markov model.

tions, and deletions, and gives these a probability, depending on the values of the various parameters in the model. The object is to adjust these parameters so that the model represents the observed variation in a group of related protein sequences. A model trained in this manner will provide a statistically probable msa of the sequences.

One problem with HMMs is that the training set has to be quite large (50 or more sequences) to produce a useful model for the sequences. A difficulty in training the HMM residues is that many different parameters must be found (the amino acid distributions, the number and positions of insert and delete states, and the state transition frequencies add up to thousands of parameters) to obtain a suitable model, and the purpose of the prior and training data is to find a suitable estimate for all these parameters. When trying to make an alignment of short sequence fragments to produce a profile HMM, this problem is worsened because the amount of data for training the model is even further reduced.

Algorithms for calculation of an HMM. As illustrated in Figure 5.16, the goal is to calculate the best HMM for a group of sequences by optimizing the transition probabilities between states and the amino acid compositions of each match state in the model. The sequences do not have to be aligned to use the method. Once a reasonable model length reflecting the expected length of the sequence alignment is chosen, the model is adjusted incrementally to predict the sequences. Several methods for training the model in this fashion have been described (Baldi et al. 1994; Krogh et al. 1994; Eddy et al. 1995; Eddy 1996; Hughey and Krogh 1996; Durbin et al. 1998). For example, the Baum–Welch algorithm, previously used in speech recognition methods, adjusts the parameters of HMMs for optimal matching to sequences, as discussed below. This HMM is developed as follows:

1. The model is initialized with estimates of transition probabilities, the probability of moving from one state to another particular state in the model (e.g., the probability of moving from one match state to the next), and the amino acid composition for each match and insert state. If an initial alignment of the sequences is known, or some other kinds of data suggest which sequence positions are the same, these data may be used in the model. For other cases, the initial distribution of

amino acids to be used in each state is described below. The initial transition probabilities that are chosen generally favor transitions from one match state, a part of the model that represents one column in an msa, to the next match state, representing the next column. The alternative of using transitions to insert and delete states, which would delete a position or add another sequence character, is less favored because this builds more uncertainty into the HMM sequence model.

2. All possible paths through the model for generating each sequence in turn are examined. There are many possible such paths for each sequence. This procedure would normally require a huge amount of time computationally. Fortunately, an algorithm, the forward–backward algorithm, reduces the number of computations to the number of steps in the model times the total length of the training sequences. This calculation provides a probability of the sequence, given all possible paths through the model, and, from this value, the probability of any particular path may be found. The Baum–Welch algorithm, referred to above, then counts the number of times a particular state-to-state transition is used and a particular amino acid is required by a particular match state to generate the corresponding sequence position.

3. A new version of the HMM is produced that uses the results found in step 2 to generate new transition probabilities and match-insert state compositions.

4. Steps 3 and 4 are repeated up to ten more times to train the model until the parameters do not change significantly.

5. The trained model is used to provide the most likely path for each sequence, as described in Figure 5.16. The algorithm used for this purpose, the Viterbi algorithm, does not have to go through all of the possible alignments of a given sequence to the HMM to find the most probable alignment, but instead can find the alignment by a dynamic programming technique very much like that used for the alignment of two sequences, as discussed in Chapter 3. The collection of paths for the sequences provides an msa of the sequences with the corresponding match, insert, and delete states for each sequence. The columns in the msa are defined by the match states in the HMM such that

amino acids from a particular match state are placed in the same column. For columns that do not correspond to a match state, a gap is added.

6. The HMM may be used to search a sequence database for additional sequences that share the same sequence variation. In this case, the sum of the probabilities of all possible sequence alignments to the model is obtained. This probability is calculated by the forward component of the forward–backward algorithm described above in step 2. This analysis gives a type of distance score of the sequence from the model, thus providing an indication of how well a new sequence fits the model and whether the sequence may be related to the sequences used to train the model. In later derivations of HMMs, the score was divided by the length of the sequence because it was found to be length dependent. A z score giving the number of standard deviations of the sequence length-corrected score from the mean length-corrected score is therefore used (Durbin et al. 1998).

Recall that for the Bayes block aligner, the initial or prior conditions were amino acid substitution matrices, block numbers, and alignments of the sequences. The sequences were then used as new data to examine the model by producing scores for every possible combination of prior conditions. By using Bayes' rule, these data provided posterior probability distributions for all combinations of prior information. Similarly, the prior conditions of the HMM are the initial values given to the transition values and amino acid compositions. The sequences then provide new data for improving the model. Finally, the model provides a posterior probability distribution for the sequences and the maximum posterior probability for each sequence represented by a particular path through the model. This path provides the alignment of the sequence in the msa; i.e., the sequence plus matches, inserts, and deletes, as described in Figure 5.16. (Bayes' rule is discussed in Chapter 4, p. 148, along with related terms of conditional probability including prior and posterior probability.)

Prior conditions for an HMM. The success of the HMM method depends on having appropriate initial or prior conditions, i.e., a good model for the sequences and a sufficient number of sequences to train the model. The prior model should attempt to capture, for example, the expected amino acid frequencies found in various types of structural and functional domains in proteins. As the distributions are modified by adding amino acid counts from the training sequences, new distributions should begin to reflect common patterns as one moves through the model and along the sequences. It is important that the model reflect not only the patterns in the training sequences, but also pattern variations that might be present in other members of the same protein family. Otherwise, the model will be overtrained and only recognize the training sequences but not other family members. Thus, some smoothing of the amino acid frequencies is desirable, but not to the extent of suppressing highly conserved pattern information from the training sequences. Such problems are avoided by using a method called regularization to avoid overfitting the data to the model. Basically, the method involves using a carefully designed amino acid distribution as the prior condition and then modifying this distribution in a manner that uses sequence information in the training sequences in a complementary manner to build a model that represents the sequences but also includes a reasonable degree of variation as found in related sequences.

Other prior conditions may be used for match states in an HMM. For example, rather than using simple amino acid composition as a prior condition for the match states in the HMM, amino acid patterns that capture some of the important features of protein structure and function have been used with considerable success (Sjölander et al. 1996). Other prior conditions include using Dayhoff PAM or BLOSUM amino acid substitution matrices modified by adding additional counts (pseudocounts) to smooth the distributions (Tatusov et al. 1994; Eddy 1996; Henikoff and Henikoff 1996; Sonnhammer et al. 1997; and see Information Content of PSSM, below).

Dirichlet mixtures used as prior information. A particular set of amino acid substitutions, called Dirichlet mixtures, have been prepared by Sjölander et al. (1996) to use as prior information in the match states of the HMM. These mixtures provide amino acid compositions that have proven to be useful for the detection of weak but significant sequence similarity. They are a method for weighting the prior distributions expected for several different multinomial distributions into a combined frequency distribution. Calculation of these mixtures is a complex mathematical procedure (Sjölander et al. 1996). Dirichlet mixtures recommended for use in aligning proteins by the HMM method have been described previously (Karplus 1995) and are available from http://www.cse.ucsc.edu/research/compbio/dirichlets/ in the open source libraries.

As an example of this approach, the amino acid frequencies that are characteristic of a particular set of nine blocks in the BLOCKS database have been determined as Dirichlet mixtures. These blocks represent amino acid frequencies that are favored in certain chemical environments such as aromatic, neutral, and polar residues and are useful for detecting such environments in test sequences. This nine-component system has been used successfully for producing an HMM for globin sequences (Hughey and Krogh 1996). To use these frequencies as prior information, they are treated as possible posterior distributions that could have generated the given amino acid frequencies as posterior probabilities. The probability of a particular amino acid distribution given a known frequency distribution, i.e., 100 A, 67 G, 5 C, etc., where pA is the probability of A given by the frequency of A, pG the probability of G, etc., and n is the total number of amino acids given by the multinomial distribution, is

$$P(100A, 67G, 5C, \ldots)$$
$$= n!\, p\mathrm{A}^{100} p\mathrm{G}^{67} p\mathrm{C}^{5} \cdots /\ 100!\ 67!\ 5! \cdots \quad (6)$$

The prior distribution for the multinomial distribution is the Dirichlet distribution (Carlin and Louis 1996), whose formulation is similar to that given in Equation 6 with a similar set of parameters but with factorials and powers reduced by 1. The idea behind using this particular distribution is that if additional sequence data with a related pattern are added, then by the Bayesian procedure of multiplying prior probabilities with the likelihood of the new data to obtain the posterior distribution, the probability of finding the correct frequency of amino acids is favored statistically. Because the amino acid frequencies in the test sequences could be any one of several alternatives, a prior distribution that reflects these several choices is necessary. After the prior amino acid frequencies are in place in the match states of the model, these are modified by training the HMM with the sequences, as described in steps 2 and 3 above. For each match state in the model, a new frequency for each amino acid is calculated by dividing the sum of all new and prior counts for that amino acid by the new total of all

amino acids. In this fashion, the new HMM (step 4 above) reflects a combination of expected distributions averaged over patterns in the Dirichlet mixture and patterns exhibited in the training sequences. A similar method is used to refashion the transition probabilities in the HMM during training following manual insertion of initial values.

Number of sequences used for training the HMM. Another consideration in using HMMs for msa's is the number of sequences being aligned. If a good prior model such as the above Dirichlet distribution is used, it should be possible to train the HMM with as few as 20 sequences (Sequence Alignment and Modeling System [SAM] manual, see http://www.cse.ucsc.edu/research/compbio/sam.html; Eddy 1996; Hughey and Krogh 1996). In general, the smaller the sequence number, the more important the prior conditions. If the number of sequences is ~50, the initial conditions play a lesser role because the training step is more effective. As with any msa method, the more sequence diversity, the more challenging the task of aligning sequences with HMMs. HMMs are also more effective if methods to inject statistical noise, e.g., simulated annealing described below, into the model are used during the training procedure. As the model is refashioned to fit the sequence data, it sometimes goes into a form that provides locally optimal instead of globally optimal alignments of the sequences. One of several noise injection methods (Baldi et al. 1994; Krogh et al. 1994; Eddy et al. 1995; Eddy 1996; Hughey and Krogh 1996) may be used in the training procedure. One method called simulated annealing is used by SAM (Hughey and Krogh 1996). A user-defined number of sequences are generated from the model at each cycle and the counts so generated are added to those from the training sequences. The noise generated in this way is reduced as the cycle number is increased. Finally, the HMM program SAM has a built-in feature of "model surgery" during training. If a match state is used by fewer than half of the sequences, it is deleted. These same sequences then have to use an insert state in the revised model. Similarly, if an insert state is used by more than half of the sequences, a number of additional match states equal to the average number of insertions is added, and the model has to be revised accordingly. These fractions may be varied in SAM to test the effect on the type of HMM model produced (Hughey and Krogh 1996).

Producing the Best HMM

In trying to produce an HMM for a set of related sequences (see Fig. 5.17), the recommended procedure is to produce several models by varying the prior conditions. Using regularization by adding prior Dirichlet mixtures to the match states produces models more representative of the protein family from which the training sequences are derived. Varying the noise and model surgery levels is another way to vary the training procedure and the HMM model. The best HMM model is the one that predicts a family of related sequences with the narrowest distribution of probability scores. An example of a portion of an HMM trained on a set of globin sequences is shown in Figure 5.17.

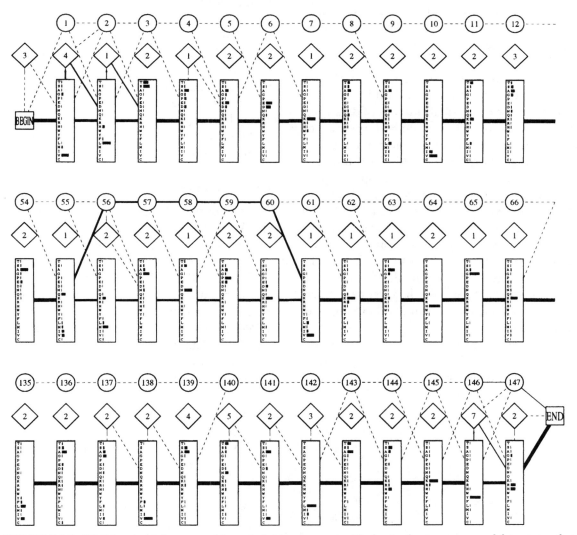

FIGURE 5.17. HMM trained for recognition of globin sequences. Circles in the top row are delete states that include the position in the alignment; the diamonds in the second row are insert states showing the average length of the insertion; and the rectangles in the bottom row show the amino acid distribution in the match states: V is common at match position 1, L at 2, and so on. The width of each transition line joining these various states indicates the extent of use of that path in the training procedure, and dotted lines indicate a rarely used path. The most-used paths are between the match states, but about one-half of the sequences use the delete states at model positions 56–60. Thus, for most of the sequences, the msa or profile will show the first two columns aligned with a V followed by an L, but at 56–60, about one-half of the sequences will have a 5-amino-acid deletion. (Reprinted, with permission, from Krogh et al. 1994 [©1994 Elsevier].)

Motif-based HMMs

The program Meta-MEME uses the HMM method to find motifs in a set of related protein sequences and the spacer regions between them (Grundy et al. 1997) and is built in part on the HMM program HMMER (Eddy et al. 1995). (A similar method was originally used to analyze prokaryotic promoters with two conserved patterns separated by a variable spacer region [Cardon and Stormo 1992].) A Meta-MEME analysis may be performed at http://www.sdsc.edu/MEME using the University of California at San Diego Supercomputing Center. The use of hidden Markov models for producing a global msa is described in the above section.

As discussed above, making an alignment of short sequence fragments to produce a profile HMM is problematic because of the small amount of data available for training the model. Two methods are used by Meta-MEME to circumvent this problem. First, another pattern-finding algorithm, the EM algorithm (discussed on p. 198), is used to locate ungapped regions that match in the majority of the sequences. Second, a simplified HMM with a much reduced number of parameters is produced. The model includes a series of match states that model the patterns located by MEME with transition probabilities of 1 between them and a single insert state between each of these patterns, as illustrated in Figure 5.18. As a result, fewer parameters need to be used, mostly for the amino acid frequencies in the match states.

Next, the most probable order and spacing of the patterns must be found by using another program, Motif Alignment and Search Tool (MAST; Bailey and

Gribskov 1997). MAST searches a sequence database for the patterns and reports the database sequences that have the best statistically significant matches. The order and spacing of the patterns found in the highest-scoring database sequences are then imported into Meta-MEME as a basis for designing the number of match and insert states and the transition probabilities for the insert states. The match states are filled with modified Dirichlet mixtures (see p. 207) (Baylor and Gribskov 1996), and the model is trained by the motif models found by MEME.

For example, for the 4Fe-4S ferredoxins, a measure of the success of the HMM for database search, the ROC_{50} score (see p. 192), was approximately 0.6–0.8 for 4 to 8 training sequences, compared to 0.95–0.96 using an evolutionary profile of 6 to 12 sequences. However, this family was one of the most difficult ones to model, and other families produced an ROC_{50} of 0.9 or better when the HMM was trained by 20 or more sequences.

Using Position-specific Scoring Matrices

Analysis of msa's for conserved blocks of sequence by the BLOCKS and MEME servers leads to production of the position-specific scoring matrix (PSSM). An example of a PSSM produced by the MEME Web site is shown in Figure 5.15. The PSSM may be used to search a sequence to obtain the most probable location or locations of the motif represented by the PSSM. Alternatively, the PSSM may be used to search an entire database to identify additional sequences that also have the same motif. Consequently, it is important to make the PSSM as representative of the

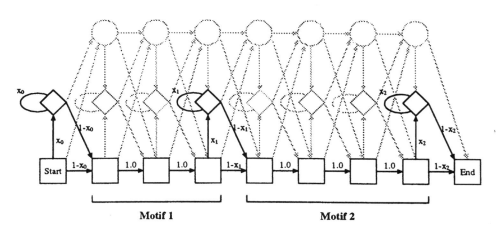

FIGURE 5.18. The HMM used by Meta-MEME to estimate motifs in sequences. (Reprinted, with permission of Oxford University Press, from Grundy et al. 1997.)

expected sites as possible. The quality and quantity of information provided by the PSSM also varies for each column in the motif, and this variation profoundly influences the matches found with sequences. This situation can be accurately described by information theory, and the results can be displayed by a colored graph called a sequence logo as shown in Figure 5.19.

Construction of a PSSM

The PSSM is constructed by a simple logarithmic transformation of a matrix giving the frequency of each amino acid in the motif. The following considerations apply in trying to produce a PSSM that represents a family of related sequences. First, if the number of sequences with the found motif is large and reasonably diverse, the sequences represent a good statistical sampling of all sequences that are ever likely to be found with that same motif. If a given column in 20 sequences has only isoleucine, it is not very likely that a different amino acid will be found in other sequences with that motif because the residue is probably important for function. In contrast, another column in the motif from the 20 sequences may have several amino acids, and some amino acids may not be represented at all. Even more variation may be expected at that position in other sequences, although the more abundant amino acids already found in that column would probably be favored. Thus, if a good sampling of sequences is available, the number of sequences is sufficiently large, and the motif structure is not too complex, it should, in principle, be possible to produce a PSSM that is highly representative of the same motif in other sequences also (Henikoff and Henikoff 1996; Sjölander et al. 1996).

However, there may be a situation in which the number of sequences for producing the motif is small, highly diverse, or complex, thus giving rise to a second level of consideration. If the data set is small, then unless the motif has almost identical amino acids in each column, the column frequencies in the motif may not be highly representative of all other occurrences of the motif. In such cases, it is desirable to improve the estimates of the amino acid frequencies by adding user-defined, extra amino acid counts, called pseudocounts, to obtain a more reasonable distribution of amino acid frequencies in the column. Knowing how many counts to add is a difficult but fortunately solvable problem. On the one hand, if too

many pseudocounts are added in comparison to real sequence counts, the pseudocounts will become the dominant influence in the amino acid frequencies, and searches using the motif will not work. On the other hand, if there are relatively few real counts, many amino acid variations may not be present because of the small sample of sequences. The resulting matrix would then only be useful for finding the sequences used to produce the motif. In such a case, the pseudocounts will broaden the evolutionary reach of the profile to variations in other sequences. Even in this case, the pseudocounts should not drown out but should serve to augment the influence of the

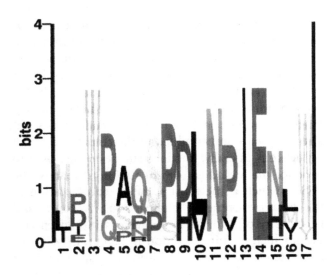

Logo of Gibbs Block D (Tc1) 9 sequences

FIGURE 5.19. A sequence logo. The logo represents the amount of information in each column of a motif corresponding to the values in PSSM of the motif discussed above. The horizontal scale represents sequential positions in the motif. The height of each column gives the decrease in uncertainty provided by the information in that column. The higher the column, the more useful that position for finding matches in sequences. In each column are shown symbols of the amino acids found at the corresponding position in the motif, with the height of the amino acid proportional to the frequency of that amino acid in the column, and the amino acids shown in decreasing order of abundance from the top of the column. From each logo, the following information may thus be found: The consensus may be read across the columns as the top amino acid in each column, the relative frequency of each amino acid in each column of the motif is given by the size of the letters in each column, and the total height of the column provides a measure of how useful that column is for reducing the level of uncertainty in a sequence matching experiment. Note that the highest values are for columns with less diversity.

real counts. Generally speaking, relatively few pseudocounts should be added when there is a good sampling of sequences, and more should be added when the data are more sparse.

The goal of adding pseudocounts is to obtain an improved estimate of the probability p_{ca} that amino acid a is in column c in all occurrences of the blocks, and not just the ones in the present sample. The current estimate of p_{ca} is f_{ca}, the frequency of counts in the data. A simplified Bayesian prediction improves the estimate of p_{ca} by adding prior information in the form of pseudocounts (Henikoff and Henikoff 1996):

$$p_{ca} = (n_{ca} + b_{ca}) / (N_c + B_c) \qquad (7)$$

where n_{ca} and b_{ca} are the real counts and pseudocounts, respectively, of amino acid a in column c, N_c and B_c are the total number of real counts and pseudocounts, respectively, in the column, and $f_{ca} = n_{ca} /N_c$. It is obvious that as b_{ca} becomes larger, the pseudocounts will have a greater infuence on p_{ca}. Furthermore, not only the types of pseudocounts, but also the total number added to the column (B_c), will influence p_{ca}. Finally, fractions such as p_{ca} are used to produce the log odds form of the motif matrix, the PSSM, which is the most suitable representation of the data for sequence comparisons. A count and probability of 0 for an amino acid a in a given column, which is quite common in blocks, may not be converted to logarithms. Addition of a small number of b_{ca} will correct this problem without producing a major change in the PSSM values. An equation similar to Equation 7 is used in the Gibbs sampler (p. 201), except that the number of sequences is $N - 1$.

Pseudocounts are added to columns of the PSSM based on simple formulas or on previous variations seen in aligned sequences. The amino acid substitution matrices, including the Dayhoff PAM and BLOSUM matrices, provide one source of information on amino acid variation. Another source is the Dirichlet mixtures (see p. 207) derived as a posterior probability distribution from the amino acid substitutions observed in the BLOCKS database (see HMMs; Sjölander et al. 1996).

One simple formula for adding pseudocounts that has worked well in some studies is to make B in Equation 7 equal to $\sqrt{N}$, where N is the number of sequences, and to allot these counts to the amino acids in proportion to their frequencies in the sequences (Lawrence et al. 1993; Tatusov et al. 1997). As N increases, the influence of pseudocounts will

decrease because $\sqrt{N}$ will increase more slowly. The main difficulties with this method are that it does not take into account known substitutions of amino acids in alignments and the observed amino acid variations from one column in the motif to the next, and it does not add enough pseudocounts when the number of sequences is small. An alternative method is to use an average sequence profile.

The information in scoring matrices may be used to produce an average sequence profile. Rather than count amino acids, the scoring table values are averaged between each possible pair of the 20 amino acids and those amino acids found in the column of the scoring matrix. Zero counts in a column are not a problem because amino acids not present are not used in the calculations. Because these averaging methods do not take into account the number of sequences in the block, they do not have the desirable effect of a reduced influence when there is a large number of sequences.

Another method of using the information from amino acid substitution matrices is to base pseudocounts on these matrices. Recall the log odds form of the matrices is derived by taking the logarithm of the frequency of substitution q_{ia} of amino acid i for amino acid a divided by the frequency of occurrence of amino acid a, p_a. Then, b_{ca} may be estimated from the total number of pseudocounts in the column by (Henikoff and Henikoff 1996):

$$b_{ca} = B_c Q_i \quad \text{where} \quad Q_i = \sum_{\text{all } i} q_{ia} \qquad (8)$$

b_{ca} in column c can also be made to depend on the observed data in that column (Tatusov et al. 1997), which is given by multiplying B_c by the following conditional probabilities:

$$b_{ca} = B_c \sum_{\text{all } i} \text{prob}(\text{amino acid } i | \text{column } c)$$

$$\times \text{prob}(\text{amino acid } a | i)$$

$$= B_c \sum_{\text{all } i} (n_{ci}/N_c \times q_{ia}/Q_i) \qquad (9)$$

where n_{ci} is the real count of amino acid i in column c.

The total number of pseudocounts in each column needs also to be estimated. As described above, one estimate is to make B_c for each column equal to $\sqrt{N}$, where N is the number of sequences, but this method does not take into account the differences between columns and, for a small number of sequences, the total number of pseudocounts is not sufficient. Allowing B_c to be a constant that can exceed N_c overcomes this limitation but still does not take into account variations in amino acid frequencies between columns, such that a column with conserved amino acids should receive fewer pseudocounts. Using the number of different amino acids in column c, R_c, as an indicator, B_c has been estimated by the formula (Henikoff and Henikoff 1996)

$$B_c = m \times R_c \qquad (10)$$

where m is a positive number derived from trial database searches and $m \leq m \times B_c \leq \min(m \times N_c, m/20)$ (the latter term meaning the minimum of the two given values). By this formula and a given value of m, when $N_c \leq m \times 20$, the total number of pseudocounts B_c is greater, and when $N_c > m \times 20$, B_c is smaller than the total number of real counts, N_c, regardless of the value of R_c. The number of pseudocounts is also reduced when $R_c = 1$. In a test search of the SwissProt and Prosite catalogs with various values of m, a value of 5–6 for m produced the most efficient PSSMs for finding known family members.

Of the several methods for making PSSMs discussed above, the one with pseudocounts derived by Equations 9 and 10 was most successful. This search was performed with PSSMs derived from blocks with amino acid counts also weighted to account for redundancy (Henikoff and Henikoff 1996). However, pseudocounts added from Dirichlet mixtures, which also vary in each column of the scoring matrix, are also very effective (Henikoff and Henikoff 1996; Tatusov et al. 1997).

Calculating the PSSM

Once pseudocounts have been added to real counts of amino acids in each column of the motif, the PSSM may be calculated. The PSSM has one column (or row) for each position in the motif and one row (or column) for each amino acid, and the entries are log odds entries. Each entry is derived by taking the logarithm to the base 2 (bit units, but sometimes also natural logarithms in nat units are used) of the total of the real counts plus pseudocounts for each amino acid, divided by the probability of that amino acid (b_{ca} / N_c). An example of a PSSM produced by MEME is shown in Figure 5.15.

As a sequence is searched with the PSSM, the value of the first amino acid in the sequence is looked up in the first column of the PSSM, then the value of the second amino acid in the matrix, and so on, until the length scanned is the same as the motif width represented by the matrix. All the log odds scores are added to produce a summed score for start position 1 in the sequence. The process is repeated starting at the second position in the sequence, and so on, until there is not enough sequence left. The highest log odds scoring sequence positions have the closest match statistically to the PSSM. Adding logarithms in this manner is the equivalent of mutiplying the probabilities of the amino acids at each sequence position. To convert each summed log odds score (S) to a likelihood or odds score of the sequence matching the PSSM, use the formula odds score = 2^S. These odds scores may be summed and each individual score divided by the sum to normalize them and to thereby produce a probability of the motif at each sequence location.

The above description and example are of using a PSSM to define motifs in protein families. PSSM are also used to define DNA sequence patterns that define regulatory sites, such as promoters or exon–intron junctions in genomic sequences. These topics are discussed in Chapter 9.

Information Content of the PSSM

The usefulness of a PSSM in distinguishing real sequence patterns from background may be measured. The unit of measure is the information content in bits. The PSSM described above gives the log odds score for finding a particular matching amino acid in a target sequence corresponding to each motif position. Variations in the scores found in each column of the table are an indication of the amino acid variation in the original training sequences that were used to produce the PSSM. In some columns, only one amino acid may have been present, whereas in others several may have been present. The columns with highly conserved positions have more information than do the variable columns and will be more definitive for locating matches in target sequences. There is a formal method known as information theory for describing the amount of information in each column that is useful for evaluating each PSSM. The information content of a given amino acid substitution matrix was previously introduced (p. 99) and is discussed in greater detail here. T. Schneider has prepared a Web site that gives excellent tutorials and a review on the topic of information theory, along with methods to produce sequence logos (Schneider and

Stephens 1990) at http://www-lmmb.ncifcrf.gov/ ~toms/sequencelogo.html .

To illustrate the concepts of information and uncertainty (see above Web site), consider 64 cups in a row with an object hidden under one of them. The goal is to find the object with as few questions as possible. The solution is quite simple. First, ask whether the object is hidden under the first or second half of the cups. If the answer is the first 32, then ask which half of that 32, the first 16 or the second 16, and so on. The sequential questions reduce the possibilities from 64 to 32 to 16 to 8 to 4 to 2 to 1, and six questions will therefore suffice to locate the object. This number is also a measure of the amount of uncertainty in the data because this number of questions must be asked to find the object. After the first question has been asked, uncertainty has been reduced by 1, so that only five questions then need to be asked to find the object. The uncertainty is zero when the object is found.

A method to calculate uncertainty (the number of questions to be asked) may be derived from the probability of finding the object under a given cup [p(object) = 1/64]. Uncertainty is found by taking the negative logarithm to the base 2 of 1/64 [$-\log_2(1/64) = 6$ bits]. A situation similar to the hidden object example is found with amino acids in the columns of a PSSM. Here, the interest is to find which amino acid belongs at a particular column in the motif. When we have no information at all, since there are 20 possible choices in all, the amount of uncertainty is $\log_2 20 = 4.32$.

The data from the PSSM provide information that reduces this uncertainty so that real occurrences of the motif can be distinguished from random matches. If only one amino acid is observed in a column of the PSSM, the uncertainty is zero because there are no other possibilities. If two amino acids are observed with equal frequency, there is still uncertainty as to which one it is, and one question must be asked to find the answer, or uncertainty = 1. The formula for finding the uncertainty in this example is the sum of the fractional information provided by each amino acid, or $-[0.5 \times \log_2 0.5 + 0.5 \times \log_2 0.5] = 1$. In general, the average amount of uncertainty (H_c) in bits per symbol for column c of the PSSM is given by

$$H_c = - \sum_{\text{all } i} p_{ic} \log_2(p_{ic}) \qquad (11)$$

where p_{ic} is the frequency of amino acid i in column c and is estimated by the frequency of occurrence of each amino acid (b_{ca}/N_c) and $\log_2(p_{ic})$ is the log odds score for each amino acid in column c. Uncertainty for the entire PSSM may then be calculated as

$$H = \sum_{\text{all columns}} H_c \qquad (12)$$

H is also known as the entropy of the PSSM position in information theory because the higher the value, the greater the uncertainty. The lower the value of the uncertainty H for the PSSM, the greater the ability of the PSSM to distinguish real occurrences of the represented sequence motif from random matches. Conversely, the higher the information content, calculated as shown below, the more useful the PSSM.

Sequence Logos

Sequence logos are graphs that illustrate the amount of information in each column of a motif. The logo is derived from sequence information in the PSSM described above. Conserved patterns in both protein and DNA sequences can be represented by sequence logos. A program for producing logos, along with several examples, is available from http://www-lmmb. ncifcrf.gov/~toms/sequencelogo.html. The Web site of S.E. Brenner at http://www.bio.cam.ac.uk/seqlogo/ will produce sequence logos from an input alignment using the Gibbs sampler method, and an implementation of an extension of the logo method for structural RNA alignment (Gorodkin et al. 1997) is at http://www.cbs.dtu.dk/~gorodkin/appl/plogo.html. A logo representation for the BLOCKS database has been implemented (Henikoff et al. 1995) and may be viewed when the information on a particular block is retrieved from the BLOCKS Web server (http:// www.blocks.fhcrc.org/). An example of a Block logo is shown in Figure 5.19. The information for a motif is also calculated by the MEME server.

Although logos are primarily used with ungapped motifs and sequence patterns, logos of alignments that include gaps in some sequence positions may also be made. If such is the case, then the height of the column with gaps is reduced by the proportion of sequence positions that are not gaps. The height of each logo position is calculated as the amount by which uncertainty has been decreased by the available data; in this case, the amino acid frequencies in each column of the motif. Thus, the higher the column, the more useful that position is for finding matching sequence. The relative heights of each amino acid within each column are calculated by determining how much each amino acid has contributed to that decrease. The height of the amino acid is proportional to the frequency of that amino acid in the column. The uncertainty at column c is given by Equation 11. Because the maximum uncertainty at a position/column when no information is available is $\log_2 20 = 4.32$, as more information about the motif is obtained by new data, the decrease in uncertainty (or increase in the amount of information) R_c is

$$R_c = \log_2 20 - (H_c + \varepsilon_n) \tag{13}$$

where H_c is given by Equation 11 and ε_n is a correction factor for a small sequence number n. R_c is used as the total height of the logo column. The height of amino acid a at position c in the motif logo is then given by $f_{ac} \times R_c$.

The above description applies to protein sequences. Sequence logos can also be produced for DNA sequences. The methodology is very similar to the above except that there are only four possible choices for each logo location. Hence, the maximum amount of uncertainty is $\log_2 4 = 2$. The above method assumes that the sequence pattern is less random than the background or expected sequence variation, and this assumption limits the ability of the method to locate subtle patterns in sequences.

An improved method for finding more subtle patterns in sequences is called the relative entropy method (Durbin et al. 1998). In this case, differences between the observed frequencies and background frequencies are used (Gorodkin et al. 1997), and the decrease in uncertainty from background to observed (or amount of information) in bits is given by

$$R_c = \sum_{\text{all } i} p_{ic} \log_2(p_{ic}/b_i) \tag{14}$$

where b_i is the background frequency of residue i in the organism and the maximum uncertainty in column c is given by $-\Sigma_{\text{all } i} [p_{ic} \log_2(1/b_i)]$. When background frequencies are taken into account, and the column frequency is less than the background frequency, it is possible for the information given by a particular residue in a logo column to be negative. To accommodate this change, the corresponding sequence character is inverted in the logo to indicate a less than expected frequency. There are also two ways used to illustrate the contribution of each character through the height of the symbol. The first method described above is to display symbol heights proportional to the frequency of the amino acid. The second method is to display them proportional to the ratio of the observed to the expected frequency, i.e., by the fraction $(p_{ic}/b_i) / (\Sigma_{\text{all } i} p_{ic}/b_i)$ for each symbol i. Gaps are included in the analysis by using $p_{\text{gap}} = 1$ and, as a result, will always give a negative contribution to the information (Gorodkin et al. 1997).

USING MULTIPLE SEQUENCE ALIGNMENT EDITORS AND FORMATTERS

Once a multiple sequence alignment has been obtained by the global msa program, it may be necessary to edit the sequence manually to obtain a more reasonable or expected alignment. Several considerations must be kept in mind when choosing a sequence editor, which should include as many of the following features as possible:

1. provision for displaying the sequence on a color monitor with residue colors to aid in a clear visual representation of the alignment

2. recognition of the multiple sequence format that was output by the msa program and maintenance of the alignment in a suitable format when the editing is completed

3. provision of a suitable windows interface, allowing use of the mouse to add, delete, or move sequence followed by an updated display of the alignment

In addition, there are other types of editing that are commonly performed on msa's such as, for example, shading conserved residues in the alignment.

The large number of multiple sequence alignment formats that are in use are discussed in Chapter 2. Two commonly encountered examples are the Genetics Computer Group's MSF format and the CLUSTALW ALN format. Because these formats follow a precise outline, one may be readily converted to another by computer programs. READSEQ by D.G. Gilbert at Indiana University at Bloomington is one such program. This program will run on almost any computer platform and may be obtained by anonymous FTP from ftp://ftp.bio.indiana.edu/molbio/readseq. There is also a Web-based interface for READSEQ from Baylor College of Medicine at http://searchlauncher. bcm.tmc.edu/seq-util/ sequtil.html. A software package SEQIO, which provides C program modules for conversion of sequence files from one format to another, is available by anonymous FTP from ftp.pasteur.fr/pub/GenSoft/unix/programming/seqio-1.2. tar.gz; documentation is available at http://bioweb. pasteur.fr/docs/doc-gensoft/seqio/.

A short list of the many available programs that have or exceed the above-listed features is discussed below. For a more comprehensive list, visit the catalog of software page at Web address http://www.ebi.ac. uk/biocat/.

Sequence Editors

1. CINEMA (Colour Interactive Editor for Multiple Alignments) at http://www.biochem.ucl.ac.uk/ bsm/dbbrowser/CINEMA2.02/kit.html is a broadly functional program for sequence editing and analysis, including dot matrix analysis (Parry-Smith et al. 1998; Lord et al. 2002). It features drag-and-drop editing, sequence shifting to left or right, viewing of different parts of an alignment using the split-screen option, multiple motif selection and manipulation, and a number of added features such as viewing of protein structures. CINEMA was developed by A.W.R. Payne, D.J. Parry-Smith, A.D. Michie, and T.K. Attwood. CINEMA is an applet that runs under a Web browser and therefore will run on almost any computer platform.

2. GDE (Genetic Data Environment) provides a general interface on UNIX machines for sequence analysis, sequence alignment editing, and display (Smith et al. 1994) and is available from several anonymous FTP sites including ftp.ebi.ac.uk/ pub/software/unix. GDE is described at http:// bimas.dcrt.nih.gov/gde_sw.html, and http:// www.tigr.org/~jeisen/GDE/GDE.html. GDE features are incorporated into the Seqlab interface for the GCG software, vers. 9. This interface requires communication with a host UNIX machine running the Genetics Computer Group software. Interface with MS-DOS or Macintosh is possible if the computer is equipped with the appropriate X-Window client software.

3. GeneDoc is an alignment editing and display editor by K. Nicholas and H. Nicholas of the Pittsburgh Supercomputing Center for MSF-formatted msa's (Fig. 5.20). It can also import files in other formats. GeneDoc can move residues by inserting or deleting gaps, and features drag-and-drop editing. As the alignment is edited, a new alignment score is calculated by the SP method or based on a phylogenetic tree. GeneDoc is available from http://www.psc.edu/ biomed/genedoc/ and runs under MS Windows.

4. MACAW is both a local multiple sequence alignment program and a sequence editing tool

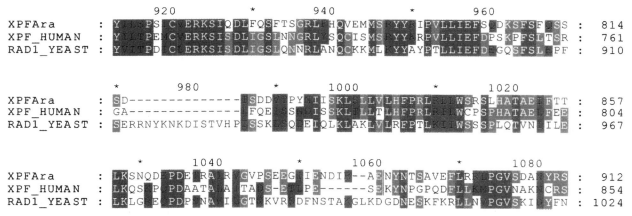

FIGURE 5.20. GeneDoc, a multiple sequence alignment editor with many useful features. Shown is an illustrative msa of three DNA repair genes similar to the *S. cerevisiae Rad1* gene. The sequences were aligned with CLUSTALW, and the FASTA-formatted alignment (Chapter 2) was imported into GeneDoc on a PC.

(Schuler et al. 1991). Given a set of sequences, the program finds ungapped blocks in the sequences and gives their statistical significance. Later versions of the program find blocks by one of three user-chosen methods: by searching for maximum segment pairs or common patterns present in the sequences scored by a scoring matrix such as PAM250 or BLOSUM matrices (the methods used by the BLAST algorithm), by using the Gibbs sampling strategy, a statistical method, or by searching for user-provided patterns provided in a particular format called a regular expression. Executable programs that run under MS Windows, Macintosh, and other computer platforms are available by anonymous FTP from ftp://ftp.ncbi.nlm.nih.gov/pub/schuler/macaw.

5. DCSE, Dedicated Comparative Sequence Editor, developed by Peter De Rijk for editing of protein, DNA, and RNA sequences using the X Windows interface is found at http://rrna.uia.ac.be/dcse/.

6. SEAVIEW, developed by Galtier et al. (1996), edits and converts multiple sequence alignment formats on a variety of computer platforms and is available from http://pbil.univ-lyon1.fr/soft ware/seaview.html.

7. SEQPUP, developed by Don Gilbert at Indiana University, is a sequence editor and sequence analysis tool that runs on most computer platforms and is available from ftp://iubio.bio.indi ana.edu/molbio/seqpup/java/.

Sequence Formatters

1. Boxshade is a formatting program by K. Hofmann for marking identical or similar residues in msa's with shaded boxes, and is available by anonymous FTP from ftp://www. isrec. isb-sib.ch/pub/. The Web server at http://www. ch.embnet.org/software/BOX_form.html takes a multiple-alignment file in either the Genetics Computer Group MSF format or CLUSTAL ALN format and can output a file in many forms, including Postscript/EPS and PICT, for editing on Macintosh and MS-DOS machines.

2. CLUSTALX is a sequence formatting tool that provides the Windows interface for a CLUSTALW msa and is available for many computer platforms, including MS-DOS and Macintosh machines, by anonymous FTP from ftp://ftp-igbmc.u-strasbg. fr/pub/ClustalX/ (Thompson et al. 1997).

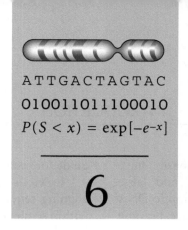

$$ATTGACTAGTAC$$
$$010011011100010$$
$$P(S < x) = \exp[-e^{-x}]$$

6

Sequence Database Searching for Similar Sequences

CONTENTS

INTRODUCTION

Similarity searches in sequence databases have become a mainstay of bioinformatics, and large sequencing projects in which all the genomic DNA sequence of an organism is obtained have become quite commonplace. Similarity searches can also be remarkably useful for finding the function of genes whose sequences have been determined in the laboratory but for which there is no biological information. In these searches, the sequence of the gene of interest is compared to every sequence in a sequence database, and the similar ones are identified. Alignments with the best-matching sequences are shown and scored. If a query sequence can be readily aligned to a database sequence of known function, structure, or biochemical activity, the query sequence is predicted to have similar properties. The strength of these predictions depends on the quality of the alignment between the sequences. As a rough rule, for searches of a protein sequence database with a query protein sequence, if more than one-half of the amino acid sequence is identical in the sequence alignments, the prediction is very strong. For searches of a nucleic acid sequence database with a nucleic acid query sequence, the sequences should be translated if they encode proteins because related protein sequences are more readily identified. If only nucleic acid sequences are compared, then most of the sequences should be identical with few gaps for a strong prediction. As the degree of similarity decreases, confidence in the prediction also decreases. The programs used for these database searches provide statistical evaluations that serve as a guide for evaluation of the alignment scores.

Previous chapters have described methods for aligning sequences or for finding common patterns within sequences. The purpose of making alignments is to discover whether sequences are homologous, i.e., likely to be derived from a common ancestor sequence. If a strong homology relationship can be established, the sequences are likely to have maintained the same function as they diverged from each other during evolution. If an alignment can be found that would rarely be observed between random sequences, the sequences can be predicted to be related with a high degree of confidence. The presence of one or more conserved patterns in a group of sequences is also useful for establishing evolutionary and structure–function relationships among them.

The methods used for establishing sequence relationships in database searches are summarized in Table 6.1. In addition to standard searches of a sequence database with a query sequence, a matrix representation of a family of related protein sequences may be used to search a sequence database for additional proteins that are in the same family, or a query protein sequence may be searched for the presence of sequence patterns that represent a protein family to determine whether the sequence belongs to that particular family. Genomic DNA sequences may also be searched for consensus regulatory patterns such as those representing transcription-factor-binding sites, promoter recognition signals, or mRNA splicing sites; these types of searches are discussed in Chapter 9.

CHAPTER GUIDE FOR BIOLOGISTS

There are few biologists and biology students who have not at some time used the BLAST server at the National Center for Biotechnology Information (NCBI), seen the large number of options available, and experienced the large volume of output. This chapter explains how to use such tools and how to evaluate the results in database sequence similarity searches. With the arrival of the genome era, it is frequently necessary to perform many such searches at a time. Automated methods that perform the same steps, sort through the results, and place selected data in a readily accessible database are needed for these searches. The relatively simple programming tools that are needed for these tasks are described briefly in this chapter and in greater detail in Chapter 12.

There are two basic kinds of sequence database similarity searches: (1) a search with a single query sequence and (2) a search with a multiple sequence alignment (msa), either of a long domain or of a series of short domains. Before performing any of these searches, DNA sequences should be converted to protein sequences as described above.

A search with a single query sequence aligns the sequence with every sequence in the database, produces a local alignment between the sequences, calculates the alignment score, and then evaluates the alignment score using the statistical methods described in Chapter 4. Because the dynamic programming algorithm normally used for such sequence alignment is slow, a much faster method based on searching for short, high-scoring alignment of short "words" is used by the FASTA and BLAST methods to screen through all of the sequence pairs to locate those that share these words. These sequences are then aligned by dynamic programming.

When performing single-query sequence searches, the parameters that restrict the output of the search to the most significant matches are important. Also, in many cases, searches confined to a well-annotated database like SwissProt or to the genome of a single organism may produce a more manageable set of results than searching through all of the predicted and experimentally derived protein sequences in GenBank. When alignments of the query sequence with the best-matching database sequence are found, the expect value (E) of the alignment score is given. E is the number of sequences that are not related to the query sequence and are predicted to produce as good an alignment score with the query sequence. The alignments should be examined for a small value of E, absence of low-complexity regions that falsely give high alignment scores, and quality of the alignment (i.e., presence of long stretches of aligned regions that do not depend on gaps to produce an alignment). It is often worthwhile to try to push the search as far as possible into lower-scoring alignments because some protein families show a great deal of divergence within the family and between species. Remotely matching sequences may be used as queries in an additional search to locate even more remote family members that can be successfully aligned.

The second database similarity search method with an msa is less commonly used. Methods for producing an msa are described in Chapter 5. Once the msa is found, the substitutions in each column of the alignment may be represented by the columns of a simple scoring matrix or by other types of models (e.g., profiles or hidden Markov models). This matrix can then be sequentially aligned with positions along a query sequence to see whether there is a region that closely matches the substitutions found in the original msa.

The advantage of the single-query sequence search is the sensitivity and accuracy of the method for finding closely related database sequences. The advantage of the msa-based search is that a series of related query sequences is used in one search. The database proteins can match any one of the sequences in the msa. Furthermore, a database sequence can also match one region of one sequence in the msa and a different region of another. Hence, this kind of msa-based search has the potential to find additional members of protein families that would not be found in the search using a single-query sequence. The msa query can also represent conserved domains in a set of related proteins and therefore can be used to locate known sequence domains in proteins. Hence, database similarity searches with a single-query sequence or with an msa each have their own advantages.

CHAPTER GUIDE FOR COMPUTATIONAL SCIENTISTS

Biologists have sequenced genes from a variety of different organisms and have made two important observations that are elaborated in detail in Chapter 1. First, many genes are highly conserved between organisms, especially closely related organisms. These genes are duplicated repeatedly and the sequence mutated to produce families of genes, thereby providing a method of diversifying biological function in an organism. Most of these genes produce proteins, but RNA-producing genes are also important. Second, gene conservation between organisms can be used to predict the function of a gene in organism a, given that biological studies have already found the function of a related gene in organism b. This analysis can be expanded to provide much information about genetic variation in b. In some cases, the structure of a protein may be predicted based on similarity with a related protein whose structure has already been solved in the laboratory. This type of observation can have significant benefits if the matching protein is a drug target, because drugs can be designed on the basis of their ability to match three-dimensional portions of a protein structure.

Relationships among a family of protein-encoding genes can be evaluated by producing an optimal local alignment of each sequence pair and evaluating each alignment score using extreme value statistics. Alternatively, related protein sequences can be aligned into a multiple sequence alignment using the methods described in Chapter 5. The alignment is then represented by a statistical model; e.g., a scoring matrix or hidden Markov model. This model can then be used to search for additional sequences in the sequence databases as best-scoring matches to the model. These models may represent a large proportion of the sequence lengths in proteins that share the same three-dimensional structure and function, or they may represent relatively short conserved domains in proteins, indicating a more distant relationship but nevertheless providing useful biological insight. These statistical models of related protein alignments are, therefore, a useful tool for sequence similarity searches.

WHAT SHOULD BE LEARNED IN THIS CHAPTER?

- Why should sequence similarity searches of a sequence database be performed?
- Why should you use protein sequences whenever possible?
- How can searches be speeded up by breaking down sequences into words?
- How are sequence alignment scores obtained and evaluated?
- What is meant by an *E* value?
- How is an alignment of a query sequence with a reported matching database sequence examined to increase confidence that there is a match?
- What are the advantages of using profiles, position-specific scoring matrices, and hidden Markov models over single-query sequences for finding similar sequences in database searches?
- How are profiles, position-specific scoring matrices, and hidden Markov models used for finding similar sequences in databases and to examine sequences for sequence domains?
- How does PSI-BLAST work?
- How are sequence databases, programs, and Web resources accessed?

Glossary Terms

Cobbler A single sequence that represents the most conserved regions in a multiple sequence alignment. The BLOCKS server uses the cobbler sequence to perform a database similarity search as a way to reach sequences that are more divergent than would be found using the single sequences in the alignment for searches.

Expect value (*E*) is the number of unrelated sequences in a similarity search of a sequence database that are expected to achieve a local alignment score as high or higher than the one obtained between the query sequence and the matching database sequence.

Extreme value distribution, as applied to sequence alignment, is the observed distribution of local alignment scores between random or unrelated sequences of approximately the same length range. The distribution has a long tail at high scores, indicating that unrelated sequences can achieve high local alignment scores. This result is found

because alignment scores do not represent a simple sum of individual scores in the alignment (e.g., to calculate the average of a set of student test scores), but instead are an optimal or best score possible (e.g., the highest test score). The statistical significance of an observed local alignment score between two sequences of interest can be evaluated on the basis of the area under the random sequence curve up to the test score value.

Extreme value statistics is a statistical analysis concerned with the maximum value that will be found in a series of observations, rather than the mean. In database similarity searches, the concern is that when the query sequence is compared to thousands of unrelated database sequences, one of them will give an alignment score that has a high enough value that they will falsely appear to be related. Extreme value statistics provides an indication as to how often this false result will occur.

Greedy algorithm is a practical computational method that follows a series of steps without considering whether it is finding the globally best solution to a problem. The solution found is a locally optimal one that may or may not also be the best or optimal one. In sequence similarity searches, for example, initially found database sequences may be used to search for more matches to the initial query sequence.

Hash is an indexed list of values in a table used in FASTA to index words in sequences by position in the sequence.

High-scoring segment pair (HSP) is a high-scoring word that is found both in a query sequence and in a database sequence. In protein searches, the HSP is usually three amino acids long. The word will be enlarged if sequence positions neighboring the word also match and provide a higher-scoring alignment. The score of the aligned amino acids, which is usually calculated using the BLOSUM62 scoring matrix values, should reach a threshold value characteristic of similar sequences.

k-tuple is the breaking down of a sequence into short sequence words or a single sequence character, where k is the length of the word.

Log odds matrix is a table of log odds scores in which the columns represent columns in a multiple sequence alignment and the rows represent sequence characters found in each column. The log odds scores for a column are found by counting the number of each character in the column and dividing the expected number based on the frequency of the character in the aligned sequences. The resulting odds score of observed/expected is converted to a logarithm to facilitate computation of a score for aligning the matrix with a sequence.

Query sequence is a sequence that is to be aligned to a large database of sequences to find the most closely matching sequences in the database. Most often the query sequence is a protein sequence or a DNA sequence translated into a protein sequence during a search.

Regular expression is a highly flexible representation of a sequence pattern that allows for a wide range of substitutions at one position, repeated occurrences of a position, variable size gaps, and many other features.

Search word is a short string (row) of sequence letters, usually 3 for proteins and 5–15 for nucleic acids, that is used to compare sequences. It is computationally simple to make a list of words in two sequences. The presence of unusual words or the occurrence of words in the same

order in both sequences is a good predictor that the sequences will align.

Selectivity (in database similarity searches) refers to the ability of a search method to locate members of a protein family without making a false-positive classification of members of other families. In general, the selectivity of a method is given by TN/(TN + FP), where TN is the number of true-negative results and FP is the number of false-positive results.

Sensitivity (in database similarity searches) refers to the ability of a search method to locate as many members of a protein family as possible, including distant members of limited sequence similarity. In general, the sensitivity of a method is given by TP/(TP + FN), where TP is the number of true-positive results and FN is the number of false-negative results.

Sequence complexity is a measure of the extent of use of all of the available sequence characters (amino acids or nucleotides) in a sequence or part of a sequence. Low-complexity sequences typically have repeats of the same character or a few characters. They pollute sequence similarity searches with high-scoring alignments although they are not related to the query sequence. In high-complexity sequences, many sequence characters are represented according to the frequencies with which they occur in sequences.

Sequence database is a collection of sequences with information about each sequence in each data entry. GenBank is an example of a database of all DNA sequences that have been collected with minimal annotation and emphasis on completeness. SwissProt is an example of a protein sequence database that has been carefully annotated with information about the proteins. Sequence databases generally include a mixture of predicted and experimentally determined gene sequences that can be difficult to tell apart. More carefully annotated sequence databases of individual genomes are very useful for these searches.

Sequence similarity is the fraction of aligned positions in a sequence alignment at which identical sequence characters or conservative substitutions are located. Positions with gaps are usually not scored.

z score is the number of standard deviations of a given data value from the mean of the set. $z = 1$ means one standard deviation above the mean, and $z = -1$ means one standard below the mean.

TABLE 6.1. *Types of database searches for proteins*

Type of search	Target database	Method	Type of query data	Examples of programs used, location (also see Tables 6.2, 6.4, 6.7, and 6.8)	Results of database search
A. Sequence similarity search with query sequence	protein sequence database (or genomic sequences[a])	search for database sequence that can be aligned with query sequence	single sequence, e.g., DAHQSNGA	FASTA (TFASTA[a]), SSEARCH http://fasta.bioch.virginia.edu/fasta/ BLASTP (TBLASTN[a]) http://www.ncbi.nlm.nih.gov/BLAST/ WU-BLAST http://blast.wustl.edu/	list of database sequences having the most significant similarity scores
B. Alignment search with profile (scoring matrix[b,d] with gap penalties)	protein sequence database	prepare profile from a multiple sequence alignment (Profilemake) and align profile with database sequence	profile representing gapped multiple sequence alignment, e.g., D-HQSNGA ESHQ-YTM EAHQSN-L EGVQSYSL	Profilesearch ftp.sdsc.edu/pub/sdsc/biology	list of database sequences that can be aligned with the profile
C. Search with position-specific scoring matrix[c,d] (PSSM) representing ungapped sequence alignment (BLOCK)	protein sequence database	prepare PSSM from ungapped region of multiple sequence alignment or search for patterns of same length in unaligned sequences;[c] then use for database search	PSSM representing ungapped alignment, e.g., DAHQSN ESHQSY EAHQSN EGVQSY	MAST http://meme.sdsc.edu/meme/ website/mast.html	list of database sequences with one or more patterns represented by PSSM but not necessarily in the same order
D. Iterative alignment search for similar sequences that starts with a query sequence, builds a gapped multiple alignment, and then uses the alignment to augment the search[d]	protein sequence database	uses initial matches to query sequence to build a type of scoring matrix and searches additional matches to the matrix by an iterative search method[d]	builds matches to query sequence, e.g., DAHQSNGA iteration 1 H-SNGA EAHQSN-L ↓ further iterations	PSI-BLAST http://www.ncbi.nlm.nih.gov/BLAST/	PSI-BLAST finds a set of sequences related to each other by the presence of common patterns (not every sequence may have same patterns)

E. Search query sequence for patterns representative of protein families[e]	database of patterns found in protein families	search for patterns represented by scoring matrix or hidden Markov model (profile HMM)[e]	single sequence, e.g., DAHQSNGA	Prosite http://www.expasy.ch/prosite INTERPRO http://www.ebi.ac.uk/interpro Pfam http://www.sanger.ac.uk/Pfam CDD/CDART http://www.ncbi.nlm.nih.gov/BLAST (also see Table 10.5)	list of sequence patterns found in query sequence

[a] Searches of this type include the use of programs that search nucleic acid databases for matches to a query protein sequence by automatically translating the nucleic acid sequences in all six possible reading frames (TFASTA, TBLASTN). These searches may be useful when only genomic sequences or partial cDNA sequences (expressed sequence tag or EST sequences) of an organism are available. Genomic sequences that encode proteins may also have been found by gene prediction programs (Chapter 9). The predicted protein is then usually entered in the protein sequence databases. Matches to these predicted proteins may be found by searches of the protein sequence databases. These gene predictions are error-prone (see Chapter 9).

[b] A multiple sequence alignment that includes gaps may be represented by a profile, a type of scoring matrix discussed in Chapter 5, page 189. The consecutive rows of the matrix represent columns of the multiple sequence alignment, and the column values represent the distribution of amino acids in each column of the alignment. The profile includes extra columns with gap opening and extension penalties. The profile is aligned to a sequence by sliding the profile along the sequence and finding the position with the best alignment score by means of a dynamic programming method. The alignment may include gaps in the database sequence. The best-scoring alignments are with database sequences that have a pattern similar to that represented by the profile.

[c] The position-specific scoring matrix (PSSM), or weight matrix as it is sometimes called, is a representation of a multiple sequence alignment that has no gaps (a BLOCK). The matrix may be made from a multiple sequence alignment or by searching for patterns of the same length in a set of sequences using pattern-finding or statistical methods, e.g., expectation maximization, Gibbs sampling, ASSET, and by aligning these patterns, as discussed in Chapter 5. The consecutive columns of the matrix represent columns of the aligned patterns and the rows represent the distribution of amino acids in each column of the alignment. The PSSM columns include log odds scores for evaluating matches with a target sequence. The matrix is used to search a sequence for comparable patterns by sliding the matrix along the sequence and, at each position in the sequence, evaluating the match at each column position using the matrix values for that column. The log odds scores for each column are added to obtain a log odds score for the alignment to that sequence position. High log odds scores represent a significant match.

[d] Using a scoring matrix instead of a single query sequence can enhance a database search because the matrix represents the greater amount of sequence variation found in a multiple sequence alignment. Amino acid representation in each column of the alignment is also reflected in the matrix scores for that column; the more common an amino acid, the higher the score for a match to that amino acid. Note also that the matrix does not store any information about correlations between sequence positions. Thus, if two amino acids are commonly found together in the sequences at two positions of the alignment, these will each be independently scored by the matrix, but there will be no information as to their co-occurrence (or covariation) in the sequences. Since this type of information is missing, the matrix can give high scores to patterns that include new combinations of amino acids not found in the original set of sequences. Scoring covariation in sequence positions is discussed further in Chapters 8, 9, and 10.

[e] Pattern databases are described in Chapter 10.

A LARGE NUMBER OF GENOMES ARE AVAILABLE FOR DATABASE SEARCHING

The genomes of a number of model organisms have been sequenced, initially including the budding yeast *Saccharomyces cerevisiae*, the bacterium *Escherichia coli*, the worm *Caenorhabditis elegans*, the fruit fly *Drosophila melanogaster*, and the human species *Homo sapiens*. These species have also been subjected to intense biological analysis to discover the functions of the genes and encoded proteins. Thus, there is a good deal of information available as to the biological function of particular sequences in model organisms that may be exploited to predict the function of similar genes in other organisms. In addition to genomic DNA sequences, complete cDNA copies of messenger RNAs that carry all the sequence information for the protein products have also been obtained for some of the expressed genes of various organisms. Translation of these cDNA copies provides a close-to-correct prediction of the sequence of the encoded proteins.

Because obtaining intact cDNA sequences is laborious and time-consuming, a common practice is to make a random library of partial cDNA sequences from the expressed genes, and then to perform high-throughput, low-accuracy sequencing of a large number of these partial sequences, known as expressed sequence tags (ESTs). The objective of an EST project is to find enough sequence of each cDNA and to have enough accuracy in the sequence that the amino acid sequence of a significant length of the encoded protein can be predicted. Overlapping ESTs can then be combined, and interesting ones can be found by database similarity searches. The full cDNA sequence of these genes of interest may then be obtained. Once all the sequence information is collected and placed in the sequence databases, the task at hand is to search through the databases to locate similar sequences in another organism that are predicted to have a similar biological function. More information regarding analysis of genomes by these methods is provided in Chapter 11.

FAST DATABASE SEARCHES ARE POSSIBLE USING FASTA AND BLAST

When database searches were first attempted, machine size and speed were limiting factors that prevented use of a full alignment program, such as the dynamic programming algorithm, for each search. Although these considerations no longer apply because of the availability of more powerful machines, the sheer number of such searches presently performed on whole genomes creates a need for faster procedures. Hence, two methods that are at least 50 times faster than dynamic programming were developed. These methods follow a heuristic (tried-and-true) method that almost always works to find related sequences in a database search but does not have the underlying guarantee of an optimal solution like the dynamic programming algorithm. The first rapid search method was FASTA, which found short common patterns in the query and database sequences and joined these into an alignment. BLAST, the second method, was similar to FASTA but gained a further increase in speed by searching only for rarer, more significant patterns in nucleic acid and protein sequences. BLAST is very popular due to availability of the program on the World Wide Web through a large server at the National Center for Biotechnology Information (NCBI) (http://ncbi.nlm.nih.gov) and at many other sites (Table 6.2). The NCBI BLAST server site receives tens of thousands of requests a day. Both FASTA and BLAST have undergone evolution to recent versions that provide very powerful search tools for the molecular biologist and are freely available to run on many computer platforms. They are discussed further below.

With the more recent increased speed and size of computers and algorithmic improvements in the Smith–Waterman dynamic programming algorithm (described in Chapter 3), database similarity searches may also be performed by a search based on a full sequence alignment. The searches are at least 50-fold

TABLE 6.2. *Web resources for performing database searches with a simple query sequence*

Server/program	Web address or FTP site	References
BLAST—Basic Local Alignment Search Tool[a]	http://www.ncbi.nlm.nih.gov/BLAST FTP to ftp.ncbi.nih.gov/blast/executables	Altschul et al. (1990, 1997); Altschul and Gish (1996)
WU-BLAST[b]	sites that run WU-BLAST 2.0 are listed at http://blast.wustl.edu programs obtainable at http://blast.wustl.edu/blast/executables with licensing agreement	Altschul et al. (1990, 1997); Altschul and Gish (1996)
FASTA[c]	http://fasta.bioch.virginia.edu/fasta FTP to ftp.virginia.edu/pub/fasta	Pearson (1995, 1996, 1998, 2000)
BCM Search Launcher (Baylor College of Medicine)	http://searchlauncher.bcm.tmc.edu/	see Web site
TIGR gene indices search	http://www.tigr.org/	see Web site

There are also many other BLAST and FASTA servers on the Web, including ones for searches in specific organisms (see Chapter 11). The TIGR site is given as an example of such a site.

[a] A stand-alone BLAST server may also be established on a local machine running Windows, UNIX, or Mac OS.

[b] Executable programs for UNIX platforms are available from the FTP site.

[c] Executable programs that run on PC, Macintosh, or UNIX platforms are available from the FTP site. The FASTA package also includes programs for performing pair-wise sequence alignments and for a statistical analysis of alignment scores (see Chapter 3). A number of Web sites offer FASTA database search, including the FASTA server and the BCM Search Launcher.

slower than FASTA and BLAST, but control experiments have revealed that more distantly related sequences will usually be found in a database search, provided that the appropriate statistical methods are used. A popular version of the Smith–Waterman program is SSEARCH (FTP to ftp.virginia.edu/pub/fasta) (see p. 259), which is also available on Web sites but usually should be established on a local computer because of the length of time required for a search. Another method for sequence alignment that has been tested in database searches is the Bayes block aligner, described in Chapter 4 (p. 152). This program has found more remotely similar sequences in protein families based on three-dimensional structure than SSEARCH, but the Bayes block aligner is a much slower method (Zhu et al. 1998).

PROTEIN SEARCHES ARE MORE SPECIFIC THAN DNA SEARCHES

One very important principle for database searches is to translate DNA sequences that encode proteins into protein sequences before performing a database search. DNA sequences comprise only four nucleotides, whereas protein sequences comprise 20 amino acids. As a result of the fivefold larger variety of sequence characters in proteins, it is much easier to detect patterns of sequence similarity between protein sequences than between DNA sequences. Pearson (1995, 1996, 2000) has proven that searches with a DNA sequence encoding a protein against a DNA sequence database yield far fewer significant matches than searches using the corresponding protein sequence against a protein sequence database. Both BLAST and FASTA provide programs that translate the query DNA sequence, the database DNA sequence, or both sequences in all six reading frames before making comparisons. There are exceptions to this rule of converting to a protein sequence. One example is the comparison of nucleic acid sequences in the same organism to locate other database entries of the same sequence; in such a search, a nucleic acid search is needed.

The sensitivity and selectivity of the methods should be considered when comparing methods used to search protein sequence databases. Sensitivity is the ability of a method to find most of the members of the protein family (sensitivity = true positives/[true positives + false negatives]) represented by the query sequence. Selectivity is the ability of the method to locate members of a protein family without making a false-positive classification of members of other families (selectivity = true negatives/[true negatives+false positives]). Ideally, both sensitivity and selectivity should be as high in quality as possible. A suitable method for describing both features is to describe a program's resulting degree of coverage of families at a given level of false positives.

The ease with which members of a protein family can be found in sequence database similarity searches depends on how much variability has occurred in the family. In such cases as a recently evolved family, all of the proteins may be readily aligned; i.e., they exhibit a high degree of sequence similarity with each other and may readily be found in database similarity searches. Members of most protein families can be readily aligned, and 50% of the aligned amino acids will be identical. In a more ancient family, there may be considerable variability within the family, with some members exhibiting strong sequence similarity with each other and others aligning less well or even very poorly with other members; i.e., exhibiting a lower degree of sequence similarity. These outlier family members may be much more difficult to find in sequence similarity searches because they will tend to produce lower-value, less-significant alignment scores, thus making it more difficult to distinguish them from non-family members.

One strategy for finding all members of such a divergent family is illustrated in Figure 6.1. An initial sequence similarity search with a query sequence finds a small number of family members that align well with the query sequence; e.g., sequence B in Figure 6.1. The initially matched family members (B in this example) are then used as queries for another similarity search, leading to more family members being identified. This second round of family hits was not found in the initial search because these sequences aligned only poorly with the original query sequence.

This search strategy is repeated using more of the initially found family members as queries until no more sequences can be found. The resulting divergent protein family will include sequences that align to different degrees and, quite possibly, some sequence pairs that cannot be aligned at all. The group as a whole does represent a divergent protein family that shares an evolutionary history. Finding protein families, including divergent ones, through sequence similarity searches is useful in protein structure prediction. When a protein of known three-dimensional structure is a member of a given protein family, the rest of that family probably has a very similar three-dimensional structure.

Two developments have greatly facilitated identifying distantly related sequences in protein sequence similarity searches: (1) amino acid substitution matrices and gap penalty scores that are most suitable for searches have been identified and (2) improved methods for establishing the statistical significance of both nucleic acid and protein sequence alignments have been developed. Use of these new tools has also greatly improved the ability to balance sensitivity of a database search with selectivity. Whether a weak alignment score between a query sequence and a database sequence is significant can now be quite readily and confidently assessed. These topics are extensively discussed in Chapter 4 and are reviewed below. Improvements in scoring matrices and gap penalties continue to be found (Muller and Vingron 2000; Reese and Pearson 2002; Muller et al. 2003; Yu et al. 2003). For example, improvements in choice of gap penalties with standard scoring matrices that improve the statistical analysis of similarity searches with short protein sequences and improve the alignment of protein sequences with repeats have been identified (Reese and Pearson 2002).

THERE ARE CHOICES FOR SCORING MATRICES USED IN DATABASE SIMILARITY SEARCHES

There are a number of choices of amino acid substitution matrices for use in similarity searches of protein sequence databases (Henikoff and Henikoff 2000). The best-performing matrices are widely used, and they often are the default choice of a particular database search so that no additional user input is needed. The most important consideration in matrix choice is that the scoring matrix should be in the log odds form so that the statistical significance of the search results can be evaluated properly. In such a log odds matrix, each matrix entry is the observed frequency of substitution of amino acids A and B for

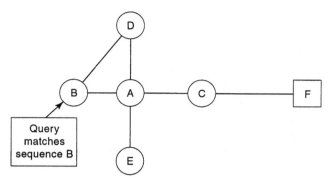

FIGURE 6.1. Structural prediction in database similarity searches. Sequences A–F refer to six members of a protein family defined by sequence similarity between some or all of the members. The sequences are represented as nodes on a graph, and similarity between sequences is represented by joining the nodes with a line (or edge). Note that not all nodes are joined. Thus, sequence A has detectable similarity to sequence B and to sequence C, but the relationship between sequence B and sequence C is not easily detectable. Suppose that sequence C can be aligned with sequence F, a protein of known three-dimensional structure. Hence, all members of this family may be predicted to have the same structure provided that the pair-wise alignments are significant and convincing. To gain further support for this prediction, improved alignments and identification of more family members to help bridge the similarity gaps may be needed.

scored and used to produce frequencies of substitutions between each amino acid pair in related proteins. These substitution frequencies are then divided by the background frequency of the amino acids in the data set to produce a table of odds scores that can be used to differentiate between related and random matches in sequence alignments. A difficulty arises in using these standard matrices to perform searches in organisms that have a biased amino acid composition. The proteins of AT-rich organisms have a greater content of amino acids encoded by AU-containing codons, and proteins of GC-rich organisms have a greater content of amino acids encoded by GC-containing codons (Yu et al. 2003). To compensate for such variations in amino acid composition in different organisms, Yu et al. (2003) have developed a procedure for adjusting the compositional background in the standard scoring matrices so that they may be used to compare organisms of diverse nucleotide richness. The BLAST server can adjust the calculation of E values in a similarity search to take into account the unusual amino acid composition of individual database sequences, thus improving the accuracy of the statistical analysis.

Matrices Based on an Evolutionary Model

For a long time, the Dayhoff PAM250 matrix was used for similarity searches of protein sequence databases by protein query sequences. This scoring matrix is based on an evolutionary model that predicts the types of amino acid changes that occur over long periods of time. The matrix is based on tallying the observed amino acid changes in a closely related group of proteins that were 85% identical. The proteins are organized into an evolutionary tree, and the predicted amino acid changes in the tree are used to estimate the frequency of substitution of each amino acid for another. These frequencies are then normalized to those expected if 1% of the sequence were to change, giving the PAM1 matrix. This level of change roughly corresponds to those amino acid changes expected over a period of 50 million years of evolutionary history. The substitution frequencies in the PAM1 matrix are then extrapolated to predict the changes occurring over longer periods of evolutionary time. For example, if D substitutes for E in the first PAM period, then in the second period, there is an additional chance that D might substitute for E. However, it is also possible that in a second PAM peri-

each other in proteins known to be related, divided by the expected frequency of a chance substitution based on the frequency of A and B in proteins; the resulting ratio is then converted to a logarithm. The score for substituting A and B for each other is simply the logarithm of the odds that a pair of aligned amino acids is found because the sequences are related to a chance alignment of the pair in an alignment between unrelated sequences. The log odds form is useful because the probabilities that successive pairs of aligned sequence characters in an alignment are related is the product of the odds of each pair. When log odds values are used, the probability may be found by addition of the log odds scores, a simpler calculation than multiplication. The choice of the best-scoring matrix for sequence alignments is discussed in detail in Chapter 3 and is reviewed in more detail below.

The same PAM or BLOSUM scoring matrices are generally used to perform protein similarity searches in databases that include sequences from a variety of organisms. In producing these scoring matrices, amino acid substitutions from columns in a multiple sequence alignment of a variety of protein families are od the initial D substitution might revert to E or change to any other amino acid.

As more time passes, the type and frequency of each substitution between the beginning and end of the time period change. PAM250 represents a period of time at which only 20% of the amino acids remain unchanged, but the expected frequencies of change are extrapolated many times from those observed in proteins that are 85% similar. Additional information concerning more recent substitution matrices that are based on an evolutionary model is discussed in Chapter 3. However, for many types of database searches, the PAM250 scoring matrix has been replaced by the BLOSUM matrices.

A new substitute set of PAM-like scoring matrices has been developed (Muller and Vingron 2000) and then compared to other matrices. The equivalent matrix to Dayhoff PAM160, designated VT160, behaves as well as the BLOSUM62 matrix in sequence similarity searches. These matrices are based on an improved method of modeling the frequencies of amino acid substitution in proteins of variable degrees of similarity. Hence, there is not a limitation of scoring amino acid substitutions only in very alike proteins, as described above, and amino acid substitutions in many more protein families can be modeled. As in the Dayhoff PAM matrices, the amino acid substitutions are modeled as a Markov process. The method used to produce the new matrices is similar to the expectation maximization method described in Chapter 5. Basically, the values in a PAM1 matrix are adjusted repeatedly, higher-order matrices are produced by multiplication, and these matrices are then tested for predicting the observed substitutions in more divergent proteins. A series of VT matrices is produced that can be used to perform similarity searches, with the advantage that searches can be made for proteins at any evolutionary distance (Muller et al. 2002).

BLOSUM62 Scoring Matrix

The BLOSUM scoring matrices were generated by S. Henikoff and J.G. Henikoff (1992), who searched for common sequence patterns (blocks) of the same length among all the related proteins in the Prosite catalog. They then added some related sequences in the current databases at the time and scored the columns in an msa of these patterns for amino acid

substitutions. In scoring the columns, some amino acid substitutions were much more common than others because many of the sequences were similar to each other and had the same amino acid. As a result, substitutions to these more common amino acids are overrepresented. To compensate for this effect, alike sequences are grouped into a smaller number of sequences. The resulting BLOSUM matrices have a number to designate to what extent these repeated occurrences are grouped, e.g., BLOSUM62.

The amino acid substitution matrix used by the BLAST programs is the BLOSUM62 scoring matrix. This matrix represents frequencies of amino acid substitutions observed in a large number of related proteins, including some quite similar and some quite different protein sequences. The observed substitutions are all lumped together to provide average frequencies of substitutions without regard to the degree of divergence between sequences. The BLOSUM62 matrix uses only 62% of the repeats in one column and thereby reduces the relative weight given to those substitutions in the matrix. BLOSUM62 is more suitable for protein database similarity searches than the Dayhoff PAM250 matrix because sequences separated by any evolutionary distance may be more readily recognized. The Dayhoff matrices are also based on a much smaller data set than the BLOSUM62 matrix.

Another scoring matrix, BLOSUM50, weights the repeated substitutions somewhat less and has been found to be more suitable for database searches by the FASTA and SSEARCH programs, which use different algorithms than BLAST. BLOSUM matrices also give the best results when the appropriate gap opening and gap extension are used (see Table 4.3, p. 142).

Other Scoring Matrices

A number of other scoring matrices have been devised in addition to the BLOSUM amino acid substitution matrices. The usefulness of various combinations of search programs and substitution matrices for identifying the largest possible number of related sequences in a database search, including remotely related sequences, has been studied in considerable detail. These studies are further discussed in Chapter 3 (p. 104).

SEQUENCE OUTPUT CAN BE LIMITED

Database similarity search programs tend to produce large volumes of output. It can become difficult to screen this volume of material and to assess whether the more remotely related sequences are really related to the query sequence. Thus, it is important to limit the sequence output; there are some relatively simple procedures that may be followed for each program, as described below. For searches of protein databases, avoid repetitive alignments with the same sequence by limiting searches to the protein sequence databases that are well curated, such as SwissProt and PIR, or to a specific genome, as opposed to the entire set of translated GenBank sequences (the GenPept database).

FLOWCHART FOR SEQUENCE SIMILARITY DATABASE SEARCHES

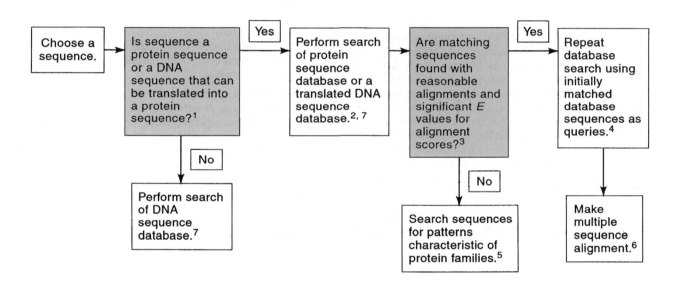

The flowchart (see above) and notes outline approaches for studies using sequence similarity database searches.

1. Translation of protein-encoding DNA sequences into protein sequences before performing sequence comparisons has been shown to be a more effective way to identify related genes than direct comparisons of untranslated DNA sequences. This method also corrects for different codon usage, base composition, and other DNA sequence variations by different organisms. However, to search for a matching DNA sequence in the same organism (e.g., a section of genomic DNA thought to encode a protein is used as a query against an EST database for the organism), a nucleic acid search is more appropriate. If the entire sequence does not encode a protein (e.g., the sequence is a genomic sequence that includes introns), the sequence can be translated in all six reading frames to locate open reading frames that may specify the amino acid sequence of a protein. The predicted translation product may then be compared to a protein sequence

database or a DNA sequence database that is translated in all six reading frames. Alternatively, a gene annotation of the genomic DNA sequence—a predicted amino acid sequence for the protein encoded by the gene that has been entered into the protein sequence database—may be used, as described in Chapter 9. If the same protein is represented in the curated protein databases, e.g., SwissProt, then in many cases the protein has probably been demonstrated biochemically and may have a predicted function. Masking low-complexity regions and sequence repeats in the query sequence is also necessary in many cases because such regions tend to give high-scoring alignments.

2. A carefully annotated protein sequence database (e.g., PIR, SwissProt) will provide a more manageable output list of matched sequences, and these proteins have probably been observed in the laboratory; i.e., the genes do produce a protein product in cells. However, investigators may also wish to expand the search to include predicted

genes from gene annotations of genomic sequences (see note 1 and Chapter 9) that are frequently entered into the DNA sequence translation databases (e.g., DNA sequences in the GenBank DNA sequence databases automatically translated into protein sequences and placed in the GenPept protein sequence database). To compare a protein or predicted protein sequence to EST sequences, the ESTs should be translated into all six possible reading frames (Pearson 2000).

3. To be a significant match, a database sequence that is listed in the program output should have a small E (expect value) and a reasonable alignment with the query sequence (or translations of protein-encoding DNA sequences should have these same features). The E of the alignment score between the sequences gives the statistical chance that an unrelated sequence in the database or a random sequence could have achieved such a score with the query sequence, given as many sequences as there are in the database. The smaller the E, the more significant the alignment. A cutoff value in the range of 0.01–0.05 may be used (Pearson 1996). In genome comparisons described in Chapter 11, a more stringent cutoff score $(10^{-100}$–$10^{-20})$ may be used to find sequences that align very well with the query sequence. However, the alignment should also be examined for absence of repeats of the same residue or residue pattern because these patterns tend to give false high alignment scores.

Filtering of low-complexity regions from the query sequence in a database search helps to reduce the number of false positives. The alignment should also be examined for reasonable amino acid substitutions and for the appearance of a believable alignment (see Chapter 3 flowchart for a summary).

To gain further confidence that the alignment between the query and database sequences are signifi-

cant, either the query sequence or the matched database sequence may be shuffled many times, and each random sequence may be realigned with the other unshuffled sequence to obtain a score distribution for a set of unrelated sequences. This distribution may then be used to evaluate the significance of the true alignment score as described in Chapter 3.

4. Including these extra steps may find additional members of a protein family that have too low a sequence similarity to the original query sequence to be detected in the first search, as illustrated in Figure 6.1.

5. These types of searches are discussed later in the chapter.

6. Methods and considerations that need to be made for producing a multiple sequence alignment are discussed in Chapter 5. Additional relationships among the matched sequences may be found by performing a phylogenetic analysis based on the multiple sequence alignments as described in Chapter 7. Such a phylogenetic analysis can reveal which sequence of several found in an organism is most closely related to a query sequence and therefore is the most likely of the group to have the same function as the query sequence, described as an ortholog of the query sequence.

7. For performing a large number of searches, there is a definite advantage to setting up the search programs on a local machine, especially since versions of the programs that run on most computer platforms are available (ftp:// ftp.ncbi.nih. gov/blast/executables/). One can then set up batch commands or scripts (shell or Perl scripts; see Chapter 12) for processing the sequences and managing the returned data. The NCBI staff provides assistance in the form of SEALS (a system for analysis of lots of sequences) at http://www.ncbi.nlm.nih.gov/CBBresearch/ Walker/SEALS/index.html (Walker and Koonin 1997).

USING A FASTA SEQUENCE DATABASE SIMILARITY SEARCH

FASTA is a program for rapid alignment of pairs of protein and DNA sequences. Rather than comparing individual residues in the two sequences, FASTA searches for matching sequence patterns or words, called k-tuples (Wilbur and Lipman 1983; Lipman and Pearson 1985; Pearson and Lipman 1988). These patterns comprise k consecutive matches of letters in both sequences. The program then attempts to build a local alignment based on these word matches. Due to the ability of the algorithm to find matching sequences in a sequence database with high speed, FASTA is useful for routine database searches of this type. Comparable methods are the BLAST program, which is faster than FASTA, of comparable sensitivity for protein queries, and also does DNA searches, and

programs that use the Smith–Waterman dynamic programming algorithm for protein and DNA searches, which are slower but more sensitive when full-length protein sequences are used as queries.

FASTA compares an input DNA or protein sequence to all of the sequences in a target sequence database and then reports the best-matched sequences and local alignments of these matched sequences with the input sequence. The input sequence is usually in the standard FASTA format, but it is also very easy to change sequence formats, as described in Chapter 2. FASTA finds sequence similarities between the query sequence and each database sequence in four steps illustrated in Figure 6.2.

In the initial stage of a search for regions of sim-

FASTA3

FASTA has gone through a series of updates and enhancements leading to version 34, which is denoted FASTA3. FASTA3 has improved methods of aligning sequences and of calculating the statistical significance of alignments. These changes result in a greatly increased ability of FASTA3 to detect distantly related sequences. The FASTA package is available by anonymous FTP from ftp.virginia.edu/pub/. Precompiled versions for Macintosh and Windows PCs are also available from this FTP site.

ilarity, FASTA uses an algorithmic method known as hashing, illustrated in Table 6.3. In this method, a lookup table showing the positions of each word of length k, or k-tuple, is constructed for each sequence. The relative positions of each word in the two sequences are then calculated by subtracting the position in the first sequence from that in the second. Words that have the same offset position are in phase and reveal a region of alignment between the two sequences. Using hashing, the number of comparisons increases linearly in proportion to average

sequence length. In contrast, using dynamic programming for a database similarity search, the number of comparisons increases in proportion to at least the square of the average sequence length. In FASTA, the k-tuple length is user-defined and is usually 1 or 2 for protein sequences (i.e., either the positions of each of the individual 20 amino acids or the positions of each of the 400 possible dipeptides are located). For nucleic acid sequences, the k-tuple is 4–6. It is much longer than for protein sequences because short k-tuples are much more common due to the four-letter alphabet

TABLE 6.3.   *Lookup method for finding an alignment*

```
position    1  2  3  4  5  6  7  8  9  10  11
sequence 1  n  c  s  p  t  a  .  .  .  .   .

position    1  2  3  4  5  6  7  8  9  10  11
sequence 2              a  c  s  p  r   k
```

amino acid	position in protein A	protein B	offset pos A - pos B
a	6	6	0
c	2	7	-5
k	–	11	
n	1	–	
p	4	9	-5
r	–	10	
s	3	8	-5
t	5	–	

```
Note the common offset for the 3 amino acids c, s, and p.
A possible alignment is thus quickly found

protein 1  n c s p t a
             | | |
protein 2  a c s p r k
```

Shown are fragments of two sequences that share a pattern c-s-p. All of the positions at which a given character is found are listed in a table. The positions of a given character in one of the sequences are then subtracted from the positions of the same character in the second sequence, giving an offset in location. When the offsets for more than one character are the same, a common word is present that includes those characters. Common words, or k-tuples, in two sequences are found by this method in a number of steps proportional to the sequence lengths.

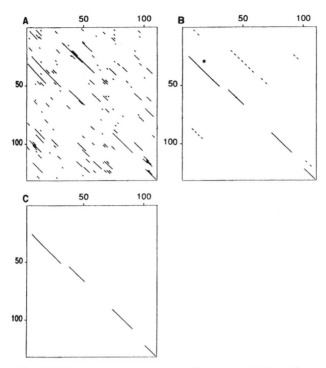

FIGURE 6.2. Methods used by FASTA to locate sequence similarities. (*A*) The 10 best-matching regions in each sequence pair are located by a rapid screen. First, all sets of *k* consecutive matches are found by a rapid method described below. For DNA sequences, *k* is usually 4–6 and for protein sequences, 1–2. Second, those matches within a certain distance of each other (for proteins, 32 for *k* = 1 and 16 for *k* = 2) are joined along with the region between them into a longer matching region without gaps. The regions with the highest density of matches are identified. The calculation is very much like a dot matrix analysis described in Chapter 3, but is calculated in fewer steps. The diagonals shown in *A* represent the locations of these common patterns initially found in the two sequences. (*B*) The highest-density regions of protein sequences identified in *A* are evaluated using an amino acid substitution matrix such as a PAM or BLOSUM scoring matrix. A corresponding matrix may also be used for DNA sequences. The highest-scoring regions, called the best initial regions (INIT1), are identified and used to rank the matches for further analysis. The best-scoring INIT1 region is shown marked by an asterisk. (*C*) Longer regions of identity of score INITN are generated by joining initial regions with scores greater than a certain threshold. The INITN score is the sum of the scores of the aligned individual regions less a constant gap penalty score for each gap introduced between the regions. Later versions of FASTA include an optimization step. When the INITN score reaches a certain threshold value, the score of the region is recalculated to produce an OPT score by performing a full local alignment of the region using the Smith–Waterman dynamic programming algorithm. By improving the score, this step increases the sensitivity but decreases the selectivity of a search (Pearson 1990). INITN and OPT scores are used to rank database matches. Finally, not shown, an optimal local alignment between the input query sequence and the best-scoring database sequences is performed based on the Smith–Waterman dynamic programming algorithm. (Reprinted, with permission, from Pearson and Lipman 1988.)

of nucleic acids. The larger the *k*-tuple chosen, the more rapid, but less thorough, a FASTA database search. An example of this initial stage of a FASTA alignment of a query and database sequences is illustrated in Figure 6.2A. The plot shown represents the positions of *k*-tuples in the two sequences. A consecutive series of *k*-tuples in the sequences produces a diagonal which represents regions in the sequences that can be aligned. Using a small *k*-tuple, e.g., 1, for protein sequences would provide considerable detail

of these aligned regions, whereas a larger *k*-tuple, e.g., 2, would provide less detail.

Detailed performance studies of programs for database similarity searches have been made, including one showing that the Smith–Waterman dynamic programming algorithm and FASTA outperformed BLAST (Pearson 1995). The FASTA programs have undergone enhancements that have improved detection of more remotely related sequences. For sequence fragments, FASTA is as good as Smith–Waterman methods. For DNA searches, FASTA is theoretically better able than BLAST to find matches because a *k*-tuple smaller than the minimum obligatory one of 7 (default size 11 for BLASTN; 3 for TBLASTN, BLASTX, TBLASTX) may be used. For reviews on using FASTA, see Pearson (1995, 1996, 1998).

Significance of FASTA Matches

The methods used by FASTA to report the significance of the results of a database search are revised in later versions, and use of the latest version, FASTA3, is strongly recommended. Revisions include the use of a detailed statistical analysis to evaluate local alignment scores between query and database sequences. The statistical distribution of scores between a query sequence and unrelated sequences found in a database search using FASTA follows the extreme value distribution, described in detail in Chapter 4 (p. 136). The statistical score gives the number of local alignment scores (the expect value, *E*) in a search between the query sequence and unrelated database sequences that are expected to be as large as the one found between the query sequence and a related database sequence (Pearson 1998). All of the local alignment scores are plotted on a curve and fitted to an extreme value distribution. The *E* of a related local alignment score is then calculated from its position under the curve as described below. Similar methods are used by the database search program SSEARCH, which is based on a slower Smith–Waterman type of alignment.

Recall that the parameters of the extreme value distribution, *u* and λ, vary with the length and composition of the sequences being compared, and also with the particular scoring system. In database searches, the expected score between the query sequence and an unrelated database sequence increases in proportion to the logarithm of the length of the database sequence. The parameters change when a different scoring system, e.g., a different scoring matrix or gap penalty, is used. FASTA calculates these parameters from the local alignment scores found with unrelated sequences during the database search. Some of the sequence scores in the database search arise from matches with related sequences and must be removed before the statistical calculations are performed. FASTA performs these tasks in the following manner:

1. The average local alignment score for database sequences in the same length range is determined.

2. The average score is plotted against the logarithm of average sequence length in each length range.

3. The plotted points are then fitted to a straight line by linear regression.

4. A *z* value, the number of standard deviations from the fitted line, is calculated for each score.

5. High-scoring, presumably related sequences, and also very low-scoring alignments that do not fit the straight line are removed from consideration.

6. Steps 1–5 are repeated one or more times.

7. The known statistical distribution of alignment scores is used to calculate the probability that a *z* value of the alignment scores between the query sequence and unrelated sequences of the same length as the query and matching database sequence (designated *Z* in the following equation) could be greater than the *z* value found for the alignment score between the query sequence and a related database sequence (designated *z* in the following equation):

$$P(Z > z) = 1 - \exp\left(-e^{-1.2825z - 0.5772}\right) \qquad (1)$$

The derivation of this equation is given in Chapter 4, page 138.

The expect value *E* of observing, in a database of *D* sequences, no unrelated alignments with scores higher than *z* is given by e^{-DP} and that of observing at least one score *z* is $E \approx 1 - e^{-DP}$. For $P < 0.1$, this relationship is approximated by $E \approx DP$ as indicated below:

$$E(Z > z) = D \times P(Z > z) \qquad (2)$$

8. Normalized similarity scores are calculated for each score by the formula $z' = 50 + 10z$. Thus, an alignment score with a standard deviation of 5 has a normalized score of 100. These normalized scores are reported in the program output.

9. The significance of an alignment score between a given sequence and a database sequence may be further analyzed by aligning a sequence with a shuffled library or a shuffled sequence with an unshuffled library (Pearson 1996) as described in Chapter 4, page 145.

An example of a database search with FASTA, version 3 is shown in Figure 6.3.

A. Score distribution

```
  z'    opt     E()
< 20    176     0:==
   22     1      0:=            one = represents 115 library sequences
   24     2      0:=
   26     2      2:*
   28    21     17:*
   30   109    102:*
   32   355    393:===*
   34  1053   1067:=========*
   36  2214   2191:===================*
   38  4004   3620:===============================*===
   40  5285   5050:=========================================*==
   42  6350   6173:===============================================*==
   44  6716   6809:===================================================*
   46  6847   6935:===================================================*
   48  6661   6640:==================================================*
   50  5777   6059:===============================================  *
   52  5015   5327:==========================================    *
   54  4344   4550:====================================    *
   56  3772   3801:==============================*
   58  3025   3120:=========================*
   60  2475   2528:===================*
   62  1968   2026:=================*
   64  1607   1612:==============*
   66  1362   1274:===========*
   68   983   1002:========*
   70   823    785:======*=
   72   597    614:=====*
   74   584    478:====*=
   76   456    372:===*
   78   306    289:==*
   80   238    225:=*=
   82   230    172:=*
   84   141    136:=*
   86   131    105:*=
   88    93     82:*          inset = represents 2 library sequences
   90    52     63:*
   92    55     49:*      :====================*===
   94    41     38:*      :=================*==
   96    37     29:*      :===============*====
   98    20     23:*      :==========  *
  100    20     17:*      :========*=
  102    16     14:*      :======*=
  104     7     10:*      :====*
  106     9      8:*      :===*=
  108     7      6:*      :==*=
  110     3      5:*      :==*
  112     6      4:*      :=*=
  114     2      3:*      :=*
  116     4      2:*      :*=
  118     3      2:*      :*=
 >120    14      1:*      :*======
```

B. Fit of data to extreme value distribution.

```
26840295 residues in 74019 sequences
 statistics extrapolated from 50000 to 73831 sequences
 Expectation_n fit: rho(ln(x))= 5.9599+/-0.000515; mu= 7.4670+/- 0.029;
 mean_var=81.3676+/-15.767, Z-trim: 42  B-trim: 68 in 1/63
 Kolmogorov-Smirnov  statistic: 0.0106 (N=29) at  42
```

C. Identification of database sequences which give high scoring alignment with probe sequence.

```
FASTA (3.14 April, 1998) function (optimized, BL50 matrix) ktup: 2
 join: 39, opt: 27, gap-pen: -12/ -2, width:  16 reg.-scaled

The best scores are:        initn  init1 opt   z-sc     E()
XPF_HUMAN   11/97 ( 905) 5893   5893  5893 6529.7    0
RA16_SCHPO  11/97 ( 892) 1569    519   749  827.2  2.1e-39
RAD1_YEAST  11/97 (1100)  975    362   619  681.7  2.7e-31
YIS2_YEAST  11/95 ( 993)   37     37   161  174.6  0.0047
YAXB_SCHPO  10/96 ( 578)   91     91   133  147.1  0.16
```

FIGURE 6.3. (*Figure continues on facing page.*)

D. Local alignments of probe sequence and high-scoring database sequences.

```
>>XPF_HUMAN  11/97  ASCII  Len Q92889 homo sapi (905 aa)
  initn: 5893 init1: 5893 opt: 5893 Z-score: 6529.7 expect()      0
Smith-Waterman score: 5893;  100.000% identity in 905 aa overlap

>XPF_HU    1- 905:--------------------------------------:

              10        20        30        40        50
gi|284 MAPLLEYERQLVLELLDTDGLVVCARGLGADRLLYHFLQLHCHPACLVLV
       :::::::::::::::::::::::::::::::::::::::::::::::::::
XPF_HU MAPLLEYERQLVLELLDTDGLVVCARGLGADRLLYHFLQLHCHPACLVLV
              10        20        30        40        50
......

>>RA16_SCHPO  11/97  ASCII  Len P36617 schizosa (892 aa)
  initn: 1569 init1: 519 opt: 749 Z-score: 827.2 expect() 2.1e-39
Smith-Waterman score: 1691;  34.056% identity in 922 aa overlap

>RA16_S    5- 896:--------------------------------------:

               10        20        30        40
gi|284    MAPLLEYERQLVLELLDTDGLVVCARGLGADRLLYHFLQLHCHPAC
          : :..:.   ::..  ::: : ::: :.            :.   :.
RA16_S METKVHLPLAYQQQVFNELIEEDGLCVIAPGLSLLQIAANVLSYFAVPGS
          10        20        30        40        50

         50        60        70        80        90
gi|284 LVLVLNTQPAEEEYFINQLKIEGVEHLPRRVTNEITSNSRYEVYTQGGVI
       :.:.:.    :: ..::  ..:         : ..  .:.. : .::..
RA16_S LLLLVGANVDDIELIQHEMESHLEKKLITVNTETMSVDKREKSYLEGGIF
         60        70        80        90       100

        100       110       120       130       140
gi|284 FATSRILVVDFLTDRIPSDLITGILVYRAHRIIESCQEAFILRLFRQKNK
       :::::::.:.:: ::..  ::::..  .:  .: .       :::.::. ::
RA16_S AITSRILVMDLLTKIIPTEKITGIVLLHADRVVSTGTVAFIMRLYRETNK
         110       120       130       140       150
......

>>RAD1_YEAST  11/97  ASCII  Len P06777 saccharo (1100 aa)
  initn: 975 init1: 362 opt: 619 Z-score: 681.7 expect() 2.7e-31
Smith-Waterman score: 1366;  30.258% identity in 1008 aa overlap

>RAD1_Y    5- 892:--------------------------------------:

                          10        20
gi|284                MAPLLEYERQLVLE-LLDTDGLVVCARGL
                      : .....:  :. . :.:.. ...:
RAD1_Y EPDDIETSKPNINDIRPVDIQLTLPLPFQQKVVENSLITEDALIIMGKGL
          70        80        90       100       110

        30        40        50        60
gi|284 GADRLLYHFLQLHCHPAC--------LVLVLNTQP--------AEEE--Y
       :  ..:..  .:           : :: .:.    :  .      : ::
RAD1_Y GLLDIVANLLHVLATPTSINGQLKRALVLVLNAKPIDNVRIKEALEELSW
          120       130       140       150       160

               70        80        90
gi|284 FINQLK------IEGVEHLPRRVTNEITSNS-----RYEVYTQGGVIFAT
       : : :       :.   : :..::   : ..:      : ::..:.:
RAD1_Y FSNTGKDDDDTAVESDDELFERPFNVVTADSLSIEKRRKLYISGGILSIT
          170       180       190       200       210

         100       110       120       130       140
gi|284 SRILVVDFLTDRIPSDLITGILVYRAHRIIESCQEAFILRLFRQKNKRGF
       :::::::.:.:: ::..  ::::. ..  :.  .:   .:::.::.
RAD1_Y SRILIVDLLSGIVHPNRVTGMLVLNADSLRHNSNESFILEIYRSKNTWGF
          220       230       240       250       260
......
```

FIGURE 6.3. Example of a FASTA, version 3 search. The SwissProt protein database was searched with the human XPF DNA repair protein on a local UNIX server with a locally written Web page interface. The recommended (default) BLOSUM50 amino acid scoring matrix and gap penalties of −12/−2 were used. Actual z scores are normalized to a z' score with a mean of 50 and a standard deviation of 10. These normalized values may be converted back to actual z scores for statistical calculations by subtracting 50 and dividing by 10. (A) Histogram of the normalized similarity scores and the expected score distribution. The first column gives the lower score in each range of scores, the second labeled "opt" is the number of optimized scores in that range, and the third labeled "$E(\ )$" is the number of alignment scores expected to be in that range for unrelated sequences based on the extreme value distribution and the calculated values of u and λ. The "=" signs outline an approximate curve for the actual score distribution and the "*" gives the same information for the expected score distribution. Note the excellent agreement between the observed and expected numbers until a normalized score >120 is reached, at which point some high-scoring alignments are revealed. (B) An evaluation of the fit of the data to the expected curve is given by the Kolmogorov–Smirnov statistic, which compares the maximum deviation between the observed and expected values. In his FASTA distribution notes, W. Pearson indicates that statistic values <0.10 (for $N = 30$) reveal excellent agreement. If this statistic is >0.2, he suggests repeating the analysis with higher gap penalties, e.g., −16, −4 rather than −12, −2. (C) Database sequences that have high normalized alignment scores are listed along with the raw init1, initn, opt, z' score, and $E(\ )$ for a z' score of that value. $E(\ )$ gives the probability that alignment of the query sequence with D database sequences unrelated to the query sequence could generate at least one such z' score. Note that the first row of scores is that for aligning the query sequence with a database copy of itself, followed by very high-scoring alignments to two yeast DNA repair genes on the next rows. (D) Smith–Waterman local alignments are shown along with additional information about the percent identity. A ":" in the alignment is an identity and "." is a conservative substitution. Included is a sketch indicating the extent to which the sequences can be locally aligned.

Versions of FASTA

There are several implementations of the FASTA algorithm (W. Pearson, release notes for FASTA vers. 3 and earlier releases; Pearson et al. 1997; Pearson 1998) using newly developed algorithms (Zhang et al. 1997):

1. FASTA compares a query protein sequence to a protein sequence library to find similar sequences. FASTA also compares a DNA sequence to a DNA sequence library.

2. TFASTA compares a query protein sequence to a DNA sequence library, after translating the DNA sequence library in all six reading frames.

3. FASTF/TFASTF and FASTS/TFASTS compare a set of short peptide fragments, as obtained from analysis of a protein, against a protein sequence database (FASTF/FASTS) or a DNA sequence database translated in all six reading frames (TFASTF/TFASTS). The FASTF programs analyze a set of amino acid sequence fragments following cleavage and sequencing of protein bands resolved by electrophoresis, and the FASTS programs analyze data from a mass spectrometry analysis of a protein. Note that a different sequence format is required to specify the separate peptides (see http://fasta.bioch.virginia.edu/fasta/).

Additional programs have been developed that are designed to align a DNA sequence translated in all three forward or reverse reading frames with a protein sequence, allowing gaps and frameshifts. If a DNA sequence has a high possibility of errors, such as EST sequences, the translated sequence may be inaccurate due to amino acid changes or frameshifts. These programs are designed to go around such errors by allowing gaps and frameshifts in the alignments. FASTX and TFASTX allow only frameshifts between codons, whereas FASTY and TFASTY allow substitutions and frameshifts within a codon. These programs have been shown to be very useful for gene panning, the search for related sequences in EST databases (Retief et al. 1999).

1. FASTX and FASTY translate a query DNA sequence in all three reading forward frames and compare all three frames to a protein sequence database.

2. TFASTX and TFASTY compare a query protein sequence to a DNA sequence database, translating each DNA sequence in all six possible reading frames.

The above FASTA suite of programs is available as executable binary files for most computer systems including Windows, Macintosh, and UNIX platforms (ftp.virginia.edu/pub/FASTA).

The FASTA algorithm has also been adapted for searching through a pattern database instead of a sequence database (Ladunga et al. 1996). FASTA-pat and FASTA-swap are accessible at the Baylor College of Medicine Web site (Table 6.2). Instead of comparing a query sequence to a sequence database, these programs compare the query sequence to a pattern database that contains patterns representative of specific protein families (see below, p. 268). A match between the query sequence and specific database patterns is an indication of a familial relationship between that sequence and the sequences from which those database patterns were generated.

Matching Regions of Low Sequence Complexity

Sequences of low complexity have a narrow representation of sequence characters such as runs of the same amino acid or repeats of amino acid patterns in protein sequences. In database similarity searches, low-complexity query sequences and database sequences align more easily and give higher alignment scores with sequences of normal complexity and hence appear to be related to the normal sequences when they actually are not. FASTA and SSEARCH in the FASTA program suite do not provide a method for avoiding low-complexity sequences or sequence repeats (Pearson 1998). An update to this feature can be anticipated. The BLAST2 programs described below filter regions of low complexity in both DNA and protein query sequences. Programs and Web sites for filtering low-complexity regions are described below in the description of BLAST. The program PRSS in the FASTA distribution package provides an alternative, straightforward way of establishing whether low complexity plays a role in the alignment score between two sequences.

In a statistical analysis primarily intended for proteins, FASTA and PRSS shuffle the matching library sequences many times and realign each of the shuffled sequences with the query sequence. Two levels of shuffling are possible, one at the level of individual amino acids and a second at the level of sequence segments of a chosen length. The first residue shuffling method explores the possibility that restricted amino acid composition plays a role in the alignment, and the second segment shuffling method that particular regions in the query sequence, such as sequence repeats, influence the score. If low complexity or other sequence features that increase align-ment scores at either level are a problem, high scores will be produced when shuffled sequences are aligned with the query sequence. This result provides a warning sign that alternative alignments are possible, thus reducing confidence that the unshuffled alignment is novel. The distribution of scores from alignments between shuffled and query sequences is used to compute the statistical significance of the actual alignment score between the sequences. An example of using PRSS is presented in Chapter 4 (p. 145).

RECOMMENDED STEPS FOR A FASTA SEARCH

The following strategy is recommended for searches with FASTA for finding the most homologous sequences in a database search while avoiding false-negative matches (Pearson 1996, 2000):

1. Look for agreement between the real and theoretical distribution of scores. If the query sequence has a low-complexity, repeated domain or if the gap penalties are set too low, there may be an excess of unrelated sequences with E less than 0.1. If there is an excess of three- to fivefold more sequences than expected in the score range of 80–110, repeat the search after removing the low-complexity regions from the query sequence or else increase the gap penalties from −12/−2 to −14/−4. Another test to apply is to examine the number of high-scoring unrelated sequences, if these are known ahead of time, with E smaller than 1.0. If there are more than 5–10 such sequences, the analysis is suspect.

2. Recall that the expect value E of a database match is the number of times that an unrelated database sequence would obtain a higher z just by chance. For a match to be significant, E should be 0.01–0.05. If the search has correctly identified homologous sequences, the corresponding E values should be much less than 0.01, whereas scores between unrelated sequences should be much greater than this value, e.g., at least 0.5. If there are no E values less than 0.1, the search has not found any sequences with significant similarity to the query sequence.

3. If there are no matches with E less than 0.1, repeat the search with FASTA with k-tuple = 1, or else use the Smith–Waterman dynamic programming method with a program such as SSEARCH. If the program now finds matches with E less than 0.02, the sequences may be homologous, if there is not a low-complexity region in the query sequence. Computer experiments with FASTA have revealed that sequences with scores of 0.2–10 may also be homologous but have marginal sequence similarity. For further study of this possibility, select some of these marginal sequences and use them as query sequences for additional database searches with FASTA. Additional family members with significant similarity may then be found.

4. Confirm homology of marginal matches by shuffling the query or database sequence many times to calculate the significance of the real alignment. The program PRSS described in Chapter 4, page 145, performs this task.

5. Protein sequence alignments with 50% identity in a short 20- to 40-amino-acid region are common in unrelated proteins. To be truly significant, the alignment should extend over a longer region.

USING THE BASIC LOCAL ALIGNMENT SEARCH TOOL (BLAST)

The BLAST algorithm was developed as a way to perform a sequence similarity search by an algorithm that is faster than FASTA but considered to be as sensitive. A powerful computer system dedicated to running BLAST was established at the NCBI. Access to this BLAST system is possible through the Web site http://www.ncbi.nlm.nih.gov/. There are also numerous other Web sites that provide a BLAST database search for specific genome sequences. In addition to the BLAST programs developed at the NCBI, an independent set of BLAST programs has been developed at Washington University (see Table 6.2). These programs perform similarity searches using the same methods as NCBI-BLAST and produce gapped local alignments. The statistical methods used to evaluate sequence similarity scores are different, and thus WU-BLAST and NCBI-BLAST can produce different results (see box below, point 11). The NCBI BLAST Web server is the most widely used one for sequence database searches and is backed up by a powerful computer system so that there is usually very little wait.

Like FASTA, the BLAST algorithm increases the speed of sequence alignment by searching first for common words or k-tuples in the query sequence and each database sequence. Whereas FASTA searches for all possible words of the same length, BLAST confines the search to the words that are the most significant. For proteins, significance is determined by evaluating these word matches using log odds scores in the BLOSUM62 amino acid substitution matrix. For the BLAST algorithm, the word length is fixed at 3 (formerly 4) for proteins and 11 for nucleic acids (3 if the sequences are translated in all six reading frames). These lengths are the minimum needed to achieve a word score that is high enough to be significant but not so long as to miss short but significant patterns. FASTA theoretically provides a more sensitive search of DNA sequence databases because a shorter word length may be used.

The BLAST algorithm has gone through several developmental stages. The most recent gapped BLAST, or BLAST2, is recommended, in part because older versions of BLAST are slower and reported to overestimate the significance of database matches (Brenner et al. 1998). The most important recent change is that BLAST reports the significance of a gapped alignment of the query and database sequences. Former versions reported several ungapped alignments, and it was more difficult to evaluate their overall significance. The statistical analysis of sequence alignments that made this change possible is discussed in detail in Chapter 4, page 141.

STEPS USED BY THE BLAST ALGORITHM

Steps for searching a protein sequence database by a query sequence include the following (Altschul et al. 1990, 1994, 1997; BLAST Web server help pages):

1. The sequence is optionally filtered to remove low-complexity regions that are not useful for producing meaningful sequence alignments.

2. A list of words of length 3 in the query protein sequence is made starting with positions 1, 2, and 3; then 2, 3, and 4, etc., until the last three available positions in the sequence are reached (word length 11 for DNA sequences, 3 for programs that translate DNA sequences).

3. Using the BLOSUM62 substitution scores, the query sequence words in step 1 are evaluated for an exact match with a word in any database sequence. The words are also evaluated for matches with any other combination of three amino acids, the object being to find the scores for aligning the query word with any other three-letter word found in a database sequence. There are a total of 20 x 20 x 20 = 8000 possible match scores for any one sequence position. For example, suppose that the three-letter word PQG occurs in a query sequence. The likelihood of a match to itself is found in the BLOSUM62 matrix as the log odds score of a P-P match, plus that for a Q-Q match, plus that for a G-G match = 7 + 5 + 6 = 18. These scores are added because the BLOSUM62 matrix is made up of logarithms of odds of finding a match in sequences. In Chapter 3, we learned that to find

the odds score for matching three consecutive amino acid pairs in an alignment, the odds scores for matching each pair must be multiplied. Using log odds scores for amino acid pairs simplifies this calculation because they can be added to give the log odds score of the alignment. Similarly, matches of PQG to PEG would score 15, to PRG 14, to PSG 13, and to PQA 12. For DNA words, a match score of +5 and a mismatch score of –4 are used, corresponding to the changes expected in sequences separated by a PAM distance of 40 (see p. 105).

4. A cutoff score called neighborhood word score threshold (T) is selected to reduce the number of possible matches to PQG to the most significant ones. For example, if this cutoff score T is 13, only the words that score above 13 are kept. In the above example, the list of possible matches to PQG will include PEG (15) but not PQA (12). The list of possible matching words is thereby shortened from 8000 of all possible to the highest scoring number of approximately 50.

5. The above procedure is repeated for each three-letter word in the query sequence. For a sequence of length 250 amino acids, the total number of words to search for is approximately 50 × 250 = 12,500.

6. The remaining high-scoring words that comprise possible matches to each three-letter position in the query sequence are organized into an efficient search tree for comparing them rapidly to the database sequences.

7. Each database sequence is scanned for an exact match to one of the 50 words corresponding to the first query sequence position, for the words to the second position, and so on. If a match is found, this match is used to seed a possible ungapped alignment between the query and database sequences.

8. (a) In the original BLAST method, an attempt was made to extend an alignment from the matching words in each direction along the sequences, continuing for as long as the score continued to increase, as illustrated below. The extension process in each direction was stopped when the accumulated score stopped increasing and had just begun to fall a small amount below the best score found for shorter extensions. At this point, a larger stretch of sequence (called the HSP or high-scoring segment pair), which has a larger score than the original word, may have been found.

```
    L P      P  Q  G   L L    QUERY sequence
    M P      P  E  G   L L    DATABASE SEQUENCE
             < WORD  >        THREE LETTER WORD FOUND
                             INITIALLY
             7  2  6          BLOSUM62 scores, word
                             score = 15

        <------'     '------>
EXTENSION TO LEFT    EXTENSION TO RIGHT
    2 7      7  2  6   4 4
    <            HSP        >
    HSP SCORE = 9 + 15 + 8 = 32
```

(b) In the more recent version of BLAST produced by NCBI, called BLAST2 or gapped BLAST, a different and much more time-efficient method is used (Altschul et al. 1997). The method starts by making a list of high-scoring matching words, as in steps 1–4 above, with the exception that a lower value of T, the word cutoff score, such as 11 in the above example of the word PQG, is used. This change results in a longer word list and matches to lower-scoring words in the database sequences. Matches between the query sequence and one database sequence are illustrated in Figure 6.4. The x's mark positions of the words with scores at least as high as the new value of T. The object is to use these short matched regions lying on the same diagonal and within distance A of each other as the starting points for a longer ungapped alignment between the words. Once found, these joined regions are then extended using the method in part (a). Usually only a few such regions are extended. Because the new matches depend on finding two contiguous words, it is necessary to use a lower value of T to maintain the same level of sensitivity for detecting sequence similarity. The newly found diagonals are then scored by summing the scores of the individually matched sequence pairs (see Fig. 6.4).

9. The next step is to determine whether each HSP score found by one of the above methods is greater in value than a cutoff score S. A suitable value for S is determined empirically by examining the range of scores found by comparing random sequences, and by choosing a value that is significantly greater. The HSPs matched in the entire database are identified and listed.

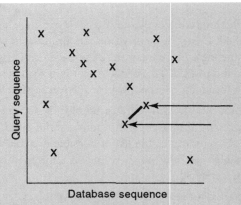

FIGURE 6.4. Scoring diagonals in BLAST2.

10. BLAST next determines the statistical significance of each HSP score. A probability that two random sequences, one the length of the query sequence and the other the entire length of the database (which is approximately equal to the sum of the lengths of all of the database sequences), could achieve the HSP score is calculated. The topic of sequence statistics is discussed in detail in Chapter 4 and therefore the procedure is only reviewed briefly here. The main problem encountered is that scores between random sequences can reach extremely high values and can become higher, then longer as the random sequences become longer. The probability p of observing a score S between random sequences is equal to or greater than x found between a query sequence, and a database sequence is given by the equation

$$p\,(S \geq x) = 1 - \exp(-\,e^{-\lambda(x-u)}) \tag{3}$$

where $u = [\log{(Km'n')}]/\lambda$ and where K and λ are parameters that are calculated by BLAST for the amino acid substitution scoring matrix, n' is the effective length of the query sequence, and m' is the effective length of the individually matched database sequence. Methods for calculating the parameters K and λ are described in Chapter 3.

The effective sequence lengths are the actual lengths of the query and database sequences less the average length of an alignment between two random sequences of the same length. m' and n' are calculated from the following relationship:

$$m' \approx m - (\ln Kmn)/H \tag{4}$$

$$n' \approx n - (\ln Kmn)/H \tag{5}$$

where H is the average expected score per aligned pair of residues in an alignment of two random sequences (Altschul and Gish 1996). H is calculated from the relationship $H = (\ln Kmn)/l$, where l is the average length of the alignment that can be achieved between random sequences of lengths m and n using the same scoring system as used in the database search. l is measured from actual alignments of random sequences. H is similar to the relative entropy of a scoring matrix described in Chapters 3 and 4, except that in this case, H is calculated from alignments of random sequences for a given scoring matrix, usually BLOSUM62. The basis for using these reduced lengths in statistical calculations is that an alignment starting near the end of one of the sequences is likely not to have enough sequence to build an optimal alignment. Using this correction also provides an improved match to statistical theory (Altschul and Gish 1996).

Note that the higher the value of H for a scoring matrix–gap penalty combination, the smaller the correction to the sequence length in Equations 4 and 5. Hence, to obtain alignments with shorter sequences, a scoring system with a higher H value is the most suitable combination. For example, for protein queries in the length range 50–85, the BLAST help pages recommend using BLOSUM80 with gap penalties (–10,–1) instead of BLOSUM62 with gap penalties (–11,–1) because the value of H is higher. To see these recommendations, click on the matrix link on the BLASTP page. For the BLOSUM62 scoring matrix and ungapped alignments, these values are $K = 0.14$ and $\lambda = 0.318$. The probability of the HSP score given by the above equation is adjusted to account for the multiple comparisons performed in the database search. The expect value E of observing a score $S \geq x$ in a database of D sequences is approximately given by the Poisson distribution,

$$E \approx 1 - e^{-p(s > x)D} \tag{6}$$

and for $p < 0.1$, E is approximately pD. The E value is the chance that a score as high as the one observed between two sequences will be found by chance in a search of a database of size D. Thus, $E = 1$ means that there is a chance that one unrelated sequence will be found in the database search. A similar expect value E is calculated by FASTA and SSEARCH. Note that the expect value E used to score HSP regions is not the same value of E reported by BLAST for the final local alignment scores. For HSP E values, K and λ derived from ungapped alignment scores between random sequences using BLOSUM62 (or the same matrix as the similarity search) are used. HSP E values determine which HSPs are significant enough to produce a local alignment of the sequences; the sequence is then evaluated using gapped alignment statistical parameters K and λ.

11. Sometimes, two or more HSP regions that can be made into a longer alignment will be found, thereby providing additional evidence that the query and database sequences are related. In such cases, a combined assessment of the significance will be made. Suppose that two sets of HSP scores are found between two query–database pairs of sequences in a similarity search; one set is 65 and 40, and the second 52 and 45. Which combination of scores is more significant, the one with the highest score (65 vs. 52) or the one with the higher of the lower score of each set (45 vs. 40)? Two methods have been used by BLAST for calculating this probability (Altschul and Gish 1996). One, the Poisson method, assumes that the probability of the multiple scores is higher when the lower score of each set is higher (45 is better than 40)? The other, the sum-of-scores method, calculates the probability of the sum of the scores. In this example, 65 + 40 = 105 is more significant than 52 + 45 = 97.

   Earlier versions of NCBI-BLAST use the Poisson method; WU-BLAST (Washington University BLAST) and gapped BLAST use the sum-of-scores method. The most recent versions of NCBI-BLAST2 perform a local gapped alignment of the sequences and calculate the expect value E of the alignment score. Such calculations became possible when it was realized that a statistical score could be calculated for gapped alignments (see Chapter 4, p. 141; Altschul and Gish 1996). To calculate the significance of the gapped alignment score, values of K and λ are determined on the basis of the alignment scores of random sequences using a combination of scoring matrix and gap penalties, and the methods described in Chapter 4 are used to calculate K and λ.

12. Smith–Waterman local alignments are shown for the query sequence with each of the matched sequences in the database. Earlier versions of BLAST produced only ungapped alignments that included the initially found HSP. If two HSPs were found, two separate alignments were produced because the two regions could not be aligned without gaps. Newer BLAST2 produces a single alignment with gaps that can include all of the initially found HSP regions. The procedure of aligning of sequences may be divided into subalignments of the sequences, one starting at some point in sequence 1 and going to the beginning of the sequences, and another starting at the distal ends of the sequences and ending at the start of the first alignment in sequence 1. A similar method is used to produce an alignment starting with the alignment between the central pair in the highest-scoring region of the HSP pattern as a seed for producing a gapped alignment of the sequences. The score of the alignment is obtained and the expect value E for that score is calculated using statistical parameters previously found for gapped alignments using the same scoring matrix and gap penalty combination used in the similarity search.

13. When the expect value E for the local alignment score of the query sequence with a database sequence satisfies the user-selectable threshold value, the match with the database sequence is reported. The results of the search are shown as a graphical representation of the sequence alignments, followed by a list of matches sorted by alignment score and E value, and then by the sequence alignments. An example of a BLASTP v2 output file is shown in Figure 6.5.

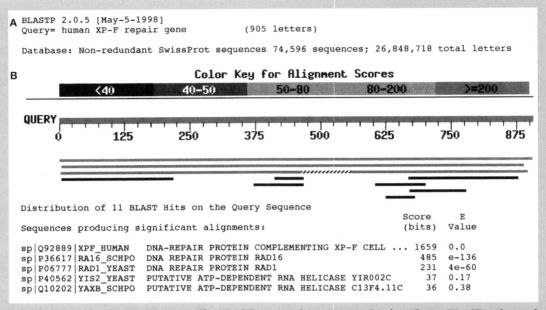

FIGURE 6.5. Example of BLASTP output. The BLAST server at http://www.ncbi.nlm.nih.gov/BLAST/, advanced version BLAST2, was given the human XP-F DNA repair sequence in FASTA format (providing the sequence accession number is another option). Program option BLASTP, database option SwissProt, and default program settings (gapped alignment, expect value $E = 10$, and low-complexity filtering), and ten descriptions and alignments were chosen. The expect value E is the number of matches expected by chance between the query sequence and random or unrelated database sequences from a database of the size used. If the program is allowed to report all of the matches that it finds, the number found will include at least this many matches with unrelated sequences in the database. The BLAST Web site has excellent help pages that should be consulted, especially when new updates of BLAST have revised Web pages. For example, one revised page did not provide an option for changing the amino acid substitution matrix from the default BLOSUM62 scoring matrix. The comments given below summarize the results of the above BLAST2 search. (A) BLAST version number, query sequence, and sequence database are identified. (B) First, a graphical representation of the extent to which database sequences match the query sequence is shown. Note that three database sequences can be aligned with the entire length of the query sequence and are therefore likely to be highly significant alignments. Other alignments found are only with portions of the query sequence. The mouse may be used to go directly to the alignments represented in the graph. The scores of the requested ten highest-scoring database sequences and the one identical database sequence are reported, one in each row. Each row includes the database sequence identifier where "sp" indicates a SwissProt match followed by SwissProt accession number and locus name, the score of the alignment in bits (see C), and the expect value (E) of the alignment. E values of 0.0 and e^{-136} (which is 10^{-36}) in the first and second rows indicate that the match is highly significant. The first match is to the query sequence itself and the next two matches are closely related to the query sequence as indicated by their low E values. (C) Gapped alignments between the query sequence and the matched database sequences are next shown (one example given). The query sequence is named as such and the database sequence is called the subject sequence. Filtering of a low-complexity region in the query sequence is indicated by the replacement of sequence by X. Gaps are indicated by a dash. Shown in each alignment are the sequence ID and length, and the score of the alignment in bits ("score" is the sum of log odds scores of each matching amino acid pair in the alignment less gap penalties; the raw score in bits is the log odds score in units of logarithms to the base 2). The score shown in the program output is in units of normalized bits = ($\lambda \times$ raw score – ln K) / ln 2. This number is independent of the scoring matrix used. The raw score in bits is also shown in parentheses. The expect value E of chance matches of unrelated sequences from a database of this size, percent identities in the alignment, percent positives in the alignment (identities plus positive scoring matches in the BLOSUM62 matrix), and percent of the alignment that is gaps are also shown. Statistical information about the search is provided at the end of the output, including the numbers found in the steps outlined above. The statistical parameters K and λ, which are different for gapped and ungapped alignments (Chapter 4), and the gap penalty scores are also shown. This information is useful as a basis for adjustment of the basic input parameters.

```
sp|Q92889|XPF_HUMAN DNA-REPAIR PROTEIN COMPLEMENTING XP-F CELL (XERODERMA PIGMENTOSUM GROUP F COMPLEMENTING PROTEIN
(DNA EXCISION REPAIR PROTEIN ERCC-4) Length = 905  Score = 1659 bits (4249), Expect = 0.0  Identities = 838/905(93%)
Positives = 838/905 (92%)


Query: 1    MAPLLEYERQLVLELLDTDGLVVCARGLGADRLLYHFLQLHCHPACLVLVLNTQPAEEEY 60
            MAPLLEYERQLVLELLDTDGLVVCARGLGADRLLYHFLQLHCHPACLVLVLNTQPAEEEY
Sbjct: 1    MAPLLEYERQLVLELLDTDGLVVCARGLGADRLLYHFLQLHCHPACLVLVLNTQPAEEEY 60
.


Query: 301  SLRATEKAFGQNSGWLFLDSSTSMFINARARVYHLPDAXXXXXXXXXXXXXXXXXXXXXXX 360
            SLRATEKAFGQNSGWLFLDSSTSMFINARARVYHLPDA
Sbjct: 301  SLRATEKAFGQNSGWLFLDSSTSMFINARARVYHLPDAKMSKKEKISEKMEIKEGEETKK 360
.

.

sp|P36617|RA16_SCHPO DNA REPAIR PROTEIN RAD16 Length = 892 Score =  485 bits (1236), Expect = e-136  Identities =
303/918 (33%), Positives = 497/918 (54%), Gaps = 76/918 (8%)


Query: 5    LEYERQLVLELLDTDGLVVCARGLGADRLLYHFLQLHCHPACLVLVLNTQPAEEEYFINQ 64
            L Y++Q+  EL++ DGL V A GL  ++  + L   P L+L++    + E   ++
Sbjct: 9    LAYQQQVFNELIEEDGLCVIAPGLSLLQIAANVLSYFAVPGSLLLLVGANVDDIELIQHE 68
.


Query: 304  -----ATEKAFGQNSGWLFLDSSTSMFINARARVYHLPDAXXXXXXXXXXXXXXXXXXXXX 358
                 ++  + Q S WL LD++  M  AR RVY  +
Sbjct: 309  LSVNVSSYPSNAQPSPWLMLDAANKMIRVARDRVYKESEGPNMDAIP------------- 355


sp|P06777|RAD1_YEAST DNA REPAIR PROTEIN RAD1 Length = 1100  Score =  231 bits (583), Expect = 4e-60  Identities =
136/369 (36%), Positives = 208/369 (55%), Gaps = 37/369 (10%)


Query: 559  LHEVEPRYVVLYDAELTFVRQLEIYRASRPGKPLRVYFLIYGGSTEEQRYLTALRKEKEA 618
            L E+ P Y+++++ +++F+RQ+E+Y+A      +VYF+ YG S EEQ +LTA+++EK+A
Sbjct: 704  LQEMMPSYIIMFEPDISFIRQIEVYKAIVKDLQPKVYFMYYGESIEEQSHLTAIKREKDA 763
.


sp|P40340|TBP7_YEAST TAT-BINDING HOMOLOG 7  Length = 1379  Score = 31.7 bits (70), Expect = 5.6   Identities = 12.
(21%), Positives = 29/55 (51%)

Query: 625  EKASMVVPEEREGRDETNLDLVRGTASADVSTDTRKAGGQEQNGTQQSIVVDMRE 679
            +K  V+PE+   +E  +L++  T ++++TD  +  +E  + S+   + E
Sbjct: 1209 DKEKAVIPEDSGANEEYTTELIQATCTSEITTDDDERARKEPKENEDSLQTQVTE 1263

D.

Database: Non-redundant SwissProt sequences
Number of letters in database: 26,848,718
Number of sequences in database:  74,596
Lambda      K        H
0.320     0.136     0.394
Gapped
Lambda      K        H
0.270    0.0470     0.230
Matrix: BLOSUM62
Gap Penalties: Existence: 11, Extension: 1
Number of Hits to DB: 42777291
Number of Sequences: 74596
Number of extensions: 1706128
Number of successful extensions: 4638
Number of sequences better than 10.0: 12
Number of HSP's better than 10.0 without gapping: 4
Number of HSP's successfully gapped in prelim test: 8
Number of HSP's that attempted gapping in prelim test: 4616
Number of HSP's gapped (non-prelim): 16
length of query: 905
length of database: 26848718
effective HSP length: 55
effective length of query: 850
effective length of database: 22745938
effective search space: 19334047300
effective search space used: 19334047300
T: 11
A: 40
X1: 16 ( 7.4 bits)
X2: 38 (14.8 bits)
X3: 64 (24.9 bits)
S1: 41 (21.8 bits)
S2: 68 (30.9 bits)
```

FIGURE 6.5. (*See facing page for legend.*)

Sequence Filtering

As discussed above, low-complexity regions have fewer sequence characters in them because of repeats of the same sequence character or pattern. These sequences produce artificially high-scoring alignments that do not accurately convey sequence relationships in sequence similarity searches. Regions of low complexity or repetitive sequences may be readily visualized in a dot matrix analysis of a sequence against itself (see Chapter 3, p. 79). Low-complexity regions with a repeat occurrence of the same residue can appear on the matrix as horizontal and vertical rows of dots representing repeated matches of one residue position in one copy of the sequence against a series of the same residue in the second copy. Repeats of a sequence pattern appear in the same matrix as short diagonals of identity that are offset from the main diagonal (see Fig. 3.6). Such sequences should be excluded from sequence similarity searches.

The BLAST programs include a feature for filtering the query sequence through programs that search for low-complexity regions. Filtering is only applied to the query sequence and not to the database sequences. Low-complexity regions are marked with an X (protein sequences) or N (nucleic acid sequences) and are then ignored by the BLAST program. Removing low-complexity and repeat sequences increases emphasis on the more significant database hits. The NCBI programs SEG and PSEG are used to mask amino acid sequences, and NSEG is used to mask nucleic acid sequences (Wootten and Federhen 1993, 1996). The SEG programs are available by anonymous FTP from ncbi.nlm.nih.gov/pub/seg, including documentation. The program DUST is also used for DNA sequences (see http://www.ncbi.nlm.nih.gov/BLAST/filtered.html). RepeatMasker (described below) is another program for this same purpose.

The compositional complexity in a window of sequence of length L is given by (Wootten and Federhen 1996):

$$K = 1/L \log_N (L!/\prod_{all\ i} n_i!) \qquad (7)$$

where N is 4 for nucleic acid sequences and 20 for protein sequences, and n_i are the numbers of each residue in the window. K will vary from 0 for very low complexity to 1 for high complexity. Thus, complexity is given by:

For the sequence GGGG,
$$L! = 4 \times 3 \times 2 \times 1 = 24$$
$$n_G = 4,\ n_C = 0,\ n_T = 0,\ n_A = 0$$

$$\prod_{All\ i} n_i = 4 \times 3 \times 2 \times 1 \times 0! \times 0! \times 0!$$
$$= 24 \times 1 \times 1 \times 1 = 24$$
$$K = 1/4 \log_4 (24/24) = 0$$

For the sequence CTGA,
$$L! = 24$$
$$n_G = n_C = n_T = n_A = 1$$

$$\prod_{All} n_i = 1$$
$$K = 1/4 \log_4 (24/1) = 0.573$$

Compositional complexities are sometimes calculated to produce K scores in bit units of logarithms to the base 2. A sliding window (usually 12 residues) is moved along the sequence, and the complexity is calculated at each position. Regions of low complexity are identified using Equation 7, neighboring low-complexity regions are then joined into longer regions, and the resulting region is then reduced to a single optimal segment by a minimization procedure. The SEG program is used for analysis of either proteins or nucleic acids by the above methods. PSEG and NSEG are similar to SEG but are set up for analysis of protein and nucleic acid sequences, respectively. These versatile programs may also be used for locating specific sequence patterns that are characteristic of exons (Chapter 9) or protein structural domains (Chapter 10). In database searches involving comparisons of genomic DNA sequences with EST sequence libraries, use of repeat masking is important for filtering output to the most significant matches because of the presence of a variety of repetitive sequences ranging from mononucleotide repeats to larger repeated elements in genomes (Claverie 1996).

In addition to low-complexity regions, BLAST will also filter out repeat elements (such as human SINE and LINE retroposons; see Chapter 11). Another filtering program for repeats of periodicity less than ten residues called XNU (Claverie and States 1993) is used by the BLAST stand-alone programs, but is not available on the NCBI server.

Another important Web server, RepeatMasker (http://ftp.genome.washington.edu/), screens sequences for interdispersed repeats known to be present in mammalian genomes and also can filter out low-complexity regions (A.F.A. Smeet and P. Green, see Web site above). A dynamic programming search program, cross-match (P. Green, see Web site), performs a search of a repeat database with the query sequence (Claverie 1996). A database of repetitive elements (Repbase) maintained at http://www.girinst. org/ by the Genetics Information Research Institute (Jurka 1998) can also be used for this purpose.

Other BLAST Programs and Options

The extensive set of BLAST resources, including programs and sequence databases, is fully described on the NCBI site map at http://www.ncbi.nlm.nih.gov/ Sitemap/index.html. There are a number of variations of the BLAST program for comparing either nucleic acid or protein query sequences with nucleic acid or protein sequence databases. If necessary, the programs translate nucleic acid sequences in all six possible reading frames to compare them to protein sequences.

These BLAST programs are shown in Table 6.4 along with the types of alignment, gapped or ungapped, that they produce. Table 6.5 lists the databases available, and Table 6.6 lists the options and parameter settings available on the BLAST server. These various options are also described on the main BLAST Web page at http://www.ncbi.nlm.nih. gov/BLAST. The results produced by a sample BLASTP version 2 output are shown and described in Figure 6.5.

1. MEGABLAST is a version of BLAST which uses a "greedy" algorithm that is optimized to search for strongly similar sequences that are ~300–100,000 bp long as in EST or genome DNA fragment comparisons. A long word length is used for searching, and the gap penalty is calculated from the match and mismatch scores.

2. RPS BLAST searches through a protein sequence for conserved domains using a domain database (type E search, Table 6.1).

3. CDART searches through a protein sequence for conserved domains using RPSBLAST and then reports GenPept sequences that share one or more of these domains.

4. BLAST CLIENT (BLASTcl3) is a network-client BLAST that may be established on a local machine and used to access the BLAST server (point browser to ftp://ftp.ncbi.nih.gov/blast/ blastcl3/) rather than using a Web browser.

5. Stand-alone BLAST. Executable versions of all of the BLAST programs for Windows, Macintosh, and UNIX platforms are available (point browser to ftp://ftp.ncbi.nih.gov/blast/executables/).

6. PSI-BLAST and PHI-BLAST perform iterative searches to locate conserved domains in a query protein sequence that are also present in database sequences or to locate domains in database sequences by starting with a regular expression pattern, respectively, and are described below in greater detail.

TABLE 6.4. *BLAST programs provided by the National Center for Biotechnology Information*

Program	Query sequence	Database	Type of alignment[a]
BLASTP	protein	protein	gapped
BLASTN	nucleic acid	nucleic acid	gapped
BLASTX	translated nucleic acid[b]	protein	each frame gapped
TBLASTN	protein	translated nucleic acid[b]	each frame gapped
TBLASTX[c]	translated nucleic acid[b]	translated nucleic acid[b]	ungapped

[a] Type of alignment available between query and database sequences in BLAST2. A gapped alignment is usually preferred, if available. BLASTX and TBLASTN generate gapped alignment for each reading frame found and may use sum statistics. TBLASTX provides only ungapped alignments and sum statistics. Ungapped alignments available as option for BLASTP and BLASTN.

[b] Nucleic acid sequence is translated in all six possible reading frames and then compared to the protein sequence.

[c] TBLASTX is a heavy user of computer resources and therefore cannot be used with the nr nucleic acid database on the BLAST Web page.

TABLE 6.5. *Databases available on BLAST Web server*

Database/Description

A. Peptide Sequence Databases

nr
All non-redundant GenBank CDS translations+RefSeq Proteins+PDB+SwissProt+PIR+PRF

swissprot
Last major release of the SwissProt protein sequence database (no updates)

pat
Proteins from the Patent division of GenPept

Yeast
Yeast (*Saccharomyces cerevisiae*) genomic CDS translations

ecoli
Escherichia coli genomic CDS translations

pdb
Sequences derived from the three-dimensional structure from Brookhaven Protein Data Bank

Drosophila genome
Drosophila genome proteins provided by Celera and Berkeley *Drosophila* Genome Project (BDGP)

month
All new or revised GenBank CDS translation+PDB+SwissProt+PIR+PRF released in the last 30 days

B. Nucleotide Sequence Databases

nr
All GenBank+RefSeq Nucleotides+EMBL+DDBJ+PDB sequences (but no EST, STS, GSS, or phase 0, 1, or 2 HTGS sequences); no longer "non-redundant"

est
Database of GenBank+EMBL+DDBJ sequences from EST Divisions

est_human
Human subset of GenBank+EMBL+DDBJ sequences from EST Divisions

est_mouse
Mouse subset of GenBank+EMBL+DDBJ sequences from EST Divisions

est_others
Non-Mouse, non-Human sequences of GenBank+EMBL+DDBJ sequences from EST Divisions

gss
Genome survey sequence, includes single-pass genomic data, exon-trapped sequences, and Alu PCR sequences

htgs
Unfinished high-throughput genomic sequences: phases 0, 1, and 2 (finished, phase 3 HTG sequences are in nr)

pat
Nucleotides from the Patent division of GenBank

yeast
Yeast (*Saccharomyces cerevisiae*) genomic nucleotide sequences

mito
Database of mitochondrial sequences

vector
Vector subset of GenBank(R), NCBI, in ftp://ftp.ncbi.nih.gov/blast/db/

E. coli
Escherichia coli genomic nucleotide sequences

pdb
Sequences derived from the three-dimensional structure from Brookhaven Protein Data Bank

***Drosophila* genome**
Drosophila genome provided by Celera and Berkeley *Drosophila* Genome Project (BDGP)

month
All new or revised GenBank+EMBL+DDBJ+PDB sequences released in the last 30 days

alu
Select Alu repeats from REPBASE, suitable for masking Alu repeats from query sequences. It is available by anonymous FTP from ftp.ncbi.nih.gov (under the /pub/jmc/alu directory). See "Alu alert" by Claverie and Makalowski (1994)

TABLE 6.5. (*Continued.*)

Database/Description

dbsts
  Database of GenBank+EMBL+DDBJ sequences from STS Divisions
chromosome
  Searches complete genomes, complete chromosome, or contigs from the NCBI Reference Sequence project

C. Human Genome Blast Databases
  genome
    Human genomic contig sequences with NT_#### accessions
  mrna
    Human RefSeq mrna with NM_#### or XM_#### accessions
  protein
    Human RefSeq proteins with NP_#### or XP_#### accessions
  gscan mrna
    Predicted mRNA sequences generated by running GenomeScan program on human genomic contigs
  gscan protein
    CDS translations from gscan mrna set

D. CDD Search
  Compares protein sequences to the conserved Domain Database. The CDD is a database containing a collection of functional and/or structural domains derived from two popular collections, Smart and Pfam, plus contributions from colleagues at NCBI. For more information, see the CDD homepage.

Source: http://www.ncbi.nlm.nih.gov/blast/html/blastcgihelp.html#protein_databases.

TABLE 6.6. *Options and parameter settings available on the BLAST server*

Parameter	Range of choices or values	Function
Limit by ENTREZ query		as used to limit ENTREZ query restricts databases search
Filter (low complexity)	yes or no	either apply filter or not; removes regions of low sequence; can apply selectively complexity or human repeats
Word size	2–3 proteins, 7–17 nucleotides	lower number increases sensitivity; higher number increases speed
Descriptions	0–500	number of matching sequences to report
Alignments	0–500	number of alignments to show
Expect value (*E*)	0.001–1000	number of matches from unrelated sequences expected by chance from the selected database; smaller values decrease chance of reporting of such matches
Genetic code	various codons	for translation of nucleic acid sequence; use tables[a]
Graphical	various choices	useful display of matches to the sequence query overview; mouse may be used to show alignment
Advanced options	—	type into space provided[b]

[a] Codon tables include standard, vertebrate mitochondrial, yeast mitochondrial, mold mitochondrial, invertebrate mitochondrial ciliate nuclear, echinoderm mitochondrial, euplotid nuclear, bacterial, alternative yeast nuclear, ascidian mitochondrial, flatworm mitochondrial, and blepharisma macronuclear. These are numbered 1–15, respectively, for E-mail access.

[b] Options include (where *n* is an integer 0,1,2, . . .): –G *n*, penalty or cost to open a gap; –E *n*, penalty to extend a gap; –*q n*. penalty for a mismatch in BLASTN; –*r n*, match score in BLASTN; –*W n*, initial word size; –*v n*, number of descriptions; *b n*, number of alignments to show; and –*E r*, expect value, where *r* is a real number such as 10.0. For example, to set the gap opening penalty to 10 and the gap extension penalty to 2, click the mouse on the advanced options form and then type –G 10 –E 2. For more advanced searches of the entire proteome of an organism using stand-alone BLAST on a local machine, additional options must be used, e.g., effective database size, to obtain reliable statistical results.

Other BLAST-related Programs

1. *BLAST-enhanced alignment utility (BEAUTY).* BEAUTY adds additional information to BLAST search results, including figures summarizing the information on the locations of HSPs and any known protein domains and sites such as PFAM domains and Prosite patterns described in Chapter 10 that are present in the matching database sequences (Worley et al. 1995). To make this enhanced type of analysis possible, a database of domains and sites was created for use with the BEAUTY program. A new database of sequence domains and sites was made showing for each sequence in Entrez the possible location of patterns in the Prosite catalog, the BLOCKS database, and the PRINTS protein fingerprint database. The BEAUTY program is accessible on the BCM Search Launcher (Table 6.2).

2. *BLAST searching with a Cobbler sequence.* The BLOCKS server (http://www.blocks.fhcrc.org) offers a variety of BLAST searches that use as a query sequence a consensus sequence derived from multiple sequence alignment of a set of related proteins. This consensus sequence, called a Cobbler sequence (Henikoff and Henikoff 1997), is used to focus the search on residues that are in the majority in each column of the multiple sequence alignment, rather than on any one particular sequence. Hence, the search may detect additional database sequences with variation unlike that found in the original sequences, yet still representing the same protein family. An example of a Cobbler sequence is shown in the BLOCKS search example on page 265.

3. *BLAST2.* This program uses the BLASTP or BLASTN algorithms for aligning two sequences and may be reached on the BLAST server site at NCBI. This program is useful for aligning very long sequences, but sequences >150 kb are not recommended.

SMITH–WATERMAN DYNAMIC PROGRAMMING METHOD PROVIDES OPTIMAL RESULTS

The objective of similarity searching in a sequence database is to discover as many sequences as possible that are similar to the query sequence. For proteins, the resulting collection of sequences may represent a sequence family. Because fewer than 20% of the amino acid pairs in a sequence alignment of some family members may be identical, finding such distant relatives is a difficult task. The aforementioned programs, FASTA and BLAST, are designed to find database sequences related to a query sequence rapidly and with high reliability. They achieve their speed by searching first for short identical patterns in the query sequence and each database sequence and then by aligning the sequences starting at these patterns. Because patterns are very often found in related sequences, the methods work most of the time. FASTA and BLAST are not based on an algorithm that guarantees the best or optimal alignment, but instead on a heuristic method that works most of the time in practice; thus, they may fail to detect some distant sequence relationships.

The Smith–Waterman dynamic programming algorithm discussed in Chapter 3 is mathematically designed to provide the best or optimal alignment between two sequences and is therefore expected to be the most reliable method for finding family members in a database search for a given scoring matrix and gap penalty combination. Several studies discussed below have shown that such is the case. The disadvantage of using dynamic programming is that it is 50–100 times slower than FASTA and BLAST, and until recently, a search could take up to several hours on a typical medium-sized computer. Myers and Durbin (2003) have determined that a set of tables based on the query sequence can be built and used to eliminate calculations for much of the dynamic programming matrix and to simplify calculations within the remainder of the matrix.

With the advent of faster and more powerful computers and improvements in the dynamic programming algorithm discussed in Chapter 3, it has become possible to perform Smith–Waterman database searches in an hour or less. Some institutions have gone so far as to establish a powerful system of several computers linked together in a parallel architecture that allows a search to be performed within minutes. Several of the sites listed in Table 6.5 offer public access through the Web. It is important to examine the site for use of up-to-date databases and use of an appropriate statistical analysis. Detection of distant sequence relationships depends on use of the statistical methods that have been developed for BLAST and FASTA.

For routine use of dynamic programming methods for database searches, installing the program SSEARCH (FTP to ftp.virginia.edu/pub/fasta; Pearson 1991; Pearson and Miller 1992) and the appropriate sequence databases on a local UNIX server is recommended. In several studies (Pearson 1995, 1996, 1998; Agarwal and States 1998; Brenner et al. 1998), it has been shown that using SSEARCH, which is based on the Smith–Waterman dynamic programming algorithm, is more suitable for identifying related proteins of limited sequence similarity than FASTA and BLAST in a database search. In several of these studies, known members of protein families are used as a query sequence searching for the remaining family members in a protein sequence database. In another study, the performance of the sequence analysis methods was determined using protein sequences of known structural relationships (Brenner et al. 1998). The results are presented in terms of the sensitivity and selectivity of the algorithm, or the ability to identify correct family members, including some that are only weakly similar, without incorrect-

ly identifying other unrelated proteins as members (Pearson 1995, 1998). The ability to discriminate true from false matches depends on the use of appropriate amino acid substitution matrices, gap opening and extension penalties that provide local alignments, and a careful statistical analysis of the search results using the extreme value distribution to predict scores from unrelated sequences (Brenner et al. 1998; Pearson 1998). The program SSEARCH has the necessary features for discriminating true from false matches and is available for database searches.

The reliability of the statistical scores reported by FASTA, BLAST2, and SSEARCH has been determined using sequences of known structural relatedness as a guide. The E values reported by FASTA and SSEARCH are reliable, with the number of false positives agreeing with the scores. BLAST2 E-value scores also appear to be reliable (see Brenner et al. 1998). Brenner et al. (1998) also compared the ability of these programs to identify proteins in the families of the SCOP structural database (Murzin et al. 1995, and see Chapter 9) and reported that, in most cases and for the most recent versions, the calculated E values reflect the correct structural relationships. The output of an SSEARCH version 3 database search is similar to that of the FASTA search shown in Figure 6.3 (p. 244). Several guest Web sites for performing a database search with the Smith–Waterman dynamic programming algorithm are listed in Table 6.7.

BAYES BLOCK ALIGNER DATABASE SEARCHES DETECT DISTANT SEQUENCE RELATIONSHIPS

From the discussion so far, it is apparent that the fastest and most convenient way to perform sequence database searches is with the FASTA and BLAST2 programs. The much slower Smith–Waterman dynamic programming programs, such as SSEARCH, may, however, be more effective in finding more distantly related sequences. The significance of the alignment scores can be accurately evaluated by all of these programs.

An even better method than Smith–Waterman searches for detection of distant sequence relationships is the Bayes block aligner (Zhu et al. 1998), which was previously discussed in Chapter 4 (p. 152). This program requires several series of computational steps roughly proportional to the product of the

sequence lengths and is therefore considerably slower than SSEARCH. As an indication of the length of time required by the Bayes block aligner for a sequence similarity search, the author of the present text estimates that a search of SwissProt with a 300-amino-acid query sequence on a 2-GHz desktop computer with 1 Gbyte of memory running the Linux operating system will take approximately 60 hours.

Programs for finding related proteins are evaluated on the basis of database searches for protein families using sequence similarity (Pearson 1998). A more difficult type of evaluation is based on the searches of structural databases (Brenner et al. 1998). In these databases, discussed in Chapter 10, sequences have been organized into families having similar three-

TABLE 6.7. *Examples of guest Web sites for performing a database search based on the Smith–Waterman dynamic programming algorithm*

Server/program	Reference	Web address
BCM Search Launcher (with programming links to several servers)	Baylor College of Medicine	http://searchlauncher.bcm.tmc.edu/ seq-search/protein-search.html
MPsrch[a]	EMBL/EBI	http://ebi.ac.uk/MPsrch
Scanps	G.Barton, European Bioinformatics Institute	http://www.compbio.dundee.ac.uk http://www.ebi.ac.uk/scanps
Swat[b]	Phil Green, University of Washington	http://www.genome.washington.edu/UWGC/ analysistools/Swat.cfm
SWsrch	DNA Databank of Japan	http://www.dna.affrc.go.jp/search

A comprehensive list of servers for these types of analyses may be found at http://restools.sdsc.edu/biotools/biotools1.html.

[a] MPSearch is an extremely fast implementation of the Smith–Waterman dynamic programming algorithm by J.F. Collins and S. Sturrock, Biocomputing Resource Unit, the University of Edinburgh, distribution rights by Oxford Molecular Ltd. Some versions of the MPsearch algorithm at this site use the same penalty for all gaps, others use gap opening and extension penalties. The former is designed to find similar sequences in which gaps are less important in the alignment, the latter the more distant sequence alignments. Current versions of these programs rank the sequences found by two kinds of scoring systems. A statistical analysis is performed but the scores do not appear to be length-normalized. Hence, the sensitivity of the program may not exceed that shown by FASTA (Pearson 1996).

[b] Includes Smith–Waterman and Needleman–Wunsch search algorithms. Calculates statistical significance using extreme value statistics (like FASTA and BLAST).

dimensional structures. Three of these databases representing groups of proteins that have less than 25%, 35%, or 45% identities (Hobohm et al. 1992) were searched using representatives of structural families in each. In each case, the Bayes block aligner slightly but significantly outperformed SSEARCH in finding structural relatives. For example, at the 1% false-positive level, the Bayes block aligner found an average of 14.4% of the proteins in the less-than-25% identity group, whereas SSEARCH with usual scoring matrix, gap penalties, and statistical score options found 12.9%, a difference of 1.5%. In addition, the Bayes block aligner can align sequences that have very little similarity, but can provide alignments that closely match those found by a careful structural analysis described in Chapter 9 using the VAST program (Madej et al. 1995).

The Bayes block aligner uses a new method for producing sequence alignments. The method, discussed in detail in Chapter 4 (p. 152), starts by finding all possible ungapped patterns (blocks) that are located in the two sequences. A large number of possible alignments between two sequences are generated by aligning combinations of blocks. Gaps will be present between the blocks, as illustrated in Figure 6.6. The sequence alignments are scored only in the regions where the sequences are aligned in blocks. There is no gap penalty as in the dynamic programming method of alignment.

Alignments are also scored differently by the Bayes block aligner than by the dynamic programming method. In the Bayes block aligner, a set of amino acid substitution matrices is used. Each scoring matrix models a different degree of substitution between the sequences, and the matrices that best represent this degree should give the highest alignment scores. When PAM-type matrices are used, the evolutionary distance between parts of sequences can be estimated knowing the best-scoring matrix. After the analysis has been completed, the choices of block number, alignments, and scoring matrices can be examined.

By using a Bayesian statistical analysis of the results, it is possible to derive block alignments in which amino acids in each sequence are most often associated, regardless of the many possible choices of block number and scoring matrix. It is these alignments that are statistically the best representation of the alignment between two sequences. The probability that the sequences are related can also be calculated from Bayesian statistical principles. An example of sequence alignment using the Bayes block aligner is given in Chapter 4 (p. 157).

```
Sequence 1 xxxxxx---o--xxxxxxxxxxxooo-oooxxxxxx
Sequence 2 xxxxxxooooooxxxxxxxxxxo-ooo-oxxxxxx
           block1       block2        block3
```

FIGURE 6.6. Alignment found by the Bayes block aligner. The alignment between two sequences includes ungapped blocks (marked by x where aligned x's may be identical or substitutions; there will be at least one identity in each block used to identify the block) and intervening unaligned regions with gaps (marked by o for unaligned residue and – for a gap). These two regions are designed to represent conserved structural alignments in the protein core and variable surface loops, respectively. A large number of alignments of this type involving different combinations of blocks are found. These alignments are then evaluated by a set of scoring matrices. The best alignment is then derived by a Bayesian statistical analysis, described in Chapter 4.

SCORING MATRIX OR PROFILE DATABASE SEARCHES ARE PERFORMED WITH MULTIPLE SEQUENCE ALIGNMENTS

The methods for database searching discussed so far in this chapter are based on using a single query sequence to search a sequence database. Another method of database searching is to use the variation found in a multiple sequence alignment of a set of related sequences to search for matching database sequences. This enhanced type of search will locate database sequences that match new combinations of sequence characters in the multiple sequence alignment. For example, if column 1 of a multiple sequence alignment includes the amino acids P and Q, and column 2 the amino acids D and E, then database sequences that match all four combinations of these two amino acids can be found, whereas only the combinations found in the original sequences would be matched if single query sequences were to be used.

Multiple sequence alignments reveal the sequence variation found in a set of three or more sequences. As described in Chapter 5, the alignment may be global, including all parts of each sequence, or local, including only aligned portions of the most similar parts of each sequence. Local alignments may be relatively short or may include long conserved stretches of sequence. Two methods for identifying local alignments in sequences were described in Chapter 5. The first method, sequence extraction, removes conserved regions of aligned sequence from a global multiple sequence alignment, previously produced by a program such as CLUSTALW. The second method, statistical pattern searching, uses pattern-finding and statistical methods to locate common sequence patterns in unaligned sequences and aligns them using these found patterns. Once a local alignment has been found by one of the above methods, it

is represented by a position-specific scoring matrix, sequence profile, or profile hidden Markov model as discussed below. Chapter 5 should be consulted for a discussion of these local alignment tools; the relevant programs and Web sites are described below.

The simplest method to search a sequence database for matches to a local alignment of sequences is to store the alignment information in a scoring matrix. The simplest scoring matrix, the position-specific scoring matrix (PSSM), represents an alignment of sequence patterns of the same length with no gaps. The production of a PSSM is also discussed in Chapter 5 (p. 211). To summarize, the sequence patterns are first aligned as a local multiple sequence alignment so that corresponding residues are in the same column. Raw amino acid counts are first found by summing the numbers in each column of the alignment, and these numbers are placed in the corresponding columns of the scoring matrix, one for each amino acid in a designated row. These counts are then adjusted by a weighting method designed to prevent overrepresentation of the amino acids in the more closely related sequences. Otherwise, the matrix would be more tuned to those sequences than to the less-alike ones in the group. To these raw scores, additional counts are added based on previously observed general types of amino acid variations in alignments of related proteins.

The idea behind this strategy is that the small number of sequences usually present in these alignments does not represent the full range of expected amino acid variations. Therefore, additional pseudocounts are added based on substitution patterns found in an amino acid substitution matrix or representative blocks in the BLOCKS database (Dirichlet mixtures). The statistical basis for adding counts is that including prior information in the form of pseudocounts should increase the sensitivity of the scoring matrix. The sum of the raw and additional counts in each column is then divided by the expected frequency of the amino acid from the sequence data or from other sources. The resulting ratio represents the odds for finding a match of another related sequence to the column divided by the chance of a random match with an unrelated sequence. For ease in multiplying probabilities by adding their logarithms, each odds score is converted to a log odds score, usually logarithms to the base 2. Each column

in the sequence alignment is placed in a corresponding column of the matrix, and there is one row of scores for each amino acid. The resulting PSSM is easy to align with a sequence, as discussed below.

PSSMs are useful for representing sequence variation on local multiple sequence alignments that does not include gaps. When gaps are present, a sequence profile method is used. Sequence profiles are similar to PSSMs but include extra rows with gap penalties. More commonly, profile hidden Markov models (Durbin et al. 1998) are used for representing gapped alignments. Sequence profiles (Gribskov and Veretnik 1996) may be used to search a sequence database for matches to the aligned sequences, as described below.

POSITION-SPECIFIC SCORING MATRIX OR SEQUENCE PROFILE DATABASE SIMILARITY SEARCHES ARE USEFUL FOR DISCOVERING PROTEIN FAMILIES

Alignment of a PSSM with a protein sequence is illustrated in Figure 6.7. Every possible sequence position is scored by sliding the matrix along the sequence one position at a time. The amino acid substitution scores in each column of the PSSM are used to evaluate each sequence position. Positions with the highest scores are the best matches of the corresponding set of sequence patterns with the sequence. In searches of a sequence database, those sequences with a region that is a close match to the pattern will produce the highest scores and may be readily identified.

The problem of finding the best alignment between a sequence profile (a PSSM with gap scores) and a given position in a sequence is similar to the problem of aligning two sequences. As with alignment of sequences, the dynamic programming algorithm is used, except that the match scores and gap penalties are site-specific and are the values given in the profile columns. Scoring matrices that correspond to a sequence profile include two extra rows for gap penalty scores (sometimes these scores may be found in extra columns if the labeling of the rows and columns is reversed). When aligning a sequence profile with a sequence, a procedure similar to that for PSSMs described above is followed in that the score for matching the profile scoring matrix to each sequence character is calculated. In addition, a gap of any length may be inserted into the sequence or profile at that position, and the gap penalties are those given in the relevant column of the profile. The gap

penalties are usually quite high with respect to the match scores, but are less when gaps were present in the original multiple sequence alignment.

Web sites and programs for finding common motifs and profiles in a set of related sequences, or for searching a protein sequence database with these patterns, are listed in Table 6.8. Also shown are sites that can be given an ambiguous pattern, called a regular expression, which is a highly flexible pattern format that includes substitutions at certain sites, repeated patterns, and intervening regions that are variable. An example is given in the box below (p. 268). The first programs available for producing profiles and for sequence searches were Profilemake for making profiles from a multiple sequence alignment, Profilegap for aligning a profile with one or more sequences, and Profilesearch for searching a protein sequence database with a profile (Gribskov and Veretnik 1996). These programs are best known as components of the Genetics Computer Group (GCG) suite of programs. Profiles produced by newer versions of these programs use evolutionary predictions of the amino acid changes in each column, which improves the ability of the profile to find related proteins in a database search. Methods for making evolutionary profiles and for using them are discussed in Chapter 5 (p. 191). Profile searches may be performed at two supercomputer centers (Table 6.8). The standard GCG multiple sequence alignment format, the MSF file (described in Chapter 2), is used as input to these programs.

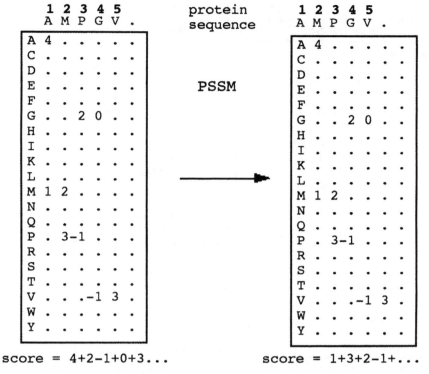

FIGURE 6.7. Scoring an alignment of a PSSM with a sequence. Only a few matrix values are shown for simplicity. The PSSM is first aligned with sequence position 1 so that columns 1–10 match positions 1–10 in the sequence. The column values that match each sequence position are then added to give a total log odds score for sequence position 1. The PSSM is then moved to sequence position 2 and a new score for matches with sequence positions 2–11 is calculated. This process if repeated until the last ten sequence positions are matched. The highest scores represent the best matches of the motif represented by the PSSM to the sequence. This type of scoring system preferentially rewards matches to columns that have a conserved amino acid more than to matches with variable columns, and penalizes mismatches to columns that have a conserved amino acid more heavily than mismatches to variable columns.

READSEQ and other sequence reformatting programs can be used to change the sequence format to the MSF format (see Chapter 2), which can then be used with Profilemake.

There is a difference in the ways PSSM and the profile matrix are generated that influence the results of a database search. The PSSM treats all amino acids as being equal, so that matching an Ala with an Ala is as significant as matching a Cys with a Cys. Scores for amino acid substitutions are based on the distribution of amino acids in each column of the alignment on which the PSSM is based. Profile scores are also based on the distribution of amino acids in each column of the alignment, but the matrix values are derived by using values from an amino acid substitution matrix, such as the Dayhoff PAM matrices, to score the substitutions found in each column of the

profile. Hence, the PSSM contains only information on sequence variation found in the multiple sequence alignment. In contrast, the profile uses the information on sequence variation in the local alignment to average global amino substitution scores found in many sequence alignments, as represented in the PAM and BLOSUM matrices. As a result, PSSM and profile scores for the same local alignment should give different results.

To illustrate these methods, the results of finding blocks by the BLOCKS server, and of using motifs found by the Multiple Expectation Maximization for Motif Elicitation (MEME) server, for a search of the SwissProt database by the Motif Alignment and Search Tool (MAST) are given below. The terms "blocks" and "motifs" are used interchangeably by these sites in that both mean a reasonably long,

TABLE 6.8. *Programs and Web sites for database similarity searches with a regular expression, motif, block, or profile*

Program	Database searched	Source or location of analysis
1. Regular expressions and motifs[a]		
EMOTIF Scan	SwissProt and Genpept	http://dna.stanford.edu/emotif/emotif-scan/html/
Prosite patterns	SwissProt and TrEMBL	http://au.expasy.org/tools/scanprosite/
ISREC pattern-finding service	SwissProt and non-redundant EMBL database	http://hits.isb-sib.ch/cgi-bin/hits_patsearch
fpat	PDB SwissProt Genpept	Web site not currently active
PHI-BLAST	BLAST databases	http://www.ncbi.nlm.nih.gov/
MOTIF	SwissProt, PDB, PIR, PRF, Genes	http://www.motif.genome.jp/
2. Blocks		
BLOCKS[b]	most databases	http://www.blocks.fhcrc.org/blocks/make_blocks.html
MAST[c]	most databases	http://meme.sdsc.edu/meme/website/
BLIMPS[d]	locally available databases	anonymous FTP ftp.ncbi.nih.gov/repository/blocks/unix/blimps
Probe[e]	BLAST databases	anonymous FTP ftp.ncbi.nih.gov/pub/neuwald/probe1.0
Genefind[f]	PIR	http://pir.georgetown.edu/gfserver
3. Profiles		
Profilesearch[g]	locally available databases	anonymous FTP ftp.sdsc.edu/pub/sdsc/biology/profile_programs
Profile-SS[h]	most databases	http://www.psc.edu/general/software/packages/profiless/profiless.html

These resources search for similarity to a sequence pattern. Resources for producing patterns from aligned or unaligned sequences are described in Chapter 4. An individual sequence may also be searched for matches to a motif database, and this procedure is discussed in Chapter 9. Additional resources for database searching are listed in Bork and Gibson (1996).

A statistical estimate of finding the site by random chance in a sequence is sometimes but not always given. Reading how these estimates are derived by the individual programs is strongly recommended. The statistical theory for sequence alignments described in Chapter 3 can be used in these types of analyses (Bailey and Gribskov 1998) but may not always be implemented.

[a] The Scan Web page shows how to compile a regular expression. Mismatches within the expression are allowed. The Prosite form of a regular expression is at http://www.expasy.ch/tools/scnpsit3.html. PHI-BLAST is a BLAST derivative that searches a given sequence for a regular expression and then searches iteratively for other sequences matching the pattern found, at each iteration including the newly found sequences to expand the search.

[b] The BLOCKS server will send a new block analysis to the MAST server.

[c] MAST is the Motif Alignment and Search Tool (Bailey and Gribskov 1998). Available protein databases are similar to those on the BLAST server. It is also possible to search translated nucleotide sequence databases.

[d] BLIMPS will prepare a PSSM from a motif and perform a database search with the PSSM (see README file on FTP site).

[e] PROBE (Neuwald et al. 1997) is described in the text.

[f] The GENEFIND site has the program MOTIFIND for Motif Identification by Neural Design (Wu et al. 1996). This motif finder uses a neural network design to generate motifs and a search strategy for those motifs. The method performed favorably in sensitivity and selectivity with others such as BLIMPS and Profilesearch and is in addition very fast. Neural networks are described in Chapters 8 and 9.

[g] Profilesearch is one of a set of programs in the GCG suite (see text). It is important to review the parameters of the program which if used inappropriately can lead to incomplete or low-efficiency searches (Bork and Gibson 1996).

[h] A version of Profilesearch running at the University of Pittsburgh Supercomputing Center.

ungapped pattern in a family of protein sequences. Once matching sequences have been found, other searches can be performed with sets of blocks or motifs of shorter or longer length and of one or more occurrences in the local alignment of the sequences. Use of the program MACAW, which runs on many computer platforms with a Windows interface, is also very useful for exploring motif size and number (see Chapter 5, p. 201). This program can find motifs either by an alignment method using an amino acid scoring matrix or by the statistical Gibbs sampling method. The relative positions of the found motifs are shown on a graphical representation of the sequences. With these methods, PSSM and profile representations of a local multiple sequence alignment may be found and then used to search protein sequence databases for members of protein families. Compared to searches with single query sequences, these methods provide a more robust kind of search that can locate more diverse family members.

EXAMPLE: BLOCKS SERVER

The BLOCKS Web server takes a set of unaligned sequences and finds blocks of sequence (matching ungapped patterns) that are present in the sequences. As an example, a request to the BLOCKS Web site to find blocks in three sequences similar to the human *XPF* DNA repair gene was made, and an example of the program output is shown in the following example (Fig. 6.8). The sequences were input into a Web form in FASTA format. The server finds blocks by two methods, a pattern-searching method called MOTIF and a statistical method called the Gibbs sampler. These two methods are described in detail in Chapter 5 (pp. 196 and 201). The blocks found by each method may not be the same because each method uses a different pattern-finding method. Examples of a representative block found by MOTIF, www.bloA, is shown in Figure 6.8A. Also shown is a portion of a Cobbler sequence of one of the input sequences, xpf95.pro, an *Arabidopsis* gene. In the Cobbler sequence, the sequence in xpf95.pro corresponding to the block location in each sequence has been replaced by a consensus sequence derived from bloA. These replaced regions are capitalized. In Figure 6.8B, an example of a Gibbs sampler block also called www.bloA is shown. A Cobbler sequence was also produced from these blocks (not shown). There are two options given after each list of blocks: (1) the Cobbler sequence may be used in a BLAST search and (2) the blocks may be sent to the MAST server to search a protein sequence database.

```
A.

**BLOCKS from MOTIF**

>www.blo xpf95.pro    957  a.a.  family
3 sequences are included in 15 blocks

            www.bloA, width = 30
gi|131810|    113 GKGLGLLDIVANLLHVLATPTSINGQLKRA
gi|2842712     27 GLGADRLLYHFLQLHCHPACLVLVLNTQPA
 xpf95.pro     27 GLSLAKLIASLLILHSPSQGTLLLLLSPAA


COBBLER sequence from MOTIF
>www.blo gi|2842712 from 1 to 905 with embedded consensus blocks
maplleyerqlvlelldtdglvvcarGLGLGRLIYHFLLLHCHPTCTVLVLQTKPAeeeyfinqlkiegvehlprrvtne
itsnsRYKLYTSGGVLFITSRILIVDLLTDRIPPNRITGILVLNAHSIRENCNEAFILRIYRSKNSWGFIKAFSDRPQAF
VTGFchvervmrnlfvrklylwprfhvavnsfleqhkpEVVEIRVSMTNTMVGIQFAIMECLNACLKELKRHNPsleved
lslenaigkpfdktirhyldPLWHRLGYKTKQLVKDLKFLRHLLQYLVQYDCVDFlnlLEALKPTEKAKYQNSPWLFVDS
SYKVFDYAKKRVYhlpdakmskkekisekmeikegeetkkEYVLEENPKWEALTEILHEIeaenkesealggpGPVLVCC
SDDRTCMQLrdyitlgaeafllrlyrktfekdskaeevwmkfrkedsskrirkshkrpkdpqNKERHVDKARCTKKKkrk
ltltqmvgkpeeleeegdveegyrreisssespescpeeikheefdvnlssdaafgilkepltiihpllgcsdpyaltrvlh
evePSYIIMYDPDLSFVRQLEVYKASNPGKPLKVYFLYYGESTEEQKYLTAIRREKEAFEKLIREKASMvvpeeregrde
tnldlvrgtasadvstdTRKAGGQQqngtqqsivvdmrefrselpslihrrgidiepvtlevgdyiltpemcverksiSD
LIGSLNNGRLYHQCEKMSRYYRYPVLLIEFDQDKSFSltsrgalfqeissndisskltlltlHFPRLRILWSPSPHATAE
IFTELKQNRDQPDaatalaitadsetlpesENYNPSPFEFLLKMPGVSKANYRSLMHKIKSFAELASLsqdeltsilgna
anakqlydfihtsfaevvskgkgkk

B.
            **BLOCKS from GIBBS**

>www.blo xpf95.pro    957  a.a.  family
3 sequences are included in 10 blocks

            www.bloA, width = 45

gi|131810|  201 EKRRKLYISGGILSITSRILIVDLLSGIVHPNRVTGMLVLNADSL
gi|2842712   84 NSRYEVYTQGGVIFATSRILVVDFLTDRIPSDLITGILVYRAHRI
 xpf95.pro   85 NQRYSLYTSGSPFFITPRILIVDLLTQRIPVSSLAGIFILNAHSI
```

FIGURE 6.8. Example of BLOCK output.

EXAMPLE: USING THE MEME TO PRODUCE A LOCAL ALIGNMENT PSSM AND SERVER TO USE THE PSSM FOR A DATABASE SIMILARITY SEARCH

The MAST server searches a protein database for best matches to a set of ungapped motifs or blocks (Bailey et al. 1997). The motifs may also be found by MEME by submitting unaligned sequences to the MEME server for analysis by a statistical method, the expectation maximization method, described in Chapter 5. The same three DNA repair sequences used above were input into a Web form in FASTA to the MEME server. To simplify output of the many possible choices, MEME was asked for one motif per sequence and for up to six different motifs of short length. Once received by E-mail, the motif messages were saved to a local file, and then this file was submitted to the MAST server (http://www.sdsc.edu/MEME) using the browse option. It is also possible to submit an already-found motif in GCG, MEME, or PSSM format (see http://www.sdsc.edu/MEME/meme/website/motif-format.html).

Another method for readily accessing MAST is through the BLOCKS server. As shown in Figure 6.8, unaligned sequences are searched for blocks by two methods, and from the BLOCKS Web page, the results may be immediately submitted to the MAST server. MAST uses the method shown in Figure 6.7 to align the blocks with each database sequence. If not specified otherwise, the output files are sent by E-mail as HTML files suitable for viewing by a Web browser. These files, which include useful graphical representations of the alignments, are first saved to a file and then opened with a Web browser.

Alternatively, the files may be requested in text format, as was done below (see Fig. 6.9). The initial list in the MAST output is of the motifs found by MEME. Note that motifs are given an ID number (1–6) that is used later in the MAST report. Section I then lists the scoring matches found in the SwissProt sequence database. The expect value (E) is the number of unrelated sequences in a database of the size of SwissProt that would achieve a score as high as the one shown with the motifs used in the search and is based on the scores of individual motifs with the sequence using the extreme value distribution described previously (Bailey and Gribskov 1998). The highest-scoring matches are with the two input DNA repair proteins, but there are also several lower-scoring matches with other proteins that interact with DNA, suggesting a common structural motif; however, caution is necessary in interpreting these kinds of matches (Bork and Gibson 1996). One of the input sequences was that of an *Arabidopsis* DNA repair protein that is not reported because it is not in the database yet.

Section II shows the locations of the motifs in each sequence. The motifs are shown in brackets and numbered as at the top of the file. Note that the order in the first three sequences is approximately the same, but that there are more and more variations going down the list, reflecting more divergence.

Finally, in section III, the matched motifs are aligned with the matched sequence. At each aligned position, the motif number, the E value of each match, the motif sequence giving the best match between sequence and motif, and a plus sign to indicate sequence letters corresponding to a positive match score in the motif column are given. A diagram shows the order of the motifs found and a combined P (combined probability for matching all matrices to an unrelated sequence) and the expect value E (number of matches expected with an unrelated sequence in a database of the size searched). The combined probabilities are calculated using the extreme value distribution as described above for FASTA and BLAST local alignment scores (Bailey and Gribskov 1998).

```
MAST - version 2.2
DATABASE swissprot contains 74596 sequences

     MOTIF WIDTH BEST POSSIBLE MATCH
     ----- ----- -------------------
       1    11    VGDYILTPDIC
       2     9    QCKMMSRYY
       3     8    YFMFYGES
       4     8    WPRFHVDV
       5     9    HFPRLRILW
       6     8    IVDMREFM
```

FIGURE 6.9. Example of MAST output. (*Continued on facing page.*)

```
SECTION I: HIGH-SCORING SEQUENCES

SEQUENCE NAME              DESCRIPTION                          E-VALUE  LENGTH
-------------             -----------                          -------  ------
sp|Q92889|XPF_HUMAN       DNA-REPAIR PROTEIN COMPLEMENTING XP... 5.3e-35    905
sp|P06777|RAD1_YEAST      DNA REPAIR PROTEIN RAD1               1.1e-31   1100
sp|P36617|RA16_SCHPO      DNA REPAIR PROTEIN RAD16              8.4e-23    892
sp|Q07864|DPOE_HUMAN      DNA POLYMERASE EPSILON, CATALYTIC S...   0.62   2257

SECTION II: MOTIF DIAGRAMS

SEQUENCE NAME             E-VALUE    MOTIF DIAGRAM
-------------             -------    -------------
sp|Q92889|XPF_HUMAN       5.3e-35    181-[4]-405-[3]-71-[6]-20-[1]-20-[2]-41-
                                     [5]-114
sp|P06777|RAD1_YEAST      1.1e-31    298-[4]-433-[3]-75-[6]-20-[1]-20-[2]-55-
                                     [5]-146
sp|P36617|RA16_SCHPO      8.4e-23    185-[4]-241-[2]-134-[3]-83-[6]-20-[1]-20-
                                     [2]-41-[5]-106
sp|Q07864|DPOE_HUMAN         0.62    190-[6]-175-[2]-381-[2]-426-[5]-34-[4]-
                                     366-[2]-478-[6]-147

SECTION III: ANNOTATED SEQUENCES

gi|548659|sp|P36617|RA16_SCHPO
  DNA REPAIR PROTEIN RAD16
    LENGTH = 892  COMBINED P-VALUE = 1.12e-27  E-VALUE =  8.4e-23
    DIAGRAM: 185-[4]-241-[2]-134-[3]-83-[6]-20-[1]-20-[2]-41-[5]-106

                                  [4]
                                  1.9e-07
                                  WPRFHVDV
                                  +++++ +
    151 TGFIKAFSDDPEQFLMGINALSHCLRCLFLRHVFIYPRFHVVVAESLEKSPANVVELNVNLSDSQKTIQSCLLTC

                                              [2]
                                              8.8e-05
                                              QCKMMSRYY
                                              +  ++++
    376 ETMLADTDAETSNNSIMIMCADERTCLQLRDYLSTVTYDNKDSLKNMNSKLVDYFQWREQYRKMSKSIKKPEPSK

                                              [3]
                                              6.8e-10
                                              YFMFYGES
                                              ++++++++
    526 NSIYIYSYNGERDELVLNNLRPRYVIMFDSDPNFIRRVEVYKATYPKRSLRVYFMYYGGSIEEQKYLFSVRREKD

                                                        [6]
                                                        3.6e-09
                                                        IVDMREF
                                                        +++ +++
    601 SFSRLIKERSNMAIVLTADSERFESQESKFLRNVNTRIAGGGQLSITNEKPRVRSLYLMFICIKTLKVIVDLREF

                    [1]                      [2]
                    6.0e-13                  8.5e-09
            M       VGDYILTPDIC              QCKMMSRYY
            +       +++++++ ++               +++ ++ ++
    676 RSSLPSILHGNNFSVIPCQLLVGDYILSPKICVERKSIRDLIQSLSNGRLYSQCEAMTEYYEIPVLLIEFEQHQS

                    [5]
                    3.7e-08
                    HFPRLRILW
                    ++ +++ +
    751 FTSPPFSDLSSEIGKNDVQSKLVLLTLSFPNLRIVWSSSAYVTSIIFQDLKAMEQEPDPASAASIGLEAGQDSTN


CPU: ghidorah
Time 68.583141 secs.
```

FIGURE 6.9. (*Continued from facing page.*)

REGULAR EXPRESSIONS

Regular expressions are a method used to search for complex patterns of letters and words in text files. The UNIX programs grep and egrep are used for this purpose. These expressions are also used to represent complex patterns in protein sequences. The Prosite catalog also uses regular expressions to describe variability in the amino acid patterns for the active sites of proteins. For example, the expression [LIVMF]-G-E-x-[GAS]-[LIVM]-x(5,11)-R-[STAQ]-A-x-[LIVMA]-x-[STACV] means that one of LIVMF in the first position, followed by G and then E, followed by any single character (indicated by x), followed by one of GAS and then by one of LIVM, followed by any 5–11 characters indicated by x(5,11), then by R, one of STAQ, then A, then any single character, then one of LIVMA, then any single character, and finally by one of STACV. More information about these patterns may be provided by the investigator in a standard file, as described on the Prosite Web site (http://www.expasy.ch/prosite/.

THERE ARE OTHER METHODS FOR COMPARING DATABASES OF SEQUENCES AND PATTERNS

One variation of the method for comparing sequences and patterns is to search a query sequence with a database of patterns (search type E, Table 6.1). If the sequence contains patterns representative of a protein family, the sequence is a candidate for membership in that same family. A large number of protein pattern databases are available (see Chapter 10), most of them offering this type of search.

FASTA-pat and FASTA-swap are versions of FASTA that may be used for comparing a query sequence to a database of patterns characteristic of protein families. They are designed to search for remotely related protein sequences by a finely tuned system of amino acid matches. The FASTA algorithm normally identifies sequence similarity very rapidly by a method for finding common patterns, or k-tuples, in the same order in two sequences. In FASTA-pat and FASTA-swap, the same rapid method is used to find common patterns. FASTA-pat performs a faster method of comparing sequences to patterns by means of a lookup table, as described above (Table 6.3). FASTA-swap performs a more rigorous search for the most significant matches of sequence to patterns.

These programs use databases of patterns found in columns of multiple sequence alignments of related protein sequences. They produce scoring matrices but, unlike PSSMs, the columns in these matrices only indicate whether a given amino acid is present; there is no score indicating frequency. Multiple sequence alignments of a large number of protein families were prepared using the PIMA program (see Chapter 5, p. 188). A large number of conserved pat-

terns were identified from these alignments, and the pattern was placed in a new type of scoring matrix.

In addition to these pattern matrices, two log odds scoring matrices, weighted-match minimum average matrix (WMM) and empirical matrix (EMMA), can be prepared from the scoring matrices. These scoring matrices are used by FASTA-pat and FASTA-swap for comparing a query sequence with a database of pattern matrices. These special scoring matrices have different types of information from the more common types of scoring matrices described previously. The scoring system for WMM and EMMA takes into account the possibility that the substitution of amino acid a for amino acid b may not be as likely as the substitution of b for a. An example from Ladunga et al. (1996) is informative. On the one hand, if an alignment column has 9 Cys and 1 Ala, the substitution of Ala for Cys in this column would be given a low substitution score because Cys is involved in disulfide bonds and this function cannot be replaced by Ala. Cys-to-Cys substitutions receive a high score for the same reason. On the other hand, if a column has 1 Cys and 9 Ala, the Cys might readily substitute for the Ala, which has no comparable specific function. The substitution of Cys for Ala is considered to be a random insertion of no particular significance and is therefore given a corresponding likelihood score of 0. When aligning a query sequence to a pattern, a single amino acid in the sequence is matched to a series of possible substitutions in the pattern. WMM uses the minimum of the scores for aligning the amino acid in the query sequence with each of the amino acids in the pattern. WMM gives significantly better results than EMMA, probably because it is more finely tuned for detecting the types of variations in related sequences. Program outputs of FASTA-pat and FASTA-swap are very similar to those of FASTA described above.

Another type of pattern database searching is to use a pattern query to search a database of patterns to discover possible relationships, such as overlapping patterns. The LAMA (Local Alignment of Multiple Alignments) server at the BLOCKS Web site, described below, performs such a search (Henikoff et al. 1995). LAMA, provided on the BLOCKS server (http://blocks.fhcrc.org/blocks-bin/LAMA_search.sh), compares a query PSSM representing a particular set of proteins with a database of such matrices to find related sets of proteins (Pietrokovski 1996). In this manner, new and larger related sets of proteins not identified previously can be discovered. Because the search is for matching sequence patterns instead of entire sequence alignments, there is an opportunity to analyze the evolution of function in different parts of a protein molecule (Henikoff et al. 1997). For example, a given group of proteins may be found to have two regions, one related to one particular group of proteins and a second related to another group. The LAMA program compares the scores found in each column of one PSSM to those in a second to discover whether there is any correlation. Examples of the procedure are given at http://www.blocks.fhcrc.org/.

PSI-BLAST, an Improved Version of BLAST for Finding Protein Families

As described above, there are advantages to using a scoring matrix that represents conserved sequence patterns in a protein family instead of a single query sequence to search a sequence database. The search of sequence databases is expanded to identify additional related sequences that might otherwise be missed. The major difficulty with such an expanded search is that an alignment of related sequences must already be available to know the variations at each position in the query sequence. A new version of BLAST called position-specific-iterated BLAST, or PSI-BLAST, has been designed to discover related sequences and build a local multiple sequence alignment starting with a BLAST search by a single query sequence. The PSI-BLAST program has been enhanced (Schäffer et al. 2001) to optimize the identification of protein family members in

sequence similarity searches using the receiver operator characteristic (ROC) test described in Chapter 5 (p. 192) as a measure of success. A similar program pattern-hit-initiated BLAST, or PHI-BLAST, performs a similar type of search starting with a specified pattern in a query sequence, which is discussed in more detail below. A more detailed discussion of the programs may be found on the NCBI BLAST Web site.

The method used by PSI-BLAST to perform extended similarity searches involves a series of repeated steps or iterations. First, a similarity search of a protein sequence database is performed using a query sequence. Second, the results of the search are presented to the user and can be assessed visually to see whether any database sequences that are significantly related to the query sequence are present. Third, if such sequences are present, a mouse-click on a decision box initiates another iteration of the search. The high-scoring sequence matches found in the first step are then aligned, and, from the alignment, a PSSM is produced based on the initial alignments found. The database is then searched again using this scoring matrix to find sequences that align with the PSSM. Thus, the search is expanded to include sequences that match the variations found in the multiple sequence alignment at each sequence position. The results are again displayed, indicating any newly discovered sequences that are significantly related to the aligned sequences in addition to those found in the previous iteration. Again, an opportunity is given to the user to go through another iteration of the program, but this time including any newly recruited sequences to refine the alignment. In this fashion, a new family of sequences that can be aligned with the original query sequence to produce alignment scores having significant E values are produced.

The PSI-BLAST method was made possible by the development of the gapped BLAST program, which increased the speed of the BLAST algorithm by over one-half so that more sophisticated search routines of PSI-BLAST could be added without an overall loss of speed. PSI-BLAST may not be as sensitive as other pattern-generating and searching programs described in Chapter 5 and above, but the simplicity and ease of use of this program are very attractive features for exploring protein family relationships. In a comparison of the ability of PSI-BLAST with the Smith–Waterman dynamic programming program SSEARCH to identify members of 11 protein families defined by sequence similarity, PSI-BLAST found more sequences and, in some cases, many more sequences, than SSEARCH and at a 40-fold greater speed (Schäffer et al. 2001).

A similar program, called MAXHOM (Sander and Schneider 1991), builds a sequence alignment in two steps. Matching sequences found in a database search are aligned by dynamic programming with a query sequence, and a profile is made from the alignment. A new round of sequences that match the updated profile are then picked from the SwissProt database (visit http://www.embl-heidelberg.de/predictprotein/predictprotein.html).

The main difficulty associated with sequence similarity searches using any search method, including PSI-BLAST, is determining the significance of the subtle sequence relationships that are found. Such subtle similarities may be evidence of structural or evolutionary relationships, but they could also occur as a result of the aligning random sequence variations that have no common evolutionary origin or function (Bork and Gibson 1996). Limited sequence similarity is not a good indicator of structural similarity. Protein structures are in general composed of a tightly packed core and outside loops. Amino acid substitutions within the core are common, but only certain substitutions will work at a given amino acid position. Thus, sequence similarity searches based on the commonly found global substitutions found in many protein families may not provide reliable information on substitutions on specific protein structures (see Chapter 10).

Another difficulty with the PSI-BLAST program is that the procedure follows a type of algorithm called a greedy algorithm. Put simply, once sequences that match the query are found, these newly found sequences are used to find more sequences like themselves. If a different, but also related, query sequence is used initially, a different group of sequences may be found. Thus, there is no guarantee that related sequences discovered by a search initiated by one query sequence will be the same as those found using a second query sequence, even though the two queries sequences are related. Nevertheless, PSI-BLAST potentially offers exciting opportunities to the curious but careful investigator. New types of relationships in the protein databases may be readily discovered and used to infer evolutionary origins of proteins (Tatusov et al. 1997).

The later steps of a PSI-BLAST search use a PSSM that represents the alignments found. PSI-BLAST has been engineered to find database matches to this matrix almost as rapidly as BLASTP finds matches to

a query sequence. However, there are some differences between the PSSM produced by PSI-BLAST and those produced by other programs: (1) The PSI-BLAST PSSM is based on a global alignment of all of the sequences, whereas PSSMs normally represent a local multiple sequence alignment; (2) the same gap penalties are used throughout the procedure and there is no position-specific penalty as in other programs; and (3) each subsequent alignment is based on using the query sequence as a master template for producing a multiple sequence alignment of the same length as the query sequence. Columns in the alignment involve varying numbers of sequences, depending on the extent of the local alignment of each sequence with the query, and columns with gaps in the query sequence are ignored. Sequences >98% similar to the query are not included to avoid biasing the matrix. Thus, the multiple sequence alignment is a compilation of the pair-wise alignments of each matching database sequence with the query sequence and is not a true multiple sequence alignment, as illustrated below. The resulting alignment provides the columns for the scoring matrix.

```
xxxxxxxxxxxxxxx    query sequence with no gaps
  xx-xxxx          alignment of sequence 1
        xxx-x              alignment of sequence 2
    xxxx-xx        alignment of sequence 3
---------------    columns of the PSI-BLAST
                   alignment
```

Once the alignment has been found, the frequencies of amino acids in each column are adjusted by weighting the sequences to reduce the influence of the more-alike sequences, and by adding more counts (pseudocounts) representing other common amino acid substitutions to broaden the search capability of the matrix. Using pseudocounts in scoring matrices is discussed in Chapter 5 (p. 211). The resulting scores in each column of the scoring matrix are scaled using the same scaling factor λ as the BLOSUM62 scoring matrix so that a threshold value T for HSPs and other statistical parameters used by BLASTP may also be used by PSI-BLAST. An example of the program output of a PSI-BLAST search is shown in Figure 6.10.

Pattern-Hit-Initiated BLAST (PHI-BLAST)

This program functions much like PSI-BLAST, except that the query sequence is first searched for a complex pattern provided by the investigator (Zhang et al.

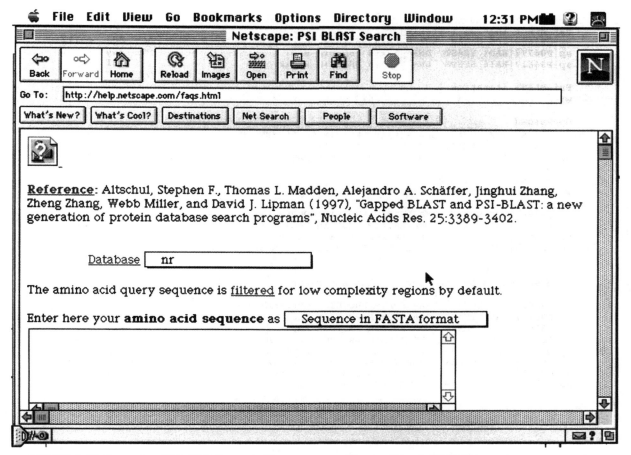

FIGURE 6.10. Example of PSI-BLAST search. The sequence of the *Arabidopsis XPF* DNA repair gene was used to query the SwissProt database, with a threshold *E* setting of 0.01, requesting ten descriptions and alignments with otherwise the recommended default program settings. The initial iteration found three matching sequences, and these were used to enter iteration 1. Iteration 1 did not produce any additional matches at the chosen level of significance, and the program indicated that the search had converged with no more sequences at the chosen level of significance. Therefore, for iteration 2 the sequences scoring worse than the threshold were used. Since only those lower-scoring sequences that have an alignment with the query could influence the result, this option could potentially find additional sequences. A yeast transport protein was then reported. With another iteration using the four sequences above threshold, another set of sequences was now pulled into the high-scoring group. This search therefore revealed that the SwissProt database has three other sequences strongly related to the query sequence but that other sequences of less-significant similarity were also present. (*Figure continues on next page.*)

1998). The subsequent search for similarity in the protein sequence database is then focused on regions containing the pattern. Thus, the method provides an opportunity to explore variations of a known pattern in the sequence database. This program is accessible from the BLAST server at http://www.ncbi.nlm.nih.gov/. The chosen query sequence is first searched for a particular pattern or class of patterns called a regular expression, which allows a wide range of pattern-matching options, and is described in the box on page 268. An iterative search for matching proteins follows using PSI-BLAST.

PROBE

PROBE is a database search tool that is similar to PSI-BLAST but performs a more complex and rigorous type of data analysis (Neuwald et al. 1997). Like PSI-BLAST, the program PROBE starts with a single query sequence and searches for family members by a BLASTP search. After removing the most-alike sequences, PROBE constructs an alignment model by means of a Bayesian statistical approach that uses both a Gibbs sampling procedure and the genetic algorithm (both methods are described in Chapter 5) to sort the

```
<Psi-BLAST output example>
Psi-BLAST initial iteration
sp|Q92889|XPF_HUMAN   DNA-REPAIR PROTEIN COMPLEMENTING XP-F CELL ...    504   e-142
sp|P06777|RAD1_YEAST  DNA REPAIR PROTEIN RAD1                          300   6e-81
sp|P36617|RA16_SCHPO  DNA REPAIR PROTEIN RAD16                         231   3e-60

Psi-BLAST iteration 1
with sequences scoring better than E threshhold

Converged
sp|Q92889|XPF_HUMAN   DNA-REPAIR PROTEIN COMPLEMENTING XP-F CELL ...   1020   0.0
sp|P06777|RAD1_YEAST  DNA REPAIR PROTEIN RAD1                          953   0.0
sp|P36617|RA16_SCHPO  DNA REPAIR PROTEIN RAD16                         897   0.0

Psi-BLAST iteration 2
with sequences scoring worse than E threshhold

sp|Q92889|XPF_HUMAN   DNA-REPAIR PROTEIN COMPLEMENTING XP-F CELL ...   1020   0.0
sp|P06777|RAD1_YEAST  DNA REPAIR PROTEIN RAD1                          967   0.0
sp|P36617|RA16_SCHPO  DNA REPAIR PROTEIN RAD16                         939   0.0
sp|P25386|USO1_YEAST  INTRACELLULAR PROTEIN TRANSPORT PROTEIN USO1      53   3e-06

Psi-BLAST iteration 3
with sequences scoring better than E threshhold

sp|Q92889|XPF_HUMAN   DNA-REPAIR PROTEIN COMPLEMENTING XP-F CELL ...   1007   0.0
sp|P06777|RAD1_YEAST  DNA REPAIR PROTEIN RAD1                          950   0.0
sp|P36617|RA16_SCHPO  DNA REPAIR PROTEIN RAD16                         884   0.0
sp|P25386|USO1_YEAST  INTRACELLULAR PROTEIN TRANSPORT PROTEIN USO1     294   5e-79
sp|Q08696|MST2_DROHY  AXONEME-ASSOCIATED PROTEIN MST101(2)              52   4e-06
sp|Q62209|SCP1_MOUSE  SYNAPTONEMAL COMPLEX PROTEIN 1 (SCP-1 PROT...     49   5e-05
sp|Q03410|SCP1_RAT    SYNAPTONEMAL COMPLEX PROTEIN 1 (SCP-1 PROTEIN)    49   5e-05
sp|Q02224|CENE_HUMAN  CENTROMERIC PROTEIN E (CENP-E PROTEIN)            45   5e-04
```

FIGURE 6.10. (*Continued.*)

patterns in all possible combinations to find the most significant set. As in PSI-BLAST, the alignment model is then used as a query for additional database sequences.

PROBE provides a new and powerful approach toward finding a sequence family and is available by anonymous FTP from ncbi.nlm.nih.gov/pub/neuwald/.

SUMMARY

As the sequence databases continue to increase in size, for the most part with genomic DNA sequences of unknown function, it is important to have a set of computational tools for predicting possible functions of these sequences. The first choice of a biologist is usually to go to the BLAST Web site because a variety of database searches are possible against regularly updated databases and can be performed with rapid turnaround time. This chapter has discussed a variety of additional resources for such searches, most available on Web sites or available for setup on a local computer system. For extensive searching, establishment of the databases and programs on a local sys-

tem is a reasonable and achievable option. It is then possible to set up batch files or scripts that automate the searches. Many searches generate large amounts of information that will then need to be organized into a database. Ways to handle these data are described in Chapter 12.

Some of the most interesting database matches to protein query sequences are those to more distantly related sequences. A short alignment region between a query and a database sequence is usually not biologically significant, even though there may be a number of identities in the alignment. If additional sequences can be found that share the same alignment, however, it is possible that the pattern represents a common structure in a family of related proteins.

Also interesting and useful are database queries against the databases of conserved patterns in protein families. It has been estimated that about one-half of these patterns can be linked to a protein structural fold. Thus, it is very worthwhile to follow the distant relationships identified further with the eventual goal of trying to discover a relationship to a protein of known structure. There are some excellent computer tools available to the molecular biologist for finding conserved patterns in protein families and for searching new sequences with these patterns, and it can be anticipated that the number will continue to grow. There are a large number of Web servers for this purpose, and these are described in Chapter 10.

As methods such as those described in this chapter are used to search for related protein sequences, it is important to keep an eye on the statistical significance of the matches and the plausibility of the observed amino acid substitutions from a structural perspective. It is quite easy to end up with a group of sequences that are related to each other but not to the query sequence. There are presently no guides as to which of the above methods is most likely to work best for a particular biological question. The best advice when performing DNA or protein database sequence searches is to go farther than the basic methods and to become familiar with the range of available methods. These include database similarity searches using individual query sequences and PSSMs or profiles that represent msa's. Individual sequences may also be searched for matches to a database of PSSMs or profile HMMs representing conserved domains in proteins and to assess whether such domains are present.

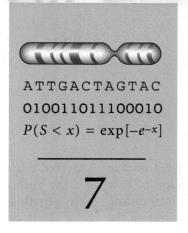

ATTGACTAGTAC
010011011100010
$P(S < x) = \exp[-e^{-x}]$

7

Phylogenetic Prediction

C O N T E N T S

INTRODUCTION

A phylogenetic analysis of a family of related nucleic acid or protein sequences is a determination of how the family members might have been derived during evolution. The evolutionary relationships among the sequences are depicted by using a graph called a tree. Sequences are placed as outer branches of the tree, and the branching relationships of the inner part of the tree then reflect the degree to which different sequences are related. For example, two sequences that are very much alike will be located at neighboring outside branches and will be joined to a common branch beneath them. Less related sequences will be on branches that are more distant from each other on the tree. The object of phylogenetic analysis is to discover the branch arrangements and branch lengths in trees that best represent the relationship among all the sequences.

Phylogenetic analysis of nucleic acid and protein sequences is an important area of sequence analysis, for example, in the study of the evolution of a family of sequences. Using this type of analysis, sequences that are the most closely related can be identified by their occupying neighboring branches on a tree. Thus, when a gene family is found in an organism or group of organisms, phylogenetic relationships among the genes can help to predict which ones might have an equivalent function that has been con-served during evolution of the corresponding organisms. Such functional predictions can then be tested by genetic experiments.

Phylogenetic analysis may also be used to follow the changes occurring in a rapidly changing species, such as a virus. Analysis of the types of changes within a population can reveal, for example, whether or not a particular gene is under selection (McDonald and Kreitman 1991; Nielsen and Yang 1998), or the timing of genetic variation in the human genome (Toomajian et al. 2003).

Procedures for phylogenetic analysis are strongly linked to those for sequence alignment, which was already discussed in Chapters 3 and 5. Similar problems are also encountered. For example, just as two very similar sequences can be easily aligned even by eye, a group of sequences that are very similar but with a small level of variation throughout can easily be organized into a tree. Conversely, as sequences become more and more different through evolutionary change, they can be much more difficult to align. A phylogenetic analysis of very different sequences is also difficult to do because there are so many possible evolutionary paths that could have been followed to produce the observed sequence variation. Because of the complexity of this problem, considerable expertise is required for difficult situations.

CHAPTER GUIDE FOR BIOLOGISTS

The goal of a phylogenetic analysis of nucleic acid and protein sequences is to analyze the relationships among a group of sequences that can be aligned into a multiple sequence alignment (msa). The sequences should have some variation from one to another, but not so much that they become difficult to align into an msa. Evolutionary biologists have used a type of graph called a tree to illustrate these differing relationships between sequences. This tree has branches, just as in plants, with the outer branches representing the sequences as they currently exist and the inner branches representing common ancestor sequences. The branches are joined through nodes that represent relationships among current and ancestor sequences. Thus, the tree represents a reasonable guess as to how the sequences might have evolved from ancestor sequences during evolutionary time and how they might be related. For example, two similar sequences will be located on two adjacent outer branches. The lower branch formed by joining these two branches represents the common ancestor sequence of these two sequences. We do not know what this sequence was and how it changed to produce the later descendant sequences. There are many possibilities. This lower branch will then be joined with other branches to ever more distant ancestor sequences. Eventually we will reach a main branch with a root. More often, however, the phylogenetic tree is left without a root (an unrooted tree) because more assumptions must be made to predict the older branches and the root of the tree.

In phylogenetic analysis, we try to predict the best trees that will generate the observed differences among the sequences and, if the sequences are not too diverse, the changes that occurred along each branch. Three methods—maximum parsimony, distance, and maximum likelihood—are used for predicting such trees. The sequence comparisons made by these methods are not particularly complicated. Trees are predicted such that the best account for the sequence variation in each column of the msa. Alternatively, the number of different characters in the alignment between each sequence pair is counted as an index of the evolutionary distance between the sequences. These methods may be computationally complex and require a day of computing on a desktop machine. More than one tree may be equally probable, or almost equally probable, for describing the observed pattern of sequence evolution. Using two or more of the above methods, including a more distantly related reference sequence, the tree can be improved or justified by examining other biological or sequence data or testing how well the data support the trees (a bootstrap analysis).

Alignments of related DNA, RNA, and protein sequences are used for phylogenetic analysis. Because RNA secondary and tertiary structure is highly conserved through evolution, phylogenetic analysis of RNA sequences uses alignments of these structural features rather than strict alignment of the sequences themselves. Unless mRNA sequences for the same gene from closely related species are being compared, there may be a large number of sequence changes from one gene to another, thus generating many possible trees that can be difficult to distinguish. However, comparing gene sequences has many useful applications in evolutionary biology, and these sequences can be aligned using the translated protein sequences as a guide for aligning codon positions. Either full-length or partial-length protein sequences encoding conserved domains are much more highly conserved between species than mRNA sequences, thus simplifying the prediction of phylogenetic relationships.

As more and more individual genes are compared in diverse genomes, particularly prokaryotes, it has been found that ancient genes, such as ribosomal genes and genes involved in transcription and translation, appear to share a common evolutionary origin. When a phylogenetic analysis is performed on these sets of genes, the resulting trees are similar. However, for many other genes, such as those for metabolic functions, the trees can be quite different. These dissimilarities suggest that the evolutionary origin of these genes is complex and probably has involved the horizontal transfer of genes between species. Evolutionary biologists frequently use particular sets of genes to analyze relationships in greater detail, e.g., mitochondrial genes for mammals and chloroplast genes for plants. A final cautionary note is a reminder that some gene sequences may vary within populations of a species, making it necessary to analyze sequence variations both within a species and between species.

CHAPTER GUIDE FOR COMPUTATIONAL SCIENTISTS

A major finding in evolutionary studies has been that the many different types of organisms found on the earth employ the same basic sets of genes for biological function. A popular evolutionary model is that these genes have been altered, copied, and rearranged from an original basic set (or sets) to create new genes and gene families that are the origin of biological variation. Three original classes of organisms have been identified on the basis of sequence variation in some of the most highly conserved genes—these include the single-celled prokaryotes (cells without a recognizable nucleus during microscopic examination), the eukaryotes (cells with a nucleus), and the more recently identified archaea, another major class of organisms with features of both prokaryotes and eukaryotes. Organisms are further classified by biologists into species generally defined by individuals having a similar biological appearance and behavior and by the reproduction of individuals of the species into more individuals of the same kind.

Reproduction occurs with transfer of the same basic set of genes acquired from the parent to the offspring. A pedigree for an individual organism can be made starting with a specific individual and listing the parental lineages. The pedigree represents vertical transmission of genetic information from one generation to the next. Genetic variation can occur within a species, particularly when there is a large population, but as long as the biological properties within the population are similar or sexual reproduction is still possible in species with males and females, the species remains the same. If a population becomes split into two parts, for example, into A and B geographically, then each part can diverge biologically and sexually, and two new species may eventually emerge. The source of variation is in the genes of the organisms. Parts A and B will have approximately the same set of genes, but part A will have a different set of sequence variations and possibly additional or fewer genes than B. The genomes will now be evolving differently, and, eventually, over long periods of evolutionary time, the number of sequence changes between them will increase so that the two species may become very different from each other. Phylogenetic analysis is designed to discover the evolutionary history of the genes as revealed in sequence variation. There is a historical record of these species variations within the genes, and phylogenetic analysis simply involves building trees that account for the observed variation in specific genes.

A more recent discovery in the analysis of genes of some species, particularly more ancient ones like the single-celled bacteria and the fruit fly *Drosophila*, is that different genes are not inherited according to a simple pattern of vertical descent. Genes for the most important biological functions, including gene copying from one generation to the next and gene reading into protein sequences, follow the same inheritance patterns in evolution. However, other genes that are more dispensable and having to do with specific types of biological function do not. The explanation is that the genetic material (DNA) for these latter genes has been transferred between species when they are in close contact, as, for example, between single-celled bacteria. Like the vertically transmitted genes, this horizontal transfer of DNA between species has been faithfully recorded in the sequence variation between species and may be discovered through variations for the individual genes in phylogenetic analysis.

This chapter describes computational methods for producing trees from msa's of related nucleic acid and protein sequences that represent the same gene or gene family in different organisms. These trees reflect the ancestor relationships among the genes. They can also reveal which genes have followed the same evolutionary path and which have not as a result of evolutionary histories.

WHAT SHOULD BE LEARNED IN THIS CHAPTER?

- What is meant by a phylogenetic analysis and how are trees interpreted based on this analysis?
- What is the relationship between an msa and a phylogenetic analysis?
- What is the universal tree of life and how was it derived?
- What is a species based on sequence analysis and how is this concept disturbed by horizontal transfer of DNA between unrelated species?
- There are three major methods for performing a phylogenetic analysis—the maximum parsimony, distance, and maximum likelihood methods. Where are the tools for performing these analyses found?
- What are the consequences of the assumption of a constant or variable mutation rate in the branches of a phylogenetic tree?
- Study the neighbor-joining algorithm for progressive clustering of sequences and appreciate its value in dealing with unequal mutation rates.
- Learn methods to correct for multiple mutations in the branches of a phylogenetic tree.

Glossary Terms

Archaea A third class of organisms shown to be different from prokaryotes and eukaryotes through phylogenetic examination of rRNA sequences and other essential genes. Together with prokaryotes, these organisms are thought to be the ancient ancestors of eukaryotes.

Binary tree In the context of this chapter, a tree whose inner branches representing ancient lines of inheritance repeated-

ly split into two new branches, eventually terminating in outer, terminal branches representing current sequences. The direction of the tree from early to late evolutionary times is from the inner branches to the outer. Branches are only joined through this ancestor–descendant network.

Bootstrap analysis A random resampling of the data used to predict an evolutionary tree, in this case the columns in the msa of the sequences, to determine how well the msa data predict the tree.

Branch-and-bound search   In phylogenetic analysis, a method that stops analyzing a particular area of branching in candidate trees when it is not possible to obtain a more parsimonious solution than has already been found.

Distance method   A phylogenetic analysis based on the number of differences between pairs of sequence in an msa.

Eukaryote   An organism comprising one or more cells that has a recognizable nucleus upon microscopic examination. This classification includes plants and animals.

Homologous genes   Genes whose sequences are so similar that they almost certainly arose from a common ancestor gene.

Homoplasy   In a phylogenetic analysis of sequences, refers to the same character arising from different genetic starting points and thus has the opposite meaning to homology. For one column of an msa, if all the identical sequences resulted from a change in one branch of the predicted tree, they are homologous and there is no homoplasy in the tree. Conversely, if they arose from changes in more than one branch, they are not homologous and there is homoplasy in the tree.

Horizontal DNA transfer (also called lateral DNA transfer)   The rare transfer of genetic information between different species. The transferred genes will have a different phylogenetic history than the original set in the recipient species.

Maximum likelihood method   A method for predicting an evolutionary tree that uses an expected pattern of mutational changes from one DNA base to another and probability calculations to find the most likely arrangement of branches that generates the set of sequences.

Maximum parsimony method   A method for predicting an evolutionary tree that best fits the sequence variation in each column of an msa. This tree will provide the minimal number of evolutionary steps to produce the sequences.

Neighbor-joining algorithm   A greedy algorithm that predicts an evolutionary tree based on progressively adding the next most-alike sequence (or set of sequences) as an additional branch to an existing tree using distances between the sequences.

Nonsynonymous mutation   A change at one of the three positions in a codon to another codon that specifies a different amino acid.

Orthologous genes   A pair of genes in two organisms whose sequences are so strikingly and uniquely similar that they are strongly predicted to have the same function.

Outgroup sequence   An additional sequence that is included in a phylogenetic analysis. The outgroup sequence is more distantly related to the other sequences than they are to each other. Including the outgroup sequence assists in correctly arranging the other sequences by acting as an external reference.

Paralogous genes   Genes that have arisen by gene duplication events in an organism and are transmitted to offspring as a gene family. Members of the resulting gene family are identified by sequence similarity, but their functions have diversified so that each has a different but necessary biological role. If the same family is present in two organisms, the most closely related pairs in these organisms are predicted to have the same function (they are orthologous genes).

Phylogenetic analysis   An investigation of the evolutionary relationships among a group of related sequences by producing a tree representation of the relationships.

Poisson distribution   A probability distribution that is used to predict the probability of the actual number of changes along a branch in an evolutionary tree when the average number of changes has been estimated. For example, if the average is three changes, then the probability of 0, 1, 2, ... changes can be calculated.

Polymorphism   In sequence analysis, refers to genetic variability across a population of the same species. Within a particular gene, there may be single or multiple base changes, and within a localized region on a chromosome, there may be insertions, deletions, and rearrangements. These changes may not affect an individual, or they may cause a biological change. They will be distributed in a population in different frequencies depending on when they occurred, their biological effects, and the reproductive history of the population.

Prokaryote   A one-celled organism that does not have a recognizable nucleus upon examination by microscopy. This classification includes many species of bacteria, including plant and animal pathogens.

Rooted tree   A tree representation of a group of related sequences in which all the sequences are descended from a common point in one of the tree branches. The path from that point through the tree to each sequence defines the predicted evolutionary path to that sequence.

Species   For the purposes of sequence analysis, a population of biologically alike individuals that share nearly the same genetic makeup. Their offspring are also like them through succeeding generations. When males and females are present in the same species, they produce biologically viable offspring having a genetic makeup that is a blend of the parents. In rare cases, genes are transferred between species, thus confounding this simplified definition of a species.

Step or cost matrix A table of scores for substituting one nucleotide for another or one amino acid for another in related nucleic acid or protein sequences.

Synonymous mutation A change at one of the three positions in a codon to another codon that specifies the same amino acid.

Transition A change of a pyrimidine base (C or T) in DNA to another pyrmidine, or of a purine base (A or G) to another purine, often considered to be more likely than changes between pyrimidines and purines (transversion).

Transversion A change of a pyrimidine base (C or T) in DNA to a purine base (A or G) or vice versa, often considered to be less likely than changes to the same base type (transition).

Tree In a molecular phylogenetics context, a graphical representation that depicts relationships among a set of sequences. Most-alike sequences are placed at the outer ends of two branches that are joined below into a lower common branch, representing their derivation from an ancestor sequence. The lower branch is then joined to other branches representing other sequences according to relationships among them. These steps are repeated to obtain a global branching pattern that reflects the relationships among all the sequences.

Tree of life An attempt to place all organisms on the same phylogenetic tree that dates back to the earliest recognizable types.

Unrooted tree A tree representation of a group of related sequences that does not indicate which of the sequences is the ancestor of the others.

Vertical transmission The transfer of genetic information from one generation to the next, from parent to offspring, within the same species.

PHYLIP AND PAUP ARE COMMONLY USED PHYLOGENETIC ANALYSIS PROGRAMS

Phylogenetic analysis programs are widely available at little or no cost. A comprehensive list has been published previously in Swofford et al. (1996). Two programs in common use are PHYLIP (phylogenetic inference package) (Felsenstein 1989, 1996) available from Dr. J. Felsenstein at http://evolution.genetics. washington.edu/phylip.html and PAUP (phylogenetic analysis using parsimony) developed by D. Swofford at http://paup.csit.fsu.edu/ and available from Sinauer Associates, Sunderland, Massachusetts. These programs have excellent documentation and offer a broad range of models of evolutionary change in sequences. Current versions of these programs provide the three main methods for phylogenetic analysis—parsimony, distance, and maximum likelihood methods (described below)—and also include many types of evolutionary models for sequence variation. Examples using these programs are given later in the chapter. Each program requires a particular type of input sequence format. PHYLIP programs automatically read in a sequence in the PHYLIP infile format and automatically produce files called outfile and treefile for the predicted tree. PAUP reads in a file in

the NEXUS format. These input formats are described in Chapter 2.

Another program, MacClade, is useful for detailed analysis of the predictions made by PHYLIP, PAUP, and other phylogenetic programs and is also available from Sinauer (also see http://phylogeny.arizona.edu/macclade/macclade.html). MacClade, as the name suggests, runs on a Macintosh computer. PHYLIP and PAUP run on practically any machine, but the user interface for PAUP has been most developed for use on the Macintosh computer. Wayne and David Maddison have also developed a suite of program modules and a programming environment for phylogenetic analysis and related types of analysis called MESQUITE at http://mesquiteproject.org/. The EMBOSS group at http://www.hgmp.mrc.ac.uk/Software/EMBOSS/ also includes programs and programming modules that function similarly to PHYLIP. Other software for more specialized types of analyses are described later in this chapter.

There are also several Web sites that provide information on phylogenetic relationships among

TABLE 7.1. *Phylogenetic relationships among organisms*

Site name	Address	Description	Reference
Entrez	http://www3.ncbi.nlm.nih.gov/ Taxonomy/taxonomyhome.html	taxonomically related structures or group of organisms	see Web page
RDP (Ribosomal database project)	http://rdp.cme.msu.edu	ribosomal RNA-derived trees	Maidak et al. (1999)
Tree of life	http://phylogeny.arizona.edu/tree/ phylogeny.html	information about phylogeny and biodiversity	Maddison and Maddison (1992)

organisms (Table 7.1). There are several excellent descriptions of phylogenetic analysis in which the methods are covered in considerable depth (Li and Graur 1991; Miyamoto and Cracraft 1991; Felsenstein 1996; Li and Gu 1996; Saitou 1996; Swofford et al. 1996; Li 1997).

HOW IS PHYLOGENETIC ANALYSIS RELATED TO MULTIPLE SEQUENCE ALIGNMENT?

When the sequences of two nucleic acid or protein molecules found in two different organisms are similar, they are likely to have been derived from a common ancestor sequence. Chapter 3 discusses sequence alignment methods used to determine sequence similarity. Chapter 5 discusses multiple sequence alignment (msa) methods that need to be applied to a set of related sequences before a phylogenetic analysis can be performed. Chapter 6 describes methods for searching through a database of sequences to locate sequences that are similar to a query sequence. A sequence alignment reveals which positions in the sequences were conserved and which diverged from a common ancestor sequence, as illustrated in Figure 7.1. When one is quite certain that two sequences share an evolutionary relationship, the sequences are referred to as being homologous.

GAATC sequence 1

GAGTT sequence 2

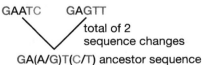

GA(A/G)T(C/T) ancestor sequence

FIGURE 7.1. Origin of similar sequences. Sequences 1 and 2 are each assumed to be derived from a common ancestor sequence. Some of the ancestor sequence can be inferred from conserved positions in the two sequences. For positions that vary, there are two possible choices at these sites in the ancestor.

Phylogenetic analysis of sequence depends on having a reliable alignment of the sequences using an msa program. The commonest method of msa (the progressive alignment method, p. 180) first aligns the most closely related pair of sequences and then sequentially adds more distantly related sequences or sets of sequences to this initial alignment. The alignment obtained is influenced by the most-alike sequences in the group and thus may not represent a reliable history of the evolutionary changes that have occurred. Other methods of msa attempt to circumvent the influence of alike sequences (see Chapter 5, p. 184). Once an msa has been obtained, each column is assumed to correspond to an individual site that has been evolving according to the observed sequence variation in the column. Most methods of phylogenetic analysis assume that each position in the protein or nucleic acid sequence changes independently of the others (analysis of RNA sequence evolution is an exception: see Chapter 8).

The analysis of sequences that are strongly similar along their entire lengths is quite straightforward. However, to align most sequences requires the positioning of gaps in the alignment. Gaps represent an insertion or deletion of one or more sequence characters during evolution. Proteins that align well are likely to have the same three-dimensional structure. In general, sequences that lie in the core structure of such proteins are not subject to insertions or deletions because any amino acid substitutions must fit into the packed hydrophobic environment of the core. Therefore, gaps should be rare in regions of msa's that represent these core sequences. In contrast, more variation, including insertions and deletions, may be found in the loop regions on the outside of the three-dimensional structure because these regions do not influence the core structure as much. Loop regions interact with the environment of small molecules, membranes, and other proteins (see Chapter 10).

Gaps in alignments can be thought of as representing quite drastic mutational changes in sequences, including insertions, deletions, or rearrangements of genetic material. The expectation that a gap of virtually any length can occur as a result of a single event introduces the problem of judging how many individual changes have occurred and in what order

they have occurred in a set of sequences. Gaps are treated in various ways by phylogenetic programs, but no clear-cut model as to how they should be treated has been devised. Many methods ignore gaps or focus on regions in an alignment that do not have any gaps. Nevertheless, gaps can be useful as phylogenetic markers in some situations.

Another approach for handling gaps is to avoid analysis of individual sites in the sequence alignment and instead to use sequence similarity scores as a basis for phylogenetic analysis. Rather than trying to decide what has happened at each sequence position in an alignment, a similarity score based on a scoring matrix with penalties for gaps is often used. As discussed below, these scores may be converted to distance scores that are suitable for phylogenetic analysis (Feng and Doolittle 1996) by distance methods (p. 311).

GENOME COMPLEXITY MUST BE CONSIDERED IN PHYLOGENETIC ANALYSIS

When performing a phylogenetic analysis, it is important to keep in mind that the genomes of most organisms have a complex origin. Some parts of the genome are passed on by vertical descent through the normal reproductive cycle. Other parts may have arisen by horizontal transfer of genetic material between species through a virus, DNA transformation, symbiosis, or some other method of horizontal transfer. Accordingly, when a particular gene is being subjected to phylogenetic analysis, the evolutionary history of that gene may not coincide with the evolutionary history of another. Many aspects of genome complexity must be considered in performing a phylogenetic analysis, and comparisons of the phylogenetic history of different genes can be used to analyze the evolutionary history of genomes (Lerat et al. 2003).

One of the most significant uses of phylogenetic analysis of sequences is to make predictions concerning the so-called tree of life. For this purpose, the gene selected for analysis should be universally present in all organisms and easily recognizable by the conservation of sequence in many species. At the same time, there should be enough sequence variation to determine which groups of organisms share the same phylogenetic origin. Ideally, the gene should also not be under selection, meaning that as variation occurs in

the sequence of the gene in populations of organisms, certain sequences are not favored with a loss of earlier variation.

Another consideration in phylogenetic analysis is the use of molecules that carry a great deal of evolutionary history in interspecies sequence variations. Examples of two such molecules are the small subunit of ribosomal RNA (rRNA) and mitochondrial DNA sequences. A large number of rRNA sequences from a variety of organisms have been aligned, and the secondary structure was deduced following methods discussed in Chapter 8. Phylogenetic predictions were then made using the distance method described below (Woese 1987). On the basis of rRNA sequence signatures, or regions within the molecule that are conserved in one group of organisms but different in another (Fig. 7.2), Woese (1987) predicted that early life diverged into three main kingdoms—Archaea, Bacteria, and Eukarya—a view that has not gone unchallenged (Mayr 1998). A more detailed analysis was used to find relationships among individual species within each group. Evidence for the presence of additional organisms in these groups has since been found by PCR amplification of environmental samples of RNA (Barns et al. 1996). The types of relationships found among the prokaryotic organisms

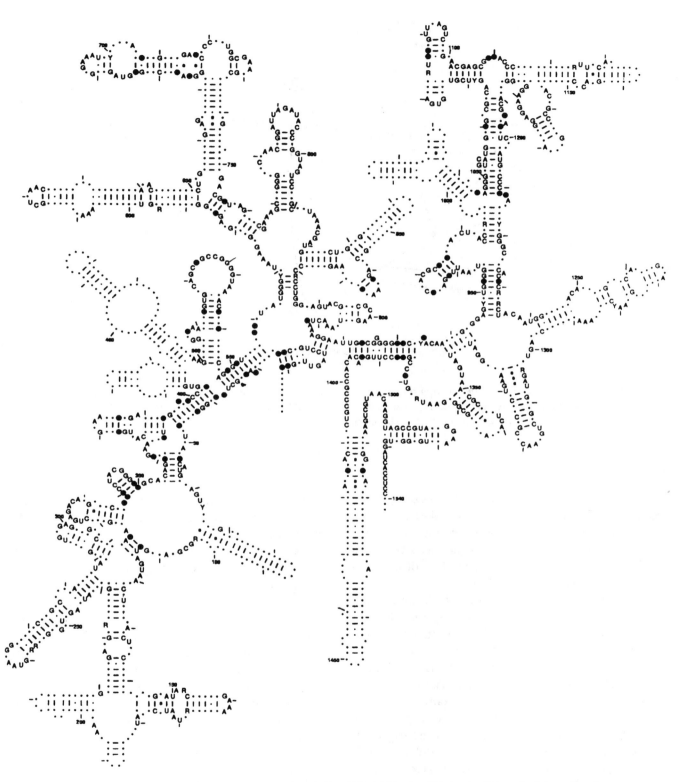

FIGURE 7.2. The signature positions in rRNA that distinguish Archaea and Bacteria. Shown is the predicted secondary structure for *E. coli* 16S ribosomal RNA with the most highly conserved sequence positions marked by the sequence character and the positions that distinguish Archaea and Bacteria shown by a black dot. Other marker positions in the sequence were used to define the third group, the Eukarya. (Reprinted, with permission, from Woese 1987 [© American Society for Microbiology].)

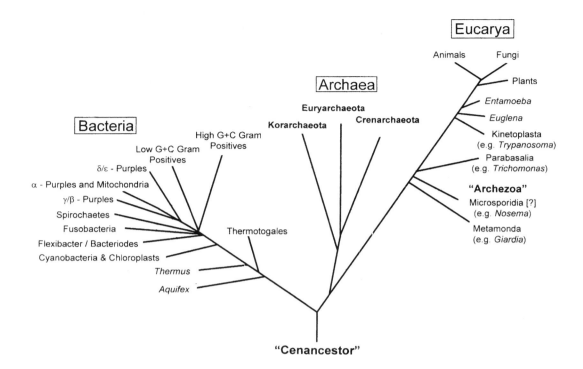

FIGURE 7.3. Rooted tree of life showing relationships among the three principal groups of organisms—Prokarya, Archaea, and Eukarya—originally based on analysis of ribosomal RNA. (Reprinted, with permission, from Brown and Doolittle 1997 [©American Society for Microbiology].)

are illustrated in Figure 7.3. The second example of a sequence molecule used for species analysis, the DNA sequence of mitochondria, has been employed in the analysis of primate evolution. An analysis is described in the section on the parsimony method of phylogenetic analysis (see below, p. 296).

Although these studies of rRNA and other sequences suggest a quite clear-cut model for the evolution of life, phylogenetic analysis of other genes and gene families has revealed that the situation is probably more complex and that a more appropriate evolutionary model might be the one shown in Figure 7.4. Thus, another consideration is the many examples of horizontal or lateral transfer of genes between species (see Fig. 3.3, p. 73) that introduce new genes and sequences into an organism (Brown and Doolittle 1997; Doolittle 1999). These types of transfers are inferred from the finding that the phylogenetic histories of different genes in an organism, such as genes for metabolic functions, are not the same. Alternatively, codon use in different genes varies, implying transfer from another

organism with a different codon use pattern (for more information, see Chapters 9 and 11). In another study, phylogenetic analysis of 205 different sets of orthologous genes (predicted to have the same function based on high-scoring sequence alignments) in a divergent group of bacteria, the γ Proteobacteria, revealed that most (203/205) shared the same phylogenetic history (Lerat et al. 2003). Because these conserved genes are more similar than most genes across species, their phylogenetic history is more readily predicted. Another type of phylogenetic analysis is based on the number of genes shared between genomes and produces a tree similar to the rRNA tree (Snel et al. 1999). These studies have revealed that gene relationships and species evolution must be taken into account and that the quality of the phylogenetic analyses is an important factor.

To track the evolutionary history of genes more carefully, closer attention has been paid to the underlying assumptions that are commonly made in phylogenetic analysis and that could be a source of error

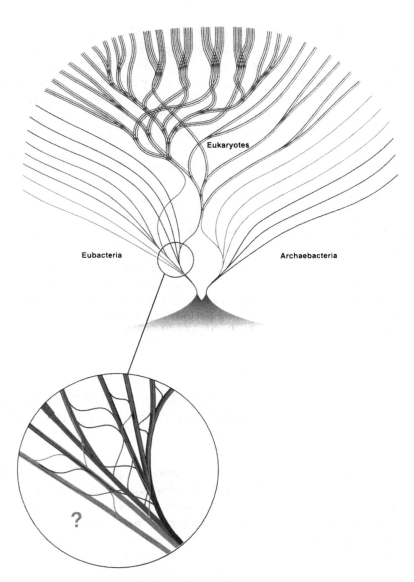

FIGURE 7.4. The reticulated or net-like form of the tree of life. Analysis of rRNA sequences originally suggested three main branches in the tree of life—Archaea, Bacteria, and Eukarya. Subsequent phylogenetic analysis of genes for some metabolic enzymes is not congruent with the rRNA tree. Hence, for these metabolic genes, the tree has a reticulated form (see magnified region) due to horizontal transfer of these genes between species. (Reprinted, with permission, from Martin 1999 [©Wiley-Liss, Inc., a subsidiary of John Wiley & Sons, Inc.].)

(Doolittle 1999). For example, one possible source of error is the assumption that the same sequence positions are evolving at the same rate in different genomes. These assumptions become particularly problematic when analyzing more distantly related sequences. Other evidence reveals that genomes undergo extensive rearrangements, placing sequences of different evolutionary origin physically next to each other in a genome and even causing rearrangements within protein-encoding genes (Henikoff et al. 1997). Consequently, sequence alignment needs to be carefully examined.

As a result of such changes and rearrangements, different regions of independent evolutionary origin in protein or DNA sequences need to be identified.

Proteins are modular with functional domains that are sometimes repeated within a protein and sometimes shared within a protein family (see Chapter 10). These regions are identified by their sharing significant sequence similarity. The remainder of the aligned regions in the protein family may have variable levels of similarity. In nucleic acid sequences, a given sequence pattern may provide a binding site for a regulatory molecule, leading to promoter function, RNA splicing, translation, or some other function. It may be difficult to decide the extent of these patterns for phylogenetic analysis; however, statistical approaches for identifying common patterns in sequences discussed in Chapter 5 may be used.

Another feature of genome complexity and evolution that should be considered in phylogenetic analysis is the occurrence of gene duplication events that can create tandem copies of a gene in a genome. An example is the olfactory receptor genes. These genes form the largest known multigene family in the human genome that is found in separate chromosomal clusters, which appear to have undergone rearrangements based on their phylogenetic history (Niimura and Nei 2003). Such duplicated gene copies may evolve along separate pathways leading to modified functions, or they may become functionally defective (pseudogenes). However, these gene copies maintain a certain level of similarity and sometimes undergo concerted evolution, a process of acquiring mutations in a coordinated way through rarely occurring gene conversion or recombination events within the duplicated region over long periods of evolutionary time. Following gene duplications within a species, separation of the species into two new species followed by additional mutations can lead to the production of two separable sets of gene sequences that include a history of the original duplication events. As discussed in Chapter 3 and illustrated in Figure 3.3, two genes in the same lineage can have different relationships. In the example shown in Figure 3.3, genes a1 and a2 have been derived from gene a. The pair is then segregated by speciation such that there is one a1 a2 pair in one species evolving along one path and a second a1 a2 pair in a second species evolving along a second path, reproductively and genetically isolated from each other. The a1 genes in the different species are orthologous to each other, as are the a2 genes, but the a1 and a2 genes are paralogous because they arose from a gene duplication event. These relationships can be determined by a careful analysis of genomes and sequence relationships (Tatusov et al. 1997; Niimura and Nei 2003).

EVOLUTIONARY TREES SHOW EVOLUTIONARY RELATIONSHIPS AMONG ORGANISMS, GENES, AND GENE FAMILIES

An evolutionary tree is a two-dimensional graph showing evolutionary relationships among organisms, or, in the case of sequences, among specific genes from separate organisms or among gene family members. The separate organisms or sequences are referred to as taxa (singular taxon), illustrated as sequences A–D in Figure 7.5. Taxa are defined as phylogenetically distinct units on the tree. The tree is composed of outer branches (or leaves) representing the taxa and nodes, and inner branches with connecting nodes (or internal vertices) representing evolutionary relationships among the taxa. The arrangement of nodes and branches is described as the topology of the tree. In Figure 7.5, sequences A and B are derived from a common ancestor sequence represented by the node below them, and C and D are similarly related. The A/B and C/D common ancestors also share a common ancestor represented by a node at the lowest level of the tree. It is important to recognize that each node in the tree represents a splitting of the evolutionary path of the gene into two different species that are isolated reproductively. Beyond that point, any further evolutionary changes in each new branch are independent of those in the other new branch.

A phylogenetic tree is also a bifurcating or binary tree—that is, only two branches emanate from each node. This situation is what one would expect during evolution—only one splitting away of a new species at a time. Trees can have more than one branch emanating from a node if the events separating taxa are so close that they cannot be resolved, or to simplify

A. Rooted tree

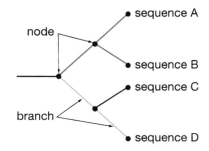

B. Unrooted tree

FIGURE 7.5. Structure of evolutionary trees.

the tree. The length of each branch to the next node represents the number of sequence changes that occurred prior to the next level of separation. In this example, note that the branch length between the A/B node and A is approximately equal to that between the A/B node and B, indicating that the species are evolving at the same rate.

The amount of evolutionary time that has transpired since the separation of A and B is usually not known. What is estimated by phylogenetic analysis is the sum of the amount of sequence change between the A/B node and A and that between the A/B node and B. Hence, judging by the branch lengths from this node to A and B, the same number of sequence changes has occurred. However, it is also likely that for some biological or environmental reason unique to each species, one taxon may have undergone more mutations since diverging from the ancestor than the other. In this case, different branch lengths would be shown on the tree. Some types of phylogenetic analyses assume that the rates of evolution in the tree branches are the same, whereas others assume that they vary, as discussed below.

The assumption of a uniform rate of mutation in the tree branches is known as the molecular clock hypothesis and is usually most suitable for closely related species (Li and Graur 1991; Li 1997). Tests for the molecular clock hypothesis have been devised as described below (see p. 302). Even if there is a common rate of evolutionary change, statistical variations from one branch to another can influence the analysis. The number of substitutions in each branch of a phylogenetic tree is generally assumed to vary according to the Poisson distribution (see Chapter 4, p. 135, for an explanation of the Poisson distribution), and

the rate of change is assumed to be equal across all sequence positions (Swofford et al. 1996).

Two different types of trees, rooted and unrooted, are depicted in Figure 7.5 A and B, respectively. In Figure 7.5A, a root or basal node has been placed at the base of the tree representing a common ancestor of all of the other sequences. A unique path leads from the root node to any other node, eventually reaching the outermost nodes and the sequences, and the direction of the path indicates the passage of evolutionary time. If the root were moved to another branch on the tree, a different tree would be produced. The root may be located by including a related taxon that we are reasonably sure branched off earlier than the other taxa under study. It is also possible to predict a root, assuming that the molecular clock hypothesis holds; i.e., that the rate of evolution in each of the branches is the same. Sometimes, the position of the root may not be of interest or may be difficult to locate, resulting in an unrooted tree as illustrated for sequences A–D in Figure 7.5B. An unrooted tree also depicts the relationships among taxa or sequences but does not provide information on a common ancestor to the group. The unrooted tree in Figure 7.5B also depicts the evolutionary relationships among sequences A–D, but it does not reveal the location of the oldest ancestry. B could be converted into A by placing another node and adjoining root to the central branch (black). A root could also be placed anywhere else in the tree. Hence, there are a great many more possibilities for rooted than for unrooted trees for a given number of taxa or sequences, as described in Table 7.2. For any set of sequences, there are usually many different trees, each predicting a different evolutionary relationship

TABLE 7.2. *Number of possible evolutionary trees to consider as a function of number of sequences*

Taxa or sequence no.	No. of rooted trees	No. of unrooted trees
3	3	1
4	15	3
5	105	15
—	—	—
7	10,395	954

among the sequences. The number of possible rooted and unrooted trees describing the evolution of a group of taxa or sequences increases very rapidly with the number of sequences or taxa being analyzed, as shown in Table 7.2. The challenge of phylogenetic analysis is to determine the tree that best accounts for the observed variation in the sequences. A useful concept in this regard is the tree length. The sum of all the branch lengths in a tree is the tree length.

THERE ARE THREE MAIN METHODS FOR PHYLOGENETIC PREDICTION

Three methods—maximum parsimony, distance, and maximum likelihood—are generally used to find the evolutionary tree or trees that best account for the observed variation in a group of sequences. Each of these methods uses a different type of analysis as described below. Programs based on distance methods are commonly used in the molecular biology laboratory because they are straightforward to use and can be used with a large number of sequences. Maximum likelihood methods are more challenging and require more understanding of the evolutionary models on which they are based. Because they involve so many computational steps and because the number of steps increases dramatically with the number of sequences, maximum likelihood programs are limited to a smaller number of sequences. They can be implemented on a supercomputer in order to analyze a greater number of sequences.

The flowchart below describes the types of considerations that need to be made in choosing a method but is not intended as a strict guide. It can be useful to try at least two of these methods, which can add confidence to the resulting analysis if the same results are obtained. These methods may find that more than one tree meets the criterion chosen for being the most likely tree. The branching patterns in these trees may be compared to find which branches are shared and therefore are more strongly supported. PAUP provides methods for finding consensus trees, and such trees are also calculated by the CONSENSE program in the PHYLIP package. Trees are stored as a tree file that shows the relationships in nested-parentheses notation; i.e., a file with the line (A,(B,(C,D))) represents the tree shown below the data in Table 7.2. Sometimes, branch lengths are also included next to the names; e.g., A:0.05. From this information, a tree-drawing program may be used to produce a tree representation of the data.

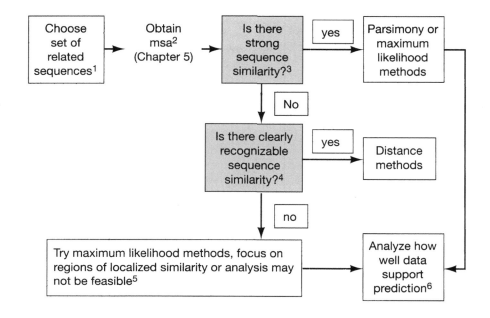

1. The sequences chosen for phylogenetic analysis can be either DNA or protein sequences. Different programs and program options are used for each type. The sequences should align with each other along their entire lengths, or else each should have a common set of patterns or domains that provides a strong indication of evolutionary relatedness. RNA sequences are analyzed by covariation methods and changes in secondary structure, as outlined in Chapter 8.

2. A phylogenetic analysis should be performed when the sequences produce an msa in which sequence similarity is apparent by the presence of conserved positions in the columns of the alignment. Some variation in these columns is necessary to produce the phylogenetic analysis, but too much makes the msa itself uncertain and the resulting phylogenetic analysis more difficult. The alignments should not include a large number of gaps that are obviously necessary to align identical or related characters (see Chapter 3 flowchart, p. 75). Some aligned regions may be better conserved than others and the analysis can then be restricted to these conserved regions. In general, phylogenetic methods analyze conserved regions that are represented in all the sequences. The more similar the sequences are to each other, the better. The simplest evolutionary models assume that the variation in each column of an msa represents single-step changes and that no reversals ($A\rightarrow T\rightarrow A$) have occurred. As the observed variation increases, more multiple-step changes ($A\rightarrow T\rightarrow G$) and reversions are likely to be present. Most phylogenetic analysis programs can correct for such variation according to predicted patterns of mutational change over time. These corrections usually assume a uniform rate of change at all sequence positions over time. Gaps in the msa are usually not scored because there is no suitable model for the evolutionary mechanisms that produce them.

3. This question is designed to select sequences in which there is a clear-cut majority of certain residues in some columns of the msa but also some variation. Some columns in the msa will have the same residue in all sequences; other columns will include variation. The more common residues in the variable columns are taken to represent an earlier group of sequences from which others were derived. If there is too much variation, there will be too many possible ancestral relationships. If the amount of variation is small but definitely present, these sequences are then suitable for maximum parsimony analysis. For parsimony analysis, the trees that best fit the observed variations in the columns of the alignment are found. The best results are obtained when the amount of variation among all pairs of sequences is similar (no very different sequences are present) and when that amount of variation is small. Because the maximum parsimony method has to attempt to fit all possible trees to the data, the method is not suitable for more than 12 sequences because there are too many trees to test. During a maximum parsimony analysis, more than one tree may be found to be equally parsimonious. A consensus tree representing the conserved features of the different trees may then be produced. A maximum likelihood

analysis that also produces a tree which best predicts the sequence variation in each alignment column may also be used.

4. The purpose of this question is to select sequences for phylogenetic analysis by distance methods. These methods do not depend on the presence of limited variation in each column of the msa like the maximum parsimony method. In distance methods, the amount of variation between each pair of sequences in the alignment is measured as the fraction of aligned characters that change (the genetic distance). As a result, the method is not as sensitive as maximum parsimony to variation in the aligned columns. Distance methods predict an evolutionary tree based on the degree of difference among pairs of sequences in the msa and can be used when the amount of variation is sufficient to distinguish the sequence pairs on this basis. As distances increase, corrections are necessary for deviations from single-step changes between sequences (see note 2). In addition, as distances increase, the uncertainty of alignments also increases (see Chapter 5), and a reassessment of the suitability of the msa method may be necessary. Sequences with this type of variation may also be suitable for phylogenetic analysis by maximum likelihood methods. Distance methods may be used with a large number of sequences and usually are not significantly affected by variations in rates of mutation over evolutionary time.

5. If there is considerable variation among the sequences in the msa, then the options given in this box should be considered. The msa itself may be improved by using programs that have been shown to produce alignments of more variable sequences (see Table 5.1, p. 172), or the more similar regions may be extracted and used for the phylogenetic analysis. Maximum likelihood methods may be used for any set of related sequences, but they are particularly useful when the sequences are more variable. These methods are computationally intense, and computational complexity increases with the number of sequences, since the probability of every possible tree must be calculated as described in the text (see p. 319). An advantage of maximum likelihood methods is that they include evolutionary models to account for the variation in the sequences.

6. This box addresses how well the sequence variation that is present in the msa supports the tree or trees predicted by the phylogenetic analysis. If the computed tree is based on variation in only a few columns of the alignment or between certain pairs of sequences, it is not representative of the variation in all of the sequences. To address this possibility, the columns in the msa are resampled randomly to produce many new alignments, and a new phylogenetic analysis is then performed on these resampled alignments — a procedure known as bootstrapping. The frequency with which a particular branch in the original tree appears in these new alignments is then given. The more often the branch appears, the better the data in the original alignment support that particular branch in the predicted tree.

Maximum Parsimony Method

Maximum parsimony, the first main method for phylogenetic analysis, predicts the evolutionary tree (or trees) that minimizes the number of steps required to generate the observed variation in the sequences from common ancestral sequences. For this reason, the method is also sometimes referred to as the minimum evolution method. An msa is required to predict which sequence positions are likely to correspond. These positions will appear in vertical columns in the msa. For each aligned position, phylogenetic trees that require the smallest number of evolutionary changes to produce the observed sequence changes from ancestral sequences are identified. This analysis is continued for every position in the sequence alignment. Finally, those trees that produce the smallest number of changes overall for all sequence positions are identified. This method is best suited for sequences that are quite similar and is limited to small numbers of sequences.

The algorithm followed in the maximum parsimony method is not particularly complicated, but it is guaranteed to find the best tree, because all possible trees relating a group of sequences are examined. For this reason, the method is quite time-consuming and is not useful for data that include a large number of sequences or sequences with a large amount of variation. One or more unrooted trees are predicted and other assumptions must be made to root the predicted tree. PAUP offers a number of options and parameter settings for a parsimony analysis in the Macintosh environment.

The main programs for maximum parsimony analysis in the PHYLIP package at http://evolution.genetics.washington.edu/phylip.html (Felsenstein 1996) are listed below. For analysis of nucleic acid sequences, the programs are:

1. DNAPARS treats gaps as a fifth nucleotide state.

2. DNAPENNY performs parsimonious phylogenies by the branch-and-bound method that limits the number of trees searched and can therefore analyze more sequences.

3. DNACOMP performs phylogenetic analysis by searching for the tree that supports the largest number of sites (compatible sites) rather than overall parsimony at all sites in the msa. This method is recommended when the rate of evolution varies among sites.

4. DNAMOVE performs parsimony and compatibility analysis interactively.

For analysis of protein sequences, the program is:

1. PROTPARS counts the minimum number of mutations to change a codon for the first amino acid into a codon for the second amino acid, but only scores those mutations in the mutational path that actually change the amino acid. Silent mutations that do not change the amino acid are not scored on the grounds that they have little evolutionary significance.

The maximum parsimony analysis is illustrated in the following example of four sequences shown in Table 7.3 and Figure 7.6 (adapted from Li and Graur 1991). An example of a parsimony analysis of mitochondrial sequences using PAUP and MacClade is then given. Note that in an msa, only certain sequence variations at a given site are useful for a parsimony analysis. In the analysis, all of the possible unrooted trees for each site or column in the msa are considered. For example, there are three unrooted trees for four sequences (Table 7.2). The sequence variations at each site in the alignment are placed at the tips of the trees, and the tree that requires the smallest number of

TABLE 7.3. *Example of phylogenetic analysis to find the correct unrooted tree from four aligned sequences by the maximum parsimony method*

Taxa	Sequence position (sites) and character								
	1	2	3	4	5	6	7	8	9
1	A	A	G	A	G	T	G	C	A
2	A	G	C	C	G	T	G	C	G
3	A	G	A	T	A	T	C	C	A
4	A	G	A	G	A	T	C	C	G

Adapted from Li and Graur (1991).

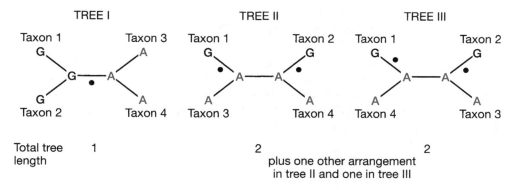

TREE I   TREE II   TREE III

Total tree   1   2   2
length   plus one other arrangement
                         in tree II and one in tree III

● is a substitution

FIGURE 7.6. Example of phylogenetic analysis based on sequence position 5 in Table 7.3, using the maximum parsimony method. (Redrawn, with permission, from Li and Graur 1991 [©Sinauer Associates].)

EXAMPLE: MAXIMUM PARSIMONY ANALYSIS OF SEQUENCES

Table 7.3 shows an example of phylogenetic analysis by maximum parsimony. This method finds the tree that changes any sequence into all of the others by the least number of steps.

Rules for analysis by maximum parsimony in this example are:

1. There are four taxa giving three possible unrooted trees.
2. Some sites are informative, i.e., they favor one tree over another (site 5 is informative, but sites 1, 6, and 8 are not).
3. To be informative, a site must have the same sequence character in at least two taxa (sites 1, 2, 3, 4, 6, and 8 are not informative; sites 5, 7, and 9 are informative).
4. Only the informative sites need to be analyzed.

The three possible trees are shown in Figure 7.6. The optimal tree is obtained by adding the number of changes at each informative site for each tree, and picking the tree requiring the least number of changes. A scoring matrix may be used instead of scoring a change as 1. Tree 1 is the more parsimonious one and the tree length will be 4 (one change at each of the positions 5 and 7 and two changes at position 9).

changes to produce this variation at that site is determined. This analysis is repeated for each informative site, and the tree (or trees) that supports the smallest number of changes overall is found. The length of the tree, defined as the sum of the number of steps in each branch of the tree, will be a minimum.

Because there were only four sequences to consider in the example above, it was necessary to consider only three possible unrooted trees. For a larger number of sequences, the number of trees becomes so large that it may not be feasible to examine all possible trees. The example of 12 sequences below took only a few seconds on a personal computer. For large numbers of sequences, the program PAUP provides an option called "heuristic," which searches for random examples of trees among all possible trees and keeps representative trees that best fit the data. The presence of common branch patterns in these trees reveals some of the broader features of the phylogenetic relationships among the sequences. The branch-and-bound options of the PAUP will analyze all possible trees, but will not examine further rearrangements in particular groups of trees when a more parsimonious tree will be found. If the sequence number is large, the program can be run on a supercomputer.

ANALYSIS OF MITOCHONDRIAL SEQUENCES BY PAUP

An example of a maximum parsimony analysis is illustrated and described in Figure 7.7. To search for the most parsimonious tree that best fits the sequence data, the trees that best fit each vertical column of sequence characters in Figure 7.7A are first determined. In some columns, the data are not informative, as in the case of all nucleotides being the same. For a nucleotide position to be informative, at least two different nucleotides must be present in at least two sequences. A tree that provides the least number of evolutionary steps to satisfy the data in all columns—the most parsimonious tree—is then found.

Problems in Using Maximum Parsimony

Parsimony can give misleading information when rates of sequence change vary in the different branches of a tree that are represented by the sequence data. These variations produce a range of branch lengths, with long ones representing more extended periods of time and short ones representing shorter times. For example, the tree shown in Figure 7.8A includes two long branches in which G has turned to A independently. Because in a parsimony analysis rates of change along all branches of the tree are assumed to be equal, the tree predicted by parsimony and shown in Figure 7.8B will not be correct.

Although other columns in the msa that show less variation may provide the correct tree, the columns representing greater variation dominate the analysis (Swofford et al. 1996). Such long branches may be broken down if additional taxa are present and are more closely related to taxa 1 and 4, thereby providing branches that intersect the long branches and give a better resolution of the changes.

Another method for identifying such long branches is called Lake's method of invariants, also called evolutionary parsimony, available in PAUP. In this method, four of the sequences are chosen at a time, and only transversions in the aligned positions are scored as changes on the grounds that they have greater evolutionary significance (see box Transitions and Transversions, p. 300). Transversions of any base to each possible derivative, e.g., A→C or T, are assumed to change at the same rate to create a balanced distribution, and the changes in each column of the alignment (each sequence position) are assumed to occur independently of each other.

Suppose that there are two long branches as in the case discussed immediately above. The correct tree is shown in Figure 7.9A, and one of the sites has changed multiply but ends up as the same base A by chance. Traditional parsimony will identify this tree incorrectly, as indicated above. If these long branches do indeed exist, then other sites should give the type of transversion events shown in Figure 7.9B. The greater the number of B-type sites, the less one can depend on the A-type sites revealed in A. The evolutionary parsimony method subtracts the number of type B from the number of type A. If, on the one hand, long branches are not present in the quartet of sequences, there will be very few type B, and type A will be taken as evidence for the correct tree. On the other hand, if many examples of type B are present, the A type will carry little weight.

FIGURE 7.7. *(See facing page for figure.)* Analysis of mitochondrial sequences using the maximum parsimony method provided by the PAUP program. *(A)* Portion of an msa of the mitochondrial sequences provided in the PAUP distribution package. PAUP will import sequences in other msa formats and convert them into the NEXUS format. The program READSEQ will reformat msa's into the NEXUS format, a format that allows additional information about the sequences, species relationship, and a scoring system for base substitution, referred to as a cost or step matrix. This format includes information about type of sequence, coding information, codon positions, differential weights for transitions and transversions, treatment of gaps, and preferred groupings (see Chapter 2). Only a portion of the NEXUS file is shown. In this analysis, branch-and-bound and otherwise default options were used. Gaps are treated as missing information. The number of sequences is indicated as ntaxa, number of alignment columns as nchar, and the interleave command allows the data to be entered in readable blocks of sequence 60 characters long. *(B)* One of the two predicted trees. The tree file of PAUP was edited in MacClade and output as a graphics file.

A. Mitochondrial sequences.

```
#NEXUS

begin taxa;
      dimensions ntax=12;
end;

begin characters;
      dimensions nchar=898;
      format missing=? gap=- matchchar=. interleave datatype=dna;
      options gapmode=missing;
      matrix

Lemur_catta        AAGCTTCATAGGAGCAACCATTCTAATAATCGCACATGGCCTTACATCATCCATATTATT
Homo_sapiens       AAGCTTCACCGGCGCAGTCATTCTCATAATCGCCCACGGGCTTACATCCTCATTACTATT
Pan                AAGCTTCACCGGCGCAATTATCCTCATAATCGCCCACGGACTTACATCCTCATTATTATT
Gorilla            AAGCTTCACCGGCGCAGTTGTTCTTATAATTGCCCACGGACTTACATCATCATTATTATT
Pongo              AAGCTTCACCGGCGCAACCACCCTCATGATTGCCCATGGACTCACATCCTCCCTACTGTT
Hylobates          AAGCTTTACAGGTGCAACCGTCCTCATAATCGCCCACGGACTAACCTCTTCCCTGCTATT
Macaca_fuscata     AAGCTTTTCCGGCGCAACCATCCTTATGATCGCTCACGGACTCACCTCTTCCATATATTT
M._mulatta         AAGCTTTTCTGGCGCAACCATCCTCATGATTGCTCACGGACTCACCTCTTCCATATATTT
M._fascicularis    AAGCTTCTCCGGCGCAACCACCCTTATAATCGCCCACGGGCTCACCTCTTCCATGTATTT
M._sylvanus        AAGCTTCTCCGGTGCAACTATCCTTATAGTTGCCCATGGACTCACCTCTTCCATATACTT
Saimiri_sciureus   AAGCTTCACCGGCGCAATGATCCTAATAATCGCTCACGGGTTTACTTCGTCTATGCTATT
Tarsius_syrichta   AAGTTTCATTGGAGCCACCACTCTTATAATTGCCCATGGCCTCACCTCCTCCCTATTATT

Lemur_catta        CTGTCTAGCCAACTCTAACTACGAACGAATCCATAGCCGTACAATACTACTAGCACGAGG
Homo_sapiens       CTGCCTAGCAAACTCAAACTACGAACGCACTCACAGTCGCATCATAATCCTCTCTCAAGG
Pan                CTGCCTAGCAAACTCAAATTATGAACGCACCCACAGTCGCATCATAATTCTCTCCCAAGG
Gorilla            CTGCCTAGCAAACTCAAACTACGAACGAACCCACAGCCGCATCATAATTCTCTCTCAAGG
Pongo              CTGCCTAGCAAACTCAAACTACGAACGAACCCACAGCCGCATCATAATCCTCTCTCAAGG
Hylobates          CTGCCTTGCAAACTCAAACTACGAACGAACTCACAGCCGCATCATAATCCTATCTCGAGG
Macaca_fuscata     CTGCCTAGCCAATTCAAACTATGAACGCACTCACAACCGTACCATACTACTGTCCCGAGG
M._mulatta         CTGCCTAGCCAATTCAAACTATGAACGCACTCACAACCGTACCATACTACTGTCCCGGGG
M._fascicularis    CTGCTTGGCCAATTCAAACTATGAGCGCACTCATAACCGTACCATACTACTATCCCGAGG
M._sylvanus        CTGCTTGGCCAACTCAAACTACGAACGCACCCACAGCCGCATCATACTACTATCCCGAGG
Saimiri_sciureus   CTGCCTAGCAAACTCAAATTACGAACGAATTCACAGCCGAACAATAACATTTACTCGAGG
Tarsius_syrichta   TTGCCTAGCAAATACAAACTACGAACGAGTCCACAGTCGAACAATAGCACTAGCCCGTGG
.
.
.
end;
```

B. Phylogenetic tree

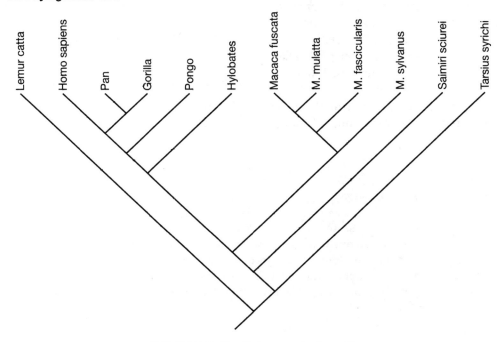

FIGURE 7.7. (*See facing page for legend.*)

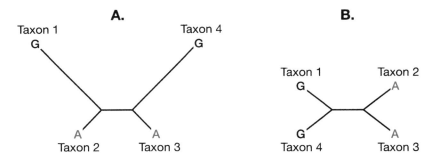

FIGURE 7.8. Type of sequence variation that leads to an incorrect prediction by the maximum parsimony method.

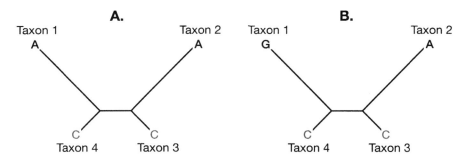

FIGURE 7.9. Type of sequence variation that, if detected, can reduce incorrect predictions by the maximum parsimony method.

These calculations are performed for all three possible unrooted trees and all possible types of transversions for the four sequences, and the tree receiving the most support is chosen. These methods and other more sophisticated methods for correcting uneven branch lengths are discussed in detail in Swofford et al. (1996). The PHYLIP program DNAINVAR computes Lake's and other phylogenetic invariants for nucleic acid sequences. PAUP also includes an option for Lake's invariant.

Compared to the above variation of the maximum parsimony method, the maximum likelihood and distance methods provide more reliable predictions when corrections are made for multiple substitutions. Distance methods such as neighbor joining discussed below have been shown generally to be better predictors than both standard and evolutionary parsimony methods when branch lengths are varying (Jin and Nei 1990; Swofford et al. 1996).

Selecting the Most Parsimonious Trees

There are options in PAUP and MacClade for selecting among the most parsimonious trees. With

TRANSITIONS AND TRANSVERSIONS

There are two ways for any base in DNA to change by transversion, e.g., A→C or T, but only one way to change by transition, e.g., A→G; however, transitions are observed to be approximately twofold more common in evolving DNA sequences than transversions (Li and Graur 1991). Transversions are also probably less common than expected because they produce a more extreme chemical change in DNA sequences, often causing a change in the amino acid sequence of proteins or other functional changes. Transversions are hence considered to be a better indicator of evolutionary change in sequences.

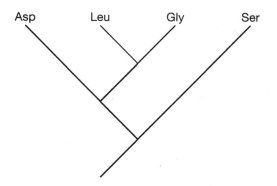

FIGURE 7.10. Tracing of sequence characters in an evolutionary tree by MacClade.

MacClade it is possible to view the changes in each sequence character in each branch of a given tree. Columns in the msa give the characters that are on the outer branches of the tree. The given tree provides the pattern of lower tree branches that leads to these sequence characters. As changes are traced from positions lower in the tree to upper positions, some nodes in the tree may be assigned an unambiguous character (shown in color in Fig. 7.10). For other nodes, the assignment may be ambiguous because the node leads to two different characters above (branch shown in black). It is possible to arrange these ambiguities optionally in two ways: one is to delay them going as far up the tree away from the root as possible (the Deltran option in the program; not shown in figure); a second is to introduce them as soon as possible and as close to the root as possible (the Acctran option in the program; not shown in figure). The effect of using Deltran is to force parallel changes in the upper branches of the trees; that of Acctran is to force reversals in the upper branches. Using these options is not recommended unless such variations are expected, as in analysis of more divergent sequences (Maddison and Maddison 1992).

Another feature of MacClade is a measurement of the degree to which the individual sites in an alignment support the predicted tree. Homoplasy refers to the occurrence of the same sequence change in more than one branch of a tree showing the variation at one site and is an indication that the site does not support the tree. For example, suppose that the sequence changes necessary for a particular position in an msa to match a given tree are examined, as illustrated in Figure 7.10. If there is a good agreement between the changes in this position and the predict-

ed tree based on all sequence positions, each change will only occur in one branch of the tree. In reality, some sequence positions will not support the tree, resulting in more than one branch producing the same sequence change to generate the observed sequence variation.

The number of sequence positions at which two changes are needed instead of one to match the sequence variation to the predicted tree increases the length of the tree and provides a measurement of homoplasy. Homoplasy is usually found for some characters for any tree. MacClade allows changing the tree to avoid homoplasy at a specific sequence position; however, the new tree length will often increase, thus making the tree a less parsimonious choice than the original. A parameter used for measuring homoplasy is the consistency index (CI), which is the minimum possible tree length originally calculated divided by the actual tree length after any site changes have been made. The more homoplasy, the greater the actual tree length, and the smaller the value of CI. The CI value provides an indication as to how well the sequence variations at each site are consistent with the predicted tree.

Parsimony methods can use information on the number of changes or steps required to change one residue into another. For example, the number of mutations required to change one amino acid into another in one branch of a tree can be taken into account. The parsimony method then attempts to minimize the number of such steps. This number of steps for interchanging characters can be incorporated into a matrix, called a step or cost matrix, for programs such as PAUP and MacClade to use.

The program PROTPARS, for protein sequences in the PHYLIP package, scores only those mutations that produce amino acid changes (Felsenstein 1996). Amino acid changes are the most likely ones to alter the function of the gene and should be a predictor of evolutionary change. This program uses an algorithm similar to one described by Sankoff (1975) for determining the minimum number of mutations in a tree needed for changing one sequence into another. Similar types of analyses for proteins are also available in PAUP and MacClade.

Distance Methods

Distance methods, the second main method for phylogenetic analysis, are based on genetic distances between sequence pairs in an msa. The genetic distance between two sequences is the fraction of aligned positions in which the sequence has been changed. In contrast, sequence identity is the fraction of the aligned positions that are identical. Gaps may be ignored in distance calculations or treated like substitutions. A scoring or substitution matrix may also be used, making the calculation slightly more complicated, but the principle is the same. Sequence pairs that have the smallest distances are "neighbors." On a tree, these sequences share a node or common ancestor position and are each joined to that node by a branch. The goal of distance methods is to identify a tree that positions the neighbors correctly and that has branch lengths which, when added up between each sequence pair, closely reproduce the original distance measurements. Finding the distances between all pairs of sequences in an msa is the first step in the distance method of phylogenetic analysis. These distances are then placed into a table. The most closely related sequence pairs with the smallest distances between them are then chosen as neighbors to be placed on adjacent branches of the phylogenetic tree. More distantly related sequences are then progressively added to the tree based on the distance table. When a scoring or substitution matrix is used, the calculation is slightly more complicated, but the principle is the same.

The distance method was pioneered by Feng and Doolittle, and a collection of programs by these authors will produce both an alignment and tree of a set of protein sequences (Feng and Doolittle 1996). The program CLUSTALW, discussed in Chapter 5, uses the neighbor-joining distance method as a guide to msa's. PAUP version 4 has options for performing a phylogenetic analysis by distance methods.

Programs of the PHYLIP package at http://evolution.genetics.washington.edu/phylip.html that perform a distance analysis include the following:

1. DNADIST computes distances among input nucleic acid sequences. There are choices given for various models of evolution as described below and a choice for the expected ratio of transitions to transversions.

2. PROTDIST computes a distance measure for protein sequences, based on the Dayhoff PAM model (see p. 95) or other models of evolutionary change in proteins (Felsenstein 1996).

Once distance matrices have been produced using one of the above options, they may be used as input to the following distance analysis programs in PHYLIP.

Distance analysis programs in PHYLIP are:

1. FITCH estimates a phylogenetic tree by assuming additivity of branch lengths using the Fitch–Margoliash method described below. It does not assume a molecular clock (i.e., it allows rates of evolution along branches to vary).

2. KITSCH estimates a phylogenetic tree using the Fitch–Margoliash method and assuming a molecular clock.

3. NEIGHBOR estimates phylogenies using the neighbor-joining or the unweighted pair group method with arithmetic mean (UPGMA) described below. The neighbor-joining method does not assume a molecular clock and produces an unrooted tree. The UPGMA method assumes a molecular clock and produces a rooted tree.

The success of distance methods depends on the degree to which the distances among a set of sequences can be made additive on a predicted evolutionary tree. Suppose there are four sequences, A–D, as shown below in Figure 7.11A, and that they were derived from evolutionary changes reflected by the tree in Figure 7.11D. The number of changes along the branches of the tree corresponds to distances between the sequences shown in Figure 7.11, B and C. In this tree, each change only occurs once, and there are no examples of the same change occurring twice (homoplasy). Although this pattern of change is idealized and most groups of sequences would have examples of the same change occurring more than once, as well as reversions of a change to the original character, this example illustrates the additivity principle for four sequences. The principle is that for four sequences predicted by this tree, $d_{AB} + d_{CD} \leq d_{AC} + d_{BD} = d_{AD} + d_{BC}$. In this example, the additivity is $3 + 3 \leq 7 + 7 = 8 + 6$. For any other tree, there would be examples of parallel changes and reversions. The additivity condition can be relaxed such that $d_{AB} + d_{CD} \leq d_{AC} + d_{BD}$ and $d_{AB} + d_{CD} \leq d_{AD} + d_{BC}$ will still hold, even for sequences in which the changes in the sequence are not fully additive. For each set of four sequences, the tree for which the above additivity condition among the distances best holds provides information as to which sequences are neighbors.

A. Sequences

sequence A A C G C G T T G G G C G A T G G C A A C
sequence B A C G C G T T G G G C G A C G G T A A T
sequence C A C G C A T T G A A T G A T G A T A A T
sequence D A C A C A T T G A G T G A T A A T A A T

B. Distances between sequences, the number of steps required to change one sequence into the other.

n_{AB} 3
n_{AC} 7
n_{AD} 8
n_{BC} 6
n_{BD} 7
n_{CD} 3

C. Distance table

	A	B	C	D
A	–	3	7	8
B	–	–	6	7
C	–	–	–	3
D	–	–	–	–

D. The assumed phylogenetic tree for the sequences A-D showing branch lengths. The sum of the branch lengths between any two sequences on the trees has the same value as the distance between the sequences.

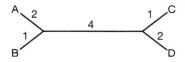

FIGURE 7.11. Set of idealized sequences for which the branch lengths of an assumed tree are additive.

This method may be used to evaluate trees and find the minimum evolution tree for four sequences and for any additional number of sequences by extending the analysis to additional groups of four sequences (Sattath and Tversky 1977; Fitch 1981; for additional references, see Swofford et al. 1996). To calculate branch lengths, distance methods assume additivity in the distances between sequences. However, real sequence data may not fit these idealized conditions. As a result, a small positive, a zero, or even a negative value may be calculated for a branch length. This result may be due to errors in the sequences or sequence alignment, statistical variation, or simply a reflection of two or more sequences diverging at approximately the same time from a common ancestor. This coincident divergence can create a short length, zero length, or negative length branch, depending on the observed distance measurements.

An even more demanding condition, rarely found in real distance data, is that the distances are ultrametric, meaning that for three taxa, $d_{AC} \leq \max(d_{AB}, d_{BC})$. If the data meet this condition, the distances between two taxa and their common ancestor are equal (Swofford et al. 1996). If the distances follow this relationship, the rates of evolution in the tree branches are approximately the same, thereby meeting the expectations of the molecular clock hypothesis. If these conditions are not met, an analysis based on the assumption of a molecular clock may give misleading results. To accommodate such variation, some distance methods are based on the molecular clock hypothesis and others are not.

One method of finding the best tree when the clock does not apply is to transform the sequences after identifying one or more sequences that are least like the rest, called an outgroup (Li and Graur 1991). The outgroup provides a reference point that should be close to the root of the tree and the analysis can be constrained to place the outgroup near the root by most phylogenetic programs.

Fitch and Margoliash Method and Related Methods

The Fitch and Margoliash (1987) method uses a distance table illustrated in Figure 7.11C. The sequences are first combined into groups of three and the combined data are used to calculate branch lengths in the tree. Additional sequences are then progressively added to the analysis. This method of averaging distances is most accurate for trees with short branches. The presence of long branches tends to decrease the reliability of the predictions (Swofford et al. 1996). The branch lengths are assumed to be additive, as described above. Example 1 describes the use of the algorithm for three sequences, and Example 2 expands the analysis to more than three sequences.

The procedure generally followed is to join all combinations of sequences in pairs to find a tree that best predicts the data in the distance table. The percent change from the actual to the predicted distance (calculated from the sums of branch lengths between the sequences) is determined for each sequence pair. These values are squared and summed over all possible pairs. This sum divided by the number of pairs = $n(n-1)/2$ less 1 (the number of degrees of freedom) provides the square of the percent standard deviation of the result.

EXAMPLE 1: USE OF THE FITCH–MARGOLIASH ALGORITHM FOR THREE SEQUENCES

Steps in the algorithm for three sequences:

1. Draw an unrooted tree with three branches emanating from a common node and label ends of branches as shown in Figure 7.12. Given the closer distance between A and B, the branch lengths between these sequences are expected to be shorter, as indicated.

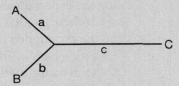

FIGURE 7.12. Tree showing relationship among three sequences A, B, and C.

2. Calculate lengths of tree branches algebraically:

Distances among sequences A, B, and C are shown in the following table:

	A	B	C
A	–	22	39
B	–	–	41
C	–	–	–

The branch lengths may be calculated algebraically using the branch labels a–c in Figure 7.12:

distance from A to B = $a + b$ = 22 (1)
distance from A to C = $a + c$ = 39 (2)
distance from B to C = $b + c$ = 41 (3)

subtract (3) from (2), $a - b = -2$ (4)
add (1) and (4), $2a = 20$, $a = 10$
from (1) and (2), $b = 12$, $c = 29$

Note that this calculation finds that the branch lengths of A and B from their common ancestor are not the same. Hence, A and B are diverging at different rates of evolution by this calculation and model. For the rates to be the same, these distances would be the same and equal to the distance from A to B divided by 2 = 22/2 = 11.

EXAMPLE 2: USE OF THE FITCH–MARGOLIASH ALGORITHM FOR FIVE SEQUENCES

	A	B	C	D	E
A	—	22	39	39	41
B	—	—	41	41	43
C	—	—	—	18	20
D	—	—	—	—	10
E	—	—	—	—	—

Distance data derived from a hypothetical msa of five sequences are shown above. The Fitch–Margoliash method may be extended from three sequences as shown in Example 1 to these five sequences by following the steps shown in this example. The steps are also summarized in the box on page 306 entitled "Steps Followed by Fitch–Margoliash Algorithm for Phylogenetic Analysis of More Than Three Sequences." The method will find the unrooted tree and branch lengths shown in Figure 7.13.

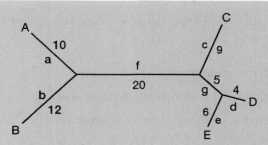

FIGURE 7.13. Tree showing relationships among sequences A–E.

1. The most closely related sequences given in the distance table are D and E. A new table is made with the remaining sequences combined.

2. The average distances from D to A, B, and C and from B to A, B, and C are calculated.

	D	E	ave. ABC
D	–	10	32.7
E	–	–	34.7
average ABC	–	–	–

3. The average distances from D to ABC and from E to ABC can also be found by averaging the sum of the appropriate branch lengths a–g.

Distance between D and E = $d + e$
Average distance between D and ABC = $d + m$, $m = g + [(c + 2f + a + b)/3]$
Average distance between E and ABC = $e + m$
By subtracting the third from the second equation and adding the result to the first equation, $d = 4$ and $e = 6$.

4. D and E are now treated as a single composite sequence (DE), and a new distance table is made. The distance from A to (DE) is the average of the distance of A to D and of A to E. The other distances to (DE) are calculated accordingly.

	A	B	C	(DE)
A	–	22	39	40
B	–	–	41	42
C	–	–	–	19
(DE)	–	–	–	–

5. The next most closely related sequences are identified, in this case C with the (DE) composite group. The new table is:

	DE	C	Ave. AB
DE	–	19	41
C	–	–	40
Ave. AB	–	–	–

By algebraic manipulations similar to those described above, $c = 9$ and the composite distance of $g + [(d + e)/2] = 10$.

6. Given the above composite distance and the previously calculated values of d and e, then $g = 10 - [(d + e)/2] = 5$. The next round of tree-building is that A and B are the next matching pair, giving $a = 10$ and $b = 12$, and a composite distance of $29.7 = [3f + c + 2g + d + e]/3$ giving $f = 29.7 - [(9 + 10 + 10)/3] = 20$. These values are precisely those given in the original tree.

7. Although by design we have generated the correct tree, normally the next step is to repeat the process starting with another sequence pair, such as A and B. We will leave this step as a student exercise to show that the correct tree will again be predicted.

STEPS FOLLOWED BY THE FITCH–MARGOLIASH ALGORITHM FOR PHYLOGENETIC ANALYSIS OF MORE THAN THREE SEQUENCES

Steps in the algorithm for more than three sequences:

1. Find the most closely related pair of sequences, for example, A and B.

2. Treat the rest of the sequences as a single composite sequence. Calculate the average distance from A to all of the other sequences, and B to all of the other sequences.

3. Use these values to calculate the distances a and b as in the above example with three sequences.

4. Now treat A and B as a single composite sequence AB, calculate the average distances between AB and each of the other sequences, and make a new distance table from these values.

5. Identify the next pair of most closely related sequences and proceed as in step 1 to calculate the next set of branch lengths.

6. When necessary, subtract extended branch lengths to calculate lengths of intermediate branches.

7. Repeat the entire procedure starting with all possible pairs of sequences A and B, A and C, A and D, etc.

8. Calculate the predicted distances between each pair of sequences for each tree to find the tree that best fits the original data.

Neighbor-joining and Related Neighbor Methods

The neighbor-joining method (Saitou and Nei 1987) is very much like the Fitch–Margoliash method except that the choice as to which sequences to pair is determined by a different algorithm. The neighbor-joining method is especially suitable when the rate of evolution of the separate lineages under consideration varies. When the branch lengths of trees of known topology are allowed to vary in a manner that simulates varying levels of evolutionary change, the neighbor-joining method and the Sattath and Taversky method, described below, are the most reliable in predicting the correct tree (Saitou and Nei 1987). If not used properly, the neighbor-joining algorithm is a method that does not guarantee finding the minimum length tree. Pearson et al. (1999) have enhanced the neighbor-joining method so that a set of trees that fit the data, rather than just a single tree, may be determined. The general neighbor-joining (GNJ) is available from ftp.virginia.edu/pub/fasta/GNJ.

Neighbor-joining chooses the sequences that should be joined to give the best least-squares estimates of the branch lengths that most closely reflect the actual distances between the sequences. It is not necessary to compare all possible trees to find the least-squares fit as in the Fitch–Margoliash method. The method pairs sequences based on the effect of the pairing on the sum of the branch lengths of the tree. These steps are followed:

1. To start, the distances between the sequences are used to calculate the sum of the branch lengths for a tree that has no preferred pairing of sequences. The star-like appearance of such a tree and the calculation of the length of the tree using the data in Example 2 above are shown in Figure 7.14.

2. The next step in the neighbor-joining algorithm is to decompose or modify the star-like tree in Figure 7.14 by combining pairs of sequences. When this step is performed for sequences A and B in Example 2, the new tree shown in Figure 7.15 will be produced. The tree has A and B paired from a common node that is joined by a new branch j to a second node to which C, D, and E are joined. The sum of the branch lengths of this new tree is calculated as shown in Figure 7.15.

3. In the neighbor-joining algorithm, each possible sequence pair is chosen and the sum of the

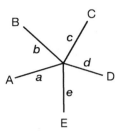

FIGURE 7.14. Tree for five sequences with no pairing of sequences. In the neighbor-joining method, the sum of the branch lengths $S_0 = a + b + c + d + e$ is calculated. The known distances from (1) A to B, $D_{AB} = a + b$; (2) A to C, $D_{AC} = a + c$; (3) B to C, $D_{BC} = b + c$; and finally (4) D to E, $D_{DE} = d + e$, for a total of $4 + 3 + 2 + 1 = 10$ combinations. In summing the ten distances $= 22 + 39 + \ldots + 10 = 314$, each branch a, b, c, etc., is counted four times. Hence, the sum of branch lengths is $314/4 = 78.5$. In general, for N sequences, $S_0 = \Sigma D_{ij}/(N-1)$, where D_{ij} represents the distances between sequences i and j, $i < j$.

branch lengths of the corresponding tree is calculated. For example, using the data of Example 2, $S_{AB} = 67.7$, $S_{BC} = 81$, $S_{CD} = 76$, and $S_{DE} = 70$, plus six other possible combinations. Of these, S_{AB} has the lowest value. Hence, A and B are chosen as neighbors on the grounds that they reduce the total branch length to the largest extent. Once the choice of neighbors has been made, the branch lengths a and b and the average distance from AB to CDE may be calculated by the FM method, as described in the last section. a is calculated by

$$a = [d_{AB}+(d_{AC}+d_{AD}+d_{AE})/3-(d_{BC}+d_{BD}+d_{DE})/3]/2$$
$$= (22+39.7-41.70)/2=10,$$

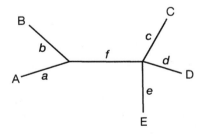

FIGURE 7.15. Tree for five sequences with pairing of A and B. The sum of the branch lengths $S_{ab} = a + b + c + d + e + f$ is calculated algebraically from the original distance data. The sum is given by $S_{ab} = [(d_{AC} + d_{AD} + d_{CE} + d_{BC} + d_{BD} + d_{BE})/6] + d_{AB}/2 + [(d_{CD} + d_{CE} + d_{DE})/3] = 244/6 + 22/2 + 48/3 = 67.7$. In general, the formula for N sequences when m and n are paired is $S_{mn} = [(\Sigma d_{im} + d_{in})/2(N-2)] + d_{mn}/2 + \Sigma d_{ij}/N - 2$, where i and j represent all sequences except m and n, and $i < j$.

and b is calculated by

$$b = [d_{AB}+(d_{BC}+d_{BD}+d_{BE})/3-(d_{AC}+d_{AD}+d_{AE})/3]/2$$
$$= (22+41.7-39.7)/2=12.$$

4. The next step of the neighbor-joining algorithm is like that of the Fitch–Margoliash method: A new distance table with A and B forming a single composite sequence is produced. The neighbor-joining algorithm is used to find the next sequence pair and the Fitch–Margoliash method is then used to find the next branch lengths. The cycle is repeated until the correctly branched tree and the branch distances on that tree have been identified.

The neighbors relation method (Sattath and Tversky 1977; Li and Graur 1991) is also a reliable predictor of trees when the rate of evolution varies between sequences. In this method, the sequences are divided into all possible groups of four. The sum of the pair-wise distances for the three possible neighbor groupings (AB/CD, AC/BD, AD/BC for four sequences) for each group of four are then compared to find which of the three neighbor groupings gives the lowest sum of distances. This procedure is repeated for all possible groups of four. The pair that appears most often in the lowest sum of pairs is selected as neighbors. The pair is then treated as a composite grouping and the entire process is repeated to find the next closest neighbor until all of the sequences have been included. An example of this method for five sequences is shown in Table 7.4.

The Unweighted Pair Group Method with Arithmetic Mean

The above distance methods provide a good estimate of an evolutionary tree and are not influenced by variations in the rates of change along the branches of the tree. The UPGMA method is a simple procedure for tree construction that assumes the rate of change along the branches of the tree is a constant; i.e., assumes that the molecular clock hypothesis holds and the distances are approximately ultrametric (see above). There are also a number of variations of this method for pairing or clustering sequences. The UPGMA method (1) calculates branch lengths between the most closely related sequences, (2) averages the distance between this pair or sequence cluster and the next sequence or sequence cluster, (3) continues until all the sequences are included in the tree, and, finally, (4) predicts a position for the root of the tree.

TABLE 7.4. *The Sattath and Tversky (1977) method for finding repeated neighbors*

Chosen set of 4	Sum of distances	Pairs chosen
ABCD	$n_{AB} + n_{CD} = 22 + 18 = 40$ $n_{AC} + n_{BD} = 39 + 41 = 80$ $n_{AD} + n_{BC} = 39 + 41 = 80$	AB, CD
ABCE	$n_{AB} + n_{CE} = 22 + 20 = 42$ $n_{AC} + n_{BE} = 39 + 43 = 82$ $n_{AE} + n_{BC} = 39 + 41 = 82$	AB, CE
ABDE	$n_{AB} + n_{DE} = 22 + 10 = 32$ $n_{AD} + n_{BE} = 39 + 43 = 82$ $n_{AE} + n_{BD} = 41 + 41 = 82$	AB, DE
ACDE	$n_{AC} + n_{DE} = 39 + 10 = 49$ $n_{AD} + n_{CE} = 39 + 20 = 59$ $n_{AE} + n_{CD} = 41 + 18 = 59$	AC, DE
BCDE	$n_{BC} + n_{DE} = 41 + 10 = 51$ $n_{BD} + n_{CE} = 41 + 20 = 61$ $n_{BE} + n_{CD} = 43 + 18 = 61$	BC, DE

Totals from Column 3 giving the number of times a pair gives the lowest score: AB (3), DE (3), CD (1), CE (1), and BC (1). AB and DE are therefore closest neighbors.

The five sequences used in the above example (see Fig. 7.13) are divided into the five possible groups of four. The sums of distances for each set of sequence pairs for the three possible groupings are then determined and the closest pairs in each grouping are determined. The closest neighbors overall are those that appear as neighbors most often. In this example, AB and DE appear most often as neighbors. These sequences are then chosen as neighbors to calculate the branch lengths on the phylogenetic tree by the method of Fitch and Margoliash.

Using Example 2 from the above analysis:

EXAMPLE: UPGMA ANALYSIS

1. Sequences D and E are the most closely related. The branch distances d and e to the node below them are calculated as $d = e = n_{de}/2 = 5$ based on the assumption of an equal rate of change in each branch of the tree. The tree is often drawn in a dendogram form (Fig. 7.16A) where only the horizontal lines indicate branch lengths, but the branches are intended to be joined to a common node as in Figure 7.16B.

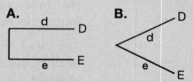

FIGURE 7.16. Branch lengths of most closely related sequences by UPGMA method.

2. Treating D and E as a composite sequence pair, find the next most related pair. The calculations will be similar to the FM method above and the distance between DE and C, $n_{DE,C} = 19$, will be the shortest one. Because we are assuming an equal rate of change in each branch of the tree, there will be two equal-length branches, one including D and E and passing to a common node for C and DE, and a second from the common node to C. Some simple arithmetic gives $c = 19/2 = 9.5$ and $g = 9.5 - 5 = 4.5$ (Fig. 7.17).

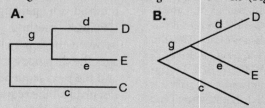

FIGURE 7.17. Inclusion of third sequence for calculation of branch lengths by UPGMA method.

3. With CDE now being treated as a composite trio of sequences, the next closest pair is A and B, giving an estimate of the distance between them and a common node in the tree of $a = b = n_{AB}/2 = 11$ (Fig. 7.18).

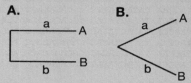

FIGURE 7.18. Inclusion of fourth and fifth sequences in UPGMA tree.

4. The final calculation is to take the average distance between the two composite sets of sequences CDE and AB. The average of n_{AC}, n_{AD}, n_{AE}, n_{BC}, n_{BD}, and n_{BE} = 39 + 39 + 41 + 41 + 41 + 43 = 40.7. One half of this distance (40.7/2 = 20.35) is included in the part of the tree that goes from the root to CDE, and the other half goes from the root to AB. Note also that the presence of the root breaks the branch between AB and CDE, previously denoted f in this example, into two components $f1$ and $f2$. Hence, $f2 + g + d = 20.35$, $f2 = 20.35 - 4.5 - 5 = 10.85$, and $f1 + a = 20.35$, $f1 = 20.35 - 11 = 9.35$ (Fig. 7.19).

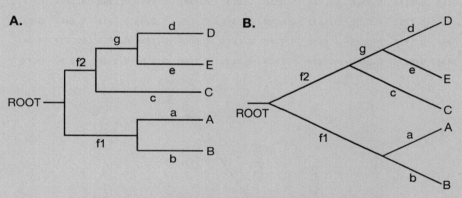

FIGURE 7.19. Final UPGMA rooted tree for five sequences.

EXAMPLE OF DISTANCE ANALYSIS: USING THE PHYLIP PROGRAMS DNADIST AND FITCH (FITCH–MARGOLIASH DISTANCE METHOD)

A set of aligned DNA sequences was converted to the PHYLIP format and placed in a text file called infile in the same folder/directory as the programs (Fig. 7.20A). READSEQ may be used to produce a file with this format from an msa. Note the required spacing of the sequences including spaces for a sequence name at the start of each row of sequence, and note that line 1 includes two numbers giving the number of sequences and the length of the alignment. Note also the presence of ambiguous sequence characters that are recognized appropriately by the program. Longer sequence alignments may be continued in additional blocks without the identifying names.

DNADIST was invoked, the program automatically read the infile, and after setting various options on a menu, an outfile was produced (Fig. 7.20B). This file was edited to remove all but the distance matrix shown. Note the required number on line 1 giving the number of taxa or sequences. Each distance is given twice as a mirror image about the upper-right to lower-left diagonal.

The predicted unrooted tree is given in the outfile and the treefile by the Fitch program. The average percent standard deviation of the predicted intersequence distance was 14, and 990 trees were analyzed to produce this result. The treefile was used as input to the program DRAWTREE, and shown in Figure 7.20C.

A. Sequences in Phylip format (partial list)

```
 20      60
MACHIERH    AACNGGCCTT CTACTAGCCA TACACTACAC CGCAGACACC ACCCTAGCCT TTTCATCTGT
CIRCUS      AACTGGCCTN CTACTAGCAA CACACTATTC CGCAGACACT ACCCTGGCTT TCTCATCCGT
LOPHICTI    AACTGGCCTC CTACTGGCCA TGCACTACAC CGCAGACACA TCACTAGCCT TCTCGTCCGT
AQUILA      AACCGGCCTC CTATTAGCCA TACACTACAC GGCAGACACC ACCCTAGCCT TCTCATCCGT
ACCIPITE    AACCGGCCTC CTCCTAGCAA TACACTACAC CGAAGACACC ACCCTAGCCT TTTCATCAGT
BUTASTUS    AACCGGCCTC CTCCTAGCAA TACACTACAC CGCAGACACC ACCCTAGCCT TTTCATCAGT
HAERAETU    AACCGGCCTC CTACTAGCCA TGCACTACAC CGCAGACACC ACCCTAGCCT TCTCGTCCGT
```

B. DNA distances (partial list)

```
   20
MACHIERH    0.0000  0.1739  0.1705  0.0899  0.0899  0.0711  0.0899  0.1496 0.1292  0.1705  0.10
0.1292  0.1496
CIRCUS      0.1739  0.0000  0.2373  0.1921  0.2144  0.1921  0.1921  0.1292 0.1496  0.1496  0.21
0.2144  0.2853
LOPHICTI    0.1705  0.2373  0.0000  0.1674  0.2326  0.2102  0.0883  0.1885 0.1674  0.2557  0.18
0.1674  0.1468
AQUILA      0.0899  0.1921  0.1674  0.0000  0.1268  0.1073  0.0698  0.1268 0.1468  0.1885  0.08
0.0698  0.1674
ACCIPITE    0.0899  0.2144  0.2326  0.1268  0.0000  0.0169  0.1268  0.1468 0.1268  0.1674  0.14
0.1885  0.2326
BUTASTUS    0.0711  0.1921  0.2102  0.1073  0.0169  0.0000  0.1073  0.1268 0.1073  0.1468  0.12
0.1674  0.2102
HAERAETU    0.0899  0.1921  0.0883  0.0698  0.1268  0.1073  0.0000  0.1268 0.1073  0.1674  0.08
0.1268  0.1468
ELANUS      0.1496  0.2853  0.1468  0.1674  0.2326  0.2102  0.1468  0.2102 0.2326  0.2795  0.21
0.1268  0.0000
```

C. Fitch tree

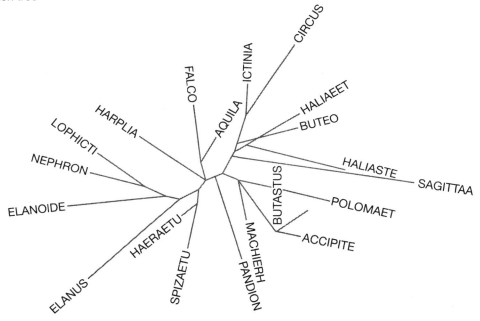

FIGURE 7.20. Tree predicted by Fitch (Fitch–Margoliash distance method) for the DNA sequences given in the example above.

Choosing an Outgroup

If there is independent information that a certain sequence or group of sequences is more distantly related to the rest of the sequences, the sequence should be separated from the others on the phylogenetic tree and, in a rooted tree, should be located at or near the root. Sequences may be designated as outgroups in phylogenetic analysis programs, and using this feature improves the prediction of trees produced by distance methods. Outgroup designations are also used by other phylogenetic analysis methods. This designation ensures that the analysis program will keep those sequences separated from the others in the tree. For example, suppose that sequences A and B are from species that are known to have separated from the others at an early evolutionary time based on the fossil record. A and B may then be treated as an outgroup. Choosing one or more outgroups with the distance method can also assist with localization of the tree root (Swofford et al. 1996). In this case, the root will be placed between the outgroup and the node that connects to the rest of the sequences.

However, there are important considerations in the use of outgroups:

1. The outgroup sequence must be closely related to the rest of the sequences, but there should also be significantly more differences between the outgroup and the other sequences than there are among the remainder of the sequences to each other.

2. Choosing too distant a sequence as the outgroup may lead to incorrect tree predictions because of the more random and complex nature of the differences between the distant outgroup and the other sequences (Li and Graur 1991; Li 1997).

3. For the same reason, using sequences that are too different in the distance method of phylogenetic prediction can lead to errors (Swofford et al. 1996). As the number of differences increases, the history of sequence changes at each site becomes more and more complex, and therefore much more difficult to predict.

4. In choosing an outgroup, one is assuming that the evolutionary history of the gene under study is the same as that provided by the external information. If this assumption is incorrect, such as if horizontal gene transfer has occurred, an incorrect analysis could result.

Including an outgroup can thus improve the predictions made in a phylogenetic analysis, but the choice of an appropriate outgroup requires careful consideration of the relationship of the outgroup sequence to the rest of the sequences.

Converting Sequence Similarity to Distance Scores

To determine phylogenetic relationships among a group of sequences, it is necessary to know the distances between the sequences. The majority of the available sequence alignment programs determine the degree of similarity between sequences rather than distances. For simple scoring systems, similarity is a measure of the number of sequence positions that match in an alignment, whereas distance is the fraction of positions that are different and must be changed to convert one sequence into the other. The difference reflects the number of changes that occurred since the sequences diverged from a common ancestor.

As outlined in Chapter 3, similarity methods provide an alignment score, and the significance of this score can be quite reliably calculated based on the probability that a score between unrelated sequences could achieve that score. What is needed is a way to convert such a score to a distance equivalent so that the appropriate phylogenetic analysis can be performed. A simple method, described and used above, is to count the number of different sequence pairs in an alignment. Another method is to convert the similarity score between two sequences to a normalized measure of similarity that varies from 0 for no similarity to 1 for full similarity. The distance can then be readily calculated.

Feng and Doolittle (1996) describe a method for calculating such a normalized score between a pair of aligned protein sequences. They calculate the similarity score between two sequences S_{real} for a given scoring matrix and gap penalty using a Needleman–Wunsch alignment algorithm (see Chapter 3). They then shuffle both sequences many times, align pairs of shuffled sequences using the same scoring system, and obtain a background average score, S_{rand}, for unrelated sequences. Finally, each protein sequence is aligned with itself to give a maximum score that could be obtained in an alignment of two identical sequences with the scoring system used, and the average of these two scores, S_{ident}, is calculated. This normalized similarity score S between the proteins is then given by

$$S = (S_{real} - S_{rand})/(S_{ident} - S_{rand}) \qquad (1)$$

A different method for calculating S_{rand} from the scoring matrix, amino acid composition, and number of gaps in an msa is also given (Feng and Doolittle 1996).

If, instead, a local alignment based on the Smith–Waterman algorithm is obtained (see Chapter 3), the statistics of local similarity scores can be used to determine distance. If λ and K have been calculated for a given scoring matrix and gap penalty combination, the standardized score of an alignment of score S_{rand} is given by

$$S' = \lambda \, S_{rand} - \log Kmn \qquad (2)$$

where m and n are the sequence lengths. Recall that S' gives approximate probability of a higher score by $e^{-S'}$ (see Chapter 4, p. 138). A conservative value of 5 for S' corresponds to a probability of 7×10^{-3}. A value of S_{rand} is then given by

$$S_{rand}(p = 0.007) = 1 / \lambda \, (5 + \log Kmn) \qquad (3)$$

An expected value for S_{ident}, $S_{ident(calc)}$ is provided by the scoring matrix as the score for a match of identical amino acids (the scores along the diagonal of the log odds form of the amino acid substitution matrix) averaged over the amino acid composition for the matrix. If s_{ii} is the score for a match and p_i is the proportion of each amino acid, the predicted score for an alignment of sequences of length m and n, $S_{ident(calc)}$, where n is the length of the shorter sequence, is given by

$$S_{ident(calc)} = n \sum_{i=1}^{20} p_i s_{ii} \qquad (4)$$

where $\Sigma \, p_i = 1$. For the PAM250 matrix, the average expected score for a matched pair of identical amino acids is 4.95. Subtracting S_{rand} from this value is not appropriate because the score is not a local alignment score but a global one that grows proportional to sequence length. With the above changes, Equation 1 becomes

$$S = (S_{real} - S_{rand(p = 0.007)}) / S_{ident(calc)} \qquad (5)$$

Once the similarity score S has been obtained, it is tempting to calculate the distance between the sequences as $1 - S$. Recall that a simple model of amino acid substitutions is a constant probability of change per site per unit of evolutionary time. Accordingly, some of the observed substitutions in a sequence alignment represent a single amino acid change between the two sequences, but others represent two or more sequential changes. The model predicts that the expected number of 0, 1, 2, . . . substitutions will follow the Poisson distribution, where D is the average number of substitutions. The calculated probability of zero changes is e^{-D}. The probability of one or more changes, which corresponds to $1 - S$, is then given by $1 - e^{-D}$ so that

$$S = e^{-D} \qquad (6)$$

Taking logarithms of both sides and rearranging then gives

$$D = - \log (S) \qquad (7)$$

which is used to calculate D.

EXAMPLE: CALCULATING GENETIC DISTANCE FROM SIMILARITY SCORES

Two sequences of length 250 have an alignment score of 700, using the PAM250 scoring matrix and gap penalties of −12, −2, which are small enough to give a long but local alignment score, then $\lambda = 0.145$ and $K = 0.012$ (Altschul and Gish 1996). Then $S_{rand(p = 0.007)} = 1 / 0.145 \, (5 + \log 0.012 \times 250 \times 250) = 80$ and $S_{ident(calc)} = 4.95 \times 250 = 1238$. Then, $S = (700 - 80) / 1238 = 0.50$, and $D = -\log 0.50 = \log 2 = 0.69$.

There are some additional points to make about the above procedure for calculating genetic distance from similarity scores:

1. The use of scoring matrices based on an evolutionary model are much preferred to matrices that are based on some other criterion. The Dayhoff PAM matrices meet this criterion, but they are based on a small data set. Two more recent sets of PAM matrices (Jones et al. 1992; Muller and Vingron 2000) are based on larger data sets and the same evolutionary model as the Dayhoff matrices. Also see later matrices on page 238.

2. A scoring matrix that models the amino acid substitutions expected for a particular distance should be used. This condition may be met by using a PAM-type matrix at the corresponding evolutionary distance between sequences. The PAM250 matrix models a separation giving only a remaining level 20% similarity. In the above example, the alignment should be rescored using the log odds PAM80 matrices, which model the expected substitution proteins that are 50% similar, and a better alignment score may be obtained. Suitable gap penalties will have to be found by trial and error, and statistical parameters will be calculated as described above. One must also be sure that the scoring system chosen provides a local alignment by demonstrating a logarithmic dependence of the growth of the alignment score on sequence length.

3. Note that Equation 7 provides an estimate of distance based on the observed similarity. The relationship only holds for sequences that are 50% or more similar. Beyond that point, so many multiple substitutions are possible that the distance essentially becomes 1.

4. When Feng and Doolittle perform distance calculations, they use msa's to assess the changes that occur in a family of related proteins. This method is a large improvement over aligning sequence pairs because the presumed evolutionary changes can be seen from the perspective of a whole related family of proteins.

Correction of Distances between Nucleic Acid Sequences for Multiple Changes and Reversions

In the above examples for genetic distance calculation, the assumption is made that each observed sequence change represents a single mutational event. This assumption may be reasonable for sequences that are very much alike, but as the number of observed changes increases, the chance increases that two or more changes actually occurred at the same site and that the same site changed in both sequences. Some of the types of changes that may have occurred are illustrated in Figure 7.21. Note that of all the possible changes, only certain classes shown cause sequence variations.

In the PAM model of evolutionary change described in Chapter 3 (p. 95), such multiple evolutionary changes and reversions are taken into account for a fixed period of evolutionary time called 1 PAM, where 1 PAM roughly equals 10 million years (my). Such tables provide a way to score a sequence alignment by taking into account all possible changes that may have occurred. If sequences are aligned by a series of PAM matrices, the highest-scoring alignments are found when the matrix that corresponds to the evolutionary distance between the sequences is used and the PAM value of this table then provides a measure of the evolutionary distance between the sequences.

There are a large number of models of evolution-

ary change of increasing complexity for correcting for the likelihood of multiple mutations and reversions in nucleic acid sequences, and only a few representative ones are listed here. A more complete list is provided in Swofford et al. (1996). These models use a normalized distance measurement that is the average degree of change per length of aligned sequence. For example, in a 20-nucleotide-long nucleic acid sequence alignment where there are three changes between two sequences A and B, $d_{AB} = n_{AB}/N = 3/20 = 0.15$.

The simplest model, called the Jukes–Cantor model, is that there is the same probability of change at each sequence position, and that once a mutation has occurred, that position is also just as likely to change again. The model also assumes that each base will eventually have the same frequency in DNA sequences (0.25) once equilibrium has been reached. It may be shown (Li and Graur 1991; Li 1997) that the average number of substitutions per site K_{AB} between two sequences A and B by this model is given by

$$K_{AB} = -3/4 \log_e [1 - 4/3 \, d_{AB}] \tag{8}$$

Thus, K_{AB} in the above example is $K_{AB} = -3/4 \log_e [1 - (4/3 \times 0.15)] = 0.17$, which is slightly greater than the observed number of changes (0.15) to compensate for some mutations that may have reverted or changed multiple times. For more different

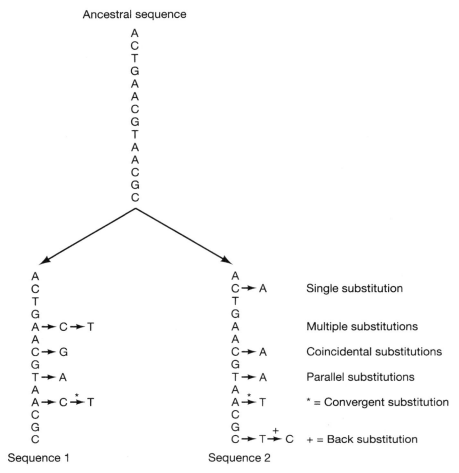

FIGURE 7.21. Types of mutational changes in nucleic acid sequences that have diverged during evolution. Note that the observed sequence changes between these homologous sequences represent only a fraction of the actual number of sequence variations that may have occurred during evolution and that multiple changes may have occurred at many sites. (Redrawn, with permission, from Li and Graur 1991 [©Sinauer Associates].)

sequences, e.g., $d_{AB} = 8/20 = 0.4$, the number of substitutions will be relatively higher than the observed number of changes. $K_{AB} = -3/4 \log_e [1 - (4/3 \times 0.4) = 0.57]$. Hence, the difference between the estimated and observed substitution rates will increase as the number of observed substitutions increases.

The Jukes–Cantor model has been modified to take into account unequal base frequencies (Swofford et al. 1996), which may be calculated from the msa of the sequences.

$$K_{AB} = - B \log_e [1 - d_{AB}/B] \qquad (9)$$

where B is given by $B = 1 - (f_A^2 + f_G^2 + f_C^2 + f_T^2)$ and f_A is the frequency of A in the set of sequences, and so forth.

A slightly more complex model of change, the so-called Kimura two-parameter model (Kimura 1980),

assumes that transition and transversion mutations occur at different frequencies. This model also assumes that the eventual frequency of each base in the two sequences will be 1/4. In this case, it is necessary to calculate the proportion of transition and transversion mutations between two sequences. If the frequencies of transitions and transversions between two sequences A and B are $d_{ABtransition}$ and $d_{ABtransversion}$, respectively, and if

$$a = 1/(1 - 2d_{ABtransition} - d_{ABtransversion})$$

and

$$b = 1/(1 - 2d_{ABtransversion})$$

the number of substitutions per site K_{AB} (Li and Graur 1991) is given by

$$K_{AB} = 1/2 \log_e (a) + 1/4 \log_e (b) \qquad (10)$$

For example, suppose that between two 20-nucleotide-long aligned sequences there are six transitions and two transversions, then $a = 1/(1 - 2 \times 0.3 - 0.1) = 3.33$, $b = 1/(1 - 2 \times 0.1) = 1.25$, and $K_{AB} = 1/2 \log_e (3.33) + 1/4 \log_e (1.25) = 0.66$. For comparison, by the Jukes–Cantor model, $K_{AB} = -3/4 \log_e [1 - 4/3 \times 8/20] = 0.57$. The larger predicted distance between A and B in the Kimura two-parameter model is due to the greater number of sequence changes in this model that could have given rise to the two observed transversion mutations.

The Jukes–Cantor and Kimura two-parameter models can be modified to take into account variations in the rates of mutation at different sites in the sequence alignment (see Swofford et al. 1996, p. 436), and there is also a Kimura three-parameter model that distinguishes between A↔T/G↔C transversions with A↔C/G↔T transversions. These various models are used in the distance methods for phylogenetic construction described above.

For distance calculations between sequences, these base-change models provide ways to improve estimates of the average mutation rate between sequences. They have less effect on phylogenetic predictions of closely related sequences and on the tree branch lengths, but a stronger effect on the more distantly related sequences.

Comparison of Protein Sequences and Protein-encoding Genes

One of the commonest types of phylogenetic comparisons made by biologists is to perform an msa of a set of proteins using the BLOSUM50 or BLOSUM62 scoring matrix and then to design a phylogenetic tree using the neighbor-joining method. The fraction of sequence positions in an alignment that match provides a similarity score. Ambiguous matches and gaps may also be included in the scoring system for similarity. The distance, 1 minus the similarity score (expressed as a fraction of the aligned positions that are identities or conservative substitutions), is calculated and used to produce a tree. CLUSTALW and other programs described in Chapter 5 provide both an alignment and a tree.

Using amino acid variations for phylogenetic predictions offers several advantages over DNA sequence variations. Amino acids confer structure and function to proteins. Therefore, amino acid substitutions in the tree may provide information concerning the influence of those amino acids on function. A limitation in performing phylogenetic analysis of proteins is that models of protein sequence variation are not commonly used. Some amino acid substitutions are much more rare than others and should therefore reflect a longer evolutionary interval. Consequently, treating amino acid substitutions equally may not provide the best phylogenetic prediction.

One method for scoring protein sequence variation is to use PAM scoring tables. Recall that as evolutionary distance between proteins increases, the expected pattern of amino acid changes varies. Rarer substitutions come into play, and the rate of increase of other changes slows down with increasing time. The Dayhoff PAM amino acid scoring matrices were designed to predict the expected substitutions for proteins separated by different evolutionary distances. Thus, the PAM score of the matrix that provides the best alignment score between two sequences reflects the evolutionary separation of the proteins, a distance of 1 PAM being approximately 10 my. Some phylogenetic programs use these original Dayhoff PAM tables, but they are based on a limited number of sequences. Updated sets of protein PAM tables are available (Jones et al. 1992; Muller and Vingron 2001).

The PAM tables have been criticized for failure to take the mutational origin of amino acid changes into account. Although useful for analyzing amino acid variation, they do not allow for the multiple mutations required for some amino acid changes (see Chapter 3, p. 100). Amino acid variation arises through mutation and natural selection acting on DNA sequences. Some amino acid changes require several mutations in codons and should therefore be more rare than amino acid mutations, which require only one mutation in a codon.

Another method for comparing protein sequences is to assess the number of nucleic acid changes that are likely to generate the amino acid differences. In the original Fitch–Margoliash method, when only amino acid sequences were available, the distance between an amino acid pair was chosen to be the minimum number of base changes that would be required to change from a codon for the first amino acid into a codon for the second.

In another approach, with the availability of the cDNA sequences that encode proteins, cDNA sequences may be compared instead of the amino acid sequences of the encoded proteins. Distance methods may be applied directly to the DNA sequence after the number of different positions in the sequences has been determined. If the protein sequences are very similar, most of the DNA changes that will be observed are silent changes that do not change the amino acid and should provide an accurate representation of the phylogenetic history without the complications of evolutionary selection. However, as the amount of variation increases, the number of silent changes will increase and multiple mutations at some of these sites will occur, whereas at other sites, other more rare types of changes will appear. It is very difficult to make accurate predictions when faced with such variation in the rate of change at different sites. One method around this difficulty is to analyze changes in only the first and second base positions in each codon, ignoring the third position, which is the source of most silent mutations (Swofford et al. 1996). A comparison of nucleic acid sequences that encode proteins for mutations that either (1) change the amino acid (synonymous) or (2) do not change the amino acid (nonsynonymous) may be made. Once these types of changes have been distinguished, phylogenetic predictions based on only one of them may be made.

A final type of correction that may be made to phylogenetic predictions is for the increase in multiple substitutions as the evolutionary distance between protein sequences increases. Although use of the PAM matrices provides this type of correction, another way is to adapt the Jukes–Cantor model for nucleic acid sequences to protein sequences. The correction to the distance is given by Equation 9, where $B = 19/20$ for the assumption of equal amino acid representation and $B = 1 - \Sigma f_{aai}$ for unequal representation of the amino acids, where f_{aai} is the frequency of amino acid i, and the sum is taken over all 20 amino acids. The second representation is, of course, much preferred, since amino acid frequencies in proteins vary.

One correction that may be applied to protein distances has been suggested by Kimura (1983). This correction is based on the Dayhoff PAM model of amino acid substitution. If K is the corrected distance and D the observed distance (number of exact matches between two sequences divided by total number of matched residues in the alignment), then

$$K = - \ln(1 - D - 0.2 D^2) \qquad (11)$$

This formula may be used up to values of $D =$

0.75. Above this value, tables based on the Dayhoff PAM model at these distances are used instead because the equation gives an inaccurate value or no value at all. This correction is applied by CLUSTALW, a commonly used program for msa and phylogenetic analysis (Higgins et al. 1996).

Comparison of Open Reading Frames by Distance Methods

When nucleic acid sequences that encode proteins first became available, the existence of synonymous substitutions that do not change the amino acid (silent changes) and nonsynonymous substitutions (replacement changes) that do change the amino acid was analyzed. Separate analyses of these two kinds of substitutions can help remove site-to-site variation in more closely related sequences and background noise of silent mutations in more distantly related sequences (Swofford et al. 1996).

One method of estimating the rates of synonymous and nonsynonymous mutations (Li et al. 1985; Li and Graur 1991; Li 1997) employs the following steps:

1. The fraction of substitutions at each codon position that can give rise to synonymous substitutions and the fraction that can give rise to nonsynonymous substitutions are counted. The first two positions of most codons count as two nonsynonymous sites because the amino acid will change regardless of the substitution. Similarly, many third-codon substitutions are synonymous. Other sites contribute synonymous and nonsynonymous substitutions. The total number of each of these two possible substitutions is determined for each sequence, and the average of these two values for the two sequences is then calculated. N_{syn} is the average number of synonymous sites and N_{nonsyn} is the average number of nonsynonymous sites in the two sequences.

2. Each pair of codons in the alignment is then compared to classify nucleotide differences into synonymous and nonsynonymous types. A single base difference can readily be designated as synonymous or nonsynonymous. When the codons differ by more than one substitution, all of the possible pathways of sequence change must be considered, and the number of synonymous and nonsynonymous changes in each pathway is identified. The average of each type of change in

the two pathways is then calculated. Weights derived from the frequency of these pathways for known codon pairs may be used to derive this average, or else the pathways may be weighted equally. These calculations give the number of synonymous differences M_{syn} and the number of nonsynonymous differences M_{nonsyn} between the sequences.

3. The fraction of synonymous differences per synonymous site ($f_{syn} = N_{syn}/M_{syn}$) and the fraction of nonsynonymous differences per nonsynonymous site ($f_{nonsyn} = N_{nonsyn}/N_{nonsyn}$) are calculated. These fractions may then be corrected for the effect of multiple changes at the same site by the Jukes–Cantor formula (Eq. 8) or by some alternative method.

Another method for estimating synonymous and nonsynonymous substitutions (Li et al. 1985; Li and Graur 1991; Li 1993, 1997) is to classify each nucleotide position in the coding sequences as nondegenerate, twofold degenerate, or fourfold degenerate. The Genetics Computer Group program DIVERGE uses this method. A site is nondegenerate if all possible changes at this site are nonsynonymous, twofold degenerate if two of the three possible changes are nonsynonymous, and fourfold degenerate if none of the possible changes are nonsynonymous. For simplification, the third position of isoleucine codons (ATA, ATC, and ATT in the universal code) is treated as a twofold degenerate site, even though in reality it is threefold degenerate. The number of each type of site in each of the two sequences is calculated and the average values for the two sequences are calculated.

Each pair of codons in the sequence alignment is then compared to classify nucleotide differences as to type of site (nondegenerate, twofold degenerate, or fourfold degenerate) and as to whether the change is a transition or a transversion.

The scored codon differences are then used to calculate the proportions of each type of site that are transitions or transversions. The proportion of synonymous substitutions per synonymous site and the corresponding proportion for transversions may then be calculated and used to determine distances between the sequences. The Kimura two-parameter model may be used to correct for multiple mutations and for differences between rates of transitions and transversions before these calculations are performed.

The Maximum Likelihood Approach

The third method for phylogenetic prediction is the maximum likelihood (ML) method. ML methods start with a simple model, in this case a model of rates of evolutionary change in nucleic acid or protein sequences and tree models that represent a pattern of evolutionary change, and then adjust the model until there is a best fit to the observed data. For phylogenetic analysis, the observed data are the observed sequence variations found within the columns of an msa. The ML method is similar to the maximum parsimony method in that the analysis is performed on each column of an msa.

The following steps are taken by ML programs:

1. A list of all possible trees for the given number of sequences is generated.

CALCULATION OF NONSYNONYMOUS AND SYNONYMOUS CHANGES

To calculate these values, note that by definition all substitutions at nondegenerate sites are nonsynonymous, and all substitutions at fourfold degenerate sites are synonymous. At twofold degenerate sites, transitions nearly always produce synonymous changes, and transversions nearly always produce nonsynonymous changes. Hence, counting transitions and transversions at these sites provides a nearly exact count of the number of synonymous and nonsynonymous substitutions, respectively. One exception to this scoring scheme in the universal genetic code is that one type of transversion in the first position of the arginine codons produces a synonymous change, whereas the other transversion and the transition produce a synonymous substitution. Another exception is in the last position of the three isoleucine codons. When the codons differ by more than one substitution, a method similar to that described above is used to evaluate each possible pathway for changing one codon into the other, and the average of each type of change in the pathways is then calculated.

2. The sequence data at the first site (the first column in the msa) are placed on the outer leaves of these trees.

3. The first tree so labeled is chosen.

4. All patterns of changes at the inner nodes (vertices) of this tree that will give rise to the observed sequence variation on the outer branches are found. Each unique pattern represents a unique path along the branches starting from the lowest branches and ending at the uppermost branches of the tree at the sequence letters.

5. The probability of change at each of the branches in the tree is looked up in a table of probabilities for the chosen evolutionary model for mutation.

6. The probability of each pattern of change is calculated for the first tree using the combinations of branch probabilities.

7. These probabilities are summed for the first tree at the first site.

8. This procedure is repeated for all other possible trees labeled with the sequences from the first site.

9. The most probable tree is found for the first site.

10. The second site in the msa is analyzed using the same list of trees, but this time sequences at the second site on the msa are placed on the outer branches of the trees.

11. This procedure is repeated for all of the sites in the msa.

12. The most likely tree will be the one that gives the highest overall probability at all of the sites found by summing the site probabilities for each tree.

13. These steps are illustrated in Figure 7.22 and described in detail in the figure legend.

Because all possible trees are considered by the ML method, the method is only feasible for a small number of sequences. When a larger number of sequences must be used, supercomputers are needed for ML programs. For each tree, the number of sequence changes or mutations that may have occurred to give the sequence variation is considered. Because the rate of appearance of new mutations is very small, the more mutations needed to fit a tree to the data, the less likely that tree is (Felsenstein 1981). The ML method resembles the maximum parsimony method in that trees with the least number of changes will be the most

likely. However, the ML method presents an additional opportunity to evaluate trees with variations in mutation rates in different lineages, or site-to-site variations within the msa, as well as to use explicit evolutionary models such as the Jukes–Cantor and Kimura models described in the above section with allowances for variations in base composition. Thus, the ML method can be used to explore relationships among more diverse sequences, conditions that are not well handled by maximum parsimony methods. The main disadvantage of ML methods is that they are computationally intense. However, with faster computers, the ML method is seeing wider use and is being used for more complex models of evolution (Schadt et al. 1998). ML has also been used for an analysis of mutations in overlapping reading frames in viruses (Hein and Støvlbæk 1996). PAUP version 4 can be used to perform an ML analysis on DNA sequences. The ML method is also used for phylogenetic analysis of amino acid changes in protein sequences.

PHYLIP at http://evolution.genetics.washington.edu/phylip.html includes two programs for this ML analysis:

1. DNAML estimates phylogenies from nucleotide sequences by the ML method, allowing for variable frequencies of the four nucleotides, for unequal rates of transitions and transversions, and for different rates of change in different categories of sites, as specified by the program.

2. DNAMLK estimates phylogenies from nucleotide sequences by the ML method in the same manner as DNAML, but assumes a molecular clock.

In addition to the PHYLIP programs, several additional programs and program suites are available for phylogenetic analysis using the ML method and include a variety of evolutionary models of sequence change. These include the PAML suite at http://abacus.gene.ucl.ac.uk/software/paml.html with excellent documentation (Yang 1997) and TREE-PUZZLE at http://www.tree-puzzle.de/ (Schmidt et al. 2002). The program MrBayes uses a Bayesian statistical approach for sampling the probability distribution of trees to identify the most likely trees and runs on most computer platforms (see http://morphbank.ebc.uu.se/mrbayes; Altekar et al. 2004).

The starting point for an ML phylogenetic analysis is an evolutionary model of sequence change that provides estimates of rates of substitution of one base for another (transitions and transversions) in a set of nucleic acid sequences, as illustrated in Table 7.5. The

FIGURE 7.22. Maximum likelihood estimation of phylogenetic tree. For the hypothetical sequences shown in *A*, one of three possible unrooted trees is shown in *B*. One column has been set aside for analysis. (*C*) One of five possible rooted derivatives of the unrooted tree is shown. The position of the root is not important since the likelihood of the tree is the same regardless of the root location. This property follows the assumption that the substitutions along each branch are considered to be a Markov chain with reversible steps (Felsenstein 1981). The bases from the marked alignment column are shown on the outer branches of this tree. Also shown are three interior nodes of the tree labeled 0, 1, and 2. The object is to consider every possible base assignment to these three nodes and then to calculate the likelihood of each choice. Since there are four possible bases for each of the three node positions of the tree, there are $4 \times 4 \times 4 = 64$ possible combinations. Also shown on the tree are six likelihood values L1–L5 for the probability of a base change per site along the respective branches of the tree, and a probability L0 for the base at node 0. These probabilities depend on the bases assigned to nodes 0, 1, and 2 and on the resulting types of base substitutions in the particular tree under consideration. The likelihood of a tree with a particular choice of bases at nodes 0, 1, and 2 is given by the product of the probability of the base at node 0 times the product of each of the substitution probabilities, or L(tree) = L0 × L1 × L2 × L3 × L4 × L5 × L6 (Felsenstein 1981). (*D*) A possible tree (tree1) with T assigned to nodes 0 and 1, and G assigned to node 2. L0 will be given by the frequency of T and will have an approximate value of 0.25. L2 will be the probability of a transversion of T to G, and L5 the probability of a transition of G to A in this tree. The remaining likelihoods will have an approximate value of unity with a small adjustment for the possibility that a mutation has occurred and then reverted to the original base so that no substitutions are observed. Assuming that the probabilities of the transition and transversion are 2×10^{-6} and 10^{-6}, respectively, the likelihood of tree1 is approximately $0.25 \times 2 \times 10^{-6} \times 10^{-6} = 5 \times 10^{-13}$. These numbers are usually very small and are therefore handled as logarithms in the computer. (*E*) Another possible arrangement

A. Sequences   ★

sequence a	A C G C G T T G G G
sequence b	A C G C G T T G G G
sequence c	A C G C A A T G A A
sequence d	A C A C A G G G A A

B. An unrooted phylogenetic tree for the sequences a–d.

C. A rooted phylogenetic tree for the sequences a–d showing the bases for one set of aligned sequence positions in A.

D. A rooted phylogenetic tree showing one set of base assignments to nodes 0, 1 and 2.

E. A rooted phylogenetic tree showing one set of base assignments to nodes 0, 1 and 2.

F. L(Tree) = L(Tree1) + L(Tree2) + ... + L(Tree64)

of base assignments in tree2. The likelihood of this tree will take into account the probability of a G to T transversion (L1) and that of a G to A transition (L5). (*F*) The likelihood of the tree in *B* or the tree in *C* is given by the sum of the likelihoods of these two trees. To this sum is added the probability of the other 62 possible arrangements of bases. This calculation is repeated for all other columns in the msa. The likelihood of the tree given the data in all of the aligned columns, that in the first column, or that in the second, etc., will be the sum of the likelihoods so calculated for each column. Each of the three possible trees for four sequences is then evaluated in this same manner, and the one with the highest likelihood score is identified. These calculations can be computationally so intense for a large number of sequences that trees for a fraction of the sequences may first be found. The data for additional sequences will then be sequentially added to refine this initial tree. The procedure may then be repeated with a different starting group of sequences with the hope that the range of trees found will give an indication of the most likely tree (Felsenstein 1981). However, this procedure is not guaranteed to find the optimal tree. Additional calculations are made in the ML method. The probability of each branch in the tree is individually adjusted by a method similar to expectation maximization (see Chapter 3) to maximize the likelihood of the tree while holding the probability of the other branches at a constant value. The rate of evolution of each site or each column in the msa is also allowed to vary. Otherwise, the method will be biased by sites that do not vary much and the information in variable sites may become lost, a problem shared with the maximum parsimony method. For an average number of mutations x over all branches, the number along an individual branch is assumed to vary according to the Poisson distribution $P(n) = e^{-x} x^n / n!$. A variable giving the equivalent probability of observing a given number of changes along a particular branch for various average values of x (or a particular mutation rate along that branch) is given by the Γ distribution. Different Γ distributions give different patterns of variation that may be user-defined by varying the α variable. A small value of α (e.g., 0.05) gives a large degree of variation whereas a large value of α (e.g., 200) gives a very narrow range of variation. Alternatively, a hidden Markov model of the variation in sites or blocks of sites in an msa may be used in the PHYLIP programs (Felsenstein and Churchill 1996). More information on the use of these models may be found in the ML program manuals. Once the range of values is determined, the range of probabilities may be used in calculations of tree likelihoods (Swofford et al. 1996).

TABLE 7.5. *General model of sequence evolution*

Base	A	C	G	T
A	$-u(a\pi_C+b\pi_G+c\pi_T)$	$ua\pi_C$	$ub\pi_G$	$uc\pi_T$
C	$ug\pi_A$	$-u(g\pi_A+d\pi_G+e\pi_T)$	$ud\pi_G$	$ue\pi_T$
G	$uh\pi_A$	$uj\pi_G$	$-u(h\pi_A+j\pi_G+f\pi_T)$	$uf\pi_T$
T	$ui\pi_A$	$uk\pi_G$	$ul\pi_T$	$-u(i\pi_A+k\pi_G+l\pi_T)$

The table gives rates for any substitution in a nucleic acid sequence or for no substitution at all (the diagonal values). Base frequencies are given by π_A, π_C, π_G, and π_T, the mutation rate by u, and the frequency of change of any base to any other by a, b, c..,l. Rates of substitutions in one direction, i.e., A→G, are generally considered to be the same as that in the reverse direction so that $a = g$, $b = h$, etc. In the JC model these frequencies are all equal, and in the Kimura two-parameter there are only two frequencies, one for transitions (α) and the other for transversions (β), and the frequency for transitions is twice that for transversions. PAUP allows these numbers to be varied. This model assumes that changes in a sequence position constitute a Markov process, with each subsequent change depending only on the current base. Furthermore, the model assumes that each base position has the same probability of change in any branch of the tree (Swofford et al. 1996).

rates of all possible substitutions are chosen so that the base composition remains the same. The set of sequences to be analyzed is first aligned, and the substitutions in each column are examined for their fit to a set of trees that describe possible phylogenetic relationships among the sequences. Each tree has a certain likelihood based on the series of mutations that are required to give the sequence data. The probability of each tree is simply the product of the mutation rates in each branch of the tree, which itself is the product of the rate of substitution in each branch times the branch length. There are multiple sets of possible base changes within each tree to consider. For each column in the aligned sequences, the probability of each set of changes is found and the probabilities are then added to produce a combined probability that a given tree will produce that column in the alignment. A simple example of this approach is shown in Figure 7.22. Once all positions in the sequence alignment have been examined, the likelihoods given by each column in the alignment for each tree are multiplied to give the likelihood of the tree. Because these likelihoods are very small numbers, their logarithms are usually added to give the logarithm likelihood of each tree. The most likely tree given the data is then identified.

Sequence Alignment Based on an Evolutionary Model

Thorne et al. (1991, 1992) have introduced a method of sequence alignment based on a model (Bishop and Thompson 1986) that predicts the manner in which DNA sequences change during evolution. Although this method has limitations and is considered by these authors to be only preliminary, it is outlined here because of its relationship to the ML method for phylogenetic analysis. The basis of this method is to devise a scheme for introducing substitutions, insertions, and gaps into sequences and to provide a probability that each of these changes occurs over periods of evolutionary time. Given each of these predicted changes, the method examines all the possible combinations of mutations to change one sequence into another. One of these combinations will be the most likely one over time. Once this combination has been determined, a sequence alignment and the distance between the sequences will be known. This method is different from the Smith–Waterman local alignment algorithm in identifying the most probable ML probability alignment based on an evolutionary model of change in sequences, as opposed to a score based on observed substitutions in related proteins and a gap scoring system. The underlying mutational theory is, however, similar to those theories used to produce the PAM matrices for predicting changes in DNA and protein sequences.

Sequences are predicted to change by a Markov process (see Chapter 3 discussion of PAM matrices, p. 96) such that each mutation in the sequence is independent of previous mutations at that site or at other sites. For example, a given nucleotide at any sequence position can mutate into another at the same rate or may not change at all during a period of evolutionary time. This model is very similar to the PAM model of evolutionary change in proteins introduced by Dayhoff and discussed earlier. In the Thorne et al. (1991) model, single insertion–deletion events between any two nucleotides are modeled by a birth–death process that leaves the sequence length roughly the same. Longer insertion–deletion events are modeled in a similar way by considering the sequence to be composed of a set of fragments, and the rate of substitution of these fragments is allowed to vary (Thorne et al. 1992).

A set of transition probabilities for changing from one nucleotide to another or for introducing an insertion or deletion into a sequence is derived mathematically from the evolutionary model. The substitution probabilities are not unlike the substitution probabilities in the protein and DNA PAM matrices. An important difference between the PAM matrices and the transition probabilities is that the insertion/deletion probabilities have been derived from the evolutionary model rather than from the ad hoc gap penalty scoring system (penalty = gap opening penalty + gap extension penalty × length) that is commonly used to produce sequence alignments by dynamic programming. Two algorithms not unlike dynamic programming are then used, one to obtain a sequence alignment and the other to calculate the likelihood that the sequences are related given the calculated set of parameters. The entries in the scoring matrices are likelihood scores (giving the highest probability of arriving at that position in the scoring matrix by a combination of mutations and gaps) and not a sum of weights for substitutions based on a scoring matrix. To estimate the likelihood of the sequences also requires that the number and types of substitutions, insertions, and deletions be optimized to find the most likely path for changing one sequence into another. This path then provides an indication of the evolutionary distance between the sequences. Methods and software for sampling DNA sequence alignments based on the above evolutionary models of gaps have been described (Metzler 2003).

HOW RELIABLE ARE PHYLOGENETIC PREDICTIONS?

As discussed earlier in this chapter, phylogenetic analysis of a set of sequences that aligns very well is straightforward because the positions that correspond in the sequences can be readily identified in an msa of the sequences. The types of changes in the aligned positions or the numbers of changes in the alignments between pairs of sequences then provide a basis for a determination of phylogenetic relationships among the sequences by the above methods of phylogenetic analysis. For sequences that have diverged considerably, a phylogenetic analysis is more challenging. A determination of the sequence changes that have occurred is more difficult because the msa may not be optimal and because multiple changes may have occurred in the aligned sequence positions. The choice of a suitable msa method depends on the degree of variation among the sequences, as discussed in Chapter 5. Once a suitable alignment has been found, one may also ask how well the predicted phylogenetic relationships are supported by the data in the msa.

In the bootstrap method, the data are resampled by randomly choosing vertical columns from the aligned sequences to produce, in effect, a new sequence alignment of the same length. Each column of data may be used more than once and some columns may not be used at all in the new alignment.

Trees are then predicted from many of these alignments of resampled sequences (Felsenstein 1988). For branches in the predicted tree topology to be significant, the resampled data sets should frequently (for example, >70%) predict the same branches. Bootstrap analysis is supported by most of the commonly used phylogenetic inference software packages and is commonly used to test tree branch reliability.

Another method of testing the reliability of one part of the tree is to collapse two branches into a common node (Maddison and Maddison 1992). The tree length is again evaluated and compared to the original length, and any increase is the decay value. The greater the decay value, the more significant the original branches. In addition to these methods, there are some additional recommendations that increase confidence in a phylogenetic prediction.

One further recommendation in performing phylogenetic predictions is to use at least two of the above methods (maximum parsimony, distance, or maximum likelihood) for the analysis. If two of these methods provide the same prediction, confidence in the prediction is much higher. Another recommendation is to pay careful attention to the evolutionary assumptions and models that are used for both sequence alignment and tree construction (Li and Graur 1991; Swofford et al. 1996; Li 1997).

HOW IS PHYLOGENETIC ANALYSIS USED?

The above methods provide a useful level of sequence analysis by predicting possible evolutionary relationships among a group of related sequences. The methods predict a tree that shows possible ancestral relationships among the sequences. A phylogenetic analysis can be performed on proteins or nucleic acid sequences using any one of the three methods described above, each of which utilizes a different type of algorithm. The reliability of the prediction can also be evaluated using bootstrapping and decay analysis.

The traditional use of phylogenetic analysis is to discover evolutionary relationships among species. In such cases, a suitable gene or DNA sequence that shows just enough, but not too much, variation among a group of organisms is selected for phylogenetic analysis. For example, analysis of mitochondrial sequences is used to discover evolutionary relation-

ships among mammals. Two more recent uses of phylogenetic analysis are to analyze gene families and to trace the evolutionary history of specific genes. For example, database similarity searches discussed in Chapter 6 may identify several proteins in a plant genome that are similar to a yeast query protein. From a phylogenetic analysis of the protein family, the plant gene most closely related to the yeast gene, and therefore most likely to have the same function, can be determined. The prediction can then be evaluated in the laboratory. Tracking the evolutionary history of individual genes in a group of species can reveal which genes have remained in a genome for a long time and which genes have been horizontally transferred between species. Thus, phylogenetic analysis can also contribute to an understanding of genome evolution, as further explored in Chapter 11.

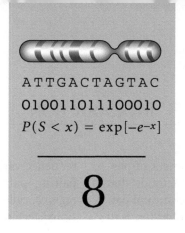

ATTGACTAGTAC
010011011100010
$P(S < x) = \exp[-e^{-x}]$

8

Prediction of RNA Secondary Structure

CONTENTS

INTRODUCTION

Chapters 3 and 5 have discussed the alignment of protein and nucleic acid sequences using methods that either align entire sequences or search for common patterns in the sequences. In either case, the objective is to locate a set of sequence characters in the same order in the sequences. Nucleic acid sequences that specify RNA molecules must be compared differently because sequence variations in RNA sequences maintain base-pairing patterns that give rise to double-stranded regions, called secondary structures, and three-dimensional, tertiary structures in the molecule that are necessary for biological function. Sequences of RNA-specifying genes may also have rows of similar sequence characters that reflect the common ancestry of the genes.

CHAPTER GUIDE FOR BIOLOGISTS

At first glance, computational analysis of RNA molecules to predict the most stable secondary structures might appear to be one of the most complex subjects in sequence analysis, but, fortunately, much of the analysis is quite intuitive and straightforward. It is quite easy to visualize two regions in the sequence of an RNA molecule that can pair to make a long, double-stranded region comprising G/C, A/U, and G/U base pairs, for example. The sequence of one strand in the double-stranded region reads from 5′ to 3′, and the sequence of the other complementary strand appears elsewhere in the sequence in the 3′ to 5′ direction. The intervening sequence is a loop or a part of some other secondary structure.

The dynamic programming algorithm used for finding best-scoring sequence alignments can also be used to find the maximum number of compatible base pairs in the RNA molecule—compatible in the sense that the RNA molecule is topologically simple with no knots and no complex interactions involving more than two regions. Dynamic programming is first used to find those regions that can potentially base-pair as the starting point for secondary structure prediction. Using estimates for stacking energies between adjacent base pairs in a double-stranded region and destabilizing energies for free strands and loops based on experimental estimates, a more detailed dynamic programming analysis is then applied to find combinations of base-paired regions that are predicted to produce the energetically most stable molecule. The result is a two-dimensional representation of the secondary structures in the molecule.

In the real world, RNA molecules have three-dimensional structures also held together by base pairs within the molecule and this aspect must also be considered in the computational analysis of RNA structure. There is a way to find both the secondary and three-dimensional, tertiary interactions between base pairs using phylogenetic comparisons of RNA sequences. Many members of one class of RNA from the same or different species are collected and subjected to multiple sequence alignment (msa). The msa is then examined to find a pair of columns anywhere in the msa in which the bases are complementary and thus capable of forming a base pair in each of the sequences in the msa. For example, the first of a pair of candidate columns might have G or U, and the second column, C or A. If these bases are located so that the same sequences have G in the first column and C in the second, or A in the first column and U in the second, then the covariance analysis predicts that these two sequence positions form a base pair in the structure that has been conserved as the sequences change over evolutionary time. Runs of self-complementary positions going forward at one location and backward in another can form a double-stranded region in the RNA molecule. Other covarying positions can also be predicted to interact, allowing a structural model with secondary and three-dimensional structural features to be produced. These base pairs include standard Watson–Crick canonical base pairs but also include a variety of other non-Watson–Crick base pairs, such as G/U wobble pairs, which play a role in protein–RNA interactions.

Given a structural model for a class of RNA molecules, e.g., transfer RNA (tRNA), one would like to be able to represent this structure with some kind of computational model that can be used to find sequences that represent other RNAs of the same class, as in a genome analysis project. When dealing with proteins, a scoring matrix or hidden Markov model is used. These types of msa models are simply a way of representing the sequence variation in an msa, going from the beginning to the end of the msa. They provide a way to search for more examples of the same protein domain given the variation found in each column of the corresponding msa. For RNA, an msa model that takes into account the covariation found in the msa is needed. When one position in the test sequence being analyzed is examined for a match to a covarying position in the model, the test sequence must also be examined for a match to the other member of the covarying pair. This goal is best accomplished by using a tree-based model of the RNA structure. The model is moved along the test sequence (a sliding tree comparable to the sliding window often used in sequence alignment) to find high-scoring regions. These regions predict a sequence that can form the RNA structure at that position. Such models representing a number of classes of small RNA molecules have been generated and used with considerable success to search newly sequenced genomes for genes that encode these classes. They have therefore become a useful tool for genome analysis projects.

Chapter Guide for Computational Scientists

The RNA molecule comprises linear chains of four molecular building blocks (bases) like those of DNA—A, G, C, and U (uracil in RNA instead of T as in DNA)—and is a special class of molecules that is believed to represent an ancient form of biological activity. RNA molecules play many important roles in modern biological systems. As messenger RNAs (mRNAs) in the cell, they are copied from one strand of DNA and their sequence is read on the ribosome and translated into the amino acid sequence of a protein. mRNAs are often processed by the cell, splicing them to remove introns and leaving behind exons that have the sequence information for proteins. The splicing machinery of the cell also requires small RNAs. The ribosome itself, the machine for making proteins, is also made up of RNA molecules complexed with proteins. Many other types of RNA molecules that are produced by cells can influence the expression of genes in diverse ways, ranging from modifications in the packaging of DNA into chromosomes to mRNA stability and translational efficiency. The function of only some of these molecules is known.

RNA has the important chemical property that it is usually single-stranded but it also contains sequences of self-complementary bases. These self-complementary regions can form double-stranded, hydrogen-bonded regions that are stable because they represent a lower energy state for the molecule. These structures are referred to as secondary structure. Through these bonding interactions, each class of RNA molecules folds into a characteristic three-dimensional (tertiary) structure that is necessary for the specific biological activity of the molecule. An easy way to predict these secondary and tertiary structures for one class of RNA is to obtain a set of known sequences, perform an msa, and then examine the columns of the msa for a pattern of covariation predictive of self-complementarity. Once the three-dimensional structure is formed, the RNA molecule can interact with other molecules, proteins, and RNAs to form more complex structures.

The exposed RNA sequences—that is, those that are not buried in the RNA structure—can interact with another RNA molecule that has a complementary sequence. If the first RNA is an mRNA for a specific gene, then the sequence on the second RNA will be antisense—that is, run in the opposite chemical direction to the protein-encoding, sense message of the mRNA. The interaction can regulate the activity of the mRNA, as for example, how many times the mRNA is translated into protein. Biologists take advantage of this sequence-specific interactive property of RNA by designing molecules that interfere with activity of specific mRNAs. These inhibitors are referred to as RNA interference (RNAi) molecules. With these molecular tools, the effects of interfering with the expression of specific genes in biological systems can be studied in the laboratory.

WHAT SHOULD BE LEARNED IN THIS CHAPTER?

- Appreciation of the great variety of types, sizes, and biological roles of RNA molecules.
- Visualization of RNA secondary structure on a dot matrix plot.
- Dynamic programming approaches for finding the minimum energy or near-minimal energy structure of RNA molecules.
- Recognition of RNA genes and prediction of their structure through covariation analysis.
- Description of a stochastic context-free grammar and how it is used to model secondary structure in RNA.

Glossary Terms

Binary tree In the context of this chapter, a tree whose branches represent the occurrence of two adjacent secondary structural elements in the RNA structure.

Covariation Coincident change at two or more sequence positions in related RNA sequences that may influence the secondary (two-dimensional) and tertiary (three-dimensional) structures of RNA. In a given structural class of RNA, if the sequence changes at one position that forms a base pair, then the sequence position of the other member of the base pair also changes to maintain complementarity. In an msa of the class of RNAs, the columns represented by these two positions will show a pattern of correlated variation.

Messenger RNA (mRNA) is copied from one strand of DNA in a region that encodes the sequence information for the amino acid sequence of a protein. After any necessary processing of the mRNA, e.g., removal of introns by RNA splicing, the sequence of three-letter codons in the processed mRNA is translated into the amino acid sequence of a protein using transfer RNAs, the ribosome, and other cellular machinery.

Pseudoknot is an RNA structure in which one strand from a hairpin structure with a double-stranded region and loop is folded back, producing a second loop and then a second double-stranded contact with the loop of the first hairpin, as illustrated in Figure 8.3A, below.

Ribosomal RNA (rRNA) Two large RNA species that are assembled along with proteins into two large ribosomal subunits and hence into ribosomes, the cellular machinery that is used to translate mRNA into proteins.

RNA secondary structure Structures formed through the formation of base pairs (C/G, A/U, and G/U) between runs of self-complementary sequences in RNA molecules. Structures include double-stranded regions and hairpins.

RNA three-dimensional, tertiary structure Structures formed by base-pairing interactions among secondary structures and unpaired sequence positions in the RNA and that are necessary for function and biological activity.

Stacking energy The energy decrease associated with formation of a second base pair adjacent to an initial one in double-stranded RNA. These pairs interact to make the double-stranded region more stable energetically.

Stochastic context-free grammar A formal linguistic method for representing the interaction of distant sequences in RNA molecules to form secondary structures. A set of these rules may be used to find new occurrences of the RNA in uncharacterized sequences. Context-free refers to the occurrence of sequence information about the grammar at distant (not localized) sites in the sequence, and stochastic refers to the expected occurrence of substitutions at these sites.

Transfer RNA (tRNA) Small RNA molecules specific for a particular amino acid. Each tRNA carries a particular amino acid to the mRNA–ribosome complex for assembly to a protein chain. Each tRNA has a sequence (the anticodon) that is aligned with the codon on the mRNA to assure the correct sequence of amino acids in the protein.

Uracil (U) A base that is found in RNA and corresponds to thymine (T) in DNA. U occurs in linear strings with G, C, and A in RNA.

WHAT ARE THE FEATURES OF RNA SECONDARY AND TERTIARY STRUCTURE?

Like protein secondary structure, RNA secondary structure can be conveniently viewed as an intermediate step in the formation of a three-dimensional structure, in this case the functional three-dimensional RNA structure (Fig. 8.1). RNA secondary structure is primarily the double-stranded regions of the molecule formed by folding the single-stranded molecule back on itself to form loops in the RNA structure. To produce these double-stranded regions, a run of bases downstream in the RNA sequence must be complementary to another upstream run so that Watson–Crick base-pairing between the complementary nucleotides G/C and A/U (analogous to the G/C and A/T base pairs in DNA) can occur. In addition, however, non-Watson–Crick G/U wobble base pairs are frequently found between the DNA strands in these double-stranded regions as they are between the third position of codons in mRNA and the corresponding anticodon in tRNA. Additional non-Watson–Crick base pairs between separate sequence positions fold the RNA into a three-dimensional structure. As in DNA, the G/C base pairs contribute

the greatest energetic stability to the molecule, with A/U base pairs contributing less stability than G/C, and G/U wobble base pairs contributing the least. From the RNA structures that have been solved, these Watson–Crick base pairs and many additional ones (see noncanonical database at http://prion.bchs.uh.edu/bp_type/ and Burkhard et al. 1999a,b) have been identified. RNA structure predictions are composed of base-paired and non-base-paired regions forming various types of loop and junction arrangements, as shown in Figure 8.2.

In addition to secondary (two-dimensional) structural interactions in RNA, there are also tertiary (three-dimensional) interactions formed by base pairs between secondary structural elements and between secondary structural elements and single-stranded regions, illustrated by the examples in Figure 8.3. Tertiary interactions are essential for biological activity of the RNA. Tertiary structures are not predictable by secondary structure prediction programs, but they can be found by careful covariance analysis as described below.

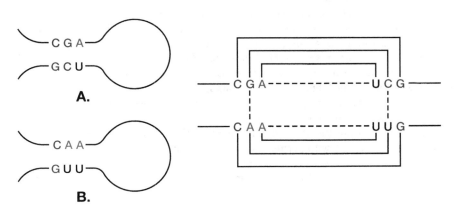

FIGURE 8.1. Complementary sequences in RNA molecules maintain RNA secondary structure. Shown is a simple stem-and-loop structure formed by the RNA strand folding back on itself. Molecule *A* depends on the presence of two complementary sequences CGA and UCG that are base-paired in the structure. In *B*, two sequence changes, G → A and C → U, which maintain the same structure, are present. Aligning RNA sequences requires locating such regions of sequence covariation that are capable of maintaining base-pairing in the corresponding structure.

A. Single-stranded RNA

B. Double-stranded RNA helix of stacked base pairs

C. Stem and loop or hairpin loop

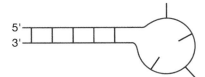

D. Bulge loop

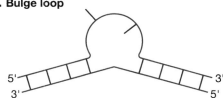

E. Interior loop

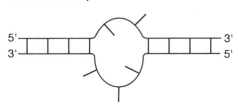

F. Junctions or multi-loops

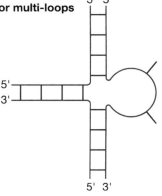

FIGURE 8.2. Types of single- and double-stranded regions in RNA secondary structures. Single-stranded RNA molecules fold back on themselves and produce double-stranded helices where complementary sequences are present. A particular base may either not be paired, as in *A*, or paired with another base, as in *B*. The double-stranded regions will most likely form where a series of bases in the sequence can pair with a complementary set elsewhere in the sequence. The stacking energy of the base pairs provides increased energetic stability. Combinations of double-stranded and single-stranded regions produce the types of structures shown in *C–F*, with the single-stranded regions destabilizing neighboring double-stranded regions. The loop of the stem and loop in *C* must generally be at least four bases long to avoid steric hindrance with base-pairing in the stem part of the structure. The stem and loop reverses the chemical direction of the RNA molecule. Interior loops, as in *D*, form when the bases in a double-stranded region cannot form base pairs, and may be asymmetric with a different number of base pairs on each side of the loop, as shown in *D* and *E*, or symmetric with the same number on each side (not shown). Junctions, as in *F*, may include two or more double-stranded regions converging to form a closed structure. The RNA backbone is red, and both unpaired and paired bases are blue. The types of loop structures can be represented mathematically, thereby aiding in the prediction of secondary structure (Sankoff et al. 1983; Zuker and Sankoff 1984). (Adapted, with permission, from Burkhard et al. 1999b.)

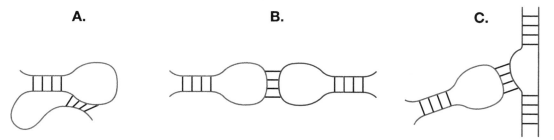

FIGURE 8.3. Examples of complex interactions between RNA secondary structural elements. (*A*) Pseudoknot. (*B*) Kissing hairpins. (*C*) Hairpin–bulge contact. (Adapted, with permission, from Burkhard et al. 1999b.)

SEQUENCE AND BASE-PAIRING PATTERNS CAN BE USED TO PREDICT RNA STRUCTURE

The most likely regions of base-pairing in an RNA molecule can be predicted computationally given only the sequence, thus providing an ab initio prediction of RNA secondary structure. From the many possible choices of complementary sequences that can potentially base-pair in an RNA molecule, the compatible sets that provide the most energetically stable molecules are chosen. Structures with energies almost as stable as the most stable one may also be produced, and regions whose predictions are the most reliable can be identified from this analysis. Variations found in related sequences may also be used to predict which base pairs are likely to be found in each of the molecules. One variation of RNA structure prediction methods predicts a set of sequences that are able to form a particular structure. Methods for predicting such three-dimensional structures from sequence are also being developed

(see M. Zuker's Web site at http://www.bioinfo.rpi. edu/).

Another type of RNA secondary structure prediction method takes into account patterns of base-pairing that are conserved during evolution of a given class of RNA molecules. Sequence positions that base-pair are found to vary at the same time during evolution of RNA molecules so that structural integrity is maintained. For example, if two different sequence positions G and C form a base pair in a given type of molecule, then sequences that have C and G reversed, or A and U or U and A at the corresponding positions, would be considered reasonable matches. These patterns of covariation in RNA molecules are a manifestation of secondary structure that leads to a structural prediction. The computational challenge is to discover these covariable positions against the background of other sequence changes.

PREDICTION OF MINIMUM-ENERGY RNA STRUCTURE IS LIMITED

In addition to using base-pairing patterns in RNA sequences for RNA structure prediction, the predicted energies of these structures can be used to find the energetically most stable molecule. In predicting minimum energy RNA secondary structure, some simplifying assumptions are usually made:

1. The most likely structure of the RNA molecule is identical or similar to the energetically most stable structure.

2. The energy associated with any position in the structure is only influenced by local sequence and structure. Thus, the energy associated with a particular base pair in a double-stranded region is assumed to be influenced only by the previous base pair and not by the base pairs farther down the double-stranded region or anywhere else in the structure. These energies can be reliably estimated by experimentation with small, synthetic RNA oligonucleotides (Tinoco et al. 1971, 1973;

Freier et al. 1986; Turner and Sugimoto 1988; SantaLucia 1998) and were recently improved to include more experimental data that reveal a dependence of the energies on neighboring sequence (Mathews et al. 1999). Energy predictions are most reliable when used for standard Watson–Crick base pairs and single G/U pairs surrounded by Watson–Crick pairs.

3. The structure is assumed to be formed by folding of the chain back on itself in a manner that does not produce any knots. The best way of representing this requirement is to draw the sequence in a circular form. The paired bases are then joined by arcs. If the total structure with all predicted base pairs is free of knots, none of the arcs must cross (Fig. 8.4). Note, however, that if a structure with interacting secondary structures (Fig. 8.3) is represented on such a diagram, the lines will cross.

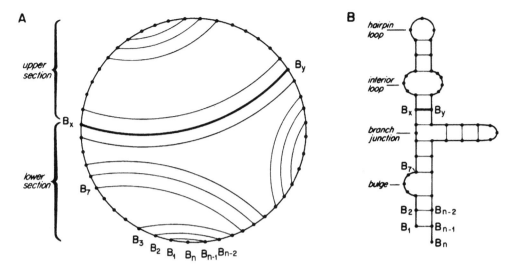

FIGURE 8.4. Display of base pairs in an RNA secondary structure by a circle plot. The predicted minimum free-energy structure shown in *B* is represented by a plot of the predicted base pairs as arcs connecting the bases in the sequence, which is drawn around the circumference of a circle, as shown in *A* (see Nussinov and Jacobson 1980). Note that none of the lines cross, indicating that the structure does not include any knots. (Reprinted, with permission, from Nussinov and Jacobson 1980.)

HOW RNA PREDICTION METHODS WERE DEVELOPED

The development of methods for predicting RNA secondary structure has been reviewed by von Heijne (1987). Development of methods progressed from early attempts to locating regions in RNA sequences that could base-pair to form secondary structures, to joining compatible secondary structures into larger structure for the RNA, and finally to predicting the most stable structure energetically.

Tinoco et al. (1971) first estimated the energy associated with regions of secondary structure by extrapolating from studies with small molecules and then attempting to predict which configurations of larger molecules were the most energetically stable. Energy estimates included the stabilizing energy associated with stacking base pairs in a double-stranded region and the destabilizing influence of regions that were not paired. Pipas and McMahon (1975) developed computer programs that listed all possible helical regions in tRNA sequences using modified Watson–Crick base-pairing rules. They developed computational methods to locate all possible secondary structures by forming permutations of compatible helical regions, and evaluated each possible structure for total free energy. Studnicka et al. (1978)

designed a computational method for generating a structure with multiple secondary structures by joining compatible double-stranded regions together that did not overlap or form knots (Fig. 8.4) and that produced the energetically most favorable structure.

Martinez (1984) made a list of possible double-stranded regions, and these regions were then given weights in proportion to their equilibrium constants, calculated by the Boltzmann function $[\exp(-\Delta G/RT)]$, where $-\Delta G$ is the free energy of the regions, R is the gas constant, and T is the temperature. The RNA molecule is folded by a Monte Carlo method in which one initial region is chosen at random from a weighted pool, similar to the method used in Gibbs sampling (see p. 201). (In the Monte Carlo method, a random drawing is made from a pool of all possible double-stranded regions, with the number of each type weighted in proportion to energetic stability.)

To illustrate this computational approach, imagine each possible double-stranded region in an RNA molecule being represented by a type of marble in a bag. The number of each type of marble in the bag is weighted by the Boltzmann probability, so that marbles corresponding to more energetically stable regions are more likely to be chosen. A secondary element is then randomly chosen as in picking a marble from the weighted collection in the bag. Additional compatible regions are then added sequentially by further selections from the weighted pool as in further drawings from the marble bag until no more can be added. This method generates a set of possible secondary structures to include in the predicted structure for the RNA molecule weighted by energy, but it does not take into account the destabilizing effect of unpaired regions. The Boltzmann probability function is used in more recent applications (described below) to find the most probable secondary structure of an RNA molecule (Hofacker et al. 1998; Wuchty et al. 1999).

Nussinov and Jacobson (1980) were the first to design a precise and efficient algorithm for predicting secondary structures in RNA. Their algorithm generates two scoring matrices—one $M(i,j)$ to keep track of the maximum number of base pairs that can be formed in any interval i to j in the sequence, and a second $K(j,k)$ to keep track of the base position k that is paired with j. From these matrices, a structure with the maximum possible number of base pairs could be deduced by a trace-back procedure similar to that used in performing sequence alignments by dynamic programming.

Zuker and Stiegler (1981) used the dynamic programming algorithm and energy rules for producing the most energetically favorable structure. Their method assumed that the most energetic, and usually longest, predicted dsRNA regions are those present in the molecule. When secondary structures are formed in RNA, there is an associated loss of energy caused by the formation of hydrogen bonds between bases that makes the structure stable. As more and more secondary structures and associated hydrogen bonds are formed in RNA, the stability increases. More hydrogen bonds are associated with some base pairs than others (e.g., a G/C base pair has three hydrogen bonds whereas an A/T has two). Hence, patterns of base-pairing that include more base pairs are expected to produce more stable molecules energetically. Because many double-stranded regions are predictable for most RNA sequences, the number of predictions is reduced by requiring that the structure be in an energetically stable configuration, and also by including known biochemical or structural information to indicate which bases should be paired or not

paired, and by enforcing topological restraints. The FOLD program of Michael Zuker (Zuker and Stiegler 1981) implemented these features for predicting the minimum energy structure.

mfold, subsequently written by Dr. Michael Zuker and colleagues, is commonly used to predict the energetically most stable structures of an RNA molecule (Jaeger et al. 1989, 1990; Zuker 1989, 1994). mfold provides a set of possible secondary structures within a user-specified energy range and provides an indication of their reliability. The program also can incorporate as an option user-provided experimental data or covariance information from phylogenetically related sequences to require pairing of specified bases in the predicted molecule (Zuker et al. 1991). mfold includes methods for graphic display of the predicted molecules. This program is very demanding on computer resources because the algorithm is of N^3 complexity, where N is the sequence length. For each doubling of sequence length, the time taken to compute a structure increases eightfold. The program also requires a large amount of memory for storing intermediate calculations of structure energies in multiple scoring matrices. As a result, mfold is most often used to predict the structure of sequences less than 1000 nucleotides in length. This method is most reliable for small molecules and becomes less reliable as the length of the molecule increases.

Access to the mfold program and to many other types of useful information on RNA are found at the Web site of Dr. Michael Zuker, at http://www.bioin fo.rpi.edu/. Details of running mfold are available in a user manual (Jaeger et al. 1990). A comparable program developed by the Vienna RNA group (Wuchty et al. 1999) uses the partition function method for finding the most probable secondary structural configuration of an RNA molecule and the most probable base pairs. This program is discussed below (p. 344).

One advance in the prediction of RNA structure came from the recognition that certain RNA sequences form specific structures and that the presence of these sequences is strongly predictive of such a structure. For example, the hairpin CUUCGG occurs in different genetic contexts and forms a very stable structure (Tuerk et al. 1988). Databases of such RNA structures and RNA sequences can greatly assist in RNA structure prediction (Table 8.1).

The genetic algorithm (see Chapter 5, p. 185) is another approach that has also been used to predict

TABLE 8.1. *Examples of RNA databases and RNA analysis Web sites*[a]

Site or resource	Web address	Reference
Comparative RNA Web site (Gutell laboratory)	http://www.rna.icmb.utexas.edu/	see Web site
mfold minimum energy RNA configuration	http://bioinfo.rpi.edu/applications/mfold/	Zuker et al. (1991)
RNA secondary structures, Group I introns, 16S rRNA, 23S rRNA, covariation analysis	http://www.rna.icmb.utexas.edu	Gutell (1994); Schnare et al. (1996 and references therein)
RNA structure database	http://www.rnabase.org/	see Web page
snoRNA database for *S. cerevisiae*	http://rna.wustl.edu/snoRNAdb/	Lowe and Eddy (1999)
tRNAscan-SE search server	http://www.genetics.wustl.edu/eddy/tRNAscan-SE/	Lowe and Eddy (1997)
Vienna RNA package for RNA	http://www.tbi.univie.ac.at/	Hofacker et al. (1998)

[a]See Web Search Terms for more sites.

secondary structure (Shapiro and Navetta 1994); for aligning RNA sequences, taking into account both sequence and secondary structure and including pseudoknots (Notredame et al. 1997); and for simulation of RNA-folding pathways (Gultyaev et al. 1995). Genetic algorithms are powerful tools, but considerable time is needed to learn to use them effectively. The program FOLDALIGN, developed by L.J. Heyer and J. Gorodkin, uses a dynamic programming algorithm to align RNAs based on sequence and secondary structure and locates the most significant motifs (Gorodkin et al. 1997). Chan et al. (1991) described another algorithm for the same purpose, and Chetouani et al. (1997) developed ESSA, a method for viewing and analyzing RNA secondary structures. Collectively, these methods offer the advantage of finding secondary structures in aligned RNAs and visualizing them. The following section describes in greater detail two methods for RNA secondary structure analysis.

METHODS FOR PREDICTING RNA STRUCTURE USE TWO MAIN APPROACHES

Methods for predicting the structure of RNA molecules include the following approaches:

1. An analysis of all possible combinations of potential double-stranded regions by energy minimization methods. Energy minimization methods have been so well refined that a series of energetically feasible models using information from multiple sequence alignments and experimental data may be computed. Thermodynamically most likely structural models are predicted. The advantages of these methods are the reliability of the prediction based on energy calculations and the built-in flexibility to accommodate experimental and alignment data. The disadvantages are that only structures with compatible secondary structures are predicted (i.e., tertiary interactions are not predicted by the method), and the computational complexi-

ty and increasing number of predictions with sequence length limits the usefulness of the method to short sequences of <1000 nucleotides.

2. Identification of base covariation that maintains secondary and tertiary structure of an RNA molecule during evolution. Sequence covariation methods are based on having a sufficient number of sequences of an RNA molecule that can be aligned and on having a conserved pattern of base-pairing in the sequences so that interacting base pairs can be identified. The advantages of these methods are their computational simplicity and that both secondary (two-dimensional) and tertiary (three-dimensional) interactions can be identified based on the observed patterns of covariation. These methods may be limited by the requirement for available information on sequence covariation.

Predictions made by either energy minimization or covariation methods serve as a guide for further experimental testing in the laboratory. These methods are discussed in detail in the following sections.

Energy Minimization

Self-complementary Regions in RNA Sequences Predict Secondary Structure

All types of RNA secondary structure analysis begin by the identification of self-complementary sequence regions. For single-stranded RNA molecules, these repeats represent regions that can potentially self-hybridize to form RNA double strands (von Heijne 1987; Rice et al. 1991). Once identified, the compatible regions may be used to predict a minimum free-energy structure. One of the simplest types of analyses that can be performed to find stretches of sequence in RNA that are self-complementary is a dot matrix sequence comparison for self-complementary regions. A more advanced type of dot matrix can be used to show the most energetic parts of the molecule (see Fig. 8.8, below).

To perform a dot matrix analysis for self-complementary regions, the sequence to be analyzed is listed in both the horizontal and vertical axes. In one method for finding such regions, the sequence is listed in the 5′ to 3′ direction across the top of the page and the sequence of the complementary strand is listed down the side of the page, also in the 5′ to 3′ direction. The matrix is then scored for identities. Self-complementary regions appear as rows of dots going from upper left to lower right. For RNA, these regions represent sequences that can potentially form A/U and G/C base pairs. G/U base pairs will not usually be included in this simple type of analysis because they play a less significant role in base-pairing. As with matching DNA sequences, there are many random matches between the four bases in RNA, and the diagonals are difficult to visualize. A long nucleotide window and a requirement for a large number of matches within this window are used to filter out these random matches.

An example of the RNA secondary structure analysis using a DNA matrix option of DNA Strider is shown in Figure 8.5. An analysis of the potato spindle tuber viroid is shown, using a window of 15 nucleotides and a required match of 11. Note the appearance of a diagonal running from the center of

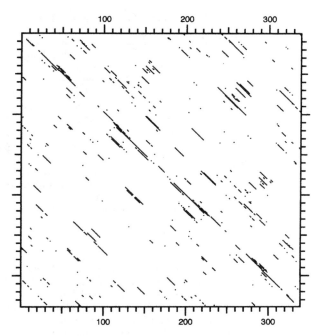

FIGURE 8.5. Dot matrix analysis of the potato tuber spindle viroid for RNA secondary structure using the MATRIX function of DNA Strider v. 1.2 on a Macintosh computer.

the matrix to the upper left, and a mirror image of this diagonal running to the lower right. The presence of this diagonal indicates the occurrence of a large self-complementary sequence such that the entire molecule can potentially fold into a hairpin structure. An alternative dot matrix method for finding RNA secondary structure is to list the given RNA sequence across the top of the page and also down the side of the page and then to score matches of complementary bases (G/C, A/U, and G/U) instead of identities as in the previous example. Diagonals indicating complementary regions will go from upper right to lower left in this type of matrix. This is the kind of matrix used to produce an energy matrix (see Fig. 8.8, below).

Minimum Free-energy Method for RNA Secondary Structure Prediction

To predict RNA secondary structure in this approach, every base is first compared for complementarity to every other base by a type of analysis very similar to the dot matrix analysis. The sequence is listed across the top and down the side of the page, and G/C, A/U, and G/U base pairs are scored as described for the second matrix example in the previous paragraph. Just as a diagonal in a two-sequence comparison indi-

TABLE 8.2. *Predicted free-energy values (kcal/mole at 37°C) for base pairs and other features of predicted RNA secondary structures*

	A. Stacking energies for base pairs					
	A/U	C/G	G/C	U/A	G/U	U/G
A/U	−0.9	−1.8	−2.3	−1.1	−1.1	−0.8
C/G	−1.7	−2.9	−3.4	−2.3	−2.1	−1.4
G/C	−2.1	−2.0	−2.9	−1.8	−1.9	−1.2
U/A	−0.9	−1.7	−2.1	−0.9	−1.0	−0.5
G/U	−0.5	−1.2	−1.4	−0.8	−0.4	−0.2
U/G	−1.0	−1.9	−2.1	−1.1	−1.5	−0.4
	B. Destabilizing energies for loops					
Number of bases	1	5	10	20	30	
Internal	–	5.3	6.6	7.0	7.4	
Bulge	3.9	4.8	5.5	6.3	6.7	
Hairpin	–	4.4	5.3	6.1	6.5	

(*A*) Stacking energy in double-stranded region when the base pair listed in left column is followed by the base pair listed in top row. C/G followed by U/A is therefore the dinucleotide 5′ CU 3′ paired to 5′ AG 3′. (*B*) Destabilizing energies associated with loops. Hairpin loops occur at the end of a double-stranded region, internal loops are unpaired regions flanked by paired regions, and a bulge loop is a bulge of one strand in an otherwise paired region (Fig. 8.2). An updated and more detailed list of energy parameters may be found at the Web site of M. Zuker (http://bioinfo.math.rpi.edu/~zuker/rna/energy/). From Turner and Sugimoto (1988); Serra and Turner (1995).

cates a range of sequence similarity, a row of matches in the RNA matrix indicates a succession of complementary nucleotides that can potentially form a double-stranded region. The energy of each predicted structure is estimated by the nearest-neighbor rule by summing the negative base-stacking energies for each pair of bases in the predicted double-stranded regions and by adding the estimated positive energies of destabilizing regions such as loops at the end of hairpins, bulges within hairpins, internal bulges, and other unpaired regions. Representative examples of the energy values that are currently used in this type of analysis are given in Table 8.2.

Several types of scoring matrices are used to evaluate all the different possible secondary structure configurations and to find the most energetically favorable ones. The complementary regions are evaluated by a dynamic programming algorithm to predict the most energetically stable molecule. The method is similar to the dynamic programming method used for sequence alignment (see Chapter 3).

An illustrative example for evaluation of energy in a double-stranded region is shown in Figure 8.6. The sequence is listed down the side of the matrix, and a portion of the same sequence is also listed across the top of the matrix; matching base pairs have been identified within the matrix. The object is to

find a diagonal row of matches that goes from upper right to lower left, and such a row is shown in color in the example. In Figure 8.6, a match of four complementary bases in a row produces a molecule of free energy −6.4 kcal/mole.

In general, each dynamic programming matrix value is obtained by considering the minimum energy values obtained by all previous complementary pairs decreased by the stacking energy of any additional complementary base pairs or increased by the destabilizing energy associated with noncomplementary bases. This calculation is repeated for the entire matrix. The increase depends on the type and length of loop that is introduced by the noncomplementary base pair, whether an internal loop, a bulge loop, or a hairpin loop, as shown in Table 8.2. This comparison of all possible matches between nucleotides in the sequence and energy values is continued until all nucleotides have been compared. The pattern followed in comparing nucleotides within the RNA molecule is illustrated in Figure 8.7. As described in the figure, several matrices are used to keep track of the base pairs and structures found. Once the most energetic structure overall is computed, the information in these matrices is used to produce the predicted secondary structure of the molecule that contributes to this minimum energy value.

A. Base comparisons

5'	A	C	G	U	3'
A					
C					
G					
U					
–					
–					
G		C/G		U/G	
C			G/C		
G		C/G		U/G	
U	A/U		G/U		
3'					

B. Free energy calculations

5'	A	C	G	U	3'
A					
C					
G					
U					
–					
–					
G				–6.4	
C			–5.2		
G		–1.8			
U					
3'					

FIGURE 8.6. Evaluation of secondary structure in RNA sequence by the matrix method. The sequence is listed down the first column of *A* and *B* in the 5'→3' orientation, and the first four bases of the sequence are also listed in the first row of the tables in the 5'→3' direction. Several complementary base pairs between the first and last four bases that could lead to secondary structure are shown in *A*. The most 5' base is listed first in each pair. The diagonal set of base pairs A/U, C/G, G/C, and U/G reveals the presence of a potential double-stranded region between the first and last four bases. The free energy associated with such a row of base pairs is shown in *B*. A C/G base pair following an A/U base pair has a base stacking energy of –1.8 kcal/mole (Turner and Sugimoto 1988). This value is placed in the corresponding position in *B*. Similarly, a C/G base pair followed by a G/C provides energy of –3.4, and a G/C followed by a U/G, –1.2 kcal/mole. Hence, the energy accumulated after stacking of these additional two base pairs is –5.2 and –6.4. The energy of this double-stranded structure will continue to decrease (become more stable) as more base pairs are added, but will be increased if the structure is interrupted by non-complementary base pairs.

Suboptimal Structure Predictions by mfold and the Use of Energy Plots

Originally, the FOLD program of M. Zuker predicted only one structure having the minimum free energy. However, there is a problem in predicting only one structure. If a structure is predicted for an RNA molecule and then a single nucleotide position is changed and the structure is recomputed, often a very different structure is predicted. A simple nucleotide change is usually not expected to produce a big change in the structure. The solution to this problem is that a series of structures, called suboptimal structures, may have an energy value that is close to the one having the minimum energy. Considering the assumptions that are made in energy calculations, these differences may not be significant. Hence, if suboptimal structures are produced, some of these may not be affected by the nucleotide change and become good candidates. Extending this analysis to a family of RNA sequences of the same molecule from different phylogenetic sources, if a particular structure among the suboptimal ones is predicted for each of these sequences, this structure has a stronger biological foundation and becomes a structural model for the whole RNA family.

mfold and the mfold Web server (Zuker 2003) provide improved prediction of energies of loop and other structural features and produce suboptimal structures within a user-specified percentage of the minimum free energy of the molecule. These predictions accurately reflect structures of related RNA molecules derived from comparative sequence analysis (Jaeger et al. 1989; Zuker 1989, 1994; Zuker et al. 1991; Zuker and Jacobson 1995).

To find suboptimal structures, the dynamic programming method was modified (Zuker 1989, 1991) to evaluate parts of a new scoring matrix in which the sequence is represented in two tandem copies on both the vertical and horizontal axes. The regions from i =

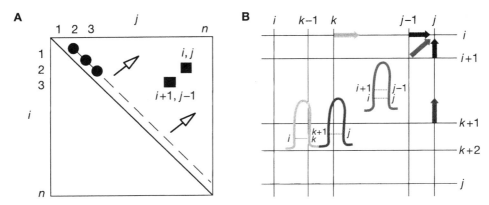

FIGURE 8.7. Method used in dynamic programming analysis for identifying the most energetically favorable configuration of a linear RNA molecule. (*A*) The sequence of an RNA molecule of length n bases is listed across the top of the page and down the side. The index of the sequence across the top is j and that down the side is i. The search only includes the upper right part of the matrix and begins at the first diagonal line for matching base pairs. First positions $i = 1$ and $j = 2$ are compared for potential base-pairing, and if pairing can occur, an energy value is placed in an energy matrix W at position 1,2. Then, $i = 2$ and $j = 3$ base are compared, and so on, until all base combinations along the dashed diagonal have been made. Then, comparisons are made along the next upper right diagonal. As each pair of bases is compared, an energy calculation is made that is the optimal one up to that point in the comparison. In the simplest case, if $i + 1$ pairs with $j - 1$, and i pairs with j, and if this structure is the most favorable up to that point, the energy of the i/j base pair will be added to that of the $i + 1/j - 1$ base pair. Other cases are illustrated in *B*. The process of obtaining the most stable energy value at each matrix position is repeated following the direction of the arrows until the last position, $i = 1$ and $j = n$, has been compared and the energy value placed at this position in matrix W, the value entered in $W(1, n)$, will be the energy of the most energetically stable structure. The structure is then found by a traceback procedure through the matrices similar to that used for sequence alignments. The method used is a combination of a search for all possible double-stranded regions and an energy calculation based on energy values similar to those in Table 8.2. The search for the most energetic structure uses an algorithm (Zuker and Stiegler 1981) similar to that for finding the structure with maximum base-pairing (Nussinov and Jacobson 1980). These authors recognized that there are three possible ways, illustrated in *B* by the colored arrows, of choosing the best energy value at position i, j in an energy matrix W. The simplest calculation (*red arrow*) is to use the energy value found up to position $i - 1, j - 1$ diagonally below i, j. If i and j can form a base pair (and if there are at least four bases between them in order to allow enough sequence for a hairpin) and $i + 1$ and $j - 1$ also pair, then the stacking energy of i/j upon $i + 1/j - 1$ will reduce the energy value at $i + 1, j - 1$, producing a more stable structure, and the new value can be considered a candidate for the energy value entered at position i, j. If i and j do not pair, then another choice for the energy at i, j is to use the values at positions $i, j - 1$ or $i + 1, j$, illustrated by the blue arrows. i and j then become parts of loop structures. Finally, i and j may each be paired with two other bases, i with k and j with $k + 1$, where k is between i and j ($i < k < j$), illustrated by the structure shown in yellow and green, reflecting the location of the paired bases. The minimum free-energy value for all values of k must be considered to locate the best choice as a candidate value at i, j. Finally, of the three possible choices for the minimum free-energy value at i, j, indicated by the four colored arrows, the best energy value is placed at position $W(i, j)$. The procedure is repeated for all values of i and j, as illustrated in *A*. Besides the main energy scoring matrix W, additional scoring matrices are used to keep track of auxiliary information such as the best energy up to i, j where i and j form a pair, and the influence of bulge loops, interior loops, and other destabilizing energies. An essential second matrix is $V(i, j)$, which keeps track of all substructures in the interval i, j in which i forms a base pair with j. Some values in the W matrix are derived from values in the V matrix and vice versa (Zuker and Stiegler 1981).

1 to n and $j = 1$ to n are used to calculate an energy $V(i, j)$ for the best structure that includes an i, j base pair and is called the included region. A second region, the excluded region, is used to calculate the energy of the best structure that includes i, j but is not derived from the structure at $i+1, j-1$ (Fig. 8.7). After energy calculations, the difference between the

included and excluded values is the most stable structure that includes the base pair i, j. All complementary base pairs can be sampled in this fashion to determine which are present in a suboptimal structure that is within a certain range of the optimal one.

An energy dot plot is produced showing the locations of alternative base pairs that produce the most

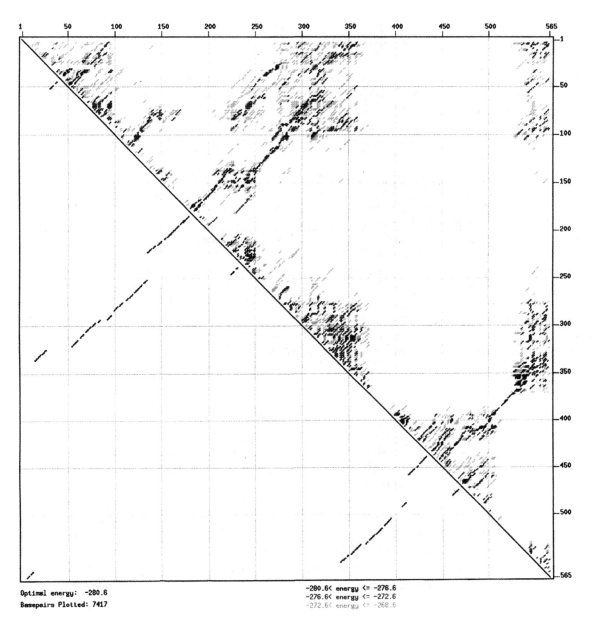

Optimal energy: -280.6
Basepairs Plotted: 7417

-280.6< energy <= -276.6
-276.6< energy <= -272.6
-272.6< energy <= -268.6

FIGURE 8.8. The energy dot plot (box plot) of alternative choices of base pairs of an RNA molecule (Jacobson and Zuker 1993). The sequence is that of a human adenovirus preterminal protein (GenBank U52533) that is given by M. Zuker as an example on his Web site at http://www.bioinfo.rpi.edu/applications/mfold/. Foldings were computed using the default parameters of the mfold program at http://www.bioinfo.rpi.edu/~zukerm/rna/energy/ (Mathews et al. 1999) using the thermodynamic values of SantaLucia (1998). The minimum energy of the molecule is −280.6 kcal/mole and the maximum energy increment is 12 kcal/mole. Black dots indicate base pairs in the minimum free-energy structure and are shown both above and the mirror image below the main diagonal. Red, green, and yellow dots are base pairs in foldings of increasing 4, 8, and 12 kcal/mole energies greater than the minimum energy, respectively. A region with very few alternative base pairs, such as the pairing of 370–395 with 530–505, is considered to be strongly predictive, whereas regions with many alternative base pairs, such as the base-pairing in the region of 340–370 with 570–530, are much less predictive.

stable or suboptimally stable structures, as illustrated in Figure 8.8. The program can be instructed to find structures within a certain percentage of the mini-

mum free energy. Parameter p provides a measure of the degree to which alternative base pairs within $p\%$ of the maximum free energy will be identified. When

RELIABILITY OF SECONDARY STRUCTURE PREDICTION

Three scores, Pnum (*i*), Hnum (*i, j*), and Snum, have been derived to assist with a determination of the reliability of a secondary structure prediction for a particular base *i* or a base pair *i, j*. Pnum(*i*) is the total number of energy dots regardless of color in the *i*th row and *i*th column of the energy dot plot as in Figure 8.8, and represents in an unfiltered dot plot the number of base pairs that the *i*th base can form with all other base pairs in structures within the defined energy range. The lower this value, the more well defined or "well determined" is the local structure because there are few competitive foldings.

Hnum(*i, j*) is the sum of Pnum(*i*) and Pnum(*j*) less 1 and is the total number of dots in the *i*th row and *j*th column. It represents the total number of base pairs with the *i*th or *j*th base in the predicted structures. The Hnum for a double-stranded region is the average Hnum value for the base pairs in that helix. The lower this number, the more well determined is the double-stranded region. In an analysis of tRNAs, 5S RNAs, ribosomal RNAs, and other published secondary structure models based on sequence variation (Jaeger et al. 1990; Zuker and Jacobson 1995), these methods correctly predict about 70% of the double-stranded regions.

Snum, also called ss-count, is the number of foldings in which base *i* is single-stranded divided by *m*, the number of foldings, and gives the probability that base *i* is single-stranded. If Snum is approximately 1, then base *i* is probably in a single-stranded region, and if Snum is approximately 0, then base *i* is probably not in such a region.

This reliability information has been used to annotate output files of mfold and other RNA display programs (Zuker and Jacobsen 1998). Plots of these values against sequence position are given by the mfold program and the Zuker RNA folding Web site at http://www.bioinfo.rpi.edu/applications/.

using mfold on a local machine with a graphic interface, clicking a part of the display will lead to program output of the corresponding structure with the RNA sequence displayed. There are various other options on the mfold Web site that allow structural restrictions, e.g., loop size, maximum distance between base pairs along the sequence, and other restrictions on forcing or rejecting base-pairing, as well as ways to limit the number of structures output by mfold. One of the predicted structures is shown in Figure 8.9.

Other Algorithms for Suboptimal Folding of RNA Molecules

A limitation of the Zuker method and other methods (Nakaya et al. 1995) for computing suboptimal RNA structures is that they do not compute all the structures within a given energy range of the minimum free-energy structure. For example, no alternative structures are produced in which each base that was paired in the original structure is now paired with another base as a test as to whether there is an alternative way to fold the RNA. Furthermore, if two substructures are joined by a stretch of unpaired bases, each structure will be computed separately without

any possible involvement between the structures or with the unpaired region. A better method is to compute structures that are globally suboptimal for both structures and the unpaired region. These types of considerations limit the number of alternative structures predicted compared to known variations based on sequence variations in tRNAs (Wuchty et al. 1999).

These limitations have been largely overcome by using an algorithm originally described by Waterman and Byers (1985) for finding sequence alignments within a certain range of the optimal one by modifications of the trace-back procedure used in dynamic programming. This method efficiently calculates alternative structures, up to a very large number, within a given energy range of the minimum free-energy structure (see Fig. 8.10, below). The method has also been used to demonstrate that natural tRNA sequences can form many alternative structures that are close to the minimum free-energy structure and that base modification plays a major role in this energetic stability (Wuchty et al. 1999).

The Waterman and Byers approach may also be used to assess the thermodynamic stability of RNA structures given expected changes in energies associ-

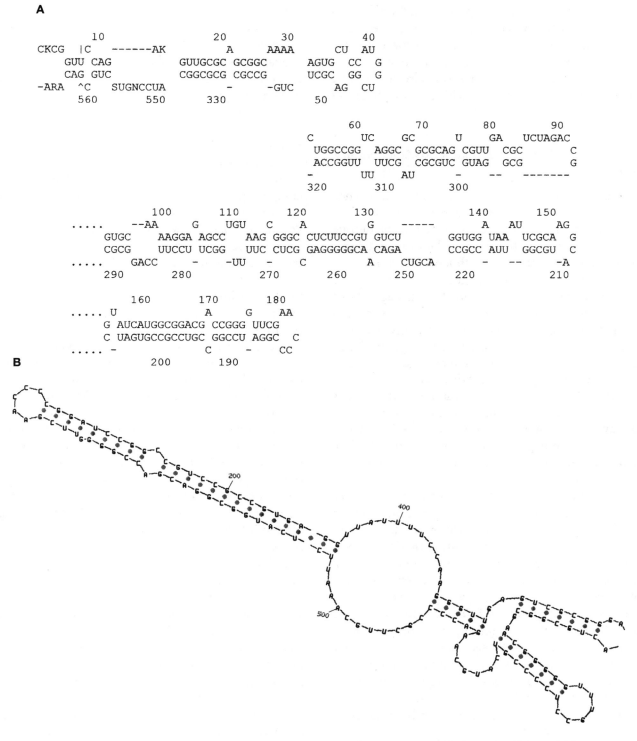

FIGURE 8.9. Model of RNA secondary structure of the human adenovirus preterminal protein. This model is one of several alternative structures represented by the above energy plot and provided as an output by the current versions of mfold. (A) Simple text representation of one of the predicted structures. Each stem-and-loop structure is shown separately and the left end of each structure is placed below the point of connection to the one above. (B) More detailed rendition of one part of the predicted structures. The structure continues beyond the right side of the page.

ated with base pairs and loops as a function of temperature. The RNA secondary structure prediction and comparison Web site at http://www.tbi.univie.ac.at/ will fold molecules of length up to 4000 bases, and the Vienna RNA Package software for folding larger molecules on a local machine is available from this site.

Prediction of the Most Probable RNA Secondary Structure

In the above types of analyses, the energies associated with predicted double-stranded regions in RNA are used to produce a secondary structure. Stabilizing energies associated with base-paired regions and destabilizing energies associated with loops are summed to produce the most stable structure or suboptimal ones of energy within a user-specified percentage of the maximum free energy. A different way of predicting the structures is to consider the probability that each base-paired region will form based on principles of thermodynamics and statistical mechanics. The probability of forming a region with free energy ΔG is expressed by the Boltzmann distribution, which states that the likelihood of finding a structure with free energy $-\Delta G$ is proportional to $[\exp(-\Delta G/kT)]$, where k is the Boltzmann gas constant and T is the absolute temperature. (The Boltzmann constant k is 8.314510 J/mole/°K.)

Note that the more stable a structure, the lower the value of ΔG. Because ΔG is a negative number, the value of $\exp(-\Delta G/kT)$ increases for more stable structures and also grows exponentially with a decrease in energy. The probability of these regions forming increases in the same manner. Conversely, the effect of destabilizing loops that have a positive ΔG is to decrease the probability of formation. By using these probability calculations and a dynamic programming method similar to that used in mfold (Zuker and Stiegler 1981) and the partition function algorithm of McCaskill (1990), it is possible to predict the most probable RNA secondary structures and to assess the probability of the base pairs that contribute energetic stability to this structure.

For a set of possible structures, the likelihood of each may be calculated using the Boltzmann distribution and the sum of the likelihoods provides a partition function that can be used to normalize each individual likelihood, providing a probability that each

will occur. Thus, probability of structure A of energy $-\Delta G_a$ is $[\exp(-\Delta G_a/kT)]$ divided by the partition function Q, where $Q = \Sigma_s [\exp(-\Delta G_s/kT)]$, the sum of probabilities of all possible structures s. This kind of analysis allows one to calculate the probability of a certain base pair forming.

The key to this analysis is the calculation of the partition function Q. A dynamic programming algorithm for calculating this function exactly for RNA secondary structure has been developed (McCaskill 1990). The algorithm is very similar to that used for computing an optimal folding by mfold. Complexity similarly increases as the cube of the sequence length, and the energy values used for base pairs and loops are also the same, except that structures with very large interior loops are ignored. Just as the minimum free-energy value is given at $W(1, n)$ in the Zuker mfold algorithm, the value of the partition function is given at matrix position $Q(1, n)$ in the corresponding partition matrix.

As indicated above, the partition function is calculated as the sum of the probabilities of each possible secondary structure. Because there are a very large possible number of structures, the calculation is simplified by calculating an auxiliary function, $Q^b(i, j)$, which is the sum of the probabilities of all structures that include the base pair i, j. The partition function $Q(i, j)$ includes both of these structures and the additional ones where i is not paired with j. An example illustrating the difference between the minimum free energy and the partition function methods should be instructive. Suppose that the bases at positions $i + 1$, $j - 1$ and i, j can both form base pairs. They then form a stack of two base pairs. In the minimum free-energy method, the energy of the i, j pair stacked on the $i + 1, j - 1$ pair will be added to $V(i + 1, j - 1)$ to give $V(i, j)$, where V is a scoring matrix that keeps track of the best structure that includes an i, j base pair. In contrast, the value for $Q^b(i, j)$ will be calculated by multiplying the matrix value $Q^b(i + 1, j - 1)$ by the probability of the base pair i, j given by the Boltzmann probability $[\exp(-\Delta G/kT)]$, where ΔG is the negative stacking energy of the i, j base pair on the $i + 1, j - 1$ base pair, and will be a large number reflecting the probability given the stability of the base-paired region.

For a hairpin structure with a row of successive base pairs, the probability of the hairpin will be the product of the Boltzmann factors associated with the stacked pair, giving a high number for the relative likelihood of formation. The procedure followed by the partition function algorithm is to calculate $Q^b(i, j)$ and $Q(i, j)$ iteratively in a scoring matrix similar to that illustrated in Figure 8.7A until $Q(1, n)$ is reached. This matrix position contains the value of the full partition function Q.

Both the partition function and the probabilities of all base pairs are computed by this algorithm, and the most probable structural model is thereby found. Information about intermediate structures, base-pair opening and slippage, and the temperature dependence of the partition function may also be determined. The latter calculation provides information about the melting behavior of the secondary structure.

A suite of RNA-folding programs available from the Vienna RNA secondary structure prediction Web site (http://www.tbi.univie.ac.at/) uses the Boltzmann/partition method to predict the most probable and alternative RNA secondary structures. An example of the predicted folding of a 300-base RNA molecule is given in Figure 8.10. The probability of forming each base pair is shown in a dot matrix display in which the dots are squares of increasing size reflecting the probability of the base pair formed by the bases in the horizontal and vertical positions of the matrix.

Identification of Base Covariation

The second major method that has been used to make RNA secondary structure predictions (Woese et al. 1983) and also tertiary structure analyses such as those shown in Figure 8.3 (Gutell et al. 1986) is RNA sequence covariation analysis. Covariance models reflecting the interactions among base pairs in RNA secondary structures were introduced.

This method examines aligned sequences of the same class of RNA molecules from different species for sequence positions that vary together in a manner that would allow them to produce a base pair in all of the molecules. The idea is quite simple. On the one hand, to conserve double-stranded regions in RNA molecules, sequence changes that take place in evolution should maintain the base-pairing for secondary structure formation. On the other hand, sequence changes in loops and single-stranded regions should not have such a constraint. The key to analysis is to look for sequence positions that are covarying in order to maintain the base-pairing properties of the RNA molecule.

The justification for this method of structure prediction is that these types of joint substitutions or covariations actually are found to occur during evolution of such genes. As shown in Figure 8.11, when one position corresponding to one-half of a base pair is changed, another position corresponding to the base-pairing partner will also change. For example, suppose that the first position in one sequence has a G and the second position a C, raising the possibility that these bases form a G/C base pair in the structure. This possibility can be tested by examining what bases are found in different sequences of the same RNA molecule at these same two positions. If the bases at these positions can consistently form a base pair, then they are covarying in a manner that maintains the base pair, providing evidence that the sequence positions do form a base pair in the RNA structure. Sequence covariability has been used to improve thermodynamic structure prediction as described in the above section (Hofacker et al. 1998). An example of using covariation analysis to decipher base-pair interactions in tRNA is shown in Figure 8.12.

One method of covariation analysis also examines which phylogenetic groups exhibit change at a given position. For each position in an RNA molecule, the base that generally predominates in one particular part of the tree is determined. This method may be used to predict phylogenetic relationships based on RNA structure variation, as described in Chapter 7. These methods have required manual examination of sequences and structures for covariation, but automatic methods have also been devised and demonstrated to produce reliable predictions (Winker et al. 1990; Han and Kim 1993; see box below).

Difficulties of Base Covariation Analysis

The difficulties that are faced in modeling the structure of RNA molecules by covariation analysis are to (1) identify covarying base pairs in a set of related RNA molecules, (2) locate secondary structures based on the covariation analysis, and (3) build a model that represents relationships among the structures. The neighboring box on page 348 describes methods to perform a covariance analysis and model

A

adeno

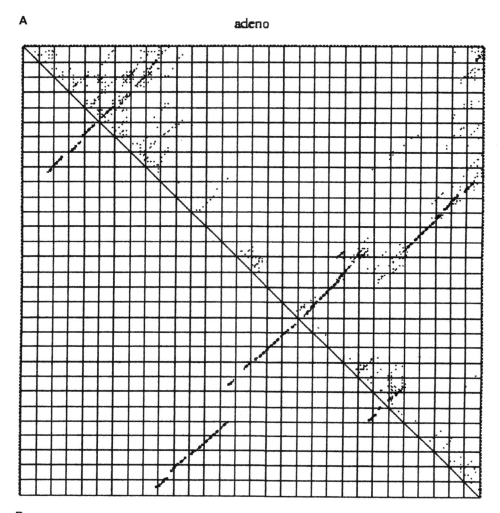

B

```
CKCGGUUCCAGAKGUUGCGCAGCGGCAAAAAGUGCUCCAUGGUCGGGACGCUCUGGCCGG
UCAGGCGCGCGCAGUCGUUGACGCUCUAGACCGUGCAAAAGGAGAGCCUGUAAGCGGGCA
CUCUUCCGUGGUCUGGUGGAUAAAUUCGCAAGGGUAUCAUGGCGGACGACCGGGGUUCGA
ACCCCGGAUCCGGCCGUCCGCCGUGAUCCAUGCGGUUACCGCCCGCGUGUCGAACCCAGG
UGUGCGACGUCAGACAACGGGGGAGCGCUCCUUUUGGCUUCCUUCCAGGCGCGGCGGAUG
. . . . . . . . . . . . . . . .((((((((((. . . . .(((((.((.(((((((((. . . .))))))))).
. . .)).)))))...)))))))).)))....(.((((((..(((((.(((((...(((.(((((.
((((((((((.((((((((((. . . . . .(((((..((.(((((((((((((((.(((((..(((.
. . . .))).)))))).)))))))))))))))).)))))........)))).(((((.((....)
)...)))))).)))))).)))))))))).)))))))))..))))))))))....))))))))....
```

FIGURE 8.10. Suboptimal foldings of an RNA sequence using probability distributions of base-pairings. The first 300 bases of the same adenovirus sequence used in Fig. 8.8 were submitted to the Vienna Web server. (*A*) The region shown represents structures within the range of bases 150–300 and may be compared to the same region in Fig. 8.8. The minimum free energy of this thermodynamic ensemble is –134.85 kcal/mole, compared to a minimum free energy of –125.46 kcal/mole. The size of the square box at highlighted matrix positions indicates the probability of the base pair and decreases in steps of tenfold; i.e., order of magnitude decreases. The size variations shown in the diagram cover a range of ~4–6 orders of magnitude. Calculations of base-pair probabilities are discussed in the text. (*B*) The minimum free-energy structure representing base pairs as pairs of nested parentheses. A low-resolution picture was also produced (not shown).

I. Sequence alignment

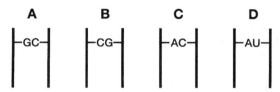

seq 1.   – – – G – – – – – C – – –
seq 2.   – – – C – – – – – G – – –
seq 3.   – – – A – – – – – C – – –
seq 4.   – – – A – – – – – U – – – .

II. Structural alignment

A   B   C   D

⊢GC⊣   ⊢CG⊣   ⊢AC⊣   ⊢AU⊣

FIGURE 8.11. Conservation of base pairs in homologous RNA molecules influences structure prediction. The predicted structure takes into account sequence covariation found at aligned sequence positions, and may also use information about conserved positions in components of a phylogenetic tree. In the example shown, sequence covariations in A, B, and D found in sequences 1, 2, and 4, respectively, permit Watson–Crick base and G-U wobble base-pairing in the corresponding structure, but variation C found in sequence 3 is not compatible. Sometimes correlations will be found that suggest other types of base interactions, or the occurrence of a common gap in an msa may be considered a match. Positions with greater covariation are given greater weight in structure prediction. Molecules with only one of the two sequence changes necessary for conservation of the base-paired position may be functionally deleterious.

A

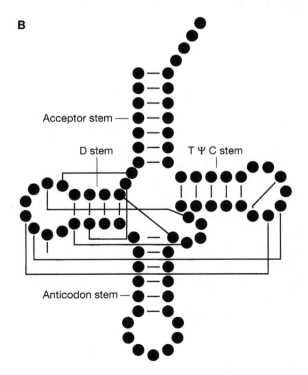

B

FIGURE 8.12. Covariation found in tRNA sequences reveals base interactions in tRNA secondary and tertiary structure. (A) Alignment of tRNA sequences showing regions of interacting base pairs. (+) Transition; (–) transversions; (│) deletion; (*) ambiguous nucleotide. (B) Diagram of tRNA structure illustrating base–base interactions revealed by a covariance analysis. (Adapted, with permission, from the Web site of R. Gutell at http://www.rna.icmb.utexas.edu.)

METHODS OF COVARIATION ANALYSIS IN RNA SEQUENCES

Secondary and tertiary features of RNA structure may be determined by analyzing a group of related sequences for covariation. Two sequence positions that covary in a manner that frequently maintains base-pairing between them provide evidence that the bases interact in the structure. Combinations of the methods in the following list have been used to locate such covarying sites in RNA sequences (see Gutell et al. 1986 for additional details and see http://www.rna.icmb.utexas.edu/).

1. Optimally align pairs of a related family of RNA molecules to locate conserved primary sequence, mark transitions and transversions from a reference sequence, and then visually examine these changes to identify complementary patterns that represent potential secondary structure, as illustrated in Figure 8.12.

2. Perform an msa using one of the programs described in Chapter 5, highlight differences using one of the sequences as a reference, and visually examine the alignment for complementary patterns.

3. Mark variable columns in the msa by numbers that mark changes (e.g., transitions or transversions) from a reference sequence; examine marked columns for a similar or identical number pattern that can represent potential secondary structure.

4. Perform a statistical analysis (χ^2 test) of the number of observations of a particular base pair in columns i and j of the msa, compared to the expected number based on the frequencies of the two bases.

5. Calculate the mutual information score (mixy) for each pair of columns in the alignment, as described in the text and illustrated in Figure 8.13.

6. Score the number of changes in each pair of columns in the alignment divided by the total number of changes (the ec score), examine the phylogenetic context of these changes to determine the number of times the changes have occurred during evolution, and choose the highest scores that are representative of multiple changes.

7. Measure the covariance of each pair of positions in the alignment by counting the numbers of all 16 possible base-pair combinations and dividing by the expected number of each combination (number of sequences × frequency of base in first position × frequency of base in second position), choose the most prevalent pair, and examine remaining combinations for additional covariation; then sum frequency of all independently covarying sites to obtain the covary score.

the structure manually. A better method is to use a computational approach that performs these steps automatically. A computational model for covariance analysis in RNA sequences has been devised by Eddy and Durbin (1994).

To relate the Eddy and Durbin model to other types of sequence models, recall from Chapter 5 that hidden Markov models (HMMs) are used for capturing the patterns of variation observed in an msa, including matches, mismatches, insertions, and deletions. HMMs are based on the assumption that each sequence can be predicted by a series of sequential states in the model that are similar to a series of independent events in a Markov chain. However, HMMs for sequence alignments do not have the features necessary to represent sequence covariation because of secondary structure in RNA molecules.

An ordered tree model is used to represent covariation and secondary structure (but not tertiary structure) in RNA sequences. A simplified tree representation of RNA secondary structure is shown in Figure 8.14. Secondary structural elements are placed at nodes on the tree and the nodes are joined in a branching pattern that reflects relationships among the secondary structures. This use of ordered trees was an important step in the development of new methods for building models of RNA structures.

The tree model in Figure 8.14 depicts a specific arrangement of secondary structures, whereas the goal is to build a model that discovers this arrangement based on covariance analysis of a set of aligned RNA sequences. This goal is achieved by starting with a general tree model and then adjusting the model by training it using covariation information in the sequences.

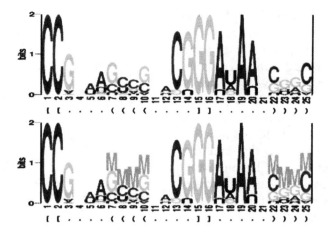

FIGURE 8.13. RNA structure logo. The top panel is the normal sequence logo showing the size of each base in proportion to the contribution of that base to the amount of information in that column of the multiple sequence alignment. The relative entropy method is used in which the frequency of bases in each column is compared to the background frequency of each base. Inverted sequence characters indicate a less than background frequency. The bottom panel includes the same information plus the mutual information content in pairs of columns. The amount of information is indicated by the letter M, and the matching columns are shown by nested sets of brackets and parentheses. All sequences have a C in column 1 and a matching G in column 16. Similar columns 2 and 15 can form a second base pair stacked upon the first. Columns 7–10 and 25–22 also can form G/C base pairs most of the time. Sequences with a G in column 7 frequently have a C in column 25, and those with a C in column 7 may have a G in column 25. Thus, there is mutual information in these two columns (Gorodkin et al. 1997, using data of Tuerk and Gold [1990]).

The model will then represent the location and arrangement of the base-paired regions. The approach is similar to training an HMM to recognize a family of protein sequences. The HMM then represents a multiple sequence alignment of the sequences. In the case of RNA secondary structure, a general tree model is trained by the sequences to produce an ordered tree model similar to that shown in Figure 8.14 that reflects the order of secondary structures found in the sequences.

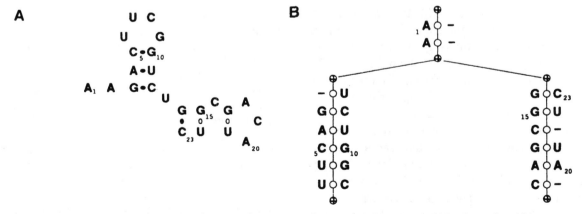

FIGURE 8.14. Tree model of RNA secondary structure. The model in *A* is represented by the ordered binary tree shown in *B*. This model attempts to capture both the sequence and the secondary structure of the RNA molecule. The tree is read like a sequence starting at the root node at the top of the model, then moving down the main branch to the bifurcation mode. Along the main trunk are nodes that represent matched or unmatched base pairs. Shown are two A's matching a "–," indicating no pairing with these bases. After the bifurcation mode, the analysis then moves down the most leftward branch to the end node. Along the branch are unmatched bases, matched base pairs, and mismatched pairs. After the end node is reached, the analysis goes back to the previous bifurcation node and follows the right branch. (Reprinted, with permission of Oxford University Press, from Eddy and Durbin [1994].)

Once available, a covariance model may be used to search newly obtained genome sequences for sequence covariation patterns that produce a high score when aligned to the model. These genomic sequences are likely to encode a similar type of RNA molecule. A covariance model for tRNA has been developed and used for scanning genomes for tRNA genes. Although very accurate when used to represent tRNA genes, the covariance model for tRNA is extremely slow and thus unsuitable for direct searches through large genome sequences. Instead, the tRNA model is used to screen through candidate tRNA genes previously identified by scanning genome sequences with less specific but faster search methods (Lowe and Eddy 1997).

In addition to representing the arrangement of secondary structures in an RNA molecule, the covariance model also needs to represent the types of sequence variations that are found in the secondary structures of each of the RNA sequences, such as insertions, deletions, and mismatches. To allow for such variations, each node in the tree is expanded to include a set of states that corresponds to all of the possible sequence variations that might be encountered in the secondary structure. These states are analogous to the match, insert, and delete states found in an HMM of an msa (see Fig. 5.16, p. 205). For RNA secondary structure, the covariance model uses states that can accommodate the types of sequence variations that are expected in the secondary structures illustrated in Figure 8.2. Examples of the types of states used in a covariance tree model are described in Figure 8.15. As in HMMs, there is a begin (BEG) state for the model and transition probabilities (arrows) give the probability of going from an earlier state to a later state in the model. The bifurcate (BIF) state is unique in representing the end of a secondary structural element and the branching of the molecule into two other secondary structures.

To obtain a model that represents the secondary structure of an aligned class of RNA sequences, the covariance pattern in the sequences is determined by calculating the mutual information content as described in the box on page 351. A covariance tree model similar to that in Figure 8.15, but based on the most significant patterns of covariation in the sequences, is then determined by a dynamic programming method. The states in the model are given initial values of base frequencies or base-pair frequencies, as appropriate for the secondary structure being represented by the model, and as described in Figure 8.15. The transition probabilities are also given

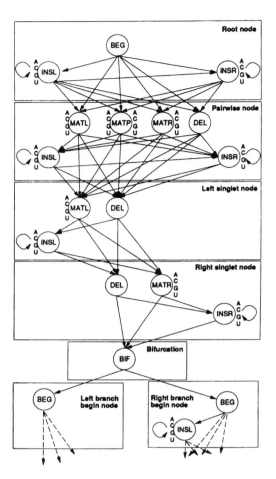

FIGURE 8.15. Details of tree model for RNA secondary structure. Each type of node in the tree shown in Fig. 8.14 is replaced by a pattern of states corresponding to the types of sequence variations expected in a family of related RNA sequences. These states each store a table of frequencies of 4 bases or of 16 possible dinucleotides. The seven different types of nodes are illustrated. BEG node includes insert states for sequence of any length on the right or left side of the node. The pair-wise node includes a state MATP for storing the 16 possible dinucleotide frequencies; MATL and MATR states for storing single-base frequencies on either the left or right side of the node, respectively; a DEL state for allowing deletions; and INSL and INSR states that allow for insertions of any length on the left or right of the node. DEL does not store information. The other five node types have the same types of states. Each state is joined to other states by a set of transition probabilities shown by the arrows. These probabilities are similar to those used in hidden Markov models. BIF is a bifurcation state with transition probabilities entering the state from above and then leaving to one or the other of two branches. (Reprinted, with permission of Oxford University Press, from Eddy and Durbin [1994].)

initial values. State frequencies and transition probabilities of the model are then adjusted by changing the frequencies and probabilities until the model gives a best fit to the sequences. The best fit is determined by measuring the probabilities of all possible paths, each through a different series of states in the model, and finding the highest scoring ones, as described for HMMs in Figure 5.16 (p. 205). These most-probable paths provide a structural alignment of the sequences. The expectation maximization method described in Chapter 5 (p. 198) is used to optimize the state frequencies and transition probabilities of the covariance model.

If a covariation model has been well trained by a sufficient number and variety of RNA sequences that represent sequence variability in a class of RNA molecules, then the model is useful for identifying other RNA molecules that have a similar pattern of base-pairing and secondary structures as that in the training sequences. This method has been useful for developing models for small RNA molecules such as tRNA (Lowe and Eddy 1997) but is not adequate for modeling the structural complexity of larger RNA molecules. The software that produces covariance models for a class of RNA sequences, predicts the structure and alignment of sequences, and performs database searches with covariance models (the COVE set of programs) is available by request from the authors (Eddy and Durbin 1994).

Stochastic Context-free Grammars for Modeling RNA Secondary Structure

Formal language theories have also been used to model base covariation. A general theory for modeling strings of symbols, such as words in sentences, has been developed by linguists. There is a hierarchy of transformational grammars originally developed by Noam Chomsky to represent how grammatical constructs are logically interpreted by the brain. The application of these grammars to sequence analysis has been extensively discussed in detail elsewhere (Durbin et al. 1998) and is only briefly introduced here.

One of these formal grammars, the context-free grammar, is suitable for finding groups of logically related symbols (words) that are separated from each other by other intervening symbols (other words or text). The words do not have to be in the same context to have grammatical meaning; they are context-free. Sequence covariation in RNA is an example of a group

MUTUAL INFORMATION CONTENT

A method used to locate covariant positions in an msa is the mutual information content, defined as the extent of evidence that two columns covary, and calculated as follows. First, for each column in the alignment, the frequency of each base is calculated. Thus, the frequencies in column m, $f_m(B_1)$, are $f_m(A)$, $f_m(U)$, $f_m(G)$, and $f_m(C)$ and those for column n, $f_n(B_2)$, are $f_n(A)$, $f_n(U)$, $f_n(G)$, and $f_n(C)$. Second, the 16 joint frequencies of two nucleotides, $f_{m,n}(B_1, B_2)$, one base B_1 in column m and the same or another base B_2 in column n, are calculated. If the base frequencies in any two columns are independent of each other, the ratio of $f_{m,n}(B_1, B_2)/[f_m(B_1) \times f_n(B_2)]$ is expected to equal 1, and if the frequencies are correlated, this ratio will be greater than 1. If they are perfectly covariant, $f_{m,n}(B_1, B_2) = f_m(B_1) = f_n(B_2)$. To calculate the mutual information content $H(m, n)$ in bits between the two columns m and n, the logarithm of this ratio is calculated and summed over all possible 16 base-pair combinations.

$$H(m, n) = \sum_{B_1, B_2} f_{m,n}(B_1, B_2) \times \log_2 \{f_{m,n}(B_1, B_2)/[f_m(B_1) f_n(B_2)]\}$$

$H(m, n)$ varies from the value of 0 bits of mutual information representing no correlation to that of 2 bits of mutual information, representing perfect correlation and implying that the bases are coevolving (Eddy and Durbin 1994).

The mutual information content may be plotted on an RNA structure logo as illustrated in Figure 8.13 (Gorodkin et al. 1997) and closely related to the sequence logo for a sequence motif, see page 214. The information content at each aligned sequence position and the contribution of each base to the information content at that position are shown in Figure 8.13 (top panel). The perfectly conserved C at positions 1 and 2 and G at positions 15 and 16 can perfectly base-pair ($G_1:C_{16}$ and $G_2:C_{15}$) and have a mutual information content of 2. Four other aligned positions in the RNA sequences (7–10 and 22–25) have mutual information content indicated by a blue M as shown in Figure 8.13 (bottom panel). These bases are predicted to base-pair (pairs 7:25, 8;24, 9:23, 10:22) as indicated by the nested parentheses.

of symbols, in this case nucleotides in RNA, that are not in the same context: A consecutive set of nucleotides in one part of RNA sequences might base-pair with another consecutive set in another part of the sequence, thus forming a secondary structure in the RNA. As a result of the similarity between context-free grammars and RNA covariation, RNA sequences can be represented and analyzed as a context-free grammar.

Another feature of context-free grammars is an ability to accommodate variations (stochasticity) in groups of symbols that are in separate contexts. Stochastic context-free grammars (SCFGs) introduce variability into the definition of each group of symbols, allowing use of alternative symbols. In the evolution of a class of RNA molecules with the same secondary structures, sequence covariation at separated sequence positions maintains the base pairs needed for conservation of secondary structure. Sequence variation at separated positions that maintains base pairs can be readily modeled by SCFGs. SCFGs thus can help define both the types of base interactions in specific classes of RNA molecules and the sequence variations at those positions. SCFGs have been used to model tRNA secondary structure (Sakakibara et al. 1994). Although SCFGs are computationally complex (Durbin et al. 1998), they play an important future role in modeling specific classes of RNA molecules and identifying new members of each class. The application of SCFGs to RNA secondary structure analysis is very similar in form to the covariance models described in the above section "Identification of Base Covariation."

For RNA, the symbols of the alphabet are A, C, G, and U. The context-free grammar establishes a set of rules called productions for generating the sequence from the alphabet, in this case an RNA molecule with sections that can base-pair and others that cannot base-pair. In addition to the sequence symbols (named terminal symbols because they end up in the sequence), another set of symbols (nonterminal symbols) designated $S_0, S_1, S_2 \ldots$, determines intermediate production stages. The initial symbol is S_0 by conven-

tion. The next terminal symbol S_1 is produced by modifying S_0 in some fashion by productions indicated by an arrow. For example, the productions $S_0 \rightarrow S_1$, $S_1 \rightarrow C\ S_2\ G$ generate the sequence C S_2 G, where S_2 has to be defined further by additional productions. The example shown in Figure 8.16 (from Sakakibara et al. 1994) shows a set of productions for generating the sequence CAUCAGGGAAGAUCUCUUG and also the secondary structure of this molecule. The productions chosen describe both features.

In this example of a context-free grammar, only one sequence is produced at each production level. In an SCFG, each production of a nonterminal symbol has an associated probability for giving rise to the resulting product, and there are a set of productions, each giving a different result. For example, the production $S_1 \rightarrow C\ S_2\ G$ could also be represented by 15 other base-pair combinations, and each of these has a corresponding probability. Thus, each production can be considered to be represented by a probability distribution over the possible outcomes. Note the identity of the SCFG representation of the predicted structure to that shown for the tree representation of the covariance model in Figure 8.14.

The use of SCFGs in RNA secondary structure production analysis is in fact very similar to that of the covariance model, with the grammatical productions resembling the nodes in the ordered binary tree. As with hidden Markov models and covariance models, the probability distribution of each production must be derived by training with known sequences. Because the programmatic requirements for aligning a sequence with the SCFG are somewhat different from those used with HMMs, and the time and memory requirements are greater (Sakakibara et al. 1994; Durbin et al 1998), different algorithms are used. The algorithms used for training the SCFG include the CYK algorithm, which is an adaptation for dynamic programming for aligning a sequence to a SCFG, and the inside–outside algorithm. which is an adaptation of the expectation maximization algorithm used for training the SCFG with the sequences (Durbin et al. 1998).

SEARCHING FOR RNA-SPECIFYING GENES

Covariance and SCFG models representing classes of RNA molecules may be used to identify further members of the same RNA class in other sequences, such as those

of newly sequenced genomes. A variety of classes of RNA play regulatory roles in cells, and the computational recognition of these classes is an important activity.

A. Productions

$P = \{ S_0 \longrightarrow S_1,$

$S_1 \longrightarrow C\, S_2\, G,$

$S_2 \longrightarrow A\, S_3\, U,$

$S_3 \longrightarrow S_4\ S_9,$

$S_4 \longrightarrow U\, S_5\, A,$

$S_5 \longrightarrow C\, S_6\, G,$

$S_6 \longrightarrow A\, S_7,$

$S_7 \longrightarrow G\, S_8,$

$S_8 \longrightarrow G,$

$S_9 \longrightarrow A\, S_{10}\, U,$

$S_{10} \longrightarrow G\, S_{11}\, C,$

$S_{11} \longrightarrow A\, S_{12}\, U,$

$S_{12} \longrightarrow U\, S_{13},$

$S_{13} \longrightarrow C \qquad \}$

B. Derivation

$S_0 \longrightarrow S_1 \longrightarrow CS_2G \longrightarrow CAS_3UG \longrightarrow CAS_4S_9UG$

$\longrightarrow CAUS_5AS_9UG \longrightarrow CAUCS_6GAS_9UG$

$\longrightarrow CAUCAS_7GAS_9UG \longrightarrow CAUCAGS_8GAS_9UG$

$\longrightarrow CAUCAGGGAS_9UG \longrightarrow CAUCAGGGAAS_{10}UUG$

$\longrightarrow CAUCAGGGAAGS_{11}CUUG$

$\longrightarrow CAUCAGGGAAGAS_{12}UCUUG$

$\longrightarrow CAUCAGGGAAGAUS_{13}UCUUG$

$\longrightarrow CAUCAGGGAAGAUCUCUUG.$

C. Parse tree

D. Secondary structure

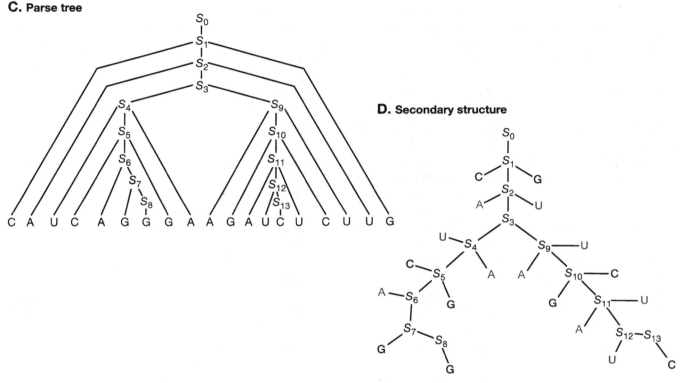

FIGURE 8.16. A set of transformation rules for generating an RNA sequence and the secondary structure of the sequence from the RNA alphabet (ACGU). (*A*) The set of production rules for producing the sequence and the secondary structure. These rules reveal which bases are paired and which are not paired. (*B*) Derivation of the sequence. (*C*) A parse tree showing another method for displaying the derivation of the sequence in *B*. (*D*) Secondary structure from applying the rules. (Redrawn, with permission of Oxford University Press, from Sakakibara et al. [1994].)

mRNAs that encode proteins may be identified computationally by the presence of exons containing runs of amino-acid-specifying codons, known as open reading frames. Ribosomal RNAs and transfer RNAs are also identifiable by patterns of complementary base pairs that produce their characteristic structures. There are also additional classes of RNA that play important structural and regulatory roles in gene expression by making chromosomal DNA (chromatin) nontranscribable and regulating translation of mRNAs into proteins (see Rivas and Eddy 2001). Some RNAs contain small coding regions and others are non-protein-coding (see

Olivas et al. 1997). Noncoding RNAs sometimes act alone and sometimes in a complex with protein. They act on specific target molecules through base complementarity with the target.

One class of RNAs, the MicroRNAs (miRNAs), are small noncoding RNA gene products about 22 nucleotides long that seem to regulate translation (Eddy 2001; Rivas and Eddy 2001). These RNAs are collected in the Rfam database of RNA families (Griffiths-Jones et al. 2003). Small RNA molecules designated RNAi with a sequence that is complementary to that of a specific mRNA sequence are routinely prepared in the laboratory and used for inhibiting expression of specific genes in target cells. (See Hannon [2003] for more information.)

One goal in RNA research has been to design methods to identify sequences in genomes that encode small RNA molecules. Larger, highly conserved molecules can simply be identified in genomic sequences on the basis of their sequence similarity with already known sequences. For smaller RNA sequences with more sequence variation, this method does not work. A number of methods for finding small RNA genes in genomic sequences have been described and are available on the Web (Table 8.1). A major problem with these methods in searches of large genomes is that a small false-positive rate becomes quite unacceptable because there are so many false positives to check out.

One of the first methods used to find tRNA genes in the context of genomic sequence was to search for sequences that are self-complementary and can fold into a hairpin like the three found in tRNAs (Staden 1980). Fichant and Burks (1991) described a program, tRNAscan, that searches a genomic sequence with a sliding window simultaneously for matches to a set of invariant bases and conserved self-complementary regions in tRNAs with an accuracy of 97.5%. Pavesi et al. (1994) derived a method for finding the RNA polymerase III transcriptional control regions of tRNA genes using a scoring matrix derived from known control regions that is also very accurate. Finally, Lowe and Eddy (1997) have devised a search algorithm, tRNAscan-SE, that uses a combination of three methods to find tRNA genes in genomic sequences— tRNAscan, the Pavesi algorithm, and the COVELS program based on sequence covariance analysis (Eddy and Durbin 1994). This method is reportedly 99–100% accurate with an extremely low rate of false positives.

Some RNA molecules are best modeled by a composite model of SCFGs to recognize base-paired regions and HMMs to recognize lengths of sequence that are not involved in base-pairing and may, for instance, interact with other molecules. An example is shown in Figure 8.17. Each SCFG component is trained to recognize one region of secondary structure and each HMM component, one region of the RNA that is not in secondary structure. The individual models are then joined into a composite model that represents the arrangement of the components in the RNA sequences. If there is more than one arrangement, a branch in the model is necessary, as shown in Figure 8.17 following component (2). The transition probabilities of going to one branch or the other are then determined from the proportion of each type of arrangement in the training sequences. Candidate sequences are aligned with the model starting at the beginning of the model, then followed by aligning with each structural component. The probabilities of all possible paths through the sequence of states in the model are calculated. If the sequence can be aligned with the model, one of these paths will generate a much higher probability than the other paths and also other sequences, a result that indicates a successful alignment.

The composite model shown in Figure 8.17 was used to identify small nucleolar (snoRNAs) in the yeast genome that methylate ribosomal RNA. The model is not used to search genomic sequences directly. Instead, a list of candidate sequences is first found by searching for sequences that are similar to the RNA sequences (Lowe and Eddy 1999). The probability model is a hybrid combination of HMMs and SCFGs trained on snoRNAs. These RNAs vary sufficiently in sequence and structure that they are not found by straightforward similarity searches. The RNAs found in the yeast genome were proven to be snoRNAs in the laboratory by insertional mutagenesis. Composite models of this type for RNAs continue to play an important role in the analysis of RNA sequences. A model of prokaryotic genes based in part on SCFGs to represent mRNA secondary structure has been developed that identifies bacterial genes (Brockhorst et al. 2003).

RNA STRUCTURAL MODELING PRODUCES IMPORTANT RESULTS

As genomic sequences of organisms become available, it is important to be able to identify the various classes of genes, including the major class of genes that encodes RNA molecules. There are a large number of Web sites, including those listed in Table 8.1, that provide programs and guest sites for RNA analysis or for access to databases of RNA molecules and sequences. These molecules perform a variety of important biochemical functions, including translation; RNA splicing, processing, and editing; cellular localization; and regulation of gene expression. Like protein-encoding genes, RNA-specifying genes may

FIGURE 8.17. Probabilistic model of snoRNAs. The numbered boxes and ovals represent conserved sequence and structural features that have been modeled by training on snoRNAs. Secondary structural features of Stem were modeled with an SCFG. Boxes with ungapped hidden Markov models, the guide sequence with a hidden Markov model, and gapped regions

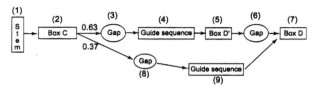

(spacers) are shown by ovals. The guide sequence interacts with methylation sites on rRNA and is targeted in each search to a complementary sequence near one of those sites. The alignment of this model produces a log odds score that provides an indication of the reliability of the match. The transition probabilities are 1, except where the model bifurcates to allow identification of two types of target sequences. The model is highly specific and seldom identifies incorrect matches in random sequences. (Reprinted, with permission, from Lowe and Eddy, 1999 [©AAAS, Washington, D.C.].)

be identified by using the unknown gene as a query sequence for DNA sequence similarity searches, as described in Chapter 6. If a significant match to the sequence of an RNA molecule of known structure and function is found, then the query molecule should have a similar role. For some small molecules, the amount of sequence variation necessitates the use of more complex search methods.

One important result already obtained using covariation analysis of RNA molecules is that of Carl Woese, who built detailed structural models for rRNAs. Woese examined the evolutionary variation in these structures and was able to predict a new model of three domains of life—the Bacteria, the Eukarya, and a newly identified Archaea. Although a large amount of horizontal DNA transfer among evolutionary lineages of other genes has added a great deal of noise to the evolutionary signal, this rRNA-based prediction is supported by other types of genome analyses.

In addition to these uses of rRNA structural analysis, excellent probabilistic models of two small RNA molecules, tRNA and snoRNA, have been built, and these models are used to search reliably through genomic sequences for genes that encode these RNA molecules. The successful analysis of these types of RNA molecules should be readily extensible to other classes of RNA molecules.

The use of laboratory-produced RNAi molecules to inhibit the expression of specific genes is playing an important role in analyzing a variety of biological processes, including gene regulation and expression, metabolic pathways, and response to environmental stress, including drugs and chemicals. RNA structure plays a significant role in the activity of these molecules.

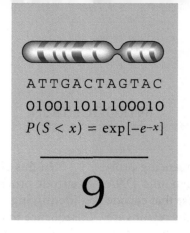

Gene Prediction and Regulation

CONTENTS

INTRODUCTION

With the advent of whole-genome sequencing projects, it has become routine to scan genomic DNA sequences to find genes, particularly those that encode proteins. The most likely protein-encoding regions are identified in a new genomic sequence and the predicted proteins are then subjected to a database similarity search to assess their function, as described in Chapter 6. This procedure is summarized in the gene prediction flowchart (p. 374). The genomic DNA sequence displayed on a Web page is then annotated with sequence positions marked to show the exon–intron structure and location of each predicted gene.

In this chapter, methods of predicting genes that encode proteins are discussed, and then methods for identifying regulatory regions in the genomic sequence that regulate the activity of protein-encoding genes are discussed. The prediction of genes that specify classes of RNA molecules is discussed in Chapter 8. A more extended review of genome analysis is presented in Chapter 11. There are a sizable number of computer programs and Web sites for gene prediction. Some representative ones are given in Table 9.1 (p. 368). General and specific Web search terms are provided at the end of the chapter.

CHAPTER GUIDE FOR BIOLOGISTS

This chapter describes several types of computational tools for gene finding and promoter analysis. Hidden Markov models (HMMs) were used previously in Chapter 5 to model multiple sequence alignments (msa's). Here, they are used to model gene structure—the expected occurrence of a series of codons in an open reading frame with an intervening spacer region and, in the case of eukaryotic genes, alternating exons and introns with splice junctions between them. You can think of these gene HMMs as representing a trip across the country in which there are alternative routes and places to stay. If we keep track of the routes taken by a lot of people in a survey, we come up with the most common route that people take, and we can even include the most common motels that are used at each stopover. We can also calculate probabilities for each choice that is made throughout the route as to which direction to go or where to stay on arrival. Similarly, if we keep track of a lot of gene sequences for a particular organism—going from one sequence position to the next and choosing a particular base at each location, skipping certain positions or adding more—we can come up with the best paths that are followed by most gene sequences, representing the conserved sequence positions, e.g., preferred codons and splice sites, in that organism. Our gene sequence survey will allow us to calculate probabilities that any new sequence is indeed a gene, just as we can predict the most likely trip that one would take across the country based on collected data from previous trips.

Another computational tool introduced in this chapter is the neural network (NN). NNs simulate the expected pattern of communication among nerve cells. In NNs representing sequences, certain computer memory positions simulating receptors receive one type of input signal from sequences. In the case of gene prediction, the signal is a series of 1s and 0s that represent some kind of sequence measurement, say GC content or presence of a sequence pattern found in genes, e.g., a hexamer that represents two commonly found adjacent codons. Those cells receiving 1s then transmit the signal to another layer of cells. Each cell in this second layer of cells receives hundreds of signals from all of the original input cells. These signals are integrated by each second layer cell, and, if there is a strong enough composite signal, the cell transmits a signal to yet another cell that is chosen to provide an output signal. After receiving signals from the previous layer of cells, the output cell integrates these signals and sends an output signal, an exon score that indicates whether or not the input sequence actually represents a gene. As with HMMs, the secret to NNs is to train the network of cells to recognize that the input actually represents a gene sequence. Training involves gradually adjusting the signals between cells until a pattern of communication is found that gives a correct answer. The trained network can then be applied to test sequences to decide whether or not there is a gene present.

A much simpler method than NNs for discriminating gene sequences from other sequences is also used. Imagine that we take all of the same information that is fed into the NN and just plot the values on a graph, say two at a time, although we can plot more axes or dimensions if we wish. We plot, first, the values for known coding sequences and, second, those for noncoding sequences. The points on the graph are labeled coding or noncoding. The graph is then examined to see which values give the best separation of the coding and noncoding points. Quite straightforward statistical methods can then be used to find the set of values that best discriminates between the coding and noncoding sequences. As before, the values found to be critical for discrimination can then be used on an unknown sequence to find any coding regions.

Additional computational tools for promoter analysis are also described in this chapter. Coordinately regulated genes, such as a set of genes expressed at a particular stage of development, in a particular tissue, or in a set of gene expression experiments using microarrays, are expected to share binding sites for transcription factors. The simplest method for representing a conserved sequence pattern in such binding sites, the position-specific scoring matrix (PSSM), was used previously for representing the variations found in msa's described in Chapter 5. The columns represent the position in the msa and the rows represent the sequence choices—A, G, C, or T. Recall that the scoring matrix has log odds scores that represent the odds of finding a particular base at that position in related sequences to that in random or unrelated sequences. For promoters, the scoring matrix represents a conserved but variable pattern, presumably the binding site of a transcription factor. If there is a similar pattern in a test sequence, a high log odds score will be found at that location using the scoring matrix.

Because promoter sequences are so variable, one cannot make an msa and then represent conserved regions of local homology by a scoring matrix or HMM as one can do for proteins. The problem is, therefore, to find patterns without having an msa. Two statistical methods, expectation maximization and Gibbs sampling, which can also be used to find common patterns in unaligned sequences (Chapter 5), are also used to find common regulatory patterns. There are some simple examples in Chapter 5 (p. 198, and problems 4 and 5 at the end of that chapter) that describe how these methods work. Another method takes every possible occurrence of a pattern (say all hexamers) in each common regulatory sequence, aligns one pattern from each sequence into a trial msa, and then evaluates many combinations for a meaningful pattern, using as a guide the principles of information theory described in Chapter 5. It is not computationally possible to evaluate all combinations, so one starts to look for a promising pair in the first two sequences, and then attempts to add the best ones sequentially from the other sequences in an approach called a greedy algorithm. Once all of the patterns have been found in a set of sequences by one of the above methods, one or more PSSMs can be made to represent them.

Another method for finding conserved patterns in the promoters of coordinately regulated genes is to do a "word count" of the sequences. The various 4-mers, 5-mers, etc., are counted. The search can include both strands, palindromes that read the same on both DNA strands (e.g., the complement of 5′-AAGCTT-3′ is 5′-AAGCTT-3′), overlapping words, and regions upstream and downstream from the promoter. The object is to find words that are overrepresented statistically in the gene set and might, therefore, represent a common regulatory signal in the promoters.

CHAPTER GUIDE FOR COMPUTATIONAL SCIENTISTS

This chapter describes methods for finding patterns in long DNA sequences that predict the presence of genes. These sequences primarily arise in large projects to sequence all of the DNA sequence—the genome—of an organism. The DNA sequence within the gene can then be translated three bases at a time (a codon) into the amino acid sequence of the encoded protein using the genetic code. The protein then folds into a three-dimensional structure and is often further modified into its biologically active form. Of the two strands in double-stranded DNA, either one can carry the sequence information for a gene, but only one strand encodes a particular gene. The other strand is a complementary sequence (A opposite T and G opposite C), which is used to make more gene copies during replication of the molecule. The DNA is only read in one chemical direction, which is referred to as 5′ to 3′.

The cell locates genes through the presence of short patterns that are present on either DNA strand, usually near the beginning of the protein-coding sequences in a region called the promoter, but regulatory patterns can also be located a considerable distance away from the start of the gene at other sites called enhancers. Proteins known as transcription factors bind to these regulatory sequences and can also bind to each other to make a specific structure, or complex, with the DNA. Another complex of proteins known as RNA polymerase II (RNAPII) then interacts with the DNA–protein structure and makes a messenger RNA (mRNA) copy of the complementary DNA strand using specific base-pairing to generate the correct sequence. The mRNA may carry extra sequences, called introns, that do not encode proteins and will have extra sequences before the start codon (the 5′ leader sequence) and after the termination codon to which a large number of A's are added in eukaryotic cells [the poly(A) tail].

Intron sequences are spliced away by processing the mRNA, thus leaving the exons that provide a continuous read of the codons for the protein. It is then the job of the ribosome and transfer RNAs described in earlier chapters to translate the mature mRNA into the amino acid sequence of the protein. In cells with a nucleus (eukaryotic cells found in plants and animals), the mRNA is produced in the nucleus and then migrates to the cytoplasm where it is translated into protein.

The computational methods used to predict genes depend on features in gene sequences not found in other sequences. The simplest is the presence of stretches of unusually long series of codons interrupted in most eukaryotes by introns. Specific patterns of five or six, and sometimes longer, bases are also abundant in the exons of a particular organism. Splice junctions between introns and exons have short conserved patterns of a few base pairs that are also diagnostic. Prediction of genes can be quite error-prone because of alternative possible intron locations. In prokaryotic cells (lacking a nucleus and introns), the prediction can be more reliable. A close match between the protein that would be encoded by the predicted gene and an already known protein sequence is often used to substantiate the prediction.

As described in the text, biologists can also isolate the mature, spliced mRNAs lacking introns from cells, copy the sequence onto DNA molecules, and then randomly sequence the DNA molecules in reads of several hundred base pairs, called expressed sequence tags or ESTs. Large collections of ESTs can then be aligned with the exons of the genome sequence and also examined for overlaps in an attempt to build a contiguous sequence for the mRNA of the gene. The contig is then translated into the protein sequence. For example, for the estimated 35,000 genes in the human genome, there are over 100,000 EST contigs in a set called the human UNIGENE set (see http://www.ncbi.nlm.nih.gov). This set is dynamically updated as more sequences are added to the EST database.

WHAT SHOULD BE LEARNED IN THIS CHAPTER?

- What are the patterns found in genomic sequences that indicate the presence of protein-encoding genes?
- How is the reliability of gene predictions assessed?
- The uses of HMMs, neural networks, and other methods that discriminate between DNA sequences that specify protein sequences and those that do not.
- The complexity of gene regulation, particularly in eukaryotic cells, and the types of DNA–protein and protein–protein interactions.
- How to discover and to use models to represent sequence patterns found in promoters.

Glossary Terms

Chromatin Nuclear DNA, usually in a compact form, complexed with proteins and organized into chromosomes that are sometimes visible.

Codon A series of three base positions in an mRNA sequence that is translated by the cell into an amino acid position in a protein.

Discriminant function A method for distinguishing objects based on a set of measurements. For gene prediction, gene sequences in the genome can be distinguished from other sequences that are not gene sequences by variation in sequence patterns and base frequencies.

Enhancer DNA sequences that act to stimulate the transcription of a gene, often from a remote location on the DNA sequence.

Eukaryotes Organisms that have a nucleus in the cell where much of the DNA is stored. The rest of the cell is cytoplasm surrounded by a membrane.

Exon DNA or mRNA sequences that include a series of codons carrying the information for a part of the amino acid sequence of a protein. In most sequences, exons are separated by introns, which are removed from mRNAs by RNA splicing.

Expectation maximization A sequence analysis method that locates a set of similar sequence patterns in a group of sequences; in this chapter, DNA sequences that have the same regulatory function. Trial patterns identified in each sequence are aligned and represented by a scoring matrix (expectation step). Each sequence is then scanned for the best matches to the matrix and the patterns and matrix are updated to improve the matches (maximization step). This process is then repeated up to hundreds of times until there is no further improvement.

Expressed sequence tag (EST) A 200- to 500-base fragment of the mRNA sequence of a gene that is sequenced from a random collection of mRNA fragments, often from the 5′ or 3′ ends.

Genomic DNA The sequence of one full haploid set of chromosomes of an organism. It includes sequences specifying the amino acid sequence of proteins (including introns if present) and sequences specifying RNA molecules. Other noncoding DNA sequences including highly repetitive sequences may also be abundant, as described in Chapter 11.

Gibbs sampling A sequence analysis method that, in this chapter, locates a set of similar sequence patterns in a group of DNA regulatory sequences. Trial patterns identified in all but one of the sequences are aligned and represented by a scoring matrix. The left-out sequence is then scanned for best matches to the matrix, a random best match is chosen and aligned with a new set of sequence patterns but one, and a new matrix is made. This process is then repeated up to hundreds of times with the goal of recruiting patterns resembling those from the left-out sequence until there is no further improvement.

Greedy algorithm A computational method that follows a series of convenient steps which do not necessarily lead to the globally best solution to a problem but rather to a locally optimal one that may or may not also be the best or optimal one. Greedy methods are used for searching regulatory DNA sequences for similar patterns, but they may not be the most similar ones in the sequences.

Hidden Markov model (HMM) For gene models, a series of states representing sequential positions in the genome sequence starting with a "BEGIN" and stopping with an "END" state are included. There may be alternative paths between states representing codons, introns, and other gene features. Each of these sequence matching states carries "hidden" information on the likelihood of each base being found, and the paths between states have probabilities that reflect the types of variation found in known gene sequences.

Interpolated Markov model (IMM) A type of hidden Markov model for locating genes in genome sequences in which DNA patterns longer than the five bases commonly used (representing two neighboring codons), but also characteristic of gene sequences for an organism, are included in the model.

Introns Sections of sequence that lack codons and are present in the genome sequence of many eukaryotic genes and initially transcribed mRNAs of those genes. Introns can be a very large fraction of the gene sequence. They are removed from mRNAs by RNA splicing.

Messenger RNA (mRNA) mRNA is copied from one strand of DNA in a region that encodes the sequence information of a protein and then is usually spliced in eukaryotic cells to remove introns, leaving behind the joined exons that include a series of codons specifying the amino acid sequence of a protein.

Neural network In bioinformatics applications in this chapter, a series of layers of "neurons" starting with an input layer that reads data from a sequence window, downstream layers that receive information from the initial layers and integrate this information, and an output layer that reports whether or not the sequence is a gene sequence. The many connections between the layers are adjusted using known gene sequences for training so that the correct output is reported for a variety of known sequences. The network can then be used to predict new gene sequences.

Open reading frame (ORF) A length of a genome sequence that is a series of amino-acid-specifying triplet codons. The ORF is commonly initiated by a codon for the amino acid methionine and ended by one or more termination codons. In any DNA sequence, there are six possible reading frames, three in the forward and three in the reverse direction, and starting at the first, second, or third base position at the ends of the sequence.

Perceptron A model of a set of DNA sequence patterns in which the individual sequence positions are read, multiplied by weighting factors, and then summed to produce a predictive score for the pattern. The weights are adjusted by using training patterns that have some function of interest. The perceptron is like a neural network without a hidden layer. Once trained, it may be used to locate similar patterns in query sequences.

Prokaryotes Organisms that do not have a nucleus in the cell, often single-celled microorganisms.

Promoter The upstream region of a protein-encoding region with short sequence patterns that are first recognized and bound by specific transcription factors. The resulting DNA–protein complex may then be recognized (or not) by RNA polymerase II, leading to transcription of the downstream sequence into mRNA. Promoters of genes

specifying RNA molecules are recognized by other RNA polymerases and have different structures.

PSSM A position-specific scoring matrix (see scoring matrix).

RNA polymerase II (RNAPII) The RNA polymerase that recognizes promoters of protein-encoding genes in eukaryotic cells.

Scoring matrix A tabular representation of the variation in a sequence pattern with the columns (or rows) representing pattern position and rows (or columns) representing the sequence variation found as an ODDS score (occurrence of each base in the pattern compared to background).

Threshold score A score that defines a boundary between two classes of objects as, for example, DNA sequences that are promoter sequences and others that are not.

Transcription factor (TF) A protein whose function is to bind to specific genomic sequences upstream (on the 5′ chemical side) from the gene sequence. The bound proteins then interact with other factors to create an environment that is either favorable for transcription (induction) or unfavorable (repression) by RNA polymerase II. TFs are themselves activated by chemical modification by other proteins, often in a cascade that starts at the surface of the cell (the cell membrane) in response to external signals such as hormones.

Untranslated region Regions at the 5′ and 3′ chemical ends of an mRNA molecule that include regulatory information but do not encode the sequence of a protein.

PROKARYOTIC AND EUKARYOTIC BIOLOGY AFFECT OPEN READING FRAME–BASED GENE PREDICTION

The purpose of gene prediction is to identify regions of genomic DNA that encode proteins. Computational methods for gene prediction work by searching through sequences to locate the most likely ones that encode proteins. These methods are trained to recognize a variety of sequence patterns in a set of known gene sequences of an organism. Once trained, the resulting gene model may then be used to search for similar patterns in newly acquired sequences from the same organism.

Genome sequencing centers often search through newly acquired sequences with gene prediction pro-

grams and then annotate the sequence database entry with this information. This annotation includes gene location, gene structure (positions of predicted exons/introns and regulatory sites), and any matches of the translated exons with the protein sequence databases. The amino acid sequence of the predicted gene may also be entered in the protein sequence databases.

Genes are also identified by isolating mRNAs from organisms in which they have been spliced; the mRNAs are reverse-translated into a cDNA copy, and then the cDNA is sequenced. The advantage of this method is that the mRNA has already been spliced by the cell to remove introns so that the only remaining prediction to be made is the start and stop codons of the open reading frame. Because it is so much trouble to keep track of every gene, methods have been devised to make cDNA copies of an entire mRNA preparation and then sequence the set randomly. The sequence fragments from individual sequence reads are called expressed sequence tags (ESTs). The task is to find enough ESTs for each expressed gene so that there is a significant degree of overlap. A consensus alignment of most or all of the mRNA for each gene can then be determined using msa methods described in Chapter 5 (see p. 170). Sometimes, mRNA goes through different types of splicing in different tissues, and sometimes mRNA is edited to produce sequence changes. EST information can greatly assist with confirming the accuracy of gene prediction models.

In prokaryotic genomes, DNA sequences that encode proteins are transcribed into mRNA, and the mRNA is usually translated directly into proteins without significant modification. Because prokaryotic organisms generally lack introns, sequences that encode proteins may be quite readily identified as the longest open reading frames (ORFs) going from a start codon (e.g., AUG in mRNA), through a series of amino acid–specifying codons and ending at one or more chain-termination codons (e.g., UAA). There may be several potential starting points for the prokaryotic gene, but the prediction can be further refined by recognizing common sequence patterns at the start of known genes (see Table 9.1 for Web sites that provide a more detailed analysis).

There are six possible reading frames in which to find ORFs in every sequence, three starting at positions 1, 2, and 3 in the 5′ to 3′ direction of a given sequence, and another three starting at positions 1, 2,

and 3 in the 5′ to 3′ direction of the complementary sequence. A reading frame of a genomic sequence that does not encode a protein will have short ORFs because of the presence of many in-frame stop codons. An example of a search of the *Escherichia coli lac* operon for ORFs is shown in Figure 9.1. In the case of *E. coli* and its phages, these predictions have to take into account the following possibilities: (1) the presence of multiple genes on mRNA and (2) sometimes the presence of overlapping genes in which two different proteins may be encoded in different reading frames of the same mRNA, either on the same or complementary DNA strands.

In eukaryotes, prediction of protein-encoding genes is a more difficult task because of their more complicated transcriptional biology. Here transcription of protein-encoding regions is initiated at specific promoter sequences and then is followed by removal of noncoding sequence (introns) from pre-mRNA by a splicing mechanism, a process that leaves the protein-encoding exons. Once the introns have been removed and certain other modifications to the mature RNA have been made, the resulting mature mRNA can be translated in the 5′ to 3′ direction, usually from the first start codon to the first stop codon. Thus, as a result of the presence of intron sequences in the genomic DNA sequences of eukaryotes, the ORF corresponding to an encoded gene will be interrupted by the presence of introns that usually generate stop codons.

Based on the lengths, composition, and characteristic sequence patterns in the introns of a particular organism, computational models may be built that can recognize introns and the boundaries of exons and introns. The intron sequences can then be removed, thus joining the exons in a long ORF that can be translated into a predicted protein sequence.

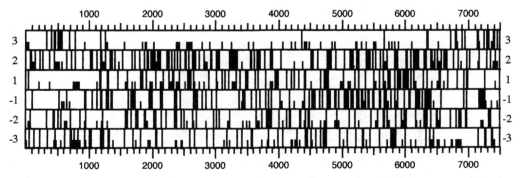

FIGURE 9.1. ORF map of a portion of the *E. coli lac* operon using the DNA STRIDER program (Marck 1988). Shown are AUG and termination codons as one-half and full vertical bars, respectively, in all six possible reading frames. The *lacZ* gene is visible as an ORF that runs from positions 1284 to 4355 in frame 3.

TABLE 9.1. *Programs and Web pages for sequence translation and related information*

Name of translation site	Web address	Reference
Arabidopsis intron splice site table	http://www.nal.usda.gov/pgdic/Probe/v2n3/codon.html	see Web site
Codon usage database	http://www.kazusa.or.jp/codon/	see Web site
EST-GENOME for alignment of EST/cDNA and genomic sequences at MRC	http://www.hgmp.mrc.ac.uk/Registered/Option/est_genome.html	see Web site; also see Florea et al. (1998)
FGENES and related programs that use linear discriminant analysis or hidden Markov models	http://searchlauncher.bcm.tmc.edu/seq-search/gene-search.html	Solovyev et al. (1995); see Web site
Gene and Promoter Analysis for plants and prokaryotes	http://mendel.cs.rhul.ac.uk/mendel.php	see Web site
FINEX—exon–intron boundary	http://www.sanger.ac.uk/cgi-bin/finex/finex_search.pl	Brown et al. (1995)
GeneFinder (hexon, fgenes, Grail, and others) collection of methods	http://searchlauncher.bcm.tmc.edu/seq-search/gene-search.html	see Web site
Genehacker for microbial genomes based on HMMs	http://www-btls.jst.go.jp/GeneHacker/	Hirosawa et al. (1997)
GeneID-3 Web server using rule-based models, and GeneID1	http://www1.imim.es/geneid.html	Guigó et al. (1992); Guigó (1998)
GeneMark and GeneMark.hmm[a] uses HMMs	http://opal.biology.gatech.edu/GeneMark/; http://www2.ebi.ac.uk/genemark/	Lukashin and Borodovsky (1998); Borodovsky and McInnch (1993)
GeneParser Web page, uses combination of neural network and dynamic programming methods	http://obesitygene.pbrc.edu/~eesnyder/geneparser.htm	Snyder and Stormo (1993, 1995)
Genetic code variations	http://www.ncbi.nlm.nih.gov/Taxonomy/taxonomyhome.html/	see Web site
Genie for finding human and *Drosophila* genes by HMMs and neural networks	http://www.cse.ucsc.edu/~dkulp/cgi-bin/genie; http://www.fruitfly.org/seq_tools/genie.html	Kulp et al. (1996); Reese et al. (1997, 2000a,b)
GenLang using linguistic methods	http://www.cbil.upenn.edu/genlang/genlang.html	Dong and Searls (1994)
GenScan based on probabilistic model of gene structure for vertebrate, *Drosophila*, and plant genes	http://genes.mit.edu/GENSCAN.html	Burge and Karlin (1997, 1998)
GenSeqr for aligning genomic and EST sequences	http://bioinformatics.iastate.edu/cgi-bin/gs.cgi	see Web site and Splicepredictor
GeneSplicer for detection of splice sites based on Markov models	http://www.tigr.org/software/	Pertea et al. (2001)

Description	URL	Reference
Glimmer uses interpolated Markov models for prokaryotic translation; also GlimmerM for eukaryotes	http://www.tigr.org/software/	Salzberg et al. (1998)
GrailII prediction by neural networks based on scores of characteristic sequence patterns and composition	http://compbio.ornl.gov/public/tools/	Uberbacher and Mural (1991); Uberbacher et al. (1996)
Human splice sites with decision tree analysis	http://industry.ebi.ac.uk/~thanaraj/splice.html	Thanaraj (1999)
INFOGENE: a database of known gene structures and predicted genes	http://genomic.sanger.ac.uk/inf/infodb.shtml (Commercial source)	Solovyev and Salamov (1999)
MZEF uses quadratic discriminant analysis for human, mouse, *Arabidopsis*, and *S. pombe* exons	http://rulai.cshl.edu/	Zhang (1997)
NetGene uses neural networks for analysis of splice sites in human, *C. elegans*, and *Arabidopsis* genes	http://www.cbs.dtu.dk/services/NetGene2/; http://www.cbs.dtu.dk/services/NetPGene/; http://www.cbs.dtu.dk/services/NetStart/	Brunak et al. (1991); Hebsgaard et al. (1996); see NetGene
Procrustes based on comparison of related genomic sequences	http://hto-13.usc.edu/software/procrustes/index.html	Gelfand et al. (1996)
Splice Predictor for plants uses trained logitlinear models	http://bioinformatics.iastate.edu/cgi-bin/sp.cgi	Brendel and Kleffe (1998); Brendel et al. (1998)
Splicing Sites by neural network at LBNL	http://www.fruitfly.org/seq_tools/splice.html	see Genie
Translate tool at ExPASy	http://us.expasy.org/tools/dna.html	see Web site
Translation machine on the Web at EBI	http://www2.ebi.ac.uk/translate/	see Web site
Veil (Viterbi exon–intron locator) uses HMMs for vertebrate DNA	http://www.tigr.org/~salzberg/veil.html	Henderson et al. (1997)
Webgene, a set of gene prediction tools and concurrent database similarity searches	http://www.itba.mi.cnr.it/webgene/	see Web site
Yeast splice site database by M. Ares Jr. laboratory	http://www.cse.ucsc.edu/research/compbio/yeast_introns.html	Spingola et al. (1999)

Abbreviations: (LBNL) Lawrence Berkeley National Laboratory, (EBI) European Bioinformatics Institute.
[a] The GeneMark.hmm program is designed to use additional information at the 5' end of bacterial sequences.

TABLE 9.2. *The universal or standard genetic code*

UUU-Phe	F	UCU-Ser	S	UAU-Tyr	Y	UGU-Cys	C
UUC-Phe	F	UCU-Ser	S	UAU-Tyr	Y	UGU-Cys	C
UUA-Leu	L	UCA-Ser	S	UAA-TER		UGA-TER	
UUG-Leu	L	UCG-Ser	S	UAG-TER		UGG-Trp	W
CUU-Leu	L	CCU-Pro	P	CAU-His	H	CGU-Arg	R
CUC-Leu	L	CCU-Pro	P	CAU-His	H	CGC-Arg	R
CUA-Leu	L	CCA-Pro	P	CAA-Gln	Q	CGA-Arg	R
CUG-Leu	L	CCG-Pro	P	CAG-Gln	Q	CGG-Arg	R
AUU-Ile	I	ACU-Thr	T	AAU-Asn	N	AGU-Ser	S
AUC-Ile	I	ACC-Thr	T	AAC-Asn	N	AGC-Ser	S
AUA-Ile	I	ACA-Thr	T	AAA-Lys	K	AGA-Arg	R
AUG-MET	M	ACG-Thr	T	AAG-Lys	K	AGG-Arg	R
GUU-Val	V	GCU-Ala	A	GAU-Asp	D	GGU-Gly	G
GUC-Val	V	GCC-Ala	A	GAC-Asp	D	GGC-Gly	G
GUA-Val	V	GCA-Ala	A	GAA-Glu	E	GGA-Gly	G
GUG-Val	V	GCG-Ala	A	GAG-Glu	E	GGG-Gly	G

Shown are each codon and the three-letter and one-letter codes for each encoded amino acid. AUG is the usual START (AUG-MET) codon and the three TER codons cause translational termination.

Four types of posttranscriptional events influence the translation of mRNA into protein and the accuracy of gene prediction:

1. The genetic code of a given genome may vary from the universal code (see Table 9.1 for reference Web sites). For the most part, the universal genetic code, shown in Table 9.2, is used.

2. One tissue may splice a given mRNA differently from another, thus creating two similar but also partially different mRNAs encoding two related but partially different proteins, thus allowing cells to use the same gene for different purposes (Lopez 1998). Understanding the molecular interactions between RNA and the RNA-binding proteins that perform these modifications is an area of active investigation. Availability of this information will assist in the prediction of such variations.

3. mRNAs may be edited, changing the sequence of the mRNA and, as a result, the encoded protein (see, e.g., Gray and Covello 1993; Paul and Bass 1998; Morse and Bass 1999). Such changes also depend on interaction of RNA with RNA-binding proteins.

4. mRNAs are subject to chemical modifications that lead to their degradation (Steiger and Parker 2002). Thus, the amount of mRNA in the cell is determined both by synthesis and degradation steps.

THE RELIABILITY OF AN OPEN READING FRAME PREDICTION CAN BE TESTED

The following characteristics of protein-encoding sequences allow testing of ORFs to verify that they are likely to encode a protein:

1. DNA sequences that encode protein are not a random chain of available codons for an amino acid, but rather an ordered list of specific codons that reflect the evolutionary origin of the gene and constraints associated with gene expression. This nonrandom property of coding sequences can be used to advantage for finding regions in DNA sequences that encode proteins (see Fickett and Tung 1992).

2. Each species also has a characteristic pattern of use of synonymous codons; i.e., codons that

TABLE 9.3. *Codon usage table*

UUU-Phe	16.6	26.0	UCU-Ser	14.5	23.6	UAU-Tyr	12.1	18.8	UGU-Cys	9.7	8.0
UUC-Phe	20.7	18.2	UCC-Ser	17.7	14.2	UAC-Tyr	16.3	14.7	UGC-Cys	12.4	4.7
UUA-Leu	7.0	26.3	UCA-Ser	11.4	18.8	UAA-TER	0.7	1.0	UGA-TER	1.3	0.6
UUG-Leu	12.0	27.1	UCG-Ser	4.5	8.6	UAG-TER	0.5	0.5	UGG-Trp	13.0	10.3
CUU-Leu	12.4	12.2	CCU-Pro	17.2	13.6	CAU-His	10.1	13.7	CGU-Arg	4.7	6.5
CUC-Leu	19.3	5.4	CCC-Pro	20.3	6.8	CAC-His	14.9	7.8	CGC-Arg	11.0	2.6
CUA-Leu	6.8	13.4	CCA-Pro	16.5	18.2	CAA-Gln	11.8	27.5	CGA-Arg	6.2	3.0
CUG-Leu	40.0	10.4	CCG-Pro	7.1	5.3	CAG-Gln	34.4	12.2	CGG-Arg	11.6	1.7
AUU-Ile	15.7	30.2	ACU-Thr	12.7	20.2	AAU-Asn	16.8	36.0	AGU-Ser	11.7	14.2
AUC-Ile	22.3	17.1	ACC-Thr	19.9	12.6	AAC-Asn	20.2	24.9	AGC-Ser	19.3	9.7
AUA-Ile	7.0	17.8	ACA-Thr	14.7	17.7	AAA-Lys	23.6	42.1	AGA-Arg	11.2	21.3
AUG-MET	22.2	20.9	ACG-Thr	6.4	8.0	AAG-Lys	33.2	30.8	AGG-Arg	11.1	9.3
GUU-Val	10.7	22.0	GCU-Ala	18.4	21.1	GAU-Asp	22.2	37.8	GGU-Gly	10.9	23.9
GUC-Val	14.8	11.6	GCC-Ala	28.6	12.6	GAC-Asp	26.5	20.4	GGC-Gly	23.1	9.7
GUA-Val	6.8	11.7	GCA-Ala	15.6	16.2	GAA-Glu	28.6	45.9	GGA-Gly	16.4	10.9
GUG-Val	29.3	10.7	GCG-Ala	7.7	6.1	GAG-Glu	40.6	19.1	GGG-Gly	16.5	6.0

Shown are frequency of each codon per 1000 codons obtained from http://www.kazusa.or.jp/codon/ for *Homo sapiens,* columns 2, 5, 8, and 11, and for *Saccharomyces cerevisiae,* columns 3, 6, 9, and 12.

stand for the same amino acid (Table 9.3) (Wada et al. 1992).

3. There are also different patterns of use of codons in strongly versus weakly expressed genes, as, for example, in *E. coli*. Also in *E. coli*, there is a strong preference for certain codons to be neighbor pairs (codon pairs) within a coding region and for certain codons to be next to the termination codon. Some of this preference is due to constraints in amino acid sequences in proteins and some to the influence of a given codon on the translation of neighboring codons (Gutman and Hatfield 1989). There is also a strong preference for codon pairs in eukaryotic exons that has been very useful for distinguishing exons and introns in eukaryotic genomic DNAs, as described later in this chapter.

4. Organisms with a high genome content of GC have a strong bias of G and C in the third codon position (for review, see Von Heijne 1987; Rice et al. 1991).

On the basis of these characteristics of protein-encoding sequences, three tests of ORFs have been devised to verify that a predicted ORF is in fact likely to encode a protein (Staden and McLachlan 1982; Staden 1990). The first test is based on an unusual type of sequence variation that is found in ORFs; namely, that every third base tends to be the same one much more often than by chance alone (Fickett 1982). This property is true for any ORF, regardless of the species. The program TESTCODE, which is available in the Genetics Computer Group (GCG) suite of programs (http://www.gcg.com), provides a plot of the nonrandomness of every third base in the sequence. An example of TESTCODE output is shown in Figure 9.2.

The second test is an analysis to determine whether the codons in the ORF correspond to those used in other genes of the same organism (Staden and McLachlan 1982). For this test, information on codon use for an organism averaged over all genes is necessary, such as shown in Table 9.3 for human and yeast genes. In addition, there may be variations in codon use by different genes of an organism providing a type of gene regulation. An example of the analysis of an *E. coli* gene for the presence of more and less frequently used *E. coli* codons is shown in Figure 9.3. A parameter that reflects the frequency of codon use may also be calculated, as in the Genetics Computer Group CODONFREQUENCY program.

Third, the ORF may be translated into an amino acid sequence and the resulting sequence then compared to the databases of existing sequences. If one or more sequences of significant similarity are found, there will be much more confidence in the predicted ORF (Gish and States 1993). The predicted protein may also be confirmed by identification in the laboratory.

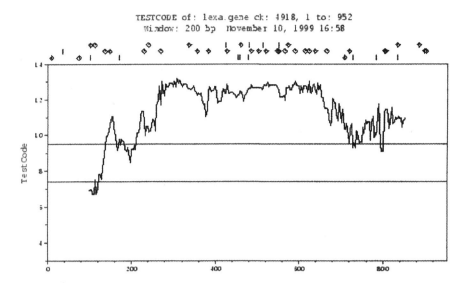

FIGURE 9.2. TESTCODE analysis of the *E. coli lexA* gene, which is known to extend from positions 102 to 707 in the sequence shown. The TESTCODE statistic (Fickett 1982; for comparison, see Staden 1990) was plotted for each base position in a sliding window of 200 nucleotides. The TESTCODE statistic is found in the following way: (1) The number of each base is counted at every third position starting at positions 1, 2, and 3, and going to the end of the sequence window; (2) the asymmetry statistic for each base is calculated as the ratio of the maximum count of the three possible reading frames divided by the minimum count for the same base plus 1; (3) the frequency of each base in the window is also calculated; (4) the resulting asymmetry and frequency scores are then converted to probabilities of being found in a codon region (found from an analysis of known coding and noncoding regions); and (5) the probabilities are multiplied by weighting factors that are summed. Weighting factors are chosen so that the resulting sum best discriminates coding from noncoding sequences. A value of >0.95 classifies the sequence as coding, and <0.74 classifies the sequence as noncoding. These cutoff values are indicated by orange horizontal lines. TESTCODE was run and displayed using TESTCODE in the Genetics Computer Group suite of programs. Above the plot, short vertical lines indicate possible start codons, and diamonds indicate possible stop codons.

EUKARYOTIC GENES HAVE REPEATED SEQUENCE ELEMENTS THAT PROBABLY REFLECT NUCLEOSOME STRUCTURE

Gene expression in eukaryotes is influenced by the availability of DNA to proteins (transcription factors) that bind to regulatory sites on DNA. Eukaryotic DNA is wrapped around histone–protein complexes called nucleosomes and forms a complex known as chromatin in the nucleus of the eukaryotic cell. The DNA helix has two grooves along the side of the molecule—one (the major groove) wider than the other (the minor groove). Some of the base pairs in the major or minor grooves of the DNA molecule face the nucleosome surface and others face the outside of the structure. Binding sites for some proteins that regulate transcription may therefore be hidden on the inside of the structure. Nucleosomes located in the promoter region are remodeled in a manner that can

influence the availability of binding sites for regulatory proteins, making them more or less available (Carey and Smale 2000).

What is interesting computationally about the nucleosome model is that repeated patterns of sequence have been found in introns and exons and near the start site of transcription of eukaryotic genes by HMM analysis (Baldi et al. 1996; for a detailed analysis, see Baldi and Brunak 1998; see also Chapter 5, p. 204) and other types of pattern-searching methods (Ioshikhes et al. 1996). These sequences appear to be strongly correlated with the position of nucleosomes and are not found in prokaryotic DNA (Stein and Bina 1999). An example of the HMM is shown in Figure 9.4. These nucleosome sequence patterns appear with a periodici-

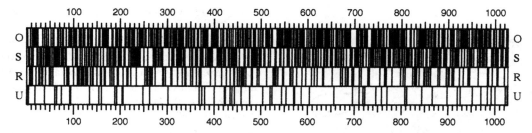

FIGURE 9.3. Analysis of *E. coli lacZ* gene for occurrence of frequent and infrequent codons using the codon adaptation analysis feature of DNA STRIDER. The positions of common (O for optimum), less common (S for suboptimal), rare (I), and unique (U, which includes the three stop codons, the AUG Met codon, and the UGG Trp codon) codons along the sequence are shown, starting at the first nucleotide in the sequence and analyzing three at a time. These first three classes correspond, respectively, to codon adaptation values (Sharp and Li 1987) of >0.9, 0.1–0.9, and <0.1. The gene is obviously represented by commonly used codons.

ty of 10; that is, the number of base pairs expected in a single turn of the DNA double-stranded helix around a nucleosome. The patterns found in promoter sequences are those that bend more easily when located in the major groove of DNA and are thought to be located on the inside of the bent molecule (Ioshikhes et al. 1996; Pederson et al. 1998), as shown in Figure 9.5. Using these observations, a model has been proposed that sequence patterns located downstream from the

transcription start site are suitable for positioning of nucleosomes, whereas upstream regions do not show the necessary patterns (Pederson et al. 1998).

Loops of chromatin are attached to the nuclear matrix by relatively short (100–1000 bp long) sequences called matrix attachment regions (MARs) or scaffold-associated regions (SARs), regions that are considered to be an indicator of the presence of expressed genes. Although the sequence of only a

FIGURE 9.4. A hidden Markov model (HMM) of eukaryotic internal exons. This HMM is designed to detect a statistically significant frequency of the same base at intervals of 10 bp in sequences. Imagine feeding an exon sequence into the part of the sequence shown by the heaviest arrow at 11 o'clock on the circle and then threading the sequence clockwise around the circle, noting the base at each subsequent position in the sequence, and recording that information in the corresponding box (the state of the HMM). If there is a small repeated pattern of a few bases at every tenth position in the sequence starting at the same position from the start of the exon sequence, the distribution of bases in some of the boxes will begin to reflect that pattern. Hence, there is a repeated pattern of not-T (i.e., A, C, or G), A or T, then a G. By a slightly more sophisticated analysis similar to that discussed in Chapter 5 (p. 205), the model can be used to show that the same pattern may start at other positions with respect to the start of the sequence (other arrows feeding into the circle) and also that some sequence positions in the circle may be skipped (arrows going around some of the states) or extra sequence may be found (loop arrow returning to same state). A similar pattern is found in introns and also around the start site of transcription. This structure is modulated by histone-modifying systems as one means of gene regulation in eukaryotes. (Redrawn, with permission, from Baldi et al. 1996 [©1996 Elsevier].)

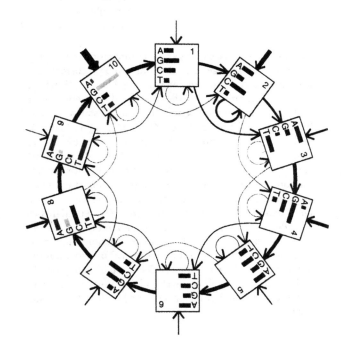

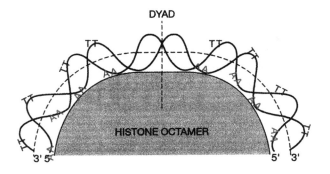

DYAD

HISTONE OCTAMER

3' 5'   5' 3'

FIGURE 9.5. A proposed role for the repeated sequence patterns in eukaryotic genes. Shown is the portion of a DNA molecule wrapped around a nucleosome. The actual length of DNA around the nucleosome will be approximately 145 bp. The repeated patterns found in eukaryotic genes (including not-T, A or G, G) and AA/TT dinucleotides influence the bendability of the DNA strand in which they are located, and hence will facilitate the folding of DNA around a nucleosome. (Redrawn, with permission, from Ioshikhes et al. 1996 [©1996 Elsevier].)

small number of such regions has been determined, several characteristic sequence patterns have been identified. The program MAR-FINDER (see Table 9.6, below, for Web site) searches for sequences that have a high representation of such sites in genomic DNA (Singh et al. 1997).

WHAT IS THE PROCEDURE FOR GENE PREDICTION?

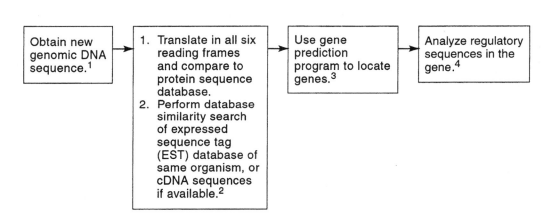

Obtain new genomic DNA sequence.[1]

1. Translate in all six reading frames and compare to protein sequence database.
2. Perform database similarity search of expressed sequence tag (EST) database of same organism, or cDNA sequences if available.[2]

Use gene prediction program to locate genes.[3]

Analyze regulatory sequences in the gene.[4]

1. Because the standards for gene identification are not uniform, and because gene predictions can be incorrect, it is a good idea to reconfirm any gene prediction of interest, perform alignments of the predicted sequence with matching database sequences to confirm statistical and biological significance (as described in Chapters 4 and 6), and confirm the predicted gene sequence by cDNA sequencing. If EST sequences are available in a sufficient coverage of the genome, these are also useful for confirmation of predicted gene sequences. For an example of the gene annotation procedure that was followed for the *Drosophila melanogaster* genome sequence, see Adams et al. (2000). The final goal of the gene annotation procedure for an organism is to produce a genome database that includes a rich supply of biological information on the function of each gene, as discussed in Chapter 11. This information will come from laboratory experimentation and manual entry of relevant published data into the genome database.

2. Database similarity searches of this type are described in the flowchart of Chapter 6. For genes of prokaryotic organisms, step 1 identifies ORFs that encode a protein similar to one found in another organism. ORFs without a similar gene in another organism may also be found, as described in the text. Genes of eukaryotic organisms often have intron and exon sequences in the genomic DNA sequence. Step 1 provides the approximate locations of exons that encode a protein similar to one in another organism. Eukaryotic genomes may also have ORFs that do not match a database sequence, and these ORFs may or may not encode a protein. In the Genome Annotation Assessment Project (GASP) of the *Drosophila* genome, one study showed that combining gene prediction methods with homology searches generally provides a reliable annotation method (Birney and Durbin 2000). Step 2 is an additional type of database similarity search that identifies protein-encoding ORFs. Because cDNA sequences and partial cDNA sequences correspond to

exons, genomic ORFs that can be aligned to these expressed gene sequences include exon sequences. This analysis can be enhanced by using databases of indexed genes in which overlapping ESTs have been identified (see flowchart, Chapter 6). EST_GENOME is a program for aligning EST and cDNA sequences to genome sequences (Table 9.1; see also Table 5.1). Collections of EST sequences for an organism are often only partial collections; thus, failure to find a matching EST is not a sufficient criterion for rejecting an ORF by this test. Searching the EST collections of related organisms, e.g., another mammal or plant, may be helpful in identifying such missing EST sequences. An additional type of gene analysis is to use an already-identified ORF as a query sequence in a database search against the entire proteome (all of the predicted proteins) of an organism to find families of paralogous genes, as described in Chapter 11.

3. There are a large number of gene prediction programs available (Table 9.1). They all have in common to varying degrees the ability to differentiate between gene sequences characteristic of exons, introns, splicing sites, and other regulatory sites in expressed genes from other non-gene sequences that lack these patterns. Because these gene sequences as well as gene structure (the number and sizes of exons and introns) vary from one organism to another (see Fig. 11.5), a program trained on one organism, e.g., the bacterium *E. coli* or the worm *Caenorhabditis elegans*, is not generally useful for another organism, e.g., another bacterial species or the fruit fly *D.*

melanogaster. Reliability tests of gene prediction programs have shown that the available methods for predicting known gene structure are, in general, error-prone. Referring to Web sites with this information (Table 9.1) or performing one's own reliability check is recommended. Some "reliability checks" should be eyed with suspicion because they are based on a comparison of new predictions with previous ones for which there is no experimental support. When gene predictions are made using gene-sized, rather than large-sized, multigene sequence genomic DNA fragments, the predictions are generally more reliable (see text).

4. In prokaryotes, the predicted genes may have conserved sequence patterns such as those for promoter recognition by RNA polymerases and transcription factors, for ribosomal binding to mRNA, or for termination of transcription, as found in the model prokaryote *E. coli*. Similarly, in eukaryotes, the region at the 5′ end of the gene may also have characteristic sequence patterns such as a high density and periodicity of putative transcription-factor-binding sites and sequence patterns characteristic of RNA polymerase II promoters. These types of analyses are enhanced by searching for similar sequence patterns in genes that are regulated by the same set of environmental conditions or that are expressed in the same tissue. Regulatory predictions are enhanced when information about conserved oligomers found in the promoters of coregulated genes is available, as described in the text.

GENE PREDICTION IS EASIER IN MICROBIAL GENOMES

As discussed earlier, predicting protein-encoding genes is generally easier in prokaryotic than in eukaryotic organisms because prokaryotes generally lack introns and because several quite highly conserved sequence patterns are found in the promoter region and around the start sites of transcription and translation, at least in the *E. coli* model of prokaryotes. When a set of different patterns characteristic of a gene are found in the same order and with the same spacing in an unknown sequence, the prediction is more reliable than if only one pattern is found.

Using Gene Regulatory Sequences

This type of information can be obtained in *E. coli*. An example of the regulatory sequences for an *E. coli*

gene, the *lexA* gene, is shown in Figure 9.6. Note the presence of the −10 and −35 regions (light brown) that mark the site of interaction with RNA polymerase, and the ribosomal binding site on the mRNA product (green) that is complementary to the ribosomal RNA. The ORF that encodes the LexA product is also indicated (blue). Also shown are three potential binding sites for the encoded LexA product to the promoter region, recognizable by searching for a consensus of known LexA-binding sites. Note that these sites are inverted repeats; i.e., the sequence on the forward and reverse sequence is approximately the same. This feature with minor variations is not uncommon in the binding sites of proteins that regulate transcription and is a reflection of the binding of a dimer of LexA protein to the two sites, which produces a stronger

DNA PATTERNS IN THE *E. coli lexA* GENE

GENE SEQUENCE	PATTERN
1 GAATTCGATAAATC TCTGGTTTATTGTGC AGTTTATGGTT	CTGN NNNNNN NNNNC AG
TT	TTGACA
41 CCAAAATCGCCTTTTGC TG TATATACTCACAGCATAACTG	CTGN NNNNNN NNNNC AG
CCAA -35 -10 TATACT >	TATAAT, > mRNA start
81 TATA TACAC CCAGGGGGCGGAATGAAAGC GTTAACGGCC A	CTGN NNNNNN NNNNC AG
+10 GGGGG Ribosomal binding site	GGAGG
121 GGCAACAAGAGGTGTTTGATCTCATCCGTGATCACATCAG	
161 CCAGACAGGTATGCCGCCGACGCGTGCGGAAATCGCGCAG	ATG
201 CGTTTGGGGTTCCGTTCCCCAAACGCGGCTGAAGAACATC	
241 TGAAGGCGCTGGCACGCAAAGGCGTTATTGAAATTGTTTC	
281 CGGCGCATCACGCGGGATTCGTCTGTTGCAGGAAGAGGAA	
321 GAAGGGTTGCCGCTGGTAGGTCGTGTGGCTGCCGGTGAAC	
361 CACTTCTGGCGCAACAGCATATTGAAGGTCATTATCAGGT	OPEN READING FRAME
401 CGATCCTTCCTTATTCAAGCCGAATGCTGATTTCCTGCTG	
441 CGCGTCAGCGGGATGTCGATGAAAGATATCGGCATTATGG	
481 ATGGTGACTTGCTGGCAGTGCATAAAACTCAGGATGTACG	
521 TAACGGTCAGGTCGTTGTCGCACGTATTGATGACGAAGTT	
561 ACCGTTAAGCGCCTGAAAAAACAGGGCAATAAAGTCGAAC	
601 TGTTGCCAGAAAATAGCGAGTTTAAACCAATTGTCGTTGA	
641 CCTTCGTCAGCAGAGCTTCACCATTGAAGGGCTGGCGGTT	
681 GGGGTTATTCGCAACGGCGACTGGCTGTAACATATCTCTG	TAA
721 AGACCGCGATGCCGCCTGGCGTCGCGGTTTGTTTTTTCATC	
761 TCTCTTCATCAGGCTTGTCTGCATGGCATTCCTCACTTCA	
801 TCTGATAAAGCACTCTGGCATCTCGCCTTACCCATGATTT	
841 TCTCCAATATCACCGTTCCGTTGCTGGGACTGGTCGATAC	
881 GGCGGTAATTGGTC ATCTTGATAGCCCGGTTTATTTGGGC	
921 GGCGTGGCGGTTGGCGCAACGGCGGACCAGCT	

Shown are matches to approximate consensus binding sites for LexA repressor (CTGNNNNNNNNNNNCAG), the -10 and -35 promoter regions relative to the start of the mRNA (TTGACA and TATAAT), the ribosomal binding site on the mRNA (GGAGG), and the open reading frame (ATG...TAA). Only the second two of the predicted LexA binding sites actually bind the repressor.

FIGURE 9.6. The promoter and open reading frame of the *E. coli lexA* gene.

interaction than binding of a single monomer to a single site. The sites in the *lexA* promoter region represent a form of down self-regulation. The two downstream sites have been shown to bind the regulatory LexA protein and to act as a repressor that prevents further transcription. The binding at two sites may be cooperative in that two dimer molecules are more effective at preventing transcription than one, possibly because the bound proteins interact, thus making the overall binding to the promoter region stronger.

LexA protein binding is a model for gene regulation in prokaryotes; however, regulation can also be up instead of down. In the case of a number of other genes, binding of a regulatory protein such as LexA to a recognizable target sequence activates transcription by stimulating the binding of RNA polymerase. The consensus patterns for these various regulatory sites may be found by sequence alignment and statistical and neural network methods. These methods are discussed in Chapters 5 and 6, and also later in this chapter. Ribosomal binding sites were the first to be modeled by a neural network with no hidden layer (or perceptron), which is discussed below (Stormo et al. 1982; Bisant and Maizel 1995).

Using Highly Conserved Gene Features

The highly conserved features of *E. coli* genes can be employed in gene identification methods, for example, use of HMMs. In this approach, an *E. coli* gene is modeled and the model then expanded to include multiple genes and the sequences between the genes. The model shown in Figure 9.7 is an example of a simple HMM of a bacterial genome as a DNA molecule that is densely packed with genes with relatively short intergenic sequences and no introns. This

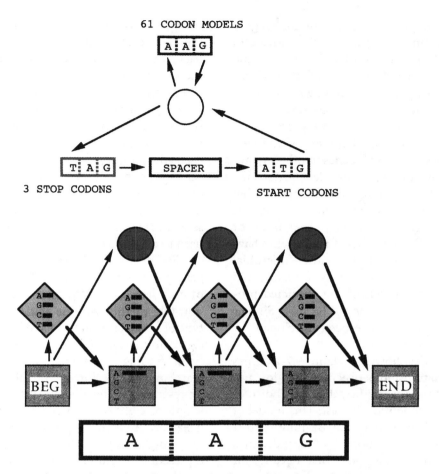

FIGURE 9.7. HMM of an *E. coli* gene (Krogh et al. 1994). This model is designed to generate a sequence of amino-acid-encoding codons of the approximate length of an *E. coli* gene starting with an ATG codon and ending with a stop codon. A set of predicted genes are separated by intergenic spacer regions of the range of lengths actually found between *E. coli* genes. Variations in this basic model are described in the text. The model is first trained on a set of known *E. coli* gene sequences with flanking sequences. The training step is performed in very much the same manner as that described in Chapter 6 for multiple sequence alignment. The trained model may then be used to find the most probable set of genes in *E. coli* genomic sequences of unknown gene composition. The model for each codon (lower part of diagram) is represented by a set of round, diagonal, or square boxes representing match, insert, and delete states, respectively. The model shown is that for the AAG codon. Each of the 61 codons has a similar structure. If a sequence were extremely accurate, only match states would be needed in the model. The insert and delete states allow an ORF with an extra or missing base to be recognized. Similarly, the inclusion of alternative bases in each match state allows for errors in base identification. Stop codons and initiation codons are assumed to be correctly represented in each sequence and no allowance for errors is made. Hence, errors in these codons would lead to an incorrect prediction. Each match and insert state has a certain probability of producing an A, another probability for producing a G, and so on. The delete state does not produce a letter but instead acts to skip a sequence position. Directional arrows (transitions) give the probability of passing from one state to another in the model. Thus, if one state generates an A with probability of 1.0, the transition probability to the next state is 0.9, and the next state generates a G with probability 0.98, then the probability of AG is 1.0 × 0.9 × 0.98 = 0.88. The model is entered at any position (upper part of diagram) and the arrows designate possible paths through the model between successive states. The central state represented by a circle does not generate a sequence position but acts as a junction between adjacent codons. For the model to generate a sequence, the probability of a codon following another codon must be quite high. Hence, the transition probability of going from the junction to a codon is much higher than for going to a stop codon. Once a stop codon has been reached, a sequence representing an intergenic spacer region is generated. Within this region is a model for sequences that are found upstream from the ATG codon for the next gene, such as the Shine-Dalgarno ribosomal binding site and other sequence information (see Hayes and Borodovsky 1998). The presence of this sequence increases the probability for a downstream gene. (*Note:* Compare this gene model with the model for protein alignments shown in Fig. 5.16, p. 205.)

model will read through a sequence of unknown gene composition and find the genes, i.e., a series of codons that specify amino acids flanked by start and stop codons, that are most like a set of known gene sequences and flanking regions that have been used to train or calibrate the model. Because codon usage and flanking sequence will probably vary from one genome to the next, the model trained with *E. coli* genes may not work for finding genes in other organisms. The reliability of the model depends on the accuracy of the gene start and stop information that is used for the training or calibration step of the HMM and on the number of such genes used for training. For *E. coli,* the positions of many genes have been accurately determined. For other microbial genomes, this information is not as available, and genes predicted by alignment of the predicted proteins with *E. coli* proteins have to be used. Similar models of gene structure have been developed for other microbial genomes.

The HMM model shown in Figure 9.7 assumes that there is no relationship between each codon and later codons in the sequence; i.e., that the choice of each codon is independent of the rest. This model of genes as a Markov chain may not be fully correct because there may be both short- and long-distance correlations between some positions because of requirements for mRNA structure or translation. However, useful gene models can be produced using this simplifying assumption. Analyses of sequential codons in genes have shown that some pairs are found at a greater frequency and others at a lesser frequency than expected by chance alone (Gutman and Hatfield 1989; Farber et al. 1992). Hence, a more appropriate choice is to design a model that uses sequence information from the previous five instead of the previous two bases to make what is called a fifth-order Markov model, which is a model that analyzes the sequence variation in five consecutive bases in DNA. In such a model, the frequency of hexamers is used to differentiate between coding and noncoding sequences. A version of GeneMark (Borodovsky and McIninch 1993) called GeneMark.HMM uses a HMM of this type to search for *E. coli* genes (Lukashin and Borodovsky 1998).

Using hexamer (6-mer) patterns instead of shorter sequence patterns to locate genes increases the ability of a gene prediction program to distinguish coding from noncoding regions. From an information perspective, as the number of consecutive sequence posi-

tions being compared in two sequences is increased, the chance of being able to find similarities above background noise increases. For example, when using the dot matrix method for comparing sequences, a sliding window in which words of length *n* are compared is used to locate the most significant matches. In comparing coding and noncoding sequences, a comparison of three consecutive positions at a time can be used to find ORFs as uninterrupted runs of amino-acid-specifying codons. Thus, extending the number of positions to a number greater than 3, such as 4–6, increases the chances of discovering higher-order sequence correlations in coding sequences that may be used to distinguish them from noncoding sequences.

For fifth-order Markov models to give accurate gene predictions, there must be many representatives of each hexameric sequence in genes, and if there is not, the method will be statistically limited. A new type of model, the interpolated Markov model (IMM; e.g., Glimmer; see Table 9.1), overcomes this difficulty of finding a sufficient number of patterns by searching for the longest possible patterns that are represented in the known gene sequences up to a length of eight bases. Thus, if there are not enough hexameric sequences, pentamers or smaller may be more highly represented, and in other cases many representative patterns even longer than six bases may be found. In general, the longer the patterns, the more accurate the prediction. The IMM combines probability estimates from the different-sized patterns, giving emphasis to longer patterns and weighting more heavily the patterns that are well represented in the training sequences (Salzberg et al. 1998).

Both GeneMark.HMM and IMM find genes in microbial genomes with an apparent high degree of accuracy, assuming that gene predictions made by other methods such as sequence similarity of the translated proteins to known *E. coli* proteins are accurate. Therefore, these methods can be expected to produce reliable predictions of genes that do not match previously identified protein sequences. A further improvement of the prediction of the bacterial start codon position has been found (Hannenhalli et al. 1999). This method sorts through a set of predictions for the start codon in a set of sequences, where the actual signal is known. These predictions depend on weighting each of the elements of a set of input sequence information. The weights are adjusted so that the predictions are made more accurate by a method called mixed integer programming.

GENE PREDICTION IN EUKARYOTES DEPENDS ON RECOGNIZING KNOWN EXONS

The presence of introns in most eukaryotic genes, and the variation in intron size and number from one organism to another, greatly increase the difficulty of gene prediction in eukaryotes. One method scans DNA sequences in all possible reading frames and predicts genes based on matches of the predicted proteins to known proteins in the sequence databases. Other computational methods are used to locate characteristic sequence patterns of genes in an organism. The resulting gene models are used to predict the sequence and location of additional genes.

Performing Sequence Database Searches

A simple method for discovering protein-encoding genes within a prokaryotic or eukaryotic genomic sequence is to perform a sequence database search by translating cloned sequences of up to several hundred kilobases in all possible reading frames and comparing the sequence to a protein sequence database using the BLASTX or FASTX programs described in Chapter 6. Alternatively, if a genomic sequence is to be scanned for a gene encoding a particular protein, the protein can be compared to a nucleic acid sequence database that includes genomic sequences and is translated in all six possible reading frames by the TBLASTN or TFASTX/ TFASTY programs. For proteins that are highly conserved, these methods can give a very good, albeit approximate, indication of the gene structure. If the proteins are not highly conserved, or if the exon structure of a gene is unusual, these methods may not work.

Additional information regarding the locations of genes in genomic DNA sequences may be found by using cDNA sequences of expressed genes (see flowchart). An enhanced method (Pachter et al. 1999) for finding eukaryotic genes rapidly is to prepare a dictionary of sequence words (4-letter words in a protein sequence database, 11-letter words in an EST database) and to use these dictionaries to compare a genomic DNA sequence to the expressed gene and protein sequence databases.

Recognizing Known Exons

The most commonly used methods for eukaryotic gene prediction depend on training a computer program to recognize sequences that are characteristic of

known introns in genomic DNA sequences. The program is then used to predict the positions of introns in the genomic sequences being analyzed, and the resulting exons are joined into a predicted gene structure. Predictions depend on analysis of a variety of sequence patterns that are characteristic of known genes in a particular organism. These include patterns characteristic of exons, intron–exon boundaries, and upstream promoter sequences. As more sequences are collected for specific organisms and the actual structures of additional genes become known, these prediction methods should become more reliable. However, patterns that specify RNA splice sites (i.e., intron–exon boundaries) are poorly conserved with only a few identical positions. Therefore, the positions of intron–exon boundaries cannot be defined precisely by simple pattern-searching methods.

Finding Complex Patterns

Neural networks (described below and in Chapter 10) provide a method of sequence analysis that has the capability of finding complex patterns and relationships among sequence positions that may not be obvious. The available methods also depend on the analysis of windows of sequence in genomic DNA to determine whether these regions are likely to be coding or noncoding. Regions that encode proteins are found to have characteristic patterns reflecting preferential codon usage and codon neighbors. These observations have led to the widely used analysis of 6-mers in DNA sequences as a basis for gene prediction.

Using RNA Polymerase II–transcribed Genes

Gene prediction methods are primarily focused on genes that encode proteins. In eukaryotes, these genes are transcribed by an enzyme complex known as RNA polymerase II (RNAPII). Transcription factors bind to the upstream promoter region of these genes and, depending on regulatory interactions among these factors, RNAPII may bind to the promoter and transcribe the gene into pre-mRNA. The pre-mRNA copy will include both exons with coding information for the amino acid sequence of the protein and intervening noncoding introns. Based on training with known exon and intron sequences, gene prediction programs give possible locations of exons that can then be joined to predict the sequence of the mature mRNA of the gene following RNA splicing. This sequence will include an upstream 5′ region (5′ untranslated region, 5′ UTR) extending from the start site of transcription to the initiation codon, the ORF for the protein ending in a translational termi-

nation codon, and the downstream region (3′ UTR) extending to the termination of transcription in the region where the signal for polyadenylation of the mRNA may be found. The initiation site for translation in eukaryotic mRNAs is usually the AUG codon nearest the 5′ end of the mRNA, but sometimes downstream AUG codons still close to the 5′ end of the mRNA may also be used (Kozak 1999).

NEURAL NETWORKS AND PATTERN DISCRIMINATION METHODS CAN BE USED TO PREDICT EUKARYOTIC GENES

As examples of the types of analyses that are available, two types of gene prediction methods—neural networks and pattern discrimination methods—are described below. Other methods and Web sites for finding genes in eukaryotic DNA are described in Table 9.1.

Neural Networks

Neural networks are a type of computational model comprising a collection of connected neuron-like elements that read sequences and pass information as signals to other neurons; the final recipient neuron makes predictions about the sequences. The signals between neurons are chosen by training with sequences of known function—in this case coding sequences—so that the network recognizes other sequences with that same function—in this case, the sequence of a eukaryotic gene that encodes a protein. (*Note:* Compare the use of neural networks for gene prediction with that for protein secondary structure prediction shown in Fig. 10.29, p. 463.)

Grail

Grail at http://grail.lsd.ornl.gov/ provides analyses of protein-coding regions, poly(A) sites, and promoters; constructs gene models; and predicts encoded protein sequences. GrailEXP is a newer version that has been used at Oak Ridge National Laboratory to annotate the human genome. This suite of programs provides EST database searching capabilities to verify gene models and explore alternative splicing models. Gene models for human, mouse, *Arabidopsis,* and *Drosophila* genome sequences have already been

developed by training the software on known gene sequences. The programs have the capability of analyzing large sequence data sets.

In Grail, a list of most likely exon candidates is first established, and these are evaluated further using the neural network described in Figure 9.8. The algorithm makes its final prediction by picking the best candidate gene. A dynamic programming approach is then used to define the most probable gene models (Uberbacher et al. 1996).

Input for Grail includes several indicators of sequence patterns. These inputs include several from different types of analyses, including a Markov model for gene recognition that, in principle, resembles the one shown in Figure 9.7, and inputs from two additional neural networks that evaluate the region for potential splice sites. One important indicator is the in-frame 6-mer preference score. Recall that the occurrence of codon pairs in coding regions is not random, whereas in noncoding regions their occurrence is random. Consequently, higher frequencies of 6-mers in genomic DNA that are more commonly found in coding regions can be an indicator of the presence of an exon. Tables have been constructed for various organisms, giving the frequency of each 6-mer (base 1 of first codon to base 3 of second, base 2 of first codon to base 1 of the third codon, and so on) of known cDNAs divided by the frequency of the 6-mer in noncoding DNA. The logarithm of this ratio gives what is called an in-frame preference value for the 6-mer. These 6-mer preference scores increase as GC composition rises, thus increasing the preference scores of a 6-mer with GC richness. Grail II automatically corrects for this increase to put predictions from GC-rich regions on an even footing with other regions.

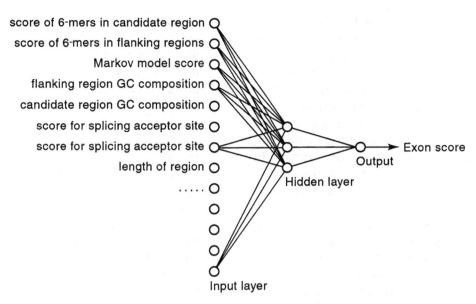

FIGURE 9.8. The Grail II system for finding exons in eukaryotic genes (Uberbacher and Mural 1991; Uberbacher et al. 1996). The method uses a neural network to identify patterns characteristic of coding sequences. The method has similarities to and differences from that used for predicting secondary structure of proteins and described in Chapter 10. Similarities include the use of three layers, an input layer for the data with the data coming from a candidate exon sequence, and a hidden layer for discerning relationships among the input data. An output layer comprising one neuron indicates whether or not the region is likely to be an exon. Each neuron receives information from a set in the layer above (some with a positive value and others with a negative value), sums these values, and then converts them to an output of approximately 0 or 1. The system is trained using a set of known coding sequences, and, as each sequence is utilized, the strengths and types of connections (positive or negative) between the neurons are adjusted, decreasing or increasing the signal to the next neuron in a manner that produces the correct output. The major difference between neural networks for exon and secondary structure prediction is that the exon prediction uses sequence pattern information as input whereas secondary structure prediction uses a window of amino acid sequence in the protein. In Grail II, a candidate sequence is evaluated by calculating pattern frequencies in the sequence and applying these values to the neural network. If the output is close to a value of 1, then the region is predicted to be an exon.

The log ratios for a potential ORF starting at base 1 in a query sequence, another for an ORF starting at base 2, and a third starting at base 3 are calculated by adding the logarithms of these individual 6-mers. These sums provide a log likelihood score for an exon starting at the first, second, or third positions in the given genomic sequence. These likelihoods are further modified by including conditional information on the likelihood of the next five bases on coding and noncoding regions, given the current 6-mer. The probability of an exon starting at base 1 is then given by a Bayesian formulation

$$P = a_1 / a + C n_1 \qquad (1)$$

where a_1 is the score for an exon starting at base 1; a is the sum of scores for base 1, base 2, and base 3; n_1 is the score for a noncoding region starting at base 1;

and C is the ratio of coding to noncoding bases in the organism. This value is used as the score of 6-mers in the candidate region (Uberbacher et al. 1996). A similar score is calculated for the regions 60 bases on each side of the candidate region. If these regions also appear to be encoding exons, the examined region will be enlarged and the prediction repeated. In this manner, a given exon candidate sequence will be enlarged until the coding signals from flanking sequences are no longer to be found.

GeneParser

This program predicts the most likely combination of exons and introns in a genomic sequence by a dynamic programming approach. Dynamic programming was introduced in Chapter 3 as a way to align sequences to obtain a most likely alignment for a given scoring system with scores for matches, mismatches, and gaps. The alignment up to a given set of sequence positions is stored in a scoring matrix, and the dynamic programming algorithm provides a method for finding the best score at that position. GeneParser uses a likelihood score for each sequence position being in an intron or exon. The intron and exon positions are then aligned with the constraint that they must alternate within a gene structure. In this manner, a combination of the most likely intron and exon regions that comprise a gene structure is found.

GeneParser includes one other novel feature, a scheme for adjusting the weights used for several types of sequence patterns that make up the intron and exon scores. A neural network is used to adjust the weights given to the sequence indicators of known exon and intron regions, including codon usage, information content (see Chapter 5, p. 213), length distribution, hexamer frequencies, and scoring matrices (see Chapter 5, p. 210) for splicing signals. The integration of the dynamic programming and neural network methods works as follows:

1. The characteristics described above for a set of intron sequences and a second set of exon sequences are determined. For example, a table of hexamer frequencies is prepared by the program.

2. For a training gene sequence, a series of indicator matrices is prepared. The sequence is listed both down the side of the matrix and across the top. Each position in one of the matrices representing positions a and b in the sequence gives the likelihood for an exon or intron that starts at position a and ends at position b. One such matrix would be the likelihood of an exon based on hexamer frequency in the a–b interval. Another matrix (or the other half of the same matrix, since only one half is needed for exon values) gives the likelihood of an intron based on the same criterion. Other sets of matrices for the sequence based on compositional complexity, length distribution or exons, or splice signals on weight matrices are also prepared.

3. The a, b values in the above indicator matrices for exons are each transformed by a weight and bias, and the sum of the weighted values is obtained. An initial arbitrary set of weights is chosen for each type of sequence information. These weights are later adjusted until they provide the correct gene structure of the training sequence. This sum (s) is then further transformed to a number (L) that is either close to 0 or 1 by using the neural network gating function $L = 1/[1 - e^{-s}]$. The transformed a, b values are then placed in another matrix L_E that gives the weighted score for exons going from position a to position b in the sequence. A similar set of transformed values for an intron at position a, b, but not necessarily weighted the same way, is placed in another matrix L_I at position a, b (which can be the other half of L_E since only half of the L_E matrix is needed). The reason for this transformation is to use the information at a later stage as input to a neural network, in the same manner as used in neural networks for prediction of protein secondary structure and discussed in Chapter 10.

4. Dynamic programming is used to predict the most compatible number and lengths of introns in the training gene up to any position j in the sequence.

5. Steps 2–4 are repeated for each training sequence.

6. The accuracy of the predictions is then determined.

7. If a certain level of accuracy in matching known gene models is not achieved, a neural network similar to that described above for Grail II is used to adjust the weights used for the input exon and intron features.

8. If the required level of accuracy is reached, the method is ready to be used for determining the structure of an unknown genomic DNA sequence.

Pattern Discrimination Methods

Discrimination methods applied to DNA sequences are statistical methods used for classifying the sequence based on one or more observed sequence patterns. For gene prediction, features of patterns found in genomic sequences are examined statistically to determine whether they are like those found in coding sequences. One such feature that is characteristic of coding sequences is the 6-mer exon preference score (EPS) described above. Another is a score for a 3′-flanking splice site (3′SS) calculated in a similar manner. In effect, the distribution of these two scores and a number of others is obtained for a large training set of known exons and also for a set of noncoding

FIGURE 9.9. Analysis of candidate sequences for exon status by a discriminant function. Up to nine different pattern features of sequences are analyzed in coding and noncoding sequence. Shown is a plot of two of these features for several exon (ex) and noncoding (nc) sequences. The object of the discriminant analysis is to define a boundary between these two groups of sequences such that they are maximally separated, or that the sum of distances from a boundary line to each point is a minimum. A linear discriminant analysis (Solovyev et al. 1994) assumes that the covariations among the data are the same for the exon and noncoding sequences and provides a straight line boundary (*dotted straight line*) between the two sets of data. Such a boundary may miss some of the data points. A quadratic discriminant analysis (Zhang 1997) is more flexible, does not assume a similar covariation in the exon and noncoding sequences, and provides a curved

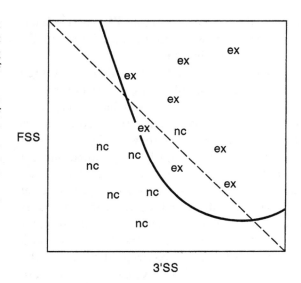

boundary formed by a quadratic equation that can, in principle, provide a better separation of the groups (*solid line*). Once these boundary lines have been calculated, the EPS and 3′SS values of a query sequence will indicate whether the sequence belongs to the exon group or noncoding group of sequences. For an actual analysis, multiple analyses are performed on a candidate sequence leading to a more complex, multidimensional type of analysis. (*Note:* Similar types of discriminant analyses are used to classify microarray data [see Figure 11.11].)

sequences. Using the EPS and 3′SS as examples, the pair of scores for each training exon or noncoding sequence is plotted on a graph, and each point is labeled as coding or noncoding, as illustrated in Figure 9.9. A line is then positioned between the two groups of sequences. A sequence of unknown coding capabil-

ity is similarly analyzed to determine whether the features of the sequence place it on one. The programs HEXON and FGENEH (combines exon prediction into a gene structure) use linear discriminant analysis (Solovyev et al. 1994), and MZEF uses quadratic discriminant analysis (see Table 9.1) (Zhang 1997).

WHAT IS THE BEST GENE PREDICTION METHOD?

The above gene prediction programs locate a contiguous set of probable exons in a genomic sequence as components for a gene model. It is important to evaluate the reliability of a given program for identifying exons; i.e., the sensitivity of the program for correctly identifying exons without making false-positive predictions. The program should also be evaluated for correctly predicting the position of exon boundaries. Information on reliability and accuracy can best be found by validating the ability of the program to predict exons correctly from experimentally derived exon sequences.

Evaluation of exon and gene prediction programs must take into account the type of analysis, whether neural network, linear discriminant, or

other; the number and types of sequences used for training and evaluation; and the method used for evaluation. In addition, choice of program variables by the user will affect the predictions that are made. As more gene sequences become known, more are becoming available for training and evaluation. The ideal method for evaluation uses a known set of gene structures for training the method and a second set that is not used in the training or similar to those used in the training for evaluation (Burset and Guigó 1996). The evaluation is usually more stringent if the evaluation set includes a gene and neighboring sequence rather than just the sequence between the first and last exons. A current evaluation of most methods is available at the Web sites for these methods listed in the footnotes to Table 9.1. These evaluations are most useful when different prediction methods are used in combination.

The method for evaluation is similar to that used for testing the reliability of scoring matrices for identifying protein family members in Chapter 5 (p. 192) or protein secondary structure prediction as described in Chapter 10 (Mathews 1975; Burset and Guigó 1996). The program, which is trained on a set of sequences from a given organism, is used to predict the exons, or set of exons, comprising a gene of a set of genomic evaluation sequences from the same organism. An evaluation is then made of the number of true positives (TP), where the length and end sequence positions are correctly predicted, and the number of overpredicted positive predictions or false positives (FP), true negative (TN), and underpredicted residues as misses or false-negative (FN) predictions. The following calculations are made:

1. The number of actual positives is AP = TP + FN.

2. The number of actual negatives is AN = TN + FP.

3. The predicted number of positives is PP = TP + FP.

4. The predicted number of negatives is PN = TN + FN.

5. The sensitivity of a method SN is given by true positives/actual positives = TP/(TP + FN).

6. The specificity by SP = true negatives/actual negatives = TN/(TN + FP).

7. The positive and negative predictive values are TP/(TP + FP) and TN/(FN + TN), respectively.

A correlation coefficient expressing the extent of agreement between the predicted and known exons, CC, is given by

$$CC = [(TP)(TN) - (FP)(FN)] / \sqrt{[(AN)(PP)(AP)(PN)]} \qquad (2)$$

By this coefficient, a method given all correct gene predictions would score 1, and the worst possible prediction would be –1. In tests of this kind on three sets of human sequences, GeneParser, GeneID, and Grail gave (1) sensitivities of 0.68–0.75, 0.65–0.67, and 0.48–0.65; (2) specificities of 0.68–0.78, 0.74–0.78, and 0.86–0.87; and (3) correlation coefficients of 0.66–0.69, 0.66–0.67, and 0.61–0.72, respectively, for the accuracy of finding the correct nucleotide ends of

exons. GeneParser was also shown to be more reliable for genes with short exons and least reliable for genes with long exons (Snyder and Stormo 1993).

A detailed evaluation of the available gene prediction programs has been performed, and the correlation coefficient was found to lie between 0.6 and 0.7, and the fraction of correctly found exons was generally less than 50%. The performance decreased when longer test sequences were used and when a 1% level of sequence noise was introduced as frameshift mutations. Programs including protein sequence database searches (GeneID+, GeneParser3, GrailEXP) showed substantially greater accuracy (Burset and Guigó 1996). Therefore, these studies indicate that gene prediction programs reliably locate genomic regions that encode genes, but they provide an only approximate indication of the gene structure. In a later similar study using the same data set as the above study, and comparing Grail, FGENEH, and MZEF, these numbers were (1) sensitivities 0.79, 0.93, and 0.95; (2) specificities 0.92, 0.93, and 0.95; and (3) correlation coefficients 0.83, 0.85, and 0.89, respectively (Zhang 1997). These numbers suggest that FGENEH and MZEF are reliable exon predictors, providing better than 85% reliable predictions. GrailEXP, an expanded Grail program, can be expected to perform considerably better than Grail because it includes database searches; e.g., EST searches to verify predictions.

In the Gene Annotation Assessment Project (GASP), exon and gene feature prediction programs from 12 groups were tested for ability to annotate a sequence contig of the *Adh* region of the *Drosophila* genome (Reese et al. 2000). Some of these programs were exon and gene predictors based on computational models as described above, and others performed database similarity searches of proteins and protein domains against the genome sequence translated in all six possible reading frames. The predictions were compared to known exon and gene assignments and promoter prediction as found by previous annotation by a panel of experts. The results were published in a series of papers in the April 2000 edition of *Genome Research*. More than 95% of the coding nucleotides (exons) in *Adh* were correctly identified by the majority of the model-based gene finders but only a small fraction of the predicted gene models were correct (Reese et al. 2000). These results suggest that using different programs to find consensus predictions is a reasonable approach.

Several groups have provided Web sites, such as GeneMachine (Makalowska et al. 2001), that will

TABLE 9.4. *Example of exons predicted in an* Arabidopsis *genomic sequence by gene prediction programs*

cDNA	Netgene[b]	GeneMark[c]	FgeneP[d]	GeneScan	Mzeff[e]
345[a]–1210	x 1210	345–1210	345–1210	530–1210	276–1210
1290–1513	1290 1513	1290–1513	x 1513	1242–1513	1290–1513
1611–1696	1611* 1696	1611–1696	x x	1611–1696	1611–1696
1880–2029	1880* 2034	1880–2029	x x	1880–2029	1880–2029
2143–2880	2143 2880	2143–2880	x 2880	2143–2880	x 2880
3143–3253	x 3253	3143–3253	x x	x x	3143–3253
3339–3599	3339* 3599	3339–3599	3339-3599	3339–3599	3339–3599
3698–3921	3698 3921	3698–3921	3698-3921	3698–3921	3698–3921
4010–4217	4010 x	4010–4220[f]	x x	4010–4220[f]	x x

This test is given as an example and should not be taken as a measure of the reliability of these programs. The Web sites were provided with the genomic sequences of the *Arabidopis UVH1* gene with approximately 250 bp upstream from the first exon and 200 bp downstream beyond the last exon. As indicated in the text, these programs are more reliable when they are presented with short genomic sequences, as was done in this example. The consensus splice sites for *Arabidopsis* may be found at http://genome-www.stanford.edu/ Arabidopsis/splice_site.html. A more detailed assessment of the reliability of gene prediction programs on Arabidopsis genomic sequences has been published (Pavy et al. 1999).

[a] Predicted.

[b] NetPlantGene was used. This program predicts intron–exon and exon–intron junctions and not most probable combinations of the two. In this case many false-positive intron–exon junctions were predicted with low probability. The highest scoring junctions are marked by *. x are actual sites not predicted. The intron–exon junctions are predicted much more reliably, and three false positives were reported.

[c] GeneMark shows a remarkably good frequency of prediction for these exons and usually joins the exons in the correct reading frame, but not always. Therefore, some parts of the predicted protein sequence are not correct.

[d] x are actual sites not predicted. Exon start sites of 1370–1513 and 2779–2880 were found illustrating a difficulty with finding exon start sites.

[e] The prior probability was set at 0.6–0.8 to obtain these results. The higher this value, the lower the level of discrimination used, the more sensitive the test, and the greater the number of exons that is predicted. x was not predicted; instead a start site of 2709 was predicted. This program predicts internal exons only.

[f] The 4220 end includes the termination codon.

accept sequences for exon and gene model prediction and automatically submit these to a series of sites running different gene prediction programs and present a useful database and display of the results for visual comparison, such as GeneComber (Shah et al. 2003). Other sophisticated computer tools have been devised for improving gene prediction using comparative nucleic acid and protein sequence information from related genes and proteins, for example, Projector (Meyer and Durbin 2004) and Combiner (Allan et al. 2004).

To illustrate the results obtained by the gene prediction programs, an *Arabidopsis* genomic sequence was submitted to several gene prediction Web servers, as shown in Table 9.4. Because the cDNA sequence was also available, the accuracy of the programs could be determined. There are computer programs designed for aligning cDNA and genomic DNA sequences of a gene (Florea et al. 1998; and see Tables 3.1 and 9.1). As shown, the results of the analyses vary considerably and the program variables must sometimes be optimized to find the correct translation. In this case, GeneMark gave a fully accurate translation of the sequence. Other programs, such as NetPlantGene, gave a large number of possible exon–intron boundaries, including some of the actual ones.

PROMOTER PREDICTION IN *E. COLI* CAN LOCATE REGULATORY PATTERNS IN DNA

Promoters are DNA sequences found at the 5′ end of genes that regulate transcription by binding regulatory proteins (transcription factors) to specific binding sites. These proteins interact in a manner that can induce or repress transcription depending on interactions among the proteins with each other and with

RNA polymerase (RNAPII in eukaryotes). The response of genes to environmental signals or to a particular transcriptional program depends on the types, numbers, and arrangements of binding sites in the promoter sequence. As a result of the shortness of the binding sites (often <10 bases) and their variability, they are difficult to identify, confirm experimentally, and locate in sequences.

The method that has most often been used to analyze *E. coli* promoters is to align a set of promoter sequences by the position that marks the known transcription start site (TSS) and then to search for conserved regions in the sequences. Following such an alignment, *E. coli* promoters are found to contain three conserved sequence features: a region approximately 6 bp long with consensus TATAAT at position −10 (the Pribnow box), a second region approximately 6 bp long with consensus TTGACA at position −35, and a distance between these regions of approximately 17 bp that is relatively constant (see Fig. 9.6 for an example). A weaker region exists around +1, the designation given to the start of transcription, and an AT-rich region is found before the −35 region (Hawley and McClure 1983; Mulligan and McClure 1986). The sequences changed to some extent as the number of sequences and the types of promoters analyzed were varied. For example, promoters that are activated by transcription factors have more variable sequences (Hertz and Stormo 1996). The RegulonDB (http://www.cifn.unam.mx/Computational_Genomics/regulondb/; Salgado et al. 1999), Dpinteract (http://arep.med.harvard.edu/dpinteract database; Robison et al. 1998), and regulatory site database (Thieffry et al. 1998; http://tula.cifn.unam.mx/~madisonp/E.coli-predictions.html) have been developed with information on the *E. coli* genome. With the availability of a large number of prokaryotic genomes (see Chapter 11 and http://www.tigr.org/tdb/mdb/mdb.html), a similar analysis of the genes and regulatory sites in these other genomes has become possible.

The aligned promoter regions provide a consensus sequence that may be used to search for matching regions as potential promoters in *E. coli* sequences. Each column in the alignment gives the variation found in that position of the promoter. Programs such as the Genetics Computer Group program FINDPATTERNS and PatScan (http://www-unix.mcs.anl.gov/compbio/PatScan/HTML/patscan.html; Dsouza et al. 1997) may be used to search for match-

es to the consensus sequence or the variation found in each column in a target DNA sequence. The difficulty with using the consensus sequence to search for new promoters is that most sequence positions in the aligned regions vary to some extent, and some regions are much less variable than others; e.g., the first, second, and sixth positions in the −10 region.

An alternative is to use the search features of FINDPATTERNS and PatScan that allow alternative symbols at one sequence position, repeats of a symbol, inverted repeats, gaps, and so on. For example, providing the pattern GAT (TG, T, G) {1,4} to FINDPATTERNS means to search for GAT followed by a TG, or a T, or a G repeated up to four times. These types of pattern expressions are similar to regular expressions that are used to specify PROSITE patterns in protein sequences and to initiate PHI-BLAST searches of protein sequence databases (see Chapter 6, p. 268, and Chapter 12 for a more detailed use of regular expressions). Although these expressions are extremely useful for locating complex regulatory patterns in DNA sequence, they do not take into account the frequency of each residue at each pattern position. What is needed is a more quantitative way to use these known sequence variations to search a target sequence. The scoring matrix method provides such an analysis.

The Scoring Matrix Method Used with Aligned Promoter Sequences

A more complex type of promoter analysis used for both prokaryotic and eukaryotic sequences is a scoring or weight matrix. This kind of matrix was previously described in Chapter 5 (p. 210) as a method for representing the variation in a set of sequence patterns in an msa, and in Chapter 6 (p. 261) as a tool for finding additional sequences with the same pattern in a database search. The scoring matrix has also been used to analyze promoters, ribosomal binding sites, and eukaryotic splice junctions (Staden 1984).

An example using a scoring matrix for representing the −10 region of *E. coli* promoters is illustrated in Table 9.5. In this example, *N* sequences have been aligned by their −10 regions, and a count of each base in each column of the alignment has been made. These counts are converted to frequencies. For example, if 79 of 100 sequences have a T in column 1, the frequency of T in column 1 of the matrix is 0.79. Similarly, a T occurs in column 2 with a frequency of 0.94. These frequencies are converted into log odds scores, as described in Table 9.5. An example of using the scoring matrix in Table 9.5 to locate the most likely 210 sites in a query sequence is shown in Figure 9.10. The matrix is moved along the query sequence one position at a time as a sliding window. At each position, the base in the sequence is noted and the corresponding score of that base in the matrix is then used. This procedure is repeated for the remaining positions. The log odds scores are then added to obtain a combined log odds score for the particular

TABLE 9.5. *A scoring matrix representing the frequency of DNA bases found in the 210 position in E. coli promoters*

A. Fraction of each base at each column of the aligned promoters in the −10 region				
Position	A	C	G	T
1	0.02	0.09	0.10	0.79
2	0.94	0.02	0.01	0.03
3..6	. . . .	. . . .	. . . .	. . . .

B. Log odds score				
Position	A	C	G	T
1	−3.80	−1.49	−1.34	1.67
2	1.92	−3.81	−4.81	−3.22
3	−0.06	−0.81	−0.66	0.81
4	1.24	−1.00	−0.72	−0.89
5	1.02	−0.35	−1.00	−0.56
6	−4.81	−3.22	−4.81	1.95

(*A*) Frequency of each base found, showing two positions as examples. (*B*) Conversion of frequencies to log odds scores. The first step is to convert the frequency of each base at each sequence position into an odds score. The odds score is simply the frequency observed in the column divided by the frequency expected, or the background frequency of the base, usually averaged over the genome. Thus, if the position frequency is 0.79 and the background 0.25, the odds score is 0.79/0.25 = 3.16. This number means that if a sequence is being examined for the presence of a promoter, and a T is present in the sequence at predicted position 1, the odds of the sequence representing a promoter (a win) to the sequence not representing a promoter (a loss) is 3.16/1. Finally, the odds score is converted to a log odds score by taking the logarithm of the odds score, usually to the base 2 (units of bits) and sometimes to the natural logarithm (units of nats). As described in Chapter 4, bit units have a special meaning in information theory. They represent the number of questions that must be asked to decide whether or not the base in the column of the scoring matrix is a match to the aligned sequence position. This number is called the information content of the matrix position. On the one hand, if all four bases are equally represented in the matrix position, the number of questions that must be asked is two. The first question might be, Is the sequence position one of A or T, or one of G and C? The second question will then find the correct base. On the other hand, if only one base is found in the matrix position, no question need be asked of the sequence position. The fewer questions that have to be asked, the more information in the matrix, and the more discriminatory it is for distinguishing real matches from random matches (Schneider et al. 1986). A set of log odds scores for the major six positions in the −10 region of *E. coli* promoters is shown (Hertz and Stormo 1996). In the actual matrices that are used, an additional 6–12 base positions that flank these major positions are also used. There is a zero occurrence of one particular base in the matrix, thus creating a problem because the logarithm of zero is infinity. In this case, a single count is substituted for the zeros and the resulting small fraction will calculate to a large negative log odds score. Alternatively, a large negative log odds score may be used at such positions in a scoring matrix.

Another formula for calculating the scoring matrix value of base i in column j, $w_{i,j}$, is given by

$$w_{i,j} = \log\left[(n_{i,j} + P_i)/\{(N + 1)P_i\}\right] \approx \ln\left(f_{i,j}/P_i\right)$$

where $n_{i,j}$ is the count of base i in column j, P_i is the background frequency of base i, N is the total number of sequences, and $f_{i,j} = n_{i,j}/N$ (Hertz and Stormo 1999). Bucher (1990) uses the formula

$$w_{i,j} = \log\left[(n_{i,j}/P_i) + (s/100)\right] + C_j$$

where s is a smoothing percentage for the column values and C_j is a column-specific constant. Bucher sometimes also uses dinucleotide composition for calculating the background base frequency to accommodate local sequence complexity (Bucher 1990). These formulas both accommodate zero occurrences of a base by adding a small value in a scoring matrix to zero positions. Another method is to add pseudocounts to these positions, as described in Chapter 5 (p. 211).

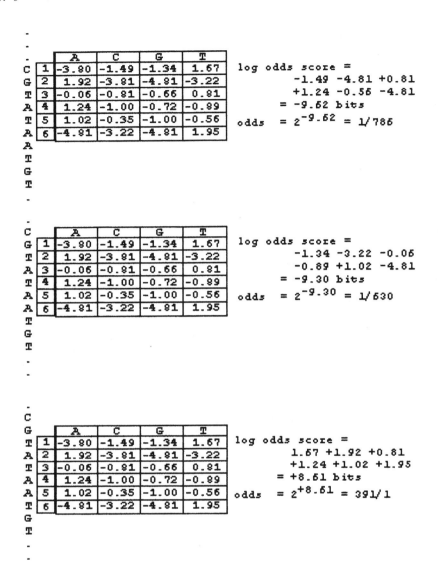

FIGURE 9.10. Locating −10 promoter sites in *E. coli* sequences using a −10 scoring matrix. The matrix is moved along a query sequence one position at a time, and at each location the sequence window is scored for a match to the matrix by summing the log odds scores. A cutoff score may be defined that permits recognition of known promoter sequences while minimizing the prediction of false-positive sites.

position in the sequence that is a −10 region in a promoter. The sum of the log odds scores in bits may be converted to odds scores by the formula odds score = $2^{(\text{log odds score})}$ or if the log odds score is in nats, by the formula odds score = $e^{(\text{log odds score})}$. These numbers vary from small fractions to large numbers reflecting variations in the likelihood of a −10 region at each sequence position.

The odds scores at every possible matching loca-

tion along the sequence may be used to find the probability of each sequence location. The odds scores are first summed to give sum S. The odds score at a particular location of six bases in the sequence divided by S then provides a probability that the location is a −10 region. To give a simple example, of the three matches in Figure 9.10, the probability of the match at the third location shown is 391/[(1/786)+(1/630)+(391)] = 1.000.

For scoring *E. coli* sequences for the presence of promoters, scoring matrices for a 35-bp region encompassing the −35 region, a 19-bp region encompassing the −10 region, and a 12-bp region encompassing the +1 region are each applied to both strands of a query DNA sequence. Each matrix will provide a distribution of odds scores that predict possible locations for matches to itself in the query sequence. These matches are then examined for spacings that are characteristic of the known promoter sequences. The region between the −10 and −35 regions varies from 15 to 21 bp but is usually 17 bp, and the region between −10 and +1 is 4–8 bp. When a suitably oriented combination of high-scoring matches is found, the log odds scores of each sequence region are added. From this sum, a penalty may be subtracted if the distance between the −10 and −35 regions is not an optimal 17 bp in length or if the distance between the −10 and +1 regions is not optimal (Hertz and Stormo 1996). The resulting log odds score represents an overall likelihood that a test sequence includes regions characteristic of *E. coli* promoters in the correct spacing. A similar application of weight matrices for identifying the start of prokaryotic genes (program ORPHEUS) has been described previously (Frishman et al. 1998).

Reliability of the Matrix Method

The reliability of a combination of scoring matrices for promoter prediction can be assessed by comparing the range of scores found in a set of known promoters versus scores in a set of random sequences. A threshold score that is achieved by most of the known promoters (true positives), but by only a small number of false-positive scores in random sequences, may then be chosen (Bucher 1990; Hertz and Stormo 1996). However, there is a tradeoff between true positives and false positives. For example, a threshold score that includes most known promoters will often be reached by random sequences. The situation can be helped somewhat by identifying the promoters that are not activated by additional transcription factors. These activated promoters are more alike than all promoters taken as a group. Random sequences rarely have the more highly conserved patterns in these activated genes. In contrast, the sequence patterns in all promoters without regard to regulation are much more variable so that random sequences are more likely to match these patterns. When an even

lower threshold is chosen to reduce this false-positive rate from random sequences, the ability to detect promoters (the true-positive rate) is decreased. To try to improve the predictive values, the lengths of the scoring matrices and the gap penalty values have been varied, but the predictive value of the matrices is not much improved above these values.

There are several reasons that matrix methods do not achieve a better prediction of *E. coli* promoters.

1. The matrix method adds the scores for each sequence position, whereas in reality, one position in the −10 region, for example, may play a role in one stage of transcription such as promoter recognition by RNA polymerase, whereas another may play a role in a subsequent stage of transcription, such as initiation of transcription or elongation of the mRNA. Matching positions with these types of functional separations are not expected to be additive, as assumed by the matrix method.

2. A second difficulty that the matrix method shares with most other methods of promoter prediction is that all promoters are treated as being in the same class, whereas different forms of RNA polymerase that are complexed with a set of transcriptional activators (σ factors) may have preference for different sequence positions in the promoter region. With the whole genome of *E. coli* now available for analysis (see http://www.genome.wisc.edu/), such additional classification may become a possibility (Hertz and Stormo 1996).

3. A third difficulty is that the promoter sequence is treated as a Markov chain, meaning that each sequence position acts independently of the others so that a match at each position may be individually scored without reference to the other positions. According to a statistical mechanical theory discussed below, the most conserved positions are thought to act independently. However, as evidenced by the fact that some weight matrices are not efficient in locating matching sites, there may be correlations between the sequence positions so that covariation of the bases at these positions occurs at frequencies greater than expected by chance. Such correlations are not easily found in a small number of training sequences. Methods include using decision trees and locating specific oligonucleotides, discussed later in the chapter.

A number of ways to improve matrix methods, including corrections for base composition, and utilizing a different number of matrix positions, have been tried, but none of these is significantly better than the basic scoring matrix described above. In addition to the matrix methods, a number of additional methods for predicting *E. coli* promoters and other regulatory sites have been developed, but without much improvement over the scoring matrix method (Hertz and Stormo 1996).

Neural Networks for Promoter Prediction

A second method for promoter prediction is the use of neural networks, which are described in Chapter 10 (p. 462). In the case of promoter prediction, a neural network is trained to distinguish *E. coli* sequences from non-promoter sequences (Horton and Kanehisa 1992; Pedersen et al. 1996). The network is like that used for prediction of protein secondary structure and is trained by similar methods. Horton and Kanehisa used a network lacking a hidden layer, called

a perceptron (see Fig. 9.11). This type of network scans the sequence to be analyzed using a sliding window and at each location reads each of the sequence positions within the window. The sliding window will scan long enough lengths of sequence to include both the −10 and −35 regions of the *E. coli* promoters. Sequence positions between these conserved regions are not scored because they do not contribute. The sequence characters are given a simple identification scheme to avoid any bias (e.g., A is 1000, G 0100, etc.), and the sum of these sequence values after weighting is used as input for a single output neuron, which produces a number close to 1 if the region is within a promoter or 0 if the region is not in a promoter. The network is trained on known promoter sequences by adjusting the weights of the input sequence positions so that the output produces the correct response. However, the perceptron method was not found to be any more effective than scoring matrices for finding *E. coli* promoters. Hence, the tendency is to use scoring matrices to represent binding sites of regulatory proteins.

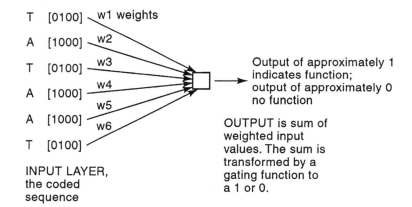

A. The perceptron

INPUT LAYER, the coded sequence

Output of approximately 1 indicates function; output of approximately 0 no function

OUTPUT is sum of weighted input values. The sum is transformed by a gating function to a 1 or 0.

B. Scoring matrix equivalent

		A	C	G	T
T	1				0.19
A	2	0.22			
T	3				0.09
A	4	0.14			
A	5	0.12			
T	6				0.24

SUM = 0.19+0.22.. approx 1 indicates function

FIGURE 9.11. The model of a perceptron used for locating *E. coli* promoters. A portion of the sequence in the −10 region that is to be scanned is shown for illustrative purposes. (*A*) A known promoter sequence is encoded and used as input into a single output neuron. The input signals are weighted, the weighted values summed, and the sum transformed into a number that is approximately 0 or 1. The network is trained by starting with an initial set of weights, then adjusting each weight by a small amount until the correct output is found for as many of the training sequences as possible. (*B*) The trained perceptron used for scanning unknown sequences for promoter-like patterns is a special type of scoring matrix. In this case, the matrix is aligned with the sequence, and the matrix values that match the sequence are added. If the sum of numbers is approximately 1, then a promoter is predicted, and, if 0, then a promoter is not predicted. The difference between this matrix and the scoring matrix described above is that in this matrix each position within the sequence window is given a different weight.

Finding Less-conserved Binding Sites for Regulatory Proteins in Unaligned Sequences

In the above example of finding consensus binding sites for RNA polymerase in *E. coli* promoters, known promoter sequences could be quite readily aligned by the transcriptional start site and the −10 and −35 regions. The binding sites for other regulatory proteins, such as the LexA protein described above, are also quite readily found because the sequence of the binding sites is conserved. However, in many other cases, particularly those for eukaryotic transcription-factor-binding sites described later in this chapter, the sites vary considerably and the surrounding regions are also variable so that it is impossible to find conserved positions in the binding site by aligning the sequences. Thus, methods are needed to find a common but degenerate pattern in sequence fragments that are expected to carry a binding site but that cannot be aligned.

The problem of representing more variable binding sites is similar to that described in Chapter 5 for finding patterns that are common to a set of related protein sequences that cannot be readily aligned. However, there is one important difference. In proteins, there are a possible 20 amino acids in each matching position of the sequence pattern, but in DNA-binding sites there are only four possible bases in each position in the pattern. Because the available alphabet is much smaller in DNA sequences, it is more difficult to detect DNA sequence patterns above background noise. The way to express this difference between DNA and protein sequences is by considering the maximum possible information content that can be present in one column of a scoring matrix. The more the information content, the better the scoring matrix is for distinguishing sequences that have the pattern represented by the scoring matrix and sequences that do not. For DNA sequences, the maximum amount of information in one column of the scoring matrix is 2 bits (for 4 bases, $\log_2 4 = 2$) whereas for protein sequences, it is 4.32 bits (for 20 amino acids, $\log_2 20 = 4.32$). Additional discussion of information content may be found in Chapter 5 (p. 213). The importance of information content is that longer, less variable patterns in DNA sequences compared to protein sequences are necessary to produce a scoring matrix that has a similar capability of discriminating patterns from background sequence variation. As a result, it is more difficult to locate significant patterns in DNA sequences than protein sequences.

Statistical Methods

Some of the statistical methods used for finding protein patterns, e.g., expectation maximization and HMMs, are also used for identifying DNA patterns in unaligned DNA sequences. The expectation maximization method is described in Chapter 5. Briefly, an initial scoring matrix of estimated length is made by a guessed alignment of the known promoter sequences (the expectation step). The scoring matrix is then used to scan each sequence in turn, and the probability of a match to each position in each sequence is calculated as discussed above. The scoring matrix is then updated by the sequence pattern found at each scanned position times the probability of a match to that position (the maximization step). The two steps are repeated until there is no improvement in probability when matching the scoring matrix to the sequences. The method has been adapted to find multiple patterns separated by a variable spacer region, to take into account the −10 and −35 regions of *E. coli* promoters (Cardon and Stormo 1992). These studies have provided useful information as to which positions in the promoter sequences provide information that enhances specificity.

HMMs such as those described in Chapter 5 (p. 204) and earlier in this chapter have also been used for prokaryotic promoter prediction (Pedersen et al. 1996). In principle, because HMM methods are based on methods that are similar to the EM method, they should be comparable in effectiveness to the EM method.

Another statistical method of finding patterns in unaligned sequences has also been used for DNA sequences. In one case, this statistical method was used with a dinucleotide analysis to reduce background noise (Ioshikhes et al. 1999). A Gibbs sampling method that takes into account additional features of DNA sequences such as inverted repeats has been described (Zhang 1999b). Align Ace is a program designed for promoter analysis that uses a Gibbs sampling strategy (http://atlas.med.harvard.edu). The inverted repeat feature is designed to identify binding sites of regulatory proteins that are inverted repeats, like LexA-binding sites in Figure 9.6.

Consensus and wconsensus

A different method has been developed for searching through a set of unaligned sequences for a common but degenerate sequence pattern (Stormo and

Hartzell 1989; Hertz et al. 1990). The program developed for this purpose, consensus, was used to produce a set of scoring matrices for eukaryotic transcription-factor-binding sites (Chen et al. 1995). Recently, a theory was developed that allows a statistical evaluation of the results (Hertz and Stormo 1999). In its simplest form, illustrated in Figure 9.12, a sliding window of sequence in each of the sequences is matched against similar windows in the remaining sequences, searching for the best scoring matrix, as judged by the information content of the matrix (p. 213). There is no allowance made for gaps, and the choice of a base at each matrix position is assumed to be independent of the other positions, although the development of methods for including such extended features has been described (Hertz and Stormo 1995). In the consensus program, parameters, such as window width, whether or not each sequence can con-

tribute at most one word, whether or not there are additional words after an initial one, whether or not words overlap, whether or not the complementary sequence is used, and the maximum number of alignments to be saved, are set by the user.

In a related program, wconsensus, the optimum window size is not set by the user. Instead, biases, values that change the range of scores obtained in using the matrix, are used and subtracted from the information content of each column in the scoring matrix to make the amount of information a smaller number, called the crude information content. The object is to reduce the average alignment score to a negative value so that an interesting alignment appears as a positive score, much like the procedure used in the Smith–Waterman algorithm for sequence alignment by dynamic programming. wconsensus finds the scoring matrix that maximizes this crude information

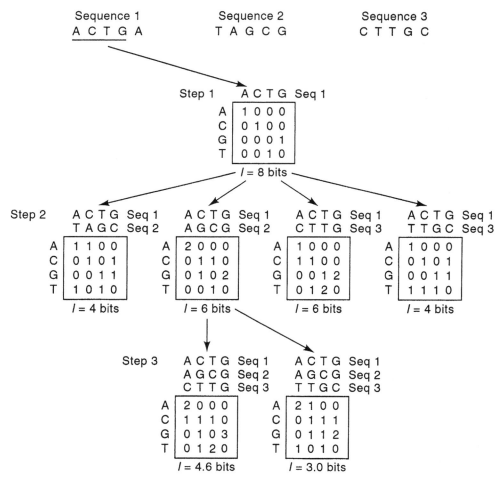

FIGURE 9.12. (*See facing page for legend.*)

content. At the same time, wconsensus also saves the flanking sequence regions from each sequence included in the matrix. As more sequences are added, these regions may also become incorporated into the alignment and help to locate additional matching regions.

The time required for computing these scoring matrices using consensus is extensive and increases as a linear function of the number of sequences being searched for patterns and as the square of the sequence lengths. The programs accept user input to reduce the computational time. These programs are not guaranteed to provide the best possible matrix, but by trying out several reasonable values for user-provided variables, there is a strong possibility of finding the best matrix. Associated with these pro-

grams is a statistical evaluation of each matrix. If I is the information content of the matrix calculated and N the number of sequences used to create the matrix, the probability of obtaining a greater product $I \times N$ from random sequences of the same length and base composition is determined. This procedure is similar in principle to the methods used to evaluate scores found in sequence alignments and database searches, except that the statistical models are quite complex (Hertz and Stormo 1999). Similar numerical methods for calculating the significance of scoring matrices and matches to scoring matrices have also been developed (Staden 1989). Thus, different matrices found by using different matrix widths, base compositions, and other variables may be evaluated for sig-

FIGURE 9.12. (*See facing page for figure.*) The Hertz, Stormo, and Hartzell method for locating common DNA-binding sites for regulatory proteins in unaligned sequences (Hertz and Stormo 1999). This example illustrates how the algorithm compares a fixed window of sequence (length 4 in this example) in a set of sequences assumed to carry one site for a DNA-binding protein that cannot be readily found by aligning the sequences. The object is to find the 4-mer in each sequence that constitutes as nearly identical a pattern as can be found in all of the sequences. The user specifies the number of matrices that can be saved by the program for further analysis. Redundant matrices are eliminated. *Step 1:* The sequence of the first four bases from sequence 1 is first chosen. An analysis of only this one window is shown in this example. Normally the program would start with all possible 4-long words in each of the sequences, thus producing a total number of six possible step 1 matrices in this example. *Step 2:* The sequence window chosen in step 1 is moved across sequence 2, then sequence 3, and so on until all possible windows in all sequences have been selected. If a sufficient number of saved matrices is specified, this procedure would be repeated for all of the six saved matrices in step 1. Only one matrix is shown for illustration purposes. At each selected position, the number of matches with sequence 1 is recorded in a scoring matrix. The amount of sequence conservation in each column is calculated as the information content (*Ic*) of the column, and the *Ic* values for each column are then added to give I of the matrix. The best-scoring matrix is chosen. Calculation of the information content of a scoring matrix is discussed in detail in Chapter 5 (p. 213). Given a position in a test sequence that is being examined for a match to a matrix column, the maximum uncertainty of a matrix column is the number of questions that must be asked to find a match to the position in a test sequence. Uncertainty is 0 if only one base is represented and 2 if all four bases are represented equally. Information content of a column is 2 minus the uncertainty of the column. For example, as each column in the first matrix in step 2 requires a single question to identify a match to a sequence (for column 1, one question must be asked: "Does the matching sequence position have an A or a T?"), then I of the matrix is $1 + 1 + 1 + 1 = 4$. The first column of the second matrix in step 2 has two As, and no other base is represented. Because no question need be asked, I is 2. A general method for calculating the amount of information in a column c is given by $Ic = \Sigma_I \{f_{ic} \log (f_{ic}/p_i)\}$, where f_{ic} is the fraction of each base in the column and p_i is the background frequency of base I in the sequences. If logarithms to the base 2 are used, then I units are in bits, and if natural logarithms are used, I units are in nats. *Step 3:* The sequence windows found in the highest-scoring matrix in step 2 are now compared to all other possible windows in the remaining sequences. In this case, only one sequence remains and the next high-scoring matrix is identified. Only one matrix is shown as an example; the maximum number that can be used for further analysis will be determined by the specified number of matrices that can be saved by the program. Additional steps (not shown) are then used to compare this best matrix with any remaining sequences until all have been included. The final matrices provide a consensus sequence by using the base in each column that has the highest score. The algorithm is greedy because the development of the highest-scoring matrix depends on matches found in ancestor matrices based on a smaller number of alignments. On the basis of this limitation and constraints provided by the user such as window size or matrix bias (see text), and on the number of matrices saved, the algorithm is not guaranteed to provide the optimal matrix for a large number of sequences.

nificance, and the best ones chosen. The consensus programs thus offer a flexible approach for finding sequence patterns in DNA sequences that are difficult to align, and produce scoring matrices that represent any patterns found. The consensus programs run under the UNIX operating system and are available by anonymous FTP from beagle.colorado.edu in the directory /pub/consensus.

Binding sites for repressors and activators of *E. coli* and other bacteria have been analyzed for conserved patterns by the above methods. One example is the set of bacterial and bacteriophage genes that is repressed by the *E. coli lexA* gene product (Lewis et al. 1994). As illustrated in Figure 9.6, these genes carry the binding site for LexA repressor, which is located in the vicinity of the promoter and transcription start site and has the consensus sequence CTGTNNNNNN NNNCAG. The more conserved positions in the binding site contribute the most to the binding of the LexA protein to these sites and, in general, the closer the binding site to consensus, the more tightly bound the protein to that site. Similar observations of several transcriptional regulators and promoters of *E. coli* have led to a statistical mechanical theory that the most conserved positions each independently contribute the most binding energy to the interaction (Berg and von Hippel 1987; Fields et al. 1997; Stormo and Fields 1998).

PROMOTER PREDICTION IN EUKARYOTES RELIES ON LOCATING BINDING SITES FOR TRANSCRIPTION FACTORS

The transcription of genes by RNAPII is carefully regulated in eukaryotic cells. Multiple transcription factors bind to the promoter region at specific sites and locations and form a complex that either favors or disfavors binding of RNAPII. The computational challenge is to build scoring matrices or develop other pattern-finding tools which represent binding sites or combinations of binding sites that are predictive for gene expression.

Transcriptional Regulation in Eukaryotes

Apart from a few short patterns described below, eukaryotic promoters show a large amount of sequence variation. The variable regions include conserved short patterns that represent specific binding sites for proteins (transcription factors). It is the presence and location of these patterns that regulate the expression of a gene. The goal of promoter prediction in eukaryotes is to identify these patterns in the promoters of a set of genes that are known to be regulated in the same manner. Once known, searching for these patterns in other eukaryotic promoters can then be used to predict their regulatory behavior.

Regulation of transcription of protein-encoding genes by RNAPII involves the interaction of a large number of proteins, called transcription factors (TFs), with DNA-binding sites in the promoter regions of eukaryotic genes. The TFs then form a complex that regulates the recruitment of RNAPII for transcribing genes into mRNA. The regions upstream from the start point of transcription, as well as those just downstream, influence the regulation and degree of expression of the gene. The region immediately upstream, the core promoter, has DNA-binding sites to which a preinitiation complex comprising RNAPII and TFIIA, B, D, E, F, and H binds (Tjian 1996).

The positions of transcription-factor-binding sites are given with reference to the TSS. A site defined as the TATA box is present in about 75% of vertebrate RNAPII promoters. A TATA box HMM trained on vertebrate promoter sequences has the consensus sequence TATAWDR (W is A or T, D is not C, R is G or A) starting at approximately –17 bp from TSS (Bucher 1990; http://www.epd.isb-sib.ch/promoter_elements/). This sequence is thought to position the initiation complex at or near the TSS. A component of TFIID, the TATA-binding protein (TBP), recognizes and binds to this sequence. INR is a loosely defined sequence around TSS that also influences the start position of transcription and may be recognized by other protein subunits of TFIID (Chalkley and Verrijzer 1999). Another conserved sequence lying upstream of the TATA box and present in about one-half of vertebrate promoters is the CCAAT box, which is thought to be the site of binding of additional proteins that influence both early and later stages of transcription. Another conserved regulatory site is the GC box. These boxes lie at variable distances from TSS and function in either orientation; i.e., pattern may be on either DNA strand. Weight matrices that describe them have been produced (Bucher 1990).

The region upstream of the core promoter of a eukaryotic gene and other enhancer sites in the neighborhood of a gene also influences gene expression. A variety of transcription factors, some affected by environmental influences such as hormone levels, bind to DNA-binding sites in these regions. These factors can also form large multiprotein complexes that interact with a preinitiation complex to induce or repress transcription. These interactions can cause remodeling of

the local nucleosome structure by histone acetylation or deacetylation, conformational changes in the transcription complex, and possibly phosphorylation of RNAPII. The independent binding of proteins to separate DNA sites in the initiation and upstream control regions is cooperative in that the binding of one protein to one site enhances the binding of another protein molecule to a second site. In this manner, a series of weak interactions between individual components is amplified by protein–protein interactions to give an overall strong binding of the transcription factor (TF) complex to the promoter.

An example of a mammalian gene with multiple regulatory elements that have been defined by experiment, the rat *pepCK* gene, is shown in Figure 9.13A. This gene encodes phosphoenol pyruvate kinase, a

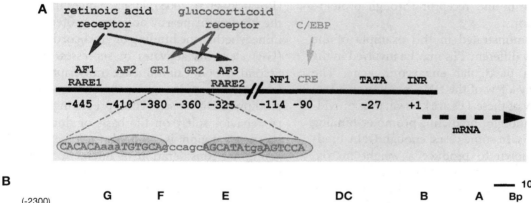

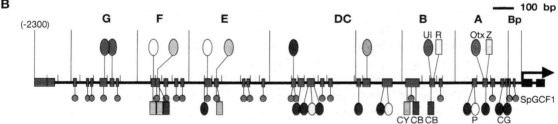

FIGURE 9.13. Regulatory elements of two promoters. (*A*) The rat *pepCK* gene. The relative positions of the TF-binding sites are illustrated (Yamada et al. 1999). The glucocorticoid response unit (GRU) includes three accessory factor–binding sites (AF1, AF2, and AF3), two glucocorticoid response elements (GR1 and GR2), and a cAMP response element (CRE). A dimer of glucocorticoid receptors bound to each GR element is depicted. The retinoic response unit (RAU) includes two retinoic acid response elements (RARE1 and RARE2) that coincide with the AF1 and AF3, respectively (Sugiyama et al. 1998). The sequences of the two GR sites and the binding of the receptor to these sites are shown. These sites deviate from the consensus sites and depend on their activity on accessory proteins bound to other sites in the GRU. This dependence on accessory proteins is reduced if a more consensus-like (canonical) GR element comprising the sequence TGTTCT is present. The CRE that binds factor C/EBP is also shown. (*B*) The 2300-bp promoter of the developmentally regulated gene *endo16* of the sea urchin (Bolouri and Davidson 2002). Different colors indicate different binding sites for distinct proteins and proteins shown above the line bind at unique locations, below the line at several locations. The regions A–G are functional modules that determine the expression of the gene in a particular tissue at a particular time of development and may either serve to induce transcription of the gene as a necessary developmental step (A, B, and G) or repress transcription (C–F) in tissues when it is not appropriate. (Reprinted, with permission, from Bolouri and Davidson 2002 [©2002 Elsevier].)

major enzyme for metabolism of glucose in mammals, and is regulated by four different hormones—glucocorticoids, glucagon, retinoic acid, and insulin—through a system of binding sites for particular transcription factors in the promoter region. The response of the cell to these agents involves binding of one of these hormones to a specific receptor protein and the subsequent binding of the hormone–receptor complex to specific sequences called response elements (REs) in the promoter region of responsive genes. The *pepCK* gene also responds to the level of cyclic AMP (cAMP) through a similar interaction. In addition, the gene has other characteristic and essential sequence features for RNAPII recognition, such as the TATA box, the initiation region (INR) that includes the TSS at +1. The REs are flanked by binding sites for other transcription factors that influence the effect of the bound receptor through protein–protein interactions.

Thus, as demonstrated in this example of the *pepK* gene, many different TFs may be involved in the regulation of a particular eukaryotic gene. The sequences of only a few of the DNA-binding sites recognized by many of these TFs are known, thus providing little information for predicting promoter-binding sites for these TFs. In some cases, enough DNA-binding sites are known to produce a weight matrix, described earlier in this chapter (Table 9.5). However, eukaryotic scoring matrices tend to be much more variable than prokaryotic matrices, so that the matrix is less useful for discriminating true binding sites from random sequence variation. If enough sequence information is known, such a matrix can be used to predict putative binding sites for a TF in a particular promoter. Because TF-binding sites may be detectable on either forward and complementary strands or present on both forward and complementary strands in a repeated configuration, both strands of the test sequence are generally searched for binding sites. Interpolated HMMs described previously for identifying prokaryotic genes have also been used for eukaryotic promoter identification (Ohler et al. 1999). This method identifies the most informative lengths of sequence in promoters and uses them for promoter prediction in test sequences.

As shown in Figure 9.13A, binding sites for TFs cluster in the promoter region of genes. This clustering is the basis of one method for promoter prediction discussed below. A search for binding sites in the Eukaryotic Promoter Database (EPD) (Table 9.6) and

a human first exon database showed that tandem binding sites for the same TF that are approximately 10 bp apart and expressing with a periodicity of 145 bp can be detected by this method. Such studies confirm that searching for multiple TF-binding sites can provide a more reasonable prediction of promoter function (Ioshikhes et al. 1999).

RNAPII Promoter Classification

Complexes of TFs bound to DNA can either activate or repress transcription through their interaction with RNAPII. Some quite remarkable variations of this theme can occur (Yamamoto et al. 1998), and the specific architecture of TF-binding sites in the promoter sequence can influence which genes are expressed or not expressed in the eukaryotic cell (Rogatsky et al. 2003). First, changes in the RE or in the binding of nearby accessory proteins can determine whether the binding of glucocorticoid response (GR) elements activates or represses transcription. Second, the GR can influence transcription simply by forming a complex with other factors and without binding to DNA itself. Thus, predicting the regulatory behavior solely on the basis of finding REs in a promoter region is probably not feasible without additional consideration of interactions among the regulatory elements and proteins themselves (Bucher et al. 1996). The correct interactions within the eukaryotic promoter are required for transcription of the gene by RNAPII at the biologically appropriate time.

Eukaryotic promoter sequences show variation not only between species, but also among genes within a species. A gene that is regulated by a certain set of signals during development will have a significantly different promoter than will a second gene that responds to a different set of signals. All of these genes are transcribed by RNAPII, but only when the required TFs for expression of a particular gene have bound to their respective binding sites and formed a suitable complex does transcription actually occur. For this reason, genes that share a regulatory response are expected to share common regulatory elements.

Developmentally regulated promoters can have modules of TF-binding sites, each of which regulates gene expression at one stage of development. A detailed laboratory and sequence analysis of the promoter of the sea urchin *endo16* gene, which is switched on and off during different stages of development, has revealed a remarkable level of organization of TF-binding sites, as illustrated in Figure 9.13B. The regions A–G are functional modules that determine the expression of the gene in a particular tissue at a particular time of development. It is informative to compare this modular level of regulation in the *endo16* gene with that of the *pepCK* gene in 9.13A. The latter gene is also regulated by a variety of hormonal signals, but in this case, the REs for these signals are clustered. These laboratory studies indicate the extent to which the binding sites for regulatory proteins can vary among different promoters. Analyzing promoters that respond to the same signals or at the same time in development to find common binding sites is therefore an approach that is likely to help discover how the promoters are regulated.

An analysis of the genes expressed in skeletal muscle for common TF-binding sites has been performed. Binding sites for TFs in skeletal muscle promoters were used to make scoring matrices, which were then used to find other muscle-regulated genes in genomic sequences. The ability of individual scoring matrices to locate signals in known muscle promoters, while at the same time not finding signals in control promoters, was then determined. The alignment scores for each matrix were then weighted in favor of the most informative matrices. The sum of these weighted scores gives a value between 0 (no promoter) and 1 (has promoter function), called the logit value of the promoter. Similar promoters from closely related species are also used to enhance the ability of the method to discriminate muscle promoters from other promoters in a method described as phylogenetic footprinting (Wasserman and Fickett 1998).

Because the usefulness of different scoring matrices for TF-binding sites is variable, other methods have been devised for weighting the scores obtained for an individual weight matrix on test sequences. One of these methods includes a new algorithm for determining the cutoff value using the background rate estimated on nonpromoters (see TFBind in Table 9.6). Scores of matches of weight matrices to test sequences follow the extreme value distribution and have also been used to evaluate matches (Claverie 1994; Claverie and Audic 1996). The application of neural networks for devising a weighting scheme as used in the gene prediction program GeneParser would be another method for weighting a group of scoring matrices to give maximum discrimination between promoter and nonpromoter sequences.

cDNA gene expression microarrays discussed in Chapter 13 (p. 636) can assist with discovering which genes are regulated in the same manner and therefore are expected to have binding sites for the same TFs (Cho et al. 1998; Eisen et al. 1998; Claverie 1999; Golub et al. 1999; Zhang 1999a,b). The promoter regions of these genes have been compared for the presence of common short words. The pentamer (5-mer) content of the promoter sequences of yeast genes that are coregulated during the cell cycle has been analyzed by the program wconsensus (described above, p. 392) and a Gibbs DNA sampler (similar to the Gibbs motif sampler described in Chapter 5 but adapted for DNA sequences) (Spellman et al. 1998; Zhang 1999b), and the results are available on a Web site (see Table 9.6).

In another study, pentamers and hexamers that are overrepresented among the upstream regions of cell-cycle-regulated genes in yeast were identified using a simple statistical sampling procedure. The sequences were divided into two sets; one set comprised cell cycle genes and the second set comprised the rest of the genome. A hexamer was then counted in both sets. The background number in the control set was used to identify overrepresented oligonucleotides in the cell cycle set. The actual number counted in the cell cycle genes was then compared to this expected value using a χ^2 test. For example, the hexamer ACGCGT is found with a variable location and orientation in the promoters of many cell cycle genes that are expressed during the late G_1 phase of the yeast cell cycle, whereas the pentamer CCCTT is located at positions −104 to −202 in one orientation in early G_1 (Wolfsberg et al. 1999). These types of analyses, which are available on Web sites (Table 9.6), demonstrate that computational analysis of the promoters of corelated genes reveals the presence of highly representative sequence patterns. Although some of these patterns correspond to the binding sites of transcription factors, others play a role that has yet to be determined. A similar method of oligomer counting has been used to identify overrepresented oligonucleotides with intron-containing genes in yeast and also to identify 3′ UTR signal for RNAs that encode proteins targeted to mitochondria (Jacobs Anderson and Parker 2000). Hence, the oligonucleotide scoring method shows considerable promise for the identification of regulatory sites in coregulated genes.

TABLE 9.6. *Promoter prediction programs, Web pages, and related information*

Name	Web address	Reference
BDNA video analysis of transcription factor–binding sites using conformational and physicochemical DNA features	see GeneExpress	Ponomarenko et al. (1999)
ConsInspector—see Transfac database[a]	http://www.gsf.de/biodv/consinspector.html	
Core-Promoter—for finding RNAPII promoters of human genes by quadratic discriminant analysis	http://rulai.cshl.edu	Zhang (1998a,b)
EPD Eukaryotic promoter database	http://www.epd.isb-sib.ch/; http://www.epd.isb-sib.ch/promoter_elements/	Bucher (1990); Périer et al. (1999, 2000)
EpoDB genes expressed during vertebrate erythropoiesis; also links to pancreas and other gene expression databases	http://www.cbil.upenn.edu/; temporarily not available	Stoeckert et al. (1999)
GeneExpress analysis of transcriptional regulations with TRRD database	http://wwwmgs.bionet.nsc.ru/mgs/systems/geneexpress/	Kolchanov et al. (1999a,b)
Genome inspector for combined analysis of multiple signals in genomes	http://www.gsf.de/biodv/genomeinspector.html	Quandt et al. (1997)
GrailII[b] prediction of TSS by neural networks based on scores of characteristic sequence patterns and composition	http://compbio.ornl.gov/Grail-1.3/	Uberbacher and Mural (1991); Uberbacher et al. (1996)
MAR-FINDER for finding matrix attachment regions	http://www.futuresoft.org/MAR-Wiz/	Kramer et al. (1997); Singh et al. (1997)
MatInd—see Transfac database		
MatInspector—see Transfac database	http://www.genomatix.de/cgi-bin/matinspector/matinspector.pl	
Mirage (Molecular Informatics Resource for the Analysis of Gene Expression)	http://www.ifti.org/	see Web page
NNPP Promoter Prediction by Neural Network for prokaryotes or eukaryotes	http://www.fruitfly.org/seq_tools/promoter.html	Reese et al. (1996)
NSITE—search for TF-binding sites or other consensus regulatory sequences	http://www.bioinformatics.vg/biolinks/bioinformatics/verbose/Promoter%2520Scan.shtml	see Web site
Nuclear (including glucocorticoid) receptor resource	http://nrr.georgetown.edu/NRR/nrrhome.htm	Martinez et al. (1997)
OOTFD Object-Oriented Transcription Factor Database	http://www.ifti.org/cgi-bin/ifti/ootfd.pl	Ghosh (1998)
PLACE plant *cis*-acting regulatory elements	http://www.dna.affrc.go.jp/htdocs/PLACE/	Higo et al. (1999)
PlantCARE plants *cis*-acting regulatory elements	http://oberon.fvms.ugent.be:8080/PlantCARE/	Rombauts et al. (1999)
Pol3scan for RNAP III/tRNA promoter sequences using pattern scoring matrices	perform Web search	Pavesi et al. (1994)
Polyadq for locating polyadenylation sites	http://rulai.cshl.edu	Tabaska and Zhang (1999)

Description	URL	Reference
Promoter element weight matrices and HMMs	http://www.epd.isb-sib.ch/promoter_elements/	Bucher (1990)
Promoter II for recognition of PolII sequences by neural networks	http://www.cbs.dtu.dk/services/promoter/	Knudsen (1999)
Promoter Inspector	http://www.gsf.de/boidv/	Klingenhoff et al. (1999)
PromoterScan[c]	http://bimas.dcrt.nih.gov/molbio/proscan/	Prestridge (1995) and see Web site
Sequence walkers for graphical viewing of the interaction of regulatory protein with DNA-binding site	uttp://www-lecb.ncifcrf.gov/~toms/walker/narcoverlogowalker.html	Schneider (1997)
Signal scan for transcriptional elements	http://bimas.dcrt.nih.gov:80/molbio/signal/	Prestridge (1991, 1996)
TESS for searching for transcription factor–binding sites	http://www.cbil.upenn.edu/tess/	Schug and Overton (1997a,b)
TFBind for transcription factor–binding sites	http://tfbind.ims.u-tokyo.ac.jp	Tsunoda and Takagi (1999)
Thyroid receptor resource[d]	http://xanadu.mgh.harvard.edu/receptor/trrfront.html	see Web page
Transfac programs providing search for TF-binding sites. MatInd for making scoring matrices and MatInspector for searching for matches to matrices	http://www.genomatix.de/cgi-bin/matinspector/matinspector.pl	see http://www.gsf.de/biodv/staff_pub.html; Knüppel et al. (1994); Quandt et al. (1995); Heinemeyer et al. (1999); Klingenhoff et al. (1999).
TRRD transcriptional regulatory region database; see GeneExpress		Kolchanov et al. (1999a)
TSSG, like TSSW but based on sequences from a different promoter database	http://genomic.sanger.ac.uk/gf/gf.shtml; http://searchlauncher.bcm.tmc.edu/seq-search/gene-search.html	see Web site
TSSW; recognition of human PolII promoter region and start of transcription by linear discriminant function analysis	http://genomic.sanger.ac.uk/gf/gf.shtml; http://searchlauncher.bcm.tmc.edu/seq-search/gene-search.html	see Web site
Yeast cell cycle gene retrieval and promoter analysis	http://www.ncbi.nlm.nih.gov/CBBresearch/Landsman/Cell_cycle_data/upstream_seq.html; http://www.ncbi.nlm.nih.gov/CBBresearch/Landsman/Cell_cycle_data/	Wolfsberg et al. (1999)
Yeast cell cycle analysis project	http://genome-www.stanford.edu/cellcycle/info	Spellman et al. (1998)

Multiple methods of analysis are offered at sites http://searchlauncher.bcm.tmc.edu/seq-search/gene-search.html and http://genomic.sanger.ac.uk/gf/gf.shtml. Lists of Web sites are given at http://linkage.rockefeller.edu/soft/linkage/ and http://biobenchelper.hypermart.net/index.html. A comparison of many of the promoter prediction programs included in this table and several additional ones on a small number of promoter-containing sequences not used in program training is available (Fickett and Hatzigeorgiou 1997).

[a] MatInspector DOS, Windows 95 and NT, and Mac versions and ConsInspector DOS and Mac versions available by FTP from ariane.gsf.de/pub/.

[b] GrailII must be given both gene and promoter sequences.

[c] Accepts one person at a time; DOS version also available.

[d] Includes links to other receptor databases.

Prediction Methods for Locating and Analyzing RNAPII Promoters

A number of methods for predicting the location of RNAPII promoters in genomic DNA have been derived. Several Web sites that offer an analysis are listed in Table 9.6. Also shown in this table are a number of Web sites that provide databases and information on TFs and their DNA-binding sites and other information related to transcriptional regulation in eukaryotes. A test analysis of these and several additional programs not listed in the table on a small number of new promoter sequences has been described (Fickett and Hatzigeorgiou 1997). The programs predicted 13–54% of the TSSs correctly, but each program also predicted a number of false-positive TSSs.

The following list briefly describes the approaches used by some of the methods of analysis and programs included in Table 9.6 (for additional information on program availability, see Fickett and Hatzigeorgiou 1997; Frech et al 1997).

1. Use of a neural network trained on the TATA and Inr sites, allowing for a variable spacing between the sites (NNPP) or a neural network–genetic algorithm approach to identify conserved patterns in RNAPII promoters and conserved spacing among the patterns (PROMOTER2.0).

2. Recognition of a TATA box using a weight matrix and an analysis of the density of TF sites. The density of TF sites at least 50 bp apart in known promoter sequences of the Eukaryotic Promoter Database (EPD) and on a set of nonpromoter primate sequences from GenBank is compared and used to produce a promoter recognition profile (PromoterScan).

3. Use of a linear discriminant function as described above for gene prediction, but in this case, used for distinguishing features of promoter sequences from nonpromoter sequences. The function is based on a TATA box score, triplet base-pair preferences around TSS, hexamer frequencies in consecutive 100-bp upstream regions, and TF-binding-site prediction (TSSD and TSSW).

4. A quadratic discriminant analysis similar to that described above for gene prediction, but in this case, applied to variable lengths of sequence in the promoter region. The frequency of pentamers in a contiguous set of thirteen 30-bp windows and also in a second set of five 45-bp windows in the same 240-bp region was compared. This double-overlapping window appeared to reduce the background noise and to enhance the transcriptional signal from the promoter region (CorePromoter).

5. Searches of weight matrices for different organisms against a test sequence (TFSearch/TESS). User-provided limits on type of weight matrix, key set of matches (core similarity) to individual matrices, and range of match scores (matrix similarity) are used, and also new matrices are generated (MatInspector and ConsInspector).

6. Evaluation of test sequences for the presence of clustered groups or modules of TF-binding sites that are characteristic of a given pattern of gene regulation (FastM).

Computational tools for gene and promoter analysis play an important role in predicting the proteins made by a particular organism and deciding how these proteins are regulated during development, in response to environmental stress or during a disease process. As more cDNA sequences from eukaryotic genomes are obtained in the laboratory, there will be a corresponding decrease in need for gene prediction. However, deciding how eukaryotic promoters function remains a challenging computational problem. The solution to understanding gene regulation in eukaryotes is going to depend on careful and detailed sequence analysis to locate the regulatory sites to which transcriptions bind and on laboratory experiments to determine how these sites relate to protein–protein interactions that lead to gene expression. Comparative genome analysis of promoter sequences (phylogenetic footprinting and genome shadowing) in closely related species, such as primates other than humans, is playing a significant role in locating the binding sites.

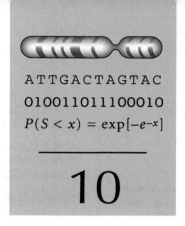

ATTGACTAGTAC
010011011100010
$P(S < x) = \exp[-e^{-x}]$

10

Protein Classification and Structure Prediction

C O N T E N T S

INTRODUCTION

One of the major goals of protein sequence analysis is to understand the relationship between amino acid sequence and three-dimensional structure in proteins. If this relationship were known, the structure of a protein could be reliably predicted from the amino acid sequence. Unfortunately, the relationship between sequence and structure is not that simple. Much progress has been made in categorizing proteins on the basis of structure or sequence, and this type of information is very useful for protein modeling. This chapter begins with a review of protein synthesis and structure, which is followed by a review of the methods used to classify proteins by structure, sequence, or both structure and sequence. Finally, the methods used to predict protein structure and evaluate the success of these predictions is reviewed.

CHAPTER GUIDE FOR BIOLOGISTS

Most methods of predicting protein secondary or tertiary structure assume that local sequence variation influences the type of local structure. The earliest methods (e.g., Chou–Fasman) scanned the composition of amino acids in a local sequence window that were typical of those found in known secondary structures and used these scans to predict the secondary structure of the central amino acid. Subsequent methods looked for amino acid patterns in these windows that were similar to those in known structures. The nearest-neighbor method simply compares the local sequence window to a database of sequence windows from known secondary elements.

Additional novel computational applications, such as neural networks (NNs), have been developed to define these patterns. In the NN application, the sequence window is input to a series of "neurons," which integrate the sequence information and produce an output signal that predicts the secondary structure of the central amino acid. This network is trained to give the correct result for a set of known structures by adjusting the signals passed between neurons. Once trained, the NNs can be used to predict the secondary structure of a protein of unknown structure. This prediction is improved by reading a window from a multiple sequence alignment (msa) of related sequences instead of one sequence.

Position-specific scoring matrices (PSSMs), described in Chapter 5 as representing a local msa of related sequences, are also used in protein structure prediction. When one of these matrices produces a strong match with an amino acid pattern in a protein of known three-dimensional structure, a similar structure may be present in other proteins matched by the PSSM.

Hidden Markov models (HMMs), used previously in Chapter 5 to model an msa, are also used in structure prediction, but in quite different ways. HMMs are designed either to match known protein structures as arrangements of secondary structural elements, sets of amino acid patterns (I-sites) that are associated with three-dimensional structures, or structural environments based on solvation energies of atoms in the environment of an amino acid.

Another prediction tool is the structural profile. The sequence profile described in Chapter 5 is a scoring matrix representing an msa and is similar to a PSSM with extra columns for gap penalties. The structural profile models the structure of a protein. For each amino acid in the structure, a column of substitution scores is given in the profile. These scores describe how well each of the 20 amino acids will substitute for a particular amino acid at one position in the three-dimensional structure. The profile may then be aligned with a new protein sequence to measure how good a fit there is to the profile.

Energy minimization is an additional method used to predict protein structures; however, the method is computationally intensive because there are so many possible ways to fold a protein. Therefore, the method is most useful for small proteins of a hundred or so amino acids.

A big leap ahead in predicting protein three-dimensional structure was made by the Rosetta method, which is based on a model of protein folding in which local interactions along the protein chain determine the eventual global folding pathway for the entire protein. In the model, a short length of protein sequence alternates among a set of possible structures that are favored by the sequence. Local regions then interact to produce additional structures that are energetically favorable until a globally favorable structure is produced. This method depends on having experimental data on the stability of local structures formed by amino acid sequences and pattern-finding methods to locate these patterns in other sequences for structure prediction purposes. Combinations of local patterns can be modeled by HMMs and used to search sequences for these combinations as a step in structure prediction.

The accuracy of a protein structure prediction can be tested by determining how well a trained model predicts a known structure. In many cases, however, these tests are unknowingly biased by sequences chosen by the investigator, and when the method is used to model completely novel sequences with their structures (i.e., the CASP project), they do not perform as well.

CHAPTER GUIDE FOR COMPUTATIONAL SCIENTISTS

Proteins are chains of 20 amino acids that are the source of most biological activity and structure in cells. They are synthesized by organisms using information encoded in genomes in which three consecutive base positions in a DNA sequence specify the amino acid of the corresponding colinear protein sequence. A protein does not have biological activity until it folds into a three-dimensional structure held together by chemical interactions between the amino acids. Folding begins by interactions between closer amino acids along the protein sequence to form helical structures known as α helices. More distant interactions then align the amino acid chains into sheets called β sheets. These initial structures are known as secondary structures. After additional amino acid interactions occur, the secondary structural elements fold into a three-dimensional structure known as the tertiary structure of the protein. The protein may also be further modified chemically by addition of additional side groups (e.g., phosphate groups or sugars), by cleavage into fragments, or by joining into complexes with other proteins or cell components.

The structure of each protein is determined primarily by its amino acid sequence. Hence, proteins with a similar sequence have a common tertiary structure, thus making identification of sequence similarity one of the most useful methods for protein structure prediction. Another important discovery about protein tertiary structure is that proteins with different sequences can fold into the same three-dimensional structure with a common arrangement of helical and strand elements known as a structural fold. More than 3000 different structural folds are found in proteins. Several structural biology research groups have produced databases of these fold families. Proteins have also been grouped into more general classes based on the content and arrangement of their secondary structural elements.

Of considerable interest and challenge is determining whether the structure of any protein can be predicted by its amino acid sequence. Local interactions in a protein that give rise to α helices are more readily predicted as are the structures of proteins when they result from interactions between amino acids that are close together in the sequence. The computational approaches range from a simple compositional analysis of amino acid sequences to the use of more complex pattern-learning tools such as NNs and HMMs based on pattern variation in sequences with the same structural fold. The most complex approach is to compute a structure that is the most energetically stable by examining how well all possible amino acid interactions provide the most favorable chemical bonds, but this method is only feasible for smaller proteins. Contests are held every 2 years in which the many methods that have been developed are tested on a set of proteins of known, but not yet published, three-dimensional structure.

WHAT SHOULD BE LEARNED IN THIS CHAPTER?

- The underlying structure of proteins.
- The relationships between amino acid sequence and structure.
- How structures are aligned to find conserved units of structure—the structural fold.
- How proteins are classified based on their three-dimensional structure.
- How proteins are clustered based on conserved sequence.
- Which computational methods are used to predict secondary and three-dimensional structure.
- An appreciation of the uncertainty associated with protein structure prediction and how reliability is assessed.

Glossary Terms

The glossary terms list is broken down into two sections: (1) computational methods used for analysis of protein sequences and structures described in this chapter, and (2) terms that describe protein structural features and classification. Some terms refer to sequence similarity, some to structural similarity, and some to both sequence and structure, and it is important not to confuse them. Other terms are used to describe protein structure and the methods for displaying and comparing protein structure.

Computational Methods Used for Protein Analysis and Classification

Back propagation is a method used in training a neural network to predict regions of protein secondary structure. The weights of signals passed down the network are adjusted slightly to produce a more correct result. The process starts at the output end of the network and works back through each connection in turn, hence the back-propagation effect.

Block is a conserved amino acid sequence pattern in a family of proteins. The pattern includes a series of possible matches at each position in the represented sequences, without any inserted or deleted positions in the pattern or in the sequences. By way of contrast, sequence profiles are a type of scoring matrix that represents a similar set of patterns, but includes insertions and deletions. Profile HMMs are hidden Markov models of such gapped patterns (see Chapters 5 and 6). There are 2290 HMM profile models in Pfam release 5.4 described below (p. 446).

Cluster (protein sequences) A cluster of protein sequences is a set of closely related protein sequences within a larger group of more loosely related sequences. The group is represented by a graph with the sequences as vertices and with edges joining related sequences. The strength of the edges is determined by the local alignment score or expect value of the sequence alignment score. A cluster is a set of the sequences that are joined by strong edges.

Contact potential method A method for describing how well each amino acid will fit into a known protein structure based on the predicted distances with neighboring atoms in the structures. These distances determine how well the atoms interact, whether functionally, not at all, or repulsively.

Discrete space model An alternative name for a hidden Markov model.

Distance matrix In the analysis of protein structures, a distance matrix is a two-dimensional plot with the sequence of one protein listed down the side and across the top. The closest amino acids in the structure are plotted, thus creating patterns of dots that represent three-dimensional arrangements of secondary structures in each protein. For example, α helices will appear as diagonal rows of dots because of the closeness of the adjacent amino acids in the helix. The distance matrix is used to find proteins that have the same arrangement of secondary structural elements.

Domain (sequence context). See Homologous domain.

Double dynamic programming A method for identifying proteins with the same three-dimensional structure by aligning the structures. A three-dimensional view is made for each amino acid by plotting the amino acid and a small number of adjacent amino acids on a three-dimensional graph (e.g., x, y, z coordinates). The same is done for each amino acid in the second sequence using the same coordinate system. The views from each amino acid are then compared in a dynamic programming matrix, the object being to find a good alignment based on having similar angular views. The results found for all amino acids in each sequence are then compared in a second dynamic programming matrix. If a particular amino acid in one sequence is in a similar structural context to one in the second sequence, a good score will be obtained. A similar alignment of the contexts of all amino acids in the sequences will demonstrate that they are similar at the three-dimensional structural level.

Energy minimization In three-dimensional protein structure prediction, energy minimization is the trial arrangement of amino acids in the structure with the goal of finding the most energetically stable interactions among the amino acids.

Family (sequence context), as defined originally by Dayhoff et al. (1978), is a group of proteins of similar biochemical function that are more than 50% identical when aligned. This same cutoff is still used by the Protein Information Resource (PIR). A protein family comprises proteins with the same function in different organisms (orthologous sequences) but may also include proteins in the same organism (paralogous sequences) derived from gene duplication and rearrangements (Henikoff et al. 1997). If a multiple sequence alignment of a protein family reveals a common level of similarity throughout the lengths of the proteins, PIR refers to the family as a homeomorphic family. The aligned region is referred to as a homeomorphic domain, and this region may comprise several smaller homology domains that are shared with other families. Families may be further subdivided into subfamilies or grouped into superfamilies based on respective higher or lower levels of sequence similarity, respectively (Barker et al. 1995; http://www-nbrf.georgetown.edu/). The SCOP database described below (p. 433) reports more than 2000 families, and the CATH database, also described below, reports more than 3000 families.

When the sequences of proteins with the same function are examined in greater detail, some are found to share high sequence similarity. They are obviously members of the same family by the above criteria. However, others are found that have very little, or even insignificant, sequence similarity with other family members. In such cases, the family relationship between two distant family members A and C can often be demonstrated by finding an additional family member B that shares significant similarity with both A and C (Pearson 1996; Park et al. 1997). Thus, B provides a connecting link between A and C. Another approach is to examine distant alignments for highly conserved matches (Patthy 1987, 1996). At a level of identity of >50%, proteins are likely to have the same three-dimensional structure, and the identical atoms in the sequence alignment will also superimpose within approximately 1 Å in the structural model (Holm and Sander 1994). Thus, if the structure of one member of a family is known, a reliable prediction may be made for a second member of the family, and the higher the identity level, the more reliable the prediction. Protein structural modeling can be performed by examining how well the amino acid substitutions fit into the core of the three-dimensional structure.

Gating function A mathematical formula for converting a range of value to binary units, usually of 0 or 1. The function is used for integrating information arriving at each collecting neuron in a neural network for predicting protein secondary structure.

Hidden layer A layer of artificial neurons that lie between the input and output layers of a neural network. The layer serves to integrate information from different amino acid positions in the input layer and passes this information along usually to the output, predictive layer.

Hidden Markov model (HMM) (as applied to structural modeling). The HMM is used in a number of different ways in protein structure prediction. First, HMMs are used to model the sequence variation that is found within families of protein sequences that have the same three-dimensional structures. As in HMMs used to model msa's, the model starts with a begin state, then includes states that match one amino acid position at a time in a three-dimensional structure, and an end state. These states are joined by transition probabilities between states that can be used to follow different paths through different sets of states that represent structural variation. Second, HMMs are used to model short amino acid patterns (I-sites) and structures that are associated with these patterns. States in these models store sequence and structural information and transition probabilities are also used to follow different paths through the model that represent structural variation. HMMs have also been used to represent transmembrane proteins and the structural environment of amino acids in known structures in order to perform structure prediction of a protein sequence of unknown structure.

Homologous domain (sequence context, also see Domain, structural context) refers to an extended sequence pattern, generally found by sequence alignment methods, that indicates a common evolutionary origin among the aligned sequences. A homology domain is generally longer than a motif (see Motif entry). The domain may include all of a given protein sequence or only a portion of the sequence. Some domains are complex and made up of several smaller homology domains that became joined to form a larger one during evolution. A domain that covers an entire sequence is called the homeomorphic domain by PIR (Barker et al. 1996; see http://www-nbrf.georgetown.edu/).

Input layer is the layer of the neural network for protein secondary structure prediction in which the sequence is located. A sliding window of sequence is examined and a structural prediction is made for the central amino acid in the window. Each amino acid in the window is represented by an individual unit of information (a neuron) in which one bit corresponding to the amino acid is turned on by giving it an "on" value of 1. Sometimes, aligned sequences are used in the input layer to include sequence variability in the network. In such cases, each aligned position provides the information for the neuron and any additional amino acid information is added by turning on additional bits. Signals pass from the input layer to one or more hidden layers, eventually reaching an output layer that makes predictions about structure.

I-sites comprise a set of sequence patterns approximately 3–15 amino acids long that correlate strongly with local protein structure and can be used to predict local structure or to determine which parts of a protein sequence fold first.

Module is a region of conserved amino acid patterns comprising one or more motifs and considered to be a fundamental unit of structure or function. The presence of a module has also been used to classify proteins into families.

Motif (sequence context) is a conserved pattern of amino acids that is found in two or more proteins. In the Prosite catalog, a motif is an amino acid pattern that is found in a group of proteins that have a similar biochemical activity and that often is near the active site of the protein. Examples of sequence motif databases are the Prosite catalog (http://www.expasy.ch/prosite) and the Stanford Motifs Database (http://dna.stanford.edu/emotif/).

Nearest-neighbor method (secondary structure prediction) A method in which a sequence of approximately 15 amino acids within a sliding window is compared to a database of sequences of the same length obtained from proteins of known structure. The closest matches of the window sequence to the database are used to predict the secondary structure of the middle amino acid in the window. The results for consecutive window positions are then averaged.

Neural network (NN) is a collection of neurons (computational units of information) organized into layers (input, hidden, and output). In protein structure prediction, the network is a feed-forward type in which information in a sliding window of sequence is placed in the input layer, which then transmits signals to the hidden layers and thence to the predictive output layer. The signals entering each neuron are weighted and then integrated to produce a binary output signal. The weights are adjusted by training the neural network to produce a good prediction for a protein of known secondary structure.

Neuron is the basic unit of a neural network, which is either labeled with an input sequence (input layer) or is a collective computational unit that receives a weighted signal from other neurons, integrates them using a gating function that generates a binary number usually either close to 0 or 1, and then transmits this signal to additional neurons. At the end of the network, the output neurons integrate the signals reaching them and produce a predictive output value.

Output layer is the layer of a neural network usually composed of three neurons, each one producing a signal that predicts one primary secondary structure type, α helix, β strand, or other structure. Information flows from the input layer (containing the protein sequence to be predicted), through one or more hidden layers, and then to the output layer of neurons.

Perceptron is a type of neural network that has only input and output layers. In structure prediction, a series of adjacent amino acid positions in a sequence window may be used to predict the secondary structure of the middle amino acid. The hidden layer provides an increased ability to recognize patterns in the window that influence this prediction.

Position-specific scoring matrix (PSSM) (sequence context, also known as weight or scoring matrix) represents a conserved region in an msa with no gaps. Each matrix column represents the variation found in one column of the msa.

Profile (sequence context) is a scoring matrix that represents an msa of a protein family. The profile is usually obtained from a well-conserved region in an msa. The profile is in the form of a matrix with each column representing a position in the alignment and each row one of the amino acids. Matrix values give the likelihood of each amino acid at the corresponding position in the alignment. The profile is moved along the target sequence to locate the best-scoring regions by a dynamic programming algorithm. Gaps are allowed during matching and a gap penalty is included as a negative score in this case when no amino acid is matched. A sequence profile may also be represented by a hidden Markov model, referred to as a profile HMM.

State is a term used in HMMs for structure prediction to represent known relationships between sequence patterns and known associations with protein structure.

Structural alignment This term refers to an alignment of proteins by their structural features rather than by sequence.

Superfamily is a group of protein families of the same or different lengths that are related by distant yet detectable sequence similarity. Members of a given superfamily thus have a common evolutionary origin. Originally, Dayhoff defined the cutoff for superfamily status as being the chance that the sequences are not related of <10⁻⁶, on the basis of an alignment score (Dayhoff et al. 1978). Proteins with few identities in an alignment of the sequences, but with a convincingly common number of structural and functional features, are placed in the same superfamily. At the level of three-dimensional structure, superfamily proteins will share common structural features such as a common fold, but there may also be differences in the number and arrangement of secondary structures. The PIR resource uses the term "homeomorphic superfamilies" to refer to superfamilies that are composed of sequences that can be aligned from end to end, representing a sharing of single sequence homology domain, a region of similarity that extends throughout the alignment. This domain may also comprise smaller homology domains that are shared with other protein families and superfamilies. Although a given protein sequence may contain domains found in several superfamilies, thus indicating a complex evolutionary history, sequences will be assigned to only one homeomorphic superfamily based on the presence of similarity throughout an msa. The superfamily alignment may also include regions that do not align either within or at the ends of the alignment (Barker et al. 1995, 1996; http://www.nbrf.georgetown.edu/). In contrast, sequences in the same family align well throughout the alignment. The SCOP Web site reports 820 superfamilies (release 1.50), and the CATH Web site (version 1.7 beta) reports 900 superfamilies (sites described below). (See also Family entry.)

Threading is the matching of a protein sequence to a set of structural models (e.g., structural profiles of certain types of HMMs) to determine if the amino acids will fit into any of the structures represented by those models.

Transition probability refers to the probability that a transition will occur between two states of an HMM.

Vector method A method used to identify proteins that have the same three-dimensional arrangement of a small number of secondary structures. The secondary structures, which comprise many atomic points, are represented by vectors that can then be aligned quite readily.

Protein Structure and Classification

Active site is a localized combination of amino acid side groups within the tertiary (three-dimensional) or quaternary (protein subunit) structure that can interact with a chemically specific substrate and that provides the protein with biological activity. Proteins of very different amino acid sequences may fold into a structure that produces the same active site.

Architecture describes the relative orientations of secondary structures in a three-dimensional structure without regard to whether they share a similar loop structure. In contrast, a fold is a type of architecture that also has a conserved loop structure. Architecture is a classification term used by the CATH database (http://www.biochem.ucl.ac.uk/bsm/cath/). SMART (Simple Modular Architecture Research Tool) is a Web tool at http://smart.embl.de/ that supports comparative studies of complex domain architectures in proteins (Letunic et al. 2004). A similar resource may be found at http://hydra.icgeb.trieste.it/sbase/ (Vlahovicek et al. 2003).

Class describes protein domains according to their secondary structural content and organization. Four classes were originally recognized by Levitt and Chothia (1976), and several others have been added in the SCOP database described below. Three classes are given in the CATH database: mainly-α, mainly-β, and α–β, with the α–β class including both alternating α/β and α+β structures. Thus, class 4 of the SCOP database is included in class 3 of the CATH database.

Contact order is a measure of the average distance between all amino acids that form contacts in a protein divided by the length of the protein. If the number is smaller, then the number of short distance interactions along the amino acid chain is higher, the protein folds more quickly, and structure prediction is easier.

Core is the portion of a folded protein molecule that comprises the hydrophobic interior of α helices and β sheets. The compact structure of the core brings side groups of amino acids into close enough proximity so that they can interact. When comparing protein structures, as in the SCOP database, core refers to the region common to most of the structures that share a common fold or that are in the same superfamily. In structure prediction, core is sometimes defined as the arrangement of secondary structures that is likely to be conserved during evolutionary change (Madej et al. 1995). A library of protein cores designated LPFC is maintained at Stanford University at http://www-camis.stanford.edu/projects/helix/LPFC/ and is based on msa's using amino acid scoring matrices based on structural substitutions.

DASEY (directional atomic solvation energy) is a measure of hydrophobicity of the structural environment of an amino acid which is obtained in the laboratory by measuring the energy to transfer atoms from a polar solvent to a nonpolar solvent.

Domain (structural context; also see Homologous domain entry) refers to a segment of a polypeptide chain that can fold into a three-dimensional structure irrespective of the presence of other segments of the chain. The separate domains of a given protein may interact extensively or may be joined only by a length of polypeptide chain. A protein with several domains may use these domains for functional interactions with different molecules. 3Dee, a database of protein domain definitions, is provided at http://barton. ebi.ac.uk/servers/3Dee.html. A structural classification of protein domains in the PDB database is maintained at http://www.ebi.ac.uk/dali/domain/ (Holm and Sander 1998). Domain databases may be found at http://hydra. icgeb.trieste.it/sbase/ (Vlahovicek et al. 2003) and http:// smart.embl.de/ (Letunic et al. 2004).

Family (structural context), as used in the FSSP (fold classification based on structure–structure alignment of proteins) database (Holm and Sander 1998) and the DALI/ FSSP Web site (see below), refers to two structures that have a significant level of structural similarity but not necessarily significant sequence similarity.

Fold is a term with similar meaning to structural motif or supersecondary structure, but in general refers to a somewhat larger combination of secondary structural units in the same configuration. Thus, proteins sharing the same fold have the same combination of secondary structures that are connected by similar loops. An example is the Rossman fold comprising several alternating α helices and parallel β strands. In the SCOP, CATH, and FSSP databases described below (p. 433), the known protein structures have been classified into hierarchical levels of structural complexity with the fold as a basic level of classification. A classification of the currently known protein structures in the Brookhaven Protein Data Bank is maintained on the DALI Web site at http://www.ebi.ac.uk/dali/ (Holm and Sander 1998) and exceeds 1375 in number. The number of distinct folds in the SCOP database is 800 (release 1.65) and the number of the equivalent topological families in the CATH database is 3300 (version 2.5.1 release). These databases are described below. Foldon is a related term that has been used to describe an independently folding unit (Panchenko et al. 1996, 1997).

Homology, structural indicates that two protein sequences were derived from a common ancestor, as evidenced by their having significant sequence similarity. Two proteins may have significant structural similarity but not detectable sequence similarity. Therefore, it may be incorrect to refer to these proteins as homologous in the absence

of evidence that they are derived from a common ancestor.

Motif (structural context) is a term with similar meaning to supersecondary structure (below).

Position-specific scoring matrix (PSSM)—three dimensional (structural context) represents the amino acid variation found in an alignment of proteins that fall into the same structural class. Matrix columns represent the amino acid variation found at one amino acid position in the aligned structures (Kelley et al. 2000).

Primary structure refers to the linear amino acid sequence of a protein, which chemically is a polypeptide chain composed of amino acids joined by peptide bonds.

Profile (structural context) is a scoring matrix that represents which amino acids should fit well and which should fit poorly at sequential positions in a known protein structure. Profile columns represent sequential positions in the structure, and profile rows represent the 20 amino acids. As with a sequence profile, the structural profile is moved along a target sequence to find the highest possible alignment score by a dynamic programming algorithm. Gaps may be included and receive a penalty. The resulting score provides an indication as to whether the target protein might adopt such a structure.

Quaternary structure is the three-dimensional configuration of a protein molecule comprising several independent polypeptide chains. A Web site for predicting quaternary structure is described at http://msd.ebi.ac.uk/Services/ Quaternary/quaternary.html. A database of experimentally identified interacting domains of protein subunits (DIP) is maintained at http://dip.doe-mbi.ucla.edu; Xenarios et al. 2000; also see Table 10.4 below).

Secondary structure refers to the interactions that occur between the C=O and NH groups on amino acids in a polypeptide chain to form α helices, β sheets, turns, loops, and other forms, and that facilitate the folding into a three-dimensional structure.

Supersecondary structure refers to a combination of several secondary structural elements produced by the folding of adjacent sections of the polypeptide chain into a specific three-dimensional configuration, such as combinations of α helices and β strands to form a loop. These elements attract each other in fold space. An example is the helix-loop-helix motif. A database of supersecondary structures is found at http://www-cryst.bioc.cam.ac.uk/~sloop/ (Burke et al. 2000).

Tertiary structure is the three-dimensional or globular structure formed by the packing together or folding of secondary structures of a polypeptide chain.

SOME PROTEIN STRUCTURES CAN BE PREDICTED

A protein's polypeptide chain is first assembled on the cell's ribosome in the cytoplasm using the nucleic acid codon sequence on mRNA as a template, as illustrated in Figure 10.1. The resulting linear chain of amino acids forms secondary structures in the nascent protein through the formation of hydrogen bonds between amino acids in the polypeptide chain. Through further interactions among amino acid side groups, these secondary structures then fold in a specific pathway into a three-dimensional structure. Chaperone proteins and cellular membranes may assist with this folding process. Further processing of the protein by cleavage or chemical modification may also be necessary for the protein to have biological activity. Therefore, protein structure is largely specified by amino acid sequence. However, it is not fully understood why one set of interactions of the many that are possible occurs in the formation of tertiary structure (Branden and Tooze 1991).

Some protein sequences have distinct amino acid motifs that always form a characteristic structure. These structures can be from sequence using available methods and information. For most proteins, however, the overall accuracy of predicting the location of secondary structures is approximately 70–75%, with the prediction for α helices being more than 90% accurate. Methods for matching sequence to three-dimensional structure have been formulated, but they are not yet very reliable. However, great forward strides have been made as more structures are known and better computational models are produced that relate sequence to structure. There is a very active community in structural biology and bioinformatics working on improvements. The rapid increases in the number of protein sequences and structures highlight the need for such an effort.

As of February, 2004, more than 22,044 protein structures had been deposited in the Brookhaven Protein Data Bank (PDB), and 144,731 protein sequence entries were in the SwissProt protein sequence database, a ratio of approximately one structure to seven sequences. The number of protein sequences can be expected to increase dramatically as more sequences are produced by research laboratories and more genome sequencing projects.

As more and more sequences and structures are found, the goals of reliable structure prediction are more within reach. It was first estimated that there were approximately 1000 protein families composed of members that share detectable sequence similarity (Dayhoff et al. 1978; Chothia 1992); this number has now been updated to more than 3000 by the CATH database (visit http://www.biochem.ucl.ac.uk/bsm/cath/). As new protein sequences are obtained, they are found to be similar to other sequences already in the databases and thus can be expected to share structural features with these proteins. Whether this low number of families represents physical restraints in folding the polypeptide chain into a three-dimensional structure or merely the selection of certain classes of three-dimensional structure by evolution has yet to be discovered (Gibrat et al. 1996). The sequence alignment, motif-finding, block-finding, and database similarity search methods described in Chapters 3, 5, and 6 may be used to discover these familial relationships. Understanding these relationships is fundamentally important because this information can greatly assist with structural predictions. As discussed below, information from amino acid substitutions at a particular sequence position as obtained from an msa has been found to increase significantly the prediction of secondary structures from

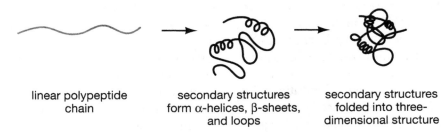

linear polypeptide   secondary structures   secondary structures
chain   form α-helices, β-sheets,   folded into three-
     and loops   dimensional structure

FIGURE 10.1. Pathway for folding a linear chain of amino acids into a three-dimensional protein structure.

protein sequences. A second major advance in protein structure analysis was the revelation that proteins adopt a limited number of three-dimensional configurations.

There are several characteristics of protein structure that affect amino acid substitution and three-dimensional configurations in protein molecules. Protein structures include a core region comprising secondary structural elements packed in close proximity in a hydrophobic environment. Specific interactions between the amino acid side chains occur within this core structure. At a given amino acid position in a given core, the amino acids that can substitute are limited by space and available contacts with other nearby amino acids. Outside of the core are loops and structural elements that are in contact with water, other proteins, and other structures. Amino acid substitutions in these regions are not as restricted as in the core. Through a close comparison of a newly generated three-dimensional structure with previously found structures, the new structure has often been found to fold into α-helical and β-sheet structural elements in the same order and spatial configuration as one or more structures already in the structural database. Proteins that show such structural similarities often do not share any detectable sequence similarity in these same regions. Hence, entirely different sequences can fold into similar three-dimensional configurations. Databases of these common structural features have been prepared and are available on Web sites described later in this chapter.

Difficulties in Using Sequence Alignment to Make Structural Predictions

The finding that only certain amino acids can be substituted at each position in a particular protein core underscores two difficulties in using sequence alignments to make structural predictions. The first difficulty is that, because a different set of substitutions applies to each position in each protein core, standard amino acid substitution matrices such as the Dayhoff PAM matrices and the BLOSUM matrices, described in Chapter 3, may not provide an alignment that has structural significance. The substitutions used to produce these tables are averaged over many sequence alignments, representing observed substitutions in both core regions and loops of sequence families.

Scoring matrices that represent a conserved region in the msa of a set of similar proteins may also be pro-

duced, as described in Chapters 5 and 6. These matrices store information on the amino acid variation found in each column of the msa. They are powerful tools for searching a new protein sequence for the presence of a sequence pattern that is similar to those in the original set of proteins. These scoring matrices include the profile, which represents gapped alignments, and the PSSM, which represents ungapped alignments. A conserved region with gaps in a multiple sequence alignment may also be represented by a profile hidden Markov model (profile HMM), which provides a probability-based model of the msa. Like the scoring matrices, the profile HMM representation of a sequence alignment can be used to identify related sequences. These methods are discussed in detail in Chapters 5 and 6.

Scoring matrices and profile HMMs can provide a direct link between sequence and structure. If one of the sequences represented by the matrix or profile model has a known three-dimensional structure, then any other sequences that match the model are also predicted to have the same structure. Conversely, if the model can be shown to match a protein of known structure, a sequence–structure link may be made. A related method is to produce an HMM (also called a discrete state-space model in the protein structure literature) for a set of proteins that belong to a structural family. These models include information on amino acid preference for positions in secondary structures. A query sequence can then be searched by a set of such models, a procedure known as threading the sequence through the structural models, to determine whether the sequence has sequence patterns that will fit the structure. A range of Web sites provide a variety of these types of analyses (see Fig. 10.30, p. 474).

A second difficulty in making sequence alignments reflect structural similarity is that gaps in the alignment should be confined to regions not in the core. Alignments that reflect structures in core regions should have few if any gaps. Some msa programs such as CLUSTALW (see Chapter 5, p. 180) and the Bayes block aligner (Chapter 4, p. 152) do provide for such variation in gap placement. These programs place alignment gaps where the alignment scores are low and, from a structural viewpoint, represent variable loops. The profile models described above also accommodate such variations of placement.

In addition to sequence-by-sequence alignment and sequence-by-structure alignment, it is also possible to perform a structure-by-structure alignment. In this type of alignment, sequential positions of the backbone carbon atoms for each amino acid in the two sequences are compared to determine whether the chain of atoms is tracing the same path in space. If two or more similar paths are found in the same relative positions and orientations, the structures corresponding to those paths are similar. From these methods, discussed below, databases of structural elements have been made and are available to the laboratory.

Goals for Protein Structure Prediction

What is a reasonable goal for protein structure prediction from the perspective of a molecular biologist? The most satisfying result is to find sequence and structural alignments of a newly identified protein with a protein of known three-dimensional structure. Even if such a prediction can be made, the positions of individual amino acids will probably not be accurately known. If the sequence identity is 50% or better, one sequence can be superimposed on the structure of the other sequence and the predicted structure will be quite accurate. If the sequence identity is greater than 30%, it may be possible to identify common structural features, but it will become more difficult to identify the precise positions of the amino acids in the structure as sequence identity decreases.

The prediction of protein structure is an active and promising area of research. As more three-dimensional structures are found and the computational tools for predicting structure are improved, structural predictions will undoubtedly improve. The existence of new groups of proteins for structural analysis is suggested by the existence of genome "ORFans," which may represent new sets of families (superfamilies) with a unique structure and function (Fischer and Eisenberg 1999). One group of investigators that works on protein classification has developed a protein structure initiative to identify new protein targets for structural analysis (http://www.structuralgenomics.org/main.html). A method for estimating the probability for a protein to have a new fold has been described (Portugaly and Linial 2000). The New York Structural Genomics Consortium (http://www.nysgrc.org/), a consortium of seven research groups, is performing high-throughput structural analysis of proteins starting with predictions from genome sequences.

The usefulness of structure prediction is increased even further with a larger set of protein models (Pennisi 1998). Many additional methods for structural classification of proteins and for displaying the structures have meanwhile been devised, and the Web has provided these resources to the research community. Formerly, special software and hardware were required to view structures. Now, there are a variety of visualization tools that work with a Web browser and allow one to view a molecule in three dimensions, to compare structures, and to perform other useful procedures. A representation of several useful Web sites for protein structure analysis is given in Table 10.1.

HOW IS PROTEIN STRUCTURE DESCRIBED?

Proteins are chains of amino acids joined by peptide bonds, as illustrated in Figure 10.2. Many conformations of the chain are possible because of the rotation of the chain about each C_α atom, and it is these conformational variations that are responsible for differences in the three-dimensional structures of proteins. Each amino acid in the chain is polar; i.e., it has separated positive and negatively charged regions with a chemically free C=O group, which can act as a hydrogen bond acceptor, and an NH group, which can act as a hydrogen bond donor. These groups interact in protein structures. The 20 amino acids found in proteins can be grouped according to the chemistry of their R groups, as depicted in Table 10.2. The R side chains also play an important structural role. Special roles are played by glycine, which does not have a side chain and can therefore increase local flexibility in structures, and cysteine, which can react with another cysteine to form a cross-link that can stabilize the protein structure. The particular chemistry of the protein chain plays the major role in determining the location of secondary structural elements, loops, supersecondary structures, folds, architecture, and overall three-dimensional structure.

TABLE 10.1. *Main Web sites for protein structural analysis*

Name of resource	Resources available	Internet address
Protein data bank (PDB) at the State University of New Jersey (Rutgers)[a]	Atomic coordinates of structures as PDB files, models, viewers, links to many other Web sites for structural analysis and classification.	http://www.rcsb.org/pdb ; also at mirror Web sites (Berman et al. 2000)
National Center for Biotechnology Information Structure Group	Molecular Modelling Database (MMDB). Vector Alignment Search Tool (VAST) for structural comparisons, viewers, threader software.	http://www.ncbi.nlm.nih.gov/Structure/
Structural Classification of Proteins at Cambridge University	SCOP database of structural relationships among known protein structures classified by superfamily, family, and fold.	http://scop.mrc-lmb.cam.ac.uk/scop ; also at Web mirror sites
Biomolecular Structure and Modelling group at the University College, London	CATH database, a hierarchical domain classification of protein structures by class, architecture, fold family and superfamily, other databases and structural analyses, threader software.	http://www.biochem.ucl.ac.uk/bsm ; also at Web mirror sites
European Bioinformatics Institute, Hinxton, Cambridge	Databases, TOPS protein structural topology cartoons, Dali domain server, and FSSP database.[b]	http://www.tops.leeds.ac.uk/
The PredictProtein server at the European Molecular Biology Laboratory at Heidelberg, Germany	Important site for secondary structure prediction by PHD, PREDATOR, TOPITS, threader.	http://cubic.bioc.columbia.edu/ predictprotein ; also at Web mirror sites[c]
Swiss Institute of Bioinformatics, Geneva	Basic types of protein analysis[d] databases, the Swiss-Model resource for prediction of protein models, Swiss-PdbViewer.	http://www.expasy.ch/

Additional sites are listed in the text. In addition to these sites, there are a number of Web sites and courses that discuss protein structure. The Swiss Institute for Bioinformatics (ISREC server) provides a tutorial on Principles of Protein Structure, Comparative Protein Modelling, and Visualisation at http://www.expasy.ch/swissmod/course/course-index.htm. There is also a Web course in protein structure at Birkbeck College http://www.cryst.bbk.ac.uk/teaching/.

[a]A summary of the PDB entries is provided at http://www.biochem.ucl.ac.uk/bsm/pdbsum/ (Laskowski et al. 1997).

[b]3Dee database of protein domains at http://barton.ebi.ac.uk/servers/3Dee.html. Dali domain server is at http://www2.embl-ebi.ac.uk/dali/domain/ and FSSP database at http://www2.embl-ebi.ac.uk/dali/fssp/fssp.html.

[c]Also performed at the structure prediction server at http://fold.doe-mbi.ucla.edu/.

[d]This site offers a series of basic types of protein analyses to assist with protein identification, including identification by amino acid composition, charge, size, and sequence fingerprint. Predictions of posttranslational modifications and oligosaccharide structures are also available.

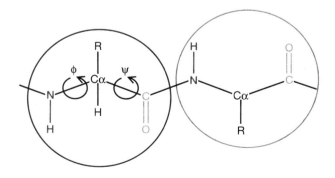

FIGURE 10.2. The structure of two amino acids in a polypeptide chain. Each amino acid is encircled by a different color ring. The R group is different for each of the 20 amino acids. Neighboring amino acids are joined by a peptide bond between the C=O and NH groups. The N-C_α-C sequence is repeated throughout the protein, forming the backbone of the three-dimensional structure. The amino acid at one end of the chain has a free NH_2 group (chain beginning) and the amino acid at the other end has a free COOH group (chain end). The bonds on each side of the C_α atom are quite free to rotate, but many combinations of angles are not possible for most amino acids because of spatial constraints from the R group and neighboring positions in the chain. The conformation of the protein backbone in space is determined by the angles of these bonds, Φ of the bond between the N and C_α atoms and Ψ of the bond between the C_α and C of the C=O group. The distribution of these two angles for the amino acids in a particular protein is often plotted on a graph called a Ramachandran plot. The angle Ω of the peptide bond joining the C=O and NH groups (not shown) is nearly always 180°. The first C atom on the R group is named C_β.

TABLE 10.2. *Chemical properties of the 20 amino acids*

Chemical group	Amino acid (one-letter code)	Name
Hydrophobic		
	A	alanine
	V	valine
	Y	phenylalanine
	P	proline
	M	methionine
	I	isoleucine
	L	leucine
Charged		
	D	aspartic acid
	E	glutamic acid
	K	lysine
	R	arginine
Polar		
	S	serine
	T	threonine
	Y	tyrosine
	H	histidine
	C	cysteine
	N	asparagine
	Q	glutamine
	W	tryptophan
Glycine		
	G	glycine
Cross-linking		
	–	cysteine + cysteine

Much of the protein core comprises regular secondary structures, α helices and β sheets, folded into a three-dimensional configuration. In these secondary structures, regular patterns of H bonds are formed between neighboring amino acids, and the amino acids have similar Φ and Ψ angles, as depicted in Figure 10.3. The formation of these structures neutralizes the polar groups on each amino acid. The secondary structures are tightly packed in the protein core in a hydrophobic environment. Each amino acid side group has a limited volume to occupy and a limited number of possible interactions with other nearby side chains, a situation that must be taken into account in molecular modeling and alignments.

α Helices

The α helix (depicted in Fig. 10.3A) is the most abundant type of secondary structure in proteins. The helix has 3.6 amino acids per turn with an H bond formed between every fourth residue; the average length is 10 amino acids (3 turns) or 10 Å but varies from 5 to 40 (1.5 to 11 turns). The alignment of the H bonds creates a dipole moment for the helix with a resulting partial positive charge at the amino end of the helix. Because this region has free NH_2 groups, it will interact with negatively charged groups such as phosphates.

The most common location for α helices is at the surface of protein cores, where they provide an interface with the aqueous environment. The inner-facing side of the helix tends to have hydrophobic amino acids and the outer-facing side hydrophilic amino acids. Thus, every third of four amino acids along the chain will tend to be hydrophobic, a pattern that can be quite readily detected. In the leucine zipper motif, a repeating pattern of leucines on the facing sides of two adjacent helices is highly predictive of the motif. A helical-wheel plot can be used to show this repeated pattern (see below). Other α helices buried in the protein core or in cellular membranes have a higher and more regular distribution of hydrophobic amino acids, and they are highly predictive of such structures. Helices exposed on the surface of the protein have a lower proportion of hydrophobic amino acids.

A. α helix

B. β-sheet configurations

parallel   antiparallel

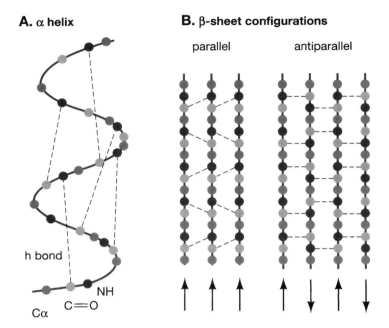

h bond

NH
C═O
Cα

FIGURE 10.3. The α helix and β sheets of protein secondary structure. The backbone of the chain is shown in red, the C$_\alpha$ atoms and the C=O and NH groups are shown in blue, yellow, and green, respectively. (*A*) In the α helix, note that each C=O group at amino acid position *n* in the sequence is hydrogen-bonded with the NH group at position *n* + 4. There are 3.6 residues per turn. The helix is usually right-handed, but short sections of 3–5 amino acids of left-handed helices occur occasionally. The average Φ and Ψ angles of the amino acids in the right-handed helix are approximately 60° and 40°, respectively. The R side chains of the amino acids are on the outside of the helix. (*B*) The β sheet is made up of β strands that are portions of the protein chain. The strands may run in the same (parallel) or opposite (antiparallel) chemical directions (or a mixture of the two), and the pattern of hydrogen bonds is different in each case and also varies in antiparallel strands.

Amino acid content can be predictive of an α-helical region. Regions richer in alanine (A), glutamic acid (E), leucine (L), and methionine (M) and poorer in proline (P), glycine (G), tyrosine (Y), and serine (S) tend to form an α helix. Proline destabilizes or breaks an α helix, but can be present in longer helices, forming a bend. There are computer programs for predicting quite reliably the general location of α helices in a new protein sequence.

β Sheets

β sheets are formed by H bonds between an average of 5–10 consecutive amino acids in one portion of the chain with another 5–10 farther down the chain, as shown in Figure 10.3B. The interacting regions may be adjacent, with a short loop in between, or far apart, with other structures in between. Every other chain may run in the same direction to form a parallel sheet, every other chain may run in the reverse chemical direction to form an antiparallel sheet, or the

chains may be parallel and antiparallel to form a mixed sheet.

As illustrated in Figure 10.3, the pattern of H bonding is different in the parallel and antiparallel configurations. Each amino acid in the interior strands of the sheet forms two H bonds with neighboring amino acids, whereas each amino acid on the outside strands forms only one bond with an interior strand. Looking across the sheet at right angles to the strands, more distant strands are rotated slightly counterclockwise to form a left-handed twist, which is apparent in some of the structures shown below. The C$_\alpha$ atoms alternate above and below the sheet in a pleated structure, and the R side groups of the amino acids alternate above and below the pleats. The Φ and Ψ angles of the amino acids in β sheets vary considerably in one region of the Ramachandran plot (see Fig. 10.2 legend). It is more difficult to predict the location of β sheets than of α helices. The situation improves somewhat when the amino acid variation in msa's is taken into account.

Loops

Loops are regions of a protein chain that are: (1) between α helices and β sheets, (2) of various lengths and three-dimensional configurations, and (3) on the surface of the structure. Hairpin loops that represent a complete turn in the polypeptide chain joining two antiparallel β strands may be as short as two amino acids in length. Loops interact with the surrounding aqueous environment and other proteins. Because amino acids in loops are not constrained by space and environment as are amino acids in the core region, and do not have an effect on the arrangement of secondary structures in the core, more substitutions, insertions, and deletions may occur. Thus, in a sequence align-ment, the presence of more substitutions and deletions may be an indication of a loop. The positions of introns in genomic DNA sometimes correspond to the locations of loops in the encoded protein. Loops also tend to have charged and polar amino acids and are frequently a component of active sites. A detailed examination of loop structures has shown that they fall into distinct families.

Coils

A region of secondary structure that is not α helix, β sheet, or a recognizable turn is commonly referred to as a coil.

PROTEINS ARE CLASSIFIED ON THE BASIS OF STRUCTURAL AND SEQUENCE SIMILARITY

Proteins can be classified according to both structural and sequence similarity. For structural classification, the sizes and spatial arrangements of secondary structures described in the above section are compared in known three-dimensional structures. For classification by sequence similarity, alignments of protein sequences are made using the methods described in Chapters 3 and 5. Historically, classification based on sequence similarity was the first to be used. Initially, similarity based on alignments of whole sequences was performed. Later, proteins were classified on the basis of the occurrence of conserved amino acid patterns that define sequence domains. Databases that classify proteins by one or more of these schemes are available. The SYSTERS database is an example of a protein family database based on protein sequence similarity, and Pfam is an example of a protein domain database. Other examples are given in Table 10.4 below.

In considering protein classification schemes, it is important to keep two principal observations in mind.

1. Two entirely different protein sequences from different evolutionary origins may fold into a similar structure. Conversely, the sequence of an ancient gene for a given structure may have diverged considerably in different species while at the same time maintaining the same basic structural features. Recognizing any remaining sequence similarity in such cases may be a very difficult task.

2. Two proteins that share a significant degree of sequence similarity either with each other or with a third sequence also share an evolutionary origin and should share some structural features also. However, gene duplication and genetic rearrangements during evolution may give rise to new gene copies, which can then evolve into proteins with new function and structure. Examples of these events are discussed at the beginning of Chapter 2 and in Chapters 7 and 11.

To make assessments of protein structure, a number of terms that describe protein similarity and structural relationships are used.

Terms Used for Classifying Protein Structures and Sequences

The more commonly used terms for describing evolutionary and structural relationships among proteins are listed in the Glossary Terms section. Many additional terms are used to describe various kinds of structural features found in proteins. Descriptions of terms used may be found at the CATH Web site (http://www.biochem.ucl.ac.uk/bsm/cath/), and the

Structural Classification of Proteins (SCOP) Web site (http://pdb.wehi.edu.au/scop/gloss.html and Web mirror sites.

Classes of Protein Structure

From the work of Levitt and Chothia (1976), four principal classes of protein structure were recognized based on the types and arrangements of secondary structural elements. These classes are described and illustrated below. In addition, several other classes recognized in the SCOP database discussed below (p. 433) (Murzin et al. 1995) are also included. Examples of this classification are taken from Branden and Tooze (1991).

1. Class α comprises a bundle of α helices connected by loops on the surface of the proteins (see Fig. 10.4).

2. Class β comprises antiparallel β sheets, usually two sheets in close contact forming a sandwich (see Fig. 10.5). Alternatively, a sheet can twist into a barrel with the first and last strands touching. Examples are enzymes, transport proteins, antibodies, and virus coat proteins such as neuraminidase.

3. Class α/β comprises mainly parallel β sheets with intervening α helices, but may also have mixed β sheets (see Fig. 10.6). In addition to forming a sheet in some proteins in this class, as illustrated below, in other proteins parallel β strands may form into a barrel structure surrounded by α

helices (not shown). This class of proteins includes many metabolic enzymes.

4. Class α + β comprises mainly segregated α helices and antiparallel β sheets (Fig. 10.7).

5. Multidomain (α and β) proteins comprise domains representing more than one of the above four classes.

6. Membrane and cell-surface proteins and peptides excluding proteins of the immune system comprise this class (see Fig. 10.8).

Protein Databases

A protein can be analyzed in the laboratory at the levels of sequence and structure. The amino acid sequence and the atomic coordinates of each atom in the structure are unique to each protein.

Sequence Analysis

The sequence of a protein is obtained in the molecular biology laboratory as a DNA sequence and translated in the computer into the amino acid sequence of the encoded protein (see Chapter 9). DNA sequences are deposited in the DNA sequence databases such as GenBank and EMBL, where they are automatically translated to produce the Genpept and TrEMBL protein databases, respectively. Sometimes protein fragments are also sequenced, and matches with DNA sequence databases are used to identify the encoding gene (Chapter 9). The encoded proteins are additionally annotated in databases such as SwissProt and PIR, as described in Chapter 2.

Another type of protein sequence analysis is a sequence alignment of protein sequences discussed in Chapter 3 or a search for similar sequences in the sequence databases, as described in Chapter 6. The alignment will reveal any significant similarity and the degree of amino acid identity between two sequences. Similarity may be present throughout the sequences or localized to certain regions. Localization of sequence similarity can best be performed by global and local sequence alignment methods, as discussed in Chapter 3. The stronger the similarity and identity, the more similar are the three-dimensional folds and other structural features of the proteins.

Another level of sequence analysis is examining a group of sequences for common amino acid patterns. Methods for finding different types of patterns,

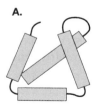
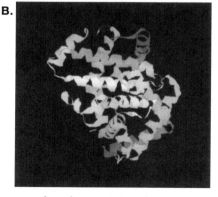

FIGURE 10.4. Structure of α class proteins. (*A*) Diagram showing α-helical pattern of this class. α Helices are orange cylinders, and black lines are loops. (*B*) Example of the class, hemoglobin, PDB file 3hhb displayed using Rasmol, using ribbons display and group color.

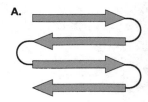
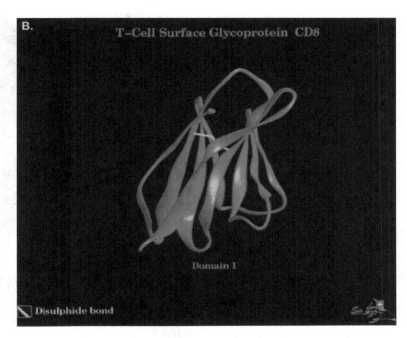

FIGURE 10.5. Structure of β class proteins. (*A*) Diagram showing typical arrangement of the antiparallel β strands (*blue arrows*) joined by loops (*black lines*) in β sheet. (*B*) Example of protein in this class, T-cell receptor CD8, PDB file 1cd8, image from http://expasy.hcuge.ch/pub/Graphics/IMAGES/.

including motifs (short gapped or ungapped patterns), blocks (ungapped patterns), and patterns with gaps (represented by profile scoring matrices and profile HMMs) are discussed in Chapter 5. These patterns may be obtained from sequences of proteins that are already known to have the same function, or they may be obtained by statistical or pattern-finding methods of any set of sequences of biological interest. Depending on the extent and significance of these

patterns and additional information about the function of the proteins, their presence may or may not represent structural similarity or an evolutionary relationship among the proteins. A combined form of sequence and structural alignments provides an additional level of analysis.

When proteins of unknown structure are similar to a protein of known structure at the sequence level, msa and pattern analysis can be used to predict the

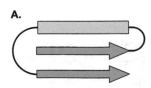
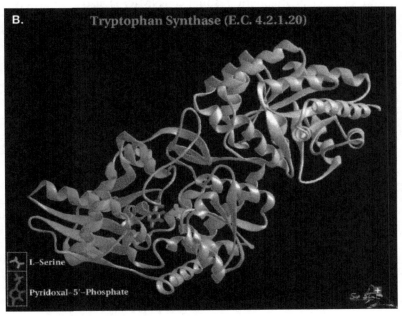

FIGURE 10.6. Structure of α/β class proteins. (*A*) Diagram showing one possible configuration of parallel β strands (*blue arrows*) in a β sheet and an intervening α helix (*orange cylinder*), joined by loops (*black lines*). (*B*) Example of protein in this class, tryptophan synthase β subunit obtained from http://expasy.hcuge.ch/pub/Graphics/IMAGES/.

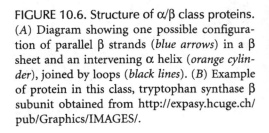

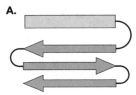

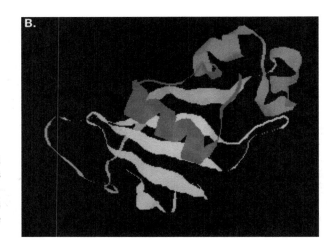

FIGURE 10.7. Structure of α + β class proteins. (*A*) Diagram showing arrangement of typical motif of antiparallel β strands (*blue arrows*) in β sheet and segregated from α helix (*orange cylinder*) and showing loops (*black lines*). (*B*) Example of protein in this class, G-specific endonuclease complex with deoxy-dinucleotide inhibitor, PDB file 1rnb viewed with Rasmol.

structures of these proteins. Databases of such related proteins are available. In another type of analysis, called threading, the sequence of amino acids in a protein of unknown structure is tested for its ability to fit into a known three-dimensional structure. The size and chemistry of each amino acid R group and proximity to other amino acids are taken into account. This analysis provides a method for aligning a sequence with a structure.

Structural Analysis

The three-dimensional structure of a protein is usually obtained by making crystals of the protein and using X-ray diffraction to determine the positions of molecules that are fixed within the crystal. The technique of nuclear magnetic resonance (NMR) is also used to obtain protein structures. Once the three-dimensional coordinates of each atom in the protein molecule have been found, a table of these coordinates is deposited with the Brookhaven Data Bank as a PDB entry. PDB entries, such as those shown in Table 10.3, give the atomic coordinates of the amino acids in proteins, protein fragments, or proteins bound to substrates or inhibitors. PDB files may be easily retrieved from the PDB Web site (http://www.rcsb.org/pdb/) and displayed with a molecular viewer such as Rasmol. Structural information may also be stored in forms other than PDB, but PDB is the most accessible for the molecular biologist. There are three different kinds of databases that provide an analysis of proteins—one kind for sequences, a second for structures, and a third for comparing sequences and structures.

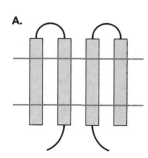

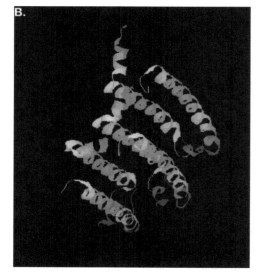

FIGURE 10.8. Structure of membrane proteins. α helices are of a particular length range and have a high content of hydrophobic amino acids traversing a membrane, features that make this class readily identifiable by scanning a sequence for these hydrophobic regions (see below). (*A*) Diagram showing typical arrangement of membrane-traversing, hydrophobic α helices (*orange*). Membrane bilayer shown as green lines. (*B*) Example of protein in this class, integral membrane light-harvesting complex, PDB file 1kzu viewed with Rasmol.

TABLE 10.3. *Brookhaven Protein Data Bank (PDB) entry 3hhb, deoxy hemoglobin for oxygen transport entered July 13, 1993*

								Occupancy	Temp. factor
ATOM	1	N	VAL A	1	5.428	17.064	5.060	1.00	41.29
ATOM	2	CA	VAL A	1	6.168	18.292	4.856	1.00	41.33
ATOM	3	C	VAL A	1	7.676	18.056	5.068	1.00	31.64
ATOM	4	O	VAL A	1	8.120	17.488	6.076	1.00	38.31
ATOM	5	CB	VAL A	1	5.644	19.268	5.884	1.00	52.26
ATOM	6	CG1	VAL A	1	6.044	20.696	5.512	1.00	52.75
ATOM	7	CG2	VAL A	1	4.124	19.120	6.000	1.00	58.75
ATOM	8	N	LEU A	2	8.444	18.512	4.116	1.00	27.63
ATOM	9	CA	LEU A	2	9.896	18.420	4.308	1.00	33.62
ATOM	10	C	LEU A	2	10.360	19.592	5.216	1.00	32.51
ATOM	11	O	LEU A	2	10.128	20.760	4.900	1.00	31.03
ATOM	12	CB	LEU A	2	10.568	18.584	2.932	1.00	34.38
ATOM	13	CG	LEU A	2	10.284	17.488	1.924	1.00	32.23
ATOM	14	CD1	LEU A	2	11.032	17.676	0.580	1.00	36.30
ATOM	15	CD2	LEU A	2	10.576	16.136	2.560	1.00	38.42

Shown is the initial part of the entry showing ATOM records that provide cartesian coordinates of all atoms in the first two amino acids Val and Leu. The last columns give the occupancy and temperature factor for each atom. The occupancy gives the frequency with which the atom is present in the crystal and is usually 1. The temperature gives a measurement of the uncertainty of the position of the atom due to the motion of the atom in the crystal. The units of temperature are Angstroms squared. A typical value of a crystal at room temperature at 2 Å resolution is 20 Å^2; the higher this value for an atom, the more uncertain the position of that atom. Structural entries sometimes provide the author's assignment of a secondary structure to each amino acid.

As more and more protein structures have been solved by X-ray crystallographic and NMR methods, these structures have been classified by various means into structural databases. This classification is based on comparison and alignment of the protein structures. The types, order, connections, and relative positions of secondary structures are compared using the known atomic coordinates of atoms in each structure and methods described below. This type of information can then be combined with sequence information to identify other proteins that might have similar structural features.

MOLECULAR VIEWERS DISPLAY PROTEIN STRUCTURES

The first major step in displaying a structure is to identify the correct PDB identification code for the structural file. Most sites provide a browser program for searching the structural database for the name of the protein, organism, or other identifying features (see below). There may be a number of choices from which to choose, including domains, folds, or protein fragments, or structures of the protein bound to a substrate or inhibitor. Some databases also include the predicted structure of mutant proteins. The available choices need to be screened carefully for the correct one.

A number of molecular viewers are freely available and run on most computer platforms and operating systems, including Microsoft Windows, Macintosh, and UNIX X-Window. These programs convert the atomic coordinates into a graphical view of the molecule. They may also recompute information to remove inconsistencies in the database or to supply missing information (Hogue and Bryant 1998a,b). Viewers also provide ways to manipulate the molecule, including rotation, zooming, and creating two images that provide a stereo view. Rotating a molecule by dragging the mouse across the image can illustrate the three-dimensional structure. Viewers can also be used to show a structural alignment of two or more structures or a predicted structure. Unless a very high-resolution view is needed, the simplest way to use a viewer is through a network browser. The browser may be readily configured to run a viewer program

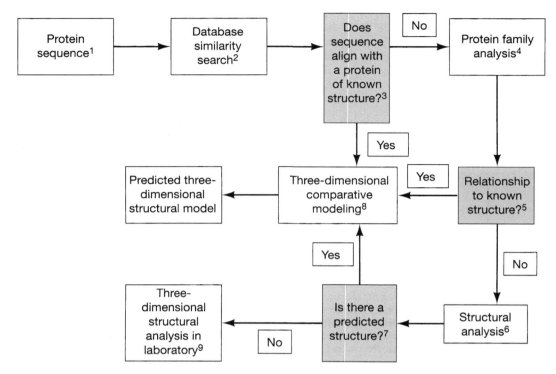

1. Amino acid sequences of proteins are derived from translation of cDNA sequences or predicted gene structures in genomic DNA sequences. Partial sequences are also derived by translation of expressed sequence tag (EST) sequences or genomic DNA sequences in all six reading frames. These predictions can be improved when genomic and EST sequences can be aligned and when overlapping EST sequences are identified by gene indexing, as described in Chapters 6 and 9.

2. The sequence is used as a query in a database similarity search against the proteins in the Protein Data Bank (PDB), all of which have a known three-dimensional structure. A significant alignment of the query sequence with a PDB sequence is evidence that the query sequence has a similar three-dimensional structure. If a relationship with a PDB protein is not found, a second database similarity search against a protein sequence database such as SwissProt can be performed. Matching sequences, including both closely related and more distantly related ones, can then be used in a search against PDB sequences. The PSI-BLAST tool described in Chapter 6 automates and enhances the process of finding related sequences in the protein database. The goal is to discover one or more database sequences that are related both to the query and to a PDB sequence, as illustrated in Figure 6.1 (p. 237).

3. If the database similarity search reveals a significant alignment between the query sequence and a PDB sequence, the alignment between the sequences can be used to position the amino acids of the query sequence in the same approximate three-dimensional structure. Testing the significance of alignment scores is discussed in Chapter 3.

4. Proteins have been classified into families on the basis of sequence similarity. The relationships are depicted in an msa of the proteins, as described in Chapter 5. Proteins of known three-dimensional structure have also been classified into fold families on the basis of a common arrangement of secondary structures. Sequences of proteins in the same fold family are often not similar, so they cannot be aligned. However, the individual proteins in a particular fold family are often members of families based on sequence similarity. Hence, these similar sequences are also predicted to have the same structural fold as the fold family. The goal of this step in the flowchart is to exploit these structure–sequence relationships. Two questions are addressed: (1) Is the new protein a member of a protein family or superfamily based on sequence similarity? (2) Does the matched family or superfamily have a predicted structural fold? The first question is usually addressed by analyzing the test sequence for patterns that represent each family using PSSMs, profile HMMs, and other tools, as described in Chapter 5. Web sites such as Interpro (Table 10.4) include a large, composite collection of patterns and will search a new sequence for matches. 3D-PSSM (Table 10.4) includes a powerful set of scoring matrices based on structural alignments for use in three-dimensional structure prediction. These Web sites usually provide links to related fold families, thus identifying a predicted structural fold for the new protein. Other Web sites employ a cluster analysis of proteins based on pair-wise alignment scores of all of the proteins in the SwissProt or other protein sequence databases. These sites offer an alternative method for finding relationships between a new sequence and all the other

sequences in SwissProt, and thus for discovering a link to a known protein structure.

5. If the family analysis reveals that the new protein is a member of a family that is predicted to have a structural fold, msa's of these proteins can be used for structural modeling.

6. This step in the flowchart includes several different types of analyses that are described below in the chapter. First, the presence of small amino acid motifs in a protein can be an indicator of a biochemical function. The Prosite catalog can be used to search a new protein sequence for motifs. Second, spacing and arrangement of specific amino acids, e.g., hydrophobic amino acids, provide important structural clues that can be used for modeling. Third, the tendency of certain amino acid combinations to occur in a given type of secondary structure provides methods for predicting where these structures are likely to occur in a new sequence. Fourth, the structural fold families described in note 4 above have been represented by PSSMs and by HMMs that capture the tendency to find each amino acid at a particular position in a structural fold and variations in the fold itself. Other models of three-dimensional structure represent the size and chemistry of amino acids or the energetic stability associated with amino acid interactions. A new protein sequence can be aligned with these models to determine whether the sequence matches one of them, a procedure known as threading a sequence into a structure.

7. The structural analysis in step 6 provides clues as to the presence of active sites, regions of secondary and three-dimensional structure, and the order of predicted secondary structures. If these predictions are convincing enough, it may be possible to identify a new protein as a member of a known structural class.

8. Sequence or structural alignments of the new protein with a protein of known structure provide a starting three-dimensional model of the protein. By using computer graphics and protein modeling software, the amino acids can be positioned to accommodate available space and interactions with neighboring amino acids.

9. Proteins that fail to show any relationship to proteins of known structure are candidates for structural analysis. There are over 3000 known fold families, and new structures are frequently found to have an already known supersecondary structure (also called structural fold). Accordingly, protein families with no relatives of known structure may represent a novel structural fold.

automatically when the particular file format used by the viewer is being downloaded from the remote computer. Most sites that provide protein structural files provide several formats allowing a choice of viewers, and they also provide Web links to other sites from which the viewer program may be downloaded. The viewer option usually appears once a particular structural file has been chosen. Some representative viewers and their features are shown in Table 10.5.

The correct processing of files with molecular structural information through the Web or through E-mail attachments is made possible by the chemical MIME (multipurpose Internet mail extension) project (http://www.ch.ic.ac.uk/chemime/iupac.html). This project acts as a repository for standard types of MIME files. As an example, if the start of the file includes the label chemical/x-pdb (MIME type chemical and subtype x-pdb), the file is a text file in the Brookhaven Protein Data Bank file format, and a viewer for a pdb file such as Rasmol or Chime is needed. Files intended for viewing by Rasmol may also be indicated by MIME type application/x-rasmol and the pdb file may also be identified by the filename extension pdb. There are also additional chemical MIME formats. For Cn3D, chemical/ncbi-asn1-binary and val are the MIME type and filename extension, respectively. Cn3D files are sent as a binary file rather than a text file, meaning that some bytes include characters other than the standard ASCII characters. For MAGE, chemical/x-kinemage and kin are used. Molecules may also be viewed by means of programs called applets written in the JAVA programming language. These programs are sent at the same time as the molecular coordinates and are run by the browser.

In addition to retrieving the three-dimensional coordinates of a molecule, already prepared graphic views of molecules may be obtained from many of the Web sites that provide pdb files. The following FTP site contains a database of stored image files: http://au.expasy.org/sw3d/. These views include two file formats commonly used on the Web, the JPEG (Joint Photographic Experts Group) format and GIF (graphics interchange format). These formats pro-

TABLE 10.4. *Databases of patterns and sequences of protein families*

Name	Web address	Description	Reference
3D-Ali	http://139.91.72.10/def2/def2.html	Aligned protein structures and related sequences using only secondary structures assigned by author of the structures.	Pascarella and Argos (1992)
3DCA	http://www.rlandgraf.med.ucla.edu/3DCA.html	Cluster Analysis integrating structural and sequence information to obtain predictions about functionally relevant clusters of residues.	see Web site
3D-PSSM	http://www.sbg.bio.ic.ac.uk/3dpssm	Uses a library of scoring matrices based on structural similarity given in the SCOP classification scheme (p. 433) for alignment with matrices based on sequence similarity.	Kelley et al. (2000)
BIND (Biomolecular Interaction Network Database)	http://www.blueprint.org/bind/bind.php	Includes information on protein–protein interactions, molecular complexes, and pathways.	see Web site
BLOCKS	http://blocks.fhcrc.org/	Ungapped blocks in families defined by the Prosite catalog.	Henikoff and Henikoff (1996); Henikoff et al. (1998)
cluSTr	http://www.ebi.ac.uk/clustr/	Clustering of all proteins in PIR, TrEMBL, and SwissProt based on pair-wise similarity.	Kriventseva et al. (2003)
COGS (Clusters of Orthologous Groups database and search site)	http://www.ncbi.nlm.nih.gov/COG	Clusters of similar proteins in at least three species collected from available genomic sequences.	Tatusov et al. (1997)
CONSURF (Protein Surface Contact Analysis)	http://consurf.tau.ac.il/	Mapping of functional regions on surface of proteins using conserved amino acid patterns.	Glaser et al. (2003)
DiffTool	http://bioweb.pasteur.fr/seqanal/difftool/	Clustering of proteins based in similarity.	Chetouani et al. (2002)
DIP (Database of Interacting Proteins)	http://dip.doe-mbi.ucla.edu	Database of interacting proteins.	Xenarios et al. (2000)
eMOTIF	http://dna.Stanford.EDU/emotif/	Common and rare amino acid motifs in the BLOCKS and HSSP databases.	Nevill-Manning et al. (1998)
HOMSTRAD	http://www-cryst.bioc.cam.ac.uk/homstrad/	Structure-based alignments organized at the level of homologous families.[a]	Mizuguchi et al. (1998a)
HSSP	http://swift.embl-heidelberg.de/hssp/	Sequences similar to proteins of known structure.	Dodge et al. (1998)
INTERPRO resource of protein domains and functional sites[b]	http://www.ebi.ac.uk/interpro	Combination of Pfam, PRINTS, Prosite, and current SwissProt/TrEMBL sequence.	see Web site

Name	URL	Description	Reference
LPFC	http://smi-web.stanford.edu/projects/helix/LPFC/	A library of protein family cores based on msa of protein cores using amino acid substitution matrices based on structure (see Chapter 3).	see Web page
NCBI	http://www.ncbi.nlm.nih.gov	Search conserved domain database (rpsblast) or for domain architecture (cdart).	see Web page
NetOGly 2.0 server	http://www.cbs.dtu.dk/services/NetOGlyc/	Predicts glycosylation sites in mammalian proteins by NN analysis.	Hansen et al. (1997)
NNPSL	http://predict.sanger.ac.uk/nnpsl/	Predicts subcellular location of proteins by NN.	see Web site
Pfam	http://www.sanger.ac.uk/Pfam	Profiles derived from alignment of protein families, each composed of similar sequence and analyzed by HMMs.	Sonnhammer et al. (1998)
PIR	http://www-nbrf.georgetown.edu/	Family and superfamily classification based on sequence alignment.	Barker et al. (1996)
PRINTS	http://www.bioinf.man.ac.uk/dbbrowser/PRINTS/	Protein fingerprints or sets of unweighted sequence motifs from aligned sequence families.	Attwood et al. (1999)
PROCLASS	http://www-nbrf.georgetown.edu/gfserver/proclass.html	Database organized by Prosite patterns and PIR super-families; NN system for protein classification into superfamily.	Wu (1996); Wu et al. (1996)
PRODOM	http://protein.toulouse.inra.fr/prodom.html	Groups of sequence segments or domains from similar sequences found in SwissProt database by BLASTP algorithm; aligned by msa.	Corpet et al. (1998)
ProtoNet	http://www.protonet.cs.huji.ac.il/	Automatic hierarchical clustering of SwissProt.	Sasson et al. (2003)
Prosite	http://www.expasy.ch/prosite	Groups of proteins of similar biochemical function on basis of amino acid patterns.	Bairoch (1991); Hofmann et al. (1999)
ProtoMap	http://protomap.cornell.edu	Classification of SwissProt and TrEMBL proteins into clusters.	Yona et al. (1999)
PSORT	http://psort.nibb.ac.jp	Predicts presence of protein localization signals in proteins.	see Web site
SignalP Web server	http://www.cbs.dtu.dk/services//SignalP-2.0/	Predicts presence and location of signal peptide cleavage sites in proteins of different organisms by NN analysis.	Nielsen et al. (1997)
SMART	http://smart.embl-heidelberg.de	Database of signaling domain sequences with accurate alignments.	Schultz et al. (1998)
SYSTERS	http://systers.molgen.mpg.de/	Classification of all sequences in the SwissProt database into clusters based on sequence similarity.	Krause et al. (2000)
TargetDB	http://targetdb.pdb.org/	Database of peptides that target proteins to cellular locations.	see Web site
Uniprot	http://www.pir.uniprot.org/	Combined protein sequence database of PIR, SwissProt, and TrEMBL.	see Web site

A list of Web sites with protein sequence/structure databases is maintained at http://www.imb-jena.de/ImgLibDoc/help/db/. Many protein family databases are accessible through the European Bioinformatics Institute (http://srs.ebi.ac.uk/). List of protein–protein interaction databases is maintained at http://www.hgmp.mrc.ac.uk/GenomeWeb/prot-interaction.html.

[a] Sequence alignments of each family shown with residues labeled by solvent accessibility, secondary structure, H bonds to main-chain amide or carbonyl group, disulfide bond, and positive Φ angle.

[b] A combination of Pfam 5.0, PRINTS 25.0, Prosite 16, and current SwissProt and TrEMBL data. Additional merges with other protein pattern databases are planned.

TABLE 10.5. *Programs for viewing protein molecules*

Viewer	Web location	Features
Chime	http://www.umass.edu/microbio/chime/	A Web browser plug-in that can be used to display and manipulate structures inside a Web page. There are many mouse-driven controls. Excellent for lecture presentations.
Cn3d[a]	http://www.ncbi.nlm.nih.gov/Structure/ (Hogue 1997)	Provides viewing of three-dimensional structures from Entrez and MMDB.[a] Cn3D runs on Windows, MacOS, and UNIX; simultaneously displays structural and sequence alignments; can show multiple superimposed images from NMR studies.
MAGE	http://kinemage.biochem.duke.edu (see Richardson and Richardson 1994)	Standard molecular viewing features with animation and kaleidoscope effects.
Rasmol[b]	http://www.umass.edu/microbio/rasmol/ (Sayle and Milner-White 1995)	Most commonly used viewer for Windows, MacOS, UNIX, and VMS operating systems. Performs many functions.
Swiss 3D viewer, Spdbv	http://www.expasy.ch/spdbv/mainpage.html (Guex and Peitsch 1997)	Protein models can be built by structural alignments; calculates atomic angles and distances, threading, energy minimization, and interacts with the Swiss Model server.

Additional viewers are accessible from the referenced Web sites. Viewer functions usually include wireframe of C_α backbone; ribbon of secondary structures; space-filling displays; color schemes to illustrate features such as residues, structures, and temperature; mouse-drag rotation; several views including stereo and zooming; and exporting to graphic file formats. Assistance with these viewers is provided at the following Web sites for obtaining molecular coordinates: Molecules R Us at NIH, http://molbio.info.nih.gov/cgi-bin/pdb, and NCBI, http://www.ncbi.nlm.nih.gov/Structure/. A large list of available graphics viewers may be found at http://www.csb.yale.edu/userguides/graphics/csb_hm_graph.html.

[a] The NCBI structure group has established a new format for databases called ASN.1 (see Chapter 2). The PDB files have been converted into this format to create another database MMDB (Molecular Modelling DataBase) that is highly suitable for structural alignments by vector methods described below. Ambiguities in PDB entries have been made explicit in the MMDB database (Hogue and Bryant 1998a,b; http://www.ncbi.nlm.nih.gov/Structure/).

[b] Rasmol and other viewers as well have many features in the molecular viewing window in addition to those described above. These additional features are accessible through a command line window that appears when the program is running.

duce images of reasonably high quality but have varying levels of detail and resolution. A higher-resolution and more detailed rendition of the molecule will have a larger file size and take longer to retrieve over the Internet. These files may be compressed to a smaller size by graphic format conversion programs.

Programs such as Raster3D (http://www.bmsc.washington.edu/raster3d/) and Molscript (http://www.avatar.se/molscript/) produce very high quality images in a number of different formats. These programs require graphics workstations and a more sophisticated level of programming experience.

PROTEIN STRUCTURE IS CLASSIFIED IN DATABASES

The following databases are accessible on the Web and provide up-to-date structural comparisons for the proteins currently in the Brookhaven PDB and access to the sequences of these proteins. The methods used to classify the protein structures in these databases vary from manual examination of structures to fully automatic computer algorithms. Hence, although one can expect to find roughly the same groupings in each database, there will be some struc-

tural relationships that are only identified by one of these methods. Each database has useful information that may be lacking in the others. The MMDB and SARF databases (see points 4 and 5 below) are based on a rapid structural alignment method that is designed to find the most significant alignments in the structural databank. The SCOP, CATH, and DALI/FSSP databases (1, 2, and 3) are based on different comparison methods and are likely to provide additional complementary information on relationships among protein structures. The folds recognized by these methods have been compared (Swindells et al. 1998) and a consensus database made (Day et al. 2003).

1. *The SCOP database.* The SCOP (*structural classification of proteins*) database (Murzin et al. 1995; Brenner et al. 1996), based on expert definition of structural similarities, is located at http://scop.mrc-lmb.cam.ac.uk/scop/. Following classification by class, SCOP additionally classifies protein structures by a number of hierarchical levels to reflect both evolutionary and structural relationships—namely, family, superfamily, and fold. Shown in Figure 10.9 is an example of the initial Web page for the lineage for the all α-class, globin-like fold, globin-like superfamily, globin, and

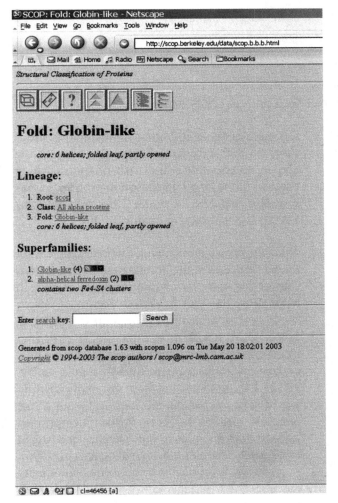

FIGURE 10.9. A portion of the SCOP structural classification showing the hierarchy of all α-class, globin, and globin-like proteins.

phycocyanin families, and finally, protein domains such as hemoglobin 1, which can be viewed by individual entry in PDB using a molecular viewer.

2. *The CATH database.* The CATH (classification by *c*lass, *a*rchitecture, *t*opology, and *h*omology) protein structure database resides at University College, London (Orengo et al. 1997; http://www.biochem.ucl.ac.uk/bsm/cath/). Proteins are classified first into hierarchical levels by class, similar to the SCOP classification except that α/β and α+β proteins are considered to be in one class. Instead of a fourth class for α+β proteins, the fourth class of CATH comprises proteins with few secondary structures. Following class, proteins are classified by architecture, fold, superfamily, and family. Similar structures are found by the program SSAP, described on page 438. The CATH Web site is shown in Figure 10.10.

3. *The DALI/FSSP database.* The FSSP (*f*old classification based on *s*tructure–*s*tructure alignment of *p*roteins) is based on a structural alignment of all pair-wise combinations of the proteins in the Brookhaven structural database by the structural alignment program DALI (Holm and Sander 1996; http://www2.embl-ebi.ac.uk/dali/fssp/fssp.html). PDB has a number of redundant structures of proteins whose sequences and structures are 25% or more identical. A subset of representative structures in PDB without these redundant entries was first produced by aligning all of the PDB structures with DALI. Each protein in the subset was then subdivided into individual domains. These domains were then aligned structurally with DALI to identify the common folds. Redundant folds were again eliminated, and a set of representative folds was chosen (Holm and Sander 1998). These fold types represent a unique configuration of secondary structural elements in the domains. For example, one fold might be composed of helix–strand–helix-6 strands joined by loops in a particular configuration.

Corresponding to each representative fold type, there is a cluster of folds that are of the same approximate structure. The domains that have a given cluster of folds are structurally related, and the cluster is represented by structural alignments of these domains. The higher the statistical score for a given domain alignment and corresponding

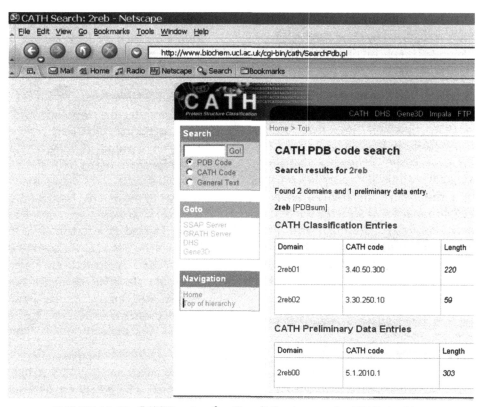

FIGURE 10.10. CATH entry for *E. coli* RecA protein (PDB 2reb).

fold (higher *Z* value), the greater the degree to which the atoms occupy similar structural positions. *Z* values >16 indicate a very good structural alignment and 8–16 a less-good alignment, down to a level of 2, which indicates the lowest level of alignment detection. Thus, fold clusters may be organized in a hierarchical fashion with folds represented by the most low-scoring alignments at the top of the hierarchy.

In addition, the sequences of the 1000 representative structures were used as probes for a sequence similarity search of the SwissProt protein sequence database. The database search program MAXHOM, which begins with a sequence similarity search and then with an expanded profile search, was used. The resulting homology-derived structures of proteins (HSSP) database (Sander and Schneider 1991; Dodge et al. 1998; http://www.sander.ebi.ac.uk/hssp/) contains lists of similar proteins, one list for each representative structure. Given the PDB database number of a known structure, the program will show the closest representative structures, and one or more may be chosen. The program will then show any significant structural alignments between the cho-

sen representative and other representative structures in FSSP. A structural alignment between the chosen representative and each of the matching proteins in the HSSP database entry for that representative may be selected. The Web site for DALI/FFSP is shown in Figure 10.11.

4. *MMDB* (*molecular modeling database*). Proteins of known structure in the Brookhaven PDB have been categorized into structurally related groups in MMDB by the VAST (*Vector Alignment Search Tool*) structural alignment program (Madej et al. 1995). VAST aligns three-dimensional structures based on a search for similar arrangements of secondary structural elements (for an example, see Fig. 10.12). This provides a method for rapidly identifying PDB structures that are statistically out of the ordinary. MMDB has been further incorporated into the Entrez sequence and reference database at http://www.ncbi.nlm.nih.gov/ (Hogue et al. 1996). Accordingly, it is possible to perform a simultaneous search for similar sequences and structures, designated neighbors, at the Entrez Web site. Structural neighbors within MMDB are based on detailed residue-by-residue alignments.

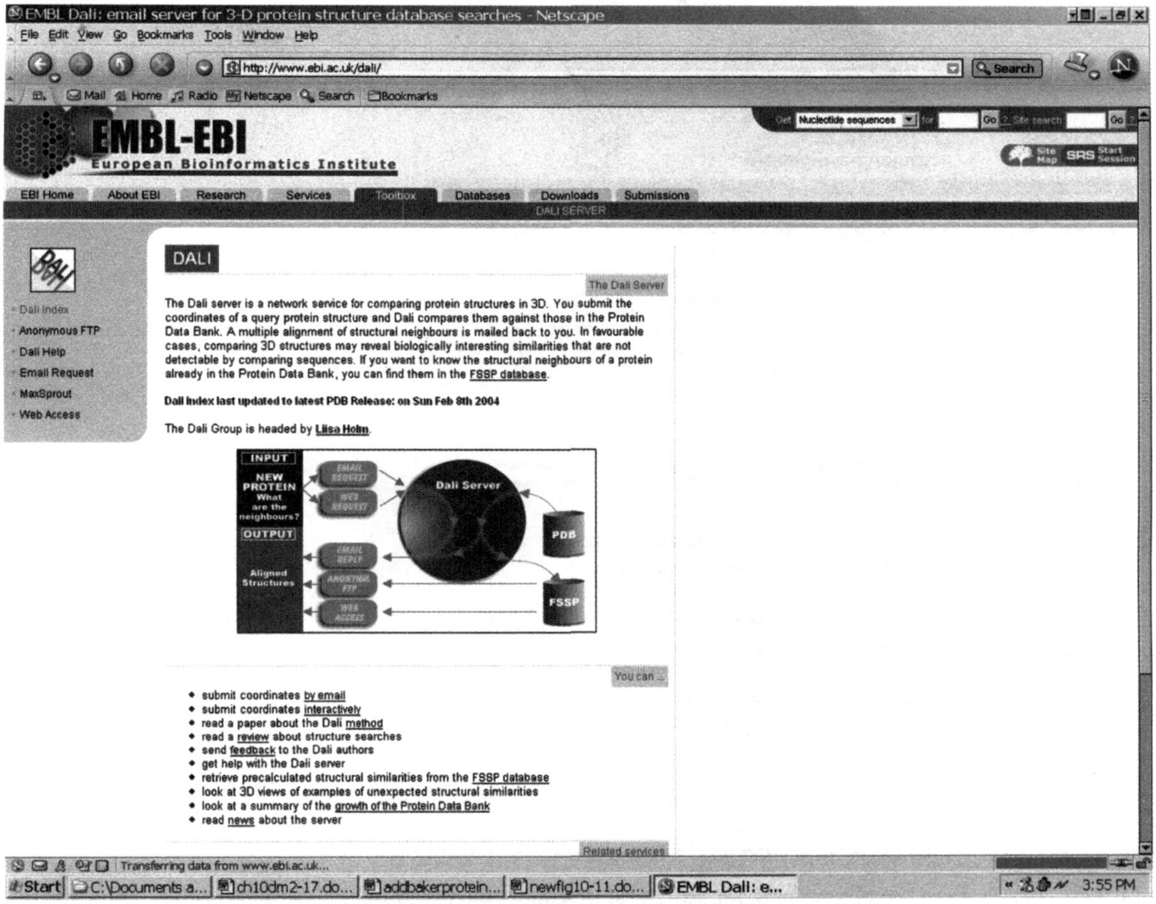

FIGURE 10.11. EBI home page for DALI, which maintains a database, formerly FSSP, of structural comparisons of all proteins in the protein databank (PDB).

5. *The SARF database.* The SARF (*s*patial *ar*rangement of backbone *f*ragments) database at http://123d.ncifcrf.gov/ (Alexandrov and Fischer 1996) also provides a protein database categorized on the basis of structural similarity. Like VAST, SARF can find structural similarity rapidly based on a search for secondary structural elements. These structural hierarchies found by this method are in good agreement with those found in the SCOP, CATH, and FSSP databases with several interesting differences. The method also found several new groupings of structural similarity. The above SARF Web site provides some excellent representations of overlaid structures.

STRUCTURAL ALIGNMENT OF PROTEINS IS MORE DIFFICULT THAN SEQUENCE ALIGNMENT

As more and more protein structures, as well as access to recently developed and rapid methods for comparing protein structures, have become available on the Web, alignment of protein structures has become a task that can be performed by laboratories not trained in the techniques of structural biology. To perform a sequence alignment, the amino acid sequence of one protein is written above the amino acid sequence of a

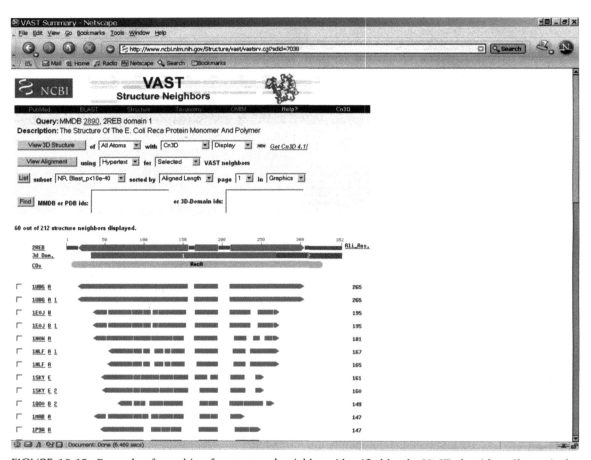

FIGURE 10.12. Example of searching for structural neighbors identified by the VAST algorithm. Shown is the result of a search for neighbors to *E. coli* RecA protein structure (PDB identifier 2reb).

second protein. Similar or identical amino acids are placed in the same columns, and gaps are placed at positions where there is no matching character. To perform a structural alignment, the three-dimensional structure of one protein domain is superimposed upon the three-dimensional structure of a second protein domain, fitting together the atoms as closely as possible so that the average spatial deviation between them is minimal.

Sequence alignments are performed to discover sequence similarity, and structural alignments to discover structural similarity (evidence that the structures share a common fold). New structural relationships are being constantly discovered. Just as a laboratory may discover a remote sequence similarity between two protein domains reflecting a family or superfamily relationship, so may the same laboratory discover a previously unknown structural relationship between two proteins.

There is one important difference between se-

quence and structural similarity. Statistically significant sequence similarity is an indicator of an evolutionary relationship between sequences. In contrast, significant structural similarity is common, even among proteins that do not share any sequence similarity or evolutionary relationship. Thus, structural similarity may or may not be an indicator of an evolutionary relationship. Further light may be shed on this question by a close examination of the structural similarity. The similarity may be quite simple, such as a common arrangement and spacing of several secondary structural elements. Alternatively, there may be a highly significant alignment of many of the proteins through the same sequence of secondary structures and loops, and many of the atoms in the two proteins may be quite superimposable. Such structural closeness may be an indication of a possible evolutionary relationship. The results of a search for remote sequence similarity in proteins with related structures by sensitive statistical methods (Gibbs sampling, expectation maximization methods, and Bayesian alignment methods discussed in Chapter 5) may provide further support for such a possibility. The ability to make such comparisons has depended on the development and availability of fast and efficient methods for performing structural comparisons.

Structural comparison methods share some of the features of methods for comparing sequences, but with additional considerations. For comparing two sequences, one searches for a row of amino acids in one sequence that matches a row in the second, allowing for substitutions and the insertion of gaps in one sequence to make up for extra characters in the other. For comparing structures, positions of atoms in two three-dimensional structures are compared. These methods initially examine the positions of secondary structural elements, α helices and β strands, within a protein domain to determine whether the number, type, and relative positions of these elements are similar or whether the proteins have a similar architecture. Distances between the C_α or C_β atoms within these structures are then examined in detail to determine the degree to which the structures may be superimposed. If a few elements can be aligned and are joined by a similar arrangement of loops, the proteins share a common fold. As the arrangement, joining, and alignment of secondary structural elements within the proteins increase, the degree of structural similarity between the proteins becomes more and more convincing and significant.

To specify a three-dimensional structure, positions of molecules are expressed as x, y, and z cartesian coordinates within a fixed frame of reference, as shown in Figure 10.13. The direction of the bond angles and the interatomic distances between amino acids along the polypeptide chain may also be represented as vectors. Secondary structures can also be represented by a vector that starts at the beginning of the secondary structural element, extends for the length of the element, and has a direction that reveals the orientation of the element in the overall structure. Comparison of these structural representations in two proteins provides a framework for comparing the structures of the proteins. In many structural comparison methods, distances between C_α atoms or between C_β atoms in two protein structures are used for comparison purposes. A more detailed comparison of the structures can be made by adding information on side chains, such as the amount of outside area of the side chain that is buried under other molecules so that the chain is not accessible to water molecules. Distances and bond

angles to other atoms in the structure may also be compared. Several of the parameters used for structural comparisons may also be used to classify the environment of a particular amino acid, e.g., a buried, hydrophobic amino acid in a β strand.

There are two reasons that it is more difficult to align structures than sequences. First, a similar structure may form by many different foldings of the amino acid C_α backbone. As a result, matched regions may not necessarily be in the same order in the two

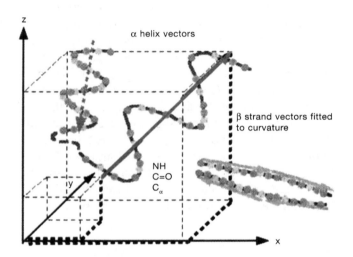

FIGURE 10.13. Alignment of the three-dimensional structure of proteins by their secondary structures. Representation of arrangement of secondary structures in three-dimensional space is shown on a two-dimensional projection. In the structural alignment programs VAST and SARF, the atoms of each secondary structural element in each protein are replaced by a vector of position, length, and direction determined by the positions of the C_α or C_β atoms along the element. Shown are projections of two α helices and two β strands and their vector representations as gray and green arrows, respectively, from a common x, y, z cartesian coordinate system. The three-dimensional cartesian coordinates of the start and end of one α-helical vector are diagrammed as wide dashed lines. Only these two sets of coordinates are needed to specify the location of the vector, whereas many such sets are required to locate the C_α or C_β atoms in the corresponding α helix. An element that is curved is approximated by two or more sequential vectors, as depicted by the two β strands, which are bent because of the twist of their composite β sheet. The joining of the helices by a short loop is also recognized by the algorithm. The vector representations of two proteins are then compared. If the type and arrangement of the elements are similar in two proteins within a reasonable margin of error and level of significance, the three-dimensional structures of the proteins are predicted to be similar.

proteins so that two matching segments are separated by unmatched segments. Second, although the local environments of many molecules in two proteins may be similar, there may also be some local differences. For example, central positions, but not the ends, of secondary structures in two proteins may match closely. For this reason, structural alignment methods often smooth out the comparisons by comparing several molecules at the same time and choosing an average result.

Structural biologists have been working on the problem of finding similar structural features in proteins for a long time, and a variety of methods have been devised for performing comparisons of protein structures (for review, see Sali and Blundell 1990; Russell and Barton 1992; Blundell and Johnson 1993; Holm and Sander 1994, 1996; Alexandrov and Fischer 1996; Falicov and Cohen 1996; Gibrat et al. 1996; Orengo and Taylor 1996; see Jewett et al. 2003 for recent heuristic method). A complete discussion of this subject is beyond the scope of this text. Programs publicly accessible at Web sites, SSAP and DALI, and two programs that utilize a fast search for common arrangements of secondary structures, VAST and SARF, are described in the following sections.

Dynamic Programming

Dynamic programming algorithms like those used for sequence alignment can be used for aligning structures. For aligning sequences, the object is to bring as many identical or similar sequence characters into vertical register in the alignment with a minimum cost of insertions and deletions. For aligning structures, the local environment of each amino acid expressed in interatomic distances, bond angles, or amino acid R group (Fig. 10.2) is given a coded value or vector representation that reflects the environment of that amino acid. Alternatively, a scoring matrix much like the amino acid scoring used for sequence alignments may be made. For protein structures, each sequential column in the scoring matrix gives a score for the fit of any of the 20 amino acids to a single position in the structure (more on matrices below). An optimal alignment between these sets of values by dynamic programming is then found.

The alignment program SSAP (*s*econdary *s*tructure *a*lignment *p*rogram) uses an algorithm called double dynamic programming to produce a structural alignment between two proteins (Taylor and Orengo 1989; Orengo et al. 1993; Orengo and Taylor 1996). A similar method has been used by Yang and Honig (2000). A local structural environment is independently defined for each residue in each sequence, and the method then matches residues by comparing these structural environments. The environment assigned to each amino acid takes into account the degree of burial in the hydrophobic core and type of secondary structure. As in sequence alignment by dynamic programming (Chapter 3), a scoring matrix is derived, and the highest-scoring regions in this matrix define the optimal structural alignment of the two proteins. One of the environmental variables that is used is a representation of the geometry of the protein by drawing a series of vectors from the C_β atoms of an amino acid to the C_β atoms of all of the other amino acids in the protein. If the resulting geometric views in two protein structures are similar, the structures must also be similar. The double dynamic programming method of aligning structures using C_β vectors is illustrated in Figure 10.14.

Because each sequential pair of amino acids is compared, an alignment will be possible only if the two protein chains follow the same approximate conformational changes throughout their lengths. If the proteins follow the same changes along some of their lengths, then diverge, then return again, it is difficult to align them through the divergent region by the above method. The problem is similar to trying to choose a gap penalty for sequence alignments, but in the case of structure, many kinds of rearrangements are possible.

Another version of SSAP (SSAP1) has been developed for identifying conserved folds/motifs, and this method circumvents the above alignment problem. This program uses all of the vector matrix values in the summary matrix and then uses a local alignment version of the dynamic programming algorithm to locate the most-alike regions in the structures. The algorithm has been greatly speeded up by comparing only pairs of amino acids with similar torsional angles (Φ and Ψ) and extent of residue burial/lack of water accessibility. SSAP is used to cluster proteins in the CATH database in a fully automated manner (Orengo et al. 1997).

Distance Matrix

The distance method uses a graphic procedure very similar to the dot matrix used for sequence alignment

FIGURE 10.14. The double dynamic programming method for structural alignment. (A) Vectors from the C_β atom of one amino acid to a set of other nearby amino acids in each of two protein segments are shown as two-dimensional projections. These vectors are given the same coordinate axes. Hence, one vector may be subtracted from the other to compare the relative positions of the C_β atoms in the two protein segments, shown in A as a vector difference. The smaller the differences, the more alike the structures. In SSAP, the vectors are subtracted (the resulting difference is δ) and the difference added to an empirically derived number, 10. The resulting value is then divided into a second empirically derived number, 500, to give a score S for the vector difference. For example, if the vector difference is 10°, then S = 500 / (10 + 10) = 25. (B) Two vector matrices that represent differences between the geometric view from one amino acid position in one protein and a view for one amino acid in the second protein. The set of vectors of one protein are listed across the top of the matrix and the set for the other are listed down the right side. The matrix is then filled with scores of vector differences. For example, if the vector from F to H in protein 1 less the vector from C to G in protein 2 is 31°, then the score placed in the upper right corner is S = 500/(31 + 10) = 12. The remaining difference scores are calculated in a similar manner. Although vectors to neighboring amino acids are shown in this example, vectors to immediate neighbor positions are actually not used to reduce the effect of local secondary structure. An optimal alignment, shown as a red path through the matrix, is then found through the vector matrix by a global form of the dynamic programming algorithm, using a constant deletion penalty of 50. For performing a structural alignment by this method, a similar set of vector differences are determined between the next amino acid V in protein A and the amino acid in protein B, as shown in the lower matrix in B, and an optimal path (blue) is obtained. This procedure is repeated until vector views between all amino acid positions have been compared. Two vector matrices are shown, comparing one position in protein A to each of two positions in protein B. (C) The resulting alignments (shown as red and blue paths) and the scores on the alignment path are transferred to a summary matrix. If two optimal alignment paths cross the same matrix position, the scores of those positions in the two alignments are summed. One part of the alignment path (black) is found in both comparisons, thereby providing corroborative evidence of vector similarity in these regions. In the example shown, the sum of the upper right positions in the two vector matrices is 12 + 16 = 28. When all of the alignments have been placed into the summary matrix, a second dynamic programming alignment is performed through this matrix. The final alignment found represents the optimal alignment between the protein structures. The logarithm of the final score is scaled such that a maximum value of 100 is possible. An adjusted score of 80 indicates a close structural relationship; one of 60–70 indicates a probable common fold. Other types of environmental variables other than the position of the C_β atoms in this example may also be aligned with this double dynamic programming method, as described in the text. (Adapted, with permission, from an example in Orengo and Taylor 1996 [©1996 Elsevier].)

A. Environmental vectors

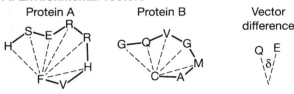

B. Vector matrices

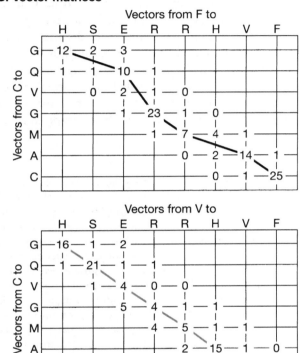

C. Summary matrix

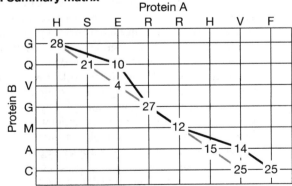

(Chapter 3) to identify the atoms that lie most closely together in the three-dimensional structure. If two proteins have a similar structure, the graphs of these structures will be superimposable. Distances between C_α atoms along the polypeptide chain and between C_α atoms within the protein structure can be compared by a two-dimensional matrix representation of the structure, as shown in Figure 10.15. Instead of aligning environmental variables of each successive amino acid in two protein structures, the distance matrix method compares geometric relationships between the structures without regard to alignment. The sequence of the protein is listed both across the top and down the side of the matrix. Each matrix position represents the distance between the corresponding C_α

atoms in the three-dimensional structure. The smallest distances represent the more closely packed atoms within secondary structures and regions of tertiary structure. Positions of closest packing are marked with a dot to highlight them, much as in a dot matrix. Distance matrices are produced for each three-dimensional structure of interest. Similar groups of secondary structural elements are superimposed as closely as possible into a common core structure by minimizing the sum of the atomic distances between the aligned C_α atoms. The method is outlined in Figure 10.15.

The program DALI (*distance alignment tool*) uses the distance method to align protein structures (Vriend and Sander 1991). The existing structures have been exhaustively compared to each other by

FIGURE 10.15. (*See facing page.*) Distance matrix of hypothetical three-helix structure. (*A*) Matrix positions that represent closest distances of approximately <12 Å between the C_α atoms in the known three-dimensional structures of the protein are marked by filling them with dots. Positions marked with black dots drawn just above the main downward-pointing diagonal (*dashed line*) from upper left to lower right represent amino acid sequential positions aa1-aa2, aa2-aa3, etc., that are close to each other because they are in the α helix. Marked regions of shortest C_α–C_α distances along this diagonal thus indicate positions of the α helices. Other marked diagonal regions (*orange and blue dots*) indicate tertiary structural interactions, including those between adjacent secondary structural elements. Helices a and b are close to each other and have opposite chemical polarities so that aa10-aa11-aa12 . . . are close to aa40-aa39-aa38 . . . on the orange surface of the helices. An upward-running diagonal (*orange dots*) from lower left to upper right reveals this spatial relationship. Helices b and c are also close to each other but have the same polarity so that aa30-aa31-aa32 . . . are close to aa50-aa51-aa52 . . . , producing a downward-directed diagonal (*blue dots*). Note that the rows of dots shown may be surrounded by other dots representative of other close amino acid pairs depending on the amino acid distance chosen for printing a dot and the arrangement of the amino acids in the structure. If another protein has a matrix pattern similar to that of the above example, then the two protein structures have the same three-helical arrangement and the loops joining the helices are of approximately the same length and conformation. The distance alignment method will find such three-helix patterns, even when the loop patterns are not similar. (*B*) Search for a common structural pattern in proteins A and B by DALI. A hypothetical example of a three-helix architecture is again used. In the top row, DALI first searches the entire distance matrix of protein A for a set of matching helices, a and b, indicated by an upward-directed diagonal whose position is the intersection of the locations of the helices in the sequence of protein A (*left column*). A similar search is performed for a corresponding pair of helices a′ and b′ in the distance matrix of protein B. In practice, the algorithm breaks down each full-sized matrix into a set of overlapping submatrices of size 6 × 6 amino acids. Distance patterns within the submatrices from each protein are then compared to locate similar structural configurations. Some matches will be longer than 6 amino acids and will therefore be found in several neighboring submatrices. A computationally sophisticated assembly step in the algorithm (see below) combines these overlaps into a complete structural alignment. Once found, individual matches are assembled. If a pair of helices is found in each structure, a beginning structural alignment of the sequences may be made (*right column*). A search for a third pair of helices c and c′ that interact with helices b and b′ in proteins A and B, respectively, is then made, as illustrated in the second row. A hypothetical pair common to A and B is shown. In this case, the order of regions b′ and c′ on the sequence of protein B is reversed from that of b and c. The composite matrices and alignment of all helices a, b, c and a′, b′, c′ are shown in the third row. Only the top one-half of the matrix is shown, leaving out the mirror image. Finally, DALI removes the insertions and deletions in the matrices and rearranges the sequence of the protein B to produce a parallel alignment of the elements in the two sequences (*bottom row*). By following these steps, an alignment of helices a, b, and c and a′, b′, and c′ in structures A and B is found by DALI, but the arrangements of sequences that produce this common architecture are different. Structural features that include β strands in proteins are found in the same manner. (Diagram derived from Holm and Sander 1993, 1996.)

DALI and the results organized into a database, the FSSP database, which may be accessed at http://www.ebi.ac.uk/dali/domain/. A newly found structure may be compared to the existing database of protein structures using DALI at the above Web site. The network version of DALI uses fast comparison methods to determine whether a new structure is similar to one already present in the FSSP database. The assembly step of the original DALI algorithm uses a Monte Carlo simulation that performs a random search strategy for submatrices that can be aligned using the similarity score defined below as a guide. The algorithm is similar to the genetic and simulated annealing algorithms (Chapter 5) in using a probabilistic method to improve previously found alignments.

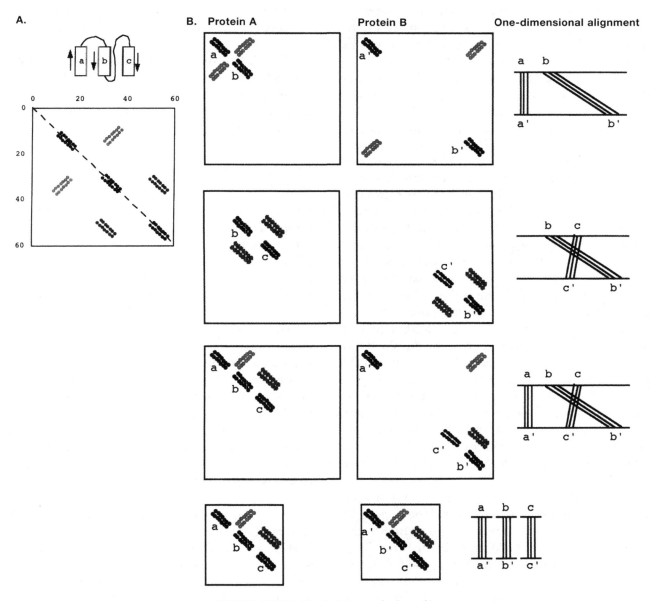

FIGURE 10.15. (*See facing page for legend.*)

ALIGNMENT IN DALI

There is no existing algorithm for direct alignment of two structures. Such an algorithm would have to find the closest alignment of two sets of points in three-dimensional space, a very difficult problem computationally, hence the need for an approximate solution. Other methods for aligning structures that are described below use simulations to find alignments. The Internet version of the DALI program uses more rapid search methods than those described above to compare new structures to existing structures in the FSSP database, but the overall analysis is very similar.

The similarity score for a structural alignment of two proteins by the distance method is based on the degree to which all of the matched elements can be superimposed. In the example shown in Figure 10.15, the score for a matching set of helices is the sum of the similarity scores of all of the atom pairs using a particular scheme for scoring each pair. Suppose that two helices a and b have been found to interact in protein A, and that a pair of helices a′ and b′ in protein B are superimposable on a and b. A certain pair of C_α atoms that are very close in the model, one in helix a (i^A) and a second in helix b (j^A), is identified. This set will correspond to a matched pair i^B in helix a′ and j^B in helix b′ of protein B. If the distance between i^A and j^A is d_{ijA} and the distance between i^B and j^B is d_{ijB}, then the similarity score for this pair of atoms is derived from the fractional deviation $|d_{ijA} - d_{ijB}|/d_{ij*}$, where d_{ij*} is the average of d_{ijA} and d_{ijB}. If two atom pairs can be superimposed, they are given a threshold similarity score of 0.20; otherwise they are given a similarity score of the threshold less the above fractional deviation. A deviation of 0.20 will correspond to adjacent β strands matching to within 1 Å and to α helices and helix strands matching to within 2–3 Å. As these scores are summed over all of the atoms in the matching helices, the contributions of more distant atoms are down-weighted by an exponential factor to allow for bending and other distortions.

The result of using this scoring system is that the similarity score for matching the two helix pairs in proteins A and B will increase in proportion to the number of superimposable atoms in the two helices. As additional matching elements are added to the structural alignment of the two proteins, the similarity scores for matching each individual pair of sec-ondary structures are added to give a higher similarity score that reflects the full alignment of the structures.

The DALI method provides one convenient method, in addition to the others described herein, to compare a new structure to existing structures in the Brookhaven structural database, and is accessible from a Web site.

Fast Structural Similarity Search Based on Secondary Structure Analysis

One class of structural alignment methods performs a comparison of the types and arrangements of α helices and β strands in one protein structure with the α helices and β strands in a second structure, as well as the ways in which these elements are connected (for review, see Gibrat et al. 1996). If the elements in two structures are similarly arranged, the corresponding three-dimensional structures are also similar. Because there are relatively few secondary structural elements in proteins and the relative positions of these elements may be quite adequately described by vectors giving their position, direction, and length, vector methods provide a fast and reliable way to align structures, as illustrated in Figure 10.13 (p. 437). It is a much simpler computational problem to compare vector representations of secondary structures than to compare the positions of all of the C_α or C_β atoms in those structures. If an element of a given type and orientation within a given tolerance level is found in the same relative position in both structures, they possess a basic level of structural similarity. Elements that do not match within the tolerance level are not considered to be structurally similar. VAST and SARF are examples of programs that are available on the Web which use this methodology (Alexandrov and Fischer 1996; Hogue et al. 1996; see http://www. ncbi.nlm.nih.gov/Entrez and http://123d.ncifcrf.gov/).

Vector methods do not use the structure authors have assigned to secondary structures in the PDB entry, but rather use automatic methods to assign secondary structure based on the molecular coordinates of atoms on the structure. Different methods are used for defining the number and extent of secondary structural elements and for the thresholds that make up an acceptable match (Bryant and Lawrence 1993; Madej et al. 1995; Alexandrov and Fischer 1996; Gibrat et al. 1996). Until one of these methods is shown to be superior, it is advisable to try all to increase the chance of a finding a biologically important match.

Once individually aligning sets of secondary structural elements have been identified, they are clustered into larger alignment groups. For example, if three matching sets of α helices have been found in two structures, a similarly oriented group of three α helices must be present in the structures. The same arrangements of a small number of secondary structural elements are commonly found in protein structures, thus this method often finds new occurrences of a previously found arrangement. An arrangement with a large number of secondary elements is less common and therefore more significant. This clustering step generates a large number of possible groups of secondary structural elements from which the most likely ones must be selected. Some methods use the clusters with the largest number of secondary structures as the most significant. Other methods perform a more detailed analysis of the aligned secondary structures. For example, the atomic coordinates of an α helix in one protein structure will be aligned with those of the matched α helix in the second structure, and the root mean square deviation (rmsd) will be calculated. The quality of this new alignment provides an indication of which secondary structure clusters are the most feasible. After starting with an alignment that includes the highest matching number of elements, the VAST algorithm examines alternative alignments that might increase the alignment score using the Gibbs sampling algorithm described in Chapter 5.

Like other structural alignment methods, VAST and SARF are available on Web pages and may be used for comparing new structures to the existing databases or for viewing structural similarities within the existing databases. An important aspect of searches for structural similarity by the vector method and other methods is the extent of the alignment found, or as Gibrat et al. (1996) state, "Is the alignment 'surprising'?"

Significance of Alignments of Secondary Structure

As in sequence alignment, it is important to estimate the reliability or statistical significance of a structural alignment. The problem is to determine the probability with which a given cluster of secondary structural elements would be expected between unrelated structures. The analogous problem with sequences is to determine whether an alignment score between two test sequences would also be found between random or unrelated sequences.

When comparing the arrangement of secondary elements in protein structures, a very large number of possible alignments are commonly found (Gibrat et al. 1996). The probability of a chance alignment of a few elements in two large but structurally unrelated proteins that have many such elements is quite high. Therefore, alignment of only a few elements in an actual comparison of two test sequences is not particularly significant. The probability of an alignment between most of the elements in large, unrelated proteins, however, is extremely low. Hence, such an alignment between structures is highly significant. The problem of significance thus boils down to assessing the number of possible ways of aligning elements in two unrelated proteins.

For calculating the probability of an alignment, the VAST algorithm uses a statistical theory very similar to that of the BLAST algorithm to calculate this probability. Recall that BLAST calculates a probability (or expect value) that a sequence alignment score at least as high as that found between a test sequence and a database sequence would also be found by alignment of random sequences. Sequence alignment scores are derived by using amino acid substitution matrices and suitable alignment gap penalties. The probabilities that alignments of random sequences could score as high as actual scores are calculated using the extreme value distribution. The equivalent VAST score is the number of superimposed secondary structural elements found in comparing two structures. The greater the number of elements that can be aligned, the more believable and significant the alignment. The statistical significance of a score is the likelihood that such a score would be seen by chance alignment of unrelated structures. This likelihood is calculated from the product of two numbers—the probability that such a score would be found by picking elements randomly from each protein domain and the number of alternative element pair combinations. Thus, if the chance of picking the number of matching elements found is 10^{-8} and the number of combinations is 10^4, the likelihood of an alignment of the same number of elements between unrelated structures is $10^{-8} \times 10^4 = 10^{-4.}$

Displaying Protein Structural Alignments

The programs and Web sites that perform a structural alignment or that provide access to databases of similar structures will transmit coordinates of the matched regions. The aligned regions may then be viewed with a number of molecular viewing programs, including Rasmol, Cn3d, and Spdbv. Cn3d also shows a second window with the matching sequence alignment, and aligned structures may be highlighted starting from this window. The program JOY provides a method for annotating sequence alignments with three-dimensional structural information (http://www-cryst.bioc.cam.ac.uk/~joy; Mizuguchi et al. 1998b).

PROTEIN STRUCTURE CAN BE PREDICTED USING AMINO ACID SEQUENCE

Use of Sequence Patterns

Although the sequences of 145,000 proteins are available, the structures of only 22,000 of these proteins are known. The increasing rate of genome sequencing can be expected to continue to outpace the rate of solving protein structures. Protein structural comparisons described above have shown that newly found protein structures often have a similar structural fold or architecture to an already-known structure. Thus, many of the ways that proteins fold into a three-dimensional structure may already be known. Structural comparisons have also revealed that many different amino acid sequences in proteins can adopt the same structural fold, and these sequences have been organized into the types of databases described above. Further examination of sequences in structures has also revealed that the same short amino acid patterns may be found in different structural contexts. Amino acid sequences present in secondary structures have been entered into databases that are useful for structure prediction. Many proteins in the sequence databases also have conserved sequence patterns upon which they may be further categorized.

If two proteins share significant sequence similarity, they should also have similar three-dimensional structures. The similarity may be present throughout the sequence lengths or in one or more localized

domains that may or may not be interrupted with gaps. When a global sequence alignment is performed, if more than 45% of the amino acid positions are identical, the amino acids should be quite superimposable in the three-dimensional structure of the proteins. Thus, if the structure of one of the aligned proteins is known, the structure of the second protein and the positions of the corresponding amino acids in this structure may be reliably predicted from the sequence alignment. If less than 45% but more than 25% of the amino acids are identical, the structures are likely to be similar, but with more variation at the lower identity levels at the corresponding three-dimensional positions.

Protein Classification Schemes

Proteins have been classified on the basis of sequence similarity or the presence of common amino acid patterns. First, they were organized into families and superfamilies on the basis of the level of sequence similarity in sequence alignments. The current method of organizing proteins by this method at the Protein Information Resource (PIR) (http://www-nbrf.georgetown.edu) is that each entry in the PIR protein sequence database is searched against the remaining entries using the FASTA algorithm. Similar sequences are then aligned with the Genetics Computer Group multiple sequence alignment program PILEUP. Families are composed of proteins that align along their entire lengths with a sequence identity of 50% or better (Barker et al. 1996).

More recent analyses of amino acid patterns in protein sequences have revealed that many proteins are made up of modular domains, short regions of similar amino acid sequence that correspond to a particular function or structure. Furthermore, sets of proteins from widely divergent biological sources may share several such domains, and the domains may not be in the same order. Hence, it has become necessary to redefine the concepts of family and superfamily. Proteins that comprise the same set of similar homology domains (extended regions of sequence similarity) in the same order are referred to as homeomorphic protein families. Protein families, members of which have the same domains in the same order, but also have dissimilar regions, are designated as a homeomorphic superfamily (Barker et al. 1996). The superfamily classification of a newly identified protein sequence may be analyzed at several Web sites (Table 10.4).

A second method of classifying proteins is based on the presence of amino acid patterns. Proteins with the same biochemical function have been examined for the presence of strongly conserved amino acid patterns that represent an active site or other important feature. The resulting database is the Prosite catalog (Table 10.4) (Bairoch 1991; Hofmann et al. 1999). Proteins have also been categorized on the basis of the occurrence of common amino acid patterns—motifs and conserved gapped and ungapped regions in msa's. These patterns are found by extracting them from msa's, by pattern-finding algorithms that search unaligned sequences for common patterns, and by several statistical methods that search through unaligned sequences. The patterns vary in length, presence of gaps, and degree of substitution. The algorithms that are used include pattern-finding algorithms, HMMs, the expectation maximization method, and the Gibbs sampling method. These methods and the computer programs and Web sites that provide them are described in Chapter 5. Listed in Table 10.4 are several databases that categorize proteins based on the occurrence of common patterns. Also shown are the locations of databases of amino acid patterns that determine cellular localization of proteins or sites of protein modification (signal or transit peptides). FSSP, a structural family database, is listed in this table because it includes links to information on sequence families and superfamilies.

A given protein sequence may be classified by using one of the resources in Table 10.4 for sequence patterns that are characteristic of a group or family of proteins. Because most of these databases are derived by quite different methods of pattern analysis, statistics, and database similarity searching, they can be expected to provide different but complementary information. Thus, a given database may include a sequence pattern that is not identified in others, and this pattern may provide an important link to structure or function for one group of proteins. Another database may provide patterns more suitable for classifying a different group of proteins. Therefore, a wise choice would be to use as many of these resources as possible for classifying a new sequence. However, note the availability of Web sites that have combined the resources of separate protein classification databases into a single database (e.g., INTERPRO, NCBI; Table 10.4). In one recent field of endeavor, protein taxonomy, genomic databases that list the entire set of proteins produced by a particular organism are searched for matches. Such searches can provide a wealth of information on protein evolution (Pellegrini et al. 1999).

Clusters

Another, more recently introduced, method for classifying proteins by sequence analysis is to use clustering methods. In these methods, every protein in a sequence database such as SwissProt is compared to every other sequence using a database search method including the BLAST, FASTA, and Smith–Waterman dynamic programming methods described in Chapter 6. Thus, each protein in the database receives a sequence similarity score with every other sequence. A similar method is used to identify families of paralogous proteins encoded by a single genome. Matching sequences are further aligned by a pair-wise alignment program like LALIGN to recalculate the significance of the alignment score as an E value (see Chapter 3 flowchart, p. 75).

In a cluster analysis, sequences are represented as vertices on a graph, and those vertices representing each pair of related sequences are joined by an edge that is weighted by the degree of similarity between the pair (see Fig. 10.4). In the first step, the clustering algorithm detects the sets of proteins that are joined in the graph by strongly weighted edges. In subsequent steps, relationships between the initial clusters found in the first step are identified on the basis of weaker, but still significant, connections between them. These related clusters are then merged in a manner that maximizes the strongest global relationships (see Web sites for ProtoMap and SYSTERS and other protein cluster data sets; Table 10.4). Clustering has been used to identify groups of proteins that lack a relative with a known structure and hence are suitable for structural analysis (Portugaly and Linial 2000). Additional information on clustering methods is provided in Chapters 11 and 13.

Proteins Comprise Motifs, Modules, and Other Sequence Elements of Structural Significance

The above analysis describes the types and distribution of motifs in proteins from the same or different organisms. A motif can represent an individual folded structure or active-site residues. Several different motifs widely separated in the same protein sequence are often found. These motifs may represent conserved regions that lie in the core of the protein structure. Hence, their presence in two sequences predicts a common structural core (for review, see Henikoff et al. 1997).

A more detailed analysis of motifs has revealed that they are components of a more fundamental unit of structure and function, the protein module. Proteins may have several modules corresponding to different units of function, and these modules may be present in a different order (Henikoff et al. 1997). These diverse arrangements suggest that a biologically important module has been employed repeatedly in protein evolution by gene duplication and rearrangement mechanisms that are discussed further in Chapter 11. The presence of modules also provides a further system of protein classification into module-based families.

An example of an important motif is the C_2H_2 (two cysteines and two histidines) zinc finger DNA-binding motif Xfin of *Xenopus laevis* illustrated in Figure 10.16. The zinc finger is one of the most commonly identified motifs, in part because of the characteristic spacing of C and H residues in the motif sequence. As indicated in Figure 10.17, the zinc atom forms bonds with these residues to create the finger-

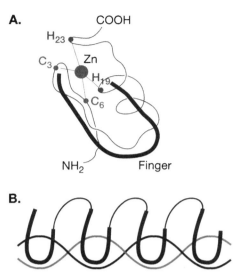

FIGURE 10.16. The zinc finger DNA-binding domain. This domain is the most commonly found due to the particular spacing of histidine and cysteine residues. (*A*) Structure found by NMR studies of a 25-residue molecule made up of a β hairpin of structural motif sheets of amino acids 1–10 followed by an α helix of amino acids 12–24. (*B*) Hypothetical binding pattern of fingers to DNA. (Adapted, with permission, from Branden and Tooze 1991 [©1991 Garland Science/Taylor & Francis Books, Inc.].)

like projection. When present in tandem copies, the finger is thought to lie in an alternating pattern in the major groove of DNA. A simple plot of the positions of C and H residues on the protein sequence as shown in Figure 10.17 provides a very simple way to locate zinc fingers in a protein sequence.

Pfam is a Web site that includes a listing of proteins that carry the zinc finger sequence motif. As shown in Figure 10.18, the zinc finger is one of the most commonly recognized motifs, and proteins that carry the motif have been classified into a family. Two other families of zinc finger proteins with four cysteine or three cysteine and one histidine residues interacting with the Zn atom, and additional variations in the basic structure of zinc fingers, have also been identified. Descriptions and alignments of these proteins are provided at the Pfam Web site, as illustrated in Figure 10.19. Other families in the Pfam classification are given a description that best reflects the extent and complexity of the conserved sequence patterns, be it a domain, module, repeat, or motif. In general, all of these patterns represent a conserved unit of structure or function.

FIGURE 10.17. Graph of the *Xenopus laevis* XFIN protein sequence which is in the Cys-Cys-His-His class of zinc finger DNA-binding proteins (Branden and Tooze 1991). The graph was produced using the AA Window, Cys + His map option of DNA STRIDER vers. 1.2 on a Macintosh computer. The bottom panel shows amino acids Y, C, F, L, and H, respectively, as bars of increasing length. The top panel shows H and C as half- and full bars, respectively. The fingers appear in the top panel as double half-bars (two Cys residues separated by two amino acids) followed by double full bars (two His residues separated by two amino acids). This type of graphic representation is extremely useful for visualizing amino acid patterns in proteins.

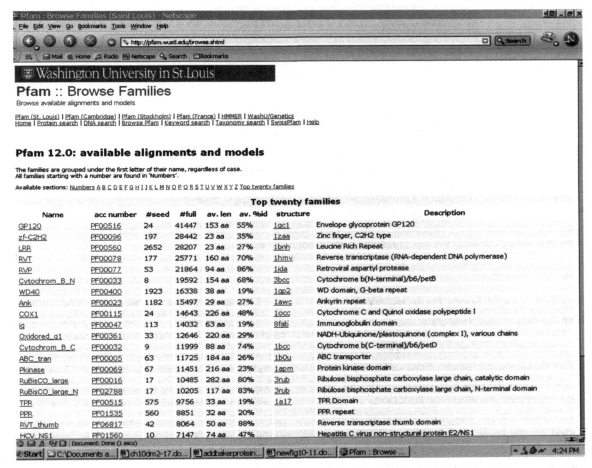

FIGURE 10.18. The Pfam Web mirror site at Washington University (http://pfam.wustl.edu/browse.shtml). Shown are the 20 most common protein families classified according to the motifs that are present. Note the presence of the Pfam entry for zf-C2H2, the name assigned to the C_2H_2 (two cysteines and two histidines) Zn finger DNA-binding motif, accession no. PF00096. Any family may be examined by clicking the mouse on the first letter of the family name. Figure 10.19 is an example of the entry for the PF00096.

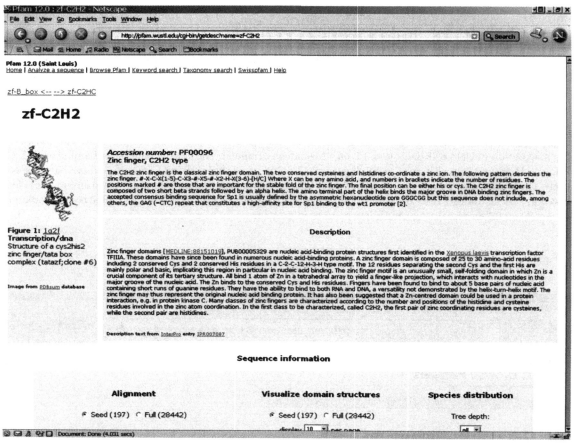

FIGURE 10.19. The Pfam entry for family zf-C2H2 (accession no. PF00096). The mouse was clicked on the entry for zf-C2H2 shown in Fig. 10.18. The Pfam database is based on a statistical analysis of sequences with the same motif using HMMs. The result is a profile of the sequences with matches, mismatches, and gaps. The entry describes how this profile was produced by the HMMER program, and also provides references and a link to a multiple sequence alignment of the sequences. As discussed in Chapter 3, this HMM of the sequences can be used to produce the msa by choosing the most probable path through the model.

Structural Features of Some Proteins Are Readily Identified by Sequence Analysis

The above section describes how a newly identified protein may be classified on the basis of the presence of sequence motifs, modules, or other sequence elements that represent structure or function. The zinc finger motif is one structural motif that may be readily identified on the basis of the order and spacing of a conserved pattern of cysteine and histidine residues in the sequence. Other classes of proteins have characteristic amino acid composition and patterns such that the structure can often be reliably predicted from the amino acid sequence. Some other examples of structure recognition on the basis of sequence are given below.

Leucine zippers and coiled coils. The leucine zipper motif is typically made up of two antiparallel α helices held together by interactions between hydrophobic leucine residues located at every seventh position in each helix, as illustrated in Figure 10.20A. The zipper holds protein subunits together. The leucines are located at approximately every two turns of the α helix. It is this repeated occurrence of leucines that makes the motif readily identifiable. In the transcription factors Gcn4, Fos, Myc, and Jun, the binding of the subunits forms a scissor-like structure with ends that lie on the major groove of DNA, as shown in Figure 10.20B. If the amino acids in each helical region are plotted as a spiral of 3.6 amino acid residues per turn, representing a view looking down the helix from the end starting at residue 1 on the inside of the spiral,

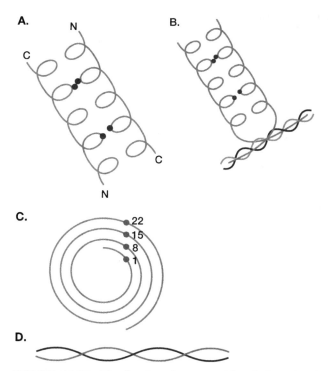

FIGURE 10.20. The leucine zipper motif and the related coiled-coil structure. (*A*) The basic motif (see text). (*B*) DNA-binding motif of transcription factors such as Gcn4. (*C*) Spiral plot of α helix in Gcn4. (*D*) A left-handed coiled-coil structure. (*Orange*) Helical backbones; (*green*) leucine residues; (*red* and *blue*) dsDNA strands. The EMBOSS program pepnet finds repeated occurrences at regular intervals of a residue such as a hydrophobic residue in a helix.

then the result shown in Figure 10.20C is found. The leucine residues are found on approximately the same side of the helix, slightly out of phase with the rotational symmetry of the helix. The predicted structure is that of a coiled coil, as shown in Figure 10.20D (Branden and Tooze 1991).

Coiled-coil structures typically comprise two to three α helices coiled around each other in a left-handed supercoil in a manner that slightly distorts the helical repeat so that it is 3.5 residues per turn instead of the usual 3.6, or an integral number of 7 residues every second turn (Lupas 1996). They occur in fibrous proteins such as keratin and fibrinogen, and are also thought to occur in leucine zippers, as there is a repeat of leucine at every seventh residue (Branden and Tooze 1991). If the spiral wheel in Figure 10.20C is plotted so that there are 7 residues every second turn instead of 7.2, then the residues align more uniformly on one face of the helix.

Consequently, the leucine zipper has been hypothesized to adopt a coiled-coil structure.

Coiled-coil regions may be predicted by searching for the 7-residue (heptad) periodicity observed in the sequence of these proteins. Naming these respective positions a, b, c, d, e, f, and g, then a and d are usually hydrophobic amino acids and the remaining amino acids are hydrophilic, because coiled coils are generally fibrous, solvent-exposed structures. As more and more of these sequential patterns are observed along a sequence, one can be more convinced that the prediction is reliable. If there are at least five of these heptads and the hydrophobicity pattern is strongly conserved, the prediction is a good one. Poorer-quality patterns come into doubt.

A program called COILS2 has been developed for predicting coiled-coil regions with greater reliability than simple pattern searching for heptad repeats (Lupas et al. 1991; Lupas 1996; program description at http://www.embl-heidelberg.de/predictprotein/). There are two Web sites for predicting the occurrence of coiled-coil regions in protein sequences using the COILS program—http://www.isrec.isb-sib.ch/software/software.html and at http://www.embl-heidelberg.de/predictprotein. The program may also be obtained from these sites for running on a local server.

Central to the COILS method is the generation of a profile scoring matrix, with each column showing the distribution of amino acids in each of the seven positions, a–g, found in all of the known coiled-coil proteins. Thus, column 1 representing residue a and column 4 representing residue d will show a high score for hydrophobic residues, but the other positions will show greater variability. These matrices may be 7, 14, 21, or 28 residues long for scoring shorter or longer sequence regions. Separate profile matrices of this kind have been prepared for two-stranded coiled coils, three-stranded coiled coils, bundles, and coiled coils with parallel or antiparallel strands. There are also two scoring matrices in use: MTK derived from the sequences of myosins, tropomyosins, and keratins (intermediate filaments type I and II), and MTIDK, a new matrix derived from a much larger number of sequences. These matrices appear to give better predictions of different sets of structures for an unknown reason. The profile scoring matrix is moved along the candidate sequence one position at a time, calculating a score at each position. This score for each window is then given a probability *P* based on the distribution of scores found in coiled-coil and globular (not coiled-coil) pro-

A.

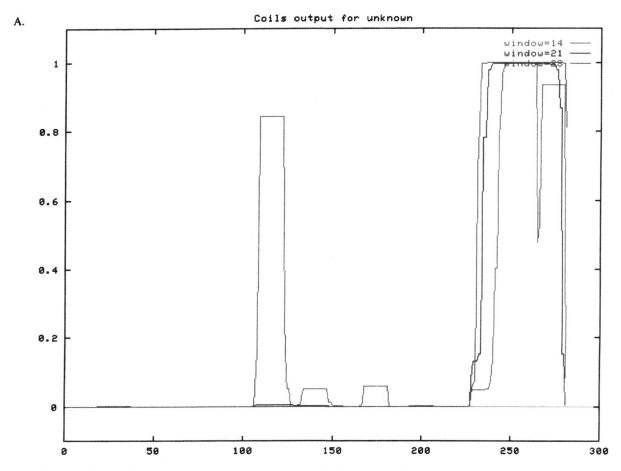

FIGURE 10.21. Prediction of coiled-coil regions by the COILS2 program. The Gcn4 protein was used as input to a Web page listed in the text. (*A*) Plot of probability of residue in coiled-coil structure versus residue number obtained from the ISREC server. (*B*) Partial list of scores by residue number. Analysis obtained from the Predict Protein server. Note that the highest probabilities are obtained with a window of 28 amino acids. The expected order of amino acids in the coiled coil is a, b, c, d, e, f, and g. This order does not start until residue 244. Amino acids found at position 243 and at lower numbered positions are characteristically found at other places in the coiled-coil heptad. The EMBOSS program pepcoil is similar to COILS2. (*Figure continues on facing page.*)

teins using the same matrix and other comparative conditions, where

$$P = G_{cc} / (R\,G_g + G_{cc}) \qquad (1)$$

and G_{cc} and G_g are probabilities derived from the statistical distribution of scores of all coiled-coil and globular proteins, respectively, and R is the predicted ratio of coiled-coil to globular residues in GenBank. These scores will vary with each window size and option chosen, and the scores may be normalized to give a better impression of their range. A false positive can occur with sequences that have a biased amino acid distribution; these false positives can be identi-

fied by the program option of weighting the two hydrophobic positions a and d the same as the five hydrophilic positions b, c, e, f, and g. Normally, these positions are weighted 2.5 times more heavily during the scoring procedure. False positives will continue to have a high score whereas true positives will not.

For candidate protein sequences, Lupas recommends using both types of weighting and both MTK and MTIDK matrices. The program reliably predicts known coiled-coil regions (Lupas 1996). An example of the program output from the ISREC Web site is shown in Figure 10.21, using as input the sequence of Gcn4 (identified as GCN4_YEAST in the SwissProt

```
B.
COILS version 2.1
using MTK matrix.
weights: a,d=2.5 and b,c,e,f,g=1.0
Input file is /home/phd/server/work/predict_h24138-21300.fasta
>prot (#) ppOld, gcn4 /home/phd/server/work/predict_h24138
    Residue      Window=14              Window=21               Window=28
             Score  Probability      Score  Probability      Score  Probability
  .
  .

    239 A    g 1.240    0.011      d 1.529    0.419      d 1.546    0.852
    240 R    a 1.240    0.011      e 1.581    0.585      e 1.546    0.852
    241 R    e 1.446    0.051      f 1.581    0.585      f 1.546    0.852
    242 S    f 1.446    0.051      g 1.581    0.585      g 1.546    0.852
    243 R    g 1.450    0.052      a 1.581    0.585      a 1.551    0.862
    244 A    b 1.529    0.093      b 1.607    0.664      b 1.643    0.968
    245 R    c 1.592    0.145      c 1.607    0.664      c 1.643    0.968
    246 K    d 1.669    0.238      d 1.607    0.664      d 1.643    0.968
    247 L    e 2.433    0.994      e 1.843    0.978      e 1.984    1.000
    248 Q    f 2.433    0.994      f 1.988    0.997      f 2.041    1.000
    249 R    g 2.433    0.994      g 2.018    0.998      g 2.052    1.000
    250 M    a 2.433    0.994      a 2.054    0.999      a 2.052    1.000
    251 K    b 2.433    0.994      b 2.054    0.999      b 2.052    1.000
    252 Q    c 2.433    0.994      c 2.054    0.999      c 2.052    1.000
    253 L    d 2.433    0.994      d 2.054    0.999      d 2.052    1.000
    254 E    e 2.433    0.994      e 2.054    0.999      e 2.052    1.000
    255 D    f 2.433    0.994      f 2.054    0.999      f 2.052    1.000
    256 K    g 2.433    0.994      g 2.054    0.999      g 2.052    1.000
    257 V    a 2.433    0.994      a 2.054    0.999      a 2.052    1.000
    258 E    b 2.433    0.994      b 2.054    0.999      b 2.052    1.000
    259 E    c 2.433    0.994      c 2.054    0.999      c 2.052    1.000

    260 L    d 2.433    0.994      d 2.054    0.999      d 2.052    1.000
    261 L    e 2.433    0.994      e 2.054    0.999      e 2.052    1.000
    262 S    f 2.421    0.993      f 2.054    0.999      f 2.052    1.000
    263 K    g 2.421    0.993      g 2.054    0.999      g 2.052    1.000
  .
  .
  .

    271 V    a 2.026    0.848      a 2.004    0.998      a 2.052    1.000
    272 A    b 2.026    0.848      b 1.968    0.996      b 2.052    1.000
    273 R    c 2.026    0.848      c 1.943    0.994      c 2.052    1.000
    274 L    d 2.026    0.848      d 1.943    0.994      d 2.052    1.000
    275 K    e 2.026    0.848      e 1.883    0.987      e 2.052    1.000
    276 K    f 2.026    0.848      f 1.883    0.987      f 2.052    1.000
    277 L    g 2.026    0.848      g 1.776    0.948      g 1.986    1.000
    278 V    a 2.026    0.848      a 1.776    0.948      a 1.949    1.000
    279 G    b 2.026    0.848      b 1.631    0.732      b 1.868    0.999
    280 E    c 2.026    0.848      c 1.631    0.732      c 1.868    0.999
    281 R    a 1.378    0.030      d 1.090    0.003      d 1.381    0.263
```

FIGURE 10.21. (*Continued; see facing page for legend.*)

database), which has a leucine zipper region. The protein is scanned for the number of occurrences of coiled coils in a sliding window of 7, 14, 21, or 28 residues.

Another method for predicting coiled coils is based on an analysis of correlations between pairs of amino acids (Berger et al. 1995), and the program is accessible at http://searchlauncher.bcm.tmc.edu/.

Transmembrane-spanning proteins. The all-α superfamily of membrane proteins (see classification of membrane proteins at the SCOP structural database at http://scop.mrc-lmb.cam.ac.uk/scop) com-

prises transmembrane (TM) proteins that traverse membranes back and forth through a series of α helices comprising amino acids with hydrophobic side chains. The typical length of 20–30 residues, and the strong hydrophobicity of these helices, provide a simple method for scanning a candidate sequence for such features. An example of such a structure is illustrated in Figure 10.22.

Membrane-spanning hydrophobic α helices can be located quite accurately by scanning for hydrophobic regions about 19 residues in length in the

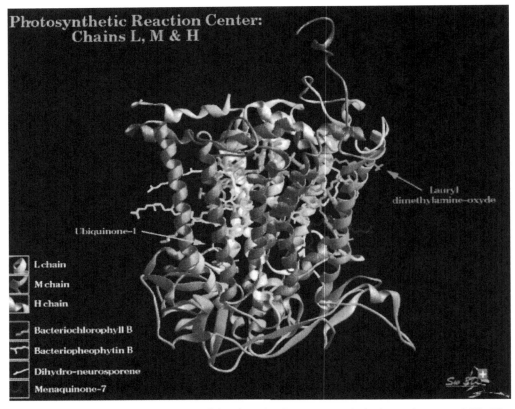

FIGURE 10.22. Three-dimensional structure of the photosynthetic center of *Rhodopseudomonas viridis*. The three subunits each cross the membranes of hollow vesicles found in these bacteria at approximately right angles, two of them back and forth multiple times. The light-harvesting pigments chlorophyll and pheophytin are bound between these helices. These membrane-spanning regions are 25–29 amino acids long and are composed of α helices. There is an abundance of hydrophobic amino acids in these helices. Hence, a hydrophobicity plot of the protein chain will show peaks centered on the position of the helices, as shown in Fig. 10.23. (Image from ftp://ftp.expasy.org/databases/swiss-3dimage/IMAGES/GIF/S3D00208.gif.)

amino acid sequence (Kyte and Doolittle 1982). The occurrence of such regions in a candidate protein of unknown structure is a good indicator that the region spans a membrane. In Figure 10.23, such an analysis is shown for subunit M of the above molecule. Membrane-spanning helices are different from α helices that are located on the surface of a protein structure. The surface helices tend to have hydrophobic residues located on the core-facing side (inside) and the hydrophilic residues on the solvent-facing side (outside) of the helix. These surface-exposed helices can be recognized by this separation of hydrophobic residues through a helical moment analysis described below. Transmembrane-spanning α helices are more like α helices that are buried in the structural core of a protein, which also have a high proportion of amino acids with hydrophobic side groups located throughout their lengths. In an effort to distinguish different classes of α helices, several methods for improving the prediction of TM regions have been devised and are available on Web sites.

One such method is one of the program choices of the PHD server for protein structure prediction at http://www.embl-heidelberg.de/predictprotein/predictprotein.html. The membrane-spanning helix predict program is named PHDhtm (protein Heidelberg predict for *h*elical *t*rans*m*embrane proteins). Briefly, a machine-learning method called a neural network (NN) (see below) is trained to recognize the sequence patterns and sequence variations of a set of α-helical TM proteins of known three-dimensional structure. A candidate sequence is then scanned for the pres-

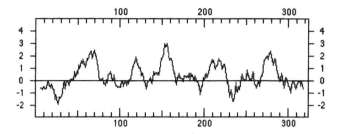

FIGURE 10.23. Hydrophobicity plot of subunit M of the photosynthetic center of *Rhodopseudomonas viridis* illustrated in Fig. 10.22. The hydrophobicity plotting program of DNA Strider 1.2 on a Macintosh computer was used with Kyte–Doolittle hydrophobicity values (Table 10.6) and a sliding window of 19, chosen to detect the approximate length of membrane-spanning α helices of 20–30 residues. The plot reliably predicts the five hydrophobic membrane-spanning helices of this protein, which are located in the three-dimensional structure at amino acid positions 52–78, 110–139, 142–167, 197–225, and 259–285. The SwissProt entry for this protein is numbered P06010 and the protein ID is RCEM_RHOVI. The program TGREASE, which is available in the FASTA suite of programs, will provide a similar plot on Macintosh or PC computers. The program tmap in the EMBOSS suite produces a similar plot.

3. The orientation of the predicted α-helical domains with respect to the inside (cytoplasmic) or outside of the membrane was also predicted based on the observed preponderance of positively charged amino acids on the cytoplasmic side of solved structures (Rost et al. 1995).

An illustrative example of a PHDhtm analysis on protein 1prc_M is shown in Figure 10.24. As shown, the program correctly predicts five transmembrane helices, but positions of the ends of these helices are not always correctly predicted, as revealed by a lack of correlation between the predicted regions (H) and the known regions (*).

A second method for prediction of TM α helices is by the TMpred server. This method scans a candidate sequence for matches to a sequence scoring matrix obtained by aligning the sequences of all of the transmembrane α-helical regions that are known from structures. These sequences have been collected into a database (TMbase) of such sequences. An example of a transmembrane analysis of 1prc_M by this method is shown in Figure 10.25. As shown, the program correct-

ence of similar sequence variations, and a prediction is made as to the occurrence and location of α-helical domains in the candidate protein.

The specific steps are as follows.

1. Each of the small number of structurally identified α-helical transmembrane proteins was used to search the SwissProt protein sequence database for additional sequences in this superfamily using the BLAST or FASTA algorithms.

2. The sequences found were assembled first into a multiple sequence alignment and then into a motif by the program MAXHOM. Sequences less than 30% identical, and therefore least likely to be in the superfamily, were not included. The most-alike sequences in the alignment were also removed to provide a representative and statistically reasonable range of amino acid substitutions in each column of the motif. The neural network was then trained to differentiate between columns in the motif representing the α-helical domains and the flanking nonhelical domains. The training method is described in greater detail below.

TABLE 10.6. *Hydrophobicity scales for the amino acids*

Residue		Value
Ala	A	1.8
Arg	R	−4.5
Asn	N	−3.5
Asp	D	−3.5
Cys	C	2.5
Gln	Q	−3.5
Glu	E	−3.5
Gly	G	−0.4
His	H	−3.2
Ile	I	4.5
Leu	L	3.8
Lys	K	−3.9
Met	M	1.9
Phe	F	2.8
Pro	P	−1.6
Ser	S	−0.8
Thr	T	−0.7
Trp	W	−0.9
Tyr	Y	−1.3
Val	V	4.2

These values are based on adjusted values derived from several sets of experimental measurements (Kyte and Doolittle 1982). The most hydrophobic amino acids are printed in green, the least hydrophobic amino acids in red. A number of additional scales are also available (von Heijne 1987).

```
       ....,....1....,....2....,....3....,....4....,....5....,....6
AA     |ADYQTIYTQIQARGPHITVSGEWGDNDRVGKPFYSYWLGKIGDAQIGPIYLGASGIAAFA|
PHD htm|                                        HHHHHHHHHHHHHHHHHHHH |
                                                ********


       ....,....7....,....8....,....9....,....10...,...11...,....12
AA     |FGSTAILIILFNMAAEVHFDPLQFFRQFFWLGLYPPKAQYGMGIPPLHDGGWWLMAGLFM|
PHD htm|HHHHHHHHHHHHHHHHHHHH                          HHHHHHHHHHH     |
        ****************                             ***********


       ....,....13...,....14...,....15...,....16...,....17...,....18
AA     |TLSLGSWWIRVYSRARALGLGTHIAWNFAAAIFFVLCIGCIHPTLVGSWSEGVPFGIWPH|
PHD htm|HHHHHHHHHHH          HHHHHHHHHHHHHHHHHHHHHHHHHHHH    HHHH    |
        *****************    **************************


       ....,....19...,....20...,....21...,....22...,....23...,....24
AA     |IDWLTAFSIRYGNFYYCPWHGFSIGFAYGCGLLFAAHGATILAVARFGGDREIEQITDRG|
PHD htm|          HHHHHHHHHHHHHHHHHHHHHHHHHHHHHHHH                    |
                  ****************************


       ....,....25...,....26...,....27...,....28...,....29...,....30
AA     |TAVERAALFWRWTIGFNATIESVHRWGWFFSLMVMVSASVGILLTGTFVDNWYLWCVKHG|
PHD htm|          HHHHHHHHHHHHHHHHHHHHHHHHHHHHHHHHHHHHHHH             |
                  *************************


       ....,....31...,....32...,....33...,....34...,....35...,....36
AA     |AAPDYPAYLPATPDPASLPGAPK|
```

FIGURE 10.24. Analysis of a known transmembrane protein by PHDhtm program at the predict protein server at Heidelberg. The same protein used for the above hydrophobicity analysis in Fig. 10.25 hydro and also the protein in Fig. 10.23 (structural name 1prc_M, SwissProt P06010) was submitted to the server at http://www.embl-heidel berg.de/predictprotein/ppDoPred.html, choosing the transmembrane prediction option on the expert page and minimizing program output. Additional program output, including probabilities, assignment of inside and out-side domain (topology), and NN details are not shown. Note that the protein 1prc_M is listed on the server as one of the proteins that was used to train the NN. Hence, using this protein is a biased test of program accuracy, which is claimed in more objective tests to identify residues in the transmembrane helices with 95% reliability and heli-cal transmembrane proteins at 86% accuracy (Rost et al. 1995, 1996; Rost 1996). The predicted helical regions are shown by an H and the known regions in the three-dimensional structure, obtained from the SwissProt entry for the protein, are shown by an asterisk.

```
2 possible models considered, only significant TM-segments used

STRONGLY preferred model: Amino-terminus inside
Five   strong transmembrane helices
 # from   to length    actual
 1   52   71 (20)      52- 78
 2  110  132 (23)     110-139
 3  146  166 (21)     142-167
 4  199  219 (21)     197-225
 5  268  289 (22)     259-285

alternative model
 five strong transmembrane helices
 # from   to length    actual
 1   53   71 (19)      52- 78
 2  113  129 (17)     110-139
 3  144  161 (18)     142-167
 4  201  225 (25)     197-225
 5  268  289 (22)     259-285
```

FIGURE 10.25. Analysis of known transmembrane protein by TMPRED. The same protein used above for the PHDhtm analysis (structural name 1prc_M, SwissProt P06010) was submitted to the server at http://www.ch.emb net.org /software/TMPRED_form.html. Shown are two predicted structural models; score and topology informa-tion is not included. Known locations of the α helices are shown in the last column for comparison. Another accu-rate predictor is TMHMM.

ly predicted five α-helical transmembrane segments. Two alternative models were predicted, the first more highly favored, but neither one matched the known ends of these regions. These examples serve to illustrate that these methods can be expected to identify membrane-spanning α-helical proteins quite reliably, but not the ends of such regions.

Another program (TMHMM) based on an HMM of the observed sequence variation in TM proteins is also available (Krogh et al. 2001; see http://www.cbs.dtu.dk/services/TMHMM/). The model predicts the location and orientation of α helices in TM-spanning proteins. TMHMM has features that match the architecture of transmembrane helices in this class of proteins. A series of consecutive states for modeling the helix core, helix caps on either side of the helix, a loop on the cytoplasmic side, two loops for the noncytoplasmic side, and a globular domain state in the middle of each loop is included. The model is trained with known TM proteins and can then be used to scan all of the predicted proteins for newly sequenced genomes for TM proteins. TMHMM has been found in comparative studies to be the best predictor (97–98% accuracy) of TM protein structures (Krogh et al. 2001; Möller et al. 2002).

A simple hydrophobicity plot may also be used as shown in Figure 10.23. The number and extent of these regions can also be predicted from the peaks in this plot. This method is unsuitable for scanning genomic sequences for possible membrane-spanning proteins; the automatic methods are much more suitable for this purpose.

Prediction of Protein Secondary Structure from the Amino Acid Sequence

Accurate prediction as to where α helices, β strands, and other secondary structures will form along the amino acid chain of proteins is one of the greatest challenges in sequence analysis. At present, it is not possible to predict these events with very high reliability. As methods have improved, prediction has reached an average accuracy of 64–75% with a higher accuracy for α helices, depending on the method used. These predictive methods can be made especially useful when combined with other types of analyses discussed in this chapter. For example, a search of a sequence database or a protein motif database for matches to a candidate sequence may

discover a family or superfamily relationship with a protein of known structure. If significant matches are found in regions of known secondary or three-dimensional structure, the candidate protein may share the three-dimensional structural features of the matched protein. Several Web sites provide such an enhanced analysis of secondary structure. These sites and others that provide secondary structure analysis of a query protein are given in the section "Web Search Terms from this Chapter" on page 485. The main methods of analyses used at these sites are described below.

Methods of structure prediction from amino acid sequence begin with an analysis of a database of known structures. These databases are examined for possible relationships between sequence and structure. When secondary structure predictions were first being made in the 1970s and 1980s, only a few dozen structures were available. This situation has now changed with present databases including more than 3000 independent structural folds. The combination of more structural and sequence information presents a new challenge to investigators who wish to develop more powerful predictive methods.

The ability to predict secondary structure also depends on identifying types of secondary structural elements in known structures and determining the location and extent of these elements. The main types of secondary structures that are examined for sequence variation are α helices and β strands. Early efforts focused on more types of structures, including other types of helices, turns, and coils. To simplify secondary structure prediction, these additional structures that are not an α helix or β strand were subsequently classified as coils. More recent methods (e.g., HMMSTR; see Fig. 10.29B below) have increased the sensitivity of detecting turns. Assignment of secondary structure to particular amino acids is sometimes included in the PDB file by the investigator who has solved the three-dimensional structure. In other cases, secondary structure must be assigned to amino acids by examination of the structural coordinates of the atoms in the PDB file.

Methods for comparing three-dimensional structures, described above, frequently assign these features automatically, but not always in the same manner. Hence, some variation is possible, and deciding which is the best method can be difficult. The DSSP database of secondary structures and solvent accessibilities is a useful and widely used resource for this purpose (Kabsch and Sander 1983; http://www.sander.ebi.ac.uk/dssp/). This database, which is based on recognition of hydrogen-bonding patterns in known structures, distinguishes eight secondary structural classes that can be grouped into α helices, β strands, and coils (Rost and Sander 1993). An automatic method makes predictions in accord with published assignments (Frishman and Argos 1995).

The assumption on which all secondary structure prediction methods are based is that there should be a correlation between amino acid sequence and secondary structure. The usual assumption is that a given short stretch of sequence may be more likely to form one kind of secondary structure than another. Laboratory analysis of proteins has revealed that short (3–15 residues) amino acid sequences called I-sites can each adopt a number of different local structures and energy conformations, leading to the prediction that folding often occurs when neighboring sites interact to form a minimum energy structure (Bystroff and Baker 1998). Many methods examine a sequence window of 13–17 residues and assume that the central amino acid in the window will adopt a conformation that is determined by the side groups of all the amino acids in the window. This window size is within the range of lengths of α helices (5–40 residues) and β strands (5–10 residues).

There is evidence that more distant interactions within the primary amino acid chain may influence local secondary structure. The same amino acid sequence up to 5 (Kabsch and Sander 1984) and 8 (Sudarsanam 1998) residues in length can be found in different secondary structures. An 11-residue-long amino acid "chameleon" sequence has been found to form an α helix when inserted into one part of a primary protein sequence and a β sheet when inserted into another part of the sequence (Minor and Kim 1996). More distant interactions may account for the observation that β strands are predicted more poorly by analysis of local regions (Garnier et al. 1996). However, the methods that have been used to predict the secondary structure of an amino acid residue all perform less well when amino acids more distant than in the small window of sequence are used.

The number of possible amino acid combinations in a sequence window of 17 amino acids is very large ($17^{20} = 4 \times 10^{24}$). If many combinations influence one type of secondary structure, examination of a large number of protein structures is required to discover the significant patterns and correlations

within this window. Earlier methods for predicting secondary structure assumed that each amino acid within the sequence window of 13–17 residues influences the local secondary structure independently of other nearby amino acids; i.e., there is no interaction between amino acids in influencing local secondary structure. Later methods assumed that interactions between amino acids within the window could play a role.

NN models described below (p. 462) have the ability to detect interactions between amino acids in a sequence window, including conditional interactions. A hypothetical example of the interactions that might be discovered illustrates the possibilities. If the central amino acid in the sequence window is Leu and if the second upstream amino acid toward the amino terminus is Asn, the Leu is in an α helix; however, if the neighboring amino acid is not Asn, the Leu is in a β strand. In another method of secondary structure prediction, the nearest-neighbor method, sequence windows in known structures that are most like the query sequence are identified. This method bypasses the need to discover complex amino acid patterns associated with secondary structure. Protein secondary structure has also been modeled by HMMs, also described as discrete state-space models, which are described below (Stultz et al. 1993; White et al. 1994; Bystroff et al. 2000).

Accuracy of Secondary Structure Prediction

One method of assessing accuracy of secondary structure prediction is to give the percentage of correctly predicted residues in sequences of known structure, called Q_3. This measure, however, is not very effective by itself, because even a random assignment of structure can achieve a high score by this test (Holley and Karplus 1991). Another measure is to report the fraction of each type of predicted structure that is correct. A third method is to calculate a correlation coefficient for each type of predicted secondary structure (Mathews 1975). The coefficient indicating success of predicting residues in the α-helical configuration, C_α, is given by

$$C_\alpha = (p_\alpha n_\alpha - u_\alpha o_\alpha) / \{\sqrt{(\ [n_\alpha + u_\alpha]\ [n_\alpha + o_\alpha] \times [p_\alpha + u_\alpha]\ [p_\alpha + o_\alpha]\ )}\} \quad (2)$$

where p_α is the number of correct positive predictions (true positives), n_α is the number of correct negative predictions (true negatives), o_α is the number of over-predicted positive predictions (false positives), and u_α is the number of underpredicted residues (false negatives). The closer this coefficient is to a value of 1, the more successful the method for predicting a helical residue. An overall level of prediction accuracy does not provide information on the accuracy of the number of predicted secondary structures, and their lengths and location in the sequence. One simple index of success is to compare the average of the predicted lengths with the known average (Rost and Sander 1993).

Another factor to consider in prediction accuracy is that some protein structures are more readily predictable than others, such that the spectrum of test proteins chosen will influence the frequency of success. A representative set of proteins that have limited similarity will provide the most objective test. Rost and Sander (1993) have chosen a set of 126 globular and 4 membrane proteins that have less than 25% pair-wise similarity, and they have used this set for training and testing NN models. A list of the structurally distinct fold types is maintained at several of the structural classification databases, including SCOP, CATH, and the DALI/FSSP (also see Holm and Sander 1998).

In the often-used jackknife test, one protein in a set of known structure is left out of a calibration or training step of the program or model being tested. The rest of the proteins are used to predict the structure of the left-out one, and the procedure is cycled through all of the sequences. The overall frequency of success of predicting the secondary structural features of the left-out sequence is used as an indicator of success. An even more comprehensive approach to the problem of accuracy is to examine the predictions for different structural classes of proteins. Because some classes are much more difficult to predict, the overall success rate with respect to protein class is an important index of success. Prediction accuracy is discussed further below.

A valuable addition to secondary structure prediction is giving the degree of reliability of the prediction at each position. Some prediction methods produce a score for each of the three types of structures (helix, strand, coil) at each residue position. If one of these scores is much higher than the other two, the score is considered to be more reliable, and a high

reliability index may be assigned that reflects high confidence in the prediction. If the scores are more similar, the index is lower. By examining predictions for known structures, as in a jackknife experiment, the accuracy of these reliability indices may be determined. What has been found is that a prediction with a high index score is much more accurate (Yi and Lander 1993; and see PHD server below), thus increasing confidence in the prediction for these residues.

Methods for Secondary Structure Prediction

Four widely used methods of protein secondary structure prediction—(1) the Chou–Fasman and GOR methods, (2) NN models, (3) nearest-neighbor methods, and (4) HMMs—are discussed below. These methods can be further enhanced by examining the distribution of hydrophobic, charged, and polar amino acids in protein sequences.

Chou–Fasman/GOR Method

The Chou–Fasman method (Chou and Fasman 1978) is based on analyzing the frequency of each of the 20 amino acids in α helices, β sheets, and turns of the then-known relatively small number of protein structures. It was found, for example, that amino acids Ala (A), Glu (E), Leu (L), and Met (M) are strong predictors of α helices, but that Pro (P) and Gly (G) are predictors of a break in α helix. A table of predictive values for each type of secondary structure was made for each of the α helices, β strands, and turns. To produce these values, the frequency of amino acid i in structure s is divided by the frequency of all residues in structure s. The resulting three structural parameters (P_α, P_β, and P_t) vary roughly from 0.5 to 1.5 for the 20 amino acids.

To predict a secondary structure, the following set of rules was used. The sequence was first scanned to find a short sequence of amino acids that has a high probability for starting a nucleation event that could form one type of structure. For α helices, a prediction is made when four of six amino acids have a high probability (>1.03) of being in an α helix. For β strands, the presence in a sequence of three of five amino acids with a probability of >1.00 of being in a β strand predicts a nucleation event for a β strand. These nucleated regions are extended along the sequence in each direction until the prediction values for four amino acids drop below 1. If both α-helical and β-strand regions are predicted, the higher-probability prediction is used.

Turns are predicted somewhat differently. Turns are modeled as a tetrapeptide, and two probabilities are calculated as follows:

1. The average of the probabilities for each of the four amino acids being in a turn is calculated as for α-helix and β-strand predictions.

2. The probabilities of amino acid combinations being present at each position in the turn tetrapeptide (i.e., the probability that a particular amino acid such as Pro is at position 1, 2, 3, or 4 in the tetrapeptide) are determined.

3. The probabilities for the four amino acids in the candidate sequence are multiplied to calculate the probability that the particular tetrapeptide is a turn.

4. A turn is predicted when the first probability value is greater than the probabilities for an α helix and a β strand in the region and when the second probability value is greater than 7.5×10^{-5}.

In practice, the Chou–Fasman method is only about 50–60% accurate in predicting secondary structural domains.

Garnier et al. (1978) developed a somewhat more involved method for protein secondary structure prediction that is based on a more sophisticated analysis. The method is called the GOR (*Garnier, Osguthorpe,* and *Robson*) method. Whereas the Chou–Fasman method is based on the assumption that each amino acid individually influences secondary structure within a window of sequence, the GOR method is based on the assumption that amino acids flanking the central amino acid residue influence the secondary structure that the central residue is likely to adopt. In addition, the GOR method uses principles of information theory to derive predictions (Garnier et al. 1996).

As in the Chou–Fasman method, known secondary structures are scanned for the occurrence of amino acids in each type of structure. However, the frequency of each type of amino acid at the next 8 amino-terminal and carboxy-terminal positions is also determined, making the total number of positions examined equal to 17, including the central one. In the original GOR method, four scoring matrices, containing in each column the probability of finding

each amino acid at one of the 17 positions, are prepared. One matrix corresponds to the central (eighth) amino acid being found in an α helix, the second for the amino acid being in a β strand, the third a coil, and the fourth, a turn. Later versions omitted the turn calculation because these were the most variable features and were consequently the most difficult to predict. A candidate sequence is analyzed by each of the three to four matrices by a sliding window of 17 residues. Each matrix is positioned along a candidate sequence, and the matrix giving the highest score predicts the structural state of the central amino acid. At least four residues in a row have to be predicted as an α helix and two in a row for a β strand for a prediction to be valid.

Matrix values are calculated in somewhat the same manner as amino acid substitution matrices (described in Chapter 3), in that matrix values are calculated as log odds units representing units of information. The information available as to the joint occurrence of secondary structural conformation S and amino acid a is given by (Garnier et al. 1996)

$$I(S; a) = \log\left[\, P(S \mid a) \,/\, P(S) \,\right] \qquad (3)$$

where $P(S \mid a)$ is the conditional probability of conformation S given residue a, and $P(S)$ is the probability of conformation S. By Bayes' rule (see Chapter 4, p. 149), the probability of conformation S given amino acid a, $P(S \mid a)$ is given by

$$P(S \mid a) = P(S, a) \,/\, P(a) \qquad (4)$$

where $P(S, a)$ is the joint probability of S and a and $P(a)$ is the probability of a. These probabilities can be estimated from the frequency of each amino acid found in each structure and the frequency of each amino acid in the structural database. Given these frequencies,

$$I(S; a) = \log\left(f_{S,a} \,/\, f_S \right) \qquad (5)$$

where $f_{S,a}$ is the frequency of amino acid a in conformation S and f_S is the frequency of all amino acid residues found to be in conformation S.

The GOR method maximizes the information available in the values of $f_{S,a}$ and avoids data size and sampling variations by calculating the information difference between the competing hypotheses that residue a is in structure S, $I(S;a)$, or that a is in a different conformation (not S), I (not $S;a$). This difference $I(\Delta S; a)$ is calculated from Equation 5 with simple substitutions by

$$
\begin{aligned}
I(\Delta S; a) &= I(S; a) - I(\text{not } S; a) \\
&= \log \{P(S,a)/[1 - P(S,a)]\} \\
&\quad + \log \{[1 - P(S)/P(S)]\} \quad (6)
\end{aligned}
$$

which is derived from the observed amino acid data as

$$
\begin{aligned}
I(\Delta S; a) &= \log [f_{S,a} / (1 - f_{S,a})] \\
&\quad + \log [(1 - f_S)/f_S] \quad (7)
\end{aligned}
$$

where the frequency of finding amino acid a not in conformation S is $1 - f_{S,a}$ and of not finding any amino acid in conformation S is $1 - f_S$. Equation 6 is used to calculate the information difference for a series of x consecutive positions flanking sequence position m,

$$
\begin{aligned}
I(\Delta S_m; a_1..a_X) \\
&= \log [P(S_m,a_1..a_X)/(1 - P(S_m,a_1..a_X)] \\
&\quad + \log [1 - P(S)/P(S)] \quad (8)
\end{aligned}
$$

from which the following ratio of the joint probability of conformation S_m given $a_1,..., a_X$ to the joint probability of any other conformation may be calculated

$$
\begin{aligned}
P(S_m,a_1..a_X)/[1 - P(S_m,a_1..a_X)] \\
= \{P(S)/[1 - P(S)]\} \, e^{-I(\Delta Sm; a1,..aX)} \quad (9)
\end{aligned}
$$

Searching for all possible patterns in the structural database would require an enormous number of proteins. Hence, three simplifying approaches have been taken. First, it was assumed in earlier versions of GOR that there is no correlation between amino acids in any of the 17 positions (both the flanking 8 positions and the central amino acid position), or that each amino acid position had a separate and independent influence on the structural conformation of the central amino acid. The steps are then:

1. Values for $I(\Delta S; a)$ in Equation 7 are calculated for each of the 17 positions;

2. these values are summed to approximate the value of $I(\Delta S_m; a1,..., a_X)$ in Equation 8;

3. the probability ratios in Equation 9 are calculated.

The second assumption used in later versions of GOR was that certain pair-wise combinations of an amino acid in the flanking region and central amino acid influence the conformation of the central amino acid. This model requires a determination of the frequency of amino acid pairs between each of the 16 flanking positions and the central one, both for when the central residue is in conformation S and when the central residue is not in conformation S. Finally, in the most recent version of GOR, the assumption is made that certain pair-wise combinations of amino acids in the flanking region, or of a flanking amino acid and the central one, influence the conformation of the central one. Thus, there are 17 × 16/2 = 136 possible pairs to use for frequency measurements and to examine for correlation with the conformation of the central residue. With the advent of a large number of protein structures, it has become possible to assess the frequencies of amino acid combinations and to use this information for secondary structural predictions. The GOR method predicts 64% of the residue conformations in known structures and quite drastically (36.5%) underpredicts the number of residues in β strands.

Use of the Chou–Fasman and GOR methods for predicting the secondary structure of the α subunit of *Salmonella typhimurium* tryptophan synthase is illustrated in Figure 10.26. In this particular case, the positions of the secondary structures predicted by either of these methods are very similar to those in the solved crystal structure (Branden and Tooze 1991). However, tests of the accuracy of these methods using sequences of other proteins whose structures are known have shown that the Chou–Fasman method is only about 50–60% accurate in predicting the structural domains.

These methods are most useful in the hands of a knowledgeable structural biologist, and have been used most successfully in polypeptide design and in analysis of motifs for organelle transport (Branden and Tooze 1991). A useful approach for enhancing these methods is to analyze each of a series of aligned amino acid sequences in an msa and then to derive a consensus structural prediction.

Patterns of Hydrophobic Amino Acids Can Aid Structure Prediction

Prediction of secondary structure can be aided by examining the periodicity of amino acids with hydrophobic side chains in the protein chain. This type of analysis was discussed above in the prediction

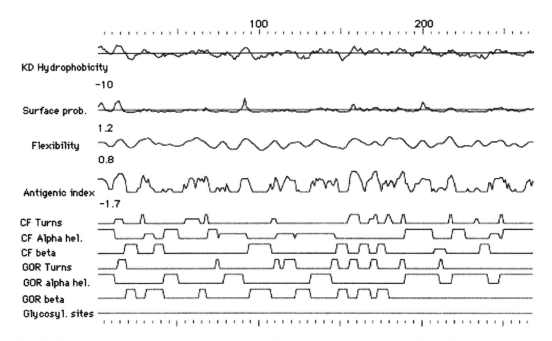

FIGURE 10.26. Example of the secondary structure predictions for the α subunit of *S. typhimurium* tryptophan synthase by the Chou–Fasman and GOR methods included in the Genetics Computer Group suite of programs. The EMBOSS garnier program also performs the GOR analysis. The predictions are shown on the lower panels, labeled as CF for the Chou–Fasman method (Chou and Fasman 1978) and GOR (referred to as GOR I) for the Garnier, Osguthorpe, and Robson method (Garnier et al. 1978). This protein is in the α-β class with an α/β barrel type of structure comprising eight parallel β strands and eight α helices in an alternating pattern and three additional α helices, illustrated in Fig. 10.6. The predicted structure is quite accurate and represents the correct pattern of secondary structure.

of transmembrane α-helical domains in proteins. Hydrophobicity tables that give hydrophobicity values for each amino acid are used to locate the most hydrophobic regions of the protein (Table 10.6) (see Lüthy and Eisenberg 1991). As for secondary structure prediction, a sliding window is moved across the sequence and the average hydrophobicity value of amino acids within the window is plotted. A hydrophobicity plot of the α subunit of *S. typhimurium* tryptophan synthase is included in the first panel of Figure 10.26.

Similar methods for predicting surface peptides including antigenic sites, chain flexibility, or glycosylation sites are also illustrated in Figure 10.26. These methods use the chemical properties of amino acid side chains to predict the location of these amino acids on the surface or buried within the core structure.

The location of hydrophobic amino acids within a predicted secondary structure can also be used to

predict the location of the structure. One type of display of this distribution is the helical wheel or spiral display of the amino acids in an α helix, as shown in Figure 10.27. The use of this display was described above as a way to visualize the location of leucine residues on one face of the helix in a leucine zipper structure. There is also a tendency of hydrophobic residues located in α helices on the surface of protein structures to face the core of the protein and for polar and charged amino acids to face the aqueous environment on the outside of the α helix. This arrangement is also revealed by the helical wheel display shown in Figure 10.27.

In another type of display, the hydrophobic moment display (Fig. 10.28), the contours show positions in the amino acid sequence where hydrophobic amino acids tend to segregate to opposite sides of a structure plotted against various angles of rotation from one residue to the next along the protein chain. For α helices, the angle of rotation is 100°, and for β

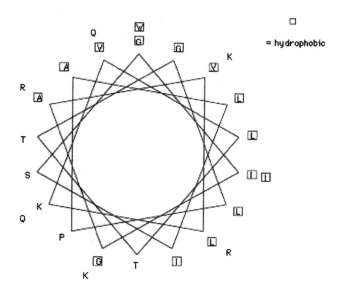

FIGURE 10.27. Helical wheel plot for the protein melittin. The plot shown was obtained using the Genetics Computer Group HELICALWHEEL program. The diagram shows the relative positions of amino acids in an end-on view of an α helix with the angle of rotation of 100° between adjacent amino acids in α helices (the angle would be 160° for β strands). The hydrophobic amino acids Leu (L), Ile (I), and Val (V) are primarily located on one side of the helix, thereby illustrating the amphiphobic nature of the helix. The pepwheel program in the EMBOSS suite produces a similar display.

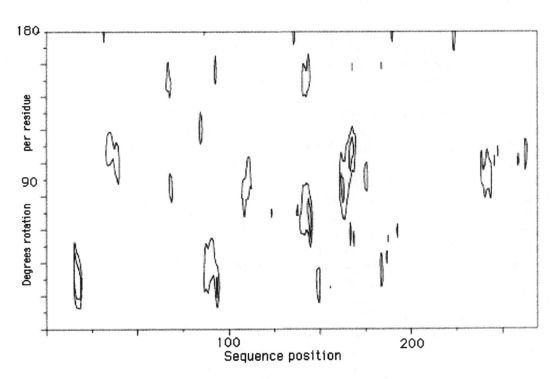

FIGURE 10.28. Hydrophobic moment plot of the sequence of the α chain of *E. coli* tryptophan synthase using the Genetics Computer Group MOMENT program. The EMBOSS hmoment program performs a similar analysis. The moment uses the hydrophobicity values shown in Table 10.6 to measure the tendency of hydrophobic residues to be located on one face of a secondary structural element (Lüthy and Eisenberg 1991). The values are normalized so that the mean value is 0 and the standard deviation is 1. The moment is calculated for a window of ten residues, five on each side of every amino acid position and for every possible rotational angle between adjacent residues. The angle is 100° for α helices and 160° for β strands. When one contour is shown, the moment values are 0.35; when two contours are shown, values are 0.35 (outer) and 0.45 (inner).

strands, 160°. The analysis in the figure predicts, for example, an α helix at approximate sequence position 165 that has segregated hydrophobic amino acids on one helix face. Helix α5 runs from positions 160 to 168 in the crystal structure of this protein.

Secondary Structure Prediction by NN Models

The most sophisticated methods that have been devised to make secondary structural predictions for proteins use artificial intelligence, or so-called NN algorithms. The earlier Chou–Fasman and GOR methods examined patterns that represent secondary structural features. However, NN methods went farther and tried to locate these patterns in a particular order that coincides with a known domain structure. Patterns typical of α/β proteins (Cohen et al. 1983), turns in globular proteins (Cohen et al. 1986), or helices in helical proteins (Presnell et al. 1992) may be located and used to predict secondary structure with increased confidence.

In the NN approach, computer programs are trained to be able to recognize amino acid patterns that are located in known secondary structures and to distinguish these patterns from other patterns not located in these structures. There are many examples of the use of this method to predict protein structures (see, e.g., Qian and Sejnowski 1988; Muggleton et al. 1992; Stolorz et al. 1992; Rost and Sander 1993), which have been reviewed (Holley and Karplus 1991; Hirst and Sternberg 1992). The early methods are reported to be up to 64% accurate. These methods have been improved to a level of over 70% for globular proteins by the use of information from msa's (Rost and Sander 1993, 1994). Two Web sites that perform an NN analysis for protein secondary structure prediction are PHD (Rost and Sander 1993; Rost 1996; http://www.embl-heidelberg.de/predictpro tein/predictprotein.html) and NNPREDICT (Kneller et al. 1990; http://www.cmpharm.ucsf.edu/~nomi/ nnpredict.html). These NN models are theoretically able to extract more information from sequences than the information theory method described above (Qian and Sejnowski 1988). NNs have also been used to model translational initiation sites and promoter sites in *E. coli*, splice junctions, and specific structural features in proteins, such as α-helical transmembrane domains. These applications are discussed elsewhere in this chapter (p. 455) and in Chapter 9.

NN models are meant to simulate the operation of the brain. The complex patterns of synaptic connections among a large number of neurons are presumed to underlie the functions of the brain. Some groups of neurons are involved in collecting data as environmental signals, others in processing data, and yet others in providing an output response to the signals. NNs are an attempt to build a similar kind of learning machine where the input is a 13–17-amino-acid length of sequence and the output is the predicted secondary structure of the central amino acid residue. The object is to train the NN to respond correctly to a set of such flanking sequence fragments when the secondary structural features of the centrally located amino acid are known. The training is designed to achieve recognition of amino acid patterns associated with secondary structure. If the NN has sufficient capacity for learning, these patterns may potentially include complex interactions among the flanking amino acids in determining secondary structures. However, two studies with NN described below have so far not found evidence for such interactions.

A typical NN model used for protein secondary structure prediction is illustrated in Figure 10.29A. A sliding window of 13–17 amino acid residues is moved along a sequence. The sequence within each window is read and used as input to an NN model previously trained to recognize the secondary structure most likely to be associated with that sequence pattern. The model then predicts the secondary structural configuration of the central amino acid as α helix, β strand, or other. Rules or another trained network are then applied that make the prediction of a series of residues reasonable. For example, at least 4 amino acids in a row should be predicted as being in an α helix if the prediction is to make structural sense.

The model comprises three layers of processing units—the input layer, the output layer, and the so-called hidden layer between these layers—as illustrated in Figure 10.29A. Signals are sent from the input layer to the hidden layer and from the hidden layer to the output layer through junctions between the units. This configuration is referred to as a feed-forward multilayer network. The input layer of units reads the sequence, one unit per amino acid residue, and transmits information on the amino acid at that location. A small window of sequence is read at a time and information is sent as signals through junctions to a

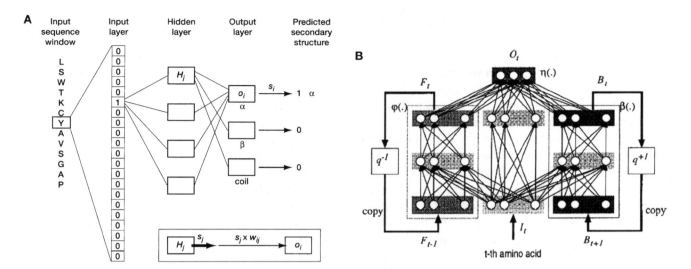

FIGURE 10.29. (*A*) A typical NN model for protein secondary structure prediction (after Rost and Sander 1993). Functions of the input (*red boxes*), hidden (*purple boxes*), and output (*green boxes*) layers are described in the text. There is one input unit for each amino acid in the sequence window of 13 (*first column*). Each input amino acid unit is made up of 21 input positions, one for each amino acid and one for a padding space when the window overlaps the end of the sequence. Other positions may be added to provide additional information. The positions each send information to the hidden unit layer. In a simple input coding system, only one of the 20 components in a given input unit has a value of 1. Shown is an example where the component for Y is turned on while the rest of the components are 0 (*second column*). When padding for the end of sequence is required, only the padding space is set to 1. When a sequence profile is used as input (not shown), each position is filled with the frequency of the amino acid in the corresponding column of the sequence profile or with a coded form of this frequency, and the numbers of insertions and deletions are added in two extra positions. Another position is used to indicate the amount of information due to the presence of conserved amino acids in the column. Signals from each position in each input unit are weighted as they proceed to units of the hidden layer. A signal from a component of one input unit will receive a different weight for each connection to a hidden unit. Each hidden unit sums the signals (s_{in}) received from the input layer and then transforms the sum using the trigger function $s_{out} = 1/(1 - e^{-ks_{in}})$ to produce an output signal that is between and close to either 0 or 1, simulating the firing of a neuron. Strong signals are transformed by this function to a number approximately equal to 1 and weak or negative values to 0. As the constant k increases, discrimination between strong and weak signals is increased. The hidden layer output signals are weighted and sent to three output units, representing prediction of an α helix, β strand, or coil (loop) for the secondary structural configuration of the central amino acid in the window. The sum of these signals is transformed to values between 0 and 1. An output signal close to 1 is a prediction for the amino acid to have the corresponding structural configuration; a weak signal close to 0 is no prediction. The example shown predicts an α-helical configuration for Y. Predictions for a series of adjacent windows are sorted out by applying rules or by additional neural networks. The insert illustrates the operation of the back-propagation algorithm that is used to train the network and is described by an example in the text (p. 465). (*B*) A new class of neural network, bidirectional recurrent neural network (BRNN). This class of NN includes three components, including a central one with a sequence-scanning window feeding signals to a hidden layer, followed by the hidden layer sending signals to an output layer, as in the more conventional feed-forward neural network shown in part *A* of this figure. In the BRNN, signals are also fed from the input layer to two outer components shown in the figure on the right and left of the feed-forward component. These two "contexts" behave as wheels, the right one starting at the beginning of the protein sequence and moving up to the current window, and the left one moving backward from the end of the sequence up to the current window. These wheels provide a "memory" of the sequence outside of the search window that can facilitate the prediction of the central amino acid in the window. Symbols shown in various parts of the model are functions that are trained to determine how signals are processed by various components of the model. This class of HMMs is normally used to model processes that change over time but in this case keeps track of information in the entire sequence as the sequence window advances. A variety of variations of the BRNN model are possible. (*B*, reprinted, with permission, from Pollastri et al. 2002 [©Wiley-Liss, Inc., a subsidiary of John Wiley & Sons, Inc.].)

number of sequential units in the hidden layer by all of the input units within the window, as shown by the lines joining units in Figure 10.29. These signals are each individually modified by a weighting factor and then added together to give a total input signal into each hidden unit. Sometimes a bias is added to this sum to influence the response of the unit. The resulting signal is then transformed by the hidden unit into a number that is very close either to a 1 or to a 0 (or sometimes to a –1). A mathematical function known as a sigmoid trigger function, simulating the firing or nonfiring states of a neuron, is used for this transformation. Signals from the hidden units are then sent to three individual output units, each output unit representing one type of secondary structure (helix, strand, or other). Each signal is again weighted, the input signals are summed, and each of the three output units then converts the combined signal into a number that is approximately a 1 or a 0. An output signal that is close to 1 represents a prediction of the secondary structural feature represented by that output unit, and a signal near to the value 0 means that the structure is not predicted.

When hidden layers are included, an NN model is capable of detecting higher levels of interaction among amino acids that influence secondary structure. For example, particular combinations of amino acids within the sliding window of sequence may produce a particular type of secondary structure. To resolve these patterns, a sufficient number of hidden units is needed (Holley and Karplus 1991); the number varies from 2 to a range of 10–40. An interesting side effect of adding more hidden units is that the NN memorizes the training set but at the same time is less accurate with test sequences. This effect is revealed by using the trained network to predict the same structures used for training. The correct number increases by over 20% as the number of hidden units increases from 0 to 10. In contrast, accuracy of prediction of test sequences not used for training decreases by 3% (Holley and Karplus 1991).

Perceptron. Without hidden layers, the NN model is known as a perceptron, and has a more limited capacity to detect such combinations. In two studies, networks with no hidden units were as successful in predicting secondary structure as those with hidden units. In addition, the number of hidden units was increased to as many as 60 in one study (Qian and Sejnowski 1988) and 20 in another (Holley and Karplus 1991) without significantly changing the level

of success. These observations imply that the influence of local sequence on secondary structure is the additive influence of individual residues and that there is no higher level of interaction among these residues. To detect such interactions, however, requires a large enough training set to provide a significant number of examples, and these conditions may not have been met. These same studies examined the effect of input window size and found that a maximum information for secondary structure prediction seems to be located within a window of 13–17 amino acids, because larger windows do not increase accuracy. However, small windows were less effective, suggesting that they have insufficient information, and below a window size of 5, success at predicting β strands was decreased.

Training the NN model is the process of adjusting the values of the weights used to modify the signals from the input layer to the hidden layer and from the hidden layer to the output layer. The object is to have these weights balance the input signals so that the model output correctly identifies the known secondary structure of the central amino acid in a sequence window of a protein of known structure. Because there may be thousands of connections between the various units in the network, a systematic method is needed to adjust these values. Initially, the weights are assigned a constant or random value (typical range –0.1 to +0.1). The sliding window is then positioned along one of the training sequences. The predicted output for a given sequence window is then compared to the known structure of the central amino acid residue. The model is adjusted to increase the chance of predicting the correct residue. The adjustment involves changing the weighting of propagated signals by a method called the back-propagation algorithm. This procedure is repeated for all windows in all of the training sequences. The better the model, the more predicted structures that will be correct. Conversely, the worse the model, the more predictions that will be incorrect. The object then becomes to minimize this incorrect number. The error E is expressed as the square of the total number of incorrect predictions by the output units.

When the back-propagation algorithm is applied, the weights are adjusted by a small amount to decrease errors. A window of a training sequence is used as input to the network, and the predicted and expected (known) structures of the central residue are compared. A set of small corrections is then made to the weights to improve an incorrect prediction, or the weights are left relatively unchanged for a correct prediction. This procedure is repeated using another training sequence until the number of errors cannot be reduced further. A large number of training cycles representing a slow training rate is an important factor for training the network to produce the smallest number of incorrect predictions. Not all of the training sequences may be used—a random input of training patterns may be used and sometimes these may be chosen from subsets of sequences that represent one type of secondary structure to balance the training for each type of structure. The back-propagation algorithm examines the contribution of each connection in the network on the subsequent levels and adjusts the weight of this connection, if needed to improve

the predictions. The example below illustrates the operation of the algorithm.

The PHDsec program in the PHD system described above in the section on prediction of transmembrane-spanning proteins (pp. 452–453) is an example of an NN program for protein secondary structure prediction (Rost and Sander 1993; Rost 1996). The Web address of this resource is http://www.embl-heidelberg.de/predictprotein/predictprotein.html. PHDsec uses a procedure similar to that used by PHDhtm. A BLAST search of the input sequence is conducted to identify similar but not closely identical sequences, and a multiple alignment of the sequences is transformed into a sequence profile. This profile is then used as input to a neural network trained to recognize correlations between a window of 13 amino

EXAMPLE: BACK-PROPAGATION ALGORITHM USED TO TRAIN THE NN (ROST AND SANDER 1993)

Consider an output unit o_i as shown in Figure 10.29A. Let us assume that this unit predicts whether or not the central residue in the scanned sequence window is an α-helical secondary structure. The output signal from this unit is s_i, which if close to 1, predicts an α-helical structure or, if close to 0, does not predict the α helix. The network has been provided with a training sequence and it is known whether the central amino acid actually is found in an α helix. If the structure is an α helix, then the output of o_i should be close to 1, and if not, then close to 0. d_i is the expected or desired output of o_i and $d_i = 1$ if a helix is expected and 0 if not. The output of o_i is determined by the sum of the inputs received from each of the hidden units with which o_i is connected. Each of the hidden units emits a signal close to 0 or 1, and each signal is separately weighted as it passes from the hidden unit to o_i. Focus on one of the hidden units H_j that is connected to o_i and emits a signal s_j that is modified by weight w_{ij}. The signal arriving at o_i is thus $s_j \times w_{ij}$, as illustrated in the insert in Figure 10.19. The problem at hand is to adjust or not to adjust w_{ij} so that the output of o_i (s_i) is close to the desired value d_i. The value of w_{ij} is adjusted according to a procedure known as gradient descent that is given by the formula

$$\Delta w_{ij} = w_{ij} - n\, \partial E/\partial w_{ij} + m \tag{10}$$

where the partial derivative of the error E with respect to w_{ij} $\partial E/\partial w_{ij}$ is calculated by

$$\partial E/\partial w_{ij} = (s_i - d_i)\, s_i\, (1 - s_i)\, s_j \tag{11}$$

where n is the rate of training (typical value 0.03) and m is a smoothing factor that allows a carryover of a fraction of previous values of w_{ij} (typical value 0.2). Suppose, for example, the s_j was sent from H_j to o_i as 0.2 and that d_i is 1, so that s_j is not contributing correct information. Then $\partial E/\partial w_{ij} = (0.2 - 1) \times 0.2 \times 0.8 \times 0.2 = -0.0256$. w_{ij} will then be increased in Equation 10 by the rate of training times this value adjusted by m for contributions from any previous value of w_{ij}. Adjusting the weights of connections between the input and hidden layers uses a more detailed formula that takes into account the effects of both the signal send from the input unit to the hidden units and that of the hidden units on each of the output units.

> ## EXAMPLE: PROGRAM OUTPUT FROM THE PHD SERVER AT
> http://www.embl-heidelberg.de/predictprotein/predictprotein.html
>
> The input sequence was the α subunit of *S. typhimurium* tryptophan synthase (SwissProt ID TRPA_SALTY, accession no. P00929), which was originally included in the training sequences. (*A*) Secondary structure prediction. (H) α helix; (E) β strand; (L) loop; (.) no prediction. Rel is the reliability of the prediction on a scale of 0–9 (highest). Rel is based on the difference between the maximal and the second largest output signals from the network. The accuracies (Q_3) to be expected for each reliability value have been calculated and are shown in *B*. For example, when a Rel score of 9 is given a prediction for a residue, that prediction is known to be 94.2% accurate. Note that for a Rel of 9 the prediction of helix residues is almost 100% accurate but that prediction of strand residues is about 70% accurate. prH, prE, and prL are probabilities of helix, strand, and loop (no prediction), respectively, based on accuracy of the predictions shown in *B*. Subset is a listing of the more reliable predictions that includes all residues for which the expected accuracy is >82%.

A. Predicted secondary structure and reliability of prediction.

```
            ....,....1....,....2....,....3....,....4....,....5....,....6
        AA  |MERYENLFAQLNDRREGAFVPFVTLGDPGIEQSLKIIDTLIDAGADALELGVPFSDPLAD|
        PHD | HHHHHHHHHH    EEEEEEE      HHHHHHHHHHHHH    EEEE         |
        Rel |934899999996348872799842489984587999999997399668944767784689|
detail:
        prH-|036899999987531100000000000016788899999998300000000001113210
        prE-|000000000000000000579886530000000000000000000000178863111000000
        prL-|963100000012368883100123689983211000000001699720036877886789
subset: SUB |L..HHHHHHHHH..LLL.EEEE...LLLL.HHHHHHHHHHHH.LLLEEE..LLLLL.LLL|


            ....,....7....,....8....,....9....,....10...,....11...,....1
        AA  |GPTIQNANLRAFAAGVTPAQCFEMLALIREKHPTIPIGLLMYANLVFNNGIDAFYARCEQ|
        PHD | HHHHHHHHHHH    HHHHHHHHHHHH      EEEEEE  HHHH HHHHHHHHHH |
        Rel |737899999999982896299999999999997289993899883422441255999999999|
detail:
        prH-|258899999999984111589999999999984100000000002356643269999999998
        prE-|000000000000000000000000000000003899885110000000000000000
        prL-|741100000000158873000000000000015899986100013554224572000000000
subset: SUB |L.HHHHHHHHHHH.LLL.HHHHHHHHHHHH.LLLL.EEEEE.......LHHHHHHHHHH|

detail:
        prH-|520000000000655799999998840000000000001899999999840000000001
        prE-|000258996200000000000000000489997200000000000000048999862
        prL-|479741002799344200000001159951000179998100000000159951000135
subset: SUB |.LLL.EEE.LLL...HHHHHHHHHH.LL.EEEEELLLLLHHHHHHHHHH.LL.EEEEE..|


            8...,....19...,....20...,....21...,....22...,....23...,....2
        AA  |GVTGAENRGALPLHHLIEKLKEYHAAPALQGFGISSPEQVSAAVRAGAAGAISGSAIVKI|
        PHD |       HHHHHHHHHHHHH    EEEE    HHHHHHHHHH    EEE HHHHHH|
        Rel |567544421256999999999981899727763668289999999953997266149999|
detail:
        prH-|111222234577899999999851000000000003899999999730000111699999
        prE-|111100100000000000000000001577732100000000000000015773100000
        prL-|677666655422000000000014899831126778510000000026997411420000
subset: SUB |LLLL......HHHHHHHHHHH.LLLL.EEE.LLL.HHHHHHHHH.LLL.EE..HHHHH|


            4...,....25...,....26...,....27
        AA  |IEKNLASPKQMLAELRSFVSAMKAASRA|
        PHD |HHH    HHHHHHHHHHHHHHHH    |
        Rel |997259918899999999999998315899|
detail:
        prH-|998420058899999999999998642100
        prE-|000000000000000000000000000000
        prL-|001569941100000000000001356899
subset: SUB |HHH.LLL.HHHHHHHHHHHHHHH..LLLL|
```

B. The accuracy of secondary structure predictions for a given reliability index. % res is the % predicted residues that receive a given reliability index, Qtot is the overall prediction accuracy, H%obs and E%obs are the observed percentage of helix and strand (known structure), H%prd and E%prd are the predicted percentages of helix and strand predictions that are associated with such an index (provided by E-mail from server).

index	0	1	2	3	4	5	6	7	8	9
%res	100.0	99.2	90.4	80.9	71.6	62.5	52.8	42.3	29.8	14.1
Qtot	72.1	72.3	74.8	77.7	80.3	82.9	85.7	88.5	91.1	94.2
H%obs	70.4	70.6	73.7	77.1	80.1	83.1	86.0	89.3	92.5	96.4
E%obs	61.5	61.7	63.7	66.6	69.1	71.7	74.6	77.0	77.8	68.1
H%prd	77.8	78.0	80.0	82.6	84.7	86.9	89.2	91.3	93.1	95.4
E%prd	64.5	64.7	67.8	71.0	74.2	77.6	81.4	85.1	89.8	93.5

acids and the secondary structure of the central amino acid in the window. The NN model is like the one shown in Figure 10.29A. Program output includes a reliability index of each estimate on a scale of 1 (low reliability) to 9 (high reliability). These reliabilities (not shown) are obtained as normalized scores derived from the output values of the three units in the output layer of the network. The highest output value is compared to the next lowest value and the difference is normalized to give the reliability index. These indices are a useful way to examine the predictions in closer detail.

Bidirectional recurrent neural networks. The neural network described in the previous section is a feed-forward type of NN used for secondary structure prediction. Signals are sent one way from an input layer that reads a window of protein sequence to a middle layer that integrates the signals and then sends them to an output layer which predicts that secondary structure of the middle amino acid in the sequence window. Another type of NN, the bidirectional recurrent neural network (BRNN) illustrated in Figure 10.29B, has also been used for secondary structure prediction (Baldi et al. 1999).

The BRNN is similar to the more commonly used NN for structure prediction in having a feed-forward component that reads a sliding sequence window and hidden and output layers. In addition, the BRNN has two other components that roll like wheels along the sequence, reading and storing information, one from the beginning of the sequence to the amino acid

being predicted and the other from the end of the sequence back to the amino acid being predicted. These components act as memories of the entire sequence and allow the network to make predictions based both on the sequence window and on sequences outside of the window. SSpro is a server for various types of protein secondary structure prediction based on BRNN technology (Baldi et al. 1999; Pollastri et al. 2002; see http://www.igb.uci.edu/tools/scratch/). Secondary structures predicted by BRNNs have an accuracy approaching 80% and are an important addition to secondary structure prediction tools. In addition to their use in predicting secondary structure, BRNNs are also used for predicting signal peptides (SIGFIND Web server), probably because they "remember" sequence information outside of a small sequence window.

Nearest-Neighbor Methods of Secondary Structure Prediction

Like NNs, nearest-neighbor methods are another type of machine-learning method. They predict the secondary structural conformation of an amino acid in the query sequence by identifying sequences of known structures that are similar to the query sequence (Levin et al. 1986; Salzberg and Cost 1992; Zhang et al. 1992; Yi and Lander 1993; Salamov and Solovyev 1995, 1997; Frishman and Argos 1996). A large list of short sequence fragments is made by sliding a window of length n (e.g., $n = 16$) along a set of approximately 100–400 training sequences of known structure but minimal sequence similarity to each other, and the secondary structure of the central amino acid in each window is recorded. A window of the same size is then selected from the query sequence and compared to each of the above sequence fragments, and the 50 best-matching fragments are identified. The frequencies of the known secondary structure of the middle amino acid in each of these matching fragments (f_α, f_β, and f_{coils}) are then used to predict the secondary structure of the middle amino acid in the query window. As with other secondary structure prediction programs, the predicted secondary structure of a series of residues in the query sequence is subjected to a set of rules or used as input to a neural network to make a final prediction for each amino acid position.

Although not implemented in the most available programs, a true estimate of probability of the above set of frequencies may be obtained by identifying sets of training sequences that give the same value of $(f_\alpha + f_\beta + f_{coils})^{1/2}$. The frequencies of the secondary structures predicted by this group then give true estimates for p_α, p_β, and p_{coils} for the targeted amino acid in the query sequence (Yi and Lander 1993). Predictions based on the highest probabilities have been shown to be the most accurate, with the top 28% of the predictions being 86% accurate and the top 43% being 81% accurate. In addition, this method of calculating probability possesses more information than single-state predictions. Using this method, therefore, a substantial proportion of protein secondary structures can be predicted with high accuracy (Yi and Lander 1993, 1996).

The several nearest-neighbor programs that have been developed for secondary structure prediction (described in "Web Search Terms from This Chapter," p. 485) differ largely in the method used to identify related sequences in the training set. Originally, an amino acid scoring matrix such as a BLOSUM scoring matrix was used (Zhang et al. 1992). Distances between sequences based on a statistical analysis of the training sequences have also been proposed (Salzberg and Cost 1992). Use of a scoring matrix (Bowie et al. 1991, 1996) based on a categorization of amino acids into local structural environments, discussed below, in conjunction with a standard amino acid scoring matrix increased the success of the predictions (Yi and Lander 1993; Salamov and Solovyev 1995, 1997). Yet further increases in success have been achieved by aligning the query sequence with the training sequences to obtain a set of nonintersecting alignments with windows of the query sequence (as described in Chapter 3), and by using a multiple sequence alignment as input with amino-terminal and carboxy-terminal positions of α helices, β strands, and β turns treated as distinctive types of secondary structure (Salamov and Solovyev 1997).

The program PREDATOR (see "Web Search Terms from This Chapter," p. 486) is based on an analysis of amino acid patterns in structures that form H-bond interactions between adjacent β strands (β bridges) and between amino acid n and $n + 4$ on α helices (Frishman and Argos 1995, 1996). The H-bond pattern between parallel and antiparallel β strands is different (Fig. 10.3), and two types of antiparallel patterns have been recognized. By using such information combined with substitutions found in sequence alignments, the prediction success of PREDATOR has been increased to 75% (Frishman and Argos 1997). Examples of the NNSSP (Salamov and Solovyev 1997) and PREDATOR (Frishman and Argos 1997) program outputs are given on page 469.

Prediction of Three-dimensional Protein Structure

Because the number of ways that a protein can fold appears to be limited, there is considerable optimism that it will be possible to predict the fold of any protein, just given its amino acid sequence. Structural alignment studies have revealed more than 3000 common structural folds in the domains of the more than 22,000 three-dimensional structures in the Brookhaven Protein Data Bank. These studies have also revealed that sequences which share no detectable similarity adopt the same fold. Thus, there are many combinations of amino acids that can fit together into the same three-dimensional conformation, filling the available space and making suitable contacts with neighboring amino acids to adopt a common three-dimensional structure. There is also a reasonable probability that a new sequence will possess an already identified fold. The object of fold recognition is to discover which fold is best matched. Considerable headway has been made toward this goal.

One approach to structure prediction has been minimization of a global energy function (Lee et al. 1999; Liwo et al. 1999). In this approach, the entire conformational space that can be occupied by the amino acids of a protein chain is explored to find a minimum energy conformation. To simplify the search, amino acids are depicted by representative

EXAMPLE: NNSSP AND PREDATOR OUTPUT

Two of the most accurate nearest-neighbor prediction programs found through a Web search are:

1. NNSSP (accuracy to 73.5%). Shown is the program output, choosing the PSSP/NNSSP option. PredSS is the predicted secondary structure by NNSSP (a = α helix; b = β strand; c = coils). The output probabilities Prob a and Prob b give a normalized score by converting the values of f_α, f_β, and f_{coils} to a scale of 0–9.

2. PREDATOR (accuracy 75%) applies the FSSP assignments of secondary structure to the training sequences. PREDATOR does not provide a normalized score. PREDATOR predictions from http://www.embl-heidelberg.de/argos/ predator_info.html/ are shown below NNSP prediction on each line (H = α helix; E = β strand). The input sequence was the α subunit of *S. typhimurium* tryptophan synthase (SwissProt ID TRPA_SALTY, accession no. P00929), which is in the training sequences because the three-dimensional structure is known.

```
nnssp  Sat Mar 13 15:49:19 CST 1999
TS_subunit_alpha
 L=  268 SS content: a-  0.56 b=  0.08 c=  0.36
                      10        20        30        40        50
 PredSS    aaaaaaaaaaaaa    bbbbbb    aaaaaaaaaaaaaaaaaaaaaaa
 AA seq    MERYESLFAQLKERKEGAFVPFVTLGDPGIEQSLKIIDTLIEAGADALEL
 Prob a    99999999999974211100000001000168889999999974578863
 Prob b    00000000000000001277788741000100000000000000001122
 Predator  ____HHHHHHHHHHHHHH_EEEEEE_____HHHHHHHHHH_____

                      60        70        80        90        100
 PredSS            aaaaaaaaaa    aaaaaaaaaaaaa        bbba
 AA seq    GIPFSDPLADGPTIQNATLRAFAAGVTPAQCFEMLALIRQKHPTIPIGLL
 Prob a    11111111100124568899887311058899999998520001111133
 Prob b    23221101100012110000001111000000000000000002335544
 Predator  _____HHHHHHHHH____HHHHHHHHHHHH____HHHH
                      110       120       130       140       150

 PredSS    aaaaaaa    aaaaaaaaaaa    bbbbb        aaaaaaa
 AA seq    MYANLVFNKGIDEFYAQCEKVGVDSVLVADVPVEESAPFRQAALRHNVAP
 Prob a    54554453447899999988400100000111222234788998731111
 Prob b    32112211000000000000001116898632211010000000000123
 Predator  HHHHH_____HHHHHHHHH____EEEEEE_____HHHHHHHH___E

                      160       170       180       190       200
 PredSS    bbb    aaaaaaaa    bbbb    aaaaaaaaaaaaaaaaa
 AA seq    IFICPPNADDDLLRQIASYGRGYTYLLSRAGVTGAENRAALPLNHLVAKL
 Prob a    00000000015899999973111121223521112555665438889999
 Prob b    89852000000000000110113677531112211100122000000000
 Predator  EEE_____HHHHHHHH____EEEEE_____HHHHH____HHHHHH

                      210       220       230       240       250
 PredSS    aaa        aaaaaaaaa        aaaaaaaaaaa    aaa
 AA seq    KEYNAAPPLQGFGISAPDQVKAAIDAGAAGAISGSAIVKIIEQHINEPEK

 Prob a    88632100111101114789999987453122226878888997542588
 Prob b    00000001334332000000000000001221110100000000000000
 Predator  HHH_____HHHHHHH_____HHHHHHHHH__HHH

                      260
 PredSS    aaaaaaaaaaaaaaaaaa
 AA seq    MLAALKVFVQPMKAATRS
 Prob a    989999998878898663
 Prob b    000000000000000011
 Predator  HHHHHHHH_____
```

atomic reference points to reduce the number of interactions that must be considered. Success has been achieved for small protein fragments of length 60–100 residues (Liwo et al. 1999), but the time required to compute a structure can be several months on a supercomputer cluster.

Using sensitive sequence alignment methods is another approach that avoids structural calculation entirely. Sequence alignment can identify a family of homologous proteins that have similar sequences and three-dimensional structures. The substitutions in an msa of a protein family reveal the amino acid substitutions that are compatible with a conserved three-dimensional structure. There are many databases that link sequence families or sequence domains to the known three-dimensional structure of a family member, e.g., Pfam, 3DPSSM. The structure of even a remote family or superfamily member can be predicted through such sequence alignment methods. The PSI-BLAST tool described in Chapter 5 and the FFAS tool (Rychlewski et al. 2000) are used to find amino acid substitutions in conserved domains that are also likely to represent structural conservation. Like PSI-BLAST, FFAS produces a position-specific scoring matrix, but the substitutions found in more divergent family members are weighted more heavily. This change increases the sensitivity of the profile for finding more remote family members, although at the expense of reduced specificity.

When the sequence of a protein of unknown structure has no detectable sequence similarity to other proteins, other methods of three-dimensional structure prediction may be employed. One such method is sequence threading. In threading, the amino acid sequence of a query protein is examined for compatibility with the structural core of a known protein structure. Recall that the protein core is made up of α helices, β strands, and other structural elements folded into a compact structure. The environment of the core is strongly hydrophobic with little room for water molecules, extra amino acids, or amino acid side chains, which are not able to fit into the available space. Side chains must also make contact with neighboring amino acid side chains in the structure, and these contacts are needed for folding and stability. Threading methods examine the sequence of a protein for compatibility of the side groups with a known protein core. The sequence is "threaded" into a database of protein cores to look for matches. If a reasonable degree of compatibility is found with a given structural core, the protein is predicted to fold into a similar three-dimensional configuration.

Threading methods have undergone a considerable degree of evolution. Their use has been extended considerably by using PSI-BLAST to locate and align conserved domains in a query sequence and represent them by scoring matrices. Variations in each column of the matrix are assumed to represent amino acid substitutions that can substitute for the original amino acid in a structural domain in the query sequence. When these substitutions are threaded into protein cores, the resulting greater flexibility in matching a position increases the chance of finding a matching domain, if one exists. An excellent description of algorithms for threading is found in Lathrop et al. (1998). Presently available methods require considerable expertise with protein structure and with programming. However, at some sites, the analysis may be performed on a Web server, as shown in Table 10.7. Additional analysis of protein structure has provided further insight into new structural prediction methods.

Contact Order

An important feature of protein structure in regard to three-dimensional structure prediction is the location in the protein sequence of the amino acids that interact in the three-dimensional structure of the protein. The contact order is a measure of the average distance between all amino acids in a protein divided by the length of the protein and is related to the time taken for the protein to fold (see Bonneau et al. 2000a for references). In some proteins, interactions occur between amino acids that are relatively close together in the sequence (i.e., they have a small-contact-order value). Because the movement of amino acids in these low-contact-order proteins is restricted in space by their close location in the protein backbone, it does not take long in the folding process for these amino acids to move into the spatial configuration needed to make the contacts necessary for three-dimensional folding. Thus, these proteins fold relatively quickly. The structure of low-contact-order proteins is also more readily predicted because there are fewer possible alternative contacts to be considered in restricting them to a smaller sequence length. For example, there are fewer possible contacts to be evaluated in trying to find a minimum energy structure for the protein.

In contrast to low-contact-order proteins, in other proteins the amino acid contacts are between more distant amino acids in the sequence (i.e., they have a high-contact-order value). As might be expected from the above discussion, a greater degree of movement and a longer time are needed for the protein chain to move into the spatial configuration necessary for the amino acid contacts to form. Thus, these proteins fold more slowly. It is more difficult to predict the structure of high-contact-order proteins

TABLE 10.7. *Threading servers and program sources*

Program	Web address	Method	Reference
123D	http://123d.ncifcrf.gov/	Contact potentials between amino acid side groups.	Alexandrov et al. (1996)
3D-PSSM	http://www.sbg.bio.ic.ac.uk/~3dpssm/	New sequence–structure using position-specific scoring matrices.	Russell et al. (1997)
GenTHREADER	http://bioinf.cs.ucl.ac.uk/psipred/	Expands a threading approach using neural neworks by adding sequence similarity in different genomes.	Jones (1999); McGuffin and Jones (2003)
Honig lab	http://honiglab.cpmc.columbia.edu/	Threading methods using biophysical properties.	see Web site
Libra I	http://www.ddbj.nig.ac.jp/search/libra_i-e.html	New target sequence and 3D profile are aligned by dynamic programming.	Ota and Nishikawa (1997)
NCBI structure site	http://www.ncbi.nlm.nih.gov/	Gibbs sampling algorithm used to align sequence and structure.[a]	Bryant (1996)
Profit	http://lore.came.sbg.ac.at/	Fold recognition by the contact potential method.	M. Sippl (see Web site)
PSIPRED	http://bioinf.cs.ucl.ac.uk/psipred/	Home of several protein analysis tools including threaders.	McGuffin and Jones (2003)
Threader 3.4	http://bioinf.cs.ucl.ac.uk/threader/threader.html	Prediction by recognition of the correct fold from a library of alternatives.	Jones et al. (1995)
TOPITS	http://www.embl-heidelberg.de/predictprotein/	Detects similar motifs of secondary structure and accessibility between a sequence of unknown structure and a known fold.	Rost (1995a,b)
TOUCHSTONE		Secondary and tertiary restraint prediction applied to ab initio folding.	Kihara et al. (2001)
UCLA-DOE structure prediction server	http://fold.doe-mbi.ucla.edu/	Already above fold-recognition using a variety of sequence similarity and structure threading methods including DASEY (see text).	Fischer and Eisenberg (1996); Mallick et al. (2002)

Information on the research groups that work on structure prediction may be found at the CASP2 Web sites accessible at http://predictioncenter.llnl.gov/.

[a]Program has to be set up on a UNIX server.

because long-range interactions, possibly over much of the protein backbone, have to be considered. The Rosetta method (see below) takes advantage of the contact order of proteins to make three-dimensional structure predictions.

The Rosetta Method Is Probably the Best Method for Three-dimensional Structure Prediction

In the CASP (*c*ritical *a*ssessment of *s*tructure *p*rediction) contests described later in this chapter, the Rosetta method gave the best prediction of structural domains from sequence (Bonneau et al. 2002b; Chivian et al. 2003). Rosetta is based on a novel com-

bination of computational modeling using amino acid pattern analysis and laboratory experiments using nuclear magnetic resonance (NMR) to estimate the energy associated with localized interactions in sequences that are the initial steps in protein folding. Experiments with proteins have suggested that, as proteins fold, a large number of partially folded, low-energy conformations form that are close to the natural structure (Shortle et al. 1998). Further research suggested that local sequence segments play an important role in this process by rapidly alternating between various structures until local structures combine to form more global structures of a favorable minimum energy state. These more extensive structures include more of the backbone sequence through interactions between these local structures. This model of protein folding focused on finding patterns of amino acids that are associated with particular three-dimensional structural features in proteins.

When msa's for proteins of known structure were examined, a large number of amino acid patterns 3–15 residues long were identified, some of which are strongly associated with particular structures (Han and Baker 1995). Subsequently, further structural analysis in the laboratory identifies a subset of the patterns, called I-sites, that corresponded strongly with local structures in proteins of predictable energy based on experimental data. I-sites were assembled into the I-sites library, which was then used to predict protein three-dimensional structures by the Rosetta method.

The Rosetta method is based on the model described above that protein folding begins with localized interactions within short sequence fragments and is then extended to include more distant sequences in the protein in a more favorable, low-energy state. To simulate this process in structure prediction, Rosetta first slides a nine-amino-acid window along a sequence searching for matches in the I-sites library at each window position. A list of matches at each window position is produced. Because I-sites are associated with local three-dimensional structures and their energies, this list provides a range of possible structure predictions for the window sequence. After obtaining predictions for all windows, Rosetta uses a Monte Carlo simulation (see genetic algorithms and simulated annealing, p. 185) of the folding process by repeatedly sampling the possible structures along the protein backbone to find a structurally compatible, minimum energy combination. The resulting combination provides a three-dimensional structure prediction for the protein.

The Robetta server (Chivian et al. 2003; see http://robetta.bakerlab.org/) combines Rosetta with a series of other search methods for sequence similarities in a query protein sequence to predict a full-chain structure. There are also a number of extensions to Rosetta including RosettaDock, which predicts the structures of protein–protein complexes, and RosettaDesign, which predicts synthetic amino acid sequences that will fold into stable structures. Structure prediction methods related to Rosetta and Robetta are not described in further detail because they are already well described and undergoing rapid development. To read about new advances in this area, visit the D. Baker Web site at http://depts.wash ingon.edu/bakerpg/. HMMs, including one based on the I-sites library, have also been developed for three-dimensional structure prediction and are described below. Reliable three-dimensional structure prediction methods such as Rosetta are greatly facilitating between-genome comparisons by providing structural and functional comparisons of the encoded proteins.

Hidden Markov Models

HMMs have been used to model alignments of three-dimensional structure in proteins (Stultz et al. 1993; Hubbard and Park 1995; Di Francesco et al. 1997, 1999; FORREST Web server at http://absalpha.dcrt.nih.gov:8008/). In one example of this approach, the models are trained on conserved amino acid patterns in α helices, β strands, tight turns, and loops in specific structural classes (Stultz et al. 1993, 1997; White et al. 1994; HMMs are referred to as discrete space models in these publications), which then may be used to provide the most probable secondary structure and structural class of a protein. The manner by which protein three-dimensional domains can be modeled is illustrated in Figure 10.30A. An example of the class prediction by the Protein Sequence Analysis (PSA) server at Boston University is shown in Figure 10.31.

HMMSTR for matching sequences to combinations of I-sites. A second HMM for three-dimensional structure prediction (HMMSTR, see Bystroff and Baker 1997, 1998) is based on the I-sites library and is described in detail in Figure 10.30B. Sequence and profile alignment HMMs discussed in Chapter 5 are characterized by a series of states that represent sequence variation within subsequent columns of an msa. These states are joined by transition probabilities that give the probability of following different paths through the model representing substitutions, insertions, and deletions within the sequences that make up the alignment. In contrast, states in HMMSTR store information about the sequence variation and structural attributes of a single position in a structural motif. HMMSTR also combines the sequence and structural information from compatible I-sites into a model that can be used to make predictions of the presence of multiple I-sites in protein sequences.

The HMM model representing one structural

motif is combined with that representing a similar structural motif to generate a new model that represents the sequence variation in both motifs, as illustrated in Figure 10.30B. Variations between the structures, such as extra positions in one structure, are represented by branches in the HMM. Transition probabilities between states in HMMSTR give the probability of following one branch or the other. When an amino acid sequence is aligned with the model in Figure 10.30B, two paths through the model are possible. The probability of each path may be found by multiplying the probability of the amino acid being found in each subsequent state times the transition probabilities along branches of the model. The most probable path will be the best alignment with the model. The structural information stored in the states of the model that are along the best-scoring path may then be used to predict structural features of each amino acid in the sequence.

This example of combining two structural motifs has been extended to include most structural motifs represented in the I-sites library. The resulting structural HMM, HMMSTR, includes divergent branches, bulges (as illustrated in Fig. 10.30B), and cycles designed to represent sequence and structural variations among the represented motifs. Additional details about HMMSTR are provided in the legend to Figure 10.30B. When the most probable alignment of a protein sequence with HMMSTR has been determined, the corresponding path through the model provides predictive structural information for each amino acid that can be used to build a structural model for the protein based on information in the I-sites library. HMMSTR and the other HMMs described are based on the conservation of amino acid patterns associated with three-dimensional structure. The following class of HMMs does not use sequence information to predict structure.

HMMs for fold prediction using solvation energies. A third type of structural HMM based on the conserved environment of each amino acid in structural folds has been described (Mallick et al. 2002). The objective was to develop HMMs with states that represent the chemical environment of each amino acid in a particular structural fold. A series of HMMs, each representing a particular fold, was produced. To use the HMMs, a query protein sequence is aligned with each of these models in turn using the algorithms described in Chapter 5 to search for a model that produces a high probability match with the query sequence. A highest-scoring alignment of the sequence with the structural fold is produced, thus providing a predictive assignment of amino acids to positions in the structural fold. This assignment can then become the basis for structural modeling of the query protein.

The states of these HMMs store information about the environment of the amino acid in a known structure, including the secondary structure. The predicted secondary structure is included because use of predictive information for the query sequence has been shown to increase the accuracy of structure prediction using structural profiles, described below (Fischer and Eisenberg 1996).

A new type of description of the chemical environment of each amino acid in a structure, called DASEY, the *directional atomic solvation energy*, is also stored in the HMM states, as described next. Each amino acid C_α atom in the protein backbone has four bonds that define four chemical directions in space as illustrated in Figure 10.2. In DASEY, a search is made along the direction of each bond for any atom that is present in a petal-shaped, normally distributed volume following for an empirically derived distance of 16 Å. The DASEY value associated with the found atoms is then looked up in a published table based on laboratory measurements, and an empirically weighted sum of these values is calculated. The result of the DASEY analysis is four energy values, one for each bond direction. These values provide a description of the environment of each amino acid as hydrophobic or hydrophilic in a particular protein structure; large DASEY values indicate a hydrophobic environment, small values a hydrophilic environment. When two proteins with the same fold but with little recognizable sequence identity have been aligned structurally, the DASEY analysis of each amino acid pair in the aligned pair gives similar results, thus indicating that the DASEY value is a reliable indicator of the expected local conservation of chemical environment in a protein structure. The DASEY method also discriminates better between random structural alignments and fold alignments than other methods, such as those used in structural profiles described in the following section (Mallick et al. 2002).

The first step in the DASEY method is to produce an HMM representing each structural fold. The HMMs have states similar to those described in

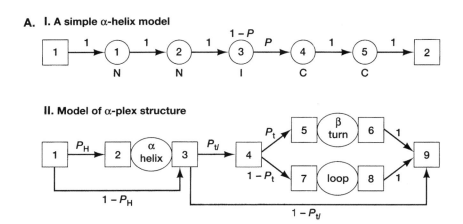

A. **I. A simple α-helix model**

II. Model of α-plex structure

FIGURE 10.30. A hidden Markov model (discrete state-space model) of protein three-dimensional structure. (*A*) Model from Stultz et al. (1993) and White et al. (1994). (*I*) Example of the model for one of the basic structural elements, an α helix. Squares indicate junctions in the model used to connect this model to models for other elements, including other α helices, β strands, β turns, and loops. Circles represent hidden states in the model. Those circles labeled N (1 and 2) are amino-terminal amino acids, I (3) is an internal amino acid, and C (4 and 5) are carboxy-terminal amino acids in the α helix. As in HMMs that represent msa's, each of these states has a distribution of amino acids as found in aligned structures. For example, the frequency of H in an amino-terminal position in an α helix is 0.015, whereas in a carboxy-terminal amino acid, it is 0.028 in the proteins modeled by Stultz et al. (1993). At portions of internal amino acids, additional amino acids may be added to the α model. Arrows indicate a path to be followed through the states of the model, starting and ending with a numbered junction. Usually, the transition probability from one state to the next is 1, but at internal positions two alternative transition probabilities are possible, one (*P*) adds only one internal amino acid, whereas a second adds an additional internal amino acid with probability $1 - P$. (*II*) Model for a particular structural complex includes elements such as the α-helix model described in *I*. This particular model is of an α-helix complex named an α plex. These plexes are then combined to produce a model of a three-dimensional structural domain. The oval α helix is a condensed representation of the finite-state α-helix model shown in *I* with functions at each end. Similar oval representations for models of a β turn and a loop are also shown in this example. The arrows between junctions indicate transitions from one type of plex to another, and in several places there are two or three alternative possibilities in this model. Note that the plex can start with an α helix with transition probability P_H, but can also skip the α helix with probability $1 - P_H$. The remaining paths in the model include the possibilities of an α turn, a loop, or no further elements. To turn domain models into a predictive tool, they are trained on a set of known proteins of that type. In this procedure, transitional probabilities between junctions, states, and plexes are adjusted using the sequences and structures of the training sequences until the model is optimal for distinguishing those domains from other types of domains. A structural prediction for a new protein sequence is then made by finding the most probable path of the sequence through a set of domain models and choosing the model that gives the best alignment with the new sequence. The procedure of calculating the probability of paths through each structural model and of finding the most probable path is similar to that used for sequence alignment models discussed in Chapter 5. Only a limited number of structural domains have been modeled by this approach. (*Continued on facing page.*)

Chapter 5 for modeling msa's. In sequence HMMs, the sequence match state includes the distribution of amino acids in one column of the alignment. In the structural HMM based on DASEY values, the equivalent state corresponds to one structurally aligned position in a structural fold, including gaps. The structural state includes probabilities of matching any amino acid substitution of a given predicted sec-

ondary structure at that position in the structural fold based on DASEY values.

To determine DASEY values for a large number of structurally aligned positions, a database of proteins that have the same fold, DAPS (database of aligned protein structures) at http://fold.doe-mbi.ucla.edu, was prepared. Each fold model is trained by a complex statistical method of modeling

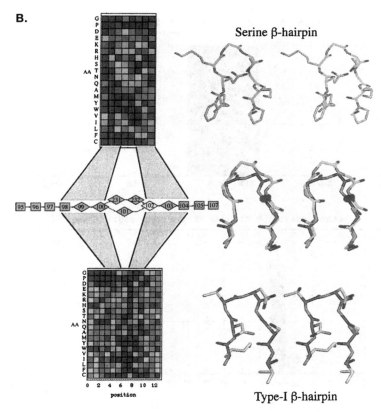

B.

Serine β-hairpin

Type-I β-hairpin

FIGURE 10.30. (*Continued from facing page.*) (*B*) HMM called HMMSTR based on I-sites, 3- to 15-amino-acid patterns that are associated with three-dimensional structural features. The two matrices with colored squares represent alignment of sets of patterns that are found to be associated with a structure, in this case the hairpin turns shown on the right. Each column in the table corresponds to the amino acid variation found for one structural position in one of the turns. (*Blue* side chains) Conserved nonpolar residues; (*green*) conserved polar residues; (*red*) conserved proline; and (*orange*) conserved glycine. The two hairpins are aligned structurally in the middle structure on the right and the observed variation in the corresponding amino acid positions is represented by the HMM between the matrices on the left. The HMM represents an alignment of the two hairpin structural motifs in three-dimensional space and an alignment of the sequences. A short mismatch in the turn is represented by splitting the model into two branches. The shaped icons represent states, each of which represents a structure and a sequence position. Each state contains probability distributions about the sequence and structural attributes of a single position in the motif, including the probability of observing a particular amino acid, secondary structure, Φ-Ψ backbone angles, and structural context, e.g., location of β strand in a β sheet. Rectangles are predominantly β-strand states, and diamonds are predominantly turns. The color of the icon indicates a sequence preference as follows: (*blue*) hydrophobic; (*green*) polar; and (*yellow*) glycine. Numbers in icons are arbitrary identification numbers for the HMM states. There is a transition probability of moving from each state in the model to the next, as in HMMs that represent msa's. This model is a small component of the main HMMSTR model that represents a merging of the entire I-sites library. Three different models, designated λ^D, λ^C, and λ^R, are included in HMMSTR, which differ in details as to how the alignment of the I-sites was obtained to design the branching patterns (topology) of the model and which structural data were used to train the model. HMMSTR may be used for a variety of different predictions, including secondary structure prediction, structural context prediction, and Φ-Ψ dihedral angle prediction. Predictions are made by aligning the model with a sequence, finding if there is a high-scoring alignment, and deciphering the highest-scoring path through the model. The HMMSTR program may be downloaded or used on a server that can be readily located by a Web search. (*B*, reprinted, with permission, from Bystroff et al. 2000 [©2000 Elsevier].)

DASEY environments of the amino acid substitutions found in all aligned proteins in the database. The set of amino acids found in one structural position of one particular fold is then used to produce one match state in an HMM that described the fold, as described in the box on page 477.

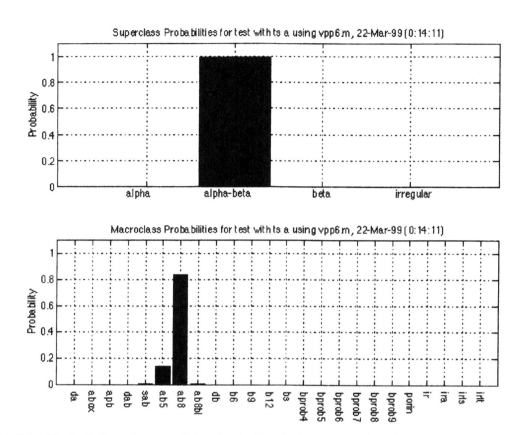

FIGURE 10.31. Prediction of structural class by the Protein Sequence Analysis (PSA) System at http://bmerc-www.bu.edu/psa/index.html. The analysis is based on the training of HMMs for each structure by the sequences that specify those structures as described in Fig. 10.30 and in the text. The input sequence was the α subunit of *S. typhimurium* tryptophan synthase (SwissProt ID TRPA_SALTY, accession no. P00929), which was correctly identified as an α-β class protein. Proteins that are homologous to the input sequence were excluded from the analysis. Hence, although the structure was available, it was not used to produce the correct model. The posterior probability of the protein's being in a particular structural class as defined by the server is given, class α-β and macroclass αβ8.

Once the HMM for a given structural fold has been trained with the aligned amino acid DASEY information, a query sequence may be aligned with the HMM, and, if there is a high-probability, best path through the model, the sequence is predicted to have the same structural fold. If no high-probability path is found, the sequence may be cycled through other HMM fold models, by searching for a high-probability alignment with these HMMs. By further analysis, this DASEY method has been shown to work well when the aligned sequences are from different sequence superfamilies, i.e., they have virtually no measurable sequence identity, in contrast to many other three-dimensional structure prediction methods that depend on sequence similarity.

Benchmark validation of the DASEY-based HMMs using data sets that were not used to train the models revealed that the DASEY-based HMMs provide a sensitive test for fold prediction, i.e., a true positive rate 0.56 at a false-positive rate of 0.1. The DASEY-based method also produced a more sensitive prediction than the sequence-derived properties (SDP) method, based on the use of a structural profile with secondary structure prediction of the query sequence described in the next section (Fischer and Eisenberg 1996), and SDP+, which uses additional sequence similarity and structural predictions (Mallick et al. 2002). DASEY is most useful for predicting fold structures of proteins that are not well represented by structural homologs in the sequence databases. DASEY works least well with proteins with a large percentage of coil regions in their structure

USING DASEY VALUES IN HMMS TO MODEL STRUCTURAL FOLDS

The following steps, illustrated in Figure 10.32, are taken to determine the DASEY values associated with amino acid substitutions in structural folds, when the secondary structure of one of the amino acids, emulating a query amino acid in a search for a structural alignment, is aligned with the other, emulating the template amino acid (Mallick et al. 2002):

1. Produce a structural alignment for each pair of proteins that lack detectable sequence similarity in the DAPS database.

2. For each structurally aligned pair of amino acids i and j, store the following information for each amino acid:

    a. the name of the amino acid residue,

    b. the secondary structure in which the amino acid is located,

    c. the predicted secondary structure in which the amino acid is located, to be used later to increase the predictive value of a search,

    d. a DASEY analysis on each amino acid in each of the four bond directions; these are combined into a four-dimensional DASEY vector.

3. Divide the results found in step 2 into 3600 ($20 \times 3 \times 20 \times 3$) bins on the basis of the following criteria, noting that there are two bin assignments for each aligned amino acid pair (i and j can be reversed). The three possible structures are helix (H), strand (E), and coil (C). The DASEY is a four-value continuous vector:

    a. residue name of template amino acid i (V in first example),

    b. secondary structure in which i is located (E in first example),

    c. the DASEY vector for amino acid i ([30,56,25,48] in first example),

    d. residue name of query amino acid j (R in first example),

    e. predicted secondary structure of amino acid j (H in first example).

4. Repeat steps 2 and 3 for every aligned amino acid pair for all structures in the database (over 800,000 pairs), and include an extra set of bins for gaps, when present. Produce a density plot of DASEY vector values for each bin composed of four component frequency distributions, one for each axis of the C_α atom in amino acid i. The density plot shown in the lower panel in Figure 10.32 is a one-dimensional projection of such a four-dimensional density plot. A different density plot is found for each bin. These plots reveal the environmental preferences, whether hydrophobic, hydrophilic, or a combination of hydrophobic and hydrophilic, for the amino acid substitution of template amino acid i of known secondary structure by query amino acid j with a predicted secondary structure.

5. To facilitate computing probability distributions for HMM states, convert these four-dimensional DASEY density plots to a mixture of normal probability density functions (not shown in figure). In one dimension, for example, this step would be the equivalent of converting a probability distribution with two distinct peaks that are approximately normal to the sum of a mixture of two separate normal distributions. This step facilitates combining two or more different probability distributions to calculate new probability distributions, as required to use the DASEY distributions to make structure predictions.

6. Use these probability density functions to determine the probability distributions in the HMM states. Each state includes a set of template residues, each with an associated secondary structure and DASEY value. The probabilities of matching any amino acid with a predicted secondary structure to this state are calculated from the aforementioned probability density functions.

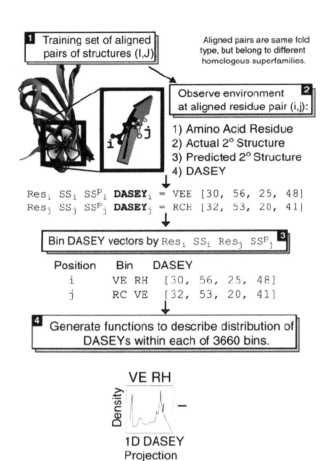

1 Training set of aligned pairs of structures (I,J)

Aligned pairs are same fold type, but belong to different homologous superfamilies.

2 Observe environment at aligned residue pair (i,j):

1) Amino Acid Residue
2) Actual 2° Structure
3) Predicted 2° Structure
4) DASEY

Res_i SS_i SS^P_i **DASEY**$_i$ = VEE [30, 56, 25, 48]
Res_j SS_j SS^P_j **DASEY**$_j$ = RCH [32, 53, 20, 41]

3 Bin DASEY vectors by Res_i SS_i Res_j SS^P_j

Position	Bin	DASEY
i	VE RH	[30, 56, 25, 48]
j	RC VE	[32, 53, 20, 41]

4 Generate functions to describe distribution of DASEYs within each of 3660 bins.

VE RH

Density

1D DASEY Projection

FIGURE 10.32. Using structurally aligned amino acids and a DASEY analysis of structural environment to produce HMMs for prediction of structural fold. The method is described in the box entitled "Using DASEY Values in HMMs to Model Structural Folds" (p. 477). (Modified, with permission, from Mallick et al. 2002 [©National Academy of Sciences, U.S.A.].)

and can incorrectly assign small proteins to a larger fold model (Mallick et al. 2002). When homologs are present, PSI-BLAST and threading combined with PSI-BLAST searches can produce better results (Mallick et al. 2002).

Environmental Template and Contact Potential Methods

The above section describes the use of HMMs for modeling protein structures and providing a method to predict the three-dimensional structure of a protein by alignment with a library of these template HMMs. This section describes the use of structural profiles, which are scoring matrices with rows representing consecutive positions in a structure and columns providing a score for substituting an amino acid at the corresponding structural position or including a gap. A library of profile templates is used for three-dimensional structure prediction in the same manner as

HMMs are used, by searching for a template profile that can be aligned with a query sequence. There are two methods in common use for evaluating amino acid substitutions in these scoring matrices, the environmental template (or structural profile) method and the contact potential method. These methods are discussed in detail in the following sections.

Environmental template or structural profile method. In the environmental template method (Bowie et al. 1991, 1996; see also Ouzounis et al. 1993; Johnson et al. 1996), the environment of each amino acid in each known structural core is determined, including the secondary structure, the area of the side chain that is buried by closeness to other atoms, types of nearby side chains, and other factors. On the basis of these descriptions at each site, the position is classified into one of 18 types, six representing increasing levels of residue burial and fraction of surface covered by polar atoms combined with three classes of secondary structure. Each amino acid is then assessed for its ability to fit into that type of site in the structure. For example, if the side group is buried, another amino acid with a hydrophobic side chain may fit best into the structure at that position. The sequence of the protein is then aligned with a series of such environmentally defined positions in the structure to see whether a series of amino acids in the sequence can be aligned with the assigned structural environments of a given protein core. The procedure is then repeated for each core in the structural database, and the best matches of the query sequence to the core are identified.

Predictions as to which amino acids might be able to fit into a given structural position are in the form of a sequence profile in the environmental template method. This method assumes that if the query protein folds the same way as a target structure, the environments of the amino acids will be in the same linear order as they are in the target. In the normal scoring matrix, it is assumed that a given amino acid substitution always has the same likelihood of every occurrence of the substitution. However, in protein three-dimensional structures, a given substitution may have quite different effects depending on where in the structure and in which structure the substitution occurs. In a loop, where there are not many chemical and physical constraints, the substitution may usually not have any deleterious effects on the overall structure of the protein. In contrast, the same substitution in protein cores, where there are many restraints, may sometimes be possible without deleterious effects, but in other cases may be extremely deleterious. Thus, a sequence profile giving values for substitutions at each amino acid position is made for each core in the PDB. These profiles, one for each core in the database, are then used to score the query sequence to be modeled for compatibility with that core.

The structural three-dimensional profile is a table of scores with one row for each amino acid position in the core and a column for each possible amino acid substitution at that position plus two columns for deletion penalties at that site, as shown in Figure 10.33. Each position in the core is assigned to one of 18 classes of structural environment. The scores in each row reflect the suitability of a given amino acid for that particular environment. The penalty at each core position reflects the acceptability of an insertion or deletion of one or more amino acids at that position in the structure. If the position is within the core,

these penalties are generally high to reflect incompatibility with the structure, but lower for positions on the surface of the core and within loop regions.

The dynamic programming algorithm is used to identify an optimal, best-scoring alignment, much as in aligning sequences by dynamic programming (discussed in Chapter 3). If a target structure is found to have a significantly high score, the new sequence is predicted to have a fold similar to that of the target core.

An entire database of sequences may be matched to a given structural profile to find the most compatible, a procedure called inverse folding. The alignment score for each protein is determined and then converted to a Z score, the number of standard deviations from the mean score for all of the sequences. The highest-scoring sequences are the most compatible with a given structure (Bowie et al. 1996).

The above three-dimensional profile provides a discrete list of scores for matching one-dimensional sequence to a three-dimensional structure. This profile undergoes sharp transitions in values as the structural environment changes. Improved performance has been achieved by smoothing the values in these transitional regions to give a more gradual change using a Fourier analysis.

Improvements in the structural profile method. Several improvements in the structural profile method that increase the sensitivity of the method for detecting matches of a sequence to a structural profile have been described. One method is the *residue pair preference profile* (R3P) method (Wilmanns and Eisenberg 1993, 1995; Bowie et al. 1996). R3P takes into account the amino acid neighbors, main-chain conformations, and secondary structure of each residue in the structure. Recall that to make amino acid scoring matrices for sequence–sequence comparisons, the frequency of amino acid substitutions in alignments is counted in sequence alignments. These frequencies are then divided by the expected frequency of finding the amino acids together in an alignment by chance. The ratio of the observed-to-expected counts is an odds score, and this score is usually converted to a log odds score for convenience in combining likelihood scores by adding their logarithms. Similarly, in the R3P method of making a three-dimensional scoring profile, the frequency of finding a particular pair of interacting amino acids, each with a particular structural feature, is calculated from the number of occurrences in known structures. For example, how often does amino

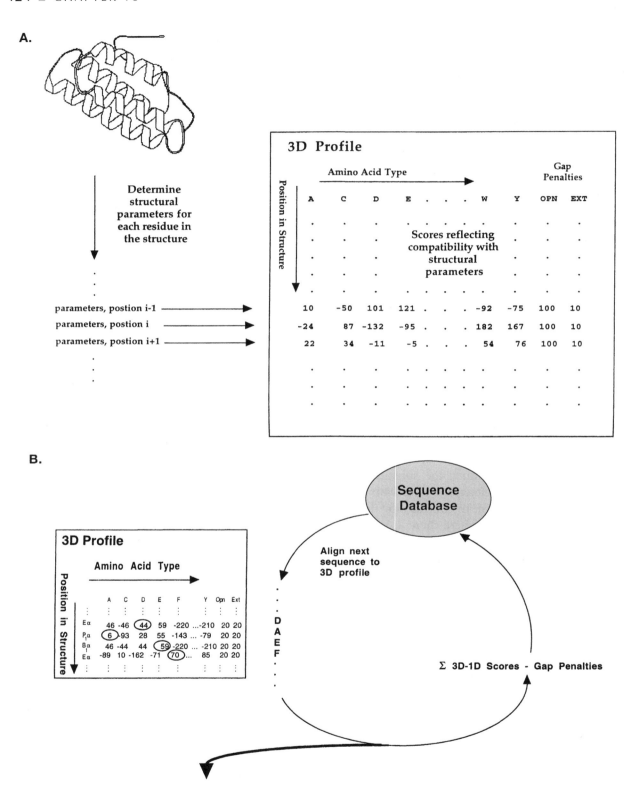

FIGURE 10.33. The structural three-dimensional profile. (*A*) Generation of a three-dimensional profile of a structural core. (*B*) Screening sequences for compatibility with 3D profile. The methods of analysis are described in the text. (*A* and *B*: Redrawn, with permission, from Bowie et al. 1996 [©1996 Elsevier].)

acid a in an α helix interact with amino acid b in a β strand? This observed frequency of interaction in a specific structural configuration is then divided by the frequency of finding a and b interacting in any configuration, and the result is converted to a log odds score.

The pair preference log odds score $S(aa_i, s_i, aa_j, s_j)$ for the amino acids aa_i and aa_j having properties c_i and c_j, respectively, is given by

$$S(aa_i,c_i,aa_j,c_j) = \ln [P(aa_i,c_i,aa_j,c_j) / P(aa_i,aa_j)] \tag{12}$$

where $P(aa_i, c_i, aa_j, c_j)$ is the frequency of amino acids aa_i and aa_j having properties c_i and c_j, respectively, and $P(aa_i, aa_j)$ is the frequency of finding an amino acid pair aa_i and aa_j. The score for position aa_i is then given by a weighted sum of all scores for the interacting pairs with aa_i:

$$S(aa_i) = \Sigma\, w_j\, S(aa_i,c_i,aa_j,c_j)/ \Sigma\, w_j \tag{13}$$

where w_j is a weight representing the compatibility of the environment residue with its own local environment.

Amino acid interactions of a given amino acid residue in a particular core are then analyzed. To determine the neighbors of a given amino acid in the structure, a sphere of radius 12 Å is drawn centered on the C_β atom (see legend to Fig. 10.2). If the C_β atom of another residue in the structure falls within this sphere, they may be interacting. A cylinder of radius 1.6 Å is then drawn between the C_β atoms and, if no H bonds or any other residue falls within this cylinder, the amino acid pair is considered to be interacting. This procedure is repeated for each amino acid that falls within the sphere, resulting in a defined list of approximately eight amino acid pairs that are close enough without barriers to prevent interaction with the given residue. The amino acid type and one of several structural properties for the residue in question and for each interacting residue are then obtained. For example, the secondary structure of the two residues (α helix, β strand, or other) may be taken into account, giving 20 X 3 possible combinations of amino acid and secondary structure. Structural properties of the interacting residue may instead include the backbone dihedral angles Φ and Ψ (see Fig. 10.2) and the number of neighboring residues.

The structural configurations of the given residue and each interacting neighbor are determined. From this information, a score for this interaction can be found from the above analysis. Scores for all of the remaining interacting residues in their particular configurations can be found and then added to give a log odds score for the given amino acid site in the core. This score represents the likelihood of finding such a set of amino acid neighbors in their respective configurations in known protein structures. A value is then determined for various amino acid substitutions or for placing an insertion or deletion at that site. A similar set of scores is then obtained for each position in the protein core to generate a three-dimensional profile matrix based on the neighboring interactions. Three such profiles have been generated, one for each type of structural property in the amino acid pairs—backbone angles, secondary structure, and the number of neighboring residues for the interacting amino acid. A combined three-dimensional profile using elements of these residue pair preference profiles and those of the neighborhood three-dimensional profiles has also been used.

Sequence–structure alignments produced by the R3P method can be improved by an iterative procedure. In the initial alignment between a sequence and the three-dimensional profile of a core, predictions are made as to which residues will interact in the modeled three-dimensional structure. This feature provides information for improving the alignment. Likelihood scores for the predicted interactions can be calculated in the same way as described above for the amino acid interactions in the core. The scores for these interactions may then be summed, as before. In this case, these scores are weighted before summing to reduce the influence of those neighboring amino acids that are not in a compatible environment (Bowie et al. 1996). In evaluations of the R3P method with known three-dimensional structures, alignments are 50% or more correct on average for sequences whose three-dimensional structure pairs superimpose with a root mean square (rms) deviation of 1.97 Å or less (Wilmanns and Eisenberg 1995).

Sequence–structure alignments may be further improved by including in the analysis the predicted secondary structure of the input query sequence, a method called sequence-derived properties (SDP). SDP results in improvements of 25% in fold assignment over standard structural profile methods (Fischer and Eisenberg 1996). Another improvement is the *motif-based assignment* method (MBA), which uses sequence motif content from the Prosite catalog of protein motifs for obtaining additional information about the structure and function of the query protein (Salwinski and Eisenberg 2001). MBA can perform as well as SDP if information about the sequence is filtered to include only that present in other protein sequence databases.

One disadvantage of the structural profile method and the use of environmental variables is that these properties are statistically associated with the original sequence. Hence, the method retains a preference for matching the original sequence of the core protein. On the other hand, the success of most present methods of three-dimensional structure prediction depends on a certain minimal level of similarity. The sequence of environmental patterns in the query sequence and the structure must also be in the same order throughout the sequence for the method to work. However, as discussed above for the SSAP alignment program, this problem may be circumvented by using local alignments.

Contact Potential Method

In the residue–residue contact potential method, the number and closeness of contacts between amino acids in the core are analyzed (Sippl 1990; Jones et al. 1992; Sippl and Weitckus 1992; Bryant and Lawrence 1993). The query sequence is evaluated for amino acid interactions that will correspond to those in the core and that will contribute to the stability of the protein. The most energetically stable conformations of the query sequence thereby provide predictions of the most likely three-dimensional structure.

In this method, each structural core is represented as a two-dimensional contact matrix. The method is very similar to that used by the distance matrix method of the program DALI and illustrated in Figure 10.15. A simple matrix is produced with the amino acids in the structure listed across the rows and down the columns. The distance between the corresponding pair of amino acids in the structure is placed in each matrix position. The amino acids in closest contact are immediately recognizable, and a group produces recognizable patterns. The object is to superimpose sets of amino acid pairs in the query sequence onto the distance matrix of the core. As shown in Figure 10.15B, sequences that fold into a similar structure should show similar contacts, although the amino acids that make up each structural feature do not have to be in the same linear order in both sequences. However, a large number of contacts must be analyzed to find the correct alignment.

To find the best combinations, the approximate conformational energies of each predicted pair are summed to predict the conformational stability of the predicted structure. Contacts have been extensively analyzed, and lookup tables with energies associated with these contacts have been produced. Hence, the energetic contributions of many possible combinations of pairs can be tested in a relatively short period of time. Computer experiments have revealed that contact energies can be used to choose the correct core in a structural database. Supporters of this method claim that the method can detect structural similarity in proteins that do not share any detectable sequence similarity. However, as shown in the next section, in truly blind experiments, the reliability of predictions drops when there is less than 25% sequence identity. A possible limitation to this analysis is that the energy associated with an isolated amino acid pair is assumed to be similar to that found in known protein structures. Recent experiments have suggested that the conformational energy of groups of amino acids larger than two may provide a more reliable prediction.

EXAMPLE: STRUCTURE PREDICTION BY WEB SERVERS THAT PROVIDE A THREADING SERVICE

These results were sent by E-mail. (*A*) Structure prediction by Libra (Table 10.7) and (*B*) UCLA-DOE structure prediction server. This server also provides a Web page for each match giving the results of other types of sequence database searches, secondary structure analyses, and the TOPITS server results. This analysis is of the α subunit of *S. typhimurium* tryptophan synthase (SwissProt ID TRPA_SALTY, accession no. P00929). The Web addresses and methods used by these servers are given in Table 10.7.

```
-----------------------------------------------------
         Forward Folding Search by LIBRA I
-----------------------------------------------------
        LIBRA I was written by M. Ota in 1994-97

gap1= 2.400 gap9= 4.800 gapE= 0.400 npdb= 1389
gap1: Gap opening penalty for exposed sites
gap9: Gap opening penalty for buried sites
gapE: Gap extension penalty
npdb: Number of the structural templates
```

[Input sequence was the *a* subunit of *S. typhimurium* tryptophan synthase, Swiss-prot ID TRPA_SALTY, accession P00929]

Compatible structures are:

```
Rk StrC  Protein                        Lsr Lal   Rsc     SD    Rs/N    ID%
 1 2tsyA  TRYPTOPHAN SYNTHASE;           262 268 -154.2  -5.57  -0.576  83.2
 2 1tlfA  TRYPTIC CORE FRAGMENT OF THE L 296 284  -99.4  -2.94  -0.350  12.0
 3 2liv-  LEUCINE(SLASH)*ISOLEUCINE(SLAS 344 275  -95.4  -2.75  -0.347   9.1
.
.
```

```
Key:
Rk  : Rank position
StrC: Structural code
Lsr : Length of the structural template
Lal : Length of the aligned region
Rsc : Raw score of the structural template
SD  : Standardized score
Rs/N: Raw score (Rsc) normalized by the alignment length (Lal)
ID% : Sequence identity
```

3D-1D alignments are:

```
 1 2tsyA structure vs your tsa sequence

3177238623 7322149997 9899888615 1775499469 6466899979
1AAAAAAAAA AAAgleeBBB BBBBelegeA AAAAAAAAAA AAAlegeBBB
MERYENLFAQ LNDRREGAFV PFVTLGDPGI EQSLKIIDTL IDAGADALEL
:!:!:! :!:!: !  !  : :!:!! :!:!:!:!:!:! :!:!:!:!:!:! :! :!:!:!:!:!
MERYESLFAQ LKERKEGAFV PFVTLGDPGI EQSLKIIDTL IEAGADALEL
.
.
```

```
Key:
1st line: Accessibility of the aligned site;
          1(exposed to water)-9(buried in protein)
2nd line: Local conformation of the aligned site;
          A(alpha), B(beta), gel(coils classified by the dihedral angles)
3rd line: Sequence of the template structure
4th line: Match site of the alignment
5th line: Query sequence
```

B.

```
Most similar fold: 1wsya
TRYPTOPHAN SYNTHASE (E.C.4.2.1.20)

RANK Z-SCORE           FOLD LENGTHALI %ID
  1   77.23           1wsya    248    85
  2    5.21           1aj0     232    24
  3    4.64           1adla    222    18.
  .
  .
```

```
LEGEND:
COL. 1: RANK. The ranks are obtained by sorting the fold library,
        by Z-SCORES, in decreasing order. Only the 15
        structures that are most compatible to your sequence
        are shown.
COL. 2: Z-SCORE. The z-scores are  computed using
        the distribution of raw scores (not shown) of all folds.
```

THE SUCCESS OF STRUCTURE PREDICTIONS IS EVALUATED BY CASP

A very large number of methods have been devised for secondary and three-dimensional protein structure prediction, and a representative sample of these methods has been described in this chapter. As the methods were developed, they were tested for their ability to predict a structure that was already known. The structure to be predicted was left out of the learning step so that the method could not be trained in any recognizable way to identify the correct structure. However, when the result is already known, there is always a possibility that the method was helped in some unintended way to identify the correct structure. Thus, a totally blind test of prediction accuracy is used to provide a more objective test. A series of contests called CASP (critical assessment of structure prediction) was conceived in which structural biologists who were about to publish a structure were asked to submit the corresponding sequence for structure prediction by the contestants. The predictions were then compared with the newly determined structures. The CASP competitions and the results obtained are described on the Protein Structure Prediction server http://predictioncenter.llnl.gov/. The results of CASP are also described in special supplements of the journal PROTEINS: Structure, Function and Genetics.

A large number of research groups enter these contests: There were 187 research groups and 72 pre-dictions servers registered for CASP5, in contrast to CASP2 held 6 years earlier, for which only 32 groups registered. The methods are also becoming more sophisticated and successful as the number and accuracy of the predictions increase. What has emerged as a first general principle of protein structure prediction is that remote sequence similarity to a known structure by domain searching using PSI-BLAST and similar tools is of paramount importance in structure prediction. A second principle is that combinations of methods based on different structural models, e.g., HMMs and structural profiles, and available on servers provided by a number of laboratories, can provide complementary predictions that aid in reliable structure prediction. Finally, new and powerful approaches, such as the Robetta server and the DASEY server, provide powerful and sophisticated new methods for structure prediction.

The CASP server provides software evaluation contests every 2 years. Other projects have been established to evaluate secondary structure predictions and three-dimensional modeling continually. The EVA server http://maple.bioc.columbia.edu/eva/doc/con cept.html downloads new PDB files on a regular basis, submits them to structure prediction servers and then reports the accuracy of the prediction. Other servers that perform the same task include http://bioinfo.pl/ LiveBench/ and http://bioinfo.pl/cafasp/.

STRUCTURAL MODELING SHOWS THE POSITIONS OF AND DISTANCES BETWEEN AMINO ACIDS

Once two- and three-dimensional structural predictions have been made for a protein, a three-dimensional model showing the positions of the amino acids and the distances between them is made. In the above section, detecting sequence similarity between a query sequence and a sequence of known structure plays an important role in successful structure prediction. Database searches as described in Chapter 6 provide alignments of a query sequence with a database of sequences, and can be used to search a database of protein sequences or sequence domains restricted to those of known structure. Alternatively, a protein sequence may be aligned to a library of structural models in a search for one that matches that amino acid sequence structurally. In either case, the alignment provides an indication as to which amino acids in the query sequence may occupy a particular position in a structure. A search of this kind may be enhanced by superimposing the query sequence onto the molecular backbone of the matched sequence to produce a PDB file suitable for analysis by a three-dimensional viewer. An example of this type of analy-

TABLE 10.8. *Web sites for predicting structural features of a query sequence*

Site	Web address	Description	Reference
Modeller	http://www.salilab.org/ modeller/modeller.html	Salilab dynamic programming alignment of sequences and structures and molecular dynamics methods.	Sali et al. (1995)
Swiss-model	http://www.expasy.ch/swissmod/ SWISS-MODEL.html	Sequence alignment of query with sequences of known structure.	Peitsch (1996)
Whatif	http://www.cmbi.kun.nl/whatif/	Flexible molecular graphics rendering of models.	Rodriguez et al. (1998)

sis is provided by the Swiss-model Web site (Table 10.8). Molecular distances, angle, and energies of the superimposed sequence may then be analyzed and manipulated by the SPDBV viewer (Table 10.4). Additional Web sites for molecular modeling are listed in Table 10.8.

SUMMARY AND FUTURE PROSPECTS

This chapter has described methods for predicting protein structure from amino acid sequence. The best approach is to locate a link by sequence analysis between a new protein and a protein of known structure. Even a marginal sequence alignment with a protein of known structure can provide a feasible structural model. Databases that organize proteins into clusters and families with links to known protein structure are also a valuable resource for structure prediction. Proteins that represent new structural folds and domains can be readily identified in these databases, and these proteins can then be targeted for structural analysis by laboratory methods. Meanwhile, the methods for secondary structure and threading analysis (fitting a sequence to a structure) can provide useful predictions, although with variable levels of reliability. Increased confidence should come when several methods give a similar prediction.

The analysis of genomes described in Chapter 11 offers an additional opportunity for protein analysis. Some genome-related projects scan new genomes with protein structure prediction tools to discover how many genes encode proteins with already known structures and how many are possibly new structures. Functions of proteins can be discovered through conserved patterns of gene regulation and organization on the chromosomes of related organisms. The function and structure of a protein in one organism can then be predicted based on the function and structure of a functionally similar protein in a second organism.

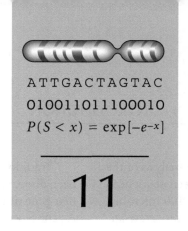

ATTGACTAGTAC
010011011100010
$P(S < x) = \exp[-e^{-x}]$

11

Genome Analysis

C O N T E N T S

INTRODUCTION

A major application of bioinformatics is analysis of the full genomes of organisms that have been sequenced starting in the late 1990s, including over 100 microbial genomes, the budding yeast *Saccharomyces cerevisiae*, the nematode *Caenorhabditis elegans*, the plant *Arabidopsis thaliana*, the fruit fly *Drosophila*, the human genome, as well as many organisms that are human pathogens. Many additional genome sequencing projects have been completed and others are planned or under way.

Traditional genetics and molecular biology have been directed toward understanding the role of a particular gene or protein in an important biological process. A gene is sequenced to predict its function or to manipulate its activity or expression. In contrast, the availability of genome sequences provides the sequences of all the genes of an organism so that important genes influencing metabolism, cellular differentiation and development, and disease processes in animals and plants can be identified and the relevant genes manipulated.

CHAPTER GUIDE FOR BIOLOGISTS

This chapter describes applications of many of the sequence analysis methods that have been introduced in previous chapters and illustrates how to use these methods for genome analysis. The methods start with genome fragment assembly into full-length genome sequences, sometimes guided with a physical map and, at other times, without a physical map. To analyze a single genome, DNA and protein sequence alignments and similarity searches are used to find and analyze repetitive DNA sequences, including highly repetitive sequences, gene families (encoding proteins that are paralogs resulting from gene duplication), gene intron structure, and gene/chromosomal duplications. Between-genome sequence similarity searches, sequence alignments, and methods to cluster genes that are similar in principle to phylogenetic analyses are used to identify strongly alike genes encoding proteins with the same function (orthologs). Protein analysis methods are also used for analysis of protein function, three-dimensional structure, and domain analysis.

Genome analysis also introduces some new computational challenges not covered previously in this book. One challenge is analyzing genome rearrangements between closely related species, e.g., human and mouse, to determine where the rearrangements occurred in the genome and in what order. In aligning DNA and protein sequences, it is assumed that the order of DNA bases or amino acids in the sequences has not changed. In aligning genomes, the gene order is often not conserved, so that new alignment methods are needed. The problem is discovering how to turn the gene order of one genome into the gene order of the other in the least (most parsimonious) number of steps. Once this number is computed, the evolutionary distance between genomes can be calculated.

A second challenge is analyzing single-nucleotide polymorphisms (SNPs) within a species population. SNPs in the human genome are linked into haplotype blocks defined by preferred recombination sites during meiosis. Genes within these blocks have been found by traditional genetic analysis to be in linkage disequilibrium; i.e., they are seldom separated by meiotic recombination events. If a human population is screened for SNPs in one chromosomal region, then they should fall into linkage groups that represent the haplotype block structure. Discovering these blocks in large sequence data sets and defining a representative set of SNPs that can be used to assess an individual's haplotype are problems of considerable interest to the computational biology community.

Developing standard formats for data storage, display, and sharing is another group of computational challenges in genome analysis. For example, as SNPs and other information bearing upon the functional organization of genomes are collected, these data need to be stored in an appropriate database format and made available for distribution on Web sites.

CHAPTER GUIDE FOR COMPUTATIONAL SCIENTISTS

Biologists have collected the genome sequence, which is the complete DNA sequence of all of an organism's chromosomes, of over 100 different organisms ranging from simple, one-celled organisms to multicellular organisms with complex developmental and life cycles. These DNA sequences include genes that specify the amino acid sequences of proteins, sequences that encode RNA molecules needed for making proteins and performing other biological functions, and highly repetitive sequences of no apparent function that are carried along in the genome. Genes that encode proteins and RNA sequences needed for basic life processes such as protein synthesis and reproduction are conserved in most of these organisms, reflecting the importance of these processes and their common evolutionary origin. Other sets of genes, including those needed for development in multicellular organisms, are also conserved as a group. These gene sequences and the amino acid sequences of the encoded proteins may be used to predict the evolutionary history of genomes. Genes frequently occur in families resulting from repeated duplication and subsequent diversification of the biological function of these genes through sequence variation.

One issue to consider in genome analysis is the content of genes and how it might change. Genome analysis has revealed that introns (sequences that interrupt the amino acid–specifying regions of messenger RNA molecules) are present in the genes of most higher organisms (eukaryotes) but are noticeably absent in most single-celled organisms (prokaryotes). The size and number of introns in particular genes can be tracked in the genomes of closely related species. The function of introns is not known, but it has been suggested that they play an important role in genome evolution of eukaryotic organisms. The birth of new introns and the death of introns in genes by processes not yet understood can potentially add or remove amino acids to the protein product of the gene, change the expression level of a gene in different tissues, and cause other changes that promote biological diversity.

There are several classes of sequences (transposable elements) that can move from one genome location to another, thus affecting gene content. Highly repetitive sequences in the genome are derived from such classes of sequences that move (transpose) from one genome location to another. These sequences fall into classes based on length, DNA sequence, and the biochemical mechanisms for producing them. Moving to new chromosomal locations potentially interrupts other sequences, but in some cases, provides a biological advantage. These sequence elements are thought to have played an important role in genome evolution by promoting genome rearrangements.

In addition to gene content, the number and arrangement of genes in genomes may also vary. Sequence similarity analysis reveals that individual genes, groups of many genes, and whole chromosomal regions can be duplicated and rearranged during the evolution of genomes. Duplications provide more diverse groups of biological functions that account for species differences; rearrangements can influence the timing and extent of gene expression. When mammalian genomes are compared, e.g., mouse and human, a large number of rearrangements are found. Modeling these rearrangements between species to discover how many there were, in what order they occurred, and how random they were is a challenging computational problem. One of the most interesting areas of genome analysis is comparing genomes of similar organisms, especially the primates that are the closest relatives of humans. These types of analyses can help to discover those sequences that are the most conserved and therefore most important for function. Human variations in these sequences can then be used as potential disease markers.

One particular benefit of the human genome project to human genetics and medicine is the information provided regarding sequence variation among individuals. Each individual has a change in DNA sequence of approximately 1 DNA base in every 500 bases compared to any other individual. These changes are called single-nucleotide polymorphisms (SNPs), and, collectively, they are responsible for most inherited differences between individuals. However, DNA sequences are transmitted from one generation to the next as haplotype blocks of sequence. Each block contains a few thousand to hundreds of thousands of DNA bases. It is these blocks that determine genetic variation between individuals by keeping groups of SNPs together in the human population, with human offspring receiving new block combinations from each parent. The challenge is to discover subsets of SNPs that indicate the block makeup of individuals. Although this analysis is primarily focused on the human genome, most other species also have sequence variations that determine biological properties in their populations, e.g., stress and disease resistance in plants.

This chapter explores the application of sequence analysis methods to these issues of genome analysis.

WHAT SHOULD BE LEARNED IN THIS CHAPTER?

- Types of sequences found in genomes and how they vary between genomes.
- Methods of sequence analysis that are used to compare genomes.
- Roles of transposable elements and gene duplication in genome evolution.
- How to discover orthologous and paralogous genes, gene families, and conserved protein domains.
- How to model genome rearrangements.
- How to collect and use information on gene function.

Glossary Terms

Annotation is the process of marking a genome sequence with information regarding function, such as, for instance, the location of exons and introns within a eukaryotic gene.

Archaea is a third class of organisms that is different from the classes Bacteria and Eukarya. It was originally determined by analysis of the evolution of ribosomal RNA structure.

Centromeres are specific chromosomal regions that serve as points of attachment between newly replicated daughter chromosomes and that are also used for pulling the chromosomes apart during cell division.

COGs (clusters of orthologous groups) analysis shows gene relationships as clusters of orthologous groups, a type of graphical representation of the sequence relationships among related genes in a group of organisms, usually prokaryotic organisms.

Computer script is a line of code in certain computer languages, e.g., Perl, that is immediately interpreted by the machine, as opposed to computer languages such as C, which require compilation to produce an executable program.

Conserved regions refers to regions of common DNA or protein sequence in two genome contexts.

Core proteome is the set of basic biological functions required by an organism for survival.

Eukaryotes are a class of organisms composed of cells that have a microscopically visible nucleus separated from the cellular cytoplasm by a nuclear membrane and containing the chromosomes of the organism.

E **value (expectation value)** is a statistical value provided by programs that perform genome analysis by sequence database similarity searches. The *E* value of the alignment score between a query sequence and a database sequence is the number of unrelated sequences in the database that are expected to achieve as good an alignment score with the query sequence.

Gene cluster is a group of related genes shown as points on a graph joined by lines (edges) that indicate close gene relationships.

Gene duplication is the process of making a new copy of an existing gene in the genome of an organism.

Genome refers to the entire DNA sequence of one set of chromosomes of an organism.

Genome comparison is a comparison between the genomes of different organisms in regard to a variety of features such as the set of encoded proteins (proteome), the order of genes on the genome, the presence of transposable elements, etc. Comparisons are also made between the genomes of individuals of the same species including sequence variation (SNPs, repeats, etc.) and association with disease, e.g., cancer.

Genome tree is a tree based on genome similarities and differences as, for example, the fraction of proteins that are similar in organisms.

Graph is a representation of relationships among a group of objects, as, for example, the genes in two genomes. The objects are represented by nodes (vertices) that may include information about the objects, and related objects are joined by edges that are labeled according to degrees of relationship. The graph may be used to find clusters of objects that are most closely related.

Haplotype block is one region of one chromosome that is transmitted to the genome of the next generation.

Haplotype map (hapmap) is a genome map that shows blocks of chromosomal sequence of variable length that are passed along from one generation to the next and maintained in the population of an organism. Discovered as conserved patterns of sequence variations (see SNPs), they are of interest in the human genome because they cause genetic variation and influence disease risk.

Horizontal gene transfer (also called lateral gene transfer) is the transfer and insertion of DNA sequences from the genome of one species into the genome of a second species.

Isochores are genome regions having a distinct level of GC-richness.

LINES are long interdispersed nuclear elements of length 4–7 kilobases found in large numbers in eukaryotic genomes including the human genome. They can promote insertion of other sequences, e.g., pseudogenes into the genome.

MITES are miniature, inverted repeat transposable elements often found in association with sequences that regulate transcription.

Orthologs are proteins of highly conserved sequence, function, and structure that are found to be the most-alike pair in whole-genome comparisons.

Paralogs are proteins that share sequence similarity and originated from gene duplication events.

Prokaryotes are single-celled organisms, such as bacteria, that lack an observable nucleus and generally have small genomes of several million DNA bases.

Protein domains are conserved amino acid sequences that are found in proteins and often represent a conserved function and structure.

Protein families are groups of proteins that are found to be at least 50% identical in sequence alignments.

Proteome is the entire complement of proteins encoded in the genome of an organism.

Pseudogene is a nonfunctional duplicate copy of a gene, usually produced by making a DNA copy of the spliced mRNA of a gene and inserting the copy into the genome.

QTLs (quantitative trait loci) are sequence variations that are so close to biologically important genes that they stay together from one generation to the next. The presence of such nearby genes is revealed by a strong association of the QTL with an important biological property, e.g., yield in plants.

Ridges are genome domains that have a high gene density, GC content, SINE repeat density, and a low LINE repeat density, as well as shorter introns than genes outside of ridges. Genes that are strongly expressed cluster into ridges. Antiridges contain clusters of weakly expressed genes and have opposite sequence characteristics to ridges.

Shotgun sequencing is the process of sequencing a genome based on assembly of random fragments based on sequence overlaps into a linear genome sequence without using a physical map as a guide.

SINES are short interdispersed nuclear elements 80–400 DNA bases long found in large numbers in eukaryotic genomes. An example is *Alu* sequences in the human.

Single-nucleotide polymorphisms (SNPs) are sequence variations at a single base position that are quite common between individuals of the same species.

Synteny is a colinearity of gene order in two species: a conserved group of genes in the same order in two genomes as a syntenic group or cluster. In genome analysis, synteny will be an extended local alignment between two genome sequences that may not necessarily be in the same chromosomal location in the species.

Telomeres are sequences composed of short repeated elements found at the ends of chromosomes and necessary for chromosomal replication.

Transposable elements are DNA sequences that move from one chromosomal location to another. Often many copies representing many transpositions will be found in a genome.

GENOME ANALYSIS PRESENTS MANY CHALLENGES

There are many challenging aspects of genome analysis. Foremost among them are gene prediction, phylogenetic history of genes, presence of repetitive sequences, and genome rearrangements. The availability of genome sequences provides an unprecedented opportunity to explore genetic variability both between organisms and within the individual organism. Thus, identification of those genes that are predicted to have a particular biological function and design of experiments to test that prediction are two of the goals of genome analysis. Such analysis depends on gene prediction using gene models for each organism followed by sequence comparisons between the predicted proteins with protein sequence databases to find a matching sequence whose function is known from biological studies.

Because a great deal of biological information is available for these model organisms, sequence comparisons between them are greatly facilitated. Many years of genetic and biochemical research of these model organisms—the bacterium *Escherichia coli*, the budding yeast *Saccharomyces cerevisiae*, the nematode *Caenorhabditis elegans*, the plant *Arabidopsis thaliana*, and the fruit fly *Drosophila melanogaster*—have led to the accumulation of a large amount of information on gene organization and function. The mouse *Mus musculus* is a genetic model for humans because the two species are both mammals that are closely related through evolution. An even closer relationship exists among primates, which include animals with similar bone and limb structure such as man, apes, lemurs, and monkeys. Sequence comparisons between primate genes can provide important biological information about sequence variation and human disease (Boffelli et al. 2003).

Thus, a newly identified gene in a newly sequenced organism can be compared to the existing databases of information to find whether it has a similar function. Genes involved in human disease, for example, are sometimes found to be similar to a fruit fly gene at the protein sequence level (for an example of how significant this kind of analysis can be, see Rubin et al. 2000). The genetic effects of mutations in the fruit fly's gene can then provide a biochemical, cellular, or developmental model for the mechanism of the human disease. However, it has not yet been possible to identify the function of all the genes found in model organisms. As a result, a similar gene or family of genes may be found in several organisms, including a model organism, but the function of the gene or genes is not known. Hence, biological analysis of model organisms in the laboratory will continue to be necessary for discovering the function of those genes that is not revealed by the tools of bioinformatics.

Tracing the phylogenetic history of uncharacterized genes, characterized genes, and gene arrangements in diverse organisms is one of the most interesting and challenging aspects of genome analysis. In addition, even though a gene that specifies an impor-

tant biological function has not been identified, the gene can be traced in individuals using sequence variations that occur among individuals in a population, called sequence polymorphisms.

Genome Sequences of Individuals Vary

Major efforts to identify the important sequence variations in humans that can be used to identify disease risk or sensitivity to drugs are under way. One is the hapmap project (http://www.hapmap.org/), an international consortium of scientists and research support agencies. In humans, for example, single-nucleotide polymorphisms (SNPs) can be found throughout the genome (Sachidanandam et al. 2001), including some that are positioned adjacent to an important disease gene. If a particular G → A polymorphism is right next to a defective tumor suppressor gene, for example, that polymorphism serves as a sequence marker for the presence of the defective gene. The genetic principle behind this use of SNP markers is that genes that are in the same chromosomal regions, which are called linked genes, tend to stay together physically from one generation to the next. Genes that are not linked in this manner can become segregated.

During meiosis, the process that makes the germ cells (egg, sperm, or pollen), each individual passes along a new set of chromosomes to potential offspring. Thus, each chromosome is a new combination of genes derived from the two parental chromosomes, as illustrated in Figure 11.1A. In this process, blocks of chromosome sequence called haplotype blocks and comprising one or more genes remain linked. The population at large includes a set of different sequences for each block called haplotypes, as shown in Figure 11.1B (Daly et al. 2001); each individual has two unique haplotype sets per chromosome. The hapmap project described above tracks the size and location of the common blocks and haplotype variants of each block in humans with the goal of finding those that represent a medical risk. Polymorphisms are also being used to study inheritance of genetic traits in other organisms.

Another example of using sequence polymorphisms to discover genes affecting important traits is in genetic analysis of crop plants. Features such as plant height and amount of seed produced are influenced by variations in sets of genes. Quantitative trait loci (QTLs) act as markers for these genes. Inheritance

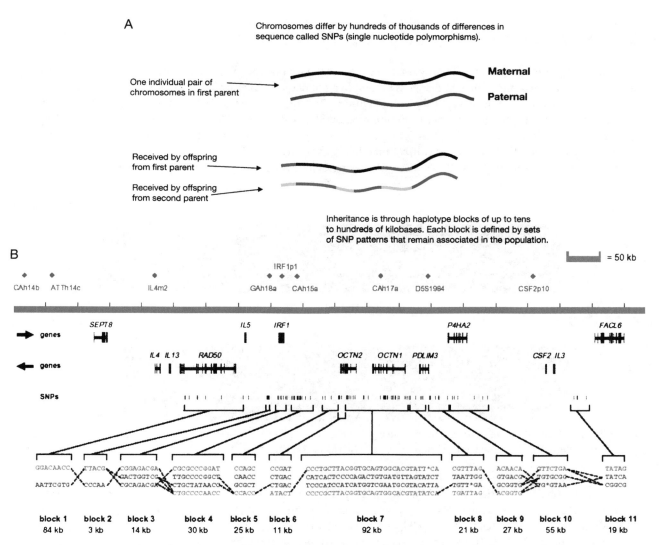

A

Chromosomes differ by hundreds of thousands of differences in sequence called SNPs (single nucleotide polymorphisms).

One individual pair of chromosomes in first parent → **Maternal** / **Paternal**

Received by offspring from first parent →
Received by offspring from second parent →

Inheritance is through haplotype blocks of up to tens to hundreds of kilobases. Each block is defined by sets of SNP patterns that remain associated in the population.

FIGURE 11.1. Source of genetic variability in the human genome. (A) The pattern of inheritance from one generation to the next. During meiosis (the process that produces germ cells), each germ cell (sperm or egg) produced by an individual receives a copy of each chromosome with a pattern of sequence blocks that is a combination of regions from the two parental chromosomes. Not all of the changes shown will occur in one cell but, in a human population, all of these regions of exchange will be observed. More widely separated blocks tend to be rearranged more often than close ones. Within a block, however, the sequences will seldom be rearranged. This lack of rearrangement within blocks is described by geneticists as linkage disequilibrium (Wall and Pritchard 2003). The pattern results from the biochemical mechanism that recombines chromosomes only acting on certain localized regions. These blocks are found by examining populations for conserved SNP patterns. Certain individuals will have the same sequence for the length of a particular block and other individuals will have a second, different sequence for the same blocks. These sequences are two haplotypes. (B) Example of blocks on human chromosome 5 position q31 of a European-derived population. Block size and number of different sequences (haplotypes) of each block are seen to vary. Dashed lines indicate some observed haplotype combinations. These haplotype combinations found in chromosomes have been modeled by a hidden Markov model (HMM) trained on the observed variations. The HMM can be used to classify the haplotype of a new chromosomal sequence (Daly et al. 2001). (B, Reprinted with permission from Daly et al. 2001 [©Nature Publishing Group].)

of QTLs can be traced from one generation to the next using sequence polymorphisms linked to genetic variation without having to wait to observe the effects on plant growth. The availability of genome sequences greatly facilitates the discovery and utilization of these sequence polymorphisms.

SNP and Haplotype Analysis

High-throughput methods are frequently used to collect sequence polymorphisms. These data need subsequent validation to confirm that the sequence change is present in a population. An example of a sequence polymorphism discovered in a gene involved in allergy is shown in Figure 11.2. Large numbers of SNPs, some validated and some not, are stored on the dbSNP (http://www.ncbi.nlm.nih.gov/SNP/) database. The human genome browser (http://genome.ucsc.edu/goldenPath/hgTracks.html) is useful for identifying possible additional SNPs within a human genomic region. The SNP Consortium (http://snp.cshl.org) has been established to identify and evaluate SNPs. There are also commercial sites that may be found in Web searches for SNP databases including proprietary SNPs.

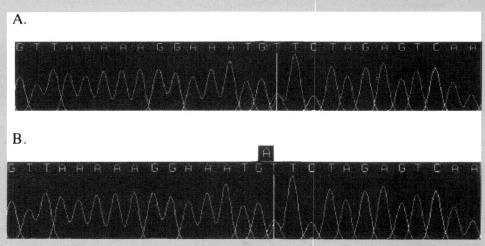

FIGURE 11.2. Example of an SNP in a human allergy gene. DNA sequences derived from human serum samples in which each chromosome has the same base (*A*) or a different base (*B*) at the same position. This analysis reveals the presence of an SNP.

Computational and statistical tools for haplotype prediction may also be readily located by Web and PubMed searches using search terms such as SNP haplotypes, linkage disequilibrium, haplotype blocks, and population genetic analysis tools. ARLEQUIN, GENEHUNTER, GOLD (Graphical Overview of Linkage Disequilibrium), HaploBlockFinder, PHASE (http://www.stats.ox.ac.uk/mathgen/software.html), SNPsim (http://www.evolgenics.com/software), and ZAPLO are examples of tools for the discovery of haplotype blocks. Once the block structure is found, the final step is to use the above tools and others, such as SNPtagger, which is a dynamic programming method (Zhang et al. 2002), to search for signature SNPs that may be used to identify presence of a particular haplotype in an individual. MutDB (http://mutdb.org/) is a tool for determining which SNPs in a gene are likely to alter the function of their associated protein product. Analysis of sequence variations in the genomes of related organisms, for example, comparing other primates with humans, can serve as a guide for predicting which SNPs may produce functional changes. These SNPs may be found through their occurrence in highly conserved regions in a multiple sequence alignment (Bofelli et al. 2003).

Genome Duplication Influences Genetic Variation

Some types of genetic variation, including those responsible for specific human diseases, are best understood at the genome-wide level. The duplication of genes, gene segments, and gene clusters pro-vides opportunities for recombination events that can cause changes in gene copy number or loss of gene function (Lupski 1998). These changes were especially apparent when the human genome was sequenced and examined for highly repetitive sequences and duplicated regions. These regions are frequently associated with known diseases, probably by causing rearrangements of the genome (Bailey et al. 2002). The analysis of genomes for gene content also helps to assess the significance of genome variation.

Analysis of the Proteome

Of the genomes that have been sequenced, one major task is to identify all of the genes that encode proteins (the proteome) for each genome and to identify the function of as many of these proteins as possible by database similarity searches. This step in genome analysis is called genome annotation. Annotation involves predicting the location of genes on the genome based on characteristic sequence variations, as described in Chapter 9. The predicted proteins are then used as query sequences in database similarity searches using the methods of Chapter 6. Significant matches are added to the genome sequence entry in the sequence database, noting the genome positions of the genes and providing a description of the protein products. More sophisticated methods of searching for protein families described in Chapters 6 and 10 are also used for annotation. Collectively, these methods provide predictions of the proteome of a new organism.

In evaluating the results of genome annotation, it is important to note the methods used, the statistical significance of the results, and the overall degree of confidence in the sequence alignments. The analysis of interesting results should be repeated if necessary. Annotation errors occur when these criteria are not followed (Kyrpides and Ouzounis 1999).

Once proteome analysis of an organism has been completed, the proteome may be compared to itself to identify paralogs, families of proteins that have arisen by gene duplication. Amplification of protein families can provide new biological functions (e.g., a new metabolic capability for a bacterium, cell-to-cell communication during development of a multicellular organism, or new cells leading to higher brain functions in primates). One proteome may also be compared to another proteome to discover genes that have kept the same function (orthologs), genes that have become fused to make a larger protein (or split into two to make two separate proteins), or new arrangements of protein domains. These types of analyses have been extended to a molecular analysis of diseases and stress responses. For example, cancer is a class of diseases in which pronounced changes in gene expression, gene amplification, and gene loss give the cell new biological properties. Understanding these changes can provide new diagnostic procedures and treatments. Finally, protein biochemists analyze all of the predicted proteins in the proteome using the structural prediction methods described in Chapter 10 in order to discover how these proteins are used for biological functions and how the corresponding genes are arranged in the genome. Detailed analyses of the predicted proteome of an organism thus play an important role in genome analysis.

Web Resources and Computational Tools for Genome Analysis

A representative set of Web pages supporting genome analysis is given in Table 11.1. At the end of the chapter, a list of search terms for finding additional sites is provided. In many cases, the large amount of data and repetitive steps involved in genome analysis requires developing one's own data analysis tools. The development of computational tools for this purpose is described in Chapter 12.

GENOME ANATOMY HAS BEEN STUDIED IN PROKARYOTES AND EUKARYOTES

Early biologists examining a particular plant, animal, or yeast cell using a microscope observed a nucleus (in a eukaryotic cell) with a specific number of chromosomes of variable length and morphology that could be seen at certain stages of cell division. The chromosomes comprised linear DNA molecules in a tightly compact form that were wrapped around protein complexes, called the nucleosome. Nuclei and chromosomes were not observed in bacteria (a prokaryotic cell), but when bacterial DNA was eventually detected, the molecule was usually circular and was also in a compacted form. The following sections outline the structure and composition of prokaryotic and eukaryotic genomes. One major accomplishment

TABLE 11.1. *Examples of Web resources for genome information and analysis*

Database	Web address
A. thaliana information resource TAIR	http://www.arabidopsis.org/
Caenorhabditis elegans (worm) database	http://www.wormbase.org/
Cold Spring Harbor Laboratory Genome Projects	http://www.cshl.org/
DOE Human Genome	http://www.doegenomes.org/
Drosophila genome project Berkeley	http://www.fruitfly.org/
Drosophila: Flybase, a genomic database	http://flybase.bio.indiana.edu/
E. coli genome project	http://www.genome.wisc.edu/
ENSEMBL *e!* genome browser (Sanger Inst., EMBL-EBI)	http://www.ensembl.org/
Genome list at NIH	http://molbio.info.nih.gov/molbio/db.html
Genome database in support of human genome project	http://www.gdb.org
Gramene—cereal grasses	http://www.gramene.org/
Human Genome Browser (Golden Path, UCSC)	http://genome.ucsc.edu/goldenPath/hgTracks.html
Mouse (*Mus musculus*) genome informatics	http://www.informatics.jax.org/
National Center for Biotechnology Information (NCBI)	http://www.ncbi.nlm.nih.gov/
NIH Human Genome Res. Initiative NHGRI	http://www.genome.gov/
Parasite genome databases	http://www.ebi.ac.uk/parasites/parasite-genome.html
PIR Georgetown University genomes	http://www-nbrf.georgetown.edu/pir/genome.html
Prokaryotic genomes P. Bork laboratory	http://www.bork.embl-heidelberg.de/Genome/
Rice (*Oryza sativa*) genome project	http://rgp.dna.affrc.go.jp/
S. cerevisiae (budding yeast) database (SGD)	http://genome-www.stanford.edu/Saccharomyces/
Stanford University genome resources	http://genome-www.stanford.edu/
The Institute of Genomics Research (TIGR)	http://www.tigr.org/
The Sanger Institute	http://www.sanger.ac.uk/
U.S. Dept. of Energy Joint Genome Initiative	http://www.jgi.doe.gov/
Washington University Genome Center	http://genome.wustl.edu/
Whitehead and Broad Institutes for Biomedical Research	http://www.broad.mit.edu/genome.html

of studies of genome anatomy has been the analysis of the ribosomal RNA molecules of prokaryotes and eukaryotes, which has led to the prediction of three main branches in the tree of life represented by the Archaea, the Bacteria, and the Eukarya.

Prokaryotic Genomes

The first bacterial genome to be sequenced was that of *Hemophilus influenzae*, a mild human pathogen (Fleischmann et al. 1995). This project was carried out at the Institute of Genomics Research (TIGR, http://www.tigr.org), in part as proof of principle for a new whole-genome sequencing method—the shotgun method. A large number of random overlapping genome fragments were sequenced and then a consensus sequence of the entire 1.8×16^6-bp chromosome of *Hemophilus* was assembled computationally based on sequence overlaps, excepting several regions that had to be assembled manually. Once the assembled genome was available, open reading frames were identified, and these were compared to known pro-

teins by similarity searches of protein sequence databases (see Chapter 6). Approximately 58% of the 1743 predicted genes matched genes of another species, the bacterial species *E. coli* K-12, which had been the subject of many years of genetic and biochemical research. The identification of these genes allowed the investigators to construct some of the biochemical pathways of the *Hemophilus* cell. The function of the other 42% of the *Hemophilus* genes could not be identified, although some of them were similar to the 38% of *E. coli* genes that were also of unknown function. Sequence repeats that may cause rearrangements and influence the production of cell-surface proteins related to pathogenesis were also found.

The success of sequencing the *Hemophilus* genome in about 6 months and with a modest budget heralded the sequencing of a large number of additional prokaryotic organisms (de Bruijn et al. 1998). The genomes of many prokaryotic species have since been sequenced. Organisms were selected for sequencing on the basis of at least three criteria:

VIRAL GENOMES

Prior to the sequencing of *H. influenzae*, the first free-living organism to be sequenced, a large number of viruses were sequenced. Viruses are unique in that they use the cell's machinery to replicate themselves and have only a small number of genes whose function is to control the cell. Many of these organisms also serve as model systems for studying replication and gene expression. As an example, the nucleotide sequence of bacteriophage λ, an important organism for understanding methods of gene regulation, was completed by Sanger et al. (1982). A further impetus to sequencing viral genomes came with the discovery that HIV, the virus that causes AIDS, undergoes rapid sequence change in infected hosts. A simple way to retrieve sequences of viral and other extrachromosomal genetic elements such as cellular organelles (e.g., the mitochondrion) is through the National Center for Biotechnology Information (NCBI) taxonomy browser at http://gov/Taxonomy/taxonomyhome.html.

(1) They had been subjected to significant biological analysis, e.g., *E. coli* and *Bacillus subtilis*, and thus were good model prokaryotic organisms; (2) they were an important human pathogen, e.g., *Mycobacterium tuberculosis* (tuberculosis) and *Mycoplasma pneumoniae* (pneumonia); or (3) they were of phylogenetic interest and likely to reveal information about the evolution of bacteria.

For genome sequencing projects, organisms have been sampled from throughout the evolutionary tree based on ribosomal RNA variation, including some that are in deeper branches of the tree and that have growth properties reminiscent of an ancient environment. A summary of the genome size and composition of a representative list of prokaryotes is given in Table 11.2. Web sites, most notably TIGR, provide a complete annotation of the prokaryotic genomes that have been sequenced thus far. Computational resources, such as GeneQuiz, facilitate the analysis of bacterial genomes.

TABLE 11.2. *Features of representative prokaryotic genomes*

Organism (reference)	Phylogenetic group	Genome size (Mbp) (number of protein-encoding genes)	Novel functions
Escherichia coli (Blattner et al. 1997)	Bacteria	4.6 (4288)	model organism
Methanococcus jannaschii (Bult et al. 1996)	Archaea	1.66 (1682)[a]	grows at high temperature and pressure and produces methane
Hemophilus influenzae (Fleischmann et al. 1995)	Bacteria	1.83 (1743)	human pathogen
Mycoplasma pneumoniae (Himmelreich et al. 1996)	Bacteria	0.82 (676)	human pathogen that grows inside cells; metabolically weak
Bacillus subtilis (Kunst et al. 1997)	Bacteria	4.2 (4098)	model organism
Aquifex aeolicus (Deckert et al. 1998)	Bacteria	1.55 (1512)[b]	ancient species; grows at high temperature, and can grow in a hydrogen, oxygen, carbon dioxide atmosphere in the presence of only mineral salts
Synechocystis sp. (Kaneko et al. 1996a,b)	Bacteria	3.57 (3168)	ancient organism that produces oxygen by light-harvesting; may have oxygenated atmosphere

The genome in each case is contained on a single circular DNA molecule except where noted. Another bacterial species, *Deinococcus radiodurans*, has two chromosomes of sizes 2.6 and 0.4 Mbp and two additional elements of size 0.17 Mb and 46 Kbp (http://www.tigr.org). Other bacterial species have linear chromosomes (for review, see Volff and Altenbuchner 2000).

[a] *M. jannaschii* has a small and a large extrachromosomal element.

[b] *A. aeolicus* has a single extrachromosomal element.

Eukaryotic Genomes

In addition to differing from prokaryotic genomes by having visible chromosomes within a nucleus, eukaryotic genomes commonly have large genomes, and tandem repeats of sequences in their genomes and their protein-coding genes include introns.

Sequence Repeats

Eukaryotic genes often have many tandem repeats of the same sequence. These repeats vary in size, number, and location in the genome and sometimes are localized in specific regions. In some cases, these sequences are on movable genetic elements that become amplified in number; in other cases, they may result from abnormal replication of DNA during copying of chromosomes that have sequence repeats or from unequal genetic exchanges between chromosomes that have sequence repeats. The presence of repetitive sequences may be demonstrated using programs for detection of low-complexity regions in sequences described in Chapters 3 and 6.

Sequence repeats and satellite DNA. Because repeats are made of the same sequence, they often skew the base composition of DNA. This skewing effect gives DNA molecules containing them a different mass per unit volume (buoyant density), allowing them to be separated as satellite DNA from the bulk DNA by density. These repetitive DNAs vary in number, length, and chromosomal location.

In nondividing eukaryotic cells, chromosomes have a mixture of lightly and darkly stained regions called heterochromatin and euchromatin, respectively. Heterochromatin occurs in a compact configuration and is not transcribed. However, heterochromatic regions play essential biological roles. Two particularly significant chromosomal features that play important roles in chromosomal replication and cell division are located near the heterochromatin. These are the centromeres, which are attachment points for the chromosomes during cell division, and the telomeres, which are located at chromosome ends and are found near heterochromatin. Most genes that are transcribed are located in the less compact and darkly staining euchromatin, to which regulatory proteins that initiate transcription have access (for review, see Brown 1999).

Three classes of satellite DNAs—satellites, minisatellites, and microsatellites—vary in their location in these chromosomal regions.

1. Satellites are composed of repeats of one thousand to several thousand bases in very large tandem arrays up to 100 million bases long and are typically near centromeres and telomeres.

2. Minisatellites are made up of repeats approximately 15 bases in length in arrays highly variable in length of a few hundred bases up to thousands of kilobases. Found throughout the genomes, they are typically located in the euchromatin. These sequences vary in length within populations, and length variation is used to identify human individuals in forensic science. They are also known as variable number tandem repeats (VNTRs). The National Institute of Standards and Technology maintains a database of short tandem repeat DNA sequences, including mini- and microsatellites.

3. Microsatellites are made up of short repeats 2–6 bases long in arrays that are highly variable in length from 10 to 100 bases. These lengths vary in populations, and the lengths are inherited from one generation to the next. The distribution of microsatellites in the genome and a variable degree of sequence variation within them make them useful as markers for both genetic analysis and evolutionary analysis of genomes. Microsatellites are also called simple sequence repeats (SSRs) and short tandem repeats (STRs). They are found in the euchromatin and the centromere and at the ends of eukaryotic chromosomes near the telomeres, which in vertebrates comprise hundreds of copies of the 6-bp repeat TTAGGG.

Sequence repeats due to transposable elements. Transposable elements (TEs) are DNA sequences that can move from one chromosomal location to another faster than the chromosome can replicate. Hence, TEs have the potential to increase in number until they comprise a large proportion of the genome sequence, a feature observed in the genomes of many plants and animals. TEs can comprise a large proportion of the repetitive sequences in the eukaryotic genome and are thought to play an important role in the evolution of genomes (Kidwell and Lisch 1997, 2000). They remain detectable in the genome until their sequences blend into the background sequence by mutation over time. The percentage of different genomes that are composed of TEs is depicted in

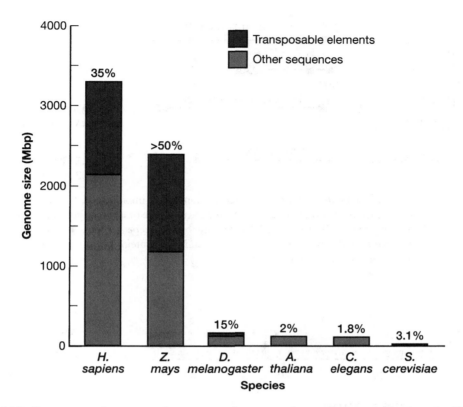

FIGURE 11.3. Percentages of representative genomes that are made up of transposable elements. The genomes include those of humans, maize, the fruit fly *Drosophila*, the model plant *Arabidopsis*, the nematode *C. elegans*, and budding yeast *S. cerevisiae*, respectively. (Adapted, with permission, from Kidwell and Lisch 2000 [©Elsevier Science].)

Figure 11.3. For example, more than one-third of the human genome consists of interspersed repetitive sequences derived from TEs.

Eukaryotic TEs fall into two main classes according to sequence similarity and the mechanism of transposition. Class I elements encode a reverse transcriptase and use RNA-mediated mechanisms of transcription. There are three main subclasses of these TEs—the long terminal repeat (LTR) retrotransposons, retroposons, and retrovirus-like elements with LTRs. The retroposons include short (80–400 bp long) interspersed nuclear elements (SINES) and long (4–7 kbp long) interspersed nuclear elements (LINES). The types of transposable elements that are present in high copy numbers in mammalian genomes are illustrated in Figure 11.4. Ten percent (1.2 million copies) of the human genome comprises one particular family of the SINE element, designated *Alu*, and 14.6% of one particular LINE designated LINE1 (593,000 copies, Smit 1996). *Alu* sequences are found in the introns, promoter regions, and untranslated intergenic regions of primates.

A second class of TEs, class II, is made up of elements that employ a DNA-based mechanism of transposition. The human genome contains about 200,000 copies of this class of elements that probably predate human evolution (Smit 1996, 1999). Class II elements also include the Activation-Dissociation (Ac-Ds) family in maize and the P element in *Drosophila*.

A third category of TEs has features of both class I and class II TEs. These miniature, inverted repeat TEs (MITES) are 400 bp in length and were discovered in diverse flowering plants where they are frequently associated with regulatory regions of genes. Hence, they could be exerting an influence on regulation of gene expression (Kidwell and Lisch 1997).

The abundance of TEs in the genomes of humans, yeast, maize, and *E. coli* is illustrated in Figure 11.5. The following features are apparent: (1) TEs are present in all of the chromosomes, ranging

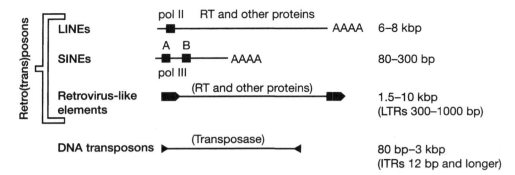

FIGURE 11.4. Transposable elements that produce high-copy-number interdispersed repeats in mammalian genomes. Shown are class of element, a representation of the structure, size of element plus, in some cases, size of terminal repeats. ■ RNA polymerase II or III promoter; ➡ long terminal repeat (LTR); ▶, ◀ inverted terminal repeats; RT, reverse transcriptase. Parentheses above elements indicate protein found in autonomous elements. (Redrawn, with permission, from Smit 1996 [©Elsevier Science].)

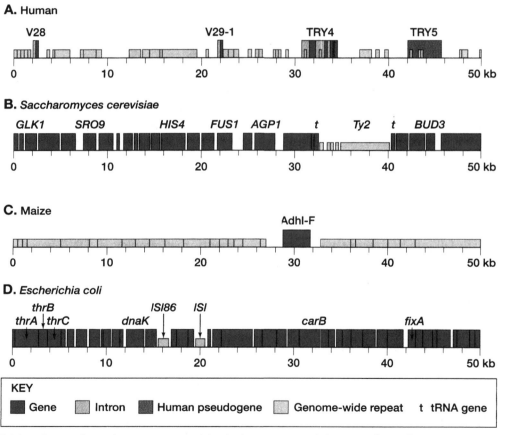

FIGURE 11.5. Comparison of genome composition in four genomes. (A) Human β T-cell receptor locus on chromosome 7. V28 and V29.1 encode parts of the β T-cell receptor proteins that are joined during development of the immune system (Rowen et al. 1996). TRY4, the gene for trypsinogen, and TRY5, a pseudogene related to the trypsinogen family, are not related to the receptor sequence. Why they are located here is not known. (B) Segment of yeast chromosome III (Oliver et al. 1992). (C, D) 50-kb fragments of the maize and E. coli chromosomes, respectively (SanMiguel et al. 1996; Blattner et al. 1997). The maize repeats are LTR retrovirus-like elements (Fig. 11.4) that have inserted within the last 3 million years (SanMiguel et al. 1998). [Redrawn, with permission, from Brown 1999 [©Wiley-Liss].)

from bacteria to humans, but their abundance varies; (2) TEs can comprise a large portion of the genomes of higher eukaryotes, both plants and animals. Thus, only a small fraction of the genome of these organisms carries gene sequences. TEs are clustered in vertebrate genomes. The presence of such clusters of sequence repeats produces localized variations in base composition.

It has been proposed that genomes are made up of distinct segments of unique base composition designated isochores. Vertebrate chromosomes have long (>300 kb) regions of distinct GC richness, repeat content, and gene density (Bernardi 1995). Human and mouse chromosomal regions that have a low density of genes are AT-rich and have more Alu or B1/B2 (SINES) than LINE1 elements. The reverse is true for regions that have a high gene density, and these regions are more GC-rich (Henikoff et al. 1997).

Gene Structure Varies in Eukaryotes

The information for the amino acid sequence of eukaryotic proteins is present in the exons, the protein-encoding sequences that are dispersed throughout the gene and interrupted by introns of varying length and number. Removal of introns by RNA splicing joins the exons, forming an open reading frame that is translated into the amino acid sequence of the proteins. In *S. cerevisiae* (budding yeast), only a small fraction of the genes contain introns; there are a total of 239 introns in the entire genome. In contrast, in individual human genes, introns may be present in numbers exceeding 100 and often comprise more than 95% of the gene. Introns can remain at the same location in a eukaryotic gene for long periods of evolutionary time. In addition, complex intron arrangements are often found. For example, the RNA of organelles can be introns containing introns (Copertino and Hallick 1993), and nuclear genes can encode genes in which one gene, including introns, is encoded within the introns of a second gene (see, e.g., Cawthon et al. 1990).

The origin of introns in eukaryotic genes is not understood, but it has been accounted for by two models. The "introns-early" view proposes that introns were used to assemble the first genes from sets of ancient conserved exons, whereas the "introns-late" view proposes that introns broke up previously continuous genes by inserting into them (Gilbert et

al. 1997). Introns are present in the earliest eukaryotes but not in any sequenced prokaryote. With the availability of whole-genome analysis of introns, additional studies support a role for introns in the evolution of eukaryotic organisms. It has been proposed that introns play this evolutionary role in multicellular organisms, including many plants and animals, that have small population sizes (Lynch and Richardson 2002; Lynch and Connery 2003). Introns are associated with the *Alu* family of sequence repeats in primates. The *Alu* family of sequence repeats has been observed to cause alternative splicing of mRNAs (Sorek et al. 2004); 5% of alternative human exons are associated with *Alu* elements, and *Alu* elements can be converted into new splice sites by mutation, resulting in human disease (Kreahling and Graveley 2004). Whether or not these splicing variations are related to genome evolution is not known. Despite these unusual variations in humans, intron number and size are usually conserved in a particular species.

The range of lengths and the number of introns in a particular eukaryote are used for predicting the location of genes in genome sequences. Other features of eukaryotic genes that are useful for gene prediction in a particular organism include the consensus sequences at exon–intron and intron–exon splice junctions, base composition, codon usage, and preference for neighboring codons. Computational methods described in Chapter 9 incorporate this information into gene models that may be used to predict the presence of genes in a genome sequence. Although not always correct, these methods provide a useful annotation of a new genome sequence, and in combination with database similarity searches, provide an indication of the gene functions that are represented in the genome of an organism. This information can be used to predict genetic and biochemical features of the organism.

The numbers of predicted genes estimated from the complete genome sequence of four model eukaryotic organisms are given in Table 11.3. The number of predicted genes in *E. coli* is also given for comparison. Because of the compact gene density in *E. coli* (see Fig. 11.5), there is about 1 gene per kilobase of genome sequence. Yeast is about twofold less compact than *E. coli*. Of the remaining genomes, *C. elegans* and *A. thaliana* have approximately the same density of genes (1 gene per 6 kb), *Drosophila* being the least dense (1 gene per 14 kb). One-sixth of the *Drosophila* sequence is composed of TEs and one-

TABLE 11.3. *Number of genes predicted to encode proteins in model organisms and humans*

Organism	Biological features	Haploid genome size (Mb)	Predicted number of genes
Arabidopsis thaliana	plant with small genome; genes for metabolism, development by hormones and cell–cell interactions and environmental responses	125	25,000
Caenorhabditis elegans	worm (nematode) genes for development by a unique cell lineage, nervous system, and reproduction	100	18,424
Drosophila melanogaster	fruit fly; model for developmental processes by hormones and cell–cell interactions	180	13,601
Escherichia coli	bacterium; genes for growth on external sources of energy, transport of molecules through cell membrane, metabolic pathways, and replication as a single cell	4.7	4,288
Homo sapiens (human)	duplicates many gene functions in other model organisms and, in addition, includes control of higher brain functions	2.9×10^3	30,000–40,000[a]
Saccharomyces cerevisiae	budding yeast; genes for existence as a single-celled organism with the basic structure and organization of the eukaryotic cell	13.5	6,241

Examples of other sequenced model organisms include the mouse (*Mus musculus*), haploid genome size 2.5×10^3 Mb, slightly smaller than the human genome, and rice (*Oryza sativa*), 430 Mb. The mouse genome is a model for the human genome with which it shares a large amount of sequence homology and local gene order. The rice genome is a model for the cereal crops such as wheat (*Triticum aestivum*, genome size 1.6×10^3 Mb). The cultivated grasses all share similar genes, and cultivation has resulted in changes in the same genes (Paterson et al. 1995). Plant genomes in general vary in genome size because of the presence of repetitive elements, including the number of copies of haploid chromosomes. Wheat, for example, has a hexaploid constitution (for review, see Devos and Gale 2000). The largest plant genomes are members of the Liliaceae family (~87×10^3 Mb, see Bennetzen 2000).

[a] The number of proteins that can be made from human genes is potentially much greater because of alternative RNA splicing of a high proportion of mRNAs of the same gene in different cells. However, there is much controversy as to whether or not alternative splicing generates functional mRNAs that can be translated (Sorek et al. 2004).

third is heterochromatic regions that do not include genes. Hence, in the euchromatic regions, the gene density in the *Drosophila* genome is 1 gene per 9 kb. Despite the fact that the lower number of predicted genes in *Drosophila* is smaller than that of the other genomes, the amount of functional diversity, as evidenced by protein family representation, is similar (Adams et al. 2000). The average gene density in the human genome is an average of about 1 gene per 80 kp, but because about one-half of the genome is sequence repeats, this number is closer to 1 gene per 40 kp. Analysis of the human genome revealed variation in gene densities.

Gene expression experiments have revealed that gene distribution is not uniform in the human genome. Genes that are strongly expressed cluster into approximately 30 genome domains, called ridges. Ridges are found to have a high gene density, GC content, and SINE repeat density, and a low LINE repeat density. Genes in ridges have significantly shorter introns than genes outside of ridges. In contrast,

weakly expressed genes cluster into a different set of domains called antiridges, which have the opposite sequence characteristics. These observations reveal higher-order structure of the human genome related to transcriptional regulation (Versteeg et al. 2003). Assessment of genome functional diversity is discussed in the following sections.

Pseudogenes

New gene functions are thought to be gained through duplication of an existing gene creating two tandem copies. Functional differentiation then occurs between the copies by mutation and selection. However, because mutations are often deleterious, and because only one gene copy may be needed for function, there is a strong tendency of one copy to accumulate mutations that render that gene nonfunctional. Accordingly, pseudogenes are DNA sequences that were derived from a functional copy of a gene but that have acquired mutations which result in loss of gene function (Li 1997). In Figure 11.5A, the pseudo-gene *TRY5* is similar to the nearby functional gene *TRY4*.

A second and more common pseudogene found in eukaryotic genomes is called a processed pseudogene. Processed pseudogenes are also derived from a functional gene, but they do not contain introns and lack a promoter; hence, they are not expressed. The origin of these pseudogenes is probably due to reverse transcrip-

tion of the mRNA of the functional gene and insertion of the cDNA copy into a new chromosomal location by a LINE1 (Fig. 11.4) reverse transcriptase (Weiner 2000). The majority of pseudogenes in mammalian genomes are processed pseudogenes and tend to be derived from highly expressed housekeeping genes such as ribosomal protein genes (Zhang et al. 2004). A database of pseudogenes is maintained at http://www.pseudogene.org.

HOW GENOME SEQUENCE IS ASSEMBLED AND GENES ARE IDENTIFIED

As discussed in Chapter 2, sequencing of genomes depends on the assembly of a large number of DNA reads into a linear, contiguous DNA sequence. The cost and efficiency of this process have been greatly improved by automatic methods of sequence assembly, first used for the sequencing of the bacterium *H. influenzae*. This same method of assembly was also used, in part, to complete the sequencing of the *Drosophila* genome (Myers et al. 2000). Sequencing of the human genome was initially guided by a physical map of the sequence fragments as described in Chapter 2. This method was used by the International Human Genome Sequencing Consortium to produce the human genome sequence. Later, an automatic sequence assembly method was used to produce the human genome sequence without using a physical map as a guide. This assembly method was used by Celera Genomics. Later comparisons revealed extremely good agreement between updated drafts of the human genome sequence by these two methods (Istrail et al. 2004). However, controversy remains as to the degree of independence of the two methods as discussed in Chapter 2. Genome assembly methods continue to be an active research area.

Genome sequences can be assembled by reducing the length of the DNA sequence fragments and using a graph-based method (Pevzner et al. 2001). The sequence fragments are nodes on the graph, and edges represent fragment overlaps due to shared sequence. The assembly process involves finding a path through the graph so that each edge is used only once, a type of graph called a Eulerian graph. This method reportedly handles sequence repeats more readily than other sequence assembly methods described above.

Searching the Genome for Protein-encoding Genes

As described in the Chapter 11 flowchart (p. 516), genome sequences are scanned for protein-encoding genes using gene models trained on known gene sequences from the same organism. Methods of gene prediction in eukaryotic genomic DNA are discussed in Chapter 8 (for RNA-encoding genes) and Chapter 9 (for protein-encoding genes). Identification of the function of protein-encoding genes is discussed in the accompanying Chapter 11 flowchart (p. 516) and in Chapter 6. For a new genome, each predicted gene is translated into a protein sequence; the collection of protein sequences encoded by the genome is the proteome of the organism. As illustrated in Figure 11.6, left panel, every protein in the proteome is then used as a query sequence in a database similarity search. Matching database sequences can be realigned with the query sequence to evaluate the extent and significance of the alignment, as described in Chapters 3 and 4.

Screening the predicted protein sequences against sequences from an expressed sequence tag (EST) library assists in predicting genes and confirms that they are expressed (see Adams et al. 2000). The collective information on proteome function can then be further analyzed by self-comparison to find duplicated genes (paralogs) and by a proteome-by-proteome comparison to identify orthologs, genes that have maintained the same function through speciation, and other sequence and evolutionary relationships that are important for metabolic, regulatory, and cellular functions. These proteome comparisons are described in the next section.

A. Types of proteome analysis

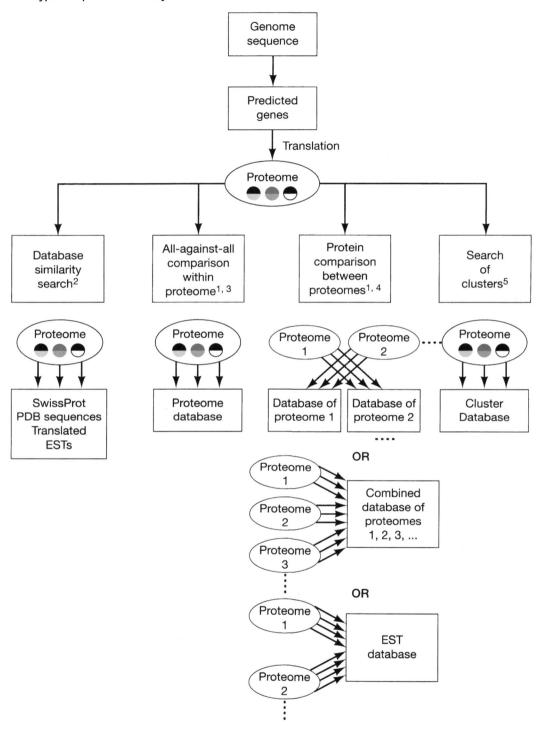

FIGURE 11.6. Analysis of the proteome encoded by genomes. (*A*) Types of proteome analyses. (*B*) Examples of database hits resulting from domain structure of proteins. (*C*) Cluster analysis of similar sequences. (*D*) Domain identification. (*Continued on following pages.*)

B. Examples of database hits resulting from domain structure of proteins[6]

Amino acid alignment	Sequence number	Typical range of P/E value[7]
(i)	1 (query)	—
	2	$<10^{-20}$
	3	$10^{-8} - 10^{-20}$
	4	$10^{-8} - 10^{-20}$
	5	$10^{-6} - 10^{-8}$
(ii)	6 (query)	—
	7	$<10^{-20}$
	8	$10^{-8} - 10^{-20}$
	9	$10^{-8} - 10^{-20}$
	10	$<10^{-20}$
(iii)	3 (query)	—
	1,2	$10^{-8} - 10^{-20}$
	5	$10^{-6} - 10^{-8}$
(iv)	1 (query)	—
	EST hits	$<10^{-4}$

Notes:

1. Because of the large number of comparisons that must be made in these types of analyses (as many as 20,000 by 20,000 sequences) and because of the volume of program output, the procedure must be automated on a local machine using Perl scripts or a similar method and a database system. For BLAST, setting an effective database size appropriate for each search and program is important for obtaining a correct statistical evaluation of alignment scores. The BioPerl project provides valuable resources for this purpose (http://www.bio perl. org). (Also see Chapter 12.)

2. Each protein encoded by the genome is used as a query in database similarity searches to identify similar database proteins, some having a known structure or function. Additional searches of EST databases can be used to identify additional relatives of the query sequence. These searches and evaluation of the alignment scores of matching sequences are described in Chapter 6.

3. An all-against-all analysis requires first making a database of the proteome. This database is then sequentially searched by each individual protein sequence of the proteome using a rapid database similarity search tool such as BLAST, WU-BLAST, or FASTA. The scoring systems of these programs vary and are described in Chapter 6. Note also that P values of WU-BLAST (Chervitz et al. 1998) are similar to E values of NCBI BLAST (Rubin et al. 2000) for values of P and $E < 0.05$. This analysis generates a matrix of alignment scores, each with an E value and corresponding alignment for each pair of proteins. Recall that the E value of an alignment score is the probability that an alignment score as good as the one found would be observed between two random or unrelated sequences in a search of a database of the same size. The lower the E value, the more

significant the alignment between a pair of matching sequences. In an all-against-all comparison within one proteome, significantly matched pairs of sequences may be paralogs that originated from a gene duplication event in this genome or the genome of an ancestor organism. Unique proteins can be identified because they do not match any other protein. A conservative cutoff E value (e.g., 10^{-6}; Rubin et al. 2000) limits the matches to the most significant ones, which are then clustered into families as described below and in the text.

4. To perform a between-proteome analysis, proteome databases are made for the known and predicted genes of two or more genomes. Both single (Chervitz et al. 1998) and combined proteome databases may be made (Rubin et al. 2000). Each protein of one proteome is then selected in turn as a query of the proteome of another organism or the combined proteome of a group of organisms. As in an all-by-all protein comparison within a proteome, a matrix of alignment scores with E values is made, and the most closely related sequences in the two organisms are identified. This analysis can predict orthologs, i.e., proteins that have an identical function attributable to descent of the respective genes from a common ancestor. The types of criteria used in bioinformatics to define orthologs include (1) reciprocal database searches with one sequence as query give a best hit of the other sequence (Tatusov et al. 1997); (2) the alignment of the sequences includes at least 80% of each sequence (Chervitz et al. 1998; Rubin et al. 2000); and (3) the sequences are clustered when all matching sequences are subjected to a cluster analysis. The likelihood of orthology is also increased if a set of orthologous pairs are linked together on the respective genomes. The types of analyses are discussed further in the text.

5. The cluster search option is most useful for prokaryotic organisms. Each protein in the proteome is used as a query of a database of protein clusters using the program COGNITOR (NCBI Web site). These clusters are composed of orthologous pairs of sequence defined by criterion 1, described in note 4. The database was made by performing an all-by-all genome comparison across a spectrum of prokaryotic organisms and a portion of the yeast proteome (Tatusov et al. 2000). Orthologous pairs of sequence were then merged with clusters of orthologous pairs (COGs) for multiple proteomes as described in the text. COGs have been linked to classes of biochemical function (Tatusov et al. 1997). Hence, matching a query sequence to the COG can potentially identify unique orthologs in another proteome that may have the same function. The COGs database is designed to provide a preliminary indication of orthologous relationships that can be tested by more detailed similarity searches, sequence alignments, and phylogenetic analysis of the matching sequences.

6. Because of the modular nature of proteins, several types of matches may be identified in the all-against-all and between-proteome comparisons. Each colored box represents a hypothetical conserved domain that is matched in the search. The dotted box (sequence 5) represents a less similar domain that will not align as well. Highest-scoring matches corresponding to matching of multiple domains present in the query and in the matched sequence (i and ii; sequence pairs 1 and 2, 6 and 7, etc.). The alignment scores of these pairs should have extremely low E values. A multidomain query protein will also match database proteins that have a single domain (as in sequences 1 and 3, 6 and 8). Because only one domain is represented by the alignment, the alignment will in general be shorter and have a poorer (higher) E value score than a multidomain alignment. The analysis will also identify matches of a query with a database protein that has two or more copies of query sequence domain (sequence 10). Query sequences with a minimal domain representation (ii) will not score particularly well with any sequence (sequence 3). Duplicate comparisons generated by the method are eliminated. When only an EST library of an organism is available, the proteome may be compared to this library. However, since these databases are generally not complete and any alignments are shorter, it is diffcult to compare these results with the full proteome comparisons. From a biological standpoint, ESTs define expressed genes, whereas proteomes are predicted genes.

7. WU-BLAST produces P scores and BLAST (NCBI) E scores where $E = -\ln(1-P)$. For values less than 0.05, $E = P$. The score ranges depicted in this column are hypothetical examples. The choice of a $<10^{-20}$ score is a conservative one for identification of orthologs that should have a similar domain structure, as do the sequences in this example (see Chervitz et al. 1998; Rubin et al. 2000). To define these groups, the distribution of hits below different thresholds should be examined, as in the above references. The higher cutoff score for EST matches is used because the search of an EST database may produce only short alignments.

8. Shown are two representations of the sequence relationships found in part B. In (i) the sequences, color coded to represent domain structure, are represented by vertices on graph. In comparing the graphic (i) and single linkage (ii) clusters, note that in (i) each sequence has multiple edges representing links to related sequences, whereas in (ii) the sequences are only connected to one branch on the outermost part of the tree.

9. The sequence alignments found above represent the presence of one or more conserved domains in each cluster or group of clusters. These clusters are next analyzed for the presence of known domains by searches of domain databases as described in Chapter 10. This analysis identifies the number and types of domains that are shared between organisms, or that have been duplicated in proteomes to produce paralogs.

C. Cluster analysis of similar sequences[8]

(i) Graphic representation

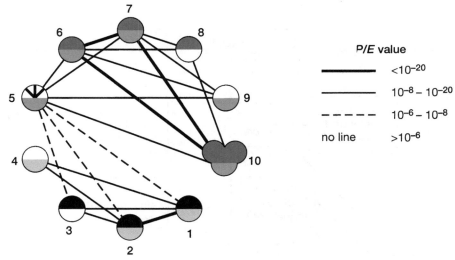

(ii) Single linkage cluster analysis

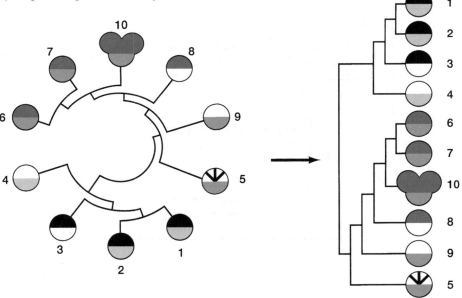

D. Domain identification[9]

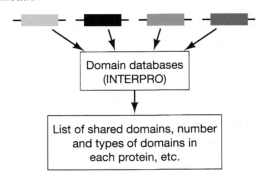

FIGURE 11.6. (*Continued.*)

HOW IS GENOME ANALYSIS PERFORMED?

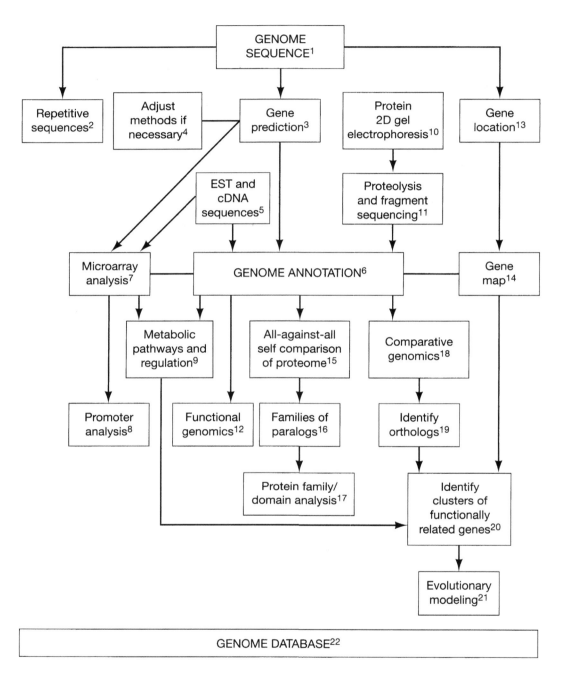

1. Genome sequences are assembled from DNA sequence fragments of approximate length 500 bp obtained using DNA sequencing machines as described in Chapter 2. Chromosomes of a target organism are purified, fragmented, and subcloned in fragments of size hundreds of 100–200 kbp in bacterial artificial chromosomes (BACs). The BAC fragments are further subcloned as smaller frag-

ments into plasmid vectors for DNA sequencing. BACs are also digested with restriction endonucleases that cut DNA at specific sequences. Those BACs that produce the same size fragments overlap in sequence and can be assembled into a sequence contig. The ends of sequences may also be sequenced to find overlapping BACs in automatic sequence assembly methods (Myers et al. 2000).

BAC contigs are often tested for the presence of particular genes to anchor them on the genetic map of an organism. Full chromosomal sequences are then assembled from the fragment order on a physical map or from the overlaps in a highly redundant set of fragments by an automatic sequence assembly method.

2. Eukaryotic genomes are partially made up of classes of repeated elements, including tandem repeats present in centromeres and telomeres, dispersed tandem repeats (satellites, minisatellites, and microsatellites), and interdispersed TEs. TEs can comprise one-half or more of the genome sequence. Analysis of sequence repeats is discussed in Chapters 3 and 6. Identification of classes of repeated elements is aided by searchable databases also discussed in Chapter 6 (p. 254).

3. Gene identification in prokaryotic organisms is simplified by their lack of introns. Once the sequence patterns characteristic of the genes in a particular prokaryotic organism (e.g,, codon usage, codon neighbor preference) have been found, gene locations in the genome sequence can be predicted quite accurately. The presence of introns in eukaryotic genomes makes gene prediction more involved because, in addition to the above features, locations of intron–exon and exon–intron splice junctions must also be predicted. Methods of gene prediction in prokaryotes and eukaryotes are discussed in Chapter 9.

4. Gene prediction methods involve training a gene model (e.g., a hidden Markov model or neural network, see Chapter 9) to recognize genes in a particular organism. Because of variations in gene codon preferences and splice junctions (see note 3 above; Fig. 11.5), a model must usually be trained for each new genome.

5. Because gene prediction methods are only partially accurate (for review, see Bork 1999; see Chapter 9), gene identification is facilitated by high-throughput sequencing of partial cDNA copies of expressed genes (called expressed sequence tags or EST sequences). Presence of ESTs confirms that a predicted gene is transcribed. Because ESTs are first-pass sequences with an error rate of about 1%, and may include sequences from pseudogenes and sequence repeats, a more thorough sequencing of full-length cDNA clones is necessary to confirm the structure of genes chosen for a more detailed analysis. The possibility of alternative splicing of mRNA in different cell types should also be considered.

6. The amino acid sequence of proteins encoded by the predicted genes is used as a query of the protein sequence databases in a database similarity search. A high-quality match of a predicted protein sequence to one or more database sequences not only serves to identify the gene function, but also validates the gene prediction. Pseudogenes (gene copies that have lost function) may also be found in this analysis. Only matches with highly significant alignment scores and alignments (see Chapter 3, p. 75) should be included. The genome sequence is annotated with the information on gene content and predicted structure, gene location, and functional predictions. The predicted set of proteins for the genome is referred to as the proteome. Accurate annotation is extremely important so that other users of the information are not misinformed. Procedures for searches starting with genome, EST, and cDNA sequence are described in Chapter 9. Usually, not all query proteins will match a database sequence. Hence, it is important to extend the analysis by searching the predicted protein sequence for characteristic domains (conserved amino acid patterns that can be aligned) that serve as a signature of a protein family or of a biochemical or structural feature (see note 17). A further extension is to identify members of protein families or domains that represent a structural fold using the computational tools described in Chapter 10. This additional information also needs to be accurately described and the significance established.

7. Gene expression analysis using microarrays provides a global picture of gene expression for the genome by revealing which genes are expressed at a particular stage of the cell cycle or developmental cycle of an organism, or genes that respond to a given environmental signal to the same extent. This type of information provides an indication as to which genes share a related biological function or may act in the same biochemical pathway and may thereby give clues that will assist in gene identification. Analysis of microarrays is discussed in Chapter 13.

8. Genes that are found to be coregulated either by a gene expression analysis or by a protein two-dimensional analysis should share sequence patterns in the promoter region that direct the activity of transcription factors. The types of analyses that are performed are discussed in Chapter 9 (p. 400).

9. As genes are identified in a genome sequence, some will be found that are known to act sequentially in a metabolic pathway or to have a known role in gene regulation in other organisms. From this information, the metabolic pathways and metabolic activities of the organism will become apparent. In some cases, the apparent absence of a gene in a well-represented pathway may lead to a more detailed search for the gene. Clustering of genes in the pathway on the genome of a related organism can provide a further hint as to where the gene may be located (see note 20).

10. Individual proteins produced by the genome can be separated to a large extent by this method and specific ones identified by various biochemical and immunological tests. Moreover, changes in levels of proteins in response to an environment signal can be monitored in much the same way as a microarray analysis is performed. Microarrays only detect untranslated mRNAs, whereas a two-dimensional gel protein analysis detects translation products, thus revealing an additional level of regulation.

Analysis of regulation by this method is discussed in Chapter 13.

11. Protein spots may be excised from a two-dimensional protein gel (see note 10) and subjected to a combination of amino acid sequencing and cleavage analyses using the techniques of mass spectrometry and high-pressure liquid chromatography. Genome regions that encode these sequences can then be identified and the corresponding gene located. A similar method may be used to identify the gene that encodes a particular protein that has been purified and characterized in the laboratory.

12. Functional genomics involves the preparation of mutant or transgenic organisms with a mutant form of a particular gene usually designed to prevent expression of the gene. The gene function is revealed by any abnormal properties of the mutant organism. This method provides a way to test a gene function that is predicted by sequence similarity to be the same as that of a gene of known function in another organism. If the other organism is very different biologically (comparing a predicted plant or animal gene to a known yeast gene), then functional genomics can also shed light on any newly acquired biological role. When two or more members of a gene family are found (see notes 16 and 17), rather than a single match to a known gene, the biological activity of these members may be analyzed by functional genomics to look for diversification of function in the family.

13. Since the entire genome sequence is available, as each gene is identified, the relative position of the gene will be known.

14. A map showing the location in the genome of each identified gene is made. These relative positions of genes can be compared to similar maps of other organisms to identify rearrangements that have occurred in the genome throughout evolution. Gene order in two related organisms reflects the order that was present in a common ancestor genome. Chromosomal breaks followed by a reassembly of fragments in a different order can produce new gene maps. These types of evolutionary changes in genomes have been modeled by computational methods that are described below. Gene order is revealed not only by the physical order of genes on the chromosome, but also by genetic analysis. Populations of an organism show sequence variations that are readily detected by DNA sequencing and other analysis methods. The inheritance of genetic diseases in humans and animals, e.g., some kinds of cancer and heart disease, and of desirable traits in plants, can be traced genetically by pedigree analysis or genetic crosses. Sequence variations (polymorphisms such as SNPs) that are close (tightly linked) to a trait may be used to trace the trait by virtue of the fact that the polymorphism and the trait are seldom separated from one generation to the next. These linked polymorphisms may then be used for mapping and identifying important genes. A more detailed analysis of the relative amount of sequence variability in a chromosomal region within populations of closely related species can reveal the presence of genes that are under selection. These regions will not have the expected amount of variability given their linkage: They are in a state of linkage disequilibrium. An example is the *BRCA1* (breast cancer 1) gene of humans and chimpanzees (Huttley et al. 2000).

15. A comparison is made in which every protein is used as a query in a similarity search against a database composed of the rest of the proteome, and the significant matches are identified by a low expect value ($E < 10^{-6}$ was used in a recent analysis by Rubin et al. [2000]). Because many proteins comprise different combinations of a common set of domains, proteins that align along most of their lengths (80% identity is a conservative choice) are chosen to select those that have a conserved domain structure.

16. A set of related proteins identified in step 15 is subjected to a cluster analysis to identify the most closely related groups of proteins. Proteins that align along most of their lengths rather than along short lengths that represent conserved domains are found by this method. A related group of proteins is derived from a gene family of paralogs that have arisen by gene duplication.

17. Each protein in the predicted proteome is again used as a query of a curated protein sequence database such as SwissProt in order to locate similar domains and sequences. The domain composition of each protein is also determined by searching for matches in domain databases such as Interpro, described in Table 10.5. The analysis reveals how many domains and domain combinations are present in the proteome, and reveals any unusual representation that might have biological significance. The number of expressed genes in each family can also be compared to the number in other organisms to determine whether there has been an expansion of the family in the genome.

18. Comparative genomics is a comparison of two or more genomes that includes a comparison of all of the predicted proteins, the relative locations of related genes in separate genomes, and any local groupings of genes that may be of functional or regulatory significance. Comparative genomics also includes a comparison of gene order, gene (intron) structure, presence and location of types of sequence repeats, sequence polymorphisms in populations, and many other features that distinguish genomes.

19. Orthologs are genes that are so highly conserved by sequence in different genomes that the proteins they encode are strongly predicted to have the same structure and function and to have arisen from a common ancestor through speciation. To identify orthologs, each protein in the proteome of an organism is used as a query in a similarity search of a database comprising the proteomes of

one or more different organisms. The best hit in each proteome is likely to be with an ortholog of the query gene. In comparing two proteomes, a common standard is to require that for each pair of orthologs, the first of the pair is the best hit when the second is used to query the proteome of the first. To find orthologs, very low E value scores ($E < 10^{-20}$) for the alignment score and an alignment that includes 60–80% of the query sequence are generally required in order to avoid matches to paralogs. Although these requirements for classification of orthologs are very stringent, a more relaxed set of conditions may lead to many more false-positive predictions. In bacteria, the possibility of horizontal transfer of genes between species also has to be considered (p. 527).

20. In related organisms, both gene content of the genome and gene order on the chromosome are likely to be conserved. As the relationship between the organisms decreases, local groups of genes remain clustered together, but chromosomal rearrangements move the clusters to other locations. In microbial genomes, genes specifying a metabolic pathway may be contiguous on the genome where they are co-regulated transcriptionally in an operon by a common promoter. In other organisms, genes that have a related function can also be clustered. Hence, the function of a particular gene can sometimes be predicted, given the known function of a neighboring, closely linked gene. Genomes are also compared at the level of gene content, predicted metabolic functions, regulation as revealed by gene expression microarray analysis, and others. These comparisons provide a basis for additional predictions as to which genes are functionally related. Gene fusion events that combine domains found in two proteins in one organism into a composite protein with both domains in a second organism are also found and provide evidence that the proteins physically interact or have a related function.

21. Evolutionary modeling can include a number of types of analyses including (1) the prediction of chromosomal rearrangements that preceded the present arrangement (e.g., a comparison of mouse and human chromosomes); (2) analysis of duplications at the protein domain, gene, chromosomal, and full genome level; and (3) search for horizontal transfer events between separate organisms.

22. Because of the magnitude of the task, the earlier stages of genome analysis including gene prediction and database similarity searches are performed automatically with little human intervention. Programming methods for automating genome analysis are described in Chapter 12. The genome sequence is then annotated with any information found without involving human judgment. The types of genome analyses in the flowchart also provide many predictions and give rise to many preliminary hypotheses regarding gene function and regulation. As more detailed information is collected by laboratory experiment and by a closer examination of the sequence data, this information needs to be linked to the genome sequence. In addition, the literature, past and present, needs to be scanned for information relevant to the genome. A carefully crafted database that takes into account the entire body of information should then be established. In addition to information on the specific genome of interest, the database should include cross-references to other genomes. To facilitate such intergenome comparisons, common gene vocabularies have been proposed. This slow, expensive, and time-consuming phase of genome analysis is of prime importance if the genome information is to be available in an accurate form for public use.

GENOMES CAN BE COMPARED FOR ORTHOLOGS, PARALOGS, AND PROTEOMES

Comparative genomics includes a comparison of all of the predicted proteins, i.e., the proteomes of two or more organisms, gene locations, and the number and location of sequence repeats, as possible influences on genome evolution. Another area of genome comparison is the degree of diversity within a species population (e.g., SNPs, microsatellite variations, gene expression levels) and the association of these sequence variations with environmental responses and disease. The availability of complete genome sequences makes possible a comparison of the proteome of one organism with that of another. Because the genome sequence provides both the sequence and the map location of each gene, both the sequence and the location of each predicted protein can be compared. As described in Chapter 10, proteins are frequently found in families that have similar biological function and three-dimensional structure. Genome analysis provides additional information on the evolutionary relationships among proteins by exploring orthologs, paralogs, and protein domain changes.

Sequence comparisons between genomes provide information on gene relationships. Genes are called orthologs when the number of genes in two organisms are so similar that they must have the same function and evolutionary history (Fitch 1970). Map locations of orthologs in two or more genomes may also be compared. If a set of genes is grouped together at a particular chromosomal location in one organism, and if a set of similar genes is also grouped together in the chromosome of another organism, these groups share an evolutionary history. Some of the group may come from a common ancestor gene whose function has been conserved over periods of evolutionary time. Others may have arisen by rare gene duplication events.

Gene families originating from rare gene duplica-

tion events over evolutionary time, called paralogs, are also found within a genome. Gene duplication events give rise to this type of gene relationship, which is found by genome comparisons. Proteins with new functions may be produced by such gene duplication events. Two tandem copies of a gene are produced (see Fig. 3.3). Through mutation and natural selection, one of the copies can develop a new function, leaving the second copy to cover for the original function. However, because most mutations are deleterious to function, often one of the copies becomes a pseudogene. Not all gene duplications are thought to have the above effects. Another scenario is that two duplicated genes both undergo change, but interactions between the proteins stabilize the original function and support the evolution of new ones (Force et al. 1999). Long chromosomal regions or even entire genomes can also be involved in genome duplications that create additional copies of genes in the duplicated regions, thus providing a similar opportunity for functional divergence of genes. As an example, extensive duplication is found in the *Arabidopsis* genome, shown in Figure 11.7. Examples

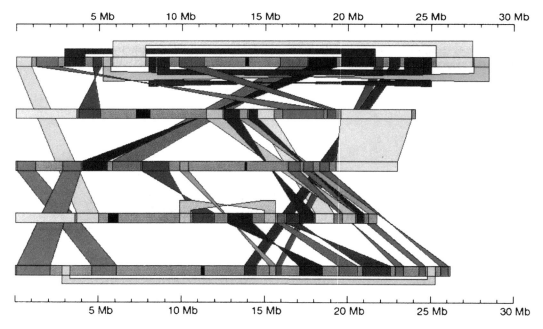

FIGURE 11.7. Presence of duplicated segments in the genome of the plant *Arabidopsis*. Multiple regions of sequence similarity are found in the same or opposite orientation by performing extensive comparisons of sections of genome sequence using the methods described in Chapter 3. These regions indicate the occurrence of genome duplications in the evolutionary history of the *Arabidopsis* genome. (Reprinted, with permission, from *Arabidopsis* Genome Initiative 2000 [©2000 Macmillan Magazines, Ltd.].) (Image by Dirk Haase.) For a similar map of duplications in the human genome, see Bailey et al. (2002).

of duplication of parts of chromosomes to other locations in the same chromosome or a different chromosome are apparent, sometimes in the same orientation and sometimes in the reverse orientation. Partial gene sequences can also be duplicated or joined with sequences of other genes, resulting in proteins with an altered sequence domain structure.

Comparisons of proteomes of different organisms can identify the types of protein domain changes over evolutionary time and provide an indication as to what biological role the domain changes may have in a particular organism. As discussed in Chapters 5 and 10, proteins are modular and comprise separate structural and functional domains. The number of protein sequences that are available is sufficient to determine that domain shuffling occurs in evolution—domains appear or disappear in particular families, become combined to make new families, or else become separated into two different proteins that are predicted to interact (Snel et al. 2000). Domain variations of these kinds are illustrated in Figure 11.6 with examples shown below in Figure 11.12. The assortment and reassortment of protein domains and domain swapping between proteins takes place in individual genomes and any beneficial results are passed along to newly derived species over evolutionary time.

The processes of domain assortment and gene duplication produce families of proteins in organisms. During speciation, a newly derived genome will inherit the families of ancestor organisms, but will also develop new ones to meet new evolutionary challenges. Comparison of each of the proteins encoded by an organism with every protein, an all-against-all comparison, reveals which protein families have been amplified and what rearrangements have occurred as steps in the evolutionary process. When two or more proteins in the proteome share a high degree of similarity because they share the same set of domains (illustrated in Fig. 11.4B), they are likely to be paralogs (Fitch 1970), genes that arose by gene duplication events. Proteins that align over shorter regions share some domains, but also may not share others. Although gene duplication events could have created such variation, other rearrangements may have also occurred, blurring the evolutionary history.

The following sections describe methods to compare prokaryotic and eukaryotic genomes for orthologs and paralogs. It is important to keep in mind the predictive nature of these types of analyses.

Decisions about gene relationships depend on careful manual inspection of sequence alignments (Huynen et al. 2000).

Proteome Analysis

All-against-All Self-comparison Reveals Numbers of Gene Families and Duplicated Genes

A comparison of each protein in the proteome with all other proteins distinguishes unique proteins from proteins that are members of families made up of paralogs resulting from gene duplication events. The analysis also reveals the number of protein families. The domain content of these proteins may also be analyzed. One type of all-against-all proteome comparison is described in Figure 11.6A (second panel). In this analysis, each protein is used as a query in a similarity search against the remaining proteome, and the similar sequences are ranked by the quality and length of the alignments found. The search is conducted in the manner described in Chapter 6, with each alignment score receiving a statistical evaluation (E value). As shown in Figure 11.6B, a match between a query sequence and another proteome sequence with the same domain structure will produce a high-scoring, highly significant alignment. These proteins are designated paralogs because they have almost certainly originated from a gene duplication event. Lower-scoring, less-significant alignments may have identified proteins that share domains but not the high degree of sequence similarity that is apparent in the best-scoring alignments. These may also be paralogs, but they may have a complicated history of domain shuffling and sequence change that is difficult to reconstruct.

Cluster analysis. In order to sort out relationships among all proteins that are found to be related in a series of searches of the types shown in Figure 11.6B, the proteins are subjected to a clustering analysis shown in Figure 11.6C. Only the relationships revealed by the hypothetical set of searches illustrated in Figure 11.6B are shown. Some of the proteins may have other relationships, which are not depicted in order to simplify the example.

Clustering organizes the proteins into groups by some objective criterion. One criterion for a matching protein pair is the statistical significance of their alignment score (the E value from BLAST searches). The lower this value, the better the alignment. There will be a cutoff E value at which the matches in the BLAST search are no longer considered significant. A value of $E > 0.01$–0.05 is usually the point at which the alignment score is no longer considered to be significant in order to focus on a more closely related group of proteins. A second criterion for clustering proteins is the distance between each pair of sequences in a multiple sequence alignment. The distance is the number of amino acid changes between each pair of aligned sequences. Deciding the criteria to be used for clustering proteins into the same or different clusters can be a difficult problem that is discussed in the box "Criteria for Clustering Protein Sequences." Two clustering methods are described below.

Clustering by making subgraphs. Figure 11.6C indicates two ways of clustering related sequences based on the above criteria. Figure 11.6C, part i, illustrates matching a protein pair according to the criterion of the statistical significance of their alignment score. In the graph shown, each sequence is a vertex and each pair of sequences that is matched with a significant alignment score is joined by an edge that is weighted according to the statistical significance of the alignment score. One way to identify the most strongly supported clusters is simply to remove the most weakly supported edges in the graph, in this case the alignments with the highest E values (dotted edges). As weaker and weaker links are removed, the remaining combinations of vertices and edges represent the most strongly linked sequences. This type of analysis was performed on an initial collection of *E. coli* genes by Labedan and Riley (1995). Their analyses revealed that *E. coli* genes clustered in this manner encode proteins already known to belong to the same broad functional category, EC number, or to have a similar physiological function. For another approach to identify orthologs in microbial genes, see Bansal (1999).

Another method for clustering similar sequences that are likely to be paralogs is described in Rubin et al. (2000). In this method, edges of E value $> 1 \times 10^{-6}$ are removed. The remaining graph is then broken down into subgraphs comprising sequences that share a significant relationship to each other but not to other sequences. The criterion chosen is that the group should mutually share at least two-thirds of all of the edges from this group to all proteins in the proteome. If two proteins A and B share a domain but do not share another domain in A, and if A shares this other

domain with a number of other sequences, the algorithm would tend not to cluster A with B (Rubin et al. 2000). Thus, the algorithm favors the selection of proteins with the same domain structure reflecting that these proteins are the most likely ones to be paralogs.

Clustering by single linkage. The method for clustering related sequences using the distance criterion for sequence relationships is shown in Figure 10.6C, part (ii). First, a group of related sequences found in the all-against-all proteome comparison is subjected to a multiple sequence alignment as described in Chapter 5. A distance matrix that shows the number of amino acid changes between each pair of sequences is then made. This matrix is then used to cluster the sequences by a neighbor-joining algorithm. This procedure and the algorithms are the same as those used to make a phylogenetic tree by the distance methods, described in Chapter 7. These methods produce a tree (Fig. 11.6C, part ii, left) or a different representation of the tree called a dendrogram (Fig. 11.6C, part ii, right) that minimizes the number of amino acid changes that would generate the group of sequences. The tree is also defined as a minimum spanning tree (Duran and Odell 1974). The tree and dendrogram cluster the sequences into the most closely related groups. Branches joining the least related sequences may be removed, thus leaving two subtrees with a small group of sequences. As smaller groups are chosen, the most strongly supported clusters are likely to be made up of paralogs. However, it is not easy to distinguish sequences that are paralogs, i.e., share several domains, from those that share domains but that also share other domains with more distantly related sequences without inspection of the alignments. GeneRage provides an automatic system for classifying protein data sets by means of an iterative refinement approach using local alignments, matrix methods, and single-linkage clustering. Tekaia et al. (1999) have used single-linkage clustering to determine the proportion of proteins in one organism that is shared with another organism to produce a genome tree. Methods of clustering data sets are further elaborated in Chapter 13.

Core proteome. The above types of all-against-all analyses provide an indication regarding the number of protein/gene families in an organism. This number represents the core proteome of the organism from which all biological functions have diversified. A representative sample is shown in Table 11.4. In *Hemophilus*, 1247 of the total number of 1709 pro-

TABLE 11.4. *Numbers of gene families and duplicated genes in model organisms*

Organism	Total number of genes	Number of gene families[a]	Number of duplicated genes[b]
Hemophilus influenzae (bacteria)	1,709	1,425[c]	284
Saccharomyces cerevisiae (yeast)	6,241	4,383	1,858
Caenorhabditis elegans (worm)	18,424	9,453	8,971
Drosophila melanogaster (fly)	13,600	8,065	5,536

Source: Rubin et al. (2000).

[a] The number of clustered groups in the all-against-all analysis using the algorithm described in the text. This number represents the core proteome of the organism.

[b] Count of number of duplicated genes within the protein family clusters.

[c] 178 families have paralogs.

teins do not have paralogs (Rubin et al. 2000). The core proteomes of the worm and fly are similar in size but with a greater number of duplicated genes in the worm. It is quite remarkable that the core proteome of the multicellular organisms (worm and fly) is only twice that of yeast.

Between-Proteome Comparisons to Identify Orthologs, Gene Families, and Domains

Comparisons between proteomes of different organisms are illustrated by the third panel in Figure 11.6A. In this analysis, each protein in the proteome is used as a query in a database similarity search against another proteome or combined set of proteomes. When the proteome of an organism is not available, an EST database may be searched for matches, but the type of search is less informative than a full-genome comparison (see below). As in the all-against-all search for paralogs, the search should identify highly conserved proteins of similar domain structure and other similar proteins that show variation in the domain structure, as illustrated in Figure 11.6B. A pair of proteins in two organisms that align along most of their lengths with a highly significant alignment score are likely to be orthologs, proteins that share a common ancestry and that have kept the same function following speciation. These proteins perform the core biological functions shared by all organisms, including DNA replication, transcription, translation, and intermediary metabolism. They do not include the proteins unique to the biology of a particular organism.

Other matching sequences in this class could be orthologs, but could also represent a match between a sequence in proteome A to a paralog of a true ortholog of the sequence in proteome B. In one method designed to identify true orthologs, the most

CRITERIA FOR CLUSTERING PROTEIN SEQUENCES

The problem of deciding which sequences to include in the same group or cluster and which to separate into different groups or clusters is a recurring one. The conservative approach is to group only very similar sequences together. However, in making a conservative multiple sequence alignment with only very-alike sequences, it is not possible to analyze the evolutionary divergence that may have occurred in a family of proteins. Furthermore, if a matrix or profile model is made from this alignment, that model will not be useful for identifying more divergent members of a family. The adventurous approach is to choose a set of marginally alignable sequences to pursue the difficult task of making a multiple sequence alignment and then to make profile models that may recognize divergence but will also give false predictions. The best method to choose is somewhere between the conservative and adventurous methods. This problem was also addressed in Chapter 5 (p. 192), where the ability of a scoring matrix or profile to distinguish known protein family members from nonmembers is analyzed. Divergence is necessary, but the sequences chosen should be clearly related based on inspection of each pair-wise alignment and a statistical analysis. Clustering analyses of the sequences can also be useful. Questionable sequences can be left out of the analysis at one stage and added in a second to determine what effect they have on the model.

closely related pairs of sequences in proteomes A and B are identified. Two proteins, X in proteome A and Y in proteome B, are predicted to be an orthologous pair if reciprocal searches of proteome A with Y and proteome B with X each produce the highest-scoring match with the other protein. Furthermore, the *E* value for each alignment should be <0.01 and the alignment should extend over 60% of each protein (Huynen and Bork 1998).

In another method to identify the mostly closely related sequences in different proteomes, Chervitz et al. (1998) kept only matched sequences with a very conservative statistical value (*E* value or equivalent statistical score) for the alignment score. The steps for identifying a group of related sequences between the yeast and worm proteomes were as follows:

1. Choose a yeast protein and perform a database similarity search of the worm proteome, a yeast-versus-worm search.

2. Make a list of the worm sequences that give a high-scoring aligning with a low statistical value (10^{-10} to 10^{-100}) and include the yeast query sequence in the list.

3. From the list in step 2, choose a worm sequence and make a search of the yeast proteome, using the same criteria as in step 2.

4. Add any matching yeast sequence to the list made in step 2.

5. Repeat steps 3 and 4 for all initially matched worm sequences.

6. Repeat steps 1–5 for every yeast protein.

7. Perform a comparable worm-versus-yeast analysis as outlined in steps 1–6.

8. Coalesce the groups of related sequences and

remove any redundancies so that every sequence is represented only once.

9. Eliminate any matched pairs in which less than 80% of each sequence is in the alignment.

The above steps locate groups of highly related sequences in two proteomes based on high-scoring alignments among the group. These groups are then subjected to the single-linkage cluster analysis described above and illustrated in Figure 11.6C. The analysis creates a multiple sequence alignment and a tree/dendrogram representation of sequence relationships very similar to that produced in a phylogenetic analysis. Orthologs appear as nearest neighbors on the tips of this tree.

The results of the above analysis with the yeast and worm proteomes are shown in Table 11.5. The numbers of sequence groups decrease about fivefold as the stringency of the statistical value of the alignment score decreases from 10^{-10} to 10^{-100}, and a similar effect is observed for the subcategories shown in the table. Given that these sequences also align to the extent of 80%, they represent highly conserved sets of genes.

Family and domain analysis. Extensive protein domain analyses have been performed for both prokaryotic and eukaryotic genomes (Chervitz et al. 1998; Huynen and Bork 1998; Rubin et al. 2000). A descriptive list of protein domain databases that may be used for such an analysis is given in Table 10.5. In a detailed analysis of the fly, worm, and yeast proteomes, 744 families and domains were common to all three organisms. More than 2000 fly and worm proteins are multidomain proteins, compared to about one-third this number in yeast (Rubin et al. 2000).

Clusters of orthologous groups. As described above, a pair of orthologous genes in two organisms shares so

TABLE 11.5. *Numbers of closely related yeast and worm sequences*

Cutoff *P* (or *E*) value	< 10^{-10}	< 10^{-20}	< 10^{-50}	< 10^{-100}
Total number of sequence groups	1171	984	552	236
Number of groups with more than two members	560	442	230	79
Number and percent of all yeast proteins (6217) represented in groups	2697 (40)	1848 (30)	888 (14)	330 (5)
Number and percent of all worm proteins represented in groups	3653 (19)	2497 (13)	1094 (6)	370 (2)

Adapted, with permission, from Chervitz et al. 1998 (©1998 AAAS).

much sequence similarity that the genes may be assumed to have arisen from a common ancestor gene. When entire proteomes of the two organisms are available, orthologs may be identified as the most-alike sequences in reciprocal proteome similarity searches, as described in the above section. Using the protein from one of the organisms to search the proteome of the other for high-scoring matches should identify the ortholog as the highest-scoring match, or best hit. However, in many cases, each of the orthologs belongs to a family composed of paralogous sequences related to each other by gene duplication events. Hence, in the above database search, the ortholog will match not only the orthologous sequence in the second proteome, but also these other paralogous sequences. The objective of the clusters of orthologous groups (COGs) approach is to identify all matching proteins in the organisms, defined as an orthologous group related by both speciation and gene duplication events. Related orthologous groups in different organisms are clustered together to form a COG that includes both orthologs and paralogs. These clusters correspond to classes of metabolic functions. A database produced by analysis of the available microbial genomes and part of the yeast genome has been made, and a newly identified microbial protein may be used as a query to search this database. Any significant matches found will provide an indication as to the metabolic function of the query protein (Tatusov et al. 1997).

To produce COGs, similarity searches were performed among the proteomes of phylogenetically distinct clades of prokaryotes. Orthologous pairs were first defined by the best hits in reciprocal searches. A cluster of three orthologs in three different species was then represented as a triangle on a diagram. Some triangles included a common side, representing the presence of the same orthologous pair in a comparison of four or more organisms. Triangles with this feature were merged into a cluster similar in appearance to Figure 11.6C, part i. Paralogs defined by sets of three matching sequences in the selected organisms were also added to these clusters. The proteins encoded by many prokaryotic organisms have been analyzed for COG relationships (Koonin et al. 1998). A COGs analysis provides an initial assessment of the genome composition of prokaryotic organisms and should be followed by a more detailed analysis as described above for the worm and yeast genomes.

Comparison of proteomes to EST databases of an

organism. For some eukaryotic organisms, the complete genome sequence is not available. What is available is a large collection of EST sequences obtained by random sequencing of cDNA copies of cell mRNA sequences. These sequences are single DNA sequence reads that contain a small fraction of incorrect base assessments, insertions, and deletions. Many sequences arise from near the 3′ end of the mRNA, although every effort is usually made to read as far 5′ as possible into the upstream portion of the cDNA. Because not all of the genes may be expressed in the cells or tissues chosen for analysis, the library will often not be complete. EST libraries are useful for preliminary identification of genes by database similarity searches as described in Chapter 6. A more detailed analysis may then be made by cloning and sequencing the intact full-length cDNA.

An EST database of an organism can be analyzed for the presence of gene families, orthologs, and paralogs. A protein from the yeast or fly proteome, for example, can be used as a query of a human EST database by translating each EST sequence in all six possible reading frames. The program TBLASTN is frequently used for this purpose. The TFASTX and TFASTY programs are designed to accommodate the errors inherent in EST sequences (p. 41). The limitations to whole-proteome searches against EST libraries are that the short length of the translated EST sequence (the equivalent of 100–150 amino acids) will only match a portion of the query protein; for example, a domain or part of domain as illustrated in Figure 11.6B. Hence, it is not possible to impose the requirement of alignment with 60–80% of the query sequence, which greatly improves the prediction of orthologs. Predictions of EST relationships can be improved by identifying overlapping EST sequences so that a longer alignment can be produced, as discussed in Chapter 6. Another method is to perform an exhaustive search for a protein family, described next.

Searching for orthologs to a protein family in an EST database. Searches of EST databases for matches to a query sequence routinely produce large amounts of output that must be searched manually for significant hits. Retief et al. (1999) have described an automatic method utilizing a computer script, FAST-PAN, that scans EST databases with multiple queries from a protein family, sorts the alignment scores, and produces charts and alignments of the matches

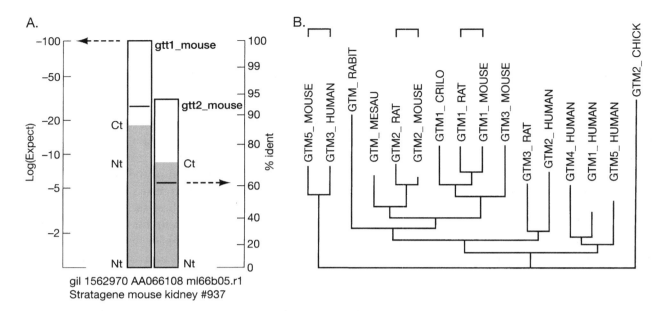

FIGURE 11.8. Prediction of paralogs and orthologs by searches of EST databases by gene panning (Retief et al. 1999). In this analysis, one class of glutathione transferase family members was used as queries to search mammalian EST databases for highly significant matches using TFASTY3 (Chapter 6). FAST_PAN is a Perl-script program (use as Web search term) that automatically searches EST databases as they are updated and compiles the results of the search. (A) Display of protein class matched (shaded), log Expect value (height of bar), length of query sequence matched (height of shaded bar), and percent identity (position of horizontal line in bar) on one graph as produced by FAST_PAN. Note that the log scales clearly reveal the lowest E value and highest identity matches. Shown are matches of two mouse ESTs to a query sequence. (B) Example of phylogenetic analysis to predict orthologs between species (bracketed). Amino acid sequences of ESTs in the matched regions were aligned, and this alignment was then used to direct an alignment of the EST codons. A phylogenetic tree was produced by the aligned EST sequences by the maximum likelihood method using the program DNAML in the PHYLIP package. As discussed by the authors, this method allows researchers to search rapidly and easily through EST databases to identify matching sequences and to examine the quality of the alignments found. In this example, a large number of glutathione transferase members were used as queries, allowing an exhaustive search of the EST database for representative family members. (Redrawn, with permission, from Retief et al. 1999.)

found. An example of using this method is shown in Figure 11.8. A chart showing the E value, percent identity, fraction of query sequence matched, and type of query matched (color coded) is shown in Figure 11.8A.

In an example by Retief et al. (1999), the large family of known glutathione transferase proteins was first subjected to multiple sequence alignment, and a phylogenetic tree was made by distance methods to identify classes of proteins within the family. These proteins represented a broad range of phylogenetic context and included classes with sometimes less than 20% identity. The object was to choose class representatives for a similarity search of mammalian EST databases for related sequences and to decide which of these sequences were orthologs and which were paralogs.

A novel feature of these searches was to use a lower-scoring PAM matrix to search for paralogs of a recently evolved group of sequences. Use of an appropriate PAM matrix that matches the expected evolutionary separation of a group of sequences provides an improved higher-scoring alignment, as described in detail in Chapter 3 (p. 98). ESTs with a high percent identity with the query sequence, a long alignment with the query sequence, and a very low E value of the alignment score represent groups of paralogous and orthologous genes. To identify orthologs as the most closely related sequence, ESTs were aligned using the amino acid alignment as a guide, and a phylogenetic tree was produced by the maximum likelihood method. This method, described in Chapter 7 (see flowchart for Chapter 7), is suitable for a divergent but recently evolved group of sequences. The predicted tree shown in Figure 11.8B predicts those pairs of sequences that are likely to be orthologous.

Ancient Conserved Regions

Phylogenetically diverse groups of organisms have been analyzed for the presence of conserved proteins and protein domains called ancient conserved regions or ACRs, domains that have been conserved over long periods of evolutionary time (Green et al. 1993). The method involves database similarity searches of the SwissProt database with human, worm, yeast, or *E. coli* genes and identification of matches with sequences from a different phylum than the query sequence. An analysis of ACRs that predate the radiation of the major animal phyla some 580–540 million years ago suggested that 20–40% of coding sequences are ACRs. For example, a search with 1916 *E. coli* proteins detected 266 ACRs found in 439 sequences, roughly one-quarter of the SwissProt database. These ACRs may represent protein present at the time of the prokaryotic–eukaryotic divergence.

With the later addition of complete genome sequences of phylogenetically diverse prokaryotic organisms, the number of ACRs could be estimated by the proportion of genes that match database sequence of known function. For the hyperthermophilic árchaeal organism *Pyrococcus hirokoshii* (Kawarabayasi et al. 1998), this proportion was 20%, perhaps representing an ancient set of prokaryotic ACRs. COGs described above represent sets of proteins that are conserved across distant phylogenetic lineages. For 11 prokaryotic genomes, the proportion of genes represented in COGs is approximately 50–70% (Koonin and Galperin 1997; Koonin et al. 1998). However, one needs to take into account that horizontal transfer of genetic material discussed below increases the sharing of genes by different lineages of prokaryotes.

Horizontal Gene Transfer

The genomes of most organisms are derived by vertical transmission, the inheritance of chromosomes from parents to offspring from one generation to the next. However, in rare instances, genomes may also be modified by horizontal (sometimes called lateral) gene transfer (HGT), the acquisition of genetic material from a different organism. (For a review of this topic, see Bushman 2002.) The transferred material then becomes a permanent addition to the recipient genome and subsequently is transmitted vertically. Although these exchanges do not occur very often on a generation-to-generation basis, a significant number of such exchanges can occur over a period of hundreds of millions of years. An extreme example is the proposed endosymbiont origin of mitochondria in eukaryotic cells and chloroplasts in plants. The endosymbiont theory proposes that these organelles were transferred from free-living bacteria to another organism with which they shared a symbiotic relationship (see Chapter 6 in Brown 1999).

HGT is a significant source of genome variation in bacteria (Ochmann et al. 2000; Bushman 2002 for overview), allowing them to exploit new environments. Such transfer is rendered possible by a variety of natural mechanisms in bacteria for transferring DNA from one species to another. Detection of HGT is made possible by the fact that each genome of each bacterial species has a unique base composition. Hence, transfer of a portion of a genome from one organism to another can generally be detected as an island of sequence of different composition in the recipient. If the amino acid composition of transferred genes is typical, these islands may be detected by a codon usage analysis as described in Chapter 9. Very ancient transfers may not be detectable because the base composition and codon usage of the transferred DNA will eventually blend into those of the recipient organism. The time of transfer of DNA may be estimated by the degree to which the composition of the HGT DNA has blended into that of the recipient genome.

Comparisons of completely sequenced bacterial genomes have revealed that they are mosaics of ancestral and horizontally transferred sequences. The proportion of the genome due to HGT sequences also varies considerably, roughly in proportion to genome size. A total of 12.8% of the genome of *E. coli* is due to HGT DNA (the highest level found), whereas it is 0.0% in *Mycoplasma genitalium*, whose genome is less than one-quarter the size of that of *E. coli*. Mycoplasma have lost many of the genes needed to be a free-living organism and instead depend on nutrients provided by the interior of the host cell. Hence, these organisms would not be expected to carry any extra unnecessary genetic baggage.

HGT DNA contributes in a major way to the disease-producing ability of pathogenic bacteria, and

GENOME ANNOTATION

Accurate annotation of genome sequences is an important step in genome analysis. An initial step in annotation is predicting the location of genes that encode proteins. Any significant alignment of the predicted protein sequences to sequences of known function is then found in database similarity searches. Matches of lesser significance provide only a tentative or hypothetical prediction and should be used as a working hypothesis of function (see Kyrpides and Ouzonis 1999). Computational tools such as MAGPIE and GENEQUIZ described below are designed to assist with accurate genome annotations. After the genes have been identified, the genome sequence can be annotated with a great deal of information about the genome sequence—the location and structure of genes (exons and introns), links to related sequences such as cDNA and encoded protein sequences, detailed information about the function of genes and the biochemistry and structure of proteins and other gene products, literature links, links to genetic maps, location of sequence repeats including transposable elements, the location of sequence tagged sites (STS) for amplification of local genome sequences by polymerase chain reaction (PCR), and sequence polymorphisms in the population (SNPs). These annotated sequences and their Web links to a variety of related information are a valuable resource to the biological community.

this DNA frequently has flanking direct repeats characteristic of transposable elements. Note that when genes are clustered on the chromosome of the donor organism (described below), the recipient organism may gain an entire metabolic pathway from another by means of HGT. Hence, clustering in combination with HGT provides an evolutionary mechanism for altering metabolic pathways in diverse organisms.

GENES CAN BE CLASSIFIED ACCORDING TO FUNCTION

An initial step in genome analysis is to identify the location of genes in the genome sequence. Once genome sequences have been annotated in this manner, a useful next step is to classify the annotated genes by function. Genes that are significantly similar in an organism, i.e., paralogs, frequently are found to have a related biological function. This discovery follows the expected origin of paralogs by gene duplication events, leaving one copy to perform the original function and producing a second copy to develop a new function not too distant from the original one under evolutionary selection. Initial classification schemes included the following approaches:

1. An early classification scheme for eight related groups of *E. coli* genes included categories for enzymes, transport elements, regulators, membranes, structural elements, protein factors, leader peptides, and carriers. Ninety percent of *E. coli* genes related by significant sequence similarity

fell into these same broad categories (Labedan and Riley 1995).

2. The Enzyme Commission numbers formulated by the Enzyme Commission of the International Union of Biochemistry and Molecular Biology provide a detailed way to classify enzymes based on the biochemical reactions they catalyze (Webb 1992; Tipton and Boyce 2000). The designation ECa.b.c.d gives the following information: (a) one of six main classes of biochemical reactions, (b) the group of substrate molecule or the nature of chemical bond that is involved in the reaction, (c) designation for acceptor molecules (cofactors), and (d) specific details of the biochemical reaction. Using this system to compare sequence-related pairs of *E. coli* genes, Labedan and Riley (1995) found that 70% of them shared the first two EC designators (a and b) in the annotation of the corresponding genes, thereby indicating

that they catalyze biochemically similar reactions.

3. A third measure of functional similarity is based on a physiological characterization of *E. coli* proteins into 118 possible categories (e.g., DNA synthesis, TCA cycle, etc.) (Riley 1993). Approximately one-quarter of *E. coli* genes fall into the same category by this scheme.

4. For genes that encode enzymes, relationships among multiple enzymes that perform the same biochemical function in the same organism are examined. Although catalyzing the same reaction, these enzymes showed variations in metabolic regulation of their activity. More than one-half of multiple enzymes in *E. coli* share significant sequence similarity; i.e., they are paralogs. However, the remainder do not share any sequence similarity. Either they were acquired by HGT from another bacterial species or the two enzymes were formed by convergent evolution from two different genetic starting points (Riley 1998). Accordingly, sequence similarity is frequently a good indicator of related biochemical function, but two enzymes that perform the same biochemical task may not share sequence similarity or evolutionary history.

5. Another functional classification scheme for genes includes a broader category for genes involved in the same biological process, e.g., a three-group scheme for energy-related, information-related, and communication-related genes has also been used. By this scheme, plants devote more than one-half of their genome to energy metabolism, whereas animals devote one-half of their genome to communication-related functions (Ouzounis et al. 1996).

6. Another scheme, described below, is to identify proteins that physically interact in a structure or biochemical pathway.

7. A system for functional annotation of genomes has also been produced (Cherry et al. 1997) and used in a comparison of the yeast and worm proteomes (Chervitz et al. 1998). *D. melanogaster* genes were classified using the Gene Ontology (GO) classification scheme (http://www.geneontology.org; Adams et al. 2000), a collaboration among yeast, fly, and mouse informatics groups to develop a general classification scheme useful for several genomes. This classification scheme provides a description of gene products based on function, biological role, and cellular location.

GENE ORDER (SYNTENY) IS CONSERVED ON CHROMOSOMES OF RELATED ORGANISMS

Synteny is a locally conserved gene order that is found in two or more genomes. Two species that have recently diverged from a common ancestor might be expected to share a similar set of genes and also similar chromosomes with these genes positioned along the chromosomes in the same order. Over evolutionary time, the sequence of each pair of genes will slowly diverge, as the species diverge and other changes such as gene duplication and gene loss change the gene content. In addition, the order of genes also changes as a result of chromosomal rearrangements. Rearrangements may be modeled by occasional, possibly random (Nadeau and Taylor 1984) chromosomal breaks and by random rejoining of the fragments by a DNA repair mechanism. Rearrangements may be analyzed by comparing the location of highly conserved DNA or protein sequences and function in prokaryotic and eukaryotic genomes from different phylogenetic lineages. As these changes become more extreme over the passage of evolutionary time, it becomes important to apply statistical tests to determine whether any observed gene clustering is a remaining order of ancestral genes or a statistical fluctuation from a random gene order. A comprehensive set of statistical tests have been devised that should be used for this purpose (Durand and Sankoff 2003).

Two additionally relevant observations have been made with regard to gene order: First, order is highly conserved in closely related species but becomes changed by rearrangements over evolutionary time.

As more and more rearrangements occur, there will no longer be any correspondence in the order of orthologous genes on the chromosome of one organism with that of a second organism. Second, groups of genes that have a similar biological function tend to remain localized in a group or cluster in many organisms. Examples of these observations and their significance are described below.

GENOMES CAN BE USED TO PREDICT EVOLUTIONARY RELATIONSHIPS

The availability of genome sequences introduces a variety of new types of analyses of the evolutionary relationships among organisms. Without genome sequences, protein sequences derived from highly conserved genes are collected from different organisms and aligned; a phylogenetic tree is then produced that reveals their phylogenetic relationships. Analysis of variations in the highly conserved and anciently derived secondary structure of ribosomal RNA also reveals phylogenetic relationships. Using whole-genome sequences, these comparisons may be extended to include all of the similar genes found between organisms, sequence repeats, and GC/AT ratios, providing more clues as to evolutionary relationships and dates of species separation.

When two species are so closely related that they share the same gene set, chromosomal rearrangments can be used for phylogenetic analysis. What is observed in such species, e.g., mouse and humans, is that short regions of genome sequence of one species can be readily aligned with regions of genome sequence of another species. This local alignment is referred to as synteny and the aligned regions as syntenic blocks. However, the alignment eventually stops, and a continued alignment of the first genome with the second genome then restarts at another location in the second genome. As this comparison is continued throughout the genome sequences, the second genome is found to be a rearrangement of fragments of the first genome. The minimum number of rearrangements required to change one genome into the other is computed as described below and used as a measure of evolutionary distance between the genomes.

Evolutionary analysis of genomes can be extended even farther to compare races and populations of one species. For this analysis, sequence variations within populations including variation in lengths and sequences of minisatellite and microsatellite sequences, and SNP haplotypes discussed earlier in this chapter (p. 506) are available.

These variations are used to track human migrations and the origins of genetic variability in our species. An important area in evolutionary comparison of genomes is computational analysis of chromosomal rearrangements.

Visualizing Chromosomal Rearrangements

A simple method of demonstrating chromosomal rearrangements between two closely related species is to draw a two-dimensional genome plot with the gene order of one species shown on the x axis and the gene order of the other species on the y axis. The syntenic blocks will be revealed as short diagonals, and rearrangements will interrupt these diagonals and move them to other locations on the plot. This method may be extended by classifying the genes by function instead of by sequence similarity. In this case, gene groupings based on function representing a level of functional organization in the genomes can be identified.

A genome plot of the positions of orthologs and paralogs on the genomes of two related bacteria, *Mycoplasma pneumoniae* and *Mycoplasma genitalium*, both human pathogens, is shown in Figure 11.9 (Himmelriech et al. 1997). This plot is very similar to the dot matrix plot used for sequence alignment (see Chapter 3), except that in this case a dot or symbol is shown at the intersection of the position of one member of an orthologous pair of sequences on genome 1 and the position of the other member of the pair on genome 2. The plot clearly shows that large sections of chromosomes are conserved but also that a number of rearrangements have occurred, making the gene order different from that of the other genome and from the common ancestor of these two organisms.

In contrast, a similar plot of orthologous genes in the genomes of the bacterial species *E. coli* and *H. influenzae* appears quite random (Tatusov et al.

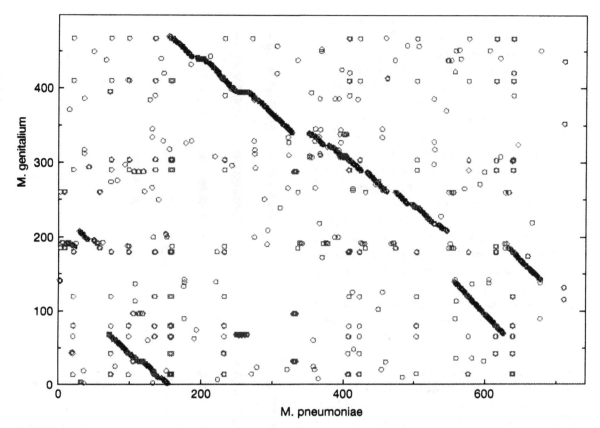

FIGURE 11.9. Genome plot of orthologous genes. Alignment of orthologous and paralogous genes in the genomes of *Mycoplasma genitalium* and *Mycoplasma pneumoniae* (Table 11.1, comparative genome analysis in P. Bork laboratory). Horizontal axis is the genome position in *M. pneumoniae*; vertical axis is the genome position of *M. genitalium*. Positions of orthologs are shown in red, paralogs in green. Orthologous genes are in the same order in both genomes except for several chromosomal rearrangements. These genes are defined by high *E* values in database searches in which one of an orthologous pair is used as query of the proteome of the other species. Proteins should also align along 60% of the length of each (Huynen and Bork 1998). Paralogs are proteins that have striking, high-scoring similarity but are not the highest scoring in reciprocal proteome searches. Note also the occurrence of paralogs within the conserved stretches of orthologs, presumably representing gene duplication in these regions. In contrast to this conserved order of gene position in the *Mycoplasma* species, the orthologous genes in two other equally related species, *E. coli* and *Hemophilus influenzae*, show no detectable conservation of order on a similar genome plot.

1996), even though the organisms are only slightly more distant in evolution than the two *Mycoplasma* species. However, on close inspection of gene function and order, similarities can be found. By classifying genes using a nine-class functional classification scheme (see above), several genes falling into the same functional category are clustered together on the chromosomes of both of these organisms, and the clusters are in a similar order (Ouzounis et al. 1996).

Comparison of the number of rearrangements in a given period of evolutionary history may vary significantly from one organism to the next. In one analysis of prokaryotic organisms of diverse phyloge-

netic origin, it has been shown that if gene A has a neighboring gene B, then if an ortholog of A occurs in another genome, there is an increased probability of an ortholog of B also occurring in the other organism. However, the B ortholog is less likely to be a neighbor of the A ortholog of the genome of the second species if the two species are more divergent (Huynen and Bork 1998).

The TIGR Web site (Table 11.1) includes a resource for comparing any two prokaryotic genomes of those available by means of a genome plot, illustrated in Figure 11.9. In general, the order of orthologs is not well conserved in prokaryotes when the genomes have diverged sufficiently that the orthologs have <50% identity (Huynen and Bork 1998).

A similar conservation of gene order also appears to be present in closely related eukaryotic genomes. The evidence is based on chromosome painting experiments in which DNA from a section of a chromosome of one organism is labeled and then hybridized to chromosomes of a second organism (fluorescent in situ hybridization, FISH). Location-specific patterns of Giemsa staining (G-banding) are also used. Regions of the second chromosome that are labeled reveal the presence of a homologous region. Although this method does not have the precision and sensitivity of sequence analysis methods, these experiments reveal that eukaryotic chromosomes also undergo rearrangements both within chromosomes and between chromosomes during evolution. An example of the differences between mouse and human chromosomes from an early, low-resolution study (Nadeau and Taylor 1984) is shown in Figure 11.10.

A much larger data collection from a variety of mammalian chromosomes suggests that each chromosome is a mosaic of a similar set of ancestral fragments (O'Brien et al. 1999). Similar studies with plant genomes have also indicated that they have a similar overall gene content but that many regional duplications and rearrangements have occurred during evolution (Bennetzen 1998, 2000; Bennetzen et al. 1998; Ilic et al. 2003). An example of genome rearrangements in the *Arabidopsis* genome is shown in Figure 11.7. The availability of genome sequences of plants and animals offers some exciting opportunities for determining the chromosomal changes that have occurred during evolution of the plant and animal kingdoms.

Computational Analysis of Gene Rearrangements

As genome-by-genome comparisons of the chromosomes of related species are made and the rearrangements are discovered, a further challenge to computational and evolutionary biologists is to estimate the number and types of rearrangements that have occurred and also to determine when they occurred. For example, a comparison of the mouse and human chromosomes reveals many more rearrangements than shown in Figure 11.10. Computational approaches to these questions are illustrated in Figure 11.11.

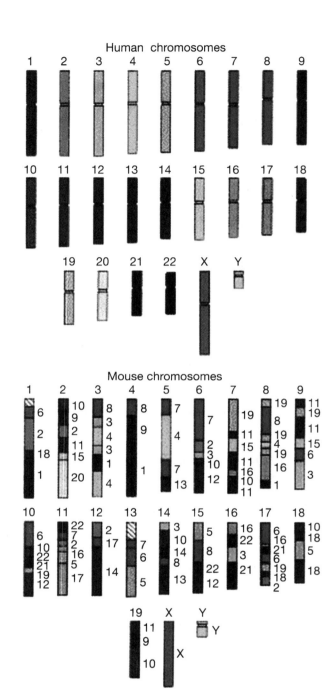

FIGURE 11.10. Similarity between local gene clusters in human and mouse chromosomes. Human chromosomes can be cut into >100 pieces and reassembled into a reasonable facsimile of the mouse chromosome. Only larger fragments are represented. Chromosomes of all mammals may share a similar relationship (O'Brien et al. 1999).

In aligning gene and protein sequences, one assumes a model in which no rearrangements have occurred so that lines can be drawn between the cor-

A. Sequence alignment

B. Genome alignments

Linear

Circular

C. Alignment reduction

Reduced alignment

Conserved segments

FIGURE 11.11. Computational analysis of genome arrangements. (*A*) In aligning two sequences, one sequence is written above the other and the highest number of consecutive matches between the sequences provides an optimal alignment as described in Chapter 3. The alignment includes matches (*solid lines*), mismatches (*dotted lines*), and insertions/deletions in order to produce an optimal number of matches. The matches are in a consecutive order in two sequences such that no rearrangements would be found. (*B*) Alignments of linear and circular chromosomes that have undergone rearrangements such as those found in mammalian chromosomes and mitochondria. In contrast to sequence alignment, lines indicating homologous positions in linear chromosomes (*left*) now cross, producing points of intersection. The more rearrangements there are, the more intersections will occur. For alignment of circular chromosomes (*right*), depending on how the chromosomes are aligned, there are two ways of showing a moved region. To go from A on the outer genome to A on the inner genome, the line joining them can go clockwise or counterclockwise and, as a result, there will be either 0 or 1 intersections with the line joining B. (*Continued on following pages.*)

D. (a)

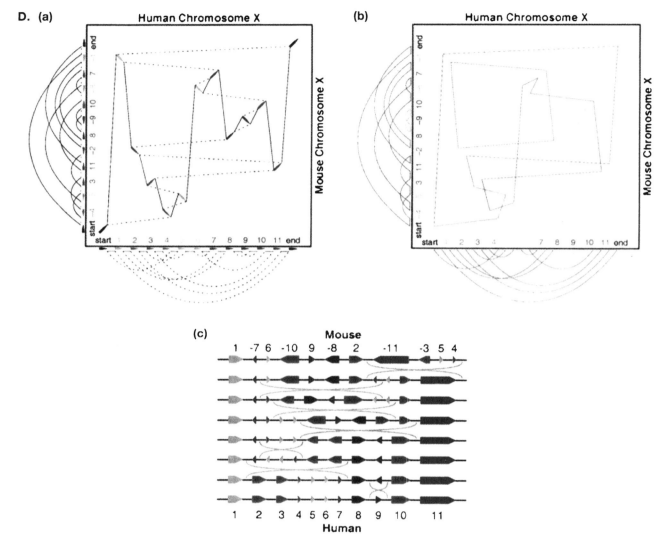

(b)

(c)

FIGURE 11.11. (*Continued.*) The complexity of alignments of circular chromosomes is reduced by limiting the joining lines to 180° of relative genome positions. Sankoff and Goldstein (1989) devised a shuffling model for estimating the number of rearrangements when the number of intersections is known. The method is analogous to shuffling an ordered deck of cards and then predicting how much order remains. Eventually, after *n* log *n* shuffles, where *n* is the number of cards, the order becomes random. Given an observed remaining order, how many shuffles have occurred? The number of observed intersections is compared to the number expected for completely shuffled genomes (Sankoff et al. 1993). (*C*) Another method for determining numbers of rearrangements is to assume that they have occurred by a number of transposition or recombination events. The object of this analysis is to try to identify the rearrangements that occurred and then to undo (or derange) the alignments accordingly. The goal is to minimize the number of rearrangements, this number then representing a genetic distance between the sequences. (*a*) Alignments of genes 1–10 in two genomes where some genes are in the same order (*red lines*) and others are inverted (*black lines*). Groups of genes such as the two joined by the black lines may be combined into a single unit representing a conserved segment since no recombination event would be required. (*b*) Alignment that can be accounted for by these inversion events. The program DERANGEII is available from the authors and FTP from ftp.ebi.ac.uk/pub/software/unix/derange2.tar.Z. These methods have been used to analyze rearrangements in mitochondrial and bacterial genomes (Sankoff et al. 1992; Blanchette et al. 1996; Sankoff and Nadeau 1996) and additional algorithms have also been developed (Kececioglu and Sankoff 1995; Kececioglu and Gusfield 1998). (Adapted from Sankoff et al. 1992, 1993.) (*D*) Method for analyzing rearrangements in the human and mouse genomes by the breakpoint graph method using the human and mouse X chromosomes as an example. (*Continued on following page.*)

responding positions in the sequences and no lines will cross or intersect, as shown in Figure 11.11A. For comparing gene orders on chromosomes that have undergone rearrangements, lines joining the corresponding genes will intersect, as shown in Figure 11.11B, and the greater the amount of rearranging, the greater the number of intersects. In the random shuffling model, genome rearrangements are modeled as cuts from an ordered set of cards. One tries to estimate the number of rearrangements that produces the observed number of intersections and to compare this number to one that would randomly shuffle the same fragments. The analysis shown in Figure 11.11C attempts to reconstruct the number and types of rearrangements (transpositions, recombination, inversions, etc.) that have given rise to the observed variation in gene order between the chromosomes. A later method of analysis is based on a mathematical analysis of permutations and reversals by P. Pevsner and S. Hannenhalli.

An example of a permutation/reversal analysis of synteny blocks in the human and mouse X chromosomes is shown in Figure 11.11D (Pevsner and Tessler 2003a,b). These blocks are defined by searching for the best sets of local alignments (anchors) with the sequences in the same or reverse orientations (for finding sequence reversals) between the DNA sequences of the human and mouse X chromosomes. Anchors within a certain minimal sequence distance are joined into a cluster using a graph-based method called GRIMM (genome rearrangements in mouse and human). Another method uses positional hashing tables similar to those used in the FASTA algorithm as a fast method to anchor sequences between genomes (Kalafus et al. 2004). After applying a minimal chromosomal fragment length requirement for clusters, the method produces a set of synteny blocks and their locations on the mouse and human X chromosomes representing genome rearrangements. These blocks represent extended chromosomal regions in the

FIGURE 11.11. (*Continued.*) (*a*) Genome dot matrix plot showing points of rearrangement between the human and mouse X chromosomes. To produce the graph, the entire X chromosome sequences are compared to locate human and mouse sequences that can align locally. Alignments in a localized region of each species are combined into synteny blocks as described in the text. These blocks represent rearrangement units between the human and mouse X chromosomes. The synteny block order on the human X chromosome is arbitrarily numbered as start, 1–11, and end and is color-coded on the *x* axis. The order of the corresponding synteny blocks on the mouse X chromosome is shown on the *y* axis. A minus sign indicates the reverse orientation of the synteny block on the mouse X chromosome relative to the human X chromosome, and no sign indicates the same orientation. Alignments of the human and mouse synteny blocks are indicated by color-coded diagonals (not to scale) going up or down to the right depending on whether the sequences in the synteny block are going in the same or reverse order on the mouse and human X chromosome. The solid lines joining the diagonals trace the location of each sequential block in the human X chromosome on the mouse X chromosome and the dotted lines trace the location of each sequential block in the mouse X chromosome on the human X chromosome. The arcs shown outside of each axis show the blocks that have to be exchanged in one X chromosome to produce the order found on the other X chromosome. (*b*) Two-dimensional breakpoint graph that represents the rearrangements as closed loops or cycles. Cycles are produced by joining the solid and dotted lines and deleting the synteny blocks (*diagonal lines*). Each cycle has a length that is the number of edges in the cycle. In this example, there are four cycles of lengths 4, 4, 8, and 8. The number of cycles is related to the minimum number of steps required to change the arrangement of synteny blocks on one X chromosome into that of the other, and is defined as the reversal distance. This distance is given by the Pevzner–Hannenhalli formula $n + 1 - c + h$ derived from a mathematical analysis of permutations and reversals, where n is the number of synteny blocks, c is the number of cycles in the breakpoint graph, and h is a constant $= 0$ in most biological situations. In this example, the reversal distance is $11 + 1 - 4 + 0 = 8$. To confirm that this number does represent the minimum number of rearrangements, the mouse X chromosome sequence is also flipped on the graph, thus representing the rearrangements from a reversed chromosomal perspective. When a new breakpoint graph is produced with the reversed X chromosome, a new set of cycles is found and the number of cycles increases by 1, reducing the reversal distance to $11 + 1 - 5 + 0 = 7$. Since 7 is less than 8, and assuming that the X chromosome rearrangements could have taken place equally as likely with the chromosomes in opposite orientations, 7 is taken to be the reversal distance. The analysis of rearrangements may be extended to include all chromosomes on the graph so that interchromosomal rearrangements may also be analyzed. Both orientations of each chromosome must be tested to find the most parsimonious solution. (*c*) A possible set of seven sequential rearrangements that could transform the synteny block order on the human X chromosome to that on the mouse X chromosome. (Adapted, with permission, from Pevsner and Tessler 2003b.)

human and mouse X chromosomes that share sequence similarity over most of the aligned regions.

The relative locations of these synteny blocks on the human and mouse X chromosomes are shown on a genome plot, with the order of the synteny blocks on the human chromosome on the x axis and the order of these blocks on the mouse chromosome on the y axis, as illustrated in Figure 11.11D, part a. This plot is similar to a dot matrix plot described in Chapter 3, except that the alignments represent long conserved patches of sequence of length up to millions of DNA base pairs. In the genome plot, two paths showing the connections between synteny blocks are traced between the diagonals, one from the perspective of the human chromosome and the other from the perspective of the mouse chromosome, shown as solid and dotted lines. The plot is used to produce a breakpoint graph, illustrated in Figure 11.11C, part b. The graph contains cycles that provide information on the number and types of rearrangements that have occurred. Using results from a mathematical analysis of permutations and reversals, the minimal number of rearrangements can be estimated as described in the figure legend. In this case, the number was 7. A series of seven rearrangements that could produce the order of the synteny blocks on the mouse X chromosome starting with their order on the human X chromosome is illustrated in Figure 11.11C, part c.

The above method has been extended to include an analysis of all the genetic rearrangements between the human and mouse genomes. To analyze the total genomes, all synteny blocks are found and a much larger genome plot that includes all of the human and mouse chromosomes is produced. As was done with the X chromosome analysis, the order and orientation of the chromosomes have to be varied to determine the minimum number of rearrangements by the breakpoint graph method (Pevzner and Tessler 2003b). This number of steps is the most parsimonious solution representing the most likely set of rearrangements over evolutionary time. On the basis of the whole-genome analysis, it has been suggested that the chromosomal breakpoints at which rearrangements start are not random as originally proposed (Nadeau and Taylor 1984), but the same breakpoints appear to be reutilized (Pevzner and Tessler 2003a). The breakpoint graph method and others for analyzing genome rearrangements will undoubtedly play a significant role in comparative

genomics, including the analysis of rearrangements in mouse versus rat and among the more closely related primate sequences.

Software tools for finding the local genome alignments necessary to perform this analysis have been described (Web search for Genome BLAT, Pattern-Hunter, Pip-Maker, and VISTA). The GRAPPA (genome rearrangement analysis under parsimony and other phylogenetic algorithms) and GRIMM methods for finding synteny blocks and analyzing genome rearrangements may be readily found by Web searches. Multiple genome rearrangements (MGR) is used to resolve when genome rearrangement has occurred (Bourque et al. 2004; http://www.cs.ucsd.edu/groups/bioinformatics/MGR).

Clusters of Metabolically Related Functions on Prokaryotic Chromosomes

In a given organism or species, genes are found in a given order that is maintained on the chromosomes from one generation to the next. Genetic analysis has revealed that genes with a related function are frequently found to be clustered at one chromosomal location. Clustering of related genes presumably provides an evolutionary advantage to a species, but the underlying biological reason is not understood. One possibility is that there is genetic variation (alleles) within each gene in a cluster of a given species and that only certain allelic combinations of different genes are compatible. Another possibility is some kind of coordinated translation of the proteins that may aid their folding. In the model bacterial species *E. coli*, genes that act sequentially in a biochemical pathway are frequently found to be adjacent to each other at one chromosomal location. For example, the genes required for synthesis of the amino acid tryptophan (*trp* genes) are clustered together on the chromosome of *E. coli*, as illustrated in Figure 11.12, where their expression is coordinately regulated by a common promoter. This coordination of expression avoids wasteful production of one enzyme when others in the same pathway are not available.

With the availability of other prokaryotic genome sequences, important metabolic genes such as *trp* can be identified in these species, and the chromosomal location of these genes can be compared with that of *E. coli*. Using the predicted tryptophan genes as an example (Fig. 11.12), the following observations were made:

1. At least some of the *trp* genes are also clustered together on the chromosomes of other species of Bacteria and Archaea.

2. The order of the genes within the cluster is conserved within the first four species listed in Figure 11.12, all of which are bacteria.

3. The order is much less conserved in the last three species, all of which are Archaea, and some of the genes have been moved to a more distant location.

4. There are multiple examples of gene fusions that give rise to a new protein that performs both biochemical functions of the single-gene, parent proteins. *trpC* has been fused independently with two other genes, *trpD* and *trpF*.

Alternatively, a composite gene may produce two smaller single-component genes by fission of a parent composite gene. Fission events have only been observed in thermophiles among prokaryotes (Snel et al. 2000). However, biochemical evidence has been presented that fission events may provide a mechanism for evolution of protein complexes (Marcotte et al. 1999b).

When a series of predicted genes in a known *E. coli* pathway is in the same order in another organism as in *E. coli*, e.g., *trpB-trpA* and *trpE-trpG*, in the

Archaea in Figure 11.12, then the same biochemical pathway is predicted also. Even if the genome annotation is based on a weak prediction of the biochemical function of two individual genes, the prediction is stronger if the two genes act in the same pathway and is strongest if the genes are clustered (Huynen et al. 2000). In the *trp* example shown in Figure 11.12, the presence of the genes in such a phylogenetically diverse group of organisms indicates that the pathway is an ancient one. Clustering of the genes further indicates that they probably originated as a group in the single chromosomal region of an ancient ancestor organism, assuming there has not been a driving force for repeated independent clustering events. What is also revealed in the *trp* example in Figure 11.12 is that some *trp* genes are found at a much more remote chromosomal location. The diverse location of the *trp* genes in *Methanococcus jannaschii* is an outstanding example. Apparently, rearrangements can break clusters and move genes to other locations, although another possibility is that the dispersed arrangement is a more ancestral state.

Two methods have been described for identifying clusters or coordinately regulated genes. In one study with three separate groups of three distantly related prokaryotes (Dandekar et al. 1998), approximately

FIGURE 11.12. Structure of tryptophan operon in different prokaryotic organisms. Numbers indicate gene number in genome; arrows indicate direction of transcription; double lines indicate a separation of more than 50 genes due to dispersion of the operon. Shown also are examples of gene fusion so examples of domain fusions (e.g., *trpD* and *trpG*) are fused in *E. coli*. Note that only the *trpA* and *trpB* genes are genetically linked and separate genes in all of the species. (Reprinted, with permission, from Dandekar et al. 1998 [©Elsevier Science].)

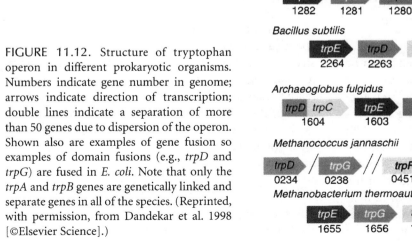
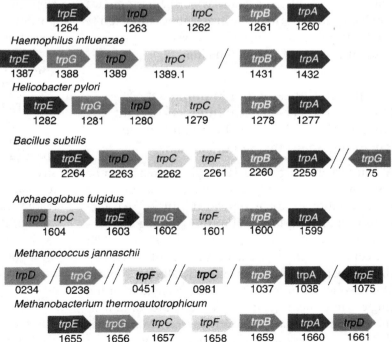

100 genes were found to be conserved as a cluster of two pairs. (Looking for a pair in three species avoided possible complications from HGT.) The direction of transcription was the same for all genes, implying a regulatory relationship as in an operon. For approximately 75% of the genes, a physical interaction between the genes had previously been demonstrated and could be predicted for almost all proteins based on additional sequence comparisons. These conserved proteins have core biological functions such as transcription, translation, and cell division.

A second method for finding coordinately regulated genes that builds upon and extends the use of reciprocal proteome similarity searches to find orthologs has been devised (Overbeek et al. 1999). Pairs of genes that show the best alignments with each other, that are located within 300 bp of each other on the chromosomes of the respective organisms, and that are transcribed from the same strand, i.e., are in a "typical" operon, are identified. A score for these pairs is formulated that is higher when the number of organisms in which the pair is observed is greater and the phylogenetic distance between the organisms is larger. Forty percent of a set of higher-scoring pairs corresponded to proteins known to act in a common metabolic pathway, as defined in metabolic function databases (Web search for BioCarta, GenMapp, KEGG). Hence, a significant proportion of the pairs of PCBBH corresponds to genes that have a related function and lie on the same pathway. This same approach could play an important role in assigning a function to uncharacterized genes in genomes based on proximity to other genes of known function.

Composite Genes with a Multiple Set of Domains Predict Physical Interactions and Functional Relationships between Protein Pairs That Share the Same Domains

As illustrated in Figure 11.12, single *trp* genes can be fused into larger composite genes. Observation of such evolutionary events provided a major step forward in understanding relationships among the proteins of diverse organisms (Enright et al. 1999; Marcotte et al. 1999b). The occurrence of a fused or composite gene in one organism is called a "Rosetta Stone sequence" because it provides evidence that the single-component genes in a separate organism encode proteins that physically interact (Marcotte et al. 1999b). For example, if a composite human gene

has two domains A and B, the analysis assumes that A and B physically interact within the protein. If two separate genes in other organisms (yeast or *E. coli*) make two proteins, one with domain A and a second with domain B, these two proteins are assumed to interact because A and B interact in the single-protein scenario. These sequence relationships may be found by sequence alignment of the composite AB protein with each of the single-component A and B proteins. However, A and B will not align with each other. If A and B do not interact in composite proteins, the prediction is a false-positive result. However, these proteins are still predicted to have related functions based on the gene fusion result.

Composite proteins were found by searching SwissProt for statistically significant matches to domains in the ProDom domain database (see Table 10.5). Six percent of the Rosetta Stone proteins were found to be represented in the DIP database of interacting proteins (see Table 10.5). Rosetta Stone predictions of interacting proteins were compared to predictions by another method for predicting related proteins, the phylogenetic profile method (Pelligrini et al. 1999; see also "bag of genes" concept in Huynen and Bork 1998). This method is based on the assumption that proteins which function together in a biochemical pathway should evolve in a correlated fashion. Databases are searched for significant matches to two proteins A and B. If A and B have related functions, they should be found together in a large proportion of genomes, whereas if they do not, they will have a random association in genomes.

Enright et al. (1999) used reciprocal searches among three complete prokaryotic proteomes, as described above in Figure 11.6, and identified related proteins that have the expected alignments for composite (AB) and component (A or B) proteins. These proteins interact functionally, act in the same biochemical pathway, or are co-regulated. Predictions are stronger when component proteins (A and B) have few paralogs, since the interacting pair can be more readily identified. Conversely, the presence of paralogs of the composite proteins increases the strength of the prediction because there are more candidates for a possible interaction (Enright et al. 1999).

Resources for Genome Analysis

The above types of analyses depend on a labor-intensive annotation of the genome and functional analy-

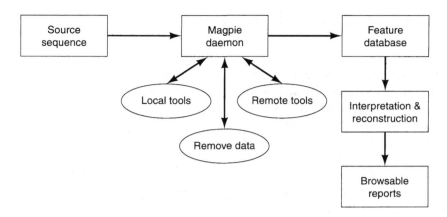

FIGURE 11.13. Automatic analysis of genome sequences with the MAGPIE system. The sequence is input into an automatic system (a daemon) that sends the sequence to local and remote resources for analysis (BLAST search, PROSITE search, etc.). The information retrieved is stored in a feature database, and the data are interpreted by a set of rules and placed in Web-browsable reports. (Redrawn, with permission, from Gaasterland and Sensen 1998 [©CRC Press].)

sis of the predicted proteins. Computational tools have been made available to automate some of these steps. Examples are MAGPIE and GeneQuiz.

MAGPIE analyzes the genome using a set of automated processes that are illustrated in Figure 11.13. Designed for high-throughput genome sequence analysis, MAGPIE automatically annotates genomic sequence data and maintains a daily up-to-date record in response to user queries about one or more genomes. The system also uses a set of rules in logic programming to make decisions that may be used to interpret information from various sources. MAGPIE has been used to locate potential promoters, termina-tors, start codons, Shine–Dalgarno sites, DNA motif sites, co-transcription units, and putative operons in microbial genomes. These sites are shown on a map display of the genome that may be edited.

GeneQuiz is an integrated system for large-scale biological sequence analysis that uses a variety of search and analysis methods using current sequence databases. By applying expert rules to the results of the different methods, GeneQuiz creates a compact summary of findings. It focuses on deriving a predicted protein function, based on a variety of available evidence, including the evaluation of the similarity to the closest homolog in a database.

MICROARRAY ANALYSIS PROVIDES INFORMATION ABOUT GLOBAL GENE REGULATION

One way to obtain useful information about a genome is to determine which genes are induced or repressed in response to a phase of the cell cycle, a developmental phase, or a response to the environment, such as treatment with a hormone. Sets of genes whose expression rises and falls under the same condition are likely to have a related function. In addition, a particular pattern of gene expression may also be an indicator of abnormal cellular regulation and is a useful tool in cancer diagnosis (see, e.g., Golub et al. 1999; Perou et al. 1999). Because genomes, especially eukaryotic genomes, are so large, a technology called microarray analysis has been developed for studying the regulation of thousands of genes on a microscope slide. These methods require complex statistical and computational types of analyses that are described in Chapter 13.

GENE FUNCTION CAN BE PREDICTED USING A COMPOSITE ANALYSIS

A considerable fraction of a genome may encode proteins whose function may be identified through similarity searches of the genomes of model organisms in which the function of the gene has been identified by biological analysis. However, in many cases, proteins predicted to be encoded by the genome are not related to another protein of known function. In the above sections, other types of evidence for a relationship between two genes are also given that are not dependent on sequence similarity. These types of evidence include:

1. Genes are closely linked on the same chromosomes and transcribed from the same DNA strand, implying coordinated regulation in an operon-like structure.

2. Gene fusions are observed between otherwise separate genes (suggesting that the encoded proteins are physically associated in a common complex).

3. Phylogenetic profiles reveal the genes are both commonly present in many organisms (implying they have interdependent metabolic functions).

Three additional types of data have been used as evidence for gene relatedness:

1. The encoded proteins each have homologs in another organism that operate in a common metabolic pathway.

2. Experimental data suggest an interaction between the proteins (stored in databases of interacting proteins; Table 10.5).

3. Patterns of mRNA expression are found to be correlated in microarray data.

The results of using the above tests for the identification of a group of related genes in yeast are shown in Figure 11.14. In an examination of the entire yeast proteome, proteins that share a relationship with the yeast Sup35 protein based on one or more of the above tests are shown as points in a two-dimensional cluster where the distances between the points are proportional to the weight of the evidence for a relationship between the protein pair, and the strength of the connection is proportional to the amount of evidence for a relationship. These types of predictions can be an important basis for hypotheses that can be tested experimentally.

FUNCTIONAL GENOMICS APPROACHES IDENTIFY GENE FUNCTION

Genome analysis depends to a large extent on sequence analysis methods that identify gene function based on similarity between proteins of unknown function and proteins of known function. Known functions are derived from experimental evidence in molecular biology and genetic studies with model organisms. Orthologous genes between biologically distinct species (e.g., yeast and fruit flies) can be identified, and the high sequence similarity between them is strong evidence for a related function. However, given the more complex multicellular biology of flies, the fly gene could have an additional function that is not predictable by the yeast model. In other cases, the occurrence of families of paralogous genes that share common domains can make a precise guess of func-

tion of one of these proteins more difficult because all match a model protein to some degree. Sequence-based methods of gene prediction can be augmented by the types of genome comparisons described above that are designed to identify related genes on the basis of common patterns of expression, evolutionary profiles, chromosomal locations, and other features. However, all of the above methods can fail to provide a precise determination of gene function. Hence, methods have been devised for directing mutations into specific genes that inactivate or modify the gene function, and the effect is then analyzed in the mutant organism.

Two general types of approaches for functional genomics experiments illustrated in Figure 11.15 are

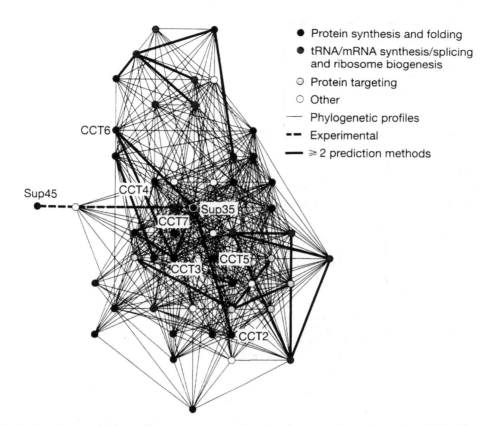

● Protein synthesis and folding
◕ tRNA/mRNA synthesis/splicing and ribosome biogenesis
◔ Protein targeting
○ Other
— Phylogenetic profiles
-- Experimental
━ ≥ 2 prediction methods

FIGURE 11.14. Genome-wide prediction of protein functions by a combinatorial method (Enright et al. 1999; Marcotte et al. 1999a). This figure shows the network of yeast proteins that are linked to the yeast prion and translation factor Sup35 (*double circle* in center of network). Each point represents a yeast protein, and branches between proteins indicate a relationship by one of several criteria indicated in the legend. Branch lengths are shorter for closely related proteins and thicker when two or more prediction methods indicate a relationship. Related to Sup35 protein are proteins involved in protein folding and targeting. The links are based on experimental data, proteins whose homologs are known to operate sequentially in metabolic pathways, proteins that evolved in a correlated fashion as evidenced by presence in fully sequenced genomes (see Snel et al. 1999), proteins whose homologs are fused into a single protein in another organism, and proteins whose mRNA expression profiles are similar under a range of cellular and environmental conditions. (Reprinted, with permission, from Marcotte et al. 1999a [©Nature Publishing Group].)

used—one in which a genetic construct is made that is targeted to interfere with the expression of a particular gene (and sometimes a set of related genes) and a second in which a large number of random mutations are generated in a population of organisms and those individuals with a mutation present in a particular gene are identified by DNA sequencing. Once mutants are obtained, the effect of the mutant genes on phenotype is determined. The gene function may then be predicted on the basis of the observed alterations. Because such extreme genetic experiments cannot be performed with humans, the mouse model

for the human genome serves the same purpose. Web sites that compare the mouse and human genomes listed in Table 11.1 provide an important basis for analyzing the human genome. An orthologous gene is identified in the mouse genome, the sequence or expression of the gene is disrupted in some fashion, and a transgenic mouse homozygous for the mutant gene is then produced. Using this technology, one can systematically go through genes that regulate cell division, for example, and determine the significance of these genes in normal versus abnormal (tumor) growth.

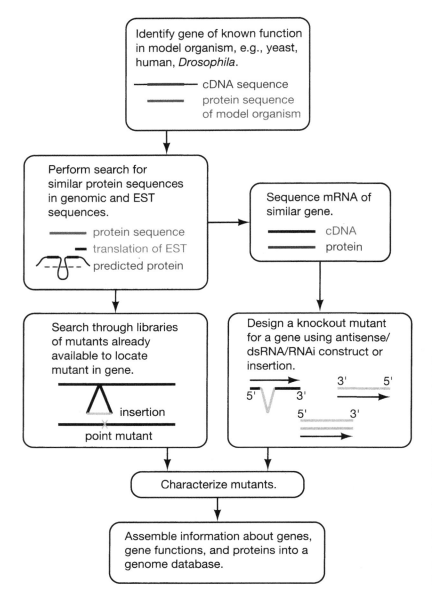

FIGURE 11.15. Reverse-genetics analysis of gene function. Steps for identification of gene function in an organism are identified. Even though a particular gene may be a highly predicted ortholog of a gene of known function in another organism, that gene may be acquired by a novel function. For example, a defect in a plant or animal gene that is a homolog of a yeast gene may have an effect on a developmental process or other biologically unique function of multicellular organisms. Information on knockout mutants in model organisms is available through the genome Web sites given in Table 11.1. Directed gene knockout and mutagenesis methods are described in Fire et al. (1998) and McCallum et al. (2000), respectively.

COLLECTING ALL OF THE INFORMATION INTO A GENOME DATABASE

The ultimate step in genome analysis is to collect the information found on gene and protein sequences, alignments, gene function and location, protein families and domains, relationships of genes to those in other organisms, chromosomal rearrangements, and so on, into a comprehensive database. This database should be logically organized so that all types of information are readily accessible and easily retrievable by users who have widely divergent knowledge of

the organism. This goal is best achieved by using controlled vocabularies that can identify the same genetic or biochemical function in different organisms without ambiguity. The genome annotation sites (e.g., the gene ontology initiative) and the maize and fly genome Web sites are developing systematic ways of defining terms and of collecting and organizing data. Other examples of database tools used to store and exchange biological information are given in

Chapters 2 and 12. The genome sites of model organisms, especially SGD and Flybase, provide examples for further study. In addition to the care needed in organizing genome databases, a great deal of human input is needed to annotate the genome manually with information about individual genes and proteins, effects of mutations in these genes, and other types of genome variations that cannot be readily incorporated into the database by automated methods. Web sites such as NCBI and Ensembl provide an invaluable service to the genome research community by providing automatic annotation of genome sequences and ready access to all of the new genome information that becomes available almost as soon as it is found in the laboratory or published in the literature. Genome analysis problems will continue to attract the interest of computational and evolutionary biologists for many years to come.

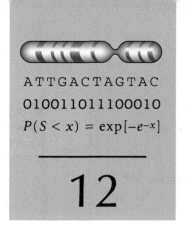

ATTGACTAGTAC
010011011100010
$P(S < x) = \exp[-e^{-x}]$

12

Bioinformatics Programming Using Perl and Perl Modules

CONTENTS

The material in this chapter was contributed by Nirav Merchant and Susan Miller of the Biotechnology Computing Facility of the University of Arizona.

INTRODUCTION

Perl (Practical Extraction and Report Language) is a high-level, yet relatively easy to use, programming language developed by Larry Wall. Since the release of version 1.0 in 1987, development has continued and expanded, with a large community of contributors who regularly add new functionality. Perl's file and text manipulation abilities make it well suited for tasks ranging from "quick" prototyping to complex and robust DNA and protein sequence analyses. Perl is a modular programming environment, comprising an exhaustive collection of contributed modules archived at the Comprehensive Perl Archive Network (CPAN). These modules facilitate tasks like database access, graphical programming, networking, and Web programming. Employing a combination of modules, one can quickly develop programs that minimize the drudgery and human intervention required with tasks associated with sequence analysis. The topics discussed in this chapter are not intended to be a definitive text on the subject matter, but rather an introduction to the advantages of using combinations of Open Source and community-evolved software and standards. Although this chapter exclusively deals with Perl and Perl modules, the reader is strongly encouraged to visit http://www.open-bio.org and examine the BioJava, BioPython, and BioRuby projects, which provide similar extensions to the traditional Java, Python, and Ruby languages. The goal in this chapter is to demonstrate the use of Perl and available modules for routinely performed bioinformatics tasks. Brief conceptual overviews and background information on related topics are also provided.

CHAPTER GUIDE FOR BIOLOGISTS

For biologists, this chapter provides an introduction to the use of Perl, a computer language widely used in bioinformatics. The need for using Perl arises when collecting and manipulating large data sets or performing the same task repeatedly, as in comparing large sets of related genes from the genomes of different organisms by repeated BLAST searches. The natural response from a biologist with little computer-related background might be to skip this chapter and rely on a programmer. Anticipating this response, biologists are encouraged to learn as much as they can about Perl and related computational information. The object should be to learn enough to be able to make informed decisions about future programming and computing needs. A well-informed biologist will appreciate more fully what can be accomplished using Perl in bioinformatics applications and thereby can lay out a program of work for trainees, programmers, and researchers. Using the more advanced applications of any programming language, including Perl, probably is not an achievable goal for most biologists, and a programmer with more extensive training and expertise must be hired for these tasks. When a well-trained individual is hired, the examples given in this chapter will point them to tools that will save a great deal of time and effort.

This chapter describes the best tools and resources available for support of computational needs in bioinformatics and how to obtain and use them. Also provided is an introduction to tried-and-true software development procedures that will help newcomers avoid pitfalls and work efficiently. Although explanations are given for setting up Perl in any operating system environment, most of the development of Perl has been accomplished using the UNIX/Linux operating systems, and Perl is used most effectively in these operating system environments. Linux is a free operating system and may be quite easily installed on any machine. Perl will usually be included in Linux or may be readily obtained. The MacOS X operating system is UNIX-based, and a Perl interpreter is included in the operating system tools. For assistance in using UNIX and Linux, a list of common UNIX/Linux commands is given in an Appendix at the end of the chapter.

This chapter will be most useful if one first reads an introductory Perl text or visits one of the many Web sites that provide an introduction to Perl. We provide a diverse set of useful Perl examples that can be downloaded from the book Web site (http://www.bioinformaticsonline.org) and run directly. These examples comprise sets of Perl instructions (Perl scripts) that can be interpreted by any machine that has a Perl interpreter installed. They can be modified to perform other tasks, and the problem set provides examples.

After providing a basic set of instructions and examples that illustrate what can be accomplished by using Perl, this chapter also explains how to use the advanced features of Perl, including modules. A module is a previously written set of programming instructions that performs a particular task; for example, translating a DNA sequence into all six possible reading frames. The module is invoked by referring to its name in a Perl script and providing the needed information for the specific problem; in the above example of a module for DNA translation, this information would be a specific DNA sequence. Within the module, the previously written instructions are followed and the result produced is sent back to the original Perl script, which can then perform further tasks with the sequence data provided by the module. In addition to describing the available modules, this chapter describes how to use Perl to place data into a database. By testing these examples and using the other information in this chapter, a biologist should be able to write useful Perl scripts with very little additional training.

CHAPTER GUIDE FOR COMPUTATIONAL SCIENTISTS

Biologists routinely handle large amounts of DNA and protein sequence data along with a variety of other types of data (e.g., gene expression, gene maps, sequence variation data). These data, along with related biological information, need to be integrated and organized into suitable database formats and displayed on interactive Web pages. Instead of starting from scratch by developing new computational tools to meet these needs, we are recommending the use of Perl. This chapter describes an extensive set of Perl programming tools that have been developed by the Open Source community to meet these needs and describes where to locate them and how to use them. Some familiarity with Perl, which is acquired easily by using the available textbooks and Web resources, is assumed. The chapter provides a number of sample scripts, including basic ones for manipulating sequences, and more advanced ones that use programming tools and modules; e.g., the BioPerl modules for automatically retrieving sequences from the existing databases, manipulating and analyzing these sequences, and placing the results in a database. We also point the way to Web resources to keep abreast of future development in this area.

WHAT SHOULD BE LEARNED IN THIS CHAPTER?

- How to choose a suitable computational environment for bioinformatics.
- Understanding useful approaches for new software development.
- How to find Perl resources and install a Perl interpreter.
- Developing the ability to write simple Perl scripts.
- Familiarity with obtaining, using, and developing Perl libraries and modules.
- How to run an external program from within a Perl script.
- Familiarity using Perl to establish a local database and perform data transactions with the database.

Glossary Terms

ANSI (American National Standards Institute) is a large voluntary organization that creates standards for the computer industry. ANSI is a member of the International Organization for Standardization (ISO).

Apache is a public-domain Open Source Web server.

Argument is a user-specified parameter passed to a program or subroutine.

Array (in programming) is a series of homogeneous objects (called elements) that are stored contiguously in memory. All elements in an array are the same size and type and are accessed by an integer indexing number. Index numbering may start at 0 or 1, depending on the programming language. Arrays may be single- or multidimensional.

bin (in UNIX) is a directory in which executable (*bin*ary) files are typically located.

BioJava is a collection of Java tools for processing biological data.

BioPerl is a collection of Perl modules for data manipulation and analysis in bioinformatics, genomics, and life science research.

BLAST database is a database comprising several files specifically formatted for BLAST search queries.

C programming language is an early portable language that was used to write the UNIX operating system, developed at Bell Labs.

C++ programming language is an enhanced version of the C language supporting type extensibility and object-oriented programming.

Compiler is a computer program that converts source code into machine language that can run directly on a particular computer architecture. The compilation step is done before program execution to optimize execution speed.

CPAN (Comprehensive Perl Archive Network) a repository for Perl modules.

CpG islands are regions of genomic sequences having a high frequency of "C" and "G" connected by a phosphodiester bond. These regions are often located near promoters of housekeeping genes and are normally not methylated.

Cross-platform Software usable on many different computers; i.e., not specific to a particular operating system.

DBD (Database Driver) is a program for connecting to a database so that the data can be manipulated. DBDs implement the DBI methods.

DBI (Database Interface) is a Perl module that provides a consistent interface between different databases.

Directory (UNIX) is a type of file that contains other files (and/or directories); UNIX directories are analogous to Windows of Mac folders.

Domain (database entry) refers to the range of values allowed in the column of a database table.

EMBOSS (the European Molecular Biology Open Software Suite) from an organization that provides a large suite of Open Software for sequence analysis.

Environment variable (UNIX) is a changeable attribute that defines an aspect of the operating environment for a user. For example, the PATH variable defines the list of directory paths in which to look for executable programs.

FAQ (Frequently Asked Questions) is a list of questions and answers provided by developers or maintainers of a project.

Flat file is a nonhierarchical file, as opposed to a structured database file.

FTP (File Transfer Protocol) is a protocol used for exchanging files on the Internet.

GCG sequence format is a file format for DNA or protein sequences, developed by the Genetics Computer Group.

GNU Project (based on the acronym Gnu is Not UNIX) provides freely redistributable software.

GPL (the GNU General Public License) which certifies that the source code licensed must be freely redistributable.

Handle (file or database) is an object that provides a connection to a file or database component.

Hash (in programming) is an unordered collection of <key, value> pairs in which a given key is used to index the data value associated with that key. Hashes are also known as "associative arrays."

Hit (in a BLAST search) is a high-scoring match between a query sequence and a subject (database) sequence. Hits may contain more than one HSP.

HSP (in a BLAST search) is a "High-scoring Segment Pair" or local alignment segment between the Query sequence and the Subject sequence.

HTML (HyperText Markup Language) is the language used to create Web documents.

Incremental development refers to developing software in stages, where later stages build on debugged and tested earlier stages.

Instance In an object-oriented software environment, when an object of a particular type is created, it is called an instance. Many independent instances of a particular object may exist.

Interpreter is a computer program that reads source code and converts it to machine code during the execution of the program, as opposed to a compiler, which compiles the source code into an executable program that can be run at a later time.

ISO (International Organization for Standardization) is an international organization composed of national standards bodies from more than 75 countries.

Java is an object-oriented programming language developed at Sun Microsystems, designed to be compiled into "byte code" that is portable across any computer architecture that runs the JVM.

JVM (Java "Virtual Machine") refers to a Java interpreter and run-time environment.

LAMP (Linux, Apache, MySQL, and PHP) is a combination of operating system, Web support server, database manager, and Web page script developer that provides an Open Source Web development standard.

Linux is a freely distributable Open Source operating system that runs on a number of hardware platforms, developed mainly by Linus Torvalds and based on the UNIX operating system.

Methods (associated with objects) In an object-oriented software environment, each type of object provides methods or functions for accessing and manipulating the object's internal information.

MySQL is an Open Source Relational Database System that relies on SQL (Structured Query Language) for data entry and retrieval.

Object (in Perl programming) is a special data type that includes particular attributes relevant to the real-life object it is modeling; e.g., a Seq object contains information about a DNA or protein sequence.

Object-oriented programming is a programming method in which data types are defined not only by a data structure, but also by the operations (methods) that can be applied to the data structure. In this way, the data structure becomes an *object* that includes both data and functions.

Open Source is a project that promotes software reliability and quality by supporting independent peer review and rapid evolution of source code. To be OSI certified, the software must be distributed under a license that guarantees the right to read, redistribute, modify, and use the software freely.

Operator (in a programming language) is a function that operates on one or more operands; e.g., the addition operator +.

Perl (Practical Extraction and Report Language) written to facilitate automated and rapid utilization of computer output.

Perl script is a series of well-defined statements in the Perl language, written to accomplish a specific set of tasks.

PHP (Hypertext Preprocessor) is an Open Source, server-side, HTML-embedded scripting language used to create dynamic Web pages.

Pipeline is a series of commands in which the output of each is fed as input to the subsequent command.

PISE (Pasteur Institute Software Environment) is a very robust program for adapting molecular biology software to Web and other human-use interfaces according to a configuration file.

Programming language refers to a precise set of syntactic constructs for manipulating data. Programming languages are developed to make it easier to interact with the underlying hardware, and can range from English-like to highly symbolic.

Programming Query (for database) is a request to retrieve specified information from database tables.

Pseudocode is the outline of a program, written in a simple English form that is both understandable and easily converted into real programming statements.

R is a language and environment for statistical computing and graphics, developed as a GNU project.

Regular expression is a representation of a type of pattern that can be recognized by a computer program.

Relational database is a type of database management system that stores data in the form of related tables.

Scalar variable (in programming) is a single data item, such as a number or string of letters, as opposed to a collection of items, such as an array or a hash.

Schema (database) provides a description of the tables and fields in a database and the relationships between them.

Scope (program variable) is the lifetime of the variable, which is dependent on how and where the variable is declared in a computer program.

Script is a file containing a series of statements that is interpreted by a shell or interpreter.

SGML (Standard Generalized Markup Language) is a system for organizing and tagging elements of a document.

Shell (UNIX) is a layer of software the runs on top of the operating system. The shell interprets user commands and dispatches them to the operating system.

SOAP (Simple Object Access Protocol) is a lightweight, XML-based messaging protocol used to encode information in Web service request and response messages before sending them over a network.

Spreadsheet refers to a table of values in rows and columns, typically represented with a flat file.

String is a series of characters treated as a single entity. Strings are often enclosed in quotes. They may consist of words, sentences, DNA/protein sequences, etc., but do not include spaces, which are used to separate one string from another.

Structured query language (SQL) is a standardized query language for requesting information from a database.

SVMs (Support Vector Machines) are learning machines that can perform binary classification (pattern recognition) and real valued function approximation (regression estimation) tasks.

Syntax refers to the precise format required in a UNIX command or programming language construct.

Tcl/Tk (Tool Command Language/Toolkit) is an easy-to-use graphical user interface package developed by John Ousterhout at the University of California, Berkeley.

UDDI (Universal Description, Discovery, and Integration) refers to a Web-based distributed directory enabling businesses to be listed on the Internet, similar to a traditional phone book's yellow and white pages.

UNIX a multiuser, multitasking operating system developed at Bell Labs in the early 1970s. UNIX was designed to be small, flexible, and, because it was written in the C language, easily portable across different architectures.

Use Case is a component of the UML software specification language, describing how a software system will be utilized by different types of users.

UML (Unified Modeling Language) is a general-purpose notational language for specifying and visualizing complex software systems.

Variable (in a program) refers to a named entity that stores data while the program is running.

Web server is a computer that delivers Web pages.

Web services are a standardized way of integrating Web-based applications using the XML, SOAP, WSDL, and UDDI standards.

WSDL (Web Services Description Language) is an XML-formatted language used to describe the capabilities of a Web service as collections of communication endpoints capable of exchanging messages, developed jointly by Microsoft and IBM.

X Window System is a network-based graphical windowing system developed at MIT. Version 11 (X11) has been adopted as an industry-standard windowing system.

XML (Extensible Markup Language) is a specification designed especially for Web documents. XML allows the creation of customized tags, enabling the definition, transmission, validation, and interpretation of data between applications.

THE RIGHT TOOLS SHOULD BE USED

In the data-rich environment of modern biology, it is necessary to choose the appropriate computer tools to keep abreast of the vast amounts of new incoming data and the new sequence and genome analysis tools. Learning how to write computer scripts that use data exchange formats and Web services, choosing suitable computer hardware and operating systems, and taking advantage of the free software available will go a long way toward keeping up. This chapter focuses on the Perl scripting language because of its popularity and adoption by the bioinformatics community and the availability of many useful extensions suited for biological data analysis. Perl is readily available for most operating systems. With some careful consideration, in most cases scripts can be developed to be cross platform. To start using these tools and resources at a proficient level, basic navigation with an operating system of choice and exposure to Perl is necessary. (Free resources available online and in print are listed at the end of this chapter.)

Perl is a relatively easy programming language to use: "It makes easy things easy and hard things possible whereas professional programming languages make all things equally difficult" (paraphrasing Perl developer Larry Wall). Perl scripts use English-like syntax, lending to its readability and ease of development. Often a series of complex tasks can be reduced to a single statement (at times making the script less readable), hence the Perl motto, "There's more than one way to do it." Today's life scientist is entrusted with analyzing and deciphering a deluge of information on a daily basis. The ability to use Perl and Perl modules to integrate and interact with databases and remote services is an essential skill, allowing one to manage knowledge effectively and efficiently. There are many Perl modules for retrieval and analysis of data.

BioPerl is a popular toolkit developed as a collection of integrated Perl modules for transforming and manipulating sequence data and annotations, accessing remote databases, and parsing output from programs such as BLAST, FASTA, etc. BioPerl also includes a collection of modules (bioPerl-run) that facilitate local execution of programs from the EMBOSS suite or remote execution via the Web using the PISE interface, a computational tool developed

for producing Web interfaces to molecular biology programs. This toolkit goes a long way to eliminate the typical "reinventing of the wheel" and common pitfalls associated with sequence analysis scripts by presenting a consistent interface and method for dealing with the multitude of available report and sequence formats. At the same time, the user is shielded from having to interact with the complex underlying methods used, providing lucid and simplified access to data.

Using Web Services versus Web Sites

Web sites provide a convenient interface for performing a variety of sequence data analyses on biological data from a desktop computer employing a Web browser. Performing sequence analysis tasks via Web sites such as those provided by National Center for Biotechnology Information (NCBI), European Molecular Biology Laboratory (EMBL), and others is quite easy to manage, especially if working with a handful of sequences or small data sets. In many cases, data can be pasted into a form and a button pressed to obtain useful results. But imagine trying to employ the same strategy when trying to analyze hundreds or thousands of sequences!

Working on this large a scale requires tools that can help automate the process and organize the resulting output. The temporal nature of the underlying data being analyzed often mandates that the analysis be repeated as new data become available, thus posing a unique problem. Ideally, the researcher must figure out how to redo the analysis with minimal effort, possibly employing some form of automation and constructing a well-designed "analysis pipeline" to prevent bottlenecks down the road.

Researchers often become reliant on Web sites for specific analyses of a few sequences, and, to avoid the manual labor associated with providing input, extracting results, or saving output, they write scripts to use in conjunction with those Web sites. The same strategy is used to circumvent limitations presented by Web-based interfaces, such as the number of sequences that can be submitted at one time. However, this practice, known as "Web scraping," should be avoided as much as possible for a variety of reasons. One major reason is that Web sites are notorious for changes in layout, often with the intention of providing updated information, which renders the script unreliable and useless. The availability of Web services circumvents this problem. Web services are provided to facilitate automation. For example, most established data providers, such as NCBI, EMBL, DDBJ, ENSEMBL, and BIND, offer bulk download of data and associated analysis programs via FTP. Recent advances in Web services have resulted in most providers offering direct access to their sites by employing standard exchange formats including XML, SOAP, and WSDL. These tools allow users to employ the programming language of their choice to interact with Web services. The use of Web services is described later in this chapter (see p. 569).

Choosing a Language and Platform

Every computer architecture has its own "machine language," represented at the lowest level by patterns of binary data or 0's and 1's. Although it is possible to read or write machine language, this is clearly not feasible because of the large number of instructions that must be given. The advent of traditional programming languages such as FORTRAN, C, or C++ enabled use of a text editor to write human-readable code called "source code." Compilers convert the source codes into the machine language of a specific computer architecture, and the resulting compiled code is called "executable." If written in a portable manner, source code can be moved from one machine architecture to another and recompiled on the second machine, whereas the compiled executable code will only work on one specific machine architecture. A disadvantage of compiled languages is that whenever a modification is made to source code, it must be recompiled before it can run. When performing the repetitive tasks required in the modern biology environment, it is often more desirable to make changes and rerun the computer code without having to recompile it. Computer scripts offer an alternative strategy.

Scripting languages use interpreters instead of compilers, so that the traditional compilation step is bypassed. The tradeoff is that the code must be converted to machine code while the program is executing; therefore, the execution time for scripts is longer than for comparable compiled programs. One must carefully consider whether execution speed is an important consideration when deciding whether to write an executable program or a script. In cases where the code is not too long or if speed is not that critical, using scripting can speed up the development of new computer software. Popular scripting languages include Perl, Python, Ruby, and PHP. Java is another popular computer language that is partially compiled into "byte code," which is portable among different architectures. There is still some interpretation being done while the Java program is running, so the execution speed is slower than that of natively compiled programs. In a sense, scripting can be thought of as a special form of programming. For the sake of simplicity, in this chapter we will use the terms "programming" and "scripting" interchangeably, although it is important to understand the distinction in practice. Scripting provides a way to automate repetitive tasks and handle large amounts of data. In addition, writing a customized script provides more flexibility than static Web interfaces, giving users the ability to customize the analysis fully for their needs.

In this chapter, we introduce some of the tools for creating powerful bioinformatics solutions. A question often asked is, "Am I writing a Perl program or a Perl script?" The apt answer from Larry Wall is, "Script is what you give an actor, but a program is what you give an audience."

Choosing a Hardware Platform

In choosing a computing environment, availability of processor (CPU) time and type of memory layout (RAM) are prime considerations. Modern small computers offer inexpensive, fast processors and large memory and storage capabilities. Clustering small computers into a supercomputer network (Beowulf clusters) has become quite common and under certain circumstances can substitute for a more powerful computer system. Recent advances in distributed computing allow harnessing of unused processor time using "grid" technology and provide options of scaling up sequence analysis applications to keep up with the pace of more incoming data.

Grid technologies are unique in that they leverage the existing infrastructure and therefore do not require up-front and ongoing investment. With this technology, a collection of a few to thousands of desktop computers running various operating systems can be operated in a coordinated manner by splitting the workload among the computer pool. The CONDOR framework, available from the University of Wisconsin (http://www.cs.wisc.edu/condor/), is an example of this technology. Similar frameworks are available from commercial vendors such as United Devices (http://www.ud.com). These systems offer Perl interfaces to submit, monitor, and manage large sequence analysis jobs on a cluster in serial or parallel configurations using Open Source standards of the Globus Toolkit (The Globus Alliance, http://www. globus. org).

Choosing an Operating System

The operating system is a layer of software that facilitates access to computer hardware by human users and program applications. Operating systems typically provide memory management, file management, task scheduling, and communication to and from input/output (I/O) devices like keyboard, network interfaces, etc. Such software is called "low-level" or "systems" software, whereas program applications are sometimes referred to as "higher-level" code. Some systems are simple single-user systems capable of executing only a few tasks at any given time. Others are multiuser, multitasking systems, meaning that many users can be logged in at once and each can have several tasks that are currently active.

Some of the more popular operating systems in use today include UNIX/Linux variants, Microsoft Windows, and MacOS. Systems such as Linux, Macintosh OS X, and FreeBSD have roots in UNIX, whereas Microsoft Windows grew out of the older DOS system. When the term UNIX is used in the remainder of this chapter, it applies to any UNIX or Linux variant. Linux has gained tremendous popularity and is widely employed for software development and providing services and resources using the Web. Many applications have been developed using the "LAMP" framework—Linux, Apache, MySQL, Perl or PHP or Python. Although working with Linux can be a challenge, new distributions of Linux do a good job of detecting hardware and are relatively easy to install.

For interested users who have not experienced Linux and prefer not to install it on their computers, an excellent starting point is the KNOPPIX project (http://www.knoppix.org). KNOPPIX is a live CD, a bootable distribution with a collection of GNU/Linux software, automatic hardware detection, and support for many graphics cards, sound cards, SCSI and USB devices, and other peripherals. KNOPPIX can be used as a Linux demo, educational CD, and even for rescuing damaged operating system installations. It is not necessary to install anything on a hard disk. Because of on-the-fly decompression, the single-installation CD can have up to 2 GB of software installed on it. KNOPPIX allows customization of installed applications. Bio-KNOPPIX is a live CD distribution developed using KNOPPIX and includes many popular sequence analysis tools and programs such as EMBOSS, BioPerl, CLUSTALW, the LAMP framework, and R statistical analysis and plotting tools, as well as other applications (http://bioknoppix. hpcf.upr.edu/).

For most practical purposes, the choice of hardware architecture, operating system, and programming language is not going to be a limiting factor for developing bioinformatics solutions. A careful analysis of the usefulness of computers that offer ease of use, such as a desktop computer, and those that offer raw computing power, such as a supercomputer, should be performed, especially for projects involving elaborate analyses. However, with the advent of Web services and grid computing technologies for data processing and storage, the distinction between operating systems and architecture is becoming blurred from an end-user perspective. A "cloud" of dedicated services whose underlying architecture and software are completely transparent to the user will eventually be used.

Using Free Software

The Free Software movement was spearheaded by Richard Stallman in the 1980s, and the GNU Public License (GPL) policy arose from his efforts. The word "free" in this case means freedom regarding what one does with the software, although GPL packages are often cost-free as well. This policy requires that software licensed under GPL be freely distributable with no restrictions by anyone who obtains the software. A related movement that has recently gained popularity among software developers and vendors is the Open Software Initiative (OSI). This initiative requires that source code be available at no cost or for only the cost of distribution, but allows certain restrictions on what can be done with the software. The http://Open Source.org Web site lists companies that currently participate in OSI. Many Internet and programming tools are OSI certified, as are many applications available at the following Web sites: http://sourceforge.net, http://freshmeat.net, and http://www.bioinformatics.org.

There is a vast repository of free software for biological data analysis, and new packages are continually being added. Many of the free software packages run on any of the popular operating systems; however, applications that have been developed under the Linux operating system and run under Linux constitute a large portion of this collection. The EMBOSS suite of programs (European Molecular Biology Open Software Suite) at http://www.emboss.org includes hundreds of programs for DNA and protein sequence analysis on UNIX/Linux and Mac OS X operating systems. NCBI (http://www.ncbi.nlm.nih.gov) provides the BLAST suite of tools for UNIX, MacOS, and Windows systems.

Some Web sites maintain lists of the available software tools (e.g., http://www.bioinformatik.de). Using a Web search engine (e.g., http://www.google.com) and entering specific keywords to describe the task of interest can also be an effective method for locating software. Searching E-mail distribution lists and newsgroups then posting succinct questions with suitable background information can also yield useful pointers and suggestions. Before developing a new sequence analysis tool, it is always a good idea to see whether there is an existing tool or one that can be adapted to your needs.

WHAT ARE THE STRATEGIES FOR WRITING SOFTWARE?

To be most effective, software development requires careful planning, a phased-in approach, close attention to style, and extensive use of documentation so that the logic of the programming steps can be followed at a later time.

Problem-solving Strategies

For life scientists who have limited exposure with programming, it helps to think of how an experiment is planned, because similar steps are followed in designing a program. The approach to problem solving is usually to partition the problem into distinct logical steps. In programming, each of the steps must be well thought out and methodically planned before writing the program. Good results cannot be expected without good planning! Furthermore, there is more than one way to solve a problem. In general, if a particular design becomes overly complicated, a different strategy should be considered.

Most fundamental computer operations are either arithmetic or logical in nature. Arithmetic operations are intuitive, but the control structures provided by a programming language include elements of the form "if condition A is true, do this, otherwise, do that" or "while condition B is true, keep doing the following." It is imperative that these types of logical flow control be understood. Before writing a script, it is helpful to sketch out the flow of logic using a mixture of English and a programming language. This step is referred to as writing "pseudocode." With pseudocode, the intent is to get the correct logic written without having to worry about the exact details of the computer language syntax. It is important to have the big picture laid out before delving into the details. Some examples of pseudocode are provided later in this chapter.

Another very important technique is called incremental development. This approach favors coding a small portion of the program and making sure that it works correctly before writing any additional portions, rather than trying to code an entire program and then trying to run it. Tackling each program component as a "proof of concept" and then proceeding to build a solution on top of that component can often be advantageous.

When designing code, it is also good practice to aim for modularity. Here, the object is to divide the programming objective into logical modules, each of which accomplishes a small task. Such modular code can easily be reused in future projects, resulting in saved time and money. An important component of reusing code is to make the code flexible. For instance, rather than writing a script that will read ten sequences and perform some sort of analysis on them, write the script so that it will work with any number of sequences. Instead of having a particular file name hard-coded into a script, have the script obtain the file name from a command line argument given when the program is first called so that the script is useful beyond the immediate needs. There are many bioinformatics problems that share common components; e.g., opening a sequence file and verifying that the sequence is formatted correctly. A repository of modules that can be mixed and matched to fit current requirements if of great value. Avoid "reinventing the wheel"!

Phases of Software Development

Software development is the process of developing software in an orderly manner. It includes not only the actual writing of code but also the preparation of requirements, objectives, and deliverables. Depending on the task at hand, one could be developing a stand-alone solution to be used by an individual or group working in close proximity or a highly integrated enterprise-wide deployed solution developed by a team that is to be widely used by the community. In either case, an orderly process is needed.

Statistics compiled by the Standish group (1994 Chaos Report, http://www.standishgroup.com/sample_research/chaos_1994_1.php) for software projects indicated that only 16% of the projects were on schedule and within budget. A total of 53% were considered as not meeting objectives because the delivered functionality was less than in the original design, cost was over budget, or the delivery was late, and 31% were considered impaired and were canceled during development. These statistics indicate the importance of careful planning in software development. An understanding of the following traditional software development phases can help one develop reliable and robust solutions regardless of project size:

- Requirement phase: Understanding, documenting, and agreeing on project requirements.

- Planning phase: The cost and duration of the project are determined.

- Design phase: Approaches and methods to solve the task at hand are chosen.

- Implementation and integration phase: Programming practices, selection of language, code reuse, and testing methods are decided.

- Maintenance and evolution phase: Methods for reporting errors and producing enhancements are included.

Documentation does not appear as a phase, but it is integral to all phases and is possibly the most underutilized and ignored aspect of software development. It is often difficult to strike a good balance between the time and effort spent in developing requirements and specifications and time and the effort spent actually programming the solution (often called "paralysis by analysis"). Unfortunately, 85% of the errors are committed during the design and requirement phases, which together generally comprise a mere 11% of the total budget, a very suggestive indicator that not enough time and effort are put into the planning phases of software development.

The development phases listed above have evolved to a point where the distinction between various phases is blurred in an effort to adopt various software life cycle models. To keep up with the pace and growing demand for software solutions, software development methods have adopted more flexible and agile approaches. The Manifesto for Agile Development discusses these principles (http://agilemanifesto.org). Extreme Programming (XP), Rational Unified Process (RUP), and Open Source are examples of this sort of approach.

Discussion on various software development methods and their benefits can be found at http://www.martinfowler.com. Glass (2002) provides good insights into the process of software development, and Rehman and Paul (2002) have a good introduction to various stages of software development, which is available in its entirety as an Open Source book at http://www.faqs.org/docs/ldev/.

Programming Style and Correct Syntax

When writing code, statements must be precisely correct. For Perl programming, the major thrust of this chapter, the Perl interpreter cannot "guess" what was intended, even if there are obvious errors in the syntax of a script. Parentheses, brackets, and braces must always be properly paired and carefully placed in the correct location in the code. Pay close attention to such details in the sample scripts in this chapter. For help using the functions and modules of Perl, documentation can be accessed via the "perldoc" command or at http://www.perldoc.com/. For those unfamiliar with Perl, the command "perldoc perldoc" is a good starting point.

Scripts should also be rigorous with respect to error checking. If the user omits a parameter or specifies something nonsensical, or if a calculation inside the script yields an unacceptable result, the script should produce a meaningful error message so that the user can tell what has gone wrong. Some things to check for are missing data, unexpected input, or operations that might result in division by 0.

In addition to correct syntax, it is vitally important to develop a programming style conducive to reading, debugging, and maintaining the code. It is good practice to use plenty of white space (lines, tabs, and blank spaces) in programs and to be consistent in the use of indentation. Often code will be passed along to another programmer. Thus, code should always be written with this in mind. Include comments to explain sufficiently what the script does and any assumptions that may have been made during development. The comments should enable anyone to read the code and understand what it does without having to scrutinize every line of the script. Style guidelines for Perl can be found by searching for "perlstyle" at http://www.perldoc.com/ or using the "perldoc perlstyle" command.

Commenting and Documenting Code

As a bare minimum, comments need to explain the purpose of a script, what the input requirements are, and any outputs generated. The programmer's name should be included, as well as a version number for the script. The version number should be increased as changes are made, and it should track archived copies of previous versions of the program. Keeping working versions of a script is an essential practice for both practical usage and debugging purposes. Code versioning systems such as Concurrent Versions System (CVS) are available free of charge and are highly recommended. Subversion is a relatively new and versatile version control system being widely adopted by the Open Source community and is described in the free online book (http://svnbook.red-bean.com or subversion.tigiris.org).

More sophisticated programs warrant more detailed comments so that subsequent code writers can thoroughly understand the strategy behind the implementation. Always write comments as if someone else will have to continue development of the program. Very often, the code writer has to revisit his or her own code after months or years. This will be a time to thank or curse oneself, depending on the usefulness of the comments.

Scripts should be adequately documented. Using the Perl POD (plain old documentation) markup language is recommended, allowing the documentation to be included in the same file as the Perl script itself. POD has the added advantage of easy conversion of the included documentation to HTML or PDF using freely available scripts.

OBTAINING AND INSTALLING PERL IS THE FIRST STEP

Perl is widely available for a variety of operating systems and computer hardware systems.

1. For UNIX-based systems (including Macintosh OS X), Perl is included with the operating system. In cases where Perl is not available, ensure the correct path to the Perl binary is used, e.g., /usr/local/bin/perl instead of /usr/bin/perl before proceeding with installation. Most vendors have a compiled binary installation or package available from their site for download.

2. Microsoft Windows users can download ActivePerl from (http://www.activestate.com/activePerl). Alternatively, users interested in working with a Linux-type environment on the Microsoft Windows platform are encouraged to install Cygwin tools from http://www.cygwin.com, which consists of many popular GNU-based tools and Linux-based utilitarian resources. This choice may be of benefit for Windows users who do not have access to Linux machines because many Linux applications can be recompiled and used on Windows in the Cygwin environment with little or no modification.

3. For pre-OS X Macintosh operating systems, visit http://www.macperl.org.

4. For other operating systems visit http://www.perl.com/CPAN/ports/.

Having established access to Perl, choose a text editor that is aware of Perl syntax, can display the statements using different colors for built-in functions and variables, and has basic bracket-matching capabilities. This choice makes error spotting easier and imparts readability to scripts. There are many editors available on the Linux platform, including emacs, vi, and nedit, that meet the above-mentioned criteria. Syn is a reasonable editor for Microsoft Windows that can be downloaded from http://syn.sourceforge.net. Make sure that the editor is configured so that it wraps text on the computer screen without inserting a new line character into the code.

Integrated Development Environments (IDEs) provide editing and testing capabilities for developing Perl software. Activestate (http://www.activestate.com) offers Komod and Visual Perl, and Perl Builder is available from Solutionsoft (http://www. solutionsoft. com). There are a few Open Source IDE packages, noteworthy among which is the Eclipse project, a cross-platform (works on Linux, Macintosh, and Windows) package that has plug-ins for many popular languages including Perl (http://www. eclipse.org). The goal of this project is to provide building blocks and a foundation for constructing and running integrated software development tools.

DEBUGGING PERL SCRIPTS LOCATES AND CORRECTS ERRORS

An important stage of developing software consists of locating and correcting bugs or errors in the programs. Error messages emitted by the Perl interpreter are often terse and must be scrutinized carefully to reveal mistakes in the command or script. In many cases, some detective work is required. The information may look cryptic at first, but if read very carefully, these error messages do provide at least a hint about the cause of the problem. The line numbers indicated in error messages may not immediately identify the offending line of code. Frequently a mistake in a script will not be detected until one or more subsequent lines are parsed and the interpreter can no longer make sense of the code. For this reason, when debugging scripts based on line numbers given in error messages, examining the neighboring lines of code can be helpful.

A debugger can be a helpful tool for discovering errors. Debuggers allow one to set breakpoints easily that interrupt execution of a script and to monitor values of variables as the program executes, easing the process of pinpointing the error. An example of the output of a debugger program is shown in Figure 12.1.

Note the capability of running the program in steps to locate an error and of showing possible errors in the programming code. For smaller or simple scripts, the use of "print" statements peppered around the source code will usually suffice to find an error. Syntax errors in scripts are easier to locate compared to errors and flaws in the logic of a script. In the latter case, closer inspection is usually required, but a debugger can still be helpful by revealing syntactic flaws in scripts.

Perl has a built-in debugger that can be invoked by using the -d option with the Perl command. This debugger is character-based and at times difficult to navigate. Ptkdb (http://world.std.com/~aep/ptkdb/) is a free cross-platform graphical debugger, implemented as a Perl module built using the Tk (toolkit) that provides a graphical interface to Perl programs. One can easily set breakpoints by clicking on the displayed line number and adding variables to monitor for changing values as the program executes. Data Display Debugger DDD (http://www.gnu.org/software/ddd/) is a versatile debugger for the UNIX operating system that may be used with many programming languages, including Perl.

HOW TO LOCATE AND INSTALL PERL MODULES

One of the most attractive features of Perl is the ability to reuse code in the form of programming modules, which are composed of lines of Perl program code that perform repetitive tasks. Modules that perform a variety of tasks may be obtained from CPAN, the Comprehensive Perl Archive Network (http://

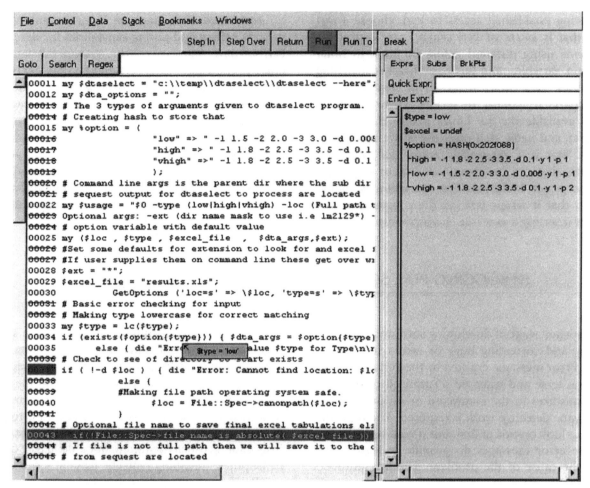

FIGURE 12.1. Screen capture of Perl debugging program ptkdb described in the text.

www.cpan.org). CPAN has a massive collection of modules (~6100 as of March 2004) that can be browsed, searched (http://search.cpan.org), and downloaded, complete with documentation and Perl source code. CPAN also includes sample Perl programs that use the module and pointers to articles, books, and a FAQ list. The modules are categorized, but the best way to locate a module is to use the text search feature provided on the Web site.

For ActivePerl users, the Perl Package Manager (PPM) is the CPAN equivalent and can be browsed at http://ppm.activestate.com. PPM does not have an easy way to search for modules. To circumvent this problem, visit CPAN and identify a module and then see if it is available as PPM. Perl even provides a module called CPAN to install new modules from the CPAN archive! This module is part of the default

CPAN installation. Some CPAN modules are dependent on other modules also being available, and these other modules also need to be obtained. A benefit from installing new modules using CPAN is that such dependencies are recognized and the additional modules are automatically downloaded and installed.

In case a needed module is not available, it can be installed in one's individual user space (e.g., UNIX account). To check if a needed module is available, type the following on the command line:

```
perl -M<module name>,
    e.g., perl -MXML::Twig
```

Note: No space after -M

```
or perldoc <module name>,
    e.g., perldoc XML::Twig
```

If the module is installed, Perl will continue without printing any errors or will display the documentation. If the module is not installed or found, this command will generate an error message because the Perl interpreter will not be able to locate it.

The following steps illustrate how one can install modules from CPAN in an individual account or user space. Note that all commands are case sensitive, as are module names.

1. Create a directory, e.g., `myperllib`, for installation of the modules. Note that the user is free to name the directory appropriately, remembering to use the full path while referring to it, e.g.,

```
/home/export/molbio/myname/
    myperllib
```

2. Type:

```
perl -MCPAN -e shell
```

If this is the first time the CPAN module has been invoked, it will ask a series of questions with multiple-choice answers to identify the closest repository and some configuration questions. (If this step fails, please see item 9 below.)

3. Once the questions are answered, the command prompt appears, i.e.,

```
cpan>
```

4. Configure cpan to install modules in the directory you created:

```
cpan> o conf makepl_arg LIB=/
    home/export/molbio/myname/
    myperllib
```

5. Now ask to obtain, compile, and install the module: e.g.,

```
cpan> install Spreadsheet::
    WriteExcel
```

6. If supporting modules are needed, there will be further prompts and the modules will be installed automatically.

7. Now that the module is installed, let Perl know where to look for it by setting the value of PERL5LIB environment variable using the command:

a. ```setenv PERL5LIB/home/export/
    molbio/myname/myperllib```

(for tcsh or csh shells)

b. ```export PERL5LIB=/home/export/
    molbio/myname/myperllib```

(for bash or sh shells)

c. The location can also be indicated by adding the following line in the start of a script:

```
use lib "/home/export/
    molbio/myname/myperllib";
```

8. If for some reason installation through the CPAN module is not possible by the above steps, use the following strategy:

a. Download the module of interest from the CPAN site.

b. Extract the files.

c. Go to the resulting directory and type the following commands:

```
>perl Makefile.PL LIB=/home/
    export/molbio/james/myperllib
>make
>make test
>make install
```

d. Refer to http://sial.org/howto/Perl/life-with-cpan/ for more tips.

For ActivePerl users, the Perl Package Manager (PPM) can be used in the following steps:

1. type ppm on the command line, resulting in the prompt:

```
ppm>
```

2. search for a module using the search command:

```
ppm> search excel
```

3. The search provides a list of matching modules.

4. Choose the appropriate module and install it:

```
ppm> install XML-Excel
```

If the name of the module is known, `install` can be called directly with the module name. (*Note:* For PPM on Windows "-" is sometimes used as the separator in module names instead of "::". For example, the statement XML-Excel on the MS Windows platform is synonymous with XML::Excel on other platforms.) Installing modules through PPM is highly recommended. Many authors provide PPM-compatible modules with instructions on how to point the PPM program to their repository for downloading and installation.

Perl Subroutines, Packages, and Modules

Subroutines

As the functionality and complexity of scripts increase, it is imperative to separate out repeated tasks and logical components into routines that can be called on when needed (in a similar manner to the built-in functions of Perl). Subroutines facilitate this task. Subroutines can be created or declared by using the "sub" keyword and can be located anywhere in the script, although good practice is to have them at the end or at the beginning of a Perl script. Alternatively, subroutines can be stored in separate files which can be loaded and used in a script by using the require, do, or use keyword:

```
sub NAME {
          ACTION;
          }
```

The subroutine NAME can then be called as

```
$out = NAME();
```

To send data to a subroutine, parameters are passed as a single list of values called scalars (integers, strings of letters, etc.). Arrays can be passed by reference by prefixing the array name with a backslash. @_ is the mysterious placeholder for parameters being passed to the subroutine and is a private array within the subroutine, for example:

```
$out = calc_len_diff($seq_in,
     $ref_seq);
sub calc_len_diff {
     my ($a,$b) = @_;
     # Now $a= $seq_in and
     # $b = $ref_seq
     my $diff = length($a) -
          length($b);
     return($diff);
}
```

A subroutine's "return" statement, shown on the last line, returns any specified values to the caller and exits the subroutine. If no return statement is present, the return value of the subroutine is the value of the last expression evaluated.

In Perl, all variables are global variables by default, meaning that, unless specifically stated otherwise, they are accessible from anywhere in the script. Creating private variables, also called lexical variables, can be accomplished using the "my" declarator. The values of lexical variables are accessible only within the block or subroutine in which they are defined. This avoids collisions with variables declared elsewhere. Constraints on the number of parameters and types of data passed to subroutines can be specified in a script using symbols that correspond to scalar ($), array (@), or other types. For example, declaring a subroutine that requires two scalars as parameters can be done as follows:

```
sub calc_len_diff($$) { ... }
```

To simpify the use of subroutines, they should reside in a file that other scripts can call on, making the code more maintainable. Thus, instead of having to modify each script individually, one can change this single file. Perl libraries are files containing collections of subroutines written for a given purpose. They consist of subroutines and must have the characters 1; as the last line. (Whenever a library, package, or a module is used, Perl has to have a true value for the last expression in the file, hence the 1; characters.) The preferred naming convention used for a library file is

```
NAMElib.pl, e.g., seqcalclib.pl
```

To use subroutines from a library file, use the require statement in a script before calling the subroutine. If the seqcalclib.pl file is not in the same directory as the script, then direct Perl to find the file using one of the following options:

1. On the command line, set the environment variable (can also be set permanently in the shells default user values in .cshrc or .bashrc):

```
setenv PERL5LIB "/export/home/
     molbio/james/myPerl/lib";
```

2. In the Perl script add either of the lines:

```
push (@INC , "/export/home/
     molbio/james/myPerl/lib");
use lib "/export/home/molbio/
     james/myPerl/lib";
```

Now, when a library is called using

```
require "seqcalclib.pl";
```

the program will find it, without having to describe a full directory path to reference the file, i.e.,

```
/export/home/molbio/myname/myPerl/
     lib/seqcalclib.pl
```

THE USE OF ":: " BY PERL TO INDICATE DIRECTORY PATHS

Conflicts among Perl program names are avoided by a nomenclature that has a similarity to file directory paths. File systems use slashes (`/home/user/myfile` or `c:\user\myfile`) as separators in paths. Instead of slashes, Perl uses the double colon (`::`). Therefore, `Bio::Factory::EMBOSS` is a directory Bio with a subdirectory Factory in which the file EMBOSS.pm is located.

Packages

A package is very similar to a library file, but it has added functionality to avoid collisions while permitting scripts to use the same variable and subroutine names. To convert `seqcalclib.pl` into a package add the following line to the top of the file:

```
package seqcalc;
    sub calc_len_diff {
        my ($a,$b) = @_;
        # Now $a= $seq_in and
            $b = $seq
        my $diff = length($a) -
            length($b);
        return($diff);
                }
1;
```

To use the package in a script, include the file using the require statement and then call upon the subroutines that are part of that package using the convention `package_name::subroutine`. For example,

```
$seq_in = "ATGC";
$ref_seq = "ATTTT";
$size = seqcalc:: calc_len_
    diff($seq_in,$ref_seq);
```

If there were another subroutine that called `calc_len_diff` and this subroutine was part of the script, or even included from another library (and possibly doing very different tasks), both of them could coexist without collisions. The default package for all scripts is the "main" package; it does not need any inclusion, so if there were a subroutine called `calc_len_diff` as a part of a script, in reality it would be `main::calc_len_diff`, which is how Perl separates names.

The use of packages becomes crucial as the functionality and size of scripts grow and as the need increases to reuse portions of the code and work with code developed by others while avoiding conflicts with the names used by others. To visualize this concept easily, think of a hard drive where one has files and directories. Files with identical names can exist as long as they are in separate folders or directories, but they cannot co-exist in the same directory. To open or work with files in separate directories, they must be referenced by their full path names, which then uniquely identify them.

Modules

A module is a reusable collection of code for carrying out specific tasks. The concept of modules is the cornerstone for reusing software in Perl and leveraging from work done by others to extend and add functionality to scripts. Perl version 4.0 was "package" based; in version 5.0, the concept of packages was extended and superceded by modules. Modules have some similarity to packages, but they offer a more robust method of interaction.

Each module has a defined set of variables and functions that can be called on. The first letter of a module name is usually capitalized. To employ a module in a script, the "use" or the "require" statement must be specified before the module is used. If a required module is missing, then the script referencing it will not proceed past the "require" statement, an important distinction between using the "require" and "use" statements.

A module provides one or more objects. An object is a collection of information about a particular type of entity, such as an annotated sequence. Objects provide functions called methods, which allow the user to retrieve and/or manipulate the data associated with the object. Every object has a method

---

EXAMPLE OF USING Getopt::Long

Create the script getopt.pl and run it on the command line

```
#!/usr/local/bin/perl -w
use Getopt::Long;
GetOptions( "name=s"=>\$in_name , "age=i"=>\$in_age);
print "Sample: $in_name is $in_age days old\n";
```

Try the options

```
getopt.pl -age 10 -name ecoli
```

or change the order and style of entry,

```
getopt.pl -n ecoli -age 10
getopt.pl -age=10 -name=ecoli
```

The conclusion is that Getopt::Long greatly facilitated reading options from the command line.

---

named new() that creates a new instance, or occurrence, of the object. The data in an object cannot be accessed directly; it is necessary to call a method to work with data in an object.

Methods are called with the "arrow" syntax, e.g., ModuleObj->new(). It may help to think of an object as a container that holds all information associated with a certain class plus the methods for accessing that information. Do not make the mistake of attempting to print the contents of an object directly by simply placing the object in a print statement. If this is done, the result appears as a nonreadable, hexadecimal HASH value, e.g., 0X10801C. Instead, look for object modules that return information about the object, and print the values returned by these modules.

Perl has a large number of standard modules included in the standard distribution. Many other modules have been contributed by different groups or individuals. There are modules for creating Web interfaces in a script, connecting to a database, creating graphics, and many more. The enormous collection of contributed modules can be viewed at http://search.cpan.org/.

One module included in the default Perl installation is Getopt::Long. This module gathers all command line options and places them into named variables that the script can then access. The ability to pass arguments and values to a script makes the script more flexible. Additionally, running a script via a single command line as opposed to prompting the user for various options makes the script more amenable for use in high-throughput, automated data-processing pipelines. Hence, Getopt::Long is an important module for bioinformatics applications.

The Getopt method GetOptions() accepts pairs of description strings and references to variables as illustrated in the example below. The description string identifies a particular option and the type of value it must have. The reference to a variable tells GetOptions where to store the value provided for the option. The Getopt::Long module allows the flexibility of providing the options on the command line in any order and even using abbreviations (e.g., -n for -name). The option description strings and variable references are connected by the => operator, which defines key value pairs.

In the example above, the description string "name=s" indicates that the "-name" option must have a value of type string (s) supplied on the command line. The variable reference \$name specifies that this value is to be stored in the variable $name. The "age=i" description and variable reference tell GetOptions that the "-age" command line option requires an integer value (i) which is to be placed in the $in_age variable.

To learn more about this module, refer to the Perl documentation by typing the command:

```
perldoc Getopt::Long
```

Perl Procedures

With this rudimentary introduction to Perl modules and CPAN, we can now locate some modules, install them, and begin the journey of code recycling. To learn more about modules not bundled with the Perl installation and ability of Perl to automate tasks, we will automate the infoseq program, which is part of the EMBOSS suite and provides a text report listing the sequences in a single file, their lengths, and %gc contents and extracts the GenBank number and other useful information. The report is plain text and hard to read. To add readability and functionality to the report, we will transform the report into the Microsoft Excel (.xls) format, link the sequence back to NCBI, and flag sequences that have %gc >70%. The Excel report can be opened in the OpenOffice package (http://www.openoffice.org) on various platforms or in the traditional MS Office Excel.

To tackle this task, let us examine the built-in Perl functions for executing external programs from within Perl scripts followed by a method to generate Excel format files easily using a Perl module.

Methods for Calling Programs from within Perl

1. `system`: This function launches the program with provided arguments, e.g.,

```
system("getopt.pl","-name",
    $name,"-age",$age);
```

and runs the `getopt.pl` program with the values stored in the scalar `$name` and `$age`, as if it were run on the command line. Depending on the task being executed, it is advisable to provide the full path, i.e., `/home/molbio/myname/bin/getopt.pl` instead of just the program name, so that Perl can correctly locate and execute the program.

2. `exec`: Similar to system with a subtle difference in how its handles execution of the external program. The `exec` function mimics the UNIX exec system call.

3. Back ticks: These are useful when output from the resulting program is to be stored in a variable in the script, e.g.,

```
$out = `getopt.pl -n $name -
    a $age`;
```

Now `$out` will contain the string output by `getopt.pl`. Again full path names are recommended for programs being executed. `@out` can be used if multiple lines are being output.

4. `open`: Typically used for opening a file for reading or writing, it can also be used for doing the same with external programs, using the pipe symbol | (located above the Enter key along with the \ on most keyboards) by appending or prepending it to the command line of the program being executed:

```
open($write_to_me, "| getopt.
    pl args");

open($read_from_me, "getopt.
    pl args |");
```

Note: If bidirectional communication is desired, then consider using `IPC::Open2` modules. For command line programs needing substantial automation, also consider the Expect module, keeping in mind that many modules used for interprocess communication (IPC) may not work across all platforms.

The system call will be used for the following example. Searching the CPAN site for the keywords "excel write" results in a few hits, one of the first is `Spreadsheet::WriteExcel`. Browsing through the description and README suggests that this module will help us with our task and is available for UNIX and

Windows! Using the steps listed above, we will install the module. Note that one does not need Excel or any other software package installed to create the XLS format output; this is a pure Perl implementation.

The following is sample script that demonstrates the execution of the EMBOSS infoseq program and conversion of the results to the Excel format with additional information and links to NCBI. This script involves a single task, but one can easily run multiple tasks. For developing an analysis pipeline or work flow, check the Biopipe project and HOWTO at http://www.bioperl.org.

Code	Comments/explanation
`#!/usr/bin/perl -w`	`#!/usr/bin/perl` (path to the perl interpreter) is always needed if script is to be directly executable.
`use strict;`	Using the strict module instructs Perl to do strict error checking and generate an error message if any unsafe constructs are found.
`use Spreadsheet::WriteExcel;`	Use the Spreadsheet::WriteExcel module.
`# Input sequence file`	
`my $file = "batchseq.tfa";`	Not a good idea to hard-code the filename, but okay for a prototype script.
	One can also read values from command line using $file = shift; or $file = $ARGV[0];
`# Execute the infoseq program from the EMBOSS suite`	
`# We are interested in name, gi, length, %gc and` `    description`	
`# Resulting output is stored in the array @out`	
`my @out = `infoseq -only -name -gi -length -pgc-` `description $file`;`	The backtick marks are used to execute the command line within them and return the command output lines as an array.
`# Create a new workbook and save as a file called` `    example.xls`	
`# and add a worksheet called Summary to that.`	
`my $workbook = Spreadsheet::WriteExcel->` `new("example.xls");`	Make a new Spreadsheet::WriteExcel object called $workbook.
`my $worksheet = $workbook->add_worksheet('Summary');`	Add worksheet to $workbook.
`# Set text bold, red, underlined for values`	
`# designated as high`	

`my $high_val = $workbook->add_format();`
`$high_val->set_bold();`
`$high_val->set_color('red');`
`$high_val->set_underline();`
	Add a format and fill in attributes.

`# NCBI URL to use for retrieving sequence`	

`my $ncbi_url =`
`"http://www.ncbi.nlm.nih.gov/entrez/query.fcgi?cmd=search`
`&doptcmdl=GenBank&db=Nucleotide&uid=";`
	This statement must be typed as one long line. It is wrapped here only to fit into our table.

`my $row = 0;`	Worksheet rows start at 0.

`# Read one line at a time and split it to get the`
`# name, gi, length, %gc and description`
`foreach my $line (@out) {`
`    my ($name,$gi,$length,$pgc,@desc) =`
`split(/\s+/,$line);`
	This line shows how to store the results of the split function into several different variables.

`    # Use gi for linking to NCBI; create the url in` `the worksheet`	
`    my $url = $ncbi_url . $gi;`	Use string concatenation to put the gi number at the end of the URL
`    $worksheet->write($row,0,$name);`	Write the name in column 0.
`    $worksheet->write($row,1,$url,$gi);`	Write the URL in column 1.
`    $worksheet->write($row,2,$length);`	Write the length in column 2.

`    # If %gc is > 70 make text red,bold,underlined`
`    if($pgc >= 70 ) {`
`        $worksheet->write($row,3,$pgc,$high_val);`
`    } else {`
`        $worksheet->write($row,3,$pgc);`
`    }`
	Write the %GC in column 3.

`    my $full_desc = join("",@desc);`	The join function connects the array elements, separated by space, into a single string.
`    $worksheet->write($row,4,$full_desc);`	Write the full description in column 4.
`# Increment $row for putting data on next line` `    $row++;`	
`} # end of foreach loop`	End of foreach loop.

PERL HAS MODULES FOR ACCESSING THE WEB

The Library for WWW access in Perl (LWP) provides a collection of modules to access resources over the Web, allowing one to perform complex tasks over the Internet with minimal effort. LWP is the foundation that most modules rely on for carrying out network access for posting and retrieving data (at times mimicking a browser). LWP is one of the default Perl modules. If it is not installed, please follow the steps listed earlier to install it. We will examine the mirror function from LWP::Simple module. The following is an example of script that checks availability of a BLAST database at the NCBI FTP site and downloads it if a more recent version than one stored locally is available. This example provides a method for updating a local sequence database automatically.

This example works with a single URL, but it can easily be modified to use multiple URLs and can be invoked to run as a scheduled task on MS Windows or using the cron or at command in UNIX. As discussed earlier, Web scraping by repeatedly running scripts should be avoided if possible; however, LWP does provide many functions to accomplish Web-scraping tasks with ease and reliability. If the need for extensive manipulation of Web pages through a script is required, consult the WWW::Mechanize module.

Code	Comments/explanation
`#!/usr/local/bin/perl -w`	Provide path to Perl interpreter on first line.
`use LWP::Simple;`	Use the `LWP::Simple` module.
`# Provide the URL for file to monitor`	
`my $url =` `"ftp://ftp.ncbi.nlm.nih.gov/blast/db/FASTA/pdbnt.gz";`	Declare and initialize lexical (non-global) variables.
`my $file = "pdbnt.gz";`	File to use for comparison with NCBI database. The "gz" indicates that the file is a compressed (g-zipped) file.
`# Invoke the LWP mirror utility which will compare` `# the files by date stamp`	
`my $rc = mirror($url, $file);`	Call the LWP `mirror()` method and save the return code in a variable.

`# If downloaded OK then code 200 is returned`
`# If file is same date as local copy, code 304`
`# is returned. For details on various codes, see`
`# ftp://ftp.rfc-editor.org/in-notes/rfc2616.txt`
	These comments are important so that the reader knows what the return codes mean.

`if ($rc == 200) {`
`# If download is OK, code 200 returned`
`# If file has some date as local copy, code 304 returned`
	For code details, see ftp://ftp.rfc-editor.org.

`    # Could send email to inform user !`
`    print "$file downloaded correctly\n";`
`}`
	See the Sendmail module.

`elsif ($rc == 304) {`	Return code 304 means file is up to date.
`    print "$file is up to date\n"` `}`	
`elsif (!is_success($rc)) {`	Call the `is_success()` method of LWP and check the result.
`    print "Problem code:$rc ", status_message($rc),` `"($url)\n";`	**status_message** is a method in LWP::Simple that provides a diagnostic message in case of failure.
`}`	

The Entrez Programming Utilities (E-utils)

LWP can be used to interact with data provided through Entrez at NCBI. The Entrez Programming Utilities (E-utils) are tools that provide access to Entrez data outside of the regular Web interface and are a good example of how one can reliably retrieve data from Entrez. E-Utils and how to use the interface are documented at http://www.ncbi.nlm.nih.gov/entrez/query/static/eutils_help.html. The utilities are

1. Einfo: Provides field index term counts, last update, and available links for each database.

2. Esearch: Searches and retrieves primary IDs (for subsequent use by EFetch, ELink, and ESummary) and term translations.

3. Epost: Posts a file containing a list of primary IDs.

4. Esummary: Retrieves document summaries from a list of primary IDs.

5. Efetch: Retrieves records in the requested format from a list of one or more primary IDs or from the user's environment.

6. Elink: Checks for the existence of an external or Related Articles link from a list of one or more primary IDs.

7. EGQuery: Provides Entrez database counts in XML for a single search using Global Query of all databases.

The E-utilities allow one to query a specific database (PubMed, Nucleotide, etc.) for a search term, similar to entering a term when using Entrez on the NCBI Web site. The result is a set of primary IDs that

USE REQUIREMENTS FOR NCBI

Before starting to use the Entrez service, take a moment and read the requirements for usage: Do not overload NCBI's systems. Users intending to send numerous queries and/or retrieve large numbers of records from Entrez should comply with the following:

1. Run retrieval scripts on weekends or between 9 p.m. and 5 a.m. ET weekdays for any series of more than 100 requests.

2. Make no more than one request every 3 seconds.

3. NCBI's Disclaimer and Copyright notice must be evident to users of your service. The National Library of Medicine (NLM) does not claim the copyright on the abstracts in PubMed; however, journal publishers or authors may. NLM provides no legal advice concerning distribution of copyrighted materials; consult your legal counsel.

One of the responsibilities for developing solutions that rely on shared resources is to respect the usage guidelines. Abusing these resources may result in the offending computer being denied access to the provider's Web sites. This result can be more than inconvenient, especially when one is behind a firewall and all traffic originating from the same organization appears as one address, resulting in the entire organization losing access simultaneously.

can be passed to other E-utils to retrieve associated data (Esummary, Efetch). The site http://www.ncbi.nlm.nih.gov/entrez/query/static/esearch_help.html provides detailed explanation on how to use this service effectively.

NCBI also provides sample programs. Users are encouraged to download and examine the programs for a better understanding of how to use the WebEnv functionality of E-utils so that the search space can be used effectively. The snippet below demonstrates the use of LWP to retrieve ten primary IDs from PubMed related to the specified search term.

The reader can also use the example URLs listed on the Esearch util help page and try those searches

Code	Comments/explanation
`#!/usr/local/bin/perl —w`	First line identifies Perl interpreter.
`use LWP::Simple;`	
`# Set base URL for all eutils`	
`my $utils =` `"http://www.ncbi.nlm.nih.gov/entrez/eutils";`	
`# Setting up to search Pubmed for the term` `  'Cancer+Prostate'`	
`# and getting back a maximum of 10 records`	

`my $db ="Pubmed";`
`my $query   ="Cancer+Prostate";`
`my $retmax = 10;`
	Set up variables for database, query, and maximum number of matches to return.

`#Build the URL to using esearch.fcgi`	
`my $esearch = "$utils/esearch.fcgi?" .` `        "db=$db&retmax=$retmax&term=";`	Construct the URL so that this script's variables are incorporated into it. The URL must be on one line and only wrapped in the screen.
`# Using LWP to post URL and get the resulting webpage`	
`# Which is saved in a variable`	
`my $esearch_result = get($esearch.$query);`	Fetch the Web page that contains the result of the query.
`# print the variable and URL that was used`	
`print "ESEARCH RESULT: $esearch_result\n";`	
`print "Using Query:\n$esearch$query\n";`	

using a Web browser. Browsing through the ouptput, it would seem desirable that one could obtain direct access to the listed primary IDs (in this case the PubMed IDs) without the accompanying information; however, the default return mode is presently in XML format, which mandates the inclusion of certain information. If correct tools are employed,

parsing XML documents to obtain specific information is a relatively simple process. Although one can rely on Perl's ability to parse text-based formats, XML documents can become quite complex. Therefore, it is worthwhile to pursue other methods of parsing XML documents. The following section discusses XML.

WHAT IS XML AND WHY USE IT?

XML is eXtensible Markup Language, a W3C (World Wide Web Consortium)-endorsed standard. Like the popular HTML (HyperText Markup Language) format for Web pages, it shares lineage with SGML (Standard Generalized Markup Language),which was developed in the early 1980s and widely used for large documentation projects. The designers of XML adopted parts of SGML and HTML to produce a regular and simple-to-use language.

The primary use for XML is structuring data such as spreadsheets for financial and genealogical data. Its strength lies in the fact that it is a meta-markup language, allowing one to create new tags to define and structure a variety of data. HTML has a limited collection of predefined tags, specifies what each tag and attribute means, and specifies how the text between the tags should appear in the Web browser, e.g., the tags `<b> </b>` will bold a value enclosed between. In contrast, XML uses tags only to delimit pieces of data, leaving the interpretation and rendering of data to the application that uses it.

This flexibility and robustness make XML an excellent candidate for exchanging information. Many sequence analysis programs like BLAST and FASTA now produce results in XML format. NCBI E-utils and similar services offered by the cancer Bioinformatics Infrastructure Objects (caBIO; http://ncicb.nci.nih.gov/core/caBIO), XEMBL (http://www.ebi.ac.uk/ xembl), and DNA Data Bank of Japan (DDBJ; http://xml.ddbj.nig.ac.jp/) make extensive use of the XML format.

The popularity of XML has given rise to and forms the basis for specialized markup formats such as MAGE-ML to describe gene expression microarray designs, manufacturing information, experimental setup, data, and data analysis results (http://www.mged.org). The Berkeley Drosophila Genome Project provides sequence annotation in GAME XML format.

The Protein Information Resources (PIR) database can be downloaded in XML, and the Gene Ontology Consortium distributes data in XML format (http://www.geneontology.org/GO.format.html#XML).

A basic understanding of XML format and how to parse it is essential for using resources available from various data providers efficiently, because XML is fast becoming the de facto standard for information interchange. The subsequent section does not do justice to the expansive abilities of XML and accompanying technologies like XSLT (eXtensible Stylesheet Language Transformations) and Xquery. A good resource for learning more about these technologies and their application is http://www.xml.com. The reader is encouraged to visit sites listed in the Appendix for in-depth explanation of this extremely powerful technology and its application in life sciences.

XML Formats and Some Basic Rules

In the following example, the first line in the XML document is the XML declaration. It defines the XML version of the document and is required. In this case, the document conforms to the 1.0 specification of XML. The next line defines the first element or the root of the document. In this example, it is `<eSearchResult>`, which is an object resulting from an Entrez search. The next four lines are the child elements: `Count`, `RetMax`, `RetStart`, and `IdList`. The `Idlist` element has a `Date` attribute and multiple `Id` subelements.

Note that in XML:

1. All tags are case sensitive.

2. All elements must have a matching closing tag.

3. All elements must be correctly nested.

```
<?xml version="1.0"?>
<eSearchResult>
        <Count>39655</Count>
        <RetMax>10</RetMax>
        <RetStart>0</RetStart>
        <IdList Date="03/03/2004">
                <Id>15045584</Id>
                <Id>15045188</Id>
                <Id>15044961</Id>
                <Id>15044850</Id>
                <Id>15044710</Id>
                <Id>15044627</Id>
                <Id>15044396</Id>
                <Id>15044326</Id>
                <Id>15043402</Id>
                <Id>15043344</Id>
        </IdList>
</eSearchResult>
```

4. All attribute values must be quoted (e.g., Date="03/03/2004" or version= "1.0").

When the above-mentioned basic rules are followed, they result in a "well-formed" XML document. XML documents can include the DTD (Document Type Definition) or a reference to it. DTD is the blueprint of how data is structured within the XML document, with information about what type of values each element can contain. The XML document can be tested for this structural integrity against the DTD specification and is considered "valid" if it meets the criteria defined in the DTD. This verification is an important step when data are being exchanged between systems, thus ensuring data integrity. As is clear from the above XML sample, defining the same text as a flat CSV (comma separated values) file is difficult because the relationships between elements are difficult to depict in CSV format.

Most Web browsers can display XML, but it is preferable to use an XML editor to work with XML. XML editors for multiple platforms can readily be located by a Web search. A free, easy-to-use, cross-platform XML editor with many advanced features is Morphon from Morphon technologies (http://www.morphon.com). An example of an XML editor is shown in Figure 12.2.

With the above basic XML knowledge, the NCBI E-utils will be revisited and the primary IDs extracted. The next goal is to parse the XML results returned by the E-utils method. The first step is to save the XML generated from the earlier script (output can be redirected to a file using > i.e., eutil.pl > pubmed.xml). Then the output will be examined using the Morphon XML editor. The elements of interested are Count and Id. The panel on the left in Figure 12.2 shows a graphical depiction and the right panel displays values of the elements. Click on a folder to see the corresponding value highlighted on the

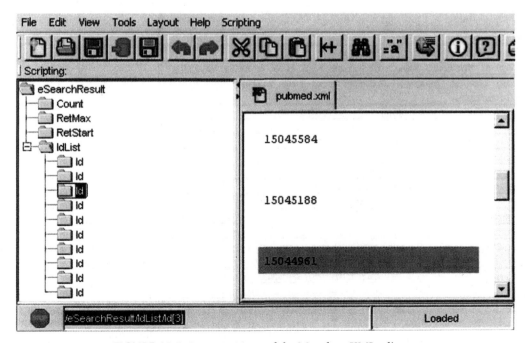

FIGURE 12.2. Screen capture of the Morphon XML editor.

opposite panel. The panel below the graphical display shows the path to that element (similar to a file path).

The path to Count is /eSearchResult/Count

The path to Id is /eSearchResult/IdList/Id[2]

Because Id has multiple values, the value in [ ] will change depending on the selected Id.

Now that the location of the elements is known, an XML parser can be used to extract the values. There are many approaches for parsing XML; most can be classified as either a "tree" or a "stream" type. A tree-based parser will typically parse the whole XML document and return a data structure made up of "nodes" representing elements, attributes, and values from the document. A stream-based parser provides a stream of "events" as the XML is being parsed. Some of the popular CPAN XML parsing modules are XML::Parser, XML::LibXML, XML::XPath, XML::Twig, XML::Simple, and XML::Smart. The Perl-XML FAQ site has good pointers and discussions about these modules (http://perl-xml. sourceforge. net/faq/). The XML Twig site also has tutorials on parsing XML with Perl http://xmltwig.com/.

In the following example, the XML::Smart module is used; it provides a lucid collection of functions to work with XML documents and can parse and generate XML documents. Based on XML:: Parser, an added benefit is that SMART relies on LWP for networking and thus it can post data to the E-util programs and parse the resulting XML document in a few simple steps. This module is available for any platform.

To parse an XML document using XML:: Smart, one has to know the path to the elements to be extracted from the document. Based on the examination in Morphon, the paths for Count and Id have been obtained:

```perl
my $count = $results->
    {eSearchResult} {Count};
```

```perl
my @Id = $results->
    {eSearchResult}{IdList}{Id}('@');
```

$results is the handle to the XML document that was returned when the request was posted to E-utils site at NCBI. The "{ }" marks separate elements. Because Count has a single value, it is stored as a scalar. Id returns multiple values and hence has ('@') at the end, thus allowing resulting values to be stored in an array. The sample code is very similar to the earlier version using LWP.

Code	Comments/explanation
`#!/usr/local/bin/perl -w`	
`use XML::Smart;`	
`# Set base URL for all eutils`	
`my $utils = "http://www.ncbi.nlm.nih.gov/entrez/eutils";`	
`# Setting up to search Pubmed for the term Cancer+Prostate`	
`# getting back a maximum of 10 records`	
`my $db ="Pubmed";`	
`my $query ="Cancer+Prostate";`	
`my $retmax = 10;`	
`#Build the URL to using esearch.fcgi`	
`my $esearch = "$utils/esearch.fcgi?" .` `"db=$db&retmax=$retmax&term=";`	
`# This module also uses LWP to post URL and get the` `# resulting XML page. Which is stored in $results`	
`my $results = XML::Smart->new($esearch.$query);`	Here is where the Smart module comes into play.
`# printing the variable and URL that was used`	
`my $count = $results->{eSearchResult}{Count};`	Parses the XML to get the Count.
`my @Id = $results->{eSearchResult}{IdList}{Id}('@');`	Parses the XML to get the list of Ids.
`my $all_Id = join("\n",@Id);`	Uses the Perl join function to put it all together.
`print "Count = $count\n";`	
`print "$all_Id\n";`	

Perl and Web Services

Traditionally, Web sites have been focused on facilitating human access. However, the growing need for applications to be able to exchange information and data with minimal human intervention has given rise to the application-centric Web. Although this concept has existed in the form of CGI programs, its implementation has been rather ad hoc. The Web services architecture formalizes this implementation and role, allowing various applications and services to interact with each other. For simplification purposes, we can categorize Web services into three major components:

1. The provider: Makes available programs and services that can be accessed to perform tasks.

2. The consumer: End-user (human or machine) that uses the programs.

3. The registry: A central repository clearinghouse where capabilities of the programs and type of data it can accept and return are registered. The formal description is known as WSDL (Web Services Description Language). The layer responsible for tying this together is called UDDI (Universal Description, Discovery, and Integration).

The language spoken among these components is XML, also known as XML messaging, where the request is encoded as an XML request and transferred using various established protocols such as HTTP (HyperText Transfer Protocol) or FTP. SOAP (Simple Object Access Protocol or Service Oriented Access Protocol) and XML-RPC (Remote Procedure Call) are two XML-based protocols for exchanging information between diverse applications and programs. The SOAP protocol allows programs written in one language to make calls to programs developed in another language. Just as Perl modules allow one to leverage from code written by others in Perl, SOAP takes the task to the next level to the point where one can use resources developed in any language, residing on a local or remote computer, as long as they provide a SOAP interface.

The following discussion focuses on SOAP from a bioinformatics data consumer point of view. caBIO provides access to programs written in Java (http://ncicb.nci.nih.gov/core/caBIO). DDBJ provides SOAP access to BLAST and retrieval of sequence data (http://xml.ddbj.nig.ac.jp/soapp.html); the European Bioinformatics Institute (EBI) provides SOAP access to FASTA, WU-BLAST, and the sequence retrieval system DBFetch (http://www. ebi.ac.uk/Tools/webser vices/); and there is a growing list of providers offering an array of services using SOAP.

Perl has a module called SOAP::Lite. The sample code below shows how to use the SOAP service and WSDL repository at DDBJ. The getFASTA_CDSEntry method retrieves coding sequence (CoDing Sequence, CDS) information from DDBJ for the specified accession number and returns it in FASTA format. This procedure is executed with a single line of Perl code!

The XMethods site has a large registry of SOAP objects one can experiment with online before implementing them as part of a program (http://services. xmethods.net/).

Code	Comments/explanation
```#!/usr/local/bin/perl -w```	
```use SOAP::Lite;```	
```# specify WSDL file```	
```my $service = SOAP::Lite->``` ```service("http://xml.nig.ac.jp/" .``` ``` "wsdl/GetEntry.wsdl");```	Pass the URL to the service method. For a list of methods available in GetEntry check: http://xml.ddbj. nig.ac.jp/doc/GetEntry.txt.
```# We will use getFASTA_CDSEntry for obtaining``` ```# the CDS in FASTA format with a given Accession number``` ```# Now call web service with Accession BN000101```	
```$result = $service->getFASTA_CDSEntry("BN000101");```	The SOAP service does all of the work.
```# print result which is in fasta format``` ```print $result;```	

BIOPERL MODULES PROCESS AND MANIPULATE BIOLOGICAL DATA

BioPerl is a large collection of modules that have been written to help process and manipulate biological data and analysis results. Using BioPerl modules and their objects can save much time and effort, although a little up-front work is required to understand the correct usage of each object. Note that BioPerl does not include executable programs, only modules that can be used in Perl scripts. Some modules provide interfaces to sequence analysis programs or databases. Others provide objects suitable for the representation of certain entities used in bioinformatics applications. Development of BioPerl is ongoing. Downloads and the latest news about the project are available at http://www.bioperl.org.

The BioPerl download includes instructions for installation on UNIX and Windows systems. To install from the Programmer's Package Manager (PPM) environment, which provides a command line interface for managing BioPerl modules and packages, simply type 'install bioperl'. Additionally, the "HOWTOs" link reveals several modules that are explained in detail with sample code showing how to use them. The reader is strongly urged to view some of the code examples in the tutorial at http://www.bioperl.org/Core/Latest/bptutorial.html. Relying on BioPerl can help one develop scripts quickly. It is easier to mix and match capabilities of existing modules to fit new tasks, thus eliminating the need to replicate functionality. Keep in mind that the BioPerl project relies on volunteers contributing their code and effort, and on occasion it may be necessary to add features to a given BioPerl module to implement a precisely required functionality. Developers are encouraged to contribute such additions to the BioPerl project.

It is likely that there are several ways to accomplish a given task using BioPerl modules. The choice of which modules to use depends on the type of results required. An earlier example illustrated the use of the XML::Smart module to parse an E-utils Esearch result, and noted that the developer must be aware of the XML element locations to extract the desired features. In contrast, one may use the module Bio::DB::GenBank to obtain data from Genbank which in turn relies on Bio::DB::NCBIHelper to obtain the results. The Bio::DB::NCBIHelper module uses the E-utils interface to obtain data from NCBI and provides methods to extract features from these data. The script developer is not exposed to the details; thus, retrieving data using accession numbers and extracting features from the resulting sequence object can be obtained with a few simple calls.

Biological Objects in BioPerl

What are some common biological objects? To name a few, there are annotated sequences and the files and databases in which they are stored, exons, alignments, microsatellites, single-nucleotide polymorphisms (SNPs) restriction enzymes, BLAST output files, and so on. All of these types of data, and many more, have representations in BioPerl modules. The following discussion takes a closer look at a few BioPerl objects and ways of using them. In the case of a sequence object, one at first may be tempted to think that a string suffices to represent a DNA or protein sequence. However, biological information about the sequences collected in a laboratory is also important. The BioPerl Seq object stores a sequence and associated attributes, such as the type of sequence (DNA/RNA/protein), the GenBank accession number, human readable identifier, sequence features, and a link to sequence annotations.

What are some methods that are useful in working with sequence objects? Certainly the ability to retrieve or modify attributes of a sequence is desirable. One may also wish to extract a portion of the sequence of bases, or the reverse-complement or translation of the sequence. All of these can be accomplished by using the methods of the BioPerl Seq object. A list of available methods appears in the BioPerl Seq Module documentation, accessible through the bioPerl.org Web site. When clicking on a BioPerl module, one sees a page that has a link to the list of methods for that module. In most cases, the page will also contain some sample code that uses the module under the Synopsis heading. Reading these code segments will greatly help one to understand how to use the module. Another thing to keep in mind is that many BioPerl objects are built as extensions of simpler objects, and sometimes the description of the method in which one is interested is in the

Methods list for the simpler object. For example, the `Bio::Seq` trunc method appears in the `Bio::PrimarySeqI` Methods list, but not on the `Bio::Seq` Methods list. A section called "Included modules" is included near the top of the module documentation page, and occasionally it is necessary to follow this link to find a method.

Recall that Perl scripts must declare which modules they want to use before trying to access a method in the module. To use the BioPerl `Seq` module, the following line should be added near the top of the script, paying special attention to the use of capital and lowercase letters in the module name:

```
use Bio::Seq;
```

Some `Seq` methods return a string result when they are called, while others create and return a new `Seq` object. It is important to not confuse a sequence string with a sequence object and to pass the appropriate type of data as required by the method. A few `Seq` methods are listed in Table 12.1.

It is often useful to name variables so that their names reflect the type of object they hold. In Table 12.1, `Bio::Seq` methods can be accessed through variables whose names end in "obj", but not through other variables. It is possible for `Bio::Seq` methods to return other types of objects as well. For instance, `$seqobj->length()` returns an integer and `$seqobj-> get_SeqFeatures` returns an array of `Bio::SeqFeatureI` objects.

The `Bio::SeqIO` object has the ability to read and write sequences in different sequence formats. Format conversions are completely transparent to the user. The `SeqIO` object is somewhat analogous to a Perl file handle—it is a connection through which file input/output (I/O) is done. A `SeqIO` object does not contain `Seq` objects; it merely provides a way to read or write `Seq` objects. On the next pages is a script that reads sequences from a GenBank format file and writes them to a FASTA format file, performing the file conversion based on attributes specified when the `SeqIO` objects are created.

TABLE 12.1. Seq *methods*

Methods that return a *Seq object*	Description
`$seqobj=Bio::Seq->new();`	Creates a new Seq object. Attribute specifications may be passed in.
`$revobj=$seqobj->revcom();`	Creates a new Seq object that has the same info as the original $seqobj except that its sequence string is the reverse-complement.
`$tseqobj=$seqobj->trunc(50,100)`	Creates a new Seq object that has the same info as the original $seqobj except that its sequence string is only bases 50–100 of the original.
`$protseqobj=$tseqobj>translate();`	Creates a new Seq object that contains the protein sequence translated from the original sequence.
Methods that return a *string*	(Simple scalar data; not an object.)
`$seqstr=$seqobj->seq();`	Returns string of bases.
`$seqidstr=$revobj->display_id();`	Returns the human readable ID.
`$accnum=$tseqobj>accession_number();`	Returns accession number as a string.
`$seqtype=$protseqobj->moltype();`	Returns "dna", "rna", or "protein".
`$subseq=$seqobj->subseq(50,100);`	Returns string of bases 50–100.

Script Example 1 (*see facing page*)—BioPerl: Reading Sequences from a GenBank Format File and Writing Them to a FASTA Format File

BioPerl concepts included

> Bio::Seq method accession_number()
> Bio::SeqIO methods new(), next_seq(), write_seq()

The pseudocode for this script is

Create a SeqIO object to handle input of GenBank file
Create another SeqIO object to handle output of FASTA file
While there is another input Seq object
    Print its accession number
    Write it to the FASTA file

Pay attention to the different ways in which `$inSeqIO` and `$outSeqIO` are used in this script. Also notice that the syntax for assigning attribute, value pairs in the BioPerl modules (e.g., `-file => "$infile"`) are consistent with the syntax used with the `Getopt::Long` module. One last detail is that when a method is called without any parameters, the `()` following the method name may be omitted. For clarity, most of the examples in this chapter use `()` to emphasize the fact that a method (i.e., a function) is being called.

How to Call EMBOSS and PISE Programs for Analysis

The EMBOSS suite of programs is an excellent source for a variety of sequence analysis tasks. Suppose there is a collection of files containing DNA sequences, and an EMBOSS utility is to be used to look for GpG Islands in each sequence. Next, the sequence will be mutated and examined to see what effect the mutations have on the CpG islands.

There is a BioPerl module that makes it easy to run external sequence analysis programs from within a Perl script. It is not part of the BioPerl core, but is an add-on called `bioPerl-run` that must be downloaded separately (see the section on downloading modules, p. 563). Having `bioPerl-run` and other collections of modules separate from the main distribution allows the user the choice of mixing and matching components to be used in addition to the BioPerl core, rather than requiring the download of a single massive software package. In the script below, the EMBOSS "`newcpgreport`" program will be called and selected results output for each DNA

sequence. Then the same thing is done with a mutated version of the sequence. Because the same thing needs to be done in two different places in the code, this is a perfect opportunity to write a subroutine and call the subroutine instead of duplicating code in the script. The desired EMBOSS programs will be run using the `Bio::Factory::EMBOSS` module. The `check_cpg` subroutine is called twice in the script. Often the main reason for creating a subroutine is so that the code can be reused. Another reason to put code into a subroutine is to make scripts more modular and easier to read, especially when the scripts are long. It often makes sense to have a subroutine for each significant "subtask" the script performs.

Script Example 2 (*see pp. 580–582*)—BioPerl: Using the `glob` Function to Obtain a List of Files Named with a Specific Extension

We will use the Perl "glob" function to obtain a list of files named with a specified extension, passed as a command line option.

Perl concepts demonstrated are

> Using the Getopt::Long module
> The "glob" function
> Perl for each loop over an array
> Perl global variables
> Perl subroutines and parameters
> Using the Bio::Factory::EMBOSS   module

The pseudocode for this script is

Get the file name extension option from the command line
If no valid extension is found, exit with an error message
Use the glob function to get the names of all files with that extension
For each of the files
    Call the check_cpg subroutine with the file name
    Call EMBOSS "msbar" to introduce mutations in the DNA sequence
    Call check_cpg with the file name of the mutated sequence

Subroutine check_cpg:
    Call EMBOSS "newcpgreport" with the sequence file name passed in
    Find the lines of output that contain the word "island" and print them

Script Example 1

Code	Comments/explanation
`#!/usr/bin/perl    -w`	Always needed if script is to be directly executable.

`# This script uses Bio::SeqIO objects`
`# to convert a Genbank format file to`
`# a FASTA format file`
`# Author:   MyName,   12 Oct 2003`
`# Version 2.0`	

`use strict;`	Using the strict module instructs Perl to do strict error checking and generate an error message if any unsafe constructs are found. Strict requires that every variable is declared with a scope specifier such as my or our.
`my ($infile, $inSeqIO, $outSeqIO, $seqobj, $accnum);`	Declare lexical (nonglobal) variables.
`use Bio::SeqIO;`	Since Bio::SeqIO includes Bio::Seq, we do not have to use Bio::Seq explicitly, although it is fine to include it.
`# Check for filename argument` `$infile = $ARGV[0];`	Always check for required arguments!
`if (!defined $infile) {`	If no argument was given, `$infile` will be undefined.
`    die "Usage: $0 inputfile\n";`	Output a Usage message that reflects the correct usage of the script.
`}`	

`# Create two SeqIO objects:`
`# one for input`
`# another for output`
	In the next two lines, notice that object methods are called using a "'single arrow'" -> whereas attribute key, value pairs are connected with a "double arrow" =>.

`$inSeqIO = Bio::SeqIO->new (-file => "$infile",` `-format => 'Genbank');`	Set up the connection to `$infile` and specify that `$inSeqIO` is to read a GenBank format file.
`$outSeqIO = Bio::SeqIO->new (-file=>">$infile.FASTA",` `-format => 'FASTA');`	Set up the connection to a file named `$infile.FASTA`, and specify that `$outSeqIO` is to write to this file in FASTA format. Notice that it is necessary to use > to indicate that the file is to be written to, as it is with the Perl `open()` function.
`# In a loop, read each Genbank# sequence, print its` `# accession number   and write it out in FASTA format`	There is very little work to do here. The BioPerl modules are doing most of it for us!
`while ($seqobj = $inSeqIO->next_seq())    {`	The `next_seq()` method returns a Seq object.
`    $accnum = $seqobj->accession_number();`	The `accession_number()` method returns a string.
`    print "Accession#    $accnum\n";` `    $outSeqIO->write_seq($seqobj);`	The `write_seq()` method requires a Seq object as a parameter.
`}    # End while next_seq()`	

Script Example 2

Code	Comments/explanation
`#!/usr/bin/perl -w`	`#!/usr/bin/perl` is always needed if the script is to be directly executable.

`# This script reads sequence files with filename`
`# extensions matching the one provided`
`# with the command line option -ext`	

`# For each file, the script calls the EMBOSS`
`# newcpgreport program on both the sequence`
`# and a mutated version and reports the`
`# CpG Islands found`	

`# Author: MyName, 16 Aug 2003` `# Version 1.01`	
`use strict;`	Using the strict module instructs Perl to do strict error checking and generate an error message if any unsafe constructs are found.
`use Getopt::Long;`	This statement is required to use the Getopt::Long module.
`use Bio::Factory::EMBOSS;`	This statement identifies the module used to call EMBOSS programs.
`my ($file_ext, $f, $mut, @files);`	Declare lexical (nonglobal) variables.
`our $emboss_fac;`	Subroutines cannot access variables external to themselves unless the variables are declared as global variables. "our" makes this variable global, meaning that it is accessible from anywhere in the file.

`# Define the check_cpg subroutine`
`# It calls newcpgreport, processes the output`
`# and prints lines containing 'island'`
	Defining the subroutine before it is called allows Perl to verify that the caller is passing the correct parameters to the subroutine.

`sub check_cpg($) {`	$ specifies that the subroutine expects one scalar parameter.
`    my ($file, $cpg, @result, $lin);` `    $file = $_[0];`	Subroutine parameters are passed in a special array named @_. In this case, there is only a single element, $_[0].
`    $cpg = $emboss_fac->program('newcpgreport');`	The "program" method of the factory object creates a representation of the EMBOSS newcpgreport program.

`    $cpg->run({'-sequence' => "$file",`
`        '-window' => '100',`
`        '-shift' => '1',`
`        '-minlen' => '200',`
`        '-minoe' => '0.6',`
`        '-minpc' => '50.0',`
`        '-outfile' => "$file.cpg"});`
	Pass all mandatory qualifiers to the newcpgreport object and run the program. The run method expects a hash as an argument, hence the { } surrounding key,value pairs.

`    # Read the newcpgreport output`	
`    open(CPGFIL, "$file.cpg") or` `    die "Could not open file $file.cpg\n";`	

Script Example 2 (Continued)

`@result = <CPGFIL>;`	Rather than reading one line at a time this statement reads all lines in the file into the @result array in a single step. Do not do this with large files!
`close (CPGFIL);`	
`foreach $lin (@result) {`	Step through array elements one at a time. $lin holds each element in turn and changes for each iteration of the foreach loop.
`    if ($lin =~ m/island/) {`	Select only lines that contain the string "island". Use a case-sensitive match.
`        print "$_[0] : $lin";`	Print the file name and the line of CpG island output.
`    }`	
`} # End of foreach result`	
`return;`	Control passes back to main script.
`} # End of subroutine check_cpg`	
`# Main script code below:`	Indicates where the script execution begins.
`# Initialize the EMBOSS factory`	
`$emboss_fac = Bio::Factory::EMBOSS->new(` `            '-verbose' => '1');`	Turn on the verbose option while debugging the script to see helpful error messages if there is a problem. The '-verbose' flag can be removed later.
`# Get the command line option`	
`GetOptions("ext=s" => \$file_ext);`	Must pass a reference to the $file_ext variable so prefix it with \.
`# Check to see if -ext was specified`	
`if (!defined $file_ext ) {`	If no extension was entered on the command line, $file_ext will not be defined.
`    die "Usage $0 -ext file_extension\n";`	
`}`	
`# Get the matching filenames`	
`@files = glob("*.$file_ext");`	The "glob" function creates an array of file names.
`# If no matching files, print message and exit`	Without a message the script would silently do nothing, leaving the user in the dark.
`if (@files < 1) {`	The array variable @files used in a scalar context gives the number of array elements.
`    die "No files with extension $file_ext\n";`	
`}`	
`foreach $f (@files) {`	No need to chomp since the @files array was created by glob.
`    check_cpg("$f");`	Call the subroutine to run newcpgreport and extract the

Script Example 2 (Continued)

	desired results. The file name $f is passed as a parameter.
`# mutate the sequence file`	
`$mut = $emboss_fac->program('msbar');`	Create a representation of the EMBOSS msbar program.

`$mut->run({'-sequence' => "$f",`
`              '-count' => "100',`
`              '-point' => "1',`
`              '-block' => "1',`
`              '-codon' => "1',`
`              '-outseq' => "$f.mut"});`
	Pass all mandatory qualifiers and run msbar. These parameter values specify 100 mutations of sort: point, block. or codon.

`# Run a CpG report on the mutated sequence` `check_cpg("$f.mut");`	Pass the file created by the msbar program to the subroutine.
`} # End of foreach @files`	

To test this script, we need sequences that contain CpG islands. One way to obtain them is to do an Entrez search (http://www.ncbi.nlm.nih.gov) for "CpG island", and find some DNA sequences that are matched. Display a sequence in FASTA format and send it to a file. Save a few sequences to files and name the files with the same extension; then try running the script with this extension. It is always a good idea to verify the correctness of scripts with sequences or data for which you know the exact result before using unknown, experimental data. Be sure to test with a sequence that has no CpG islands also.

When this script runs, it creates ".cpg" and ".mut" files for each sequence file processed. In general any incidental files created by a script should be removed before the script exits. The Perl function that removes files is unlink(), and it takes the file name as a parameter. For example,

```
unlink("$filename");
```

Of course it is necessary to exercise great care when calling unlink(). Leave it out only until the script has been thoroughly tested and debugged. It is not polite to leave extra files lying around!

Script Example 3— BioPerl: Searching IO to Read Full BLAST Output

The Bio::DB::GenBank module can be used to establish a connection to GenBank and retrieve information over the Web using a Perl script. The GenBank object has methods that allow a set of sequences to be specified by accession numbers, gi numbers, or even GenBank query strings. Some of these methods return Seq objects while others return SeqIO streams. In the latter case, the SeqIO next_seq() method is used to retrieve the individual Seq objects.

For parsing BLAST output files, there is a BioPerl module called SearchIO. Results for each BLAST query are retrieved using the next_result() method, which returns a Bio::Search::Result::ResultI object. Using the ResultI objects methods, one can access each Hit (or Subject match) as a HitI object and from the HitI, retrieve the HSP (high scoring pair) objects. If you work with large BLAST outputs, the SearchIO module is definitely one to become familiar with.

The next script will use the Bio::SearchIO module, the Bio::DB::GenBank module, and the methods of the Bio::SeqIO object. An earlier sample script processed a Tab-delimited BLAST output file that was created by specifying the output of an Alignment View "Hit Table" on the NCBI BLAST Web page, or by adding the "-m 9" option to command line BLAST. The default BLAST output includes a lot more information about the alignments, and therefore results in a much larger output file. The next script uses Bio::SearchIO to read a full BLAST output, select hits whose descriptions match a string, and retrieve the full GenBank records for those hits. The script outputs two files: a Tab-delimited file listing information for each HSP of each matching hit and the FASTA file containing the sequences retrieved from GenBank.

BioPerl concepts included are

> Bio::SearchIO object and methods
> Bio::DB::GenBank object and methods
> Bio::SeqIO methods
> Perl array function push

The pseudocode for this script is

Get command line options that specify a BLAST output file name and a pattern-matching string
> Open an output file for writing (or die)
> Create a SearchIO object read the BLAST file
> While the SearchIO object returns another Result object
>> Save the BLAST query name in a variable
>> While the Result object returns another Hit object
>>> Save the BLAST subject name in a variable
>>> Save the BLAST subject description in a variable
>>> If the subject description does NOT match the pattern string
>>>> Skip ahead to the next Hit
>>> Split subject name into an array, using | as the split delimiter
>>> If the first split element is not "gi"
>>>> Skip ahead to the next Hit
>>> Push the second split element (GI number) onto an array
>>> Save the Hit length in a variable
>>> While the Hit object returns another HSP
>>>> Save strand, percent identity, and *E* value of the HSP
>>>> Print to the output file delimited by Tabs:
>>>>> Query name, Subject Name, Subject Description, Hit Length, HSP strand, percent ID, and *E* value
># End of while next Result
>Close the output file
>Create a SeqIO object to output a FASTA file
>Create a GenBank object for retrieving sequences of GI numbers in array (Specify FASTA format and retrieval type as tempfile to avoid running out of memory)
>If there are any GI numbers in the array
>>Call GenBank get_Stream_by_id and pass a reference to the array
>>While the Input SeqIO returned has a next sequence
>>>Write it out using the Output SeqIO object
>Else
>>Print a message stating that no matches to the pattern were found

Script Example 3

Code	Comments/explanation
`#!/usr/bin/perl    -w`	

`# This script uses the Bio::SearchIO module`
`# to parse a BLAST output, extracts GI numbers`
`# from the Hit descriptions, and retrieves the`
`# sequences with those GI numbers from GenBank,`
`# using the Bio::DB::GenBank module.   It saves`
`# these sequences as a FASTA format file.`	

`# Author: My name, 03 Jun 2003` `# Version 1`	

`use Bio::SearchIO;`
`use Bio::DB::GenBank;`
`use Getopt::Long;`
	Specify modules that are needed.

`my ($blastfil, $pattern, $SrchIO, $result, $sname),`
`my ($sdesc, @part, @gi, $ngi, $hit, $hitlen);`
`my ($hsp, $hspPct, $hspEval, $inSeqIO, $outSeqIO);`
`my ($gb, $seqobj);`
	Declare lexical (nonglobal) variables.

Script Example 3 (Continued)

```GetOptions("file=s" => \$blastfil,``` ``` "pat=s" => \$pattern);```	Must pass references to these variables so prefix them with \.
`# Check to see if -file and -pat were specified`	If either option was not entered on the command line, the corresponding variable will not be defined.
`if (!defined $blastfil   || !defined $pattern)   {`	The || operator is equivalent to the "'or'" operator.
`    die "Usage $0 -file BLASTfile -pat pattern\n";` `}`	
`# Open an output file for writing Hit info`	Create a named file handle.
`open(OUTFILE, ">$blastfil.hit_info") or` `  die "Cannot open $blastfil.hit_info for writing\n";`	Keep the output file name similar to the BLAST file so it is easy to see that the files are related.
`# Create a SearchIO object to get BLAST results`	
```$SrchIO = Bio::SearchIO->new('-format' => 'blast',``` ```'-file' => "$blastfil");```	Tell the SearchIO object we want to read a BLAST format file named $blastfil.
`# Loop through each BLAST result`	
`while ($result = $SrchIO->next_result()) {`	The next_result method provides a way to get info about each query sequence.
`    $qname = $result->query_name();`	query_name() is a method of the ResultI object.
`    # Loop through each hit for this query`	
`    while ($hit = $result->next_hit()) {`	next_hit() is a method of the Hit object.
`      $sname = $hit->name();` `      $sdesc = $hit->description();`	
`      if ($sdesc !~ /$pattern/) { next; }`	!~ checks to see if the pattern is not matched.
`      @part = split   /|/,    $sname;`	Split the subject name apart using | as the element separator.
`      if ($part[0] ne 'gi') {`	If no GI in description, skip to next Hit. Notice the use of "ne" to check inequality of strings.

`        print "No gi in $sname\n";`
`        next;`
`      }`	

`      push @gi, "$part[1]";`	Push the GI number onto array @gi for processing later.

`        $hitlen = $hit->length();`
`        # Get info for each HSP within the Hit`
`        while ($hsp = $hit->next_hsp) {`
`          $strand = $hsp->hit->strand;`
`          $hspPct = $hsp->frac_identical();`
`          $hspEval = $hsp->evalue();`	

`          print OUTFILE "$qname\t$sname\t";`
`          print OUTFILE "$sdesc\t$hitlen\t";`
`          print OUTFILE "$hspPct\t$strand\t";`
`          print OUTFILE "$hspEval\n";`
	Write tab-delimited items to the output file.

`        } # End of while next HSP`
`    } # End   of while next hit`
`  } # End of while next result`
`close OUTFILE;`	

Script Example 3 (Continued)

`# Create a SeqIO object to write a FASTA file`	
`$outSeqIO = Bio::SeqIO->new('-format' => 'fasta',` `'-file' => "$blasftil.fasta");`	
`# Create a GenBank object that will give us a` `# SeqIO object to use for input`	The `get_Stream_by_ID` method returns a SeqIO object.

`$gb = Bio::DB::GenBank->new('-format' => 'fasta',`
`                           '-retrievaltype' =>`
`'tempfile');`
	We use the `retrievaltype` `tempfile` because the retrieval could result in too much info to fit into memory.

`$ngi = @gi;`	Get the number of elements in the `@gi` array.
`if ($ngi > 0) {`	
`    print "Retrieving $ngi seqs from GenBank... \n";`	
`        $inSeqio = $gb->get_Stream_by_id(\@gi);`	Must pass a reference to the `@gi` array so prefix the array name with `\`.
`    while ($seqobj = $inSeqio->next_seq()) {`	
`        $outSeqio->write_seq($seqobj);`	
`    }`	
`} else { print "No subjects matching $pat found\n"; }`	

There are many more useful BioPerl modules, such as `Bio::Graphics`, which provides a way to render sequence data and features graphically. The reader is encouraged to visit the HOWTOs link on http://bioperl.org to learn more about what BioPerl has to offer.

PERL CAN INTERACT WITH DATABASES

Sequence analysis programs typically generate text reports as output. Sifting, aggregating, and collating relevant information from these reports can be a daunting task, especially in a medium- to large-scale analysis. Perl and relevant BioPerl modules can be employed for parsing reports to carry out further analysis. Storing the processed output in a database can minimize the complexity and at the same time leverage from the optimized routines for searching, retrieval, and aggregation of data made available by virtue of the data residing in the database. There are obvious benefits in using Perl and BioPerl in conjunction with databases. Readers with limited prior database experience are encouraged to experiment with the database systems without the use of Perl, and employ Perl once they have attained basic understanding and familiarity with the database system.

Database Structures

A database is a collection of data arranged for ease and speed of search and retrieval, often known jokingly as a marvelously efficient way of putting large amounts of information into one central repository, never to be seen again. Popular biological databases have evolved from flat-file systems, in which the traditional plain text file is indexed (akin to a printed book) for easier retrieval of information. Sequence

Retrieval System (SRS) developed in the early 1990s at EMBL is an example of such a system. With the growing need to associate and integrate heterogeneous sources of information many of these systems now employ commercial or Open Source databases exclusively or alongside their traditional indexed flat-file methods for serving information.

Databases can be broadly categorized as

1. Desktop for single or a limited number of users (Microsoft Access, Borland, Paradox),

2. Workgroup for multiple users (MySQL, DB2, Oracle, Postgres, Sybase).

Database Management Systems (DBMS) are a set of programs that provide the interface between the end-user and the underlying data to define, create, administer, and process the data. DBMS allow one or more users to create, access, and manipulate data without having to understand the physical layout on the storage media (hard disk). At the same time the DBMS ensure the integrity and security of data while efficiently handling multiple requests to manipulate the data.

The concept of relational databases was invented by E.F. Codd at IBM in 1970. A relational database is a container of data items organized as a set of formally described tables from which data can be accessed or reassembled in different ways without having to reorganize the underlying database tables. A table can be generalized as a spreadsheet. Tables are composed of rows (records) and each row is composed of columns (fields). Each table contains one or more data categories in columns and each row contains an instance of data for that category defined by the column type. Tables can be related to each other by the column categories.

Database Structured Query Language

To interact with a database the standard interactive and programming language is the structured query language (SQL), which is an ISO and ANSI standard. A SQL statement is often referred to as an "SQL query." Most databases will support a subset of this standard along with proprietary additions. The SQL syntax itself is case insensitive, but capitalization is preferred for readability. The three major components of SQL are

1. Data Definition Language (DDL): Deals with structural aspects of the database—creation, modification, deletion of tables:

```
CREATE TABLE mydata (seqname
    CHAR(10) , length INT)
```

2. Data Manipulation Language (DML): Allows modification of the data contained in the tables—insertion, deletion, selection, modification, aggregation (e.g., count, sum, average):

```
INSERT INTO mydata
    VALUES("ecoli-10", 1675)
SELECT seqname FROM mydata WHERE
    length > 1600
UPDATE mydata SET length = 1677
    WHERE seqname = "ecoli-10"
```

3. Data Control Language (DCL): Deals with maintaining the integrity of the database using permissions, transactions, etc.:

```
GRANT ALL ON mydata TO myuser
REVOKE UPDATE,INSERT TO mydata FROM
myuser
```

SQL is very powerful; see http://sqlzoo.net/ and http://www.w3schools.com/sql/ to learn more about SQL statements.

MySQL and Postgres are two popular Open Source relational database systems employed extensively in bioinformatics projects (e.g., ENSEMBL uses MySQL). Popular commercial databases include Oracle, DB2 from IBM, and Microsoft SQL server. Selecting appropriate databases from Open Source or commercial vendors is not an easy task, as they all have strengths and weaknesses and there is no clear winner across the board. A key factor to keep in mind is that, as the available functionality and features of the data increase, the responsibility and complexity associated with administering and maintaining the database increase rapidly. Database vendors are trying to provide more domain-specific functionality wrapped with the database engine, which, when leveraged appropriately, can reduce the complexity of one's script. For example, Oracle 10G offers SVM (Support Vector Machine) and sequence searching integral to the database. Many vendors offer native storage and retrieval in XML format.

For individual and group projects, employing embedded database engines can offer many of the features of a full-fledged database without the overhead of having to run and administer a Relational Database Management System (RDBMS). The ease of having data stored in a single file that is readily identifiable for backups and other operations simplifies the task of relying on databases even for trivial tasks. When there is a need for connecting remotely (between two machines) or managing concurrent activities, the use of a more feature-rich solution should be employed. Most enterprise-level projects mandate such robust solutions. In this chapter we focus on SQLite, an embeddable database (http://www.sqlite.org/) and MySQL (http://www.mysql.com). Both are freely available as binaries for UNIX and Windows along with source code and Perl DBI, DBD modules.

SQLite

The single executable program SQLite will permit manual entry and execution of SQL commands against an SQLite database. The database created can be copied from Windows to UNIX or vice versa and can also be accessed from Windows applications such as OpenOffice, Excel, or Access, using ODBC (Open Database Connectivity Drivers).

EXAMPLE WITH SQLITE: A QUICK TOUR

The following example works on Windows and UNIX platforms. To create a new database mydb in the current directory (give the full path name to the database to save it in a different location)

```
sqlite mydb
SQLite version 2.8.13
Enter ".help" for instructions
sqlite>
```

Note: This will result in the creation of the database mydb and provides the prompt sqlite> at which SQL statements may be entered. If the database already exists, sqlite will connect to the database. All SQL statements must be terminated by a semicolon (;).

The SQLite utility has a help screen that can be invoked by typing .help at the sqlite> prompt. Some of the frequently used commands are

Code	Comments/explanation
dump table_name	Dump content of a table in text format.
exit	Exit the program.
output FILENAME	Send output to FILENAME (output visible only after you exit sqlite).
quit	Exit the program.
read FILENAME	Execute SQL statements from FILENAME.
schema table_name	Show the CREATE statement used for table_name.
tables	List of tables matching a pattern.

For detailed explanation visit http://www.sqlite.org/sqlite.html.

MySQL Example

Code	Comments/explanation
CREATE TABLE mydata (seqname CHAR(10) , length INT);	Create a table called mydata to hold sequence name and sequence length.
INSERT INTO mydata VALUES("ecoli-10", 1675); INSERT INTO mydata VALUES("azotov-12", 2567);	Insert data on the table mydata Note: the use of " " around character data.
SELECT seqname FROM mydata WHERE length > 1600;	Select specific data from mydata.
SELECT COUNT(seqname) FROM mydata WHERE length > 11;	Count occurrence of condition in mydata.
UPDATE mydata SET length = 1677 WHERE seqname = "ecoli-10";	Update specific records in mydata.
DELETE FROM mydata WHERE length >= 2500;	Delete specific records in mydata.

The above steps can be carried out in MySQL using the mysql command line utility. Creating a database is a special privilege and cannot be done by all users. The administrator for the MySQL database can use the mysqladmin utility to create the database:

```
mysqladmin create mydb
```

The administrator can also create a user with permission to enter data into mydb (the user must first log into mysql using the mysql command line utility)

```
GRANT ALL PRIVILEGES ON mydb.* TO 'myuser'@'localhost' IDENTIFIED BY 'secret';
```

This allows the user myuser with the password secret to connect from the localhost (same machine running the mysql database) and create, update, and modify tables in the mydb database (the mydb.* means all tables in mydb). Note that a mysql database user name and UNIX user name are two distinctly separate entities.

Once an individual has a MySQL user account with the necessary permissions, he or she can interact with it using the mysql utility:

```
mysql -u myuser -p mydb
Enter password:
Welcome to the MySQL monitor. Commands end with ; or \g.
Your MySQL connection id is 3707 to server version: 4.1.12-max
```

Type "help;" or "\h" for help. Type "\c" to clear the buffer.
```
mysql>
```

Typing help; will bring up the common commands and their usage, which are similar to SQLite.

Using the Perl DBI

As can be seen above, entering or retrieving data using the command line utility is time consuming and is not conducive to complex iterative operations. By using Perl DBI to connect and interact with the database, one can carry out fairly complex iterative operations with relative ease in a Perl script. Perl provides a set of modules (namely, DBI Database Interface and DBD Database Driver) for specific databases like MySQL, Postgres, SQLite, and Oracle. These modules are used for connecting to databases from within Perl scripts. Relying on DBI and DBD can ease the maintenance of scripts, making them more portable when the underlying database changes. DBD are available for not only databases but also for flat files in CSV or tab-delimited formats, making the task for manipulating data in these file formats more modular using SQL-like syntax. Just like other Perl modules, DBI is invoked by the `use DBI;` statement. To use the functionality of DBI we will examine some basic concepts. Retrieval of data from a database is a five-stage process:

1. Establishing the connection,

2. Preparing the SQL statement,

3. Executing the statement,

4. Fetching the resultant output,

5. Cleaning up, where memory allocated to various stages is freed.

The first and last stages are usually performed once per script, while the remaining stages are performed multiple times depending on the task. SQL operations like DELETE, INSERT, and UPDATE do not return data, in which case the fourth stage of fetching data is skipped.

To work through the above-mentioned stages, DBI provides three primary handles:

1. Driver handle, which deals with initialization and loading of drivers for specific databases to which a connection is to be established;

2. Database handle, which deals with establishing the connection using the loaded drivers (common naming convention in scripts is `$dbh`);

3. Statement handles, which are the workhorses and deal with preparing the SQL statements and executing and presenting the returned data. `$sth` is a common naming convention for statement handles in scripts.

Establishing Connection with a Database

To establish a database connection, the following information is required:

1. Data Source Name (DSN) is a combination of DBI followed by the database driver (DBD) type and name of the database to which the connection is to be established. Keep in mind that DBD names are case sensitive.

2. Login name.

3. Password.

4. Optional attributes that can be set for diagnostic messages in case of failure, provided as a hash.

Once a successful connection is made, a database handle is available, to which statements can be sent for processing.

Code	Comments/explanation
`$dbh = DBI->connect("DBI:mysql:mydb","myuser", "secret") or die "Cannot connect" . $dbh->errstr ;`	The statement below attempts to connect to the mysql database `mydb` using login `myuser` and password `secret`. If the connection is unsuccessful, an error message is stored as a string in the variable `$dbh->errstr`. The die statement will result in the script terminating and printing the error message.
`$dbh = DBI->connect("DBI:SQLite:/home/molbio/myname/db-home/my_db","", "") or die "Cannot connect" . $dbh->errstr;`	In contrast, here is a similar connection to the SQLite database `mydb`. Because it does not use login or passwords those fields are empty, but need to be specified nonetheless.
`$dbh = DBI->connect("DBI:mysql:mydb","myuser", "secret", {RaiseError=>1}) or die "Cannot connect" . $dbh->errstr ;`	Enabling the optional error checking is highly recommended.
`$dbh->disconnect();`	To disconnect from the database.

Preparing Statements

After establishing the connection and obtaining a database handle ($dbh), the next step is to issue queries to manage and manipulate data stored in the database (see above).

By using a ? in the query statement, a value can be provided during execution of the statement handles. A statement can contain multiple placeholders. Another benefit of using the placeholder is its ability to deal with occurrence of single quotes ' in variables. Database engines will produce a error if a matching ' is not found; this method escapes the single quote automatically.

Code	Comments/explanation

`$sth = $dbh->prepare("SELECT seqname FROM mydata WHERE length > 1600");`
                 `or die "Could not prepare`
`statement: " . $dbh->errstr;`
	The preparation entails taking the SQL statement and creating a statement handle. A good practice is to test the SQL and make sure it works using the utility program provided by the database (e.g., sqlite or mysql) before entering it into a Perl script. Note the use of $dbh->errstr.

`$sth = $dbh->prepare("SELECT seqname FROM mydata WHERE length > ? ");`
                 `or die "Could not prepare`
`statement: " . $dbh->errstr;`
	The resulting statement handle $sth is now ready to be executed. In the statement shown above, we have put in a hard value of 1600 for the length qualifier. In most cases, this would need to be a variable. For statements that will be repeatedly executed using different values for a given parameter one can use ? as a placeholder for a variable as shown here.

Executing Statements

Once a statement handle has been prepared it can be executed.

Code	Comments/explanation
`$sth->execute($length) or die "Couldn't execute` `statement: " . $sth->errstr;`	This example shows how to pass the variable $length to be substituted for the placeholder.
`$rows = $dbh->do("DELETE FROM mydata WHERE length > 20000 ") or die "Cannot execute: "` `$dbh->errstr;`	Statements other than SELECT do not return data and can be prepared and executed in one step using the $dbh->do method. *Note:* Some databases will return the number of rows affected. This operation is performed using the database handle.

Fetching Data

Records are returned as rows on execution of a statement. The records returned are often called the "resultant set" and can be accessed via the statement handle using a loop. There are multiple methods of accessing the resultant set; we discuss a few of them as follows.

1. An easy way to print a resultant set is to use the `dump_results` method. This method is useful for debugging and simple outputs. It takes four optional arguments of field length, line separator, field separator, and output file handle. The defaults are 35, newline, comma, and STDOUT:

```
$out = $sth->dump_results();
```

2. The simplest general method is using `fetchrow_array`. If a SELECT statement is returning two values (say, name and length), the code below will retrieve the results one row at a time. Keep in mind that the returned values are in the order as requested by the SQL statement. A simple `while` loop can be employed to work through the set:

```
while( ($name,$length) = $sth-
    >fetchrow_array) {
print "The name is: $name and
    length is $length\n";
}
```

3. Using the `fetchrow_hashref` method, one can step through the set using a hash with names of the fields being the columns requested from the database in SQL statement. A downside to this method is that, depending on the database, the return field names can be in lowercase, uppercase, or mixed case, i.e., different than expected, making the reference invalid when changing the underlying database:

```
while( $hash = $sth-
    >fetchrow_hashref) {
print "The name is: $hash-
    >{name} and length is $hash-
    >{length}\n";
}
```

4. To keep the code readable and to efficiently manage resources when the number of columns being returned is large, the output column names can be bound directly to a Perl variable instead of creating an array and storing data in it. This speeds up access when working with larger datasets.

```
$sth->bind_col(1,\$name);
$sth->bind_col(2,\$length);
while( $sth->fetch) {
print "The name is: $name and
length is $length\n";
}
```

5. To fetch resultant data as an entire set one can use the `fetchall_arrayref` method, which provides a Perl data structure.

Cleanup

The last stage of interaction with a statement handle is usually automatic; i.e., when the scope of the variable expires, Perl frees up associated memory. Depending on how the handles are defined, this occurs when a loop is completed or connection to the database is closed via the database handle. Attention should be paid to these details, especially when working with large sets of data. For details on using DBI, refer to Descartes and Bunce (2000).

The following examples demonstrate reading data from a file and populating the mydata table in the mydb database.

Script Example 4—Perl: DBI Database Interface Script

The script below demonstrates Perl's DBI database interface. The script will load data from a BLAST tab-delimited output file into a database table and/or perform a database query based on the command line options provided. *Note:* Do not confuse SQL "query" statements and BLAST query sequences. This script uses a database query statement to retrieve information pertaining to a particular BLAST query sequence. The work involved in retrieving records stored in the database table and the provided search criterion is substantially less compared to implementing the same using the BLAST file and series of Perl "if" statements. At the same time, one can easily modify the database query by adding more conditions without having to make significant changes to the script.

The pseudocode for this script is

Get options from the command line specifying database operations (database table, load, and query options)

If invalid options are given, output a Usage message and exit

Connect to the MySQL or SQLite database

If connection fails, output error message and exit

If the 'load' option was specified,

 Create the database table if necessary

Open the input file

Read data from the input file and insert into the database table

If the 'query' options were specified,

 Construct the database query statement according to the options given

 Execute the query and fetch the results

 Print each result

Script Example 4

Code	Comments/explanation
`#!/usr/local/bin/perl -w`	

`# This script takes arguments that specify the desired`
`# database interactions.  The user may load data`
`# from a tab-delimited BLAST output file into the`
`# specified table and/or query the table based on`
`# e-value, start and stop positions delimiting a region`
`# of interest in a given query sequence`	

`# If loading from the BLAST file, create and/or insert`
`# into a database table having the following fields:`
`# (q_id char, s_id char, identity int, length int,`
`#   mismatch int, gap int, q_start int, q_end int,`
`#   s_start int, s_end int, e_val int, bit_score int);`
`#`
`# If performing a query, find which table entries`
`# have e-values smaller than or equal to the e-value`
`# argument and match the specified query sequence`
`# within the range of bases passed in as arguments`	

`use strict;`	Force strict error checking.
`use Getopt::Long;`	Use the Get Options module.
`use DBI;`	Use the Database Interface module.
`my ($fname, $e_val, $qs, $qe, $load, $tbl, $qid);` `my ($dbh, @tables, $test, $sth);`	Declare all lexical (nonglobal) variables as required by "strict."

`my $usage = "Usage $0 -t table [-load -f file]`
`   [-qid query_id -qs start -qe end -e_val`
`e_value]\n";`
	Define the usage message. The string within double quotes must be on a single line! It is split here to fit into the table. The $0 variable is the script name.

`# Get the command line options`	
`$load = $e_val = $qid = $qs = $qe = "";    #Set to false`	The empty string " " evaluates to false in a conditional test.

`GetOptions("file=s" => \$fname,`
`           "table=s" => \$tbl,`
`           "load" => \$load,`
`           "e_val=s" => \$e_val,`
`           "qid=s" => \$qid,`
`           "qstart=i" => \$qs,`
`           "qend=i" => \$qe);`
	GetOptions requires *references* to variables, hence the back slash before each variable name. The -load option is the only one that does not require a corresponding value.

`# Check to see if required options were specified:` `# Must have database table name.   If load option`	

Script Example 4 (Continued)

Code	Commentary

`# is given, must have filename.   If any of e_value,`
`# qid, qstart, or qend options are specified, all`
`# of them must be specified`	

`if (!defined $tbl) {`
    `die $usage;`
`}`
	Must specify a database table.

`if ($load and !defined $fname)   {`
    `die $usage;`
`}`
	If -load specified, need a file name.

`if ((defined $qid or defined $qs or defined $qe or`
    `defined $e_val) and !(defined $qid and defined $qs`
    `and defined $qe and defined $e_val)) {`
  `die $usage;`
`}`
	If *any* query options are given, must have all of them.

`# General DBI handle for connection to SQLite of MySQL`
`# For SQLite the login and password are empty`
`#       as SQLite does not need them`
`# For MySQL the login and password are needed`
`# (Comment out the two lines for the database`
`# type that is not being used)`	

`# We print the error statement using $dbh->errstr`	The errstr method returns a string.

`# $dbh = DBI->connect(`
`# "DBI:SQLite:/home/molbio/james/db/mydb","",`
`# "",{RaiseError=>1}) or`
`# die "Cannot connect" . $dbh->errstr;`
	Because these lines are commented out, a connection to a MySQL database is established on the next line. However, the script works if these lines are uncommented and the MySQL connection lines are commented out.

`$dbh = DBI->connect('DBI:mysql:mydb','myuser',`
                            `'secret',`
`{RaiseError=>1}) or`
    `die "Cannot connect" . $dbh->errstr;`
	Note the use of single quotes around the password to prevent evaluation of the string by Perl. Concatenate a prefix to $dbh->errstr.

`if ($load) {`
  `# If the results table does not exist, create it`
  `@tables = $dbh->tables();`	

`  if(grep{$_ eq "$tbl"} @tables) {`	Perl's "grep" function looks through an array to find a string or pattern. In this statement if table "$tbl" is found, the matching array element is placed in the special variable $_.
`    print "Table $tbl exists; inserting values\n";`	
`  } else {`	

`    $test = $dbh->do("create table $tbl`
        `(q_id char(100),s_id char(100),identity float,`
            `length int,mismatch int,gap int,`
                `q_start int,q_end int,s_start int,`
                `s_end int,e_val double,bit_score`
`int)") or`
        `die "Cannot create table $tbl";`
  `}`
	The quoted string must be all on one long line! It is split here to fit into the table. This statement sends a query directly to the database to create the results table.

525

Script Example 4 (Continued)

```# Prepare an SQL statement which is a template # We will fill in data from the file, loading it into the # table line by line```	
```$sth = $dbh->prepare("INSERT INTO results VALUES (?,?,?,?,?,?,?,?,?,?,?,?)") or die "Could not prepare statement: " . $dbh->errstr;```	Prepare the query statement with placeholders, denoted by ?, whose values are passed in when the statement is executed.
```# Now read data from the BLAST file open(FH,"$fname") or die "Cannot open $fname for reading\n";```	Open a file handle for use in reading the file.
```while(my $line = <FH>) {```	Read one line at a time from the file.
``` # Remove newline chomp $line; if ($line =~ /^#/) { next; }```	If the line begins with #, skip it.
``` my($q_id,$s_id,$identity,$length,$mismatch,$gap, $q_start,$q_end,$s_start,$s_end,$e_val, $bit_score) = split(/\t/,$line) ;```	*This assignment needs to be a single line!* It is split to fit into the table. Values from the split function are sequentially assigned to the variables in the list.
``` # Execute the query $sth->execute($q_id,$s_id,$identity,$length, $mismatch,$gap,$q_start,$q_end, $s_start,$s_end,$e_val,$bit_score) or die "Couldn't execute: " . $sth->errstr; }```	Pass in values to the query statement template and execute the query.
``` $sth->finish;```	Finished with this statement handle.
```} # End if load```	
```# If no e_value, no query is needed if (!$e_val) { exit; }```	
```# Prepare the query to find the matches that satisfy # the criteria defined by $qid, $e_val, $qs, and $qe $sth = $dbh->prepare("SELECT s_id, e_val, length FROM $tbl WHERE q_id = '$qid' AND e_val <= $e_val AND $qs <= s_end AND $qe >= s_start");```	*The quoted string must be all on one long line!* It is split here to fit into the table. This statement prepares a query that will select (subject_id, *E* value, and length) of matches to the query $qid having *E* values <= $e_val AND overlapping the $qs, $qe range.
```# Execute query and fetch results $sth->execute;```	
```while (my @result = $sth->fetchrow_array) {```	The fetchrow_array function returns each match triplet (subject_id, *E* value, length) as an array.
``` print "@result\n";```	Print the entire array.
```}```	
```$sth->finish;```	Finished with this statement handle.
```$dbh->disconnect;```	Close the database connection.

PERL EXAMPLES

The example Perl scripts below are meant to be easily understood, yet useful for some basic analysis tasks. The scripts can be downloaded from the book's Web site.

Script Example 5—Perl: Reading Files and Looking for Text Patterns

The following Perl script shows how to prompt for and read keyboard input (stdin), read a text file, and check for a valid DNA sequence. The user is prompted to enter the name of a file, and the script attempts to open the file. Each line of the file is read in and the sequence is validated. Lines containing invalid DNA bases are reported. Notice that when a line is read from a file or from stdin, the last character is a "newline", represented in Perl as "\n". It is often necessary to remove these trailing newline characters from lines, as this example illustrates.

The Perl concepts included are

Perl comments and interpreter specification
The newline character \n and its presence in input and output
Opening a file for reading, file handles, and the "die" function
The "print" function
The "chomp" function
Reading from standard input (STDIN) and using a file handle
Writing to standard output (STDOUT)
Simple pattern matching
The "while" looping construct
Closing a file

The pseudocode for this script is

Print a query to stdout asking for a file name
Read the user's response from stdin
Open the file or exit with an error message if unsuccessful
Set the line number to 0
While there is another line in the file, read it and:
    Add 1 to the line number
    Chomp the newline character off the line
    Check to see if the line has any invalid DNA bases
    If so, report the line number and the invalid base
Close the file

The table below contains the script and a column of explanatory text. Compare the Perl code with the pseudocode. The code may be typed into an editor. Configure the editor to display line numbers so that any errors reported by the Perl interpreter can be located. Save the file, make it executable, and run the script. *Note:* Make sure that the row separators are preserved and that extra rows are not added between pages. This must be done for all examples of scripts in this chapter.

Script Example 5

Code	Comments/explanation
`#!/usr/local/bin/perl -w`	*Very important:* If the path to the Perl interpreter is not provided with precisely this syntax (very first line begins with # ! / with no intervening spaces), the script will not be directly executable on the command line. The -w option tells Perl to issue warnings as needed, which helps in debugging the script.
	Include white space for readability.

`#    This script will ask for the name of a file`
`#    containing DNA sequence and validate each`
`#    line of sequence.   If invalid bases are found,`
`#    the offending line number(s) are reported.`
	Begin with comments describing the purpose of the script.

`#    Author:   MyName`
`#      Version 1.2    21 Feb 04    output line numbers`
`#      Version 1.1    24 Dec 03    output invalid base`
`#      Version 1.0    15 Sep 03    initial version`
	Include the author's name in the comments. Versioning is essential for tracking purposes. Assign a version number and date each change.

	More white space for readability.

Script Example 5 (Continued)

```use strict;```	Using the strict module instructs Perl to do strict error checking and generate an error message if any unsafe constructs are found. strict requires that every variable be declared with a scope specifier such as my or our. Notice the ; at the end of all statements.
my ```($filename, $line, $lnum);```	Declare lexical (nonglobal) variables. The $ sign before the variable name indicates a scalar variable, a number of string of letters.

```#   Notice that when we print a question we do```
```#   not include a newline since we want the user's```
```#   typed response to be entered on the same line```
	Add comments that explain the next section of code.

```print "Enter DNA sequence file name: ";```	Print query to the screen (stdout).
```$filename = <STDIN>;```	The "diamond operator" <> reads one line of input from the file associated with the enclosed file handle. The entire line is stored in the variable $filename.
```# Open the file, or die with an error message```	Comment explaining the following line.
```open(SEQFIL, "$filename") or``` ``` die "Cannot open file $filename\n";```	The open function associates a file handle with a filename. Here, SEQFIL is the file handle that will be used to read from the file. File handles are normally written in uppercase. The "'or'" clause is executed only if the open fails. The "'die'" function takes an error string (series of characters and/or perl variables enclosed in double quotes) as a parameter, prints the string, and exits the script.
```# Count line numbers for error reporting```	
```$lnum = 0;```	$lnum is initialized to 0 since no lines have been read yet.
while ```($line = <SEQFIL>) {```	The <> around the file handle tells Perl to read the next line from the file. This line is assigned to the $line variable. If there are no more lines to read, $line is undefined and the while loop is exited. The statement block in the while loop is delimited with curly braces { }. These must be exactly matched. The statement block is indented for readability.
``` $lnum++;;```	The ++ operator increments the value of $lnum by 1.
``` chomp($line);```	Chomps the \n off the end of $line

Script Example 5 (Continued)

	so that the \n is not detected as an invalid DNA base.
`if ( $line   =~   /([^ACGTacgt])/ ) {`	The =~ operator searches $line for the specified pattern, which is located in between the / / characters. The pattern class [ACGTacgt] matches any of these valid DNA bases, so to check for invalid bases, the class is negated by inserting a caret ^ at the beginning: [^ACGTacgt]. The result is that any character other than ACGTacgt will be matched. To capture the invalid base that is matched, the pattern is surrounded by parentheses ( ) and the match is placed in the temporary variable $1.
`    print "Invalid DNA sequence ($1) on line $lnum\n";`	Print the offending character and the line number.
`    }`	Closing } for if statement block.
`}    # end of while <SEQFIL>`	Closing } for while loop. The comment is useful for visually matching the opening brace of the while statement block.
`close SEQFIL;`	It is good practice to close any files that have been opened. When a file is opened for writing, it must be closed to guarantee the correct output.

Script Example 6—Perl: Command Line Arguments and Text Manipulation

Running the previous script repeatedly becomes tedious because the input file name has to be manually typed each time. A more elegant script would read the file name from the command line, so that on systems with command history, the command line can be reused rather than typing it again. In general, it is best to avoid interactive prompting. Scripts that run via a single command line can more easily be integrated into a series of manipulations or "processing pipeline." All subsequent scripts presented will be suitable for use in a pipeline.

The script below reads a command line argument that specifies the name of a tab-delimited BLAST output file. This type of output can be generated by choosing the "Hit Table" Alignment View on the NCBI BLAST format Web page, or by using the "-m 9" command line option with the stand-alone BLAST program. In this output file, there are a few "header lines," followed by a series of lines, each representing an alignment between the Query sequence and a sequence in the BLAST database, or Subject sequence. The script opens the file and reads one line at a time, looking at the columns and deciding whether to output the BLAST line based on the *E* value of the alignment. The columns in the BLAST output file are Query ID, Subject Description, Percent Identity, Length of Alignment, Number of Mismatches, Number of Gaps, start and end positions of the alignment in the Query and Subject Sequences, *E* value, and Score. Here are abbreviations for the 12 columns of tab-delimited BLAST output:

Qry	Sbj	%	Align	MM	Gaps	Qry	Qry	Sbj	Sbj	E-value	Score
ID	Desc	ID	Len			Start	End	Start	End		

E values should be checked carefully because there are two possible formats: a decimal number or an exponential notation. Perl's pattern-matching makes it very easy to do this.

The following Perl features are illustrated:

Using command line arguments
Perl arrays
Perl while loops and next;
Perl for loops
File I/O (Input and Output) and
File Handles

The "open" and "close" functions
The "exit" and "die" functions
The "split" function
Simple pattern matching

The pseudocode for this script is

Get a file name from the first command line argument
If no argument is given, print an error message and exit
Attempt to open the file for input; exit with an error
if unsuccessful
Read one alignment line of the file at a time and for
each line

If it is a header line, skip it
Find the *E* value of the alignment
If *E* value is higher than 1e-20, skip to the next
line
If the *E* value is 0, print "**" then the align-
ment line,
Otherwise print " " then the alignment line

Note: When this script is run, the user must provide a
file name as an argument to the script, e.g.,

scriptname ./rdblast_tab.pl myBLAST.out

Script Example 6

Code	Comments/explanation
`#!/usr/bin/perl     -w`	Identifies the location of the Perl interpreter and enables warning messages.

`# This script reads a Tab-delimited BLAST`
`# output file, skips lines with E-values`
`# greater than 1e-20 and prints other lines,`
`# adding  '**'  for E-values of    zero`
	Comments describing the purpose of the script and its inputs and outputs.

`# Author: MyName`
`# Version 2.01    26 Oct 2003`
`#                 added exit(-1)`
`# Version 2.0     18 Aug 2003`
`#                 get filename as argument`
`#                 pattern match to skip comments`
`# Version 1.0     01 Feb 2003`
`#                 initial beta version`	

`use strict;`	Using the strict module instructs Perl to do strict error checking and generate an error message if any unsafe constructs are found. Strict requires that every variable is declared with a scope specifier such as my or our.
`my  ($blastfile, $line, $e_value, @cols);`	Declare lexical (nonglobal) variables. The @ sign indicates an array variable comprised of lines of data.
`$blastfile = $ARGV[0];`	@ARGV is a special array that holds arguments typed on the command line when the script is run. Like all Perl arrays, the first element is number 0.
`# Verify that an argument was provided`	Always check that needed arguments have been provided.
`if   (!defined   $blastfile)   {`	If no argument was given on the command line, the variable $blastfile will be undefined. Notice the negation of defined with the exclamation point !.
`     print   "Missing filename argument\n";` `     print   "Usage: $0   BLAST file\n";`	Always give the user a clue about what they did wrong and show them how to run the script correctly. $0 is a special Perl variable that holds the name of the script.

Script Example 6 (Continued)

`exit(-1);`	By convention, a nonzero exit status indicates an error. This could be useful if the script were called from another program.
`}`	Do not forget to end the if statement block!
`# Open the BLAST file`	Create named file handles. Always use "`or die`" with open.
`open(INFILE, "<$blastfile") or` `die "Cannot open file $blastfile\n";`	Notice the < that specifies that the file is being opened for reading. That is the default behavior of open, so the < may be omitted.
`# Read BLAST results, one line at a time`	
`while ($line = <INFILE>)    {`	This `while` loop will repeatedly read lines from the file until the end of file is reached. Each time a line is read, it is assigned to the variable `$line`.
`# Skip comment lines`	
`if ( $line =~   /^#/ ) { next; }`	This statement uses pattern matching to skip over comment lines. The pattern is anchored to the beginning of the line by the caret (^), so only lines with # as the first character are matched.
`@cols = split /\t/, $line;`	The `/\t/` pattern matches the tab characters. Items between the tabs are placed in the `@cols` array. The whole line is read.
`$e_value = $cols[10];`	*E* value is in the 11th column. Perl arrays are indexed from 0, so get `$cols[10]`.

`# If e-value begins with 'e', prepend a '1'`
   `# so Perl can recognize it as an exponential`
   `# number`
	Perl recognizes exponential numbers of the form 1e-20, but NCBI BLAST output can include *E* values without a leading digit, i.e., e-20.

`if ( $e_value =~   /^e/ ) {`	
`$e_value = "1" . $e_value;`	The . is Perl string concatenation.
`}`	
`# If e-value is higher than 1e-20, skip`	
`if ( $e_value > 1e-20 ) { next; }`	
`if ( $e_value   ==   0.0) {`	Excellent alignment! Note the double equal sign = = used for numeric comparison.
`print "** $line";`	It is not necessary to add \n to this print since the input line was not chomped.
`} else {`	The *E* value is greater than 0 but less than or equal to 0.01.
`print "    $line";`	
`}`	Closing } for else block.
`} # End while <INFILE>`	Closing } for while block. Adding the comment makes it easier to visually identify matching braces.
`close(INFILE);`	We are finished with the file handle so close it.

Script Example 7—Perl: Pattern Substitution and Incremental Development of Scripts

The next script will find the number of occurrences of "CAG" in DNA sequences read from a file and converted to uppercase. The file name will be passed to the script as a command line argument. The file format is a simplified version of the FASTA format, and it contains alternating lines of sequence description and actual sequence. In other words, the sequences are each on one long line. Here is an example:

```
>Sample_544 Dr. Smith 10 March 1999
ATTACGCGATCAGCAGCAGCAGCAGCAGTCGCTTTCAGCGAGT...
>Sample_626 Dr. Who 6 Aug 2000
CGCGAGCGTCGACAGCAACAGCAGCAGCAGCAGCAGCAG...
```

Files contain an arbitrary number of sequences, and we assume that each has a description line followed by a sequence line. Incremental development is illustrated by dividing the pseudocode into four parts. When coding, get Part I working completely correctly before adding code for Part II, etc. The Perl features demonstrated are

> Using command line arguments
> The "chomp" function
> The "uc" function
> The "length" function
> Pattern matching and substitution

The pseudocode for this script is

Part I:

Obtain a file name from the first command line argument

If no argument is given, print an error message and exit
Attempt to open the file for input; exit with an error if unsuccessful
Attempt to open another file for output; exit with an error if unsuccessful

Part II:

While the end of the input file has not been reached, read the next description line
Chomp the newline from the description
Remove the > symbol from the description
Print a message to STDOUT indicating that processing is occurring for the sequence identified by the description
Read the sequence line and chomp it
Convert the sequence to uppercase

Part III:

Add a check (or test) *inside the while loop* to see if the DNA sequence is valid
Find the length of the sequence
Use pattern substitution to count the number of occurrences of "CAG"

Part IV:

Add a check *inside the while loop* to see if the number of CAG repeats is greater than 50, and if so, print a message that tells the number of CAG's found and the length of the sequence
After the while loop, close the input and output file handles

Script Example 7

Code	Comments/explanation
`#!/usr/bin/perl    -w`	This directive is always needed if the script is to be directly executable.

`# The purpose of this script is to learn Perl`
`# command line arguments, file I/O, and`
`# pattern matching`
`# Author: MyName`
`# Version 0.1, 27 Sep 03`	

`use strict;`	Using the strict module instructs Perl to do strict error checking and generate an error message if any unsafe constructs are found. Strict requires that every variable is declared with a scope specifier such as my or our.
`my ($infile, $def, $seq, $len, $n_found);`	Declare lexical (nonglobal) variables.
`# Get filename from command line argument` `$infile = $ARGV[0];`	@ARGV is a special array that holds arguments typed on the command line

Script Example 7 (Continued)

	when the script is run. Like all Perl arrays, the first element is number 0.
`# Verify that an argument was provided`	Always check for required arguments.
`if    (!defined    $infile)    {`	Notice that "not defined" is written with an exclamation point in front of "`defined`"
`    print    "Missing filename argument\n";` `    print    "Usage: $0    FASTAfile\n";`	Always give the user a clue about what they did wrong and show them how to run the script correctly. The $0 is a special Perl variable that holds the name of the script. Use $0 in error messages so that the messages are relevant even if the script is renamed.
`    exit(-1);`	By convention, a nonzero exit status indicates an error. This might be checked by another program that uses the script.
`}`	Do not forget to end the "'if'" statement block!
`# Open the input and output files`	Create named file handles.
`open(INFILE, "<$infile") or` `    die "Cannot open file $infile\n";`	Notice the < that specifies that the file is being opened for reading. That is the default behavior of open, so the < can be omitted.
`# Make the output filename related to the` `# input filename`	This helps to easily identify related files.
`open(INFILE, ">$infile.out") or` `    die "Cannot open $infile.out for writing\n";`	To open for writing the > is mandatory. Notice that we can add .out as a suffix to $infile. Perl variable names cannot include ".", so Perl knows where the variable name ends within the string.
`# End of Part I`	Write this much and test it. You will see warnings that the INFILE and OUTFILE handles are not used (for now), but no other error messages should appear.
`# Read sequences from file and look for CAGs`	First definition line, then sequence line.
`while    ($def    =    <INFILE>)    {`	Notice similarity to reading from STDIN—just a different file handle.
`    chomp($def);`	Always chomp unless the newline is needed—most often it is not and it can cause problems if you forget it
`    # Remove > from the description and print` `    # progress message to STDOUT`	
`    $def    =~ s/>//;`	Replace the > in the line with nothing!
`    print "Sequence $def being processed...\n";`	It is nice to indicate progress to reassure the user that the script is doing something.
`    # Now read the sequence from the file`	
`    $seq    =    <INFILE>;`	Reads one line from the file, including the trailing newline.
`    chomp($seq);`	Remove the newline character.

Script Example 7 (Continued)

`# Convert sequence to upper case`	
`$seq = uc($seq);`	Must assign the results of "uc" to `$seq`. The "uc" function itself does not change `$seq`.
	End of Part II. To test this much, the closing } for the while loop must be here. Test Part II before inserting Parts III and IV before the } .
`# Check that the DNA sequence is valid`	
`if ($seq =~ m/[^ACGT]/) {`	If any characters that are not A, C, G, or T are matched, this condition is True. "m" is the match operator.
`print "Invalid DNA bases in $infile : $def \n";` `next;`	Skip to top of while loop to begin processing the next sequence.
`}`	Close "'if'" statement block.
`# Find the length of the sequence`	
`$len = length($seq);`	Use the built-in Perl function "length".
`# Count number of CAG's`	Use pattern substitution since it returns a count of the number of substitutions made.
`$n_found = ($seq =~ s/CAG/CAG/g);`	The sequence remains unchanged because CAG is substituted for CAG. "s" is the substitution operator. The "g" modifier indicates that the substitution is applied globally (across the entire sequence). A substitution does not occur but the number of CAGs is found.
	End of Part III. To test this much the closing } for the while loop must be here. Insert Part IV before the } after testing Part III.

`# Check to see whether number found > 50`
`# or number found >= 0`
`# If either of these is true, print msg to outfile`
`if ($n_found >= 50) {`	

`print OUTFILE "$def: found $n_found "` `print OUTFILE "CAG in $len bases \n";`	Notice the use of the file handle. Two print statements are used only to fit in this table—these can be combined into one.
`} elsif ($n_found <= 0) {`	The "elsif" keyword must be used here. The interpreter will not recognize "else if".
`print OUTFILE "$def: CAG not found!\n";` `}`	
`} # End of while <INFILE>`	The comment helps us to visually match the { } braces.
`close (INFILE);` `close (OUTFILE);`	Remember that if an output file is not closed it may not be written properly.

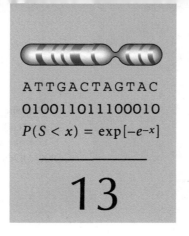

ATTGACTAGTAC
010011011100010
$P(S < x) = \exp[-e^{-x}]$

13

Analysis of Microarrays

C O N T E N T S

The material in this chapter was contributed by David A. Henderson, Department of Animal Sciences, and Division of Epidemiology and Biostatistics, University of Arizona.

INTRODUCTION

Useful biological information about the genome of an organism can be determined by finding which genes are induced or repressed in a phase of the cell cycle, a developmental phase, or as an environmental response, such as in response to treatment with a hormone or exposure to high temperature. Sets of genes whose expression rises and falls under the same condition are likely to have a related biological function and perhaps a common regulatory relationship; e.g., they may have similar promoter sequences for binding transcription factors. In addition, a pattern of gene expression may be an indicator of abnormal cellular regulation and is, therefore, a useful tool in disease diagnosis. Because eukaryotic genomes are so large, microarray technology has been developed for studying regulation or sequence variation of thousands of genes or synthesized proteins at the same time. The major use of microarray technology has been to study gene expression by detecting mRNA levels in cells and tissues. In this type of analysis, many or most of the genes of an organism are represented by oligonucleotide sequences spread out in a high-density array on glass microscope slides or some other solid support medium. mRNA, extracted from cells or tissues, is copied into DNA and labeled with a fluorescent dye or some other means. The labeled mRNA or an mRNA mixture is then hybridized to the slide and the slide is scanned with a microscope to measure the amount of label over each spot. The amount of label is assumed to be proportional to the amount of each mRNA in the original biological sample. The data are then analyzed to identify genes whose expression is changing. This chapter provides a basis for planning experiments and for the subsequent analysis of microarray data. It also describes the most common techniques of microarray technology and the advantages and disadvantages of these methods.

Rapid changes are occurring in microarray technology and analysis. Supplementary material addressing these changes will be placed on the book Web site.

CHAPTER GUIDE FOR BIOLOGISTS

This chapter addresses the analysis of large data sets, including microarrays for measuring gene expression. It is important to note that similar analysis should also apply to proteomics experiments. Almost every molecular biology laboratory performs a microarray experiment to measure gene expression in a biological sample at some time. Because of the expense, a microarray experiment usually includes only a few slides with perhaps a duplicate experiment. A commercial product or downloadable software is used to analyze the data. Most often, this involves making plots of all values, choosing a rough normalizing procedure, and deciding upon a twofold cutoff value for a change between samples. A set of clustering analyses may also be performed. The genes that are expressed differently between samples or clusters are then taken as an indicator of biological changes in the experiment. Often the experiment has many variations so that long lists of genes unique for each treatment are found. Some colleagues obtain enough data so that they can use important information (e.g., identify a strongly overexpressed gene in many cancer tissues); but, in many cases, it is a big struggle to find useful information. Many colleagues are discouraged by the lack of useful information. Some investigators are abandoning microarrays because of their associated uncertainties and are switching to proteomics. It is likely, however, that the problem with microarrays will be repeated with proteomics because of the lack of experience of most biologists in using these technologies and handling large data sets. The goal of this chapter is to show that if careful attention is paid to experimental detail and data analysis, the chance of recovering useful information from a microarray experiment is greatly increased.

Biologists first need to be aware of the many sources of variation in microarray data at all stages of the experiment, starting with sample preparation as well as variations in cDNA labeling of genes, dye choice, hybridization to individual slides, reproducibility of measurement of particular genes, and signal measurement. The experiment must be planned in a manner that provides measurements of the expected variation in the expression values of a particular gene. The amount that this expression value on a particular slide deviates from what is expected can then be determined and a statistical analysis performed on these deviations to obtain the most significantly varying genes. This result is only attainable if there is replication of samples; i.e., the same experiment and hybridizations should be repeated, usually at least three times, to measure variability adequately.

A second reason why efficient use of microarray data is important is to have a clearly defined objective in the experiment. A clearly chosen experimental goal will help with both the experimental design and the methods of data analysis. For example, one goal might to identify altered gene expression in cells treated with a drug at some time point after administration of the drug as compared to untreated cells; a second goal might be to find a set of genes that distinguish between gene expression patterns in a set of normal tissues and those in cancer tissues. The design of these two types of experiments and the data analysis will be very different. In the first experiment, there should be enough replication to identify sources of experimental variation for genes, slides, dye for cDNA arrays, and drug treatments so that a list of genes that vary significantly can be found. In the second example, enough normal and diseased samples must be used so that a small set of genes that varies significantly between the classes can be identified. This latter set of genes makes up a classifier, usually a straight line that separates the two sets of samples on a plot of expression values for one gene on the x axis and expression values for the other gene on the y axis. It is important to validate the classifier by resampling the data and showing that the classifier still separates all of the samples when some of the samples are not used to produce (or train) the classifier.

A large number of different computational methods support the above types of analyses. Statisticians have produced and support many statistical tools for analysis. Many of these tools are written in R, a high-level Open Source computing language like Perl (S+ is the equivalent commercial product) that supports statistical analysis and graphical displays. Familiarity with these methods is important so that the appropriate tool congruent with the experimental goals may be used. Examples of using R tools to analyze microarray data are given in this chapter.

Beyond these considerations, additional advice to consider is: (1) become familiar with statistical methods as much as possible; (2) obtain a copy of Snedecor and Cochran (1989) and read about analysis of variance, which is explained with many examples; (3) seek advice from a statistician at all stages of planning and analyzing an experiment; and (4) note that clustering, which is used to discover classes in microarray data, can be based on different criteria. There are methods that help one to visualize where the variation is in the data, and others that show which genes or samples are showing the most-alike patterns of variation and therefore likely to share a common regulatory relationship. These different methods are described in this chapter.

CHAPTER GUIDE FOR COMPUTATIONAL SCIENTISTS

As discussed previously, most genes are not always transcribed into mRNA and then translated into proteins. Transcription occurs when the function of a gene is biologically necessary. For many years, biologists have biochemically extracted mRNA from cells and tissues and determined the expression of a particular gene by assessing the amount that binds (hybridizes) to a DNA sequence of the same gene through hydrogen bonding. With the availability of genome sequences, biologists have access to the sequences of all the genes for an organism, a number that varies from a few thousand to more than 35,000 genes. A technology called microarray analysis has been developed to spot one or more sequences of thousands of genes in a high-density array on a solid medium such as a glass slide. For some arrays, the gene sequence is synthesized directly onto the slide. In this approach, a DNA copy is made of each mRNA in a test tube and the mixture is then hybridized to the slide. The amount of label attached to each gene position represents the amount of mRNA in the biological sample for the corresponding gene. This technology allows biologists to examine changes in gene expression that occur in many situations including, for example, changes following treatment of cells with a drug, differences between normal and diseased cells, and changes that occur during specific stages of development. The only limitations to the technology itself are the ability of the researcher to collect a pure enough example of the cells or tissue and a certain minimum amount of mRNA for labeling.

The sequences that are located on a microarray chip also are linked to biological information about each gene. This information assists biologists in making a biological interpretation of gene expression changes. Patterns of gene expression changes have also attracted the interest of computational scientists who attempt to build models designed to interpret the patterns of gene expression in terms of gene interactions and regulation.

There are several computational problems related to microarray data. First, microarray technology generates large amounts of data that need appropriate data management. Data models that support the collection and storage of information have been designed and are described in this chapter. Second, many sources of variability occur in microarray data at almost every level of the technology. In experimental design and data analysis, these variations need to be accounted for to identify the genes that vary significantly. The goal of this chapter is to help biologists get the most out of microarray experiments by using this additional information about microarray variation. Biologists need advice from a experienced computational scientist trained in statistical analysis.

Many tools have been developed for analyzing microarray data, but it is important to notice whether these methods are based on sound statistical foundations or are ad hoc, possibly misleading the researcher or failing to find a significantly varying gene. This chapter emphasizes R (http://www.r-project.org/), a computing language and environment for statistical computing and graphics. Just as Perl is useful for bioinformatics and has many useful modules for analyzing sequence information, R is useful for statistical analysis and display of microarray data. Statisticians have developed excellent Open Source tools for analysis of microarray data under the GNU license. These tools and R scripts that use them are described and used throughout this book.

WHAT SHOULD BE LEARNED IN THIS CHAPTER?

- Familiarity with devising an appropriate experimental design for a microarray experiment.
- How to normalize microarray data when necessary.
- How to develop a statistical model and use that model to identify significantly varying genes by analysis of variance and *t*-tests.
- How to cluster and group genes by various methods.
- Familiarity with methods to identify and validate gene classifiers.
- How to interpret results biologically.
- How to use results to infer regulatory relationships.

Glossary Terms

Analysis of variance (ANOVA) is a statistical procedure that assesses inferences on differences in treatment means from multiple treatments using a ratio of variation associated with the treatments to an estimate of an appropriate error variance.

Average linkage clustering is a form of hierarchical clustering in which the merging of nodes is decided using the average distance between the members of each node.

Bayesian probability is a degree of belief in the probability of occurrence of a specific outcome as opposed to a frequentist probability, which is based solely on frequency of events. A Bayesian analysis uses prior probabilities and a data model to produce posterior conditional probabilities based on the model.

Block design is an experimental design in which sources of external variation (blocks) are averaged over to remove their influence on statistical inference of treatment effects.

Bonferroni correction is a type of multiple comparison adjustment that strongly controls the family-wise type I error rate. A type I error is the rejection of a true null hypothesis. The correction is commonly approximated by the equation $p \text{ value}_{adjusted} = p \text{ value}_{raw} / \text{\# comparisons}$.

Conditional probability is the probability of an event that depends on the state or value of a second event.

Correlation coefficient is the standardized linear relationship between two continuous variables. By definition, correlation coefficients are restricted to the range of values $[-1,1]$. The Pearson correlation coefficient is calculated based on the assumption that the variables are sampled from a normal distribution, whereas the Spearman correlation coefficient is calculated from the rank of the observations.

Covariance is a measure of the relationship between two variables. It represents the unscaled deviation from independence between two variables such that their covariance increases as the two variables become more dependent. The sign of the covariance determines the direction of dependence.

Dimensional reduction is the practice of representing multidimensional data in a smaller number of dimensions created so as to minimize the loss of information contained in the entire data set.

Euclidean distance is the linear distance between two points on a graph.

F-test is a statistical test based on the ratio of variation in a variable of interest to the variation in an appropriate error term. A large F-value indicates significantly larger variation due to the experimental variable versus that in the error. The interpretation of a significant F-value is often that at least one of the levels of the numerator variable differs from the others.

Factor analysis is a multivariate statistical technique that groups variables such that the covariance explained by the retained factors is maximized.

False discovery rate is the expected proportion of false positives, or the expectation of the type I error rate, when performing multiple hypothesis testing.

Fuzzy C-means clustering clusters a data set into a specified number of clusters in which the assignment of a sample to a cluster is not hard, as in k-means clustering, but has a certain probability of assignment, making the cluster borders fuzzy. The object is to cluster the most-alike samples based on a measure of difference between them.

Hierarchical clustering is a clustering technique that progressively groups variables of small distance together and keeps distant variables separate.

k-means clustering is a method of clustering that, once a value k is specified, will minimize the variation within each of k clusters, but maximize the variance between clusters.

Loess normalization is a polynomial version of Lowess normalization.

Loop design is an experimental design with t treatments applied to blocks of size b where $b < t$ and each treatment is measured an equal number of times. Loop design is also called a partially balanced incomplete block design.

Lowess normalization is a data analysis technique for producing a "smooth" set of values from a noisy scatter plot by moving a window across the plot and producing a linear model fit to each point using the data within the window.

M versus A plot represents the relationship between the variance (M) and the mean (A) of gene expression values in microarrays.

Maximal linkage clustering, also called complete linkage clustering, is a type of hierarchical clustering in which genes are evaluated for cluster membership based on the maximal distance between that gene and all the genes within the potential new cluster.

Maximum likelihood estimation gives the probability of obtaining a particular set of data, given an expected probability distribution or underlying probability-based model for generating the data.

Minimal linkage clustering, also called single linkage clustering, is a type of hierarchical clustering in which genes are evaluated for cluster membership based on the minimal distance between that gene and all the genes within the potential new cluster.

Mixed models is a linear statistical model that includes additional random effects in the model beside the usual random error term.

Normalization is the process of removing nonlinear trends between arrays or between dye channels within a spotted array.

Posterior probability, in a Bayesian analysis, is the conditional probability of the variables of interest, given the observed data.

Principal components analysis (PCA) is a multivariate method that partitions the data into a set of components that explain decreasing amounts of variation in linear combinations of variables such that the first component explains the largest amount of variation, the second component the second largest, and so on. The linear combinations of variables are described as a collection of vectors called eigenvectors, and the variance explained by that component is the corresponding eigenvalue. Among multivariate techniques, principal components analysis is a member of the class of variance maximization methods.

Prior probability, in a Bayesian analysis, is the prior belief, or degree of confidence, of the variables of interest.

Quantile to quantile (QQ) plot compares actual data values to those expected for an assumed distribution—usually the normal distribution. A straight line indicates support for the assumed distribution.

Self-organizing maps is a clustering method closely related to k-means, with the exception that the k clusters are now oriented in a two- or three-dimensional grid.

Singular-value decomposition (SVD) is a method similar to principal components, but not limited to square matrices. Interpretations of the components are similar with the exception that singular-value decomposition produces two sets of eigenvectors.

Supervised learning describes a statistical or machine-learning procedure that does require annotated training data to perform on new, unannotated data.

Support vector machines is a classification tool that identifies class membership between two classes by means of a user-defined kernel function. The function provides a data transformation method that creates additional dimensions to facilitate finding a classifier (e.g., x, y in two-dimensional space becomes x^2, $\sqrt{(2\ xy)}$, y^2 in three-dimensional space). SVMs are a supervised learning procedure.

***t*-test** is a test statistic used to test means and differences between means when the variances are estimated.

Unsupervised learning describes a statistical or machine-learning procedure that does not require annotated training data to perform on new, unannotated data.

Variance is a measure of variability in a data set, specifically the average squared deviation from the mean.

Volcano plot is a plot of p value versus treatment contrast size, used to demonstrate the relationship between significance and estimated treatment difference.

THE COMPLEXITIES OF MICROARRAYS AFFECT EXPERIMENTAL DESIGN AND ANALYSIS

Gene expression microarrays have seen a great deal of use in the biology community since their reported invention by the Pat Brown laboratory in 1995 (Schena et al. 1995) and by Affymetrix in 1996 (see Lockhart et al. 1996). DNA microarrays are a multi-use technology—different technologies are employed to produce the microarray chips and different technical approaches are used for analyzing microarray data, ranging from statistical models that closely resemble those in agricultural research to machine-learning methods for identifying class predictors (genes that can distinguish classes of biological samples). The underlying technology is extremely complex. There are many possible sources of errors in producing the microarray slides, such as variations within and between oligonucleotide spots, efficiency of dye incorporation, efficiency of hybridization of labeled cDNA to each slide, and accuracy and fluctuations in scanning the fluorescent signals on the slide. Because of all these types of variations, it is important to plan a gene expression microarray carefully and to process and analyze the data, keeping these sources of variations in mind.

One of the major goals in this chapter is to provide the statistical tools needed for planning microarray experiments and analyzing the data. The most

TABLE 13.1. *Resources for statistical analysis of microarrays*

Resource	Developer	URL
ANOVA	G. Churchill	http://www.jax.org/staff/churchill/labsite/software/anova/index.html
GeneTS	K. Strimmer	http://www.stat.uni-muenchen.de/~strimmer/ .
Oligonucleotide chip analysis	B. Bolstad	http://www.stat.berkeley.edu/users/bolstad/index.html
R library functions for stat, analysis	G. Smyth	http://www.statsci.org/smyth/
R statistical packages	R. Irizarry	http://biosun01.biostat.jhsph.edu/~ririzarr/
SAM	R. Tibshirani	http://www-stat.stanford.edu/~tibs/SAM/
Smartpred, etc.	T. Hastie	http://stat.stanford.edu/~hastie/
Statistical analysis software	D. Allison	http://www.soph.uab.edu/ssg_content.asp?id=1290
Statistical packages	T. Speed	http://stat-www.berkeley.edu/users/terry/zarray/Html/soft.html

common techniques are described in this chapter, as are their advantages and disadvantages. A list of Web sites of statisticians who work on microarray analysis and who provide useful software and advice is given in Table 13.1. A representative list of laboratories that have developed microarray analysis tools is shown in Table 13.2. Many of the examples given in this chapter are the Open Source collection of data analysis tools called BioConductor. Finally, examples of microarray databases that store previous experiments and provide them to the research community for further analysis are listed in Table 13.3. These databases provide excellent resources that may help to provide data and information for new microarray projects.

When analyzing gene expression microarrays, two considerations should be kept in mind. First, as Pierre Baldi has pointed out (Baldi and Hatfield 2002), the underlying biological processes being investigated are often not understood and are almost certainly complex. There may be a certain level of "biological noise" between biological samples because the cells or tissues are not following an identical pattern due to sample, environmental, and genetic programming. Second, gene expression microarrays are measuring the steady-state level of an unstable molecule, mRNA. The steady-state level in the cell of each mRNA that is being measured in a microarray experiment also depends on the rates of transcription and degradation of the mRNA. The method also assumes that copying of each mRNA into a labeled cDNA copy is equally efficient, whereas the copying and labeling of mRNAs are likely to be variable.

TABLE 13.2. *Representative data analysis tools for microarrays*

Tool	Lab	URL
ArrayMaker	Joseph DeRisi lab	http://www.microarrays.org/
BioConductor, Open Source collection		http://www.bioconductor.org/
BRB array tools	National Cancer Institute	http://linus.nci.nih.gov/BRB-ArrayTools.html
DCHIP	W.H. Wong lab	http://www.dchip.org
GENECLUSTER		http://www.genome.wi.mit.edu/MPR/software.h\ml
GeneRage		http://www.ebi.ac.uk/research/cgg/services/rage/
GeneX		http://www.ncgr.org/research/genex/
Microarray guide	Pat Brown lab	http://cmgm.stanford.edu/pbrown/
Microarray software	Michael Eisen lab	http://rana.lbl.gov/
microarrays.org. software resource		http://www.microarrays.org/software.html
NIH Microarray		http://www.nhgri.nih.gov/DIR/LCG/15K/HTML/
TIGR functional genomics		http://www.tigr.org/microarray/

TABLE 13.3. *Examples of microarray databases*

Database	URL
Arrayexpress, European Bioinformatics Institute	http://www.ebi.ac.uk/arrayexpress/
GEO, Gene expression repository	http://www.ncbi.nlm.nih.gov/geo/
MAD, Jackson Labs.	http://mad.jax.org/
MGED data sharing group	http://www.mged.org/
SMD, Stanford Microarray Database	http://genome-www5.stanford.edu/
YMD, Yale Microarray Database	http://info.med.yale.edu/microarray/

Limitations in Measuring mRNA Levels to Study Biological Regulation

There are many steps after mRNA synthesis in the regulation of gene expression, including translation of mRNA into protein, cellular transport of proteins, and posttranslational modification of proteins into an active form. Individual protein levels and protein modifications may be detected by using proteomics techniques. An example is two-dimensional gel elec-trophoresis in which proteins from a sample are labeled and separated based on their size and charge. Protein modifications may then be identified and the proteins may be identified by MALDI-TOF mass spectrometry. For this reason, proteomics analysis provides a second important method of analysis of gene expression in biological samples. Many of the methods of experimental design and analysis used for expression microarrays that are described in this chapter can also be used for proteomics data.

THERE ARE TWO MAIN TYPES OF MICROARRAYS

The two types of microarrays used for measuring gene expression are the cDNA, or spotted, array and the high-density oligonucleotide array. Use of these arrays requires different experimental approaches and types of analyses as described below. A useful review of many of the procedures, along with pitfalls to avoid, is available in Bowtell and Sambrook (2003).

cDNA, or Spotted, Arrays

The production and use of spotted array microarrays are illustrated in Figure 13.1 (left panel). Basically, DNA fragments representing the exons of most genes of an organism are amplified by polymerase chain reaction (PCR) and then spotted in a high-density grid pattern on a glass slide. mRNA samples are obtained from two biological samples to be compared. DNA copies (complementary or cDNA) of these mRNA preparations are then fluorescently labeled with Cy3 or Cy5 dyes, respectively. A mixture of the labeled cDNAs is then hybridized to the slide and the slide is scanned to measure the amount of label over each DNA spot. The ratio of the labels is then used to indicate the ratio of the mRNA levels in the original samples. The ratio is usually indicated by a color. An advantage of cDNA arrays is that they can be used in experimental designs for determining sources of variation in the data, as described below (see p. 625). An example of a good-quality cDNA array is shown in Figure 13.2.

High-density Oligonucleotide Arrays

An example of this type of microarray experiment is illustrated in Figure 13.1 (right panel). Instead of spotting DNA samples onto the array, a series of 11–16 short oligonucleotides approximately 25 bp in length are synthesized at very high densities onto a solid supporting medium for each expressed gene, as shown in the upper part of the right panel in Figure 13.1. The sequences of these sets represent perfectly matched exon sequences at different positions along the sequence of the represented gene. Another set of mismatching oligonucleotides is produced that has the same sequence as the matching set except for a mismatched base in the middle. A similar set of

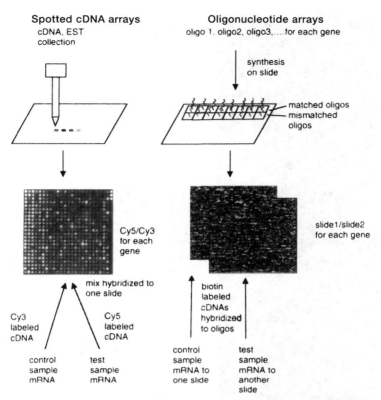

FIGURE 13.1. The production and use of spotted or cDNA arrays and high-density oligonucleotide arrays for detecting global gene expression. In the cDNA array method (*left panel*), mRNA from two biological samples is copied to cDNA, each cDNA is labeled with a different fluorescent label, and a mixture of the two cDNAs is hybridized to an array that has a single DNA spot for each gene on the array. Each spot is scanned and the ratio of the two labels is determined. In the oligonucleotide array method (*right panel*), mRNA from a single biological sample is copied to labeled cDNAs, and then hybridized to a set of short 25-mer matching oligonucleotides for each gene and also to another set of the same oligonucleotides with a mismatch to serve as controls for the hybridization specificity for the gene. To compare two biological samples using the oligonucleotide method, two hybridizations to two separate arrays are needed.

matching and mismatching oligonucleotide probes is then produced for each expressed gene in the target organism. As illustrated in the lower part of Figure 13.1 (right), mRNA is extracted from the biological sample and the sequence is copied to create a cDNA labeled with biotin. The labeled cDNA from one sample is then hybridized to the oligonucleotide array and the amount of DNA at each oligonucleotide position is measured. If two or more samples are to be compared, a different slide is used for each sample. The signal for each gene is determined by measuring the average amount of label affixed to the matching oligonucleotides while correcting for less specific binding to the mismatching oligonucleotides. Many of the uses of oligonucleotide arrays may be found on the Web site of Affymetrix, the company that developed the oligonucleotide array technology.

Other Types of Microarrays

Although this chapter focuses on the use of microarrays for measuring gene expression, microarrays have also been used for a number of other purposes, including the detection of genome loss and amplification, sequence variation (e.g., single-nucleotide polymorphisms [SNPs]), DNA methylation, DNA–protein interactions, and tissue biomarkers. An entire supplementary issue of *Nature Genetics*, volume 32 is devoted to uses of microarray technologies (especially see Pollack and Iyer 2002). Two other uses of spotted DNA arrays are illustrated in Figures 13.3 and 13.4. In Figure 13.3, DNA samples from two tissues are compared to determine if a particular gene or piece of chromosome is over- or underrepresented in one of the tissues. This method is called cancer

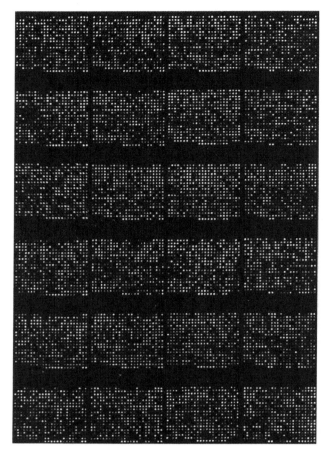

FIGURE 13.2. Example of a high-quality DNA spotted microarray for measuring global gene expression from two biological samples labeled with Cy3 and Cy5. Note that every DNA spot gives a clearly readable signal. The colors shown represent the ratio of the Cy3 and Cy5 labels at each spot. The image has a dynamic range in excess of two orders of magnitude (spot intensities ranging from 200 to 50,000 counts); minimal background fluorescence (<200 intensity counts in both Cy3 and Cy5 channels); strong spatial uniformity in both dynamic range and background intensity (i.e., from one side of the image to the other); round, evenly spaced spots; and virtually no dust, scratches, speckles, or other artifacts. (Reprinted from Bowtell and Sambrook 2003.)

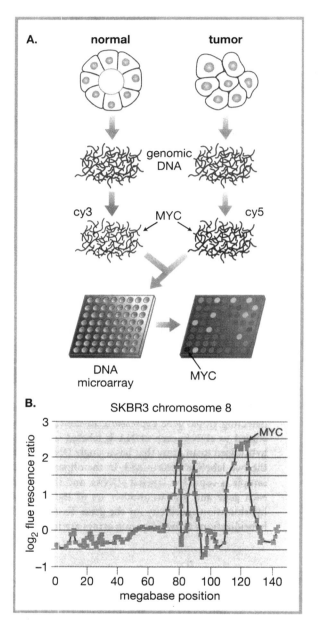

FIGURE 13.3. Example of a cancer genome hybridization (CGH) assay in which DNA spotted arrays are used for detecting gene copy number changes in cancer cells. Extracted DNAs from each cell type are labeled with Cy3 or Cy5, respectively, mixed, and hybridized to a spotted DNA array. The ratio of signals at each spot gives the gene copy number in the cancer cell relative to that in the normal cell. (Reprinted, with permission, from Pollack and Iyer 2002 [©Nature Publishing Group].)

genome hybridization (CGH) because it is used to determine gain or loss of gene function in cancer cells and tissues (see Cai et al. 2002; Pollack and Iyer 2002; Schadt et al. 2003). The two DNA samples are copied into labeled DNA as in a gene expression experiment, and hybridized to a spotted microarray slide. The ratio of the Cy3 and Cy5 fluorescent dyes at each gene position on the slide provides a measure of the relative amount of that gene in the original samples. Arrays that are spotted with 100-kb pieces of DNA

(bacterial artificial chromosome or BAC sequences) are sometimes used.

The second example of using spotted arrays shown in Figure 13.4 is for detecting DNA methyla-

**Cytosine Methylation
Analysis by CGI Microarray**

Genomic DNA from samples

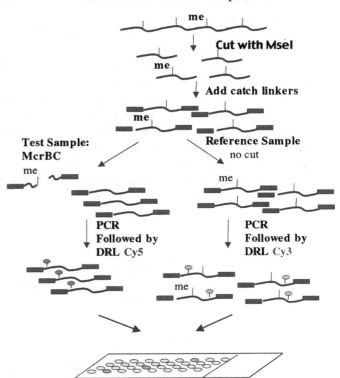

FIGURE 13.4. DNA methylation assay used for detecting methylation of CpG-rich promoter regions (CpG islands) in cancer cells. A spotted CpG island array is produced. Genomic DNA from a biological sample is fragmented into small pieces that can be amplified by polymerase chain reaction (PCR). The fragments ends are first spliced to primer sequences that may be subsequently used to amplify them by PCR. One-half of the DNA fragment mixture is then treated with an enzyme that cuts only those sequences that have been methylated, thus rendering them incapable of PCR amplification. The treated and untreated mixtures are amplified in the presence of Cy3 or Cy5 label, mixed, and hybridized to the CpG island spotted array. A loss of signal for one DNA spot following enzyme treatment indicates methylation of the corresponding promoter region, possibly leading to loss of transcription. (This figure was kindly provided by Damian Junk and Bernie Futscher, Arizona Cancer Center.)

tion. The rationale behind this assay is that methylation of the DNA base cytosine at CpG sequences occurs as a regulatory mechanism for preventing transcription. Normally, methylation of CpG sequences in the promoters of expressed genes in eukaryotic promoter sequences does not occur because it prevents transcription. In cancer cells, aberrant methylation events prevent transcription of genes needed to regulate cellular activity and contribute to the development of the disease. The assay in Figure 13.4 describes a method to detect these abnormal methylation events.

To detect methylation of promoter sequences, a special spotted DNA microarray is made from promoter sequences that are enriched for CpG sequences. The genomic DNA samples are digested with a restriction endonuclease that cuts the DNA into small fragments, ligated to end linkers sequences for PCR amplification and labeling, and then divided into two samples. One sample is digested with another enzyme that preferentially cuts methylated sequences and the other sample is left undigested. If a

DNA fragment has a methylated CpG sequence, it will then be cut by the enzyme and will no longer be capable of subsequent PCR amplification. The undigested fragments are then labeled with fluorescent probes (Cy3 or Cy5) by label incorporation during PCR amplification, mixed, and hybridized to the array of CpG island probes. The slides are scanned and the ratios of the probe signals are compared. The ratio of digested to undigested fragments reflects the degree of methylation for each CpG island on the CGH microarray. A ratio close to 0 will indicate a methylated CpG island; a ratio approaching 1 will indicate an unmethylated island. This method is used to detect loss of gene function in cancer cells due to promoter methylation. This assay depends on inheritance of the methylation pattern by most or all of the cells in the sample. If a pattern of methylation occurs in a single cell, that pattern is inherited by all daughter cells of the methylated one, such that they all acquire the loss of promoter function. The usefulness of this type of microarray analysis depends on having a suitable experimental design.

BIOINFORMATICS MEETS STATISTICS

Throughout this text, statistical methods developed for use in bioinformatics applications have been described. In some cases (e.g., testing the significance of a local alignment score described in Chapter 4), these methods are based on sound statistical methods that have been proven to give a reliable estimate of statistical significance. In other cases, often found in tests of significance and other types of analysis of microarray data in the published literature, the tests have been ad hoc, not based on sound statistical theory, and never proven to provide a reliable statistical estimate. In this chapter, an attempt is made to reveal how statistical methods developed for other applications, and based on a sound theoretical basis, can be used for the design of microarray experiments and the analysis of microarray data. Not using these proven methods can have severe consequences in the form of failing to find important information or to report incorrect information. There are important examples of incorrect analysis of microarray and proteomics data in the published literature. In cases involving the comparison and mining of large data sets to identify complex patterns and relationships in bioinformatics, it may be difficult to find an established statistical design or principle that applies. One goal of this chapter is to advocate the development and use of sound statistical approaches to meet these needs.

PROPER MICROARRAY EXPERIMENTAL DESIGN
MAXIMIZES USEFUL DATA

Before actually performing a gene expression microarray experiment, there are some important decisions that need to be made.

1. Because microarray data points are subject to considerable statistical variation from a number of different sources, it is important to determine the purpose of the experiment, for example, to find genes in tissue culture cells that are up-regulated in response to a new drug.

2. The number of genes to be sampled, the number of probes of each gene on a slide, the presence of housekeeping genes as controls of known response, and the number of slides also need to be addressed.

3. The methods of data processing and analysis should be identified. These methods assess data quality; normalize the data so that different genes, slides, time points, treatments, and so forth can be compared; and provide statistical tests to identify the most reliable data points.

The next section begins with a general introduction to experimental design when there are many sources or variation, as found in gene expression microarray data sets, and then discusses experimental designs.

An Early Experiment

Designing scientific experiments that incorporate statistical methods to facilitate inference on sets of hypotheses dates back to the early 1900s with R.A. Fisher. Legend has it that Fisher offered to get a cup of tea from an urn for a Dr. Muriel Bristol (Box 1978). Dr. Bristol declined the offer saying she preferred tea when the milk was poured in first. There was some debate over whether Dr. Bristol could differentiate between cups of tea with milk poured into the cup before the tea or after the tea, and so Fisher designed an experiment to discern Dr. Bristol's tea-tasting abilities. This experiment is depicted in Table 13.4.

This tea-tasting example is most likely the first designed experiment from a statistical perspective. The outcome of this experiment was never recorded, although the details leading up to the moments of truth (eight moments, or cups of tea, as it turns out) were recorded for posterity. Rumor has it that Dr. Bristol correctly identified all eight cups of tea (Salsburg 2002).

There are several aspects of this simple experiment that are relevant not only to microarray data analysis, but also to all scientific experiments. The principles presented are:

TABLE 13.4. *Tea-tasting experiment as designed by R.A. Fisher*

Dr. Bristol's order	True order		
	Milk first	Tea first	Total
Milk first	a	b	a+b
Tea first	c	d	c+d
Totals	a+c	b+d	n

The identifiers in each cell represent the number of cups identified by Dr. Bristol as belonging to that row (i.e., either Milk first or Tea first) and their true labels are identified by column. The total number of cups is n. Cups identified as belonging to cells b and c are incorrectly labeled by Dr. Bristol. Fisher decided to use four cups of tea with milk poured into the cup first and four cups with tea poured in first. Under this design, the expected values in the table are $a = 4$, $b = 0$, $c = 0$, and $d = 4$. This means that $a + b = a + c = 4$ and $d + b = d + c = 4$. To protect against fluctuations in sweetness, temperature, etc., in cups of tea, Fisher employed a statistical technique called randomization. The technique, as its name implies, consists of presenting Dr. Bristol with cups of tea of either milk or tea first in random order.

1. *Randomization:* As its name implies, randomization is the process of randomly assigning treatments to experimental units. Experimental units are the most basic unit to which a statistical treatment is applied. In the tea-tasting experiment, the experimental unit is a cup of tea and the randomization is the random order in which the cups of tea were presented to Dr. Bristol. The practical purpose of randomization is to remove the influence of forces external to the experiment and outside of the control of the experimenter. We will revisit this practice as applied to microarray experiments shortly.

2. *Presentation and number of the experiment choices:* In the tea-tasting experiment, cups of tea were presented in random order to allay the effects of variations in temperature and sweetness between cups of tea and to prevent bias in the experiment arising from cups of tea with identical treatments being adjacent to one another and presented in that order. Another aspect of this principle is the number of experimental units. In the tea example, eight cups of tea were used and this number was enough to provide a satisfactory test. The number of dimensions is 2, one treatment level for each order of addition of milk and tea. For a microarray experiment, every gene on a microarray chip adds another dimension (see Schwarz 1978). The number of experimental units then becomes difficult to determine, and practical, objective procedures often used in statistics to assure obtaining a

reliable answer (e.g., a power analysis to limit the risk of an incorrect answer, as described in Chapter 4) are often not helpful because of this high dimensionality or number of variables in microarray experiments. It is not that we cannot calculate the power measure, but that the number of arrays needed by the procedure to assure a reliable outcome is often impractically large, requiring additional planning and consideration.

3. *How to interpret the results:* In the tea-tasting experiment, the outcomes follow a hypergeometric distribution (which applies when a random selection is made between objects of two distinct types without repetition of the same individual) and the probability of each of 70 possible combinations found (the number of possible combinations of choosing two sets of four cups from a total of eight cups, four of which are to be labeled as milk poured first, is found by the combinatorial formula $8!/[4!(8 - 4)!] = 70$). Since only 1 of the 70 combinations is the correct one, the chance of choosing correctly all the two sets of four just by chance is $1/70 = 0.014$. If one error was made in the tea experiment, the probability of making one error can be found by summing the number of 70 combinations in which there is a deviation of 1 from the correct result. Thus, from these probability outcomes, we can make inferences regarding Dr. Bristol's tea-tasting ability. There are additional statistical procedures that can followed to produce a more reliable test.

Statistical Planning for Microarray Experiments

A useful statistical procedure not used in the tea-tasting experiment is the concept of blocking. In agricultural experiments, a field used for growing crops is often divided into blocks, each of which receives a different treatment. Using these blocks is referred to as a blocking factor in the analysis. Blocking involves varying the experimental variables, such as treatments or plant variety, across these blocks. For example, suppose an experiment is devised to examine the effects of two fertilizers on wheat yield in four available fields. In this scenario, "field" is the blocking factor and both fertilizers will be applied to two or more different blocks in each field. In this way, the fertilizer effects are averaged over the effects of each individual field, effectively removing any effect of "field" from the fertilizer yield estimates. In statistical designs of experiments using cDNA or spotted microarrays, slides or arrays are considered blocks. Hybridization of mixtures of Cy3- and Cy5-labeled cDNAs from different treatments (e.g., a different tissue, time point, or dye label) can be varied across blocks (slides or arrays) to arrive at estimates of those treatments unbiased by block effects.

The above discussion of block design leads to several other important statistical considerations that apply to both cDNA spotted/printed and oligonucleotide arrays, even though only spotted microarrays can be used in a block design (since two samples can be compared on one slide). Oligonucleotide microarrays cannot be used in a block design because each slide or array is used to assay only one sample. Thus, first, an appropriate definition of this type of microarray experiment is in order. It is common to refer to the application of molecular biology laboratory techniques to a single biological sample as an experiment. However, this choice of words is not an experiment in a statistical sense, especially when discussing a microarray experiment. Statistically speaking, an experiment is the application of measuring all samples (utilizing all laboratory procedures required to collect sample tissue, process all samples, and measure gene expression in all samples), applying a statistical analysis procedure to those samples, and drawing inferences from the results of the statistical analysis. Thus, by definition, a single experiment will involve multiple samples (experimental units) and a resulting single analysis of those samples.

It is convenient to define two types of repeated or multiple experiments for microarray analysis: technical replication and biological replication. In technical replication, for example, the same sample (or experimental unit) is hybridized to multiple slides or arrays. An example of biological replication is hybridizing multiple samples (or experimental units) to multiple slides or arrays. These replications may be used for statistical analysis and inference drawing as described below.

Technical versus Biological Replication and Estimation of Variance

Experimental replication, whether technical or biological, is important for estimating variance, the measure of reliability in a data set. The degree of experimental replication determines the precision of

variance estimation. Variance estimation is crucial to statistical inference. For example, variance is used in microarray data analysis using F-statistics:

$$F = \frac{\text{Mean Square}_{\text{Treatment}}}{\text{Mean Square}_{\text{error}}}$$

where mean squares are sums of squared deviations from the mean divided by the degrees of freedom, the effective number of independent observations. Variance is also used for analyzing microarray data using the t-statistics:

$$t = \frac{x - \bar{x}}{\sqrt{\sigma^2}}$$

in which x and $\bar{x}$ are the means at the expression levels of a gene in two sample data sets and the variance of the expression levels from the mean is σ^2. These uses will be described later in this chapter (see p. 634).

The type of experimental replication necessary for estimating variance depends on the experimental objective, which usually is to draw an inference from the results (e.g., a specific gene appears to be overexpressed). Technical replication of the same sample, such as using a ruler to measure the length of a pencil three times to produce three technical replicate measurements of that pencil, plays an important role in controlling for measurement errors in technology. The practice of spotting the same gene multiple times on an array produces one form of technical replication; splitting the same biological sample or the same pool of samples for subsequent hybridization on separate arrays produces another form of technical replication.

Biological replication is the practice of measuring multiple samples of different experimental units; for example, using a ruler to measure the length of three pencils from a box of pencils. In microarrays, the practice of using multiple samples from different experimental units or multiple pools comprising different samples is sufficient to produce biological replication. Biological replication plays an important role in statistical inference involving populations.

It should be clear from the example using the pencil that if technical replication is used, a considerable amount of information about one pencil and the reliability of using a ruler to measure the length of that one pencil is obtained. However, very little is learned about the length of pencils in a box of pencils; although a variance of the one pencil length measurement can be calculated, that variance only applies to one pencil. Thus, technical replication only measures the variability of technology, in this case using a ruler to measure pencils. If three pencils are measured, there is now information about the variability of all pencils in the box of pencils; i.e., using biological replication, inferences about the population of pencils in that box can then be made. Ideally, both types of replication should be used, but we often cannot afford that luxury. In that case, biological replication often wins out as being the more important one. The cases where this is not true mainly correspond to sample classification, such as classifying a tissue sample as belonging to a specific class of tumor. For this reason, when discussing replication in the following sections, we will refer to biological replication unless otherwise noted.

How do replication and sample size play into calculations of variance? First, consider the statistical formulas for both a sample variance and a sample mean.

$$\text{Sample mean: } \bar{X}_n = \frac{\sum_{i=1}^{n} x_i}{n} \tag{1}$$

$$\text{Sample variance: } s^2 = \frac{\sum_{i=1}^{n} (x_i - \bar{X}_n)^2}{n-1} \tag{2}$$

where x_i is the value of sample i and n is the number of samples. The degrees of freedom is the denominator of the formula for sample variance. Roughly translated, degrees of freedom is the effective number of observations used to calculate that statistic. (This is not the correct statistical definition but rather an intuitive interpretation of the term degrees of freedom.) For the variance, we effectively have one less observation $(n - 1)$ to calculate that variance than we do for the mean.

The variance of a contrast, the statistic needed to determine if two treatments are statistically different from one another, is given by

$$s^2_{t_1-t_2} = \frac{(n_1 - 1)s_1^2 + (n_2 - 1)s_2^2}{n_1 + n_2 - 2} \tag{3}$$

where n_1 is the number of samples given treatment 1 and n_2 is the number of samples given treatment 2, and s_1^2 and s_2^2 are the sample variances for the treatments. Here, the effective number of observations used to calculate the contrast variance is 2 less than the sum of the number of observations in each treatment. It should be obvious that if there are no biolog-

ical replicates, there are 0 degrees of freedom with which to calculate the variance and the test statistic is undefined; i.e., there is no information from which to make a statistical inference. Also, if there is only one biological replicate per treatment (i.e., two biologically independent observations per treatment), there are effectively two observations from which to calculate the contrast variance. It should be apparent that the greater the number of biological replicates per treatment, the better the estimate of the contrast variance and the better the statistical inference.

Principles of Design

The difference in basic experimental design between cDNA or spotted microarrays and oligonucleotide arrays is mainly due to the hybridization technology used. cDNA or spotted microarrays are inherently a dual-channel, or two-dye labeling, technology (although more dyes do exist), whereas oligonucleotide microarrays use only one channel or dye. Good experimental design recognizes the differences between these array technologies.

Spotted Microarrays

A common practice in experimental design is to block on factors external to the experiment. By blocking on external factors, each blocking factor level is assigned to all treatment levels. An example in microarrays is dividing each sample into two, labeling each half with either Cy3 or Cy5, and combining these samples with other labeled samples on a cDNA array. Making this choice allows the statistician to average over the block effects when assessing the treatment effects. In microarray studies, microarray slides are natural blocking factors because they are external to the experiment; i.e., arrays are integral to the experiment, but are not of immediate interest. The limitation for spotted microarrays is that there are usually only two dyes available and therefore only two treatments can be applied to each array. The simplest way around this problem is to use a reference-type design in which one "treatment" on each array is fixed and the other is allowed to vary. Other designs described below can also circumvent this problem.

In the reference design, a common reference sample is hybridized to each array and treatments are applied to each array, as depicted in Figure 13.5. A "dye swap" is also included in this design. It is well

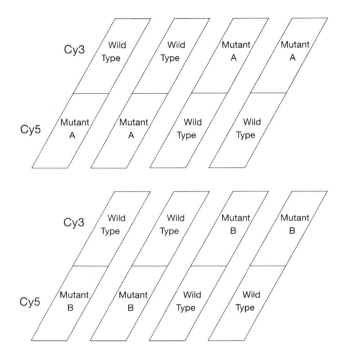

FIGURE 13.5. Reference design. The squares depict slides with different combinations of samples and dye labeling. The Wild-type sample is acting as a reference sample and the Mutant samples are the samples of interest. This experiment includes a "dye swap" and two biological replicates. The "dye swap" arrangement provides a separation of the dye effect, which can potentially introduce bias into the expression measures from other effects on the gene expression values. In this class of experimental design, treatments are not compared with each other directly, but indirectly through the use of a common reference sample. In a microarray experiment with this design, each treatment sample is hybridized to a slide along with cDNA from a common reference sample. The ratio of treatment to reference or the logarithm of the ratio is used as the observed intensity value. One benefit of using the reference design is that the experiment resembles a factorial experimental design in which a table with two rows and two columns can be used to analyze the effects of two different kinds of treatments or samples, e.g., dye effect versus sample variation. This particular design is a factorial design on Mutant dye. Reference designs can be readily analyzed using statistical software packages. A drawback to the reference design is that one-half of all hybridizations are devoted to the reference sample, which is not of experimental interest. Note how eight arrays are needed to obtain the same number of measurements per treatment (or mutant) that is required in the corresponding loop design (Figure 13.6).

known that the two dyes used in spotted arrays, Cy3 and Cy5, are not as efficiently incorporated into cDNAs (Kerr et al. 2001; Kerr and Churchill 2001a,b). Reversing the assignment of dye to the reference and

treatment samples—a "dye swap"—helps to correct this problem. A common criticism of the reference design is the extensive use of one sample, i.e., the reference itself. As can be seen in Figure 13.5, the reference clearly consumes one-half of the available array resources. Although it is often not as efficient as other designs, the reference design does have its place in the collection of microarray experimental methods.

A more efficient design for microarray experiments is the so-called loop design (Kerr et al. 2001; Kerr and Churchill 2001a,b), also known to statisticians as a partially balanced, incomplete block design and illustrated in Figure 13.6 (Lentner and Bishop 1993). The most discernible difference between loop designs and reference designs is the lack of a reference sample, as depicted in Figure 13.6. By completely removing the reference sample, greater resources are allocated to estimating effects of interest, in this case the difference between two mutants. This design also includes reversing dyes to samples as in the reference design.

Two properties of the loop design need further explanation. The first is that each array should involve

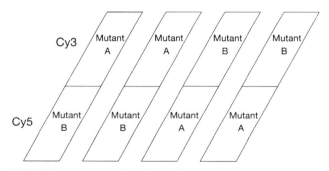

FIGURE 13.6. Partially balanced incomplete block, or loop, design. In this example, the same Mutant samples as used in Figure 13.5 are arranged in a loop design. The expression measure is the intensity values for Mutant A or Mutant B, not the ratios of these values as in the reference design in Figure 13.5. Note that there are the same number of observations per mutant as in the reference design in Figure 13.5, but one-half the number of arrays is used. The design also includes a "dye swap" feature. Also note that all samples used are of interest, making this design highly efficient. Partially balanced incomplete block or loop designs are a type of experimental design suited for situations where there are more treatments investigated than allowable space within blocks. Loop designs usually include some degree of balance in the application of treatments to blocks. Once a loop design has been constructed, it should be checked for conformity with the requirements for this type of design (Lentner and Bishop 1993).

a different pair of biological samples. In this way, loop designs have the built-in biological replication necessary for proper statistical inference. The second property involves the number of treatments or samples. In the example shown in Figure 13.6, each array involves a distinct combination of all possible treatments (i.e., mutant samples) and blocking factors (i.e., reversing Cy3 versus Cy5 dye incorporation). In general, as the number of combinations of treatments and blocking factors becomes large, the number of arrays required to form a loop design becomes larger than the number required for the corresponding reference design; i.e., the reference design becomes more efficient for large numbers of treatment by blocking factor combinations. The number of treatment-blocking factor combinations where the switch in efficiency usually occurs is 10 (Kerr and Churchill 2001a).

Judicious pruning of arrays from a complete loop design may produce an experimental design more efficient in the application of microarray resources than either the reference or loop design (Kerr and Churchill 2001a). An example of the resulting partial loop or incomplete block design is shown in Figure 13.7. The example shown includes 16 different samples in which only the combinations that are of greatest biological interest have been compared.

Although it would be impractical to present a single experimental design appropriate for all spotted microarray experiments, the following list points out some general guidelines for good design.

- Create a list of treatments (e.g., different samples, time points, drug treatments, etc.) and any external factors (e.g., variations in Cy3/Cy5 incorporation or hybridization to different slides) that might influence gene expression during the experiment.

- If the product of the number of treatments and the number of blocking factor combinations exceeds 10, a reference design may be ideal and any experimental design textbook can aid in the construction of an appropriate factorial design (the production of a table of treatments by predicted blocking factors) to separate treatment effects from external factor effects.

- Remember that an appropriate incomplete block design may be constructed that is more efficient than the reference design.

- If the number of combination levels is less than

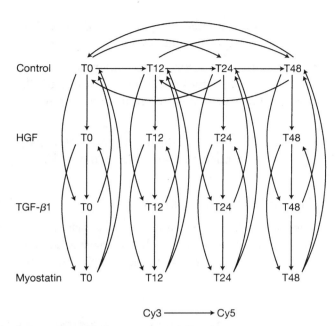

FIGURE 13.7. Partial loop, or incomplete block, design. In this example, cells have been treated with three different agents or left untreated, and then sampled at time 0 and at three subsequent times. There are four time points and four treatments making 16 total treatment combinations. The arrows indicate that the sample at the start point of the arrow is labeled with Cy3 and then mixed with the sample at the arrow end, which is labeled with Cy5, and then the mixture is hybridized to a slide. The total number of slides will be 40. If all of the samples were compared, there would be 16 x 15 = 240 combinations! Hence, this example is referred to as a partial loop design because some combinations are missing. This particular design examines in detail the changes in the control cells over time and also the variations between the untreated and treated samples at each time point. These choices reflect the types of inferences being investigated. The arrays that would have contained treatment combinations that were not of interest have been removed from the design (e.g., T0—transforming growth factor-β [TGF-β] x T24—Control). Partial loop designs share many of the properties of loop designs. One difference is that some treatments may be represented more often than others. The analysis is fairly easy to perform, provided the block-treatment patterns (e.g., the 40 sample comparison choices shown in the figure) have been constructed appropriately.

10, a looped design is the most efficient and either Kerr and Churchill (2001a) or another text on experimental design can assist in creating the appropriate design.

- Remember the lessons learned in the tea-tasting experiment, specifically randomization and presentation of the tea cups. These features aid in removing the effect of potential biases due to improper experimental design.

- Remember the lesson on the difference between technical and biological replication and the importance of biological replication in statistical inference.

- Consult a statistician during this exercise, a highly recommended and often essential practice.

Oligonucleotide Microarrays

Oligonucleotide microarrays use a single-channel, or single-dye, technology and will thus have a different blocking structure for comparing samples than that presented above for spotted arrays. Technical replication, i.e., the hybridization of samples to multiple slides, is needed to detect when the variation between treatments or samples (sample variance) is greater than that between different microarray slides (microarray variance). A potential problem would arise if estimates of both array and sample variance are needed. Experiments using oligonucleotide arrays are generally factorial in nature (see also Fig. 13.5)—designed to distinguish treatments or samples versus array variation—and most factorial designs presented in statistics textbooks will suffice. As in spotted arrays, the advice of an experienced statistician is valuable in constructing an appropriate experimental design using oligonucleotide microarrays.

STATISTICAL CONSIDERATIONS ARE ESSENTIAL FOR ANALYSIS OF MICROARRAY DATA

Microarrays are used to analyze the differential regulation of genes under alternative treatments, to identify genes that are coregulated and that therefore may have a related biological function, and to identify gene expression patterns useful for medical diagnosis by making possible the classification of biological samples. Each one of these functions requires a different statistical approach in both experimental design and analysis. First, however, data quality, background, and data normalization issues must be considered.

Data Quality and Background Subtraction

The data should be analyzed for quality, and poorer-quality or uninformative data removed. For example, dirt on the slide or variation within the signal within one DNA position leads to unreliable data points. The signal for each array position with DNA is examined microscopically by breaking the signal area into pixels and examining the variation among these pixels to determine signal uniformity within the spot. Software for scanning array slides and analyzing pixel variation within each array spot and making this determination is widely available and described in detail elsewhere (Bowtell and Sambrook 2003). Second, background signal subtraction is generally performed by simply subtracting the observed background signal based on measurements made in the neighborhood of the DNA spot from the observed foreground signal in the DNA spot. There can be problems with this procedure because it is not uncommon for some background values to be greater than their foreground counterparts; the resulting net negative intensity measurements are of no value and suggest further that the appropriate background measurement is not being used.

There are three main procedures for dealing with net negative intensity measurements. The most common is to remove those genes with negative intensity values following background correction from the analysis. If there is a replicate of the DNA spot, then those values will be used. A less common approach is to not use the signal intensity of spot for those genes that give net negative intensity values but to replace the observed intensity in those cases with the lowest observed spot intensity value on the array. A more sophisticated approach is that of Kooperberg et al. (2002), who use a Bayesian statistical approach to estimate the foreground measure based on the assumption that the foreground measure should always be greater than the background. This third method has the advantage of providing an expression value for the more weakly expressed genes without loss of information on the rest of the genes. Following the removal of unreliable data and a suitable background choice, the better-quality data can then be normalized to correct for additional sources of variation.

Data Normalization

The need for data normalization, also called data standardization, arises when microarray data points vary in a way that suggests uneven variation between two labels on one slide or between two slides. For example, if a cell line is being treated by a drug that is expected to affect only a small number of genes, and the expressed genes from untreated and drug-treated cells are labeled with Cy3 and Cy5, respectively, then a plot of Cy3 versus Cy5 values for each gene should be scattered around a straight line roughly following a 45° slope. More strongly expressed genes will be represented in the upper part of the line and weakly expressed genes in the lower part of the line. Similarly, if two related samples are labeled and assayed on two separate microarray slides, then a plot of the values on one slide against those on the other should also be scattered about a 45° line. In many cases, however, because of variations in the sample preparation, labeling, or signal measurement, the scattered points drift away from the hypothetical linear relationship in one direction or another to produce a nonlinear relationship.

An even better method for viewing the variation between slides or dyes than the simple intensity plot is the M/A plot, which is a plot of average dye intensity between samples on the x axis (A) versus the variance in the sample on the y axis (M). An example of an M/A plot is shown in Figure 13.9B on page 633. The data are also plotted on a simple intensity plot in Figure 13.9A, below. Most microarray analysis tools that produce an M/A plot show any deviation from linearity in the data. As shown in Figure 13.9B, the deviation of the red line from 0 reveals that the relationship between the samples is nonlinear; i.e., there are more genes labeled with one dye than with the other. As shown in Figure 13.9C, normalizing the data reduces the deviation of the red line from 0, producing a linear relationship between the intensity values for the two dye samples. On a simple intensity plot like that shown in Figure 13.9A, the normalized values will cluster more closely to the 45° line.

Normalization is a procedure designed to remove these nonlinear trends in intensity values, both between dye channels within an array and between arrays, and to restore a standardized, more linear relationship. The normalized data set can then be more readily examined for variations in line with the original experimental objectives with less interference from background variation in the data.

When to Normalize

It is important to know when to normalize microarray data. As described in the Principles of Design section, experiments should be designed to examine the influence of one or more variables (e.g., choice of drug, type of cell, dose of drug, etc.) by a suitable statistical model. Some of the variation in the data points will be due to the influence of each one of these variables. Other variations, such as between dyes or between slides, will also be present. Because normalization procedures remove variation in the data regardless of which variable is involved, they tend to reduce the influence of one or more variables. Incorrect use of normalization of microarray data thus runs the risk of removing potentially interesting information from the data. Hence, normalizing data should be used when necessary to remove the influence of certain factors, such as dye variations, that interfere with making the observations needed to achieve the objectives of the experiment, and not in all situations.

Normalization Based on Statistical Models

The normalization methods presented in this chapter are, for the most part, based on linear statistical models, which are algebraic equations that include as terms the variables affecting the measured outcome. In microarrays, the gene expression values are the outcome. Each observed intensity value is assumed to be a linear sum of contributions from variables, such as slide-to-slide variation, dye variation, variation, and so forth. The list of contributing terms to be included in the equation is determined by the experimental design. The expression values found in different biological samples may then be used to estimate the effects of the experimental treatments, from which statistical inferences can be drawn. There are examples in which the linear statistical model is determined after conducting the experiment, but this is a less desirable approach that generally limits the recovery of information from the experiment.

If the overall mean expression on one microarray or dye/channel in a spotted microarray is higher than for another, linear statistical methods will correctly account for these differences. Even if there is only slight nonlinearity in the expression values between different arrays or dyes/channels, the linear model methods will still perform adequately. Only in the case of moderate to severe nonlinearity will these linear model methods begin to perform poorly. By far the most commonly applied normalization techniques in use are Lowess, Loess, and quantile normalization.

Lowess Normalization

Lowess is a normalization method in which the data within a small window of expression values are fitted to a straight line by linear regression weighted as described below. This window is then moved to the next set of values and the data fitted to another line. The process is repeated across the range of values in the data set. Lowess normalization fits these multiple linear regressions, each representing a small portion of the data, to any nonlinearities in the data by connecting several small straight lines as illustrated in Figure 13.8. The red lines represent localized fits to the data and the green line represents a connection between the red lines. The green line represents a nonlinear regression to model the curved relationship between the two dyes in a spotted array or the intensity values between two oligonucleotide arrays. By fitting a series of linear regressions, any potential

bias that may arise by using other methods (e.g., fitting local constants or using local averages) is avoided. There is a risk, however, of overfitting or overcurving the true trend in the data (Hastie et al. 2001).

The mechanics of how this local weighted regression is performed involves choosing the appropriate proportion of the data (or window size) to include in each local regression. The size of the window depends on the type of kernel (a function describing the expected distribution of the data, e.g., normal distribution) used in producing the weights. As the analysis moves from one point to the next in the data, data within the window size adjacent to the current point are used to estimate the parameters of a linear model. It is important to remember that only the fit at the current point in the window center is being considered, even though other data assist in the estimation of the model parameters. Use of weights is illustrated in the following equation of the linear regression:

$$\min_{\alpha(x_0),\beta(x_0)} \sum_{i=1}^{n} K_\lambda(x_0,x_i)[y_i - \alpha(x_0) - \beta(x_0)x_i]^2 \quad (4)$$

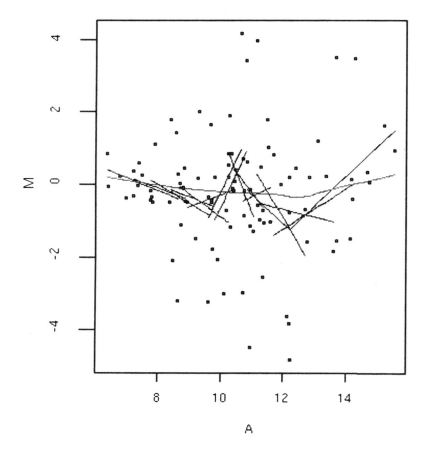

Lowess Plot

FIGURE 13.8. Lowess normalization. These data are derived from four spotted microarrays of 95 genes measured on four plant root and four plant leaf tissue samples and arranged in a loop design. The short red lines consist of a window size that includes 20 observations and, for clarity, only every fifth window is plotted. The green line displays the Lowess fit through the data. One can see how the Lowess fit follows the average of the red lines, but is less affected by extreme, outlying points.

where $K_\lambda(x_0,x_i)$ describes the weight of observation i with a window size of λ. x_0 is the center position of the window, β is the slope, and α is a factor that determines the y intercept. The straight red lines shown in Figure 13.8 are examples of the localized linear regression.

Loess Normalization

Sometimes local data are not appropriately modeled by linear regression, thus making the data unsuitable for Lowess normalization. Loess normalization is another method of data normalization that fits a more complex polynomial function, such as a quadratic or cubic function, to the local data. The polynomial curve is then fitted at each point using neighboring data in the same manner as Lowess normalization.

Quantile Normalization

The nonlinearity observed between channels or microarrays often arises from one channel, or microarray, conforming to a different statistical distribution than the other channels; here one channel arises from a normal distribution with a different mean and variance than the other. If the rankings of genes within a channel or microarray are correct, a simple manipulation of the intensity values is all that is needed to normalize the microarrays arrays. By performing a quantile normalization, which involves ranking the intensity values, the resulting intensity values for each dye within a microarray, or the intensity values between microarrays, have the same quantile values.

In quantile normalization, intensity values from each microarray, or from the two channels within a spotted microarray, are ranked from smallest to largest, preserving the gene names. The intensity value from the smallest gene value within an array called the master array is used to replace the smallest-intensity value on the other microarrays or other channels. This procedure is repeated for all the remaining gene values. Because there will be variation in gene ranking between arrays or channels, there will still be variation in intensity values for a specific gene. However, now all arrays, or both channels within an array, will have the same distribution of intensity values (i.e., same mean, median, quantiles, and variance).

Normalization of Multiple Slides in an Experiment

When designing a microarray experiment, multiple slides must allow for replication and measurement of different sources of variation. There are different recommended normalization approaches to be used with spotted or oligonucleotide arrays and with different experimental designs. For example, in Figure 13.4 spotted microarray slides are used to determine differences between two mutants. One recommended approach for normalization in this case is to produce an M/A plot for each slide. If there is a linear relationship between dyes (the red line in Fig. 13.9B does not drift from the $y = 0$ value), then the data may be used for further analysis without normalization. If the red line deviates from 0, then normalization should help to produce a linear relationship between the dye values for each gene. In some cases, of course, the M/A plot reveals that something is wrong with the slide and the data should not be used. This procedure is repeated for each slide in the spotted array experiment. Sometimes additional normalization steps would be made between slides in this example, but, in general, these extra steps are not needed or recommended. Normalization of oligonucleotide microarray slides is done somewhat differently and does requires multiple comparisons between slides.

One recommended normalization method when using a set of oligonucleotide microarray slides is first to choose a high-quality slide as a master slide. Every other slide in the experiment is then compared to this slide on an M/A plot and, if necessary, normalized against the master slide. A method-normalization method that is popular in this type of procedure is the quantile method described above. Briefly, the genes on the master slide are ranked by intensity value. The genes on each of the other slides are also ranked in this same manner. Finally, the intensity values on the master slide are then copied to the gene lists of the other slides. This method assures linearity among all of the slide combinations. The normalized data are then used for data analysis.

Detection of Differential Gene Expression

By far the most common use of microarrays is for the detection of differential gene expression. A statistical analysis is recommended as the optimal way to obtain this information. The details of any statistical analysis are dictated by the experimental design. An outline of approaches will be provided below. The types of statistical analysis described may be performed using tools from the groups listed in Table 13.1. Commercial software is also used (SAS, and S+, the commercial version of the R statistical package, for example).

EXAMPLE: SUMMARIZING DATA NORMALIZATION

To summarize data normalization, an example using the statistical and maPlot functions of the BioConductor package (found by Web search) to produce typical plots of microarray data is shown in Figure 13.9. A standard Cy3/Cy5 plot of log intensities from a spotted array is shown in Figure 13.9A. The data vary over a wide range of values, making it difficult to observe any trends in the data and to appreciate when one value is diverging from another. The data in Figure 13.9A are replotted in Figure 13.9B in a plot of M (y axis, logarithm of the ratio of red to green intensities representing the variation in the data points) versus A (x axis, average of the logarithms of the red and green intensities representing the means of the data points), called an M/A plot. The M/A plot gives a much better display of the data variability than the simple Cy3/Cy5 plot of log intensities. Note the red line, which indicates the Lowess fit, can be interpreted as the average amount of each label represented in each part of the plot. Values above or below 0 indicate that one dye, on the average, showed much higher values in samples in that region of the plot than did the other dye. This result could be due to variations in dye incorporation or in measurement of each dye signal on the microarray. Normalizing the data by Lowess smoothing is seen to reduce the amount of variability in the data and to compensate for the dye variation, as shown in Figure 13.9C. The adjusted ratios are expected to represent the true signal ratios more accurately.

The effects of background subtraction on a different data set are shown in Figure 13.9D (with background subtraction) and Figure 13.9E (without background subtraction). Background based on local intensity measurements near each DNA spot was subtracted from each spot foreground measurement. When the background value was greater than the spot value for a particular gene, thus creating a net negative intensity value for that spot, then the lowest intensity value of any DNA spot on the array was used as the intensity measurement for that particular gene. As may be seen by comparing parts D and E of Figure 13.9, background subtraction using a minimum estimate of expression for weakly expressed genes greatly inflates the variability. This example illustrates the point that the use of these simple methods may impair, rather than improve, statistical inference. Therefore, a useful recommendation is to create two data sets, one with and one without background adjustment, and then to create M/A plots for each data set and observe the impact of background adjustment on the variability of weakly expressed genes, as shown in Figure 13.9, D and E. If the impact is minimal, or the background adjustment reduces the variability across expression levels, then use the background-adjusted data set. Otherwise, use the data set that ignores background adjustment.

FIGURE 13.9. (*See facing page.*) Effects of data normalization and background subtraction on cDNA microarray data. These displays were all produced using BioConductor modules. (*A*) Plot of Cy3 versus Cy5 log intensity values from a standard cDNA array hybridization. (*B*) Plot of the same data as in *A* in which the average of each set of Cy3 and Cy5 values (*A*) is plotted on the abscissa and the log of the ratio of the Cy3 and Cy5 values (*M*) is plotted on the ordinate. This M/A plot of data variation versus average intensity value has the advantage of showing more clearly where the data variation is occurring as a function of the expression value. The red line indicates where an average ratio of 1 (e.g., equal labeling of samples) would be located. The drift of the red line below $M = 0$ indicates that more of one label is present in most samples, and is particularly noticeable at high values. This result indicates that the ratios for each gene need to be normalized in order to correct for the overrepresentation of one of the dye values. (*C*) The same M/A plot in *B* following data normalization by Lowess smoothing to reduce the influence of widely diverging data points and correct for dye value inequities. The red line indicates the path of the curve generated by the normalization curve and represents a more detailed smoothing of the red lines shown in Figure 13.9. Note that the drift of the red line below $M = 0$ is much reduced, indicating that the dye inequity has been largely corrected. (*D*) M/A plot of a different data set with local background subtraction, employing the minimum estimate of background for weakly expressed genes that is described in the text. However, there were few negative values. (*E*) M/A plot of the same data as in *D* without any background subtraction. Note the clustering of data points near the origin with little variation. Variations among low expression data points are quite extreme. Note change of scale between *D* and *E*.

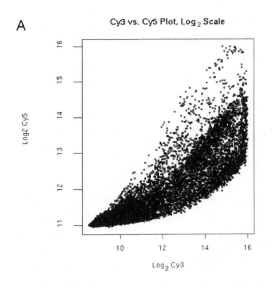

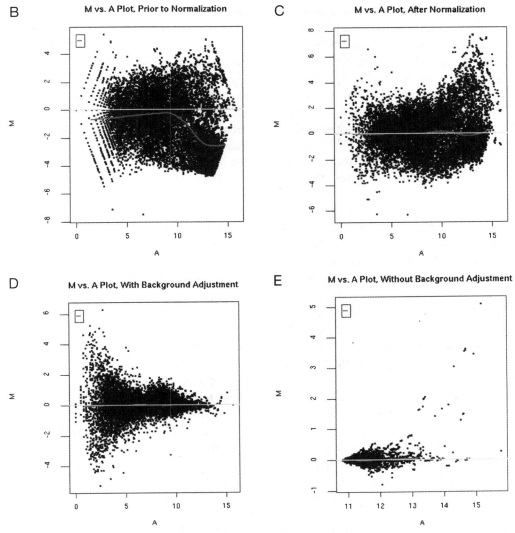

FIGURE 13.9. (*See facing page for legend.*)

A statistical model of the variation of intensity values within any microarray experiment can be broken down into two major components:

1. A component common to all genes.

2. A component specific to each gene.

The common practice of simply assessing statistical significance between two gene intensity values using a *t*-test based on the data variation between those two genes (the *t*-statistic is the difference between two mean values divided by the standard error of the difference between the means) is not appropriate because this test ignores the other major component of intensity variation in the first component.

To capture and use the information in the common component for all genes, a linear statistical model is often employed. A linear statistical model is an algebraic equation containing the variables whose parameters are to be estimated from the mean gene expression values. An example of using such a model is shown in Figure 13.10, based on the experimental model in Figure 13.6. This model contains terms estimated for slide, dye, treatment (mutant in the example in Fig. 13.6), and the individual gene on each slide (Wolfinger et al. 2001; Chu et al. 2002). These terms are called global because they are common to all genes and thus represent the variation given above in component 1.

Once this global model has been fitted to the data, residuals (the part of the intensity values that are not explained by the model) can be used as globally "normalized" intensity values for the component 2 gene-specific models. The example in Figure 13.10 explains how the statistical model can be used to sort out dye and mutant effects for each gene. This analysis allows testing for gene and treatment (mutant in the example) interactions. An assumption underlying the use of these gene statistical models is that there is no interaction of one gene with another, a useful starting assumption. However, if gene interactions are thought to be occurring, then more complex models may be needed.

Once the individual gene model has been obtained, two types of statistical hypothesis tests can be performed. The first involves testing the hypothesis, "Is at least one of the treatment levels different within this gene data set?" The second test involves testing for changes in expression levels of each gene in response to the specific individual treatments. Each method is discussed below.

Variance Test for Difference in at Least One of the Treatments

The question of testing whether at least one treatment is different from the others is statistically beneficial. As will be seen later in the section on multiple hypothesis testing, this test helps protect the experimenter from the accumulation of a gene-specific type I error (rejecting a true null hypothesis of no difference between treatments). The test statistic used to assess the risk in accepting the alternative hypothesis (the alternative to the null hypothesis) is the *F*-statistic (Snedecor and Cochran 1989). The *F*-test in this case involves an analysis of variance (ANOVA) of the data for each gene model. ANOVA involves estimating the mean and standard deviation for different groupings of data including all of the gene values and different subsets (associated with dyes, treatments, slides, etc.). Mean square values (mean square = sum of squares of deviations of each value from mean / degrees of freedom) are then calculated for each group. If the mean square value for one of the groups (e.g., treatment) is greater than the residual error term for the gene (for the origin of this term, see Fig. 13.10), then the treatment is having a greater overall effect on intensity than differences between slides. The *F*-test estimates the probability that such a variation within treatment effects would be observed by chance given the degrees of freedom associated with each mean square value (for more information on ANOVA and the *F*-test, see Snedecor and Cochran 1989). Because the number of data points for each gene model is quite small (eight data points in the example described in Fig. 13.10), statisticians will often employ additional information from the larger global data set, a technique called data shrinkage, to enhance the analysis of each gene. A discussion of statistical shrinkage is beyond the scope of this chapter, but may be found in statistical texts or by a Web search.

If the results of the ANOVA support the rejection of the null hypothesis of no difference between treatments, one is stating that the risk of declaring at least one treatment level within this gene as different than the others is acceptable. Individual treatment contrasts can be inspected and can then be protected at the level of significance corresponding to the *F*-statistic, but only when considering this one gene. If additional genes are to be considered, a different ANOVA analysis of those genes is needed to protect against the accumulation of type I errors.

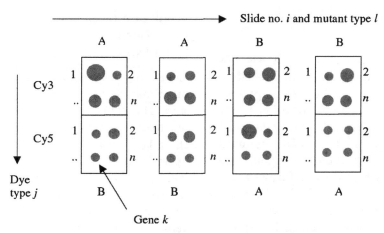

FIGURE 13.10. Employing statistical models for analyzing difference in gene expression. Shown are four microarray slides from a hypothetical loop design experiment for two mutants as previously described in Figure 13.6. There are two biological samples of each of two mutants, A and B, with the first set of samples hybridized to the first and third slide and the second set of samples hybridized to the second and fourth slides. The Cy3 and Cy5 labels are reversed in the last two slides. Shown are genes 1–4 as circles, with the diameter of the circle indicating the log intensity expression value of the gene on each slide. The goal of the experiment is to identify genes that are expressed differently in the mutants. Genes 1 and 2 represent two genes that are varying in the two mutants, with gene 1 generally expressed more strongly in mutant A and gene 2 expressed more strongly in mutant B. The last two genes shown in each half-slide position are not varying in the mutants. There are also two additional sources of variation. First, any gene labeled with Cy5 is showing a lower intensity than when labeled with Cy3. Second, there are variations between duplicate biological samples. Gene 1 shows considerable variation between the first and second slides and also between the third and fourth slides. This extreme variation in gene 1 makes an important point about replication. This gene may be chosen by an inexperienced statistical eye based on a large value in one of the samples, yet the variation brings uncertainty into the value of this gene. By contrast, gene 2 is consistently expressed more strongly in B than A, yet might not be chosen because the difference is small. The small variation between slides and experiments makes the observations of gene 2 in mutant B more significant than those of gene 1 in mutant A. Failure to recognize these differences can lead the microarray analyst badly astray. This comparison also reveals the uncertainty of choosing a cutoff score (e.g., 1.5- or 2-fold for picking a list of genes that are varying in a sample), a very common practice in the biology laboratory. There is a straightforward statistical method, however, to determine which genes are varying significantly. These sources of variation in the experiment are subjected to an analysis of variance (ANOVA) to discover the most significantly varying genes. The object of the ANOVA analysis in this particular experiment is to determine which genes are expressed differently in the two mutants. Three steps are followed to achieve this objective:

1. A statistical model that contains all of the global factors in the experiment that influence each gene value is chosen. Each gene value is assumed to be the sum of a series of average values due to slide, mutant, and dye plus an extra term due to experimental variation in the particular gene; i.e., expression of gene k on slide i, labeled with dye j, in mutant type l = an overall mean (u) + slide i mean + dye j mean + mutant type l mean + residual term ε_{ijkl}. The needed calculations are done to find each group mean (e.g., the mean of all gene values for mutant A, dye Cy3, etc.) and the residual value for each gene in each array is then calculated using the above formula. Note that with other experimental designs, a different set of terms is needed, but the principle is the same.

2. The residual ε_{ijkl} values calculated in step 1 for each i, j, k, l combination are renamed γ_{ijkl} and are then organized into tables, one for each gene. In the present example, each table will have eight values corresponding to two values on each of four slides, four for each mutant. The following analysis is performed on each gene. The value for each gene is again modeled by a reduced set of linear terms; i.e., expression of gene k on slide i, labeled with dye j, in mutant type l = dye j mean + mutant type l mean + residual term ε_{ijkl}. These calculations are performed on the small table of values for each gene. Following the calculations for each gene in turn using the values in each gene table, the residual terms ε_{ijkl} now represent gene–dye and gene–mutant interactions.

3. The specific ε_{ijkl} residual or error values for each gene found in step 2 may now be used in two statistical tests that are described in the text. First, the variance between treatments, dyes, etc., for each gene may be compared to the error value by an ANOVA test to see if there is an effect of treatment. Second, the means and standard deviations of residual values for one treatment may be compared with those for another by means of a t-test.

A set of genes from an experiment using the above model is shown in Table 13.6.

Specific t-test for Changes in Expression Levels with Treatment

The above statistical test is based on an ANOVA across genes, treatments, dyes, and other variables present. The test shows that a difference is present but does not reveal which treatment or other factor is responsible for the difference. In the *t*-test, pairwise comparisons between the means and standard deviations resulting from different treatments are compared. These values are used to perform a *t*-statistic (difference between means divided by standard error of difference between means). In the example in Figure 13.10, the differences in expression between mutants A and B can be compared by this test. The ANOVA test in the section above will help to strengthen the inference of any difference found. The R script used to identify the genes that are differentially expressed in this microarray experiment is shown in Table 13.5 and the list of genes that are showing a significant difference in expression values between mutants A and B is shown in Table 13.6.

Prediction of Genes That Are Functionally Related by Differential Gene Expression Analysis

Genes that respond the same way to a given treatment may share a regulatory or functional relationship. The concept of functionally related groups is one that is not well defined. In one context, functionally related groups may represent all genes that are up-regulated (or down-regulated) under one treatment condition. In another context, the concept may represent all genes involved in a single metabolic pathway or signal transduction cascade. This plural definition of "functionally related" means one has to be careful to describe which context is being used. For present purposes, "functionally related" refers to those genes that are coregulated in response to a given stimulus. To identify such genes, a method is needed to compare intensity changes between genes and group together those genes that are responding in the same manner. The principal methods for making these comparisons are distance-based methods and sample classification.

TABLE 13.5. *R script used to obtain the list of genes with expression values changed significantly in the two mutants from the experiment discussed in Figure 13.9*

R Command	Description
`library(limma)`	Load the limma library.
`targets <- readTargets("rootleaf.txt")`	Read in the description of the hybridization scheme.
`RG <-read.maimages(targets$File.Name,columns=list(Rf="Rf",Gf="Gf",Rb="Rb",Gb="Gb"))`	Read in the individual array data files. The labels for the Cy3 and Cy5 data are Gf and Rf and the corresponding backgrounds are Gb and Rb.
`RG <- backgroundCorrect(RG, method="none")`	Here, we are not adjusting for background effects, although there are several methods that could be specified.
`MA <- normalizeWithinArrays(RG, method="loess", span=0.4)`	Use Loess to normalize the data within arrays.
`fit <- lmFit(MA, c(-1,1,-1,1))`	Fit a global ANOVA model that will remove the effect of global variables for further processing.
`fit2 <- eBayes(fit)`	Perform an empirical Bayes shrinkage estimator for individual gene variances and use that estimate in a *t*-test. This method corresponds to the individual gene ANOVA models.
`topTable(fit2, adjust="fdr", n=29)`	Display the top 29 significant genes and adjust their *p* values to reflect the false discovery rate (FDR).
`plotMA(RG,cex=0.5)`	MA plot of raw intensities.
`plotMA(MA,cex=0.5)`	MA plot of normalized values.

The resulting list of genes is shown in Table 13.6. The script, which uses the LIMMA package in R (Smyth et al. 2003), is available on the book Web site.

TABLE 13.6. *List of genes showing a significant variation*

Gene	t-value	FDR p value
55	−34.4954	0
7	−33.2082	0
27	23.7111	1e-04
53	−22.8911	1e-04
42	−22.805	1e-04
51	22.4563	1e-04
6	21.8251	1e-04
15	−21.0525	1e-04
40	−20.3001	1e-04
93	18.8301	1e-04
45	18.4794	1e-04
23	16.5359	1e-04
16	−16.5261	1e-04
85	13.7867	3e-04
63	13.6448	3e-04
34	−12.9279	3e-04
92	−9.7432	0.0012
11	9.6559	0.0012
78	−10.5206	0.0022
57	−8.8568	0.0016
35	8.801	0.0016
21	−8.2769	0.0020
81	7.895	0.0023
1	7.6256	0.0026
84	−7.4629	0.0028
3	10.2018	0.0066
87	7.1316	0.0033
24	6.6163	0.0044
89	6.4959	0.0046
46	−12.0652	0.0167
56	−5.9726	0.0063
86	5.9709	0.0063
73	−5.8084	0.0066
69	−5.6605	0.0072
48	5.5118	0.0079
18	5.1481	0.0102
44	−4.9928	0.0113
79	4.8221	0.0127
82	4.793	0.0127
29	−4.7349	0.0128
60	4.7277	0.0128
83	−4.6866	0.0129
70	4.6684	0.0129
94	−4.4691	0.0150

The data set is from an actual data set using the experimental model in Figure 13.10. These genes are varying significantly between mutants A and B following appropriate statistical modeling as described in Figure 13.10. The BiocConductor script for generating this list is shown in Table 13.5. FDR is the false discovery rate or the probability that these significant values of t represent a false-positive rejection of the null hypothesis for this set of genes (the null hypothesis is that there is no difference between the mutants).

Distance-based Methods

One method of comparing genes is to use a distance measurement (metric) that is based on differences between intensity values. Distance-based methods use a distance rule in the formation or exclusion of group members. A distance rule is simply a mathematical formula that translates the observed data values into a numerical two-dimensional distance. For example, suppose a correlation coefficient is used as a distance rule and a correlation of 0.8 between genes A and B is determined. Traditionally, the distance d_{AB} would be calculated as $d_{AB} = 1 - r_{AB} = 1 - 0.8 = 0.2$, a relatively small distance. Thus, these two genes would likely be grouped together; i.e., they respond similarly to the application of some treatment. What if the correlation were now $r_{AB} = -0.8$? These genes obviously both respond to the same stimulus, but they respond differently and thus produce a negative correlation coefficient close to −1. The distance would be $d_{AB} = 1 - r_{AB} = 1 - (-0.8) = 1.8$, a large distance, and these two genes would likely not be grouped together. To group all responders, whether positive or negative, an alternative would be to use the distance measure $d_{AB} = 1 - |r_{AB}|$, which in the last case would produce the distance value $d_{AB} = 1 - r_{AB} = 1 - |-0.8| = 0.2$, making these two genes closely related because they do respond to the same treatment. This method of calculating distance is used quite widely because of the increased number of gene relationships represented.

Table 13.7 shows the calculation of three common distance measures between genes x and y: correlation coefficient, absolute value of correlation coefficient, and Euclidean distance. The correlation coefficient formula shown is the Pearson correlation coefficient, which is based on the assumption of a normal distribution of variables. As described above, the correlation coefficient may be used directly to calculate the distance measure. However, a more informative procedure is to use the absolute value of the correlation coefficient. The Euclidean distance is simply the distance between two points in a graph. For example, suppose there are two biological samples, each with expression values for three genes A, B, and C. If these values for sample 1 are 2, 3, and 4 for A, B, and C, respectively, and the corresponding values for sample 2 are 1, 2, and 3, then the Euclidean distance between the samples is $\sqrt{(2-1)^2 + (3-2)^2 + (4-3)^2} = \sqrt{3}$. Once distance measures between samples or between

TABLE 13.7. *Distance measures*

Type	Formula	Distance measure
Correlation coefficient	$$r_{x,y} = \frac{\sum\limits_{i=1}^{n} x_i y_i - \dfrac{\left(\sum\limits_{i=1}^{n} x_i\right)\left(\sum\limits_{i=1}^{n} y_i\right)}{n}}{\sqrt{\left(\sum\limits_{i=1}^{n} x_i^2 - \dfrac{\left(\sum\limits_{i=1}^{n} x_i\right)^2}{n}\right)\left(\sum\limits_{i=1}^{n} y_i^2 - \dfrac{\left(\sum\limits_{i=1}^{n} y_i\right)^2}{n}\right)}}$$	$d_{xy} = 1 - r_{xy}$
Absolute value of correlation coefficient	$$\left|r_{x,y}\right| = \frac{\left|\sum\limits_{i=1}^{n} x_i y_i - \dfrac{\left(\sum\limits_{i=1}^{n} x_i\right)\left(\sum\limits_{i=1}^{n} y_i\right)}{n}\right|}{\sqrt{\left(\sum\limits_{i=1}^{n} x_i^2 - \dfrac{\left(\sum\limits_{i=1}^{n} x_i\right)^2}{n}\right)\left(\sum\limits_{i=1}^{n} y_i^2 - \dfrac{\left(\sum\limits_{i=1}^{n} y_i\right)^2}{n}\right)}}$$	$d_{xy} = 1 - \left|r_{xy}\right|$
Euclidean distance	$$Ed_{xy} = \sqrt{\sum_{i=1}^{n}(x_i - y_i)^2}$$	$d_{xy} = Ed_{xy}$

The concept of a distance measure is central to many clustering methods. The *r* distance measure quantifies the strength of the relationship between two variables (genes or samples). A good distance measure should have the following properties.

1. The distance between any two variables must be greater than 0.

2. The distance between a single variable and itself must be 0.

3. If the distance between any two variables is 0, then these two variables must be identical.

4. The distance between any two variables must be symmetric; i.e., the distance between variable A and variable B must be the same as the distance between variable B and variable A.

5. The distance between variables A and C must be equal to or less than the sum of the distance between variables A and B and that between variables B and C (also known as the triangle equality).

The correlation coefficient, as presented, is the Pearson correlation coefficient. The Pearson correlation coefficient is sensitive to outliers, especially when low numbers of observations are used. A nonparametric version of the Pearson correlation is the Spearman correlation coefficient. The computation of the Spearman correlation coefficient is quite simple. First, record the ranks of each variable and store them as new measurements on each of the variables. Then, use the ranks in the correlation coefficient formula given in the table. The Euclidean distance measure depends on units of measurement, thus creating problems in comparing values that are scaled differently. A solution is to work with variables in standardized units when using Euclidean distances as a distance metric.

genes have been calculated, they may be used for discovering samples or genes related by gene expression.

The following sections describe some popular distance-based analysis methods.

Hierarchical clustering. Hierarchical clustering (HC) is a multivariate tool often used in phylogenetics (see Chapter 7; Li 1997), comparative genomics (see Chapter 11), and anthropology (Johnson and Wichern 1998) to relate the evolution of species (or languages by evolutionary time distances). The concept of HC is to group variables together according to different distance rules so that variables with a small distance are grouped together. There is an ordered fashion, a hierarchy, by which the variables are grouped together such that the clusters are in a binary branching tree. There are three types of HCs: complete-, single-, and average-linkage clustering. Note that each clustering method produces a distinct result but that gene 46 is always in a unique cluster.

1. *Complete linkage:* When joining a new gene to an existing node, the distance of the new gene from the node is the largest distance found from the new gene to all genes contained within that node. This method provides the greatest separation of clusters. An example of complete linkage analysis is shown in

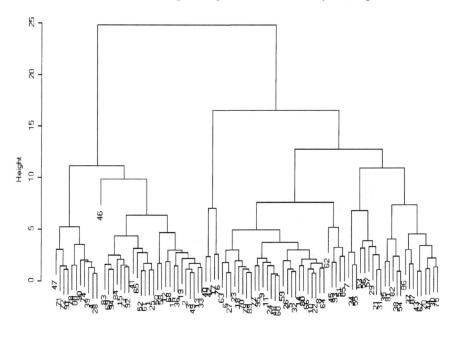

Cluster Dendrogram Using Euclidean Distance. Complete Linkage.

Gene Name

FIGURE 13.11. Hierarchical clustering by complete-linkage analysis. These data are from the same loop design experiment used in Figure 13.8. A mixed linear model excluding the treatment by gene effect appropriate for the experimental design was fitted to the data, and the residuals from this model were used to generate a distance table based on Euclidean distance, from which this dendrogram was produced. There appear to be two distinct clusters in the data, one from gene 47 to 33 on the left and the other from 40 to 75 on the right.

Figure 13.11. This experiment involved analyzing the expression of 95 genes in eight plant-tissue samples—four root and four leaf tissues. The experiment was based on a loop design model and the data were analyzed using a statistical model and subjected to an ANOVA as described above (Detection of Differential Gene Expression). The clustering reveals which genes are showing the most-alike variations across the eight tissue samples.

2. *Single linkage:* When joining a new gene to an existing node, the distance of the new gene from the node is the smallest distance found from the new gene to all genes contained within that node. Single linkage reveals the most similar clusters and is illustrated in Figure 13.12 using the same data as used in Figure 13.11. Note isolation of gene 46.

3. *Average linkage:* When joining a new gene to an existing node, the distance of the new gene from the node is the average distance from the new

gene to all genes contained within that node. Average linkage provides a balance between revealing similar clusters and separating dissimilar clusters. An example is shown in Figure 13.13.

For microarray data, these types of hierarchical clusterings differ in how genes are grouped together once nodes are formed. The R script that was used to produce the dendrograms in Figures 13.11–13.13 is listed in Table 13.8.

The use of clustering for microarray data was first made popular by Eisen et al. (1998), who developed the Cluster software (Eisen 1999) exclusively for this purpose. Common distance measures are the Pearson correlation coefficient, Spearman correlation coefficient, absolute value of the Pearson correlation coefficient, absolute value of the Spearman correlation coefficient, and the Euclidean distance (see Table 13.7). The typical output from a cluster analysis is a tree-like figure called a dendrogram. The distance between the tips of the branches represents the dis-

Cluster Dendrogram Using Euclidean Distance. Single Linkage.

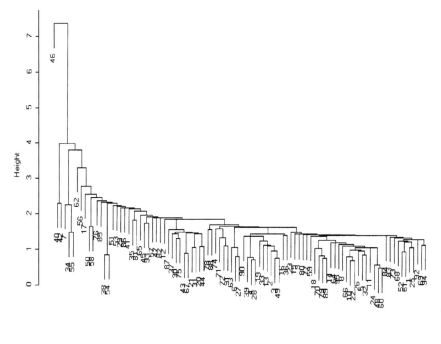

Gene Name

FIGURE 13.12. Hierarchical clustering by single-linkage analysis. These data are from the same loop design experiment used in Figure 13.8. There appears to be no significant clustering, with the exception of gene 46, which is in an outlier cluster.

Cluster Dendrogram Using Euclidean Distance. Average Linkage.

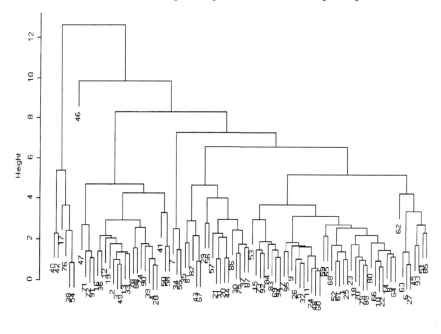

Gene Name

FIGURE 13.13. Hierarchical clustering by average-linkage analysis. These data are from the same loop design experiment used in Figure 13.8. There appear to be three clusters, the first cluster containing genes 40, 42, 17, 76, 38, and 54. The other two groups consist of gene 46 for the first of the two and the remaining genes for the other.

TABLE 13.8. *R script for producing the hierarchical clustering dendrograms shown in Figures 13.11–13.13*

R Command	Description
`library( nlme )`	Load mixed model functions.
`x <- read.csv( "rootleaf.csv" )`	Read in data.
`tlme.out <- lme( quantile ~ dye*trt + dye*oligo , data = x , random = list( slide = pdIdent( ~ -1 ) ) )`	Fit a mixed linear model that includes all ANOVA model effects except for the treatment by gene effect.
`x$resid <- resid( tlme.out )`	Get residuals from the mixed model.
`y <- matrix( nrow = 8 , ncol = 95 )`	Create a matrix to hold the residuals in a multivariate data set form.

`y[2,] <- x$resid[ x$slide == "s91" & x$dye == "cy5" ]`
`y[3,] <- x$resid[ x$slide == "s92" & x$dye == "cy3" ]`
`y[4,] <- x$resid[ x$slide == "s92" & x$dye == "cy5" ]`
`y[5,] <- x$resid[ x$slide == "s93" & x$dye == "cy3" ]`
`y[6,] <- x$resid[ x$slide == "s93" & x$dye == "cy5" ]`
`y[7,] <- x$resid[ x$slide == "s94" & x$dye == "cy3" ]`
`y[8,] <- x$resid[ x$slide == "s94" & x$dye == "cy5" ]`
	Fill the new matrix.

`gene.labels <- as.character( unique( x$oligo ) )`	Create a vector of gene names.
`d.x <- dist( y ) dist.x <- abs( cor( t( y ) ) )`	Create some distance matrices (the first one is a Euclidean distance).

`hx <- hclust( d.x )`
`hx.s <- hclust( d.x , method = "single" )`
`hx.a <- hclust( d.x , method = "average" )`
	Create a complete-, single-, and average-linkage hierarchical clustering object.

`plot( hx , main = "Cluster Dendrogram Using Euclidean Distance. Complete Linkage." , xlab = "Gene Name" , sub = "" )`
`plot( hs.x , main = "Cluster Dendrogram Using Euclidean Distance. Single Linkage." , xlab = "Gene Name" , sub = "" )`
`plot( hx.a , main = "Cluster Dendrogram Using Euclidean Distance. Average Linkage." , xlab = "Gene Name" , sub = "" )`
	Plot dendrograms for each clustering method.

The script is available on the book Web site.

tance between the variables associated with those branch tips.

An algorithm for clustering is essentially:

1. Create a distance matrix using the chosen distance metric function.

2. Locate the nearest genes in the matrix (may be a cluster of genes with common distance).

3. Join these entries to form a new cluster in the tree, where the height of the tree represents the distance between cluster members.

4. Create a new distance matrix for the distance between the new cluster, the existing clusters, and the remaining unclustered genes.

5. Return to step 2 and repeat until all genes and nodes are clustered.

A common question regarding HC is, "Which clusters are significantly different from the others?" There is no standard test that can be performed to evaluate such hypotheses. However, a bootstrap method may be used to assess the significance of each individual cluster or group if the sample size is large enough (Kerr and Churchill 2001c). Both parametric data (based on a probability model, e.g., if normal or binomial is used) and nonparametric (not based on a probability model) data may be used. The bootstrap method involves resampling the gene expression data many times, reconstructing a binary tree, and examining the tree for conservation of clusters. The clusters that remain intact throughout the bootstrap procedure then demonstrate which cluster structures in the tree are reliable. The bootstrap method is also used in Chapter 7 to test how well sequence variation in a multiple sequence alignment supports the branches of a phylogenetic tree (see p. 321).

k-means clustering. This method is similar to HC, but requires the experimenter to prespecify k, the number of clusters. One difference to HC is that the k-means method does not assume a hierarchy of clusters. Clusters are groups of similar genes based on expression and the distance metric and genes are mapped to the k clusters using the chosen distance function. Genes are initially randomly assigned to one of the k clusters. Different runs of the algorithm can result in slightly different clusterings because of this random assignment.

An algorithm to perform k-means clustering is:

1. Choose k, the number of clusters, and d_{xy}, the distance function.

2. Randomly assign each gene to one of the k clusters.

3. Calculate the centroid (middle or average) of each cluster.

4. For each gene, calculate the distance to all k centroids.

5. Move the gene to the cluster containing the closest centroid.

6. Repeat steps 4 and 5 until genes no longer change cluster membership.

This method is very sensitive to the value of k. The bootstrap method described above may be used to validate clusters. During reclustering of the bootstrap data with the same value of k, the clusters whose membership remains constant are the most reliable.

Self-organizing maps. Self-organizing maps (SOM; Tamayo et al. 1999) is a multivariate data mining tool similar to k-means clustering. For a prespecified grid (two-dimensional or three-dimensional) of data collection nodes, the data are mapped iteratively to the nearest node, one data point at a time. The resulting map of the nodes no longer conforms to the specified grid, but has nodes separated by the average distance between data observations within the respective nodes. Some nodes may also be empty at the end of the analysis. Two disadvantages of this method are that the number of nodes must be prespecified and

Self Organizing Maps, 2 by 2 grid. Expression Patterns.

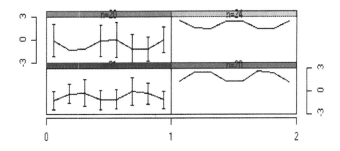

Self Organizing Maps, 2 by 2 Grid. Gene number is colored by node membership

FIGURE 13.14. Self-organizing maps (SOMs). These data are from the same loop design experiment used in Figure 13.8. The data were run through two SOM procedures, each specifying a two-by-two grid topology. (*Upper*) This figure was generated using the geneSOM option in the R statistical softwear package. The curves in each grid represent the range of standardized expression values for the genes in that grid in each of of the tissue samples. The error bars are 99% confidence intervals and, where no error bar is shown, the confidence interval is too small to draw. There appear to be four distinct clusters in the data. Note the distinct difference in the pattern between the upper two and lower two panels, yet the left and right columns appear to differ only in their relative magnitude. The values 0, 1, 2 on the abscissa are of no significance. (*Lower*) This panel depicts the gene names colored by their cluster membership. This figure was generated using the batchSOM function in R. Points within node are jittered in an attempt to improve readability.

that test statistics cannot be compared across different node topologies. An example of an SOM analysis is shown in Figure 13.14. This particular analysis revealed that based on the pattern of gene expression in the leaf and root samples, the genes could be grouped into four grids in the SOM.

Principal components analysis. In principal components analysis (PCA), the goal is to identify linear combinations of p variables (in our case genes) that explain large portions of variation in N rows of data. These linear combinations of variables are called components, and they are ordered by their rank in the portion of variability accounted for. The first component explains the largest portion of variance in the data, the second component the second largest, and so on. For this reason, the PCA method is called a variance maximization technique. PCA and related methods are used to achieve dimensional reduction, meaning that instead of having to consider a large number of dimensions to find relationships between genes or samples, such as one dimension for each gene in a microarray experiment, the relationships may be captured by a much smaller number of dimensions. When each sample or gene is compared to another using this reduced set of dimensions, the relationships among the samples or genes will be preserved and closely approximate the relationships when all genes were used. Dimensional reduction occurs in that the first n components, where $n << p$, may be sufficient to explain the majority of the variation seen in the data. For example, if the first three components are used to produce a three-dimensional plot of the gene expression values based on their component scores, then the plot will give a representative picture of any clustering of gene relationships.

PCA produces gene clusters based on finding the sources of greatest variation in the data. The method uses a distance matrix that shows distances between all pairs of genes to find these clusters. The principal components are found by matrix algebra. First, a square, symmetric matrix of correlations or covariances of expression values from an experiment is decomposed into two matrices, $\mathbf{V}$ and $\mathbf{W}$, i.e., $\text{Eigen}(\text{Cor}(\mathbf{Y})) = \mathbf{VWV}'$, where $\mathbf{V}'$ is the transpose of matrix $\mathbf{V}$. Here, $\mathbf{V}$ is a matrix of eigenvectors and $\mathbf{W}$ is a matrix of eigenvalues. The first eigenvector, i.e., the first column of $\mathbf{V}$, is associated with the first eigenvalue of $\mathbf{W}$, i.e., the first diagonal element of $\mathbf{W}$. Then, each column of $\mathbf{V}$ is multiplied to each row of $\mathbf{Y}$ to produce a vector $\mathbf{z}$ that contains the actual principal components. Clustering comes from entries in each column of $\mathbf{V}$ that are far from 0, indicating that a specific gene, specified by the row label, is clustered with other genes that have entries far from 0 within that column.

Singular value decomposition. Singular value decomposition (SVD) is a matrix method related to principal components, but is not restricted to square matrices. Whereas the principal components method uses a distance matrix to find clusters, the SVD method uses the gene expression values. The clusters produced

by SVD will be different because SVD is not restricted to Euclidean distances between genes or samples, which reduce the number of expression variables (e.g., an x, y value is reduced to a single value of $\sqrt{(x^2 + y^2)}$ for principal components analysis). SVD is like PCA in that both methods find the combination of genes or samples that maximizes the variance within components by capturing the greatest amount of variation in the data. SVD attempts to decompose an observed intensity matrix $\mathbf{Y}$ of N rows (or genes) and P columns (or samples) into three matrices, i.e., $\mathbf{Y}=\mathbf{UWV}'$. $\mathbf{U}$ and $\mathbf{V}'$ are matrices of eigenvectors that describe the linear combinations of genes ($\mathbf{U}$) or samples ($\mathbf{V}'$) that collectively explain large portions of variation (as described by the diagonal entries of $\mathbf{W}$). The matrices $\mathbf{U}$ and $\mathbf{V}'$ are orthogonal in that their product produces a matrix with nonzero diagonal elements and 0s elsewhere. This method has been used to identify genes involved in cell cycle development (Alter et al. 2000).

Independent component analysis. Unlike PCA and SVD, which build clusters by maximizing variation, independent component analysis (ICA) builds clusters of genes and samples based on the amount of similarity between them, i.e., a high degree of covariation. The object of ICA is to build clusters with high correlation among the samples, but with little correlation between clusters, thus making the clusters statistically independent. The clusters are also independent in the sense that there is no apparent causal or regulatory relationship between clusters as indicated by the same pattern of gene expression across samples between clusters. Thus, these clusters are more likely to be biologically independent, i.e., each represents the same regulatory network.

Independent component analysis has been used for analysis of microarrays (Liebermeister 2002) and is closely related to SVD in that a matrix $\mathbf{Y}$ is decomposed into a matrix product, $\mathbf{Y}=\mathbf{XA}$. This step is similar to SVD ($\mathbf{Y}=(\mathbf{UW})\mathbf{V}'$), but is not restricted to orthogonal columns of $\mathbf{UW}=\mathbf{X}$ and $\mathbf{V}'=\mathbf{A}$. Information on the "clustering" of genes to expression modes is through the nonzero entries in the columns of $\mathbf{X}$. This result means that each column is a cluster and the magnitude of the entry for a gene within that column roughly corresponds to the probability that this gene belongs in this cluster. Although the columns of $\mathbf{X}$ are not orthogonal (meaning that the principal axes for each column are perpendicular to each other), they are statistically independent. Should the columns of $\mathbf{X}$ not line up with the x and y axes in a two-dimensional plot, interpreta-

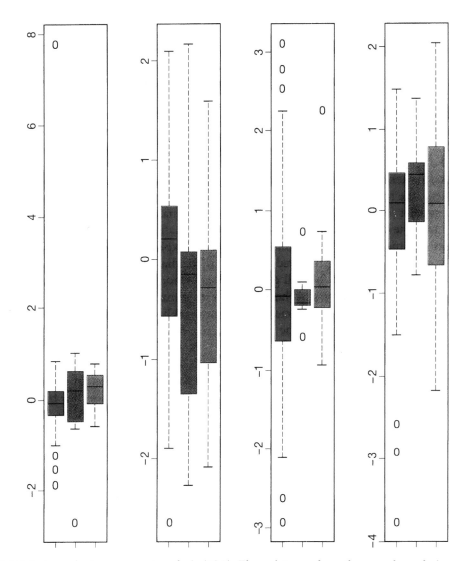

FIGURE 13.15. Independent component analysis (ICA). These data are from the same loop design experiment used in Figure 13.8. A mixed linear model excluding the treatment by gene effect appropriate for the experimental design was fitted and the residuals from this model were used to generate these graphs. The data were run through the fastICA package in R with four factors specified. There does not appear to be a difference within factors for the three groups of genes in the data set (Root Morphs [*red*], Shoot Morphs [*green*], and Photosynthetic [*blue*] genes), although there is a difference between the factors.

tion of the nonzero entries is obscured and there is no formal basis for rotation back to the grid axes. Additionally, the ICA model has no formal basis for modeling either additional covariates or a residual error term. A method called factor analysis does have formal arguments for factor rotation, modeling of additional covariates, and a residual error term. This method is explained and illustrated in Figure 13.15. Factor analysis is described below (see p. 645).

Model-based cluster analysis. Model-based cluster analysis (MBCA) is a method to cluster variables based on a mixture of distributions. A mixture of distributions simply means that the observed data contain observations that belong to one or more of a collection of possible distributions. Each distribution could differ in the mean, the variance, or both. The method simply assigns a probability for a variable to belong to each proposed distribution. Each distribution corresponds to a different cluster and the cluster with the highest probability is assumed to be the correct cluster for that variable. Fraley and Raftery (2002a,b) presented an MBCA algorithm based on these mixture distributions and implemented the method in the statistical package R (Fraley and Raftery 2002b). MBCA assumes that the observed data are produced from a mixture of several normal distributions and allows variables to be associated with multiple-component distributions (i.e., clusters). Genes can be assigned to a single cluster of the mixture based on their posterior probabilities of belonging to any of the components. Yeung et al. (2001) used MBCA to cluster gene expression data.

Factor analysis. Like the ICA and MBCA methods described above, factor analysis (FA) (Johnson and Wichern 1998) attempts to build clusters of genes whose expression is highly correlated with little, if any, correlation between clusters. In general, FA may be described as a method that describes the covariance relationships among many variables in a multivariate data set in terms of a few underlying, but unobservable, random quantities called factors. In microarray FA, genes or sample expression values are evaluated as linear combinations of such factors. Genes or samples having similar combinations of these variables are placed in the same cluster. Using these factor combinations, the correlation coefficient between genes or samples in different clusters is 0. In FA, each factor refers to a specific cluster of variables. There are two types of FA models, orthogonal and oblique. Generally, most analyses use the orthogonal FA model with the assumption that the factors are orthogonal to, or independent of, each other.

The general factor model is linear in the common factors **f**:

$$\mathbf{y} - \mu = \mathbf{Lf} + \varepsilon \qquad (5)$$

In this model, the vector **y** represents a multivariate observation, μ a vector of means, **L** a matrix of factor loadings, **f** a vector of common factors, and ε a vector of specific, or residual, factors. The vectors **f** and ε are generally not observed and assumed independent. Equations 6–11 help to define the orthogonal factor model further:

$$E[\mathbf{f}] = \mathbf{0} \qquad (6)$$

$$\text{Cov}(\mathbf{f}) = \mathbf{I} \qquad (7)$$

$$E[\varepsilon] = \mathbf{0} \qquad (8)$$

$$\text{Cov}(\varepsilon) = \psi \qquad (9)$$

$$\text{Cov}(\mathbf{y}) = \mathbf{L'L} + \psi \qquad (10)$$

$$\text{Cov}(\mathbf{y},\mathbf{f'}) = \mathbf{L} \qquad (11)$$

where $E[\ ]$ denotes the expected value and $\text{Cov}(\ )$ denotes the covariance. The matrix ψ is assumed to be a diagonal matrix of specific, or residual, variances (Johnson and Wichern 1998). This assumption implies that all covariation in the data (i.e., between the genes) is explained by the retained factors. Dimension reduction occurs by retaining a number m of factors, where m is much less than the number p of variables. Because each of the m retained factors corresponds to a unique cluster, entries in the factor-loading matrix **L** indicate which genes belong to which cluster. The factor-loading matrix has dimension p variables by m factors, such that the columns correspond to clusters. Nonzero entries in a column indicate that this gene belongs to this cluster. Note that like MBCA, a gene can belong to more than one cluster.

An interesting property of the factor-loading matrix is that it can be rotated by an orthogonal matrix without loss of information (i.e., the covariances or correlations between variables are maintained following rotation). The portion of the variance explained by the retained factors for a specific variable is termed the communality and the portion not explained by the retained factors is termed the specific variance. There are four common methods used to perform factor analysis.

1. *Principal components factor analysis (PCFA):* This method is by far the simplest of all five methods presented and is related to PCA only in that both involve an eigenvalue decomposition. PCA and PCFA are otherwise based on different methods of clustering.

2. *Principal factor FA (PFFA):* Principal factor FA is a variant of PCFA based on the observation that the specific variances are invariant to rotation by an orthogonal matrix. In this method, the specific variances are estimated first and the factor loadings afterward.

3. *Iterated principal factor:* This method is identical to PFFA, except that an iterative strategy is used to estimate the factor loadings **L** and specific variances ψ.

4. *Maximum likelihood FA (MLFA):* Estimates of **L** and ψ are obtained from the likelihood in Equation 12 given that $\Sigma = \mathbf{LL}' + \psi$, where Σ is Cov(**y**):

$$L(\mu,\Sigma) = (2\pi)^{-n/2}\, |\Sigma|^{-n/2}$$
$$\times \exp\left(-\tfrac{1}{2}\, \mathrm{tr}\left[\Sigma^{-1}(\sum_{j=1}^{n}(\mathbf{y}-\overline{\mathbf{y}})(\mathbf{y}-\overline{\mathbf{y}})'\right.\right.$$
$$\left.\left. + n(\overline{\mathbf{y}}-\mu)(\overline{\mathbf{y}}-\mu)')\right]\right) \qquad (12)$$

The method uses an expectation maximization (EM) algorithm developed by Rubin and Thayer (1982). A detailed explanation of the application of the EM algorithm for finding conserved patterns in sequence data is provided in Chapter 4.

5. *Confirmatory and exploratory Bayesian FA:* This model was initially developed for confirmatory factor analysis (Rowe and Press 1998), but has been extended to exploratory factor analysis by Henderson (2001). Both models have heavy parametric assumptions, but can accommodate the estimation of highest posterior density intervals (a Bayesian analog of confidence intervals).

Table 13.9 demonstrates how a Bayesian exploratory factor analysis can be applied to gene expression data. Like PCA, all of the above methods produce factor loadings that do not line up with the *x* and *y* axes in a two-dimensional plot. Fortunately,

factor analysis has a way to rotate the loadings to line up with the *x* and *y* axes called factor rotation. These rotations often assist in clarifying the gene membership status within each putative cluster. Factor rotation is often performed subject to an objective criterion, with the Varimax procedure (Kaiser 1958) being the most popular. Another rotation criterion that has shown promise is the minimum entropy (Inagaki 1994) factor rotation criterion, which follows from the assumption that the factor loadings within a given factor should be either near 0 or near ±1.

Sample Classification

The concept behind sample classification is that a small set of genes with expression profiles that are predictive of sample origin can be identified. The primary interest in developing this technology is in creating diagnostic tools. Once a sample can be classified, an appropriate treatment for that sample class can be applied. Important issues involve not only creation of the classifier but also identification of which genes comprise the classifier.

There are a number of types of classifiers, four of which are described below: discriminant classifiers, nearest-neighbor classifiers (NNCs), support vector machines (SVMs), and stochastic search variable

TABLE 13.9. *Bayesian exploratory factor analysis (BEFA)*

Gene	Factor 1 Value	Lower	Upper	Factor 2 Value	Lower	Upper	Factor 3 Value	Lower	Upper	Factor 4 Value	Lower	Upper	Factor 5 Value	Lower	Upper
Gene 14	**0.33**	0.02	0.61	−0.14	−0.45	0.21	−0.03	−0.36	0.35	−0.08	−0.48	0.31	0.14	−0.28	0.55
Gene 22	0.02	−0.20	0.24	**0.22**	0.03	0.37	−0.01	−0.23	0.26	0.02	−0.22	0.23	0.12	−0.12	0.33
Gene 23	−0.10	−0.36	0.14	**0.26**	0.06	0.43	0.06	−0.16	0.31	−0.03	−0.28	0.24	−0.09	−0.36	0.22
Gene 26	−0.03	−0.44	0.36	**0.34**	0.01	0.68	−0.07	−0.48	0.28	0.06	−0.36	0.48	−0.17	−0.63	0.28
Gene 27	−0.07	−0.28	0.14	**0.19**	0.01	0.33	0.08	−0.09	0.28	−0.03	−0.23	0.18	0.03	−0.17	0.27
Gene 35	−0.10	−0.32	0.11	**0.22**	0.06	0.37	0.00	−0.21	0.26	−0.01	−0.23	0.22	0.04	−0.22	0.27
Gene 38	**0.32**	0.05	0.57	0.06	−0.21	0.40	−0.14	−0.47	0.18	0.16	−0.21	0.51	0.06	−0.30	0.42
Gene 44	0.04	−0.28	0.37	−0.09	−0.39	0.18	0.00	−0.33	0.30	**0.34**	0.08	0.57	0.05	−0.29	0.40
Gene 52	−0.11	−0.40	0.18	0.05	−0.25	0.33	−0.05	−0.33	0.23	**0.28**	0.04	0.52	−0.07	−0.37	0.25
Gene 54	0.03	−0.24	0.31	0.01	−0.25	0.27	−0.05	−0.30	0.20	**0.27**	0.03	0.46	−0.06	−0.30	0.21
Gene 56	**0.36**	0.05	0.66	−0.12	−0.47	0.23	0.17	−0.21	0.57	−0.08	−0.48	0.31	−0.05	−0.51	0.45
Gene 57	**0.27**	0.03	0.47	−0.13	−0.39	0.12	−0.10	−0.37	0.18	−0.06	−0.36	0.25	0.02	−0.30	0.34
Gene 58	0.01	−0.41	0.45	0.01	−0.41	0.40	**0.36**	0.02	0.70	0.18	−0.19	0.61	0.04	−0.48	0.53
Gene 60	0.01	−0.21	0.25	0.10	−0.16	0.32	−0.05	−0.34	0.27	0.01	−0.25	0.28	**0.25**	0.01	0.45
Gene 61	**0.29**	0.06	0.50	−0.02	-0.31	0.27	−0.18	−0.45	0.10	0.02	−0.33	0.33	0.04	−0.31	0.37

These data are from the same loop design experiment used in Figure 13.8. A mixed linear model excluding the treatment by gene effect appropriate for the experimental design was fitted and the residuals from this model were used to generate these graphs. The data were run through the SG_Nirvana program (written by Dave Henderson, University of Arizona) with five factors specified. The genes in bold within a column are significant for loading on that factor (notice how the upper and lower bounds for the factor loading do not include 0). Genes that do not have a significant loading for any factor were omitted from the table.

selection (SSVS) with logistic regression. All of the methods attempt to identify a set of genes whose expression profile uniquely classifies sample type. As with the clustering methods described in the above section, it is strongly recommended that gene expression values used to identify classifiers should be based on the following:

1. A suitable experimental design.

2. A normalization option applied.

3. Application of a linear data model used as described previously in this chapter.

4. The resulting adjusted gene expression values are then used for the classification analysis.

Using incorrect, inefficient, or ad hoc methods to find classifiers is one of the commonest mistakes made in analyzing microarray data sets (Simon et al. 2003). An excellent example is the difficulty in finding reliable classifiers (Baggerly et al. 2004) for ovarian cancer in serum SELDI-TOF data as originally reported (Petricoin et al. 2002), probably based on insufficient statistical analysis of the data.

The following methods are designed to find classifiers on the basis of expression data that has been modeled as described above. In the first three methods (discriminant classifiers, NNC, SVMs), a set of predictive genes must first be identified using data reduction methods to make these methods feasible. Common methods to do so include identifying genes that have the largest variance across classes, genes that have the smallest variance within the classes, and genes that are significantly differentially expressed between samples. The last method (variable search with logistic regression) is unique because it simultaneously creates the classification rule and identifies which genes should be involved in the classification. Functions found on the R and SAS statistical packages can be used for most of these analyses or they may be obtained from authors of publications.

Discriminant classifiers. Discriminant classifiers are a group of classification tools that seek to identify a group of genes whose expression profile within a class has a smaller variance than when the variance is calculated across classes. There are two main types of discriminant classifiers: Fisher linear discriminant analysis (FLDA) and maximum likelihood discriminant analysis (MLLDA). In FLDA, the classifier rule seeks to identify groups of genes with a large ratio of between-group to within-group sums of squares. In MLLDA, given a known number of classes k and a

vector of expression values $\mathbf{x}$, the unknown sample belongs to the class where

$$\text{Class}(\mathbf{x}) = \text{argmax}_k \left[P(\mathbf{x}|y=k) \right]$$

meaning that the predicted class is the one whose expression profile produces the highest probability of belonging to class k based on the ratio of within-group to between-group sums of squares. Both the FLDA and MLLDA methods are examples of supervised learning in that training sample data sets are required where the samples belong to known classes. These data sets are used to generate parameters for the Class($\mathbf{x}$) descriptive statistics, e.g., the mean and variance for each class. Both methods are suitable when the data can be separated by a straight line or plane. LDA methods are not suitable for nonlinear data and are not easily extensible to more than two classes. Figure 13.16A provides an example of discriminant analysis to find a linear classifier between two classes. Note that the straight line is positioned so that there is a maximum separation of sample classes in the training data set.

Nearest-neighbor classifier. The NNC is a distance-based classifier and is based on several straightforward calculations. First, d, the measure of distance between samples, is defined. For example, d might be chosen as the Euclidean distance between two samples on an expression plot, i.e., $d = \sqrt{[(x_i - x_j)^2 + (y_i - y_j)^2]}$, where x_i, y_i and x_j, y_j are the expression values of the classifier genes for samples i and j on the plot. Second, divide the samples into a specified number of classes k such that the distances between the members of each class are a minimum. The method for producing these classes is very similar to that used for the k-means clustering algorithm described on page 642. Once these classes have been defined, the class prediction for a sample is decided by majority vote, meaning that the class to which the sample belongs is decided by which class has more genes close to the putative sample. To perform class prediction by the NNC method, the number of neighbors to investigate (g) and the number of similar neighbors (l) out of the g selected required to declare success are chosen. Once these numbers are set, the classification is quite simple.

NNC is not robust regarding outliers, individual samples that lie far away from most examples on a gene expression plot. An advantage of NNC is that, unlike discriminant classifiers, NNC is easily extended to more than two classes. An example of testing four gene pairs as classifiers for the same data used in Figure 13.16A by the NNC method is shown in the

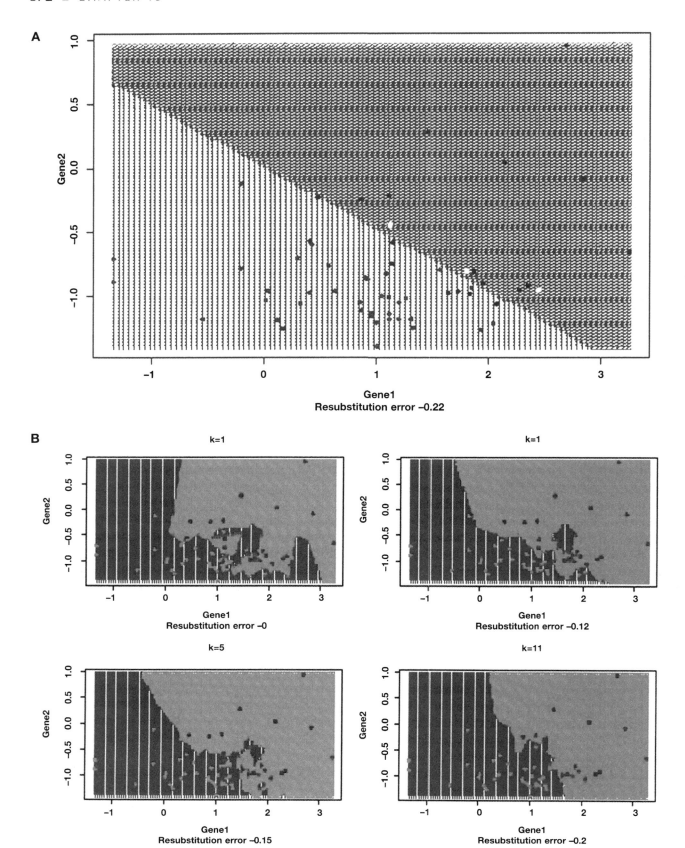

A

Gene1
Resubstitution error –0.22

B

k=1

Gene1
Resubstitution error –0

k=1

Gene1
Resubstitution error –0.12

k=5

Gene1
Resubstitution error –0.15

k=11

Gene1
Resubstitution error –0.2

FIGURE 13.16. (*See facing page for legend.*)

C

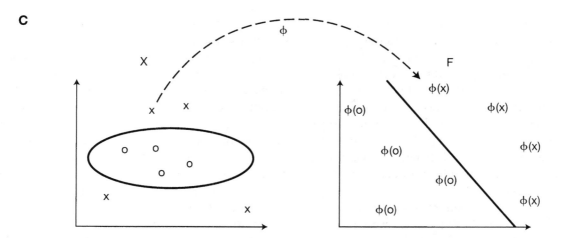

FIGURE 13.16. Methods for identifying classifiers. (*A*) Linear discriminant analysis. The *x* and *y* axes correspond to two different genes used to classify the sample. The blue points correspond to a different sample class than the red points. The red-shaded area denotes samples predicted to be of one class and the blue-shaded area denotes samples predicted to be of the other class. Blue points within the red-shaded area are correctly classified, as are red points within the blue-shaded area. Otherwise the sample class was predicted incorrectly. The incorrectly classified red points on the red background have been changed to white by us to facilitate seeing them. Note that the two-gene classifier works best and reliably for identifying blue class membership correctly. These data are from a brain tumor data set and are an attempt to identify gene expression predictors for survival. The data were fit to a statistical model and the residual terms for genes were used to test genes for ability to discriminate between treatments (known survivals) by *t*-tests, as described in Figure 13.10 and the text. This classifier is a linear separator (a line or plane) between the classes. If there are more dimensions (genes) used as classifiers, then the separator is a hyperplane. (Reprinted, with permission, from Dr. Sandrine Dudoit, University of California at Berkeley.) (*B*) Nearest-neighbor classifiers. The data set is the same one used in A. The four plots correspond to using *k* different nearest neighbors in assessing class prediction. The *x* and *y* axes of each plot correspond to two different sets of genes used to classify the sample. The blue points correspond to a different sample class than the red points. The red-shaded area denotes samples predicted to be of one class and the blue-shaded area denotes samples predicted to be of the other class. Blue points within the red-shaded area are correctly classified, as are red points within the blue-shaded area. The incorrectly classified points are not visible. The main point of the slide is that instead of using a straight line as in A to discriminate between classes, a curved line based on sample classes (see text) may be used as found by nearest-neighbor analysis. (Reprinted, with permission, from Dr. Sandrine Dudoit, University of California at Berkeley.) (*C*) Support vector machines. The plot on the left indicates a two-dimensional plot of sample points (x and o) for two genes. Note that it would be difficult to find a linear separator between the two sample classes as was done in A. The arrow connecting the two plots indicates the use of the kernel function φ. The sample values on the left are substituted in this function, which returns a new set of two or more values (called fitness values) for each sample. When these new values are plotted, a better linear separation of the two samples may be produced, resulting in a better classifier. A two-dimensional plot of a hypothetical set of two new values is shown on the right. A much better separation has been produced by using the kernel function. The straight line indicates the separating plane between the two classes.

four panels of Figure 13.16B. The main point of this figure is to illustrate that the classifier using the NNC method is not a straight line as in LDA methods. Note that each pair of genes provides a separation of the two sets of samples but the line of separation is a curve that depends on the location of sample classes on the plot. The resolution of the plot depends on the number of classes (*k*) chosen for a sample similarity analysis.

Support vector machines. The SVM method is a supervised method that uses training data of known

classes to train the kernel function for each gene in order to predict class. This training data set is used to generate the parameters for the kernel function. SVMs are a machine-learning type of classifier. The method uses a mathematical expression (a kernel function) to convert appropriately adjusted gene expression values into a new set of values that provides a better separation of the samples. Figure 13.16C demonstrates how SVMs can be used for this purpose. The original expression values for the two groups of samples shown on the right cannot be separated by a straight line. However, after applying a kernel function φ specified in the analysis to the expression values, a new set of values is produced. The samples can then be readily separated by a linear classifier. An advantage of SVMs is that nonlinear separable data as well as linear separable data can be classified by this method. However, the method is not easily extended to more than two sample classes. Another disadvantage of SVMs is the effort needed to choose and optimize a kernel function for producing the best classification efficiency.

Variable search with logistic regression. Stochastic search variable selection (SSVS) is the Bayesian analog to stepwise regression in that it seeks to identify which set of variables (adjusted gene expression values) collectively explains the largest proportion of variation in the dependent variable (the probability of a class given a set of gene expression values). In general, logistic regression is a statistical model for binary data. For classifying samples based on microarray data, the dependent variable is the probability of an outcome of 0 or 1, representing one sample class or the other, and the independent variables are the adjusted gene expression values that are tested as class predictors. SSVS linearizes what is basically a nonlinear statistical model.

Linearization is achieved by using log odds scores in favor of one class or the other—the probability of one class $P(Y=1)$ is the odds score for one class divided by the quantity 1 minus the odds score for that class. The association of each gene with one class can be expressed as the odds of one class to the other. If multiple genes are affecting one class, then the odds score for each must be multiplied to give the overall odds for one class. If the odds scores for each gene are converted to log odds scores, then probability of class is now a function of the linear sum of the log odds scores for each gene, thus generating a linear statistical model that can be used in a regression analysis. The reader may recall that a similar method is used to calculate a log odds score for a sequence alignment in Chapter 3.

The method to find the best combination of genes that define a classifier utilizes a Bayesian statistical analysis similar in concept to the sequence alignment methods described in Chapters 4 and 5. Every gene is initially given a prior probability of being included in a classifier model. A Gibbs sampling method is then used to find optimal combinations of gene expression values that best separate the samples. Recall that in

searching for a conserved sequence pattern in a set of sequences, one sequence is left out and a log odds scoring matrix representing a trial alignment in the remainder of the sequences is used to produce a probability distribution of the trial pattern in the left-out sequence (see p. 201). From this distribution, a probability-weighted pattern is chosen on which to base a new scoring matrix. This matching of a trial scoring matrix to sequences is repeated hundreds of times until a conserved sequence pattern is found and there is no better pattern to be found by further iterations. The object of Gibbs sampling is to recruit patterns present in more than one sequence by a method of the iterative matching of trial sequence patterns. Similarly, Gibbs sampling is used in gene classification to find a sum of log odds score of specific genes in the microarray data (a matrix of sample identification versus gene expression values) that most accurately classifies samples correctly. This combination of genes may then be used as a classifier for unknown samples.

Because SSVS uses logistic regression based on a choice of two sample classes, it is currently limited to this number of classes. If there are more than two classes, the method must be revised. SSVS is statistically a more rigorous method than linear discriminant analysis. An example of the results of an SSVS analysis is shown in Table 13.10. Four different models for predicting class are shown and the probability of an incorrect classification and number of misclassifications is shown in each case. The best models may be identified based on these latter values. SSVS produces a classifier that usually requires measurement of the adjusted expression values of a small number of genes to determine sample class.

Using sample classifiers. All of the classifier methods described above are based on the availability of training data sets of known sample class. Once the analysis has been performed using these known samples, the resulting classifier may be used to identify an unknown sample. The classifier also needs to be validated to assure that the prediction of unknown samples is reasonably accurate. A straightforward validation method is to test the classifier for reliability using an independent data set for which the classification is known. Various other methods of dividing a data set into training and validation sets are described below.

Classifier validation. When samples of two classes are used to build a classifier, some concerns should immediately arise. The first is, "How well do the individual samples represent the underlying populations

TABLE 13.10. *Stochastic search variable selection*

Y	Model 1 $\Pr(Y = 1 \mid X)$	Model 2 $\Pr(Y = 1 \mid X)$	Model 3 $\Pr(Y = 1 \mid X)$	Model 4 $\Pr(Y = 1 \mid X)$
1	1	1	0.9993	0.9998
1	1	1	1	0.9969
1	1	1	0.9999	1
1	1	1	0.9999	0.8605
1	1	1	0.9999	0.7766
1	1	1	0.9998	1
0	0	0	0	0
0	0	0	0	0
0	0	0	0	0.0002
0	0	0	0	0
0	0	0	0	0
0	0	0	0	0
0	0	0	0	0.0002
0	0	0	0.0018	0.0867
0	0	0	0.0005	0.007
0	0	0	0	0
0	0	0	0	0.2864
1	1	1	1	1
0	0	0	0	0
0	0	0	0	0
0	0	0	0	0
0	0	0	0	0
Deviance				
	1.2683e −12	3.1464e −7	0.0071	1.6843
Number of misclassifications				
	0	0	0	1

The SSVS procedure generates a list of possible models that could explain large proportions of variation in the dependent variable. The table shows four such models. The Y column indicates the actual sample class (designated as 0 or 1) and the columns for each model give the probability that the sample class is in class 1 ($Y = 1$) for each sample using that model and given the gene expression values for that sample (X). Based on a measure of deviation of predicted class from actual class and the number of misclassifications, models 1 and 2 are best at predicting sample class. (Reprinted, with permission of Oxford University Press, from Lee et al. 2003.)

of each class?" For example, if two or three tissue samples out of 20 chosen comprise a subtype that is different from the rest of the samples in the population, they could bias the classifier so that it is useful for finding the same subtype, but not necessarily other tissue types represented in that class and perhaps those more representative of the sampled tissue population. This concern can best be addressed by choosing a large enough number of samples and avoiding any bias for any particular subclass. The concerns regarding class variation can also be addressed by using the best methods to validate the classifier. If done carefully and adequately, validation can help to determine that the entire sample collection supports the classifier. The methods described in this section explain how to perform the validation. Simon et al. (2003) and Dudoit et al. (2002) have discussed the importance of obtaining a reasonable and most conservative, worst-case estimate for the error rate for class prediction.

There are three methods for testing the specificity of a classifier: leave-one-out cross validation, test/train data set cross validation, and the bootstrap method. Each method has its merits, although some of these methods may not be possible for most microarray experiments.

LEAVE-ONE-OUT CROSS VALIDATION. Leave-one-out cross validation is by far the most commonly used classifier validation method. The operation of the method proceeds according to the following steps:

1. Repeat the prediction of the classifier on all but one of the samples.

2. Use the new classifier rules to predict the class for the left-out sample.

3. Repeat this procedure for all samples in the data, one at a time.

To evaluate the classifier, a mechanism is needed to tabulate the loss, i.e., a loss function $L(y_i, \hat{f}^{-k(i)}(x_i))$, where the notation $-k(i)$ denotes the full data minus observation i. The loss using the loss function for all possible data sets is evaluated with one observation missing. The total loss is then calculated as the average of the leave-one-out loss function values:

$$\text{CV} = \frac{1}{N} \sum_{i=1}^{N} L\left(y_i, \hat{f}^{-k(i)}(x_i)\right) \qquad (13)$$

TEST/TRAIN DATA SET CROSS VALIDATION. The test/train cross-validation method is almost identical to the leave-one-out cross-validation method. The only real difference is in the number of observations left out. Instead of leaving out just one sample as above, k observations are left out. Using the same loss-function equations as before, $-k(i)$ is denoted as the full data set minus k observations. The data minus k observations are used to train the classifier and the validity of the classifier tested on the remaining k samples. The average loss is evaluated in the same way as the leave-one-out method.

BOOTSTRAP METHOD. Both of the above cross-validation methods require data sets of substantial size, greater than 30–40 samples, with at least five samples in each class. Many microarray data sets are small, and some classes may have fewer than five samples. In these cases, the cross-validation methods are likely to bias the classification error downward, i.e., the classifier will appear to be more accurate than is really the case. A bootstrap estimate of the classification error is likely to be more accurate for these small sample size data sets. The bootstrap method is also the most likely method to identify subclasses in the classes that bias the classifier, as described above in the introduction to this section.

The bootstrap estimator works by finding the fit for a classifier from a training data set. Then, this data set is randomly sampled with replacements to create new data sets of the same size and the classification error is estimated from these bootstrap data sets. The method is identical to the bootstrapping used to demonstrate support for branches in a phylogenetic tree by resampling the columns of a multiple sequence alignment to build new alignments and new trees based on them (see Chapter 7).

Avoiding underestimation of the error rate of a classifier. What is needed in validation is the most conservative estimate of the error rate when using a classifier. Unfortunately, there are biases that arise when using the same data set for building and validating a classifier. Thus, the bootstrap approach is likely to produce an underestimate of the error rate. One method of avoiding this problem is to divide the sample data set into two parts and use one part for training and the other part for validation. Although often done, this practice results in a loss of information for building a better classifier. Another method is to try to correct the bias by a computational method.

To avoid potential bias from using the same data points as the training data in the bootstrap data sets, the so-called 0.632+ bootstrap approach of Hastie et al. (2001) was developed. The 0.632+ bootstrap adjusts for this underestimate of the error rate by the bootstrap method. This validation measure has the form

$$Err(0.632+) = (1 - (0.632 / (1 - 0.368 R))) err \\ + (0.632 / (1 - 0.368 R)) err1 \quad (14)$$

where err is the training error, *err1* is the leave-one-out bootstrap error, and R is $(err1 - err)/(\gamma - err)$. The variable γ is a tuning parameter dependent on the number of predictors used. Hastie et al. (2001) provide a detailed use of this method.

RESULTS OF MICROARRAY ANALYSIS CAN BE INTERPRETED BY STATISTICAL INFERENCE

The above sections describe the different types of analyses but do not explain how to interpret the results. In the following section, a procedure for statistical inference of microarray data is presented. In statistical inference, conclusions are drawn on the basis of statistical analysis of the data.

Differential Expression

A set of significantly varying genes found using the LIMMA package in R and shown in Table 13.6 was described previously. There are 44 genes in this list, identified by *p* values smaller than 0.05. The *p* values presented in Table 13.6 have been adjusted for the fact that 95 hypotheses are being tested, or one hypothesis per gene, and the list of significant genes is 22 genes smaller than the corresponding unadjusted list of significant genes. The adjustment performed here is based on a method developed by Benjamini and Yekutieli (2001) called the false discovery rate (FDR). This procedure was made popular for gene expression analyses by Storey and others (Storey and Tibshirani 2001; Storey 2002; Storey et al. 2004). The FDR rate is based on certain statistical considerations.

In statistical inference, there is often concern for minimizing the risk associated with choosing one hypothesis over the other. The risk generally minimized is the type I error, or the probability of falsely rejecting the null hypothesis when the null hypothesis is actually true. The 0.05 level for *p* values has historical significance and corresponds to the probability of making a type I error, that is, in 5% of all experiments of the same size and type performed by the experimenter, the null hypothesis will at least once be falsely rejected when in reality the null hypothesis is true. Each time a hypothesis test is performed, there is the risk of making a type I error, and the probability of making a type I error accumulates as more hypotheses are tested. This means that the type I error rate for the overall experiment utilizing all genes could be much higher than 5%, even though each test was controlling the type I error at the 5% level.

Therefore, to control for this accumulation of type I error when testing multiple hypotheses, several procedures have been developed. The most common procedure is the Bonferroni correction (Snedecor and Cochran 1989), often approximated by p threshold/number of hypotheses tested. For most microarray experiments, this correction is often too severe, given the number of genes on most arrays is greater than 3000. The false discovery rate previously mentioned minimizes risk in a different way and is less severe in its effect on inference. The method corresponds to determining the probability of a falsely rejected null hypothesis given a list of rejected null hypotheses, i.e., controlling the FDR at the 5% level means that, in a list of rejected null hypotheses, 5% are falsely rejected null hypotheses. Thus, the FDR method provides a more liberal way to control type I errors than the Bonferroni correction and yet still limits the accumulation of these errors in a microarray experiment.

Cluster Analysis

The interpretation of cluster analysis, regardless of the method, is quite simple. The genes within a cluster are related in some way. How the genes are related depends on two factors: (1) what distance measure, if any, is used and (2) what principle is used to construct clusters. For example, in HCA, if one uses a Euclidean distance, the resulting clusters correspond to genes that are similarly regulated or respond to a stimulus in the same way. This prediction is easy to see in that genes whose expression patterns are similar will be close together in a Euclidean sense, and thus will be clustered together. For FA and ICA, the clusters that are produced typically correspond to functionally related genes that may respond to stimuli in different manners. This prediction is made because both FA and ICA group genes together that have high covariance or correlation.

In HCA, genes are said to be clustered if they are within the same clade, or group, in a dendrogram. For example, in Figure 13.13, the genes 76, 17, 42, and 40 are clustered together because they reside within the same clade. The height of the line connecting the clade to the rest of the tree represents the "distance" between that clade and the other clades. The same comparisons may be made to find other gene relationships.

For SOMs, the genes associated with each node are said to be clustered together, and the corresponding expression pattern is intended to be representative of all genes within that cluster. Similar reasoning applies to the clusters generated by model-based clustering.

For both FA and ICA, the magnitude of the factor loadings are intended to represent the strength of relationship between the gene and that factor. Large loadings, regardless of sign, are indicative of cluster membership. For example, in Table 13.9, genes 44, 52, and 54 are clustered together on Factor 4.

MICROARRAY DATA SHOULD BE STORED IN A STANDARDIZED MANNER

Microarray data collected by different laboratories can be very difficult to compare because of variations in the number and types of array genes, the processing and labeling of cDNA samples, and other variations in equipment and technique. To facilitate data sharing, the MIAME (Minimum Information about Microarray Experiment) guidelines have been set up by the MGED (Microarray Gene Expression Data, visit http://www.mged.org) society, an international organization of biologists, computer scientists, and data analysts whose aim is to facilitate the sharing of microarray data generated by functional genomics and proteomics experiments. These guidelines are designed so that a laboratory can repeat a microarray experiment previously performed by another laboratory and attempt to reproduce any significant findings. The MAGE workgroup in this society is developing a data exchange model to facilitate the exchange of microarray information between different microarray data storage systems. This model is based on an examination of all of the stages of microarray research. A markup language in XML format called MAGE-ML (for description and use of XML, see Chapter 12, p. 572) is used for storing this information (Spellman et al. 2002). A similar initiative for sharing proteomics data is also being undertaken by the HUPO (Human Proteome Organization; visit http://psidev.source forge.net/). It is important for microarray and proteomics groups to take advantage of these standards for efficient data storage and exchange.

TOOLS ARE BEING DEVELOPED FOR FUNCTIONAL ANALYSIS OF MICROARRAY DATA

The data analysis and statistical methods described above can be used to identify patterns of gene expression associated with biological samples. The next task is to verify these patterns and identify their biological significance. The significance can be further explored by promoter analysis to search for evidence of transcriptional regulation and by examining the biological roles of the genes. If microarray experiments have been planned with clear-cut objectives in mind, as described earlier in this chapter, then the functional analysis of the data will be more useful.

A number of tools that have been developed to assist in the biological intepretation of microarray results are shown in Table 13.11. These tools provide a variety of information about the functions of the genes that are showing altered gene expression. One useful resource for functional descriptions of genes is provided by the Gene Ontology Consortium at http://www. geneontology.org. Gene relationships are described by a set of controlled vocabularies representing molecular function, biological process, and cellular component of each gene and by representing gene relationships in the form of directed acyclic graphs. GoMiner (Table 13.11) will map a set of genes to this information. Another tool is Pathway Miner, illustrated in Figure 13.17 and Table 13.11. This Web-based resource locates the human and mouse genes on known metabolic pathways and shows a metabolic chart with the genes highlighted and color coded by expression level. Another useful type of analysis of coregulated genes is promoter analysis.

Promoter Analysis of Coregulated Genes

Promoter analysis of genes that are coregulated in a microarray experiment is performed to determine if the genes share common transcription binding sites.

Usually, the first task at hand is to collect the sequences of the promoter regions for these genes on the basis of gene models. The Perl scripts and programming modules described in Chapter 12 can facilitate this task. The next task is to search for known transcription-factor-binding sites and other conserved patterns that suggest conserved regulatory sites in the promoters of the coregulated genes. (Promoter analysis is described extensively in Chapter 9.) The main points to keep in mind are that (1) the binding sites are predicted using position-specific scoring matrices, which often produce false-negative predictions; (2) a common location of the binding site with respect to the start of transcription in two coregulated genes may be of regulatory significance; and (3) clustering of the binding sites for the same transcription factors in two promoters is highly significant. Resources for performing these types of analysis are listed in Table 9.6, page 398.

TABLE 13.11. *Examples of computational tools for biological mining of microarray data*

Tool	Reference
ChipInfo	Zhong et al. (2003)
DRAGON	Bouton and Pevsner (2002)
GenMapp	Dahlquist et al. (2002)
GoFish	Berriz et al. (2003)
GoMiner	Zeeberg et al. (2003)
KnowledgeEditor	Toyoda and Konagaya (2003)
Pathway Miner	Pandey et al. (2004)
Pathway processor	Grosu et al. (2002)

The Web sites for these tools may be found by a Web search. For a discussion of Web-based resources for functional annotation and visualization of DNA microarray data, see Guffanti et al. (2002).

FIGURE 13.17. Analysis of the location of genes showing altered expression in microarray data in biochemical pathways by Pathway Miner. This tool was developed by Ritu Pandey, Raghavendra Guru, and David W. Mount (Pandey et al. 2004) to assist with a biological interpretation of gene expression data from microarray experiments. A provided list of genes is mapped on to the pathways in the BioCarta, Kegg, or GenMapp databases (these resources may be located by a Web search). Shown is one of the displays provided by Pathway Miner, a graphical display of the pathway relationships among the genes of interest. Genes are indicated as nodes and genes with a pathway relationship are joined by edges whose width reflects the number of pathway interactions. The amount of information provided can be filtered in various ways to reduce the output. Clicking the mouse on the genes or edges provides a detailed list of information about the pathways, genes, and interactions based on locally maintained databases, including the mouse and human genomes.

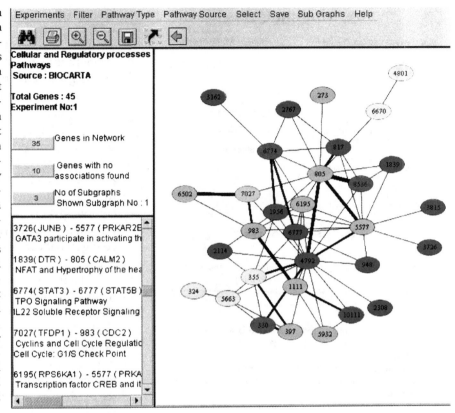

MICROARRAYS HAVE MANY APPLICATIONS

Locating Regulatory Changes by QTL Analysis

One goal in using microarray experiments is to identify the underlying genetic changes that cause a distinct phenotypic change in plants or animals, such as size, stress response, or fat metabolism. These genes are usually located by finding a nearby sequence marker (called a quantitative trait locus or QTL) that is easily followed in genetic crosses. The keys to discovery of these genes are (1) identify genes that cause the difference in phenotype (size, etc.) usually achieved by developing two inbred strains that differ in phenotype followed by genetic crosses between them and scoring QTLs to locate genes; (2) identify a set of genes that is significantly changing expression in each inbred strain; and (3) perform further genetic crosses and search for particular genes (using QTL analysis) that appear to be influencing the expression of a particular microarray

gene. These regulatory QTLs are designated eQTLs (expression QTLs). This combined genetic-microarray analysis methodology is illustrated by two published examples.

In the first experiment shown in Figure 13.18A (Schadt et al. 2003), two inbred strains of mice with altered fat metabolism (one normal and one with larger foot fat pads) were obtained. As shown in the figure, approximately 100 sequence variations between the two strains and spanning the mouse genome were identified. The mice were then crossed once and then the descendant mice (genetically the F_1 progeny) were crossed again with each other to produce about 100 inbred F_2 progeny in a second generation. The chromosomes derived from the original parent strains are reassorted in these F_2 progeny mice. The object is to locate the mice in this F_2 group that have large fat pads and then identify the chromosomal regions these mice have inherited from the abnormal parent. This analy-

A. Finding QTLs

Inbred Mouse strain A -(fat) ✕ Inbred Mouse strain B - (normal)

F₁ generation
F₂ generation

+ 100 other assortments

- Find 100 different sequence variations between A and B to serve as red, blue chromosomal markers.
- Collect 100 progeny F₂ mice
- Some are have large fat pads fat and some normal fat pads.
- Identify chromosomal regions (red) that are associated with large fat pads in these mice; these are QTLs.

B. Finding eQTLs

Inbred Mouse strain A -(fat) ✕ Inbred Mouse strain B - (normal)

F₁ generation
F₂ generation

- Perform microarray analysis on A and B to find set of genes that are varying between A and B
- Collect 100 F₂ progeny mice
- Measure gene expression in all 100 mice
- Find what chromosomal regions from A and B are in these mice as in part A.
- Find correlations between regions present and gene expression.
- Infer regulatory relationships

FIGURE 13.18. Analysis of quantitative trait loci (QTLs) for regulation of gene expression. (*A*) A genetic cross between two inbred strains of mice to identify chromosomal regions that are varying between the strains and are associated with altered fat metabolism. (*B*) An experiment similar to that in *A*, but with the inclusion of gene expression analysis using microarrays. The analysis of the data to find eQTLs that regulate gene expression is described in the text.

sis utilizes a maximum likelihood approach in which the most likely regions are identified. The programs MapMaker (http://www.broad.mit.edu/genome_software/) and QTL cartographer (http://statgen.ncsu.edu/qtlcart/) are examples of software that can be used for this purpose. Two excellent references for QTL analysis are Lynch and Walsh (1998) and Weller (2001).

The next step for analyzing regulatory changes is to combine the QTL analysis with a microarray analysis as illustrated in Figure 13.18B. First, a set of genes that was significantly varying between the normal and abnormal parents was found. About 100 F₂ progeny mice were obtained and a gene expression analysis performed on each mouse. For each over- or underexpressed gene in the abnormal F₂ progeny mice, the QTLs that influenced the expression of that gene (eQTLs) were then identified by maximum likelihood analysis. Two types of eQTLs were found. One type strongly influenced expression and was located close by the gene being regulated. These changes appear to be promoter mutations in the abnormal mouse strain that geneticists refer to as *cis*-acting. The second type of eQTLS had a smaller effect and were distantly located from the gene being regulated; these are *trans*-acting QTLs. These eQTLs appeared to be regulatory genes, possibly encoding one or more transcription factors that had changed in the abnormal mouse. This kind of combined genetic–microarray analysis thus has the potential of identifying promoter and regulatory gene changes that can produce a particular type of heritable change.

The results of a similar analysis of a second experiment to find eQTLs (Pomp et al. 2004) are shown in Figure 13.19. This figure demonstrates a graphical method for displaying regulatory effects of QTLs. The map location of each microarray gene is plotted on the *x* axis against the map position of the QTL that explains the most variation in expression levels of that gene in the expression array on the *y* axis. The chromosomes are listed one after the other by numbers on these axes. Four scenarios are depicted. First, levels of expressed genes are controlled by *trans*-acting eQTLs (scattered yellow circles). Second, *cis*-acting eQTLs (diagonal red circles) may represent genetic variation within the regulatory or coding regions of the expressed genes themselves. Third, a single eQTL may result in changes in expression levels of many unlinked genes (horizontal blue triangles), either directly by gene product interaction-promoter or by multiple changes in a regulatory cascade resulting from changes of expression in a single key gene. Finally, clusters (small green dots) of gene expression changes may result from changes due to linkage of multiple expressed genes to a single regulatory QTL, or, alternatively, as a result of coordinate regulation in specific chromosomal regions. Thus, these types of experiments provide a variety of regulatory information not previously available to the biologist.

Modeling Genetic Regulatory Networks

Instead of genetically analyzing gene regulation, another approach for analyzing regulatory networks using gene expression data is to apply computational models (for review, see Gat-Viks and Shamir 2003). Information from protein–protein and protein–DNA interactions has also been combined with microarray gene

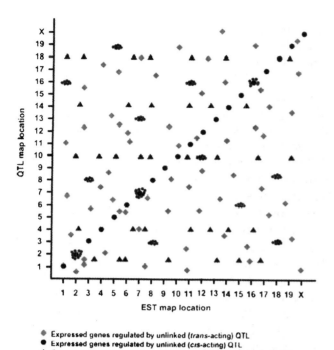

FIGURE 13.19. Types of gene regulation observed in the analysis of QTLs affecting gene expression using microarray data. The experiment is similar to that described in Figure 13.18. The location of microarray genes is shown on the x axis and that of QTLs on the y axis. Color spots indicate four types of observed regulation and the approximate location of regulatory changes. (Reprinted, with permission, from Pomp et al. 2004.)

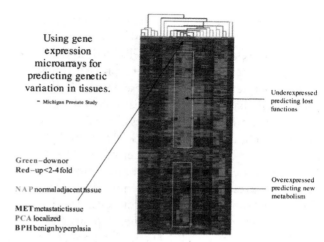

FIGURE 13.20. Example of expression microarray application in pharmacogenomics. Shown is a hierarchical clustering analysis of gene expression data from 50 different prostate tissue samples by Dhanasekaran et al. (2001). The tissue samples have been classified pathologically, but this information is not used for clustering. The expression values are shown relative to a combined pool of normal prostate tissues, which should help to control genetic variation in gene expression between different individuals. The data have been clustered with respect to both genes in the vertical axis and samples on the horizontal axis, resulting in a remarkable grouping of overexpressed and underexpressed genes that is highly conserved in each class of tissue. The two white boxes indicate a group of underexpressed genes (coded green) and overexpressed genes (coded red). This information is used to identify new targets for anti-cancer drugs as described in the text. (Modified, with permission of Nature, from Dhanasekaran et al. 2001 [©Macmillan Magazines Limited].)

expression data to facilitate prediction of regulatory networks by computational models (Ideker et al. 2002; Vert and Kanehisa 2003). A new type of analysis, network component analysis, is designed for uncovering regulatory signals in microarray data (Laio et al. 2003). These methods play an important role in the regulatory analysis of microarray data.

Applications of Expression Microarrays in Pharmacogenomics

Analysis of gene expression using microarrays presents a novel opportunity to identify new drug targets and to evaluate drugs for their overall effects on gene expression. An example that illustrates how microarray data may be used for identifying new drug targets in metastatic prostate cancer cells is shown in Figure 13.20. The altered gene expression in these cells enables them to invade surrounding tissues, establish new tumors at sites remote from the original one, and induce the growth of new blood vessels at these sites. Thus, identifying gene expression changes that make

these processes possible is a high priority so that new therapeutic agents (e.g., small RNAs that interfere with gene expression and drugs that inhibit the protein product) may be directed against these cells.

Figure 13.20 shows a hierarchical clustering of the microarray data from over 50 tumor samples. The clustering has been performed in two dimensions: First, genes whose expressions are most closely coordinated across the samples are shown on the vertical axis and, second, tissue samples whose expressions are most closely coordinated are shown on the horizontal axis. The two-dimensional plot of the combined clusterings reveals that tissue samples of one type have conserved patterns of gene expression and also shows which genes show a similar expression in related samples. For example, the data in the plot indicate that the metastatic cancer cells have a common pattern of gene expression that provides a starting point for finding therapeutic targets.

Two boxes are shown in the plot—one around a set of genes that is overexpressed in the metastatic cancer cells relative to normal cells (red) and another around a larger set that is underexpressed (green). A biological interpretation of these results is that the cancer cells have undergone many genetic changes, including the loss of the expression of many genes due to deletion, mutation, CpG island methylation, etc., and to amplification of other genes. The underexpressed genes are not needed any longer, whereas the overexpressed genes are required to meet the altered metabolism of the cancer cells. The overexpressed genes provide a first line of pharmaceutical attack on the cancer cells. The underexpressed genes also provide information on a more subtle, second line of attack according to the following line of reasoning.

Normal cells often have multiple overlapping gene functions that provide a backup in case a gene is accidentally lost by mutation or other means. Pairs of genes that provide backup or alternate functions have been found in yeast experiments. If the function of one of the genes in the pair is knocked out by an insertion of extra DNA, the cell survives. The same is true if the second gene of the pair is knocked out. However, if both genes are knocked out, the cell dies because the function of at least one of the gene pairs is essential. These lethal combinations are known as synthetic lethals by geneticists. The second line of attack in the prostate metastatic cells is to identify pairs of synthetic lethal genes, one of which is underexpressed (green region of gene expression data) and the other overexpressed (red region) in the cancer cells. A therapeutic agent directed against the overexpressed gene should be very effective since there is no backup function in the cancer cells, but the normal cells will not be affected because they should still have the backup function (see Kamb 2003a,b). This example illustrates the role of microarray experiments in the development of new drugs and disease treatments.

CONCLUDING COMMENTS

When image data are returned to the Earth by a satellite or space craft, there is often a fuzziness to the image that must be corrected. The information for a much clearer image exists in the data, but first the statistical noise from a number of sources must be removed. Once this correction is applied, the image is much clearer. Such statistical variation also exists in the data when microarrays are used to measure gene expression, and this variation needs to be removed to obtain a clearer picture of the changes in gene expression. The factors that must be considered for achieving this clear picture are careful planning of experiments, use of repeated experiments to identify noise, and use of appropriate data models and statistical analyses. These considerations also apply to proteomics data.

The extensive use of microarrays for measuring gene expression in the biology community indicates recognition of their potential. Well-designed experiments with a clearly defined objective and with data stored in sufficient detail in standard formats should provide a valuable resource for discovering many different aspects of gene interactions, metabolic pathways, and regulatory effects. By combining these data sets with other types of genomic and proteomic data, the computational biology community can build models of major biological processes. These models will lead to predictions that could not be reached by traditional research methods, and these predictions can be tested in the laboratory and the models refined as needed. Thus, the future role of these large data sets lies in the resulting cycles of modeling and experiment.